Recovery Improvement

Oil and Gas Chemistry Management Series

Recovery Improvement

Volume III

Edited by

Qiwei Wang

Saudi Aramco, Dhahran, Saudi Arabia

ELSEVIER

Gulf Professional Publishing
An imprint of Elsevier

Gulf Professional Publishing is an imprint of Elsevier
50 Hampshire Street, 5th Floor, Cambridge, MA 02139, United States
The Boulevard, Langford Lane, Kidlington, Oxford, OX5 1GB, United Kingdom

ISBN: 978-0-12-823363-4

For Information on all Gulf Professional Publishing publications
visit our website at https://www.elsevier.com/books-and-journals

Publisher: Charlotte Cockle
Acquisitions Editor: Katie Hammon
Editorial Project Manager: Sara Valentino
Production Project Manager: Manju Paramasivam
Cover Designer: Miles Hitchen

Typeset by MPS Limited, Chennai, India

Contents

List of contributors

Hakan Alkan
TU Bergakademie Freiberg, Freiberg, Germany

Baojun Bai
Geosciences and Geological and Petroleum Engineering Department, Missouri University of Science and Technology, Rolla, MO, United States

Birol Dindoruk
Department of Petroleum Engineering and Chemical and Biomolecular Engineering, University of Houston, Houston, TX, United States

Ian D. Gates
Department of Chemical and Petroleum Engineering, University of Calgary, Calgary, AB, Canada

Tara D. Gates
Department of Chemical and Petroleum Engineering, University of Calgary, Calgary, AB, Canada

B. Todd Hoffman
Montana Technological University, Butte, MT, United States

Russell Johns
Department of Energy and Mineral Engineering, Pennsylvania State University, University Park, PA, United States

Ranjani Kannaiyan
Department of Chemical and Petroleum Engineering, University of Calgary, Calgary, AB, Canada

Mahdi Kazempour
ChampionX, Sugar Land, TX, United States

Mojtaba Kiani
ChampionX, Sugar Land, TX, United States

Felix Kögler
University of Strasbourg, Strasbourg, France

Ran Luo
Department of Chemical and Petroleum Engineering, University of Calgary, Calgary, AB, Canada

Qisheng Ma
Power Environmental Energy Research Institute, Covina, CA, United States; ChemEOR Inc., Covina, CA, United States

Hassan Mahani
Department of Chemical and Petroleum Engineering, Sharif University of Technology, Tehran, Iran

Aleksandr Mamonov
The National IOR Centre of Norway, University of Stavanger, Stavanger, Norway

Soujatya Mukherjee
Wintershall Dea AG, Kassel, Germany

Iván Darío Torrijos Piñerez
The National IOR Centre of Norway, University of Stavanger, Stavanger, Norway

Tina Puntervold
The National IOR Centre of Norway, University of Stavanger, Stavanger, Norway

Alireza Roostapour
ChampionX, Sugar Land, TX, United States

Skule Strand
The National IOR Centre of Norway, University of Stavanger, Stavanger, Norway

Yi Su
Department of Chemical and Petroleum Engineering, University of Calgary, Calgary, AB, Canada

Xindi Sun
Department of Physics and Engineering, Slippery Rock University of Pennsylvania, Slippery Rock, PA, United States

Yongchun Tang
Power Environmental Energy Research Institute, Covina, CA, United States

Geoffrey Thyne
ESal, Laramie, WY, United States

Jingyi Wang
Department of Chemical and Petroleum Engineering, University of Calgary, Calgary, AB, Canada

Wei Wang
Aramco Americas: Aramco Research Center-Boston, Cambridge, MA, United States

CHAPTER

Conformance control and water shut-off

1

Mahdi Kazempour, Mojtaba Kiani and Alireza Roostapour
ChampionX, Sugar Land, TX, United States

1.1 Introduction

Excess water production creates lots of operational challenges and shortens the well (field) life affecting the return on investment. Excess water production occurs due to the reservoir characteristics and fluid properties (conformance issue) and improper operation methods. High immature water cuts will reduce the effective well (reservoir) life and will impact the final estimated ultimate recovery. This means that the well or the reservoir might not meet the economic targets. On the other hand, due to the high-water production rates, a high liquid lifting cost is expected, which again will increase the OPEX. Due to the high volumes of water present in the wellbore and producing formation, severe well and formation damage happens, which again costs the well life. Thus finding a method, such as water shut-off (WSO) techniques, to improve the conformance and control the high water cut will help with the economy of the operation and will increase the estimated ultimate recovery (returned on the investment).

The term conformance, in its general form, relates to the level of volumetric sweep efficiency achieved during an oil recovery process by most often a waterflood program. The term conformance has also been used as a measure of, and the treatment of, excessive water production from hydrocarbon-bearing reservoirs. The aforementioned is also known as WSO in which producing wells are targeted for placing the treatment [1–6].

The term volumetric sweep efficiency (E_V) is defined as the fraction of the pore volume within the flooded zone contacted by the injected fluid and it is expressed by:

$$E_V = E_A \cdot E_v \tag{1.1}$$

where E_A is the areal sweep efficiency, an areal fraction of the flooded pattern that is swept by the injected (or displacing) fluid and E_v is the vertical sweep efficiency, a fraction of the pay zone interval that is swept by the displacing fluid. Lower volumetric sweep efficiency caused either by lower areal or by vertical sweep efficiency results in lower oil recovery with such flooding program. Thus

Recovery Improvement. DOI: https://doi.org/10.1016/B978-0-12-823363-4.00001-7

maximizing volumetric sweep efficiency while keeping the cost down is always desirable.

In other words, volumetric sweep efficiency is a measure of how well the injected fluid (often water) would be coming in contact with the accessible pore space in the oil-bearing zone. Gross heterogeneities in the rock matrix lead to low sweep efficiencies. Fractures (natural or induced), high-permeability streaks, and faults (sealing and transmissible) are examples of gross heterogeneities. Homogenous rock formation provides the optimum setting for high sweep efficiencies if the mobility ratio is favorable ($M_r < 1$).

By definition, the mobility ratio is the ratio of the displacing fluid mobility to the displaced fluid (fluids) mobility (mobilities), as indicated in the following equation:

$$M_r = \frac{\text{mobility of displacing fluid}}{\text{mobility of displaced fluid}} \tag{1.2}$$

Mobility is the ratio of effective permeability to viscosity. If M_r is one, then the mobility of displacing and displaced phases becomes equal. Where the M_r is less than one, then the mobility ratio is believed to be favorable in which the areal sweep efficiency is usually high with minimal fingering. When the mobility ratio is significantly high (> 100), the cause is the low mobility of the displaced phase and high mobility of the displacing phase, the areal sweep efficiency becomes very poor ($<45\%$), and severe fingering is expected. A good example of such high and unfavorable mobility ratio would be a high-viscous oil recovery process by CO_2 injection. Here, the injection of gas under immiscible conditions would result in low recoveries since the high mobility gas phase would not be able to effectively push the heavy oil out of the reservoir and it fingers through it very rapidly.

1.2 Root causes for conformance problems

1.2.1 Mobility-driven viscous fingering

As was described earlier, mobility-induced fingering happens when the mobility ratio is above one, and the higher mobility ratio results in more severe viscous-fingering issue. Fig. 1.1 is a simplified schematic illustrating the viscous fingering caused by the high mobility ratio. This type of conformance problem is more pronounced in reservoirs, which contain relatively viscous oils. The impact of viscous fingering can be intensified by reservoir heterogeneities as described by Sydansk and Romero-Zeron [5].

A correlation between areal sweep efficiency and the mobility ratio has been readily determined by Craig [2]. In his study, they used a horizontal laboratory model mimicking a quadrant of a five-spot pattern in which they used immiscible fluids with gravity and capillary effects scaled. The areal sweep efficiencies

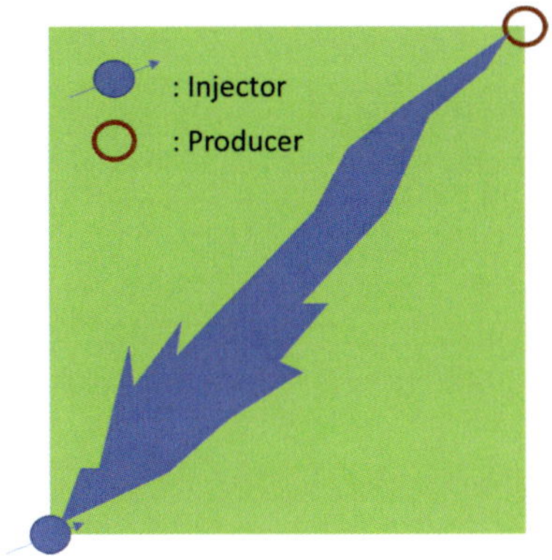

FIGURE 1.1

Viscous fingering under unfavorable mobility ratio (quarter of a five-spot pattern).

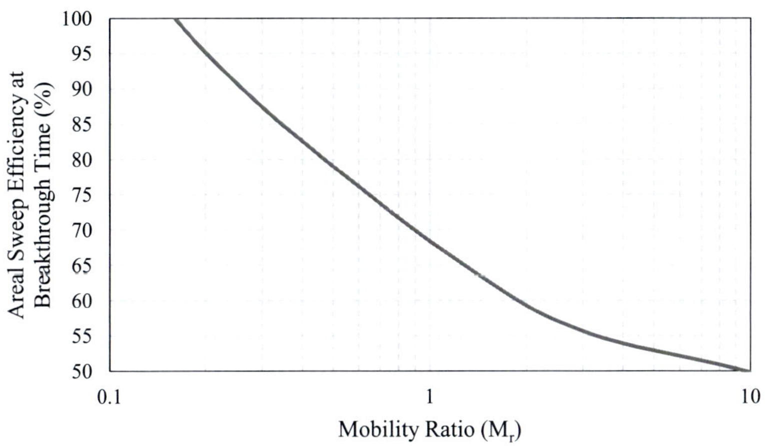

FIGURE 1.2

The areal sweep efficiency (%) as a function of mobility ratio for a five-spot pattern.

throughout the displacement stages were obtained using X-ray shadow graphs taken under different mobility ratio conditions, and the results are illustrated in Fig. 1.2.

1.2.2 Reservoir heterogeneity

As it was described, reservoir heterogeneity includes both areal and vertical heterogeneities. Areal heterogeneity (sometimes known as lateral or directional heterogeneity) is caused by anomalies in permeability distribution in horizontal direction. Vertical heterogeneities (the most common type of heterogenies in oil reservoirs) are caused by the existence of layers, channels, or strata with much higher permeability values compared to the reset of layers within the pay zone. These high-permeability channels (also known as thief zones) establish rapid communications between injectors-producers resulting in poor seep efficiency, premature water breakthrough, and undesirable water cut increase at the offset producers. Fig. 1.3 shows a thief zone in the middle of a formation (layer 4). The injected fluid (here water) follows the path of the least resistance (here layer 4 that has the maximum permeability value among all layers included) and enters this layer preferentially. This leads to a very early water breakthrough and leaves behind a substantial amount of oil in the reservoir unswept.

If the permeability values of all layers are the same ($K_1 = K_2 = \ldots = K_7$), the bottom layers would be the ones, which are swept more by water due to gravity effects, also recognized as gravity under running. The latter is more pronounced if the perforated interval is larger.

1.2.2.1 Vertical heterogeneity: the effects of crossflow

There are some formations in which there is no pressure communication and fluid crossflow among various reservoir strata. In these formations, there is usually at least one continuous impermeable layer (shale barrier) that resides between the geological layers of varying permeabilities (Fig. 1.4A). In this situation, and due to the lack of vertical communication, addressing the conformance issue is relatively simpler than the other scenarios, and the treatment volumes are smaller. If crossflow exists between the layers (Fig. 1.4B), commonly, the vertical permeability value is a fraction of horizontal permeability (10–20%). Depending on

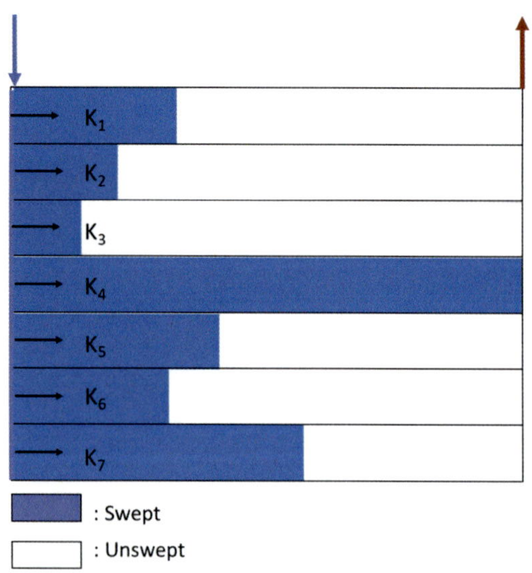

FIGURE 1.3

Heterogeneity in vertical permeabilities and its impacts on displacement efficiency.

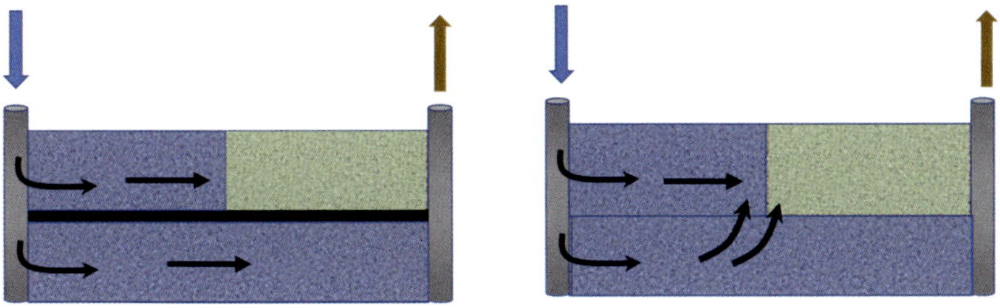

FIGURE 1.4

(A) Channeling without crossflow. (B) Channeling with crossflow.

well spacing, pay zone thickness, and vertical to horizontal permeability ratio (K_v/K_h), the volume of the conformance treatment may need to be increased, or in-depth conformance technologies must be used. If the injectivity is not a concern and the contrast in permeability values of the layers is not significant, injecting a high viscosity solution (such as polymer flooding) could also be helpful to improve vertical sweep efficiencies.

1.2.2.2 The impacts of fractures and vuggy features within the reservoir body

Conventional carbonate reservoirs are known to be naturally fractured. In some cases, those conductive features become large enough to form vuggy networks that could contain extremely permeable flow conduits. Based on fracture intensities, their orientations, continuities, and dimensions, they can cause both areal and vertical conformance issues (Fig. 1.5). Due to high contrast in permeability values between matrix and fracture network and considering the fact that the water saturation is usually higher in the fracture networks, these reservoirs are good candidates for gel-based conformance technologies. In vuggy reservoirs in which the transit time between injectors and offset producers could be very short, stronger gel-based technologies with accelerant may be considered [5,6].

1.2.3 Water conning

Sometimes the source of excessive water at producers is not from the injection side, but rather from either an underlying bottom or an edge aquifer. Presence of an aquifer is helpful to maintain reservoir pressure as production is ongoing, but it has its own inherent problems as well. The oil—water contact is rising over time, and at some point, water starts flowing to the wellbore through its own path (Fig. 1.6A). If this water encroachment is in radial fashion and taking place gradually, it would be hard to using conformance techniques and the treatment may not last long. If the water invasion is taking place through vertically oriented

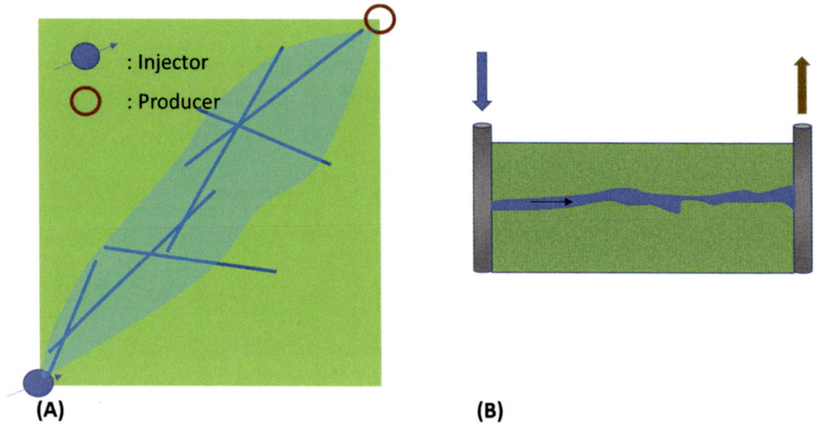

FIGURE 1.5

(A) Fracture channeling causing areal conformance issues. (B) Vugular channeling establishing strong communication between the injector and its offset producer.

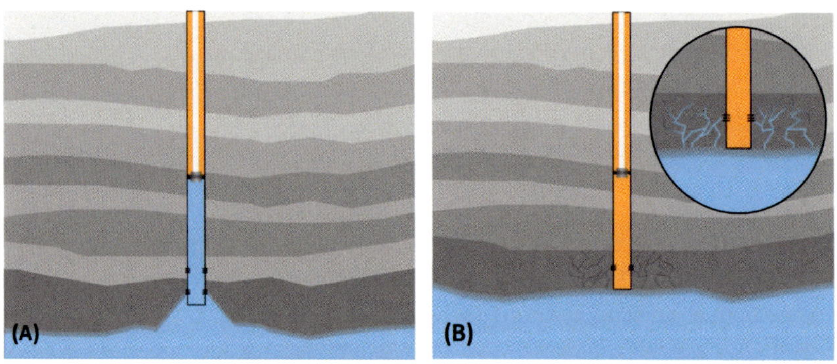

FIGURE 1.6

(A) Radial water conning. (B) Water encroachment from bottom aquifer to the wellbore through vertically oriented fractures.

fractures or it is due to some vertical heterogeneities, then conformance techniques would be well-suited to address the issue (Fig. 1.6B).

1.2.4 Wellbore integrity

Behind the pipe channeling and casing leaks may also be the causes of excessive water production. In both cases, problematic water from a water-bearing formation (above or below the oil-bearing zone) finds its way toward the wellbore (Fig. 1.7). Poor cement jobs and casing corrosions are the main reasons for such problems. The most common ways to fix these problems are through mechanical

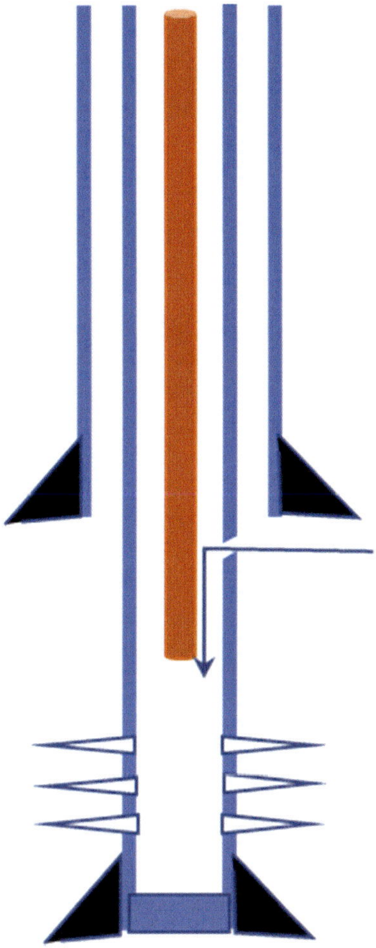

FIGURE 1.7

Casing leak example.

solutions including straddle packers, cement squeeze jobs, and gel/resin-based chemical treatments are considered only when mechanical isolation is extremely difficult or expensive to perform. In these cases, typically a very high strength gel or resin is injected in isolation in a limited volume ($<$15 bbls).

1.3 Bulk gel technology

This section provides more detailed information on existing technologies addressing conformance issues.

Bulk gel is one of the common solutions to conformance and excess water production issues with more than 35 years history. This technology has been applied in sandstone and carbonate reservoirs for horizontal, vertical, and deviated wells to address conformance issues or rejuvenate old producers. Multiple types of bulk gel system exist depending on the type of polymer and crosslinker used; however, the most popular system, called MARCIT Gel, was first introduced by Marathon Oil Company in mid-1980s [7]. This technology was developed for low to medium range reservoir temperature and has been applied in a variety of applications all over the world with reasonable levels of success.

High concentration polymer with no crosslinker added has been tested as an alternative to bulk gel on occasions by different operators mainly due to simplicity in execution, compared to bulk gel that needs specialized mixing and pumping equipment. The idea is to take advantage of polymer retention/entrapment in porous media to provide resistance to water flowing in high conductivity features. However, limited success has been observed as high concentration polymer does not get set, and it is still flowing under differential pressure under reservoir conditions. Although polymer does increase the resistance to flow by increasing the viscosity of the displacing brine, the impact is minimal unless the permeability of the feature being addressed is very low.

A series of studies and field trials were done in Argentina, where they have implemented polymer injection in a fluvial reservoir system to improve oil recovery [8–10]. In these studies, numerical modeling was applied to decouple the advantages of using polymer to improve the adverse mobility ratio and reservoir heterogeneity. It was shown that the negative impact of reservoir heterogeneity on recovery factor is less in case of polymer injection compare to straight water injection.

1.3.1 Mechanism of action

Bulk gel is a fluid-based system with a 3D-shaped structure, made by mixing a polymer with a crosslinker. Originally, bulk gel was introduced by Sydansk [7] using partially hydrolyzed polyacrylamide (HPAM) polymer and chromium III as the crosslinking agent. However, other types of polymers, such as ATBS, and crosslinking agents, such as aluminum and zirconium, have been used historically to create bulk gel as well depending on circumstances [11]. For instance, in high temperature environment organic crosslinkers are preferred as the tend to have a slower reaction with crosslinker which will result in a more stable polymer gel. Another example is preference on using a high molecular weight polymer when the matrix permeability is high and we would like to minimize the leak off to the matrix while treating a highly conductive feature.

The 3D structure of bulk gel is the reason why once formed, it can block flow in a high permeability feature and withstand high differential pressure.

The major factors impacting the strength and stability of the gels are salinity and hardness of the water used to make up the gel, reservoir temperature, presence of H_2S, and pH and bicarbonate content. For instance, high salinity and hardness may accelerate the reaction between HPAM polymer and crosslinker which will eventually impact the stability of the gel. As a result, the original HPAM-based gel that was introduced in 1980s has been modified through the years by changing the polymer and crosslinking system to address these issues.

Being a water-based fluid, bulk gel tends to travel with the water phase and follow the path of least resistance, where formation has a higher permeability and higher water saturation. In fact, this is the main reason bulk gel has the advantage of being implemented with limited need for intervention. Crosslinking reaction starts as soon as polymer and crosslinker are mixed on the surface and the mixture gains strength and viscosity over time, as it is flowing downhole and into the formation. The high resistance factor after bulk gel reaches its full strength will help in blocking the preferential flow path to water and divert the injection fluid to other areas, resulting in sweep efficiency improvement (Fig. 1.8).

1.3.2 Applications

Bulk gel is used when signs of communication exist between an injector and their offset producers. On the injector side, a sign of communication could be an increase in injectivity of the well or the well going on vacuum if communication is severe. On producer side, an increasing trend of water production and decline in oil production rates are signs of communication.

Bulk gel can also be used to address excess water production in a producer if the source of water is an aquifer or a water zone that has been connected to the well through features, such as faults, fractures, or any high permeability feature.

Candidate selection plays a very important role in the success of a conformance improvement or WSO application. The following are some of the criteria to be considered:

- Permeability of the target formation: $k > 300$ mD.
- Formation temperature: $T < 300°$F.

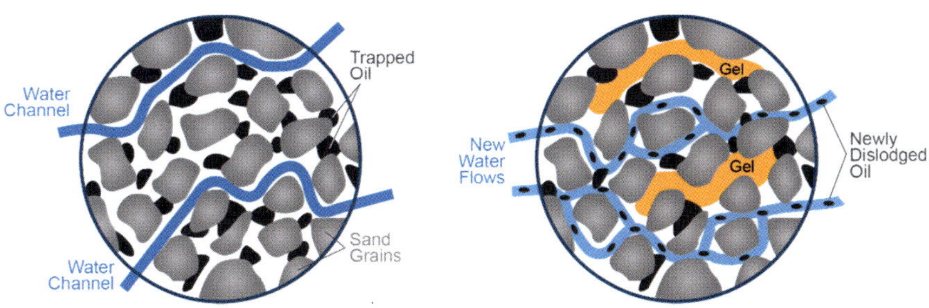

FIGURE 1.8

Before (left) and after gel injection and gel set up (right).

- Injection water temperature: 60–70°F is a perfect temperature for the source water. High source water temperature may result in premature gel formation.
- Pressure margin: when considering a treatment, it is prudent to make sure there is enough pressure margin to place the desired volume of gel without hitting fracturing pressure of the formation/well.
- Source water composition and pH: compatibility testing needs to be done to make sure strong and stable gel can be formed using the source water.

1.3.3 Case studies

1.3.3.1 Conformance improvement on the North Slope of Alaska

The HPAM-based bulk gel with chromium III as the crosslinking system was applied to a dual-lateral horizontal injector that was communicating with three offset producers. The well was injecting into Kuparuk C sand unit, which is a layered reservoir with high heterogeneity and high fracture density and major faults. The area was under miscible water alternating gas (WAG) process, and tracer testing conformed communication between injector and the offset producers (Fig. 1.9) which resulted in water production increasing while the oil rate dropped significantly.

The injector was treated with roughly 20,000 bbls of bulk gel volume starting at 3000 ppm polymer concentration and stepwise increase in concentration through the job to maximum of 10,000 ppm at 40:1 polymer to crosslinker ratio. A high molecular weight polymer was used for this treatment to minimize the polymer leak off to the matrix.

Posttreatment analysis shows that the treatment had an immediate impact on the injectivity of the well indicating that communication paths have been addressed to a degree. As a result of the treatment, 460,000 bbls of incremental oil was recovered in 2 years [12].

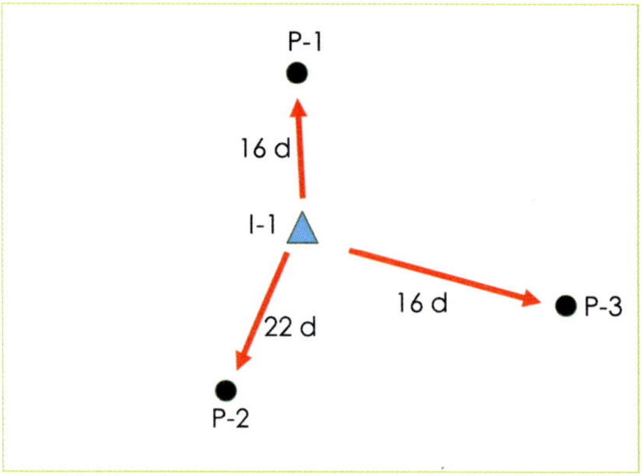

FIGURE 1.9

Schematic representation showing the injector, the offset producers and transit time to offset producers.

1.3.3.2 Conformance improvement in Canadian Bakken

Canadian Bakken, a tight oil reservoir with average permeability of less than 0.1 mD, is under waterflood to tackle the sharp decline in oil production and improve recovery. Due to tight well spacing and presence of induced fractures, signs of communication were observed between some of the injectors and offset producers. Interestingly, in most of these cases, the injector went on vacuum as soon as communication was established with an offset producer.

A series of HPAM-based bulk gel treatments were executed to address the communication. At the reservoir temperature of 65°C, a medium molecular weight polymer and chromium III was the bulk gel system applied in these treatments. However, due to the tight nature of the reservoir and small completion size, a different approach, in terms of treatment design and execution, was taken.

Posttreatment analysis indicated that the injector showed an immediate reduction in injectivity, and the wells on vacuum started showing pressure, indicating that the communication paths have been addressed. On the other hand, offset producers responded differently in different campaigns as execution approach was changed through the years to avoid significant losses in injectivity. Fig. 1.10 shows the immediate change in oil decline rate after the treatment (a group of four offset producers) [13].

1.3.3.3 Water shut-off in Manitoba

The Lodgepole Formation, located in Manitoba, Canada, is a conventional carbonate reservoir with a bottom water drive. The Lodge Pole Formation, originally discovered in the 1950s, has been the focus of horizontal programs in recent years with success. However, once penetrated by horizontal wells, the significant

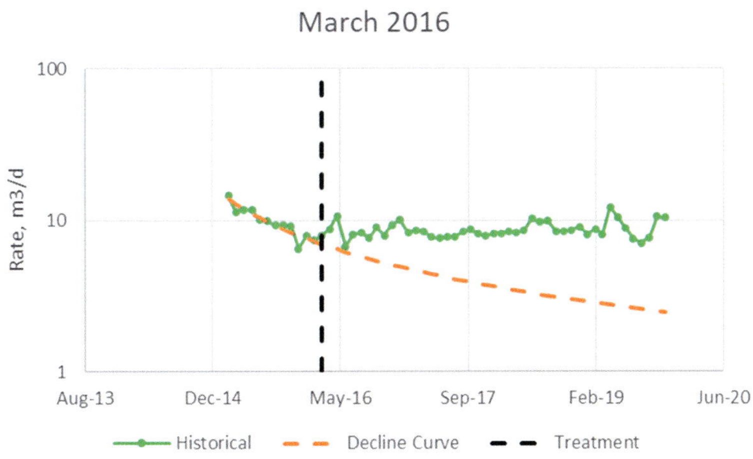

FIGURE 1.10

Conformance treatments resulting in change in decline curve.

drawdown from the well created uneconomic issues of preferential influx of water production over the desirable oil production from the Lodgepole member.

In this case, the lower leg of a dual-lateral horizontal producer is believed to be impacted from the bottom aquifer at the interval highlighted in Fig. 1.11 (965–990 m KB), which resulted in a dramatic increase in water and a significant reduction in oil production in this well. By isolating the lower leg, the operator would be able to reduce the water influx, reduce operating costs, and significantly improve oil production. With a well workover defined utilizing a mechanical

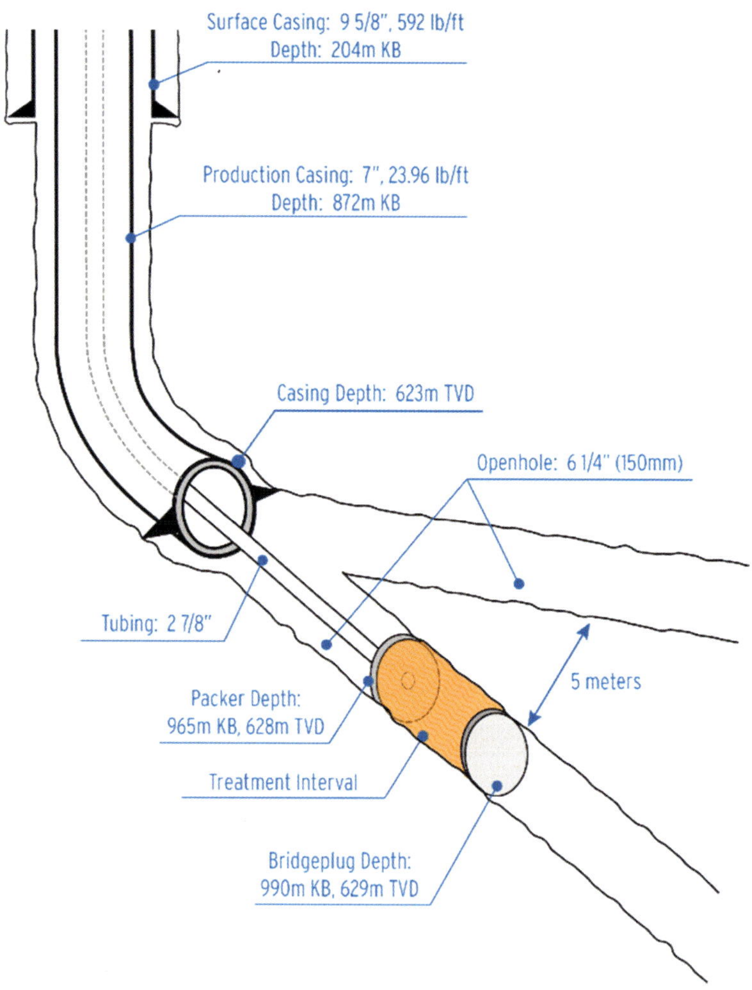

FIGURE 1.11

Schematic representation showing the dual-lateral well configuration and treated interval in the lower lateral.

bridge plug placed in open hole of the lower leg, the customer re-evaluated this workover to consider a solution that could enhance the partial isolation of the open hole.

A low molecular weight polymer gel technology was recommended to shut-off flow from the interval with high water influx. Use of low molecular weight crosslinked polymer was specifically suggested to be pumped and squeezed below a retrievable packer and above the mechanical plug. Total volume of 3.2 m^3 TIORCO Gel was injected at 50,000 ppm polymer concentration and a crosslinker ratio of 40:1 (polymer to crosslinker ratio). The well was shut in for a week to let the gel reach its final strength. After bringing the well back on production, water production was reduced by 360 bpd, and oil production saw an increase of 120 bpd in 2 months. Four subsequent treatments were implemented on well with similar conditions and positive results were observed.

1.4 Relative permeability modifiers

The idea of a technology that can impact the relative permeability of a rock to water without affecting the flow of oil has always been attractive and the topic of discussion among scientists for many years. Relative permeability modifiers (RPMs) were first introduced in 1964 [14], discussing the ability of acrylamide polymers to reduce the excess water production in producing wells.

RPMs are water-soluble polymers or aqueous gels that are believed to have the ability to impart fluid flow in the matrix by adsorption to the surface of the rock. Three generations of RPMs have been developed through the years. Modifications to the backbone and branches of polymers are the main differences between different generations of this technology. Below are the specifics of each generation regarding to backbone and branches charges of polymer structure:

- 1st generation: hydrophilic backbone and hydrophilic branches
- 2nd generation: hydrophilic backbone and hydrophobic branches
- 3rd generation: hydrophobic backbone and hydrophilic branches.

1.4.1 Mechanism of action

RPMs function based on their adsorption to the surface of the rock and create a layer on that surface that prevents the water from passing through the pore throat. The level of adsorption, and hence the impact of RPMs on fluid flow, is a function of many different parameters, such as polymer type and structure, rock minerology (clay content, quartz, or calcite rich), wettability status, and ionic strength and composition of the brine.

The 3rd generation of RPMs are designed such that their hydrophilic branches hold the water molecules close to the wall of the matrix allowing oil to flow in the middle of pore throat. Two major phenomena happening at the same time are the RPM adsorption to the surface of the rock will reduce the flow in the pore network and, at the same time, it stays selective to allowing the oil phase while reducing the flow of water phase to the charges. This is the reason RPMs claim not to reduce the relative permeability to the oil phase.

In fact, the performance of this technology depends on ionic bonding between the rock/polymer and polymer/water/oil molecules. Hence, change in minerology can impact the adsorption while change in brine composition can impact the selective performance of RPMs amongst other parameters.

1.4.2 Applications

The ideal conditions for the RPMS technology to perform best are [15]:

- Vertical wells with radial flow through matrix where there are multiple zones including one zone responsible for excess water production and no crossflow between zones
- Fractured vertical fractured wells with a hydraulic fracture intersecting a fracture
- Vertical wells in reservoirs with limited/single natural fracture network
- Horizontal well with fractures connected to a water zone/aquifer.

Finding candidates that fit these criterial perfectly is very difficult, and even if a candidate with the above criteria is found, there are still limitations to this technology, such as:

- Due to adsorption, the change in residual resistance factor will be a function of permeability of the matrix, meaning the tighter the rock, the higher the impact to flow [16].
- RPMs have shown to always reduce the relative permeability to oil [17,18].
- Polymer only RPMs are applicable to formations with certain range of permeabilities [19,20]
- On occasions when RPMs are placed where the differential pressure is high, there is a possibility of them being washed out, especially for polymer only RPMs [18]

1.4.3 Deployment history

There have been extensive applications of this technology all over the world, and the results have been erratic. Historically, numerous ineffective and underperforming applications of RPMs have been reported [17,19,21]. The reason behind so many unimpressive applications of this technology is that there are a lot of factors that must be met for the technology to reduce water production without affecting the oil and be economic at the same time. Some of these factors are reservoir heterogeneity, presence of crossflow between layers, production constraints, limitations of the technology, origin of the excess water. Below is a case

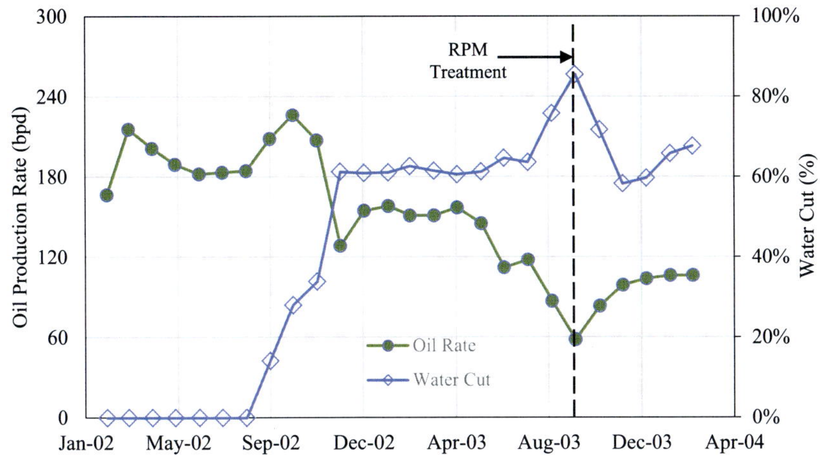

FIGURE 1.12

Positive results showing gradual reduction in water production.

Reproduced with modifications from G.A. Farrera Romo, H. Hernández Leyva, R.B. Aguilar, et al., Advanced technology to reduce water cut: case studies from the Pemex southern region, SPE Production & Operations, 25 (2010) 139–144.

study of a successful application of specific generation of RPMs called hydrophobically modified water-soluble polymer (HMP).

A HMP was used to address the excess water production from the vertical well drilled and completed in Cinco President field in southern Mexico. The Encanto formation is sandstone with an average permeability of 50 mD, porosity of 22% and initial water saturation of 17%. The well started producing 166 bpd of oil with no water production in February 2002. By August 2002, the water cut of this well increased to 65%, and the oil rate dropped down to 57 bpd [22].

Further investigation showed coning as the root cause of increase in water production in this well, and the well was treated with 250 bbls (10 ft penetration into the formation) of HMP at 0.2 wt% to address the excess water production. Posttreatment results showed a gradual decrease in the water cut with a 30% decrease in the first 2 months and a drop of 68% in 6 months, while the oil production rate went from 57 to 104 bpd (Fig. 1.12).

1.5 Sodium silicate gels

The application of inorganic-based gels has been documented since 1922. However, the technology was not appreciated for a long time due to lack of understanding of its behavior in the reservoir.

Recently, the use of inorganic material, such as fly ash, silica, magnetic materials, and kaolin, to generate composite hydrogels has gained a lot of attention and lots of experimental work has been conducted to understand the technology better [23,24].

The main reason behind the popularity of inorganics in gel systems is the discovery that thermal stability and strength of gels can be improved by adding inorganic material to the mix [25–27]. Researchers discovered that gelation time can also be modified using these components [27]. It is also considered green and low harm which is a huge advantage compared to very popular technologies, such as polymer gel. However, the technology has some disadvantages, such as sensitivity to water salinity and being susceptible to fracture and losing strength under pressure. It is also prone to syneresis which needs to be taken into consideration when designing the application and selecting the technology. A lot of research has been dedicated to investigating the conditions (concentrations, volumes, activators type, etc.) under which various technologies can perform best [28,29].

1.5.1 Mechanism of action

Sodium silicate gels can be generated using an internal activator, such as urea, that could start decomposing at certain conditions and start the gelation of silica. External activators, such as calcium and magnesium chloride, could also be used that precipitate silicate once mixed with sodium silicate solution. The two components are mixed on the surface in liquid form creating a water-like solution. This gives the technology enough time to get to the desired location in the reservoir before activation.

The silicate gelation process and its strength are function of several parameters including (but not limited to) pH value, temperature, initial concentration of monomers, reactants type and their concentrations as well as water composition. The gelation process occurs through spontaneous hydrolysis and polycondensation reactions. The condensation process is triggered by the formation of small oligomers. These oligomers act as nucleation sites to form stable particles that can aggregate to form a 3D gel-like network.

1.5.2 Applications

Different activators can be used to adjust the technology for certain application. For example, sodium silicate with urea as activator can be applied in high temperature applications. Urea starts to decompose at high temperatures and initiate the gelation process. This mechanism provides the advantage over other technologies, such as polymer gel, which tend to form very fast at high temperature resulting in stability issues.

According to Lakatos and Lakatos-Szabo [30], 80 jobs were completed using silicate gels until 2012. 60%–65% of these jobs were successful with 40% of them being economic.

Pham and Hatzignatiou [31] describe two major applications of silicate gels in the Snorre field. Both were deemed successful with one addressing an in-depth conformance issue 40 m away from the wellbore, capable of reducing the permeability 100 times.

In addition to silicate-based gels, other inorganic conformance techniques had been proposed and tested in the field, such as iron-hydroxide-based gel. Lakatos et al. [32] realized that some inorganic compounds, specially Fe (III) have the capabilities to transform into a gel-like precipitate by in-situ hydrolysis. At the conclusion of hydrolysis, they can be immobilized further by adding a flocculant agent or through spontaneous aging. They ran more than seventeen trails in which seven injection wells and ten producing wells were treated with the success rate more than 60%.

1.6 Colloidal dispersion gels

Colloidal dispersion gels (CDGs) have been used for EOR applications since the late 1980s. These CDG systems are based on specific HPAM polymers mixed with aluminum or chromium salts to create cross-links between polymer chains. Despite a significant body of literature supporting the successful application of CDG technology in different fields [33–35], there are reports concerning key aspects of the CDG technology, namely the injectability of CDG and its propagation through porous media, which were solely based on lab-scale testing and observations [36,37]. CDGs are made of low concentrations of polymer and cross-linker. Polymer concentrations in CDG's normally range from 100 to 1200 mg/l. In this range of concentration, there is not enough polymer to form a continuous and firm 3D network, so a conventional bulk gel cannot form. Instead, a solution of separate gel bundles forms, in which a mixture of predominantly intramolecular and minimal intermolecular cross-links connect relatively small numbers of molecules (Fig. 1.13).

During injection in field projects, CDG formation appears to take longer than laboratory tests indicate. Large volumes of gellant are injected over time frames of weeks to months [38–42] with slow pressure buildup at the injection well, suggesting slow, gradual development of CDG. Research has shown that CDG's tend to show significantly delayed formation when in a high shear regime [8], and it is theorized that this phenomena accounts for more delayed CDG formation in the lab than the field. Once the CDG develops in the formation, it is more resistant to flow than uncrosslinked polymer. It is also adsorbed more strongly in the formation and provides several times the residual resistance factors of uncrosslinked polymer. The enhanced properties of CDG's over uncrosslinked polymer are particularly useful in heterogeneous formations, where nonuniform rock results in an uneven flood front and incomplete volumetric sweep. In this type of rock, uncrosslinked polymer is not strong enough to overcome these adverse effects.

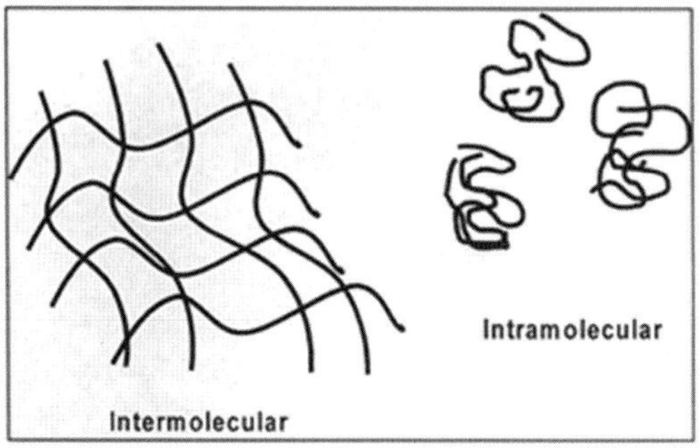

FIGURE 1.13

Gel crosslinking mechanisms—CDGs have predominantly intramolecular cross-links due to low polymer concentration.

1.6.1 Candidate selection for colloidal dispersion gels

The best candidates for CDGs are those reservoirs with high heterogeneity, where the heterogeneity is due mainly to highly permeable matrix rock. CDGs work best where tortuous flow paths exist in the high perm channels, so all but fine fractures are generally excluded. The gels have been used in a few reservoirs with fracturing, but generally if a strong fracture system exists, stronger conventional gels should be considered. CDGs have been used in reservoirs with Dykstra Parson coefficient of 0.5 and higher with good results [34,42–46]. Mean permeabilities have ranged from 10 to more than 300 mD. The CDG's strength can be adjusted based on reservoir conditions. For example, if an unfavorable mobility ratio condition exists in addition to the heterogeneity, then stronger CDG would be placed first, and later, it would be followed by a weaker CDG or even uncrosslinked polymer for mobility control.

An important consideration in candidate selection is that there be a large target of recoverable oil still in place. From this standpoint, the most successful CDG projects are those that are started early in the life of the waterflood.

1.6.2 Expected results

Table 1.1 provides a means of estimating performance of a field CDG project for budgeting purposes. The ranges given are based on experience from over 40 CDG field projects. The range should be adjusted as specified to account for the point in the waterflood where the CDGs are applied. The time frames depend on the well spacing in addition to the point in the waterflood life, and the time frames indicated assume 40-acre well spacing. If larger well spacing is used, the time frame for injection and response may be longer, while for tighter spacing, the time frame may be shorter. As with most oil field processes, there is a risk associated with applying CDGs. The success rate of past projects shown in Table 1.1 is

Table 1.1 Summary of expected results from colloidal dispersion gels.

Variable	Range	Early in WF	Later in WF
PV injected	2%–12%	TBD	TBD
Time to inject	6 months to 2 years	TBD	TBD
Response time	6 months to 1.5 years	Sooner	Later
Incremental oil recovered, % OOIP	1%–25%	Higher	Lower
Chemical cost, $/incremental bbl oil recovered	50¢ to $4.00 ($1.09 average)	Less	More
Success rate	78%	Higher	Lower

Notes: This information is useful for budgeting purposes and is based on experience from more than 40 field CDG projects. TBD, *to be determined;* WF, *waterflooding.*

one way to risk a project. Another method that is more rigorous involves the use of Monte Carlo techniques to obtain cumulative frequency distributions for potential processes, given the error associated with variables that affect performance. Norman et al. [43] reported an economic analysis of waterflooding, polymer flooding, and CDG flooding in the same reservoir, given a range of injection efficiencies and permeability variations. The results show that oil recovery and net present value for the CDG flood are significantly improved compared to the polymer flood or the waterflood, over the entire range of probabilities. The worst case for the CDG flood, which represents the highest K_v and lowest injection efficiency, has a higher net present value than the best case for the waterflood. The results imply that CDG's decrease the risk of running a straight waterflood in heterogeneous rock significantly.

1.6.3 Case studies

Two case studies are described here. The first one used CDG at early stage of a waterflood to improve oil recovery, while the second one used CDG in a mature waterflooded field.

1.6.3.1 North Rainbow Ranch

This is an example of CDGs applied later in the waterflood life [40]. In this project, CDGs were applied 2 years after the waterflood was started. Water had already broken through, and the water-oil-ratio was almost 3 when CDGs were applied. About 6 months after CDGs were started, the water–oil ratio (WOR) decreased significantly, and the oil decline curve increased slightly, indicating incremental oil from CDGs. After 300,000 bbls of incremental oil were produced at decreasing water-oil-ratios, a casing collapse in a producing well resulted in a pattern change. Beyond this point, incremental oil could not be defined accurately. Another interesting aspect of North Rainbow Ranch is that later in the life of the flood, after CDGs were placed, bulk gels were used to correct more focused channeling.

1.6.3.2 Dina field

This is one of the most recent and the largest CDG field trial implementation which was carried out in Dina field in Colombia [45]. In this case, CDG injection started after 15 years of water injection in very heterogenous sandstone reservoir with Dykstra Parson coefficient of 0.7 and the mobility ratio >4. The permeability range in this field was from 50 to 200 mD with the average porosity of 16%. From mid-2011 to mid-2015, about 3.5 MM bbls of CDG were injected over four injectors with two injectors taking almost one million bbls of CDG individually. In this case, the average polymer concentration was kept at 400 ppm with polymer to crosslinker ratio of 40:1. During and after CDG injection, neither polymer nor crosslinker was detected at any offset producers. Some of the offset producers showed very early (~3 months) and positive response to CDG injection. The oil rate increased (doubled), the water cut dropped, and the response was sustained for more than 5 years.

1.7 Thermally active polymers

Mechanical isolation, polymer bulk gels, cementing, selective completions are amongst the methods that have been used frequently to overcome the water channeling and manage the water production in both onshore and offshore environments. As discussed earlier, the use of polymer-based technologies in the form of bulk gels has been practiced as a common method for near wellbore treatments. However, it has been desired to develop technologies that could address the heterogeneity deep into the formation with no commonly observed skin damage and loss of injectivity in the wellbore. In 1996 an industry consortium (BP, NALCO, ChevronTexaco) conducted the BRIGHTWATER development efforts to develop an in-depth reservoir triggered product that could improve sweep efficiency of waterflood. This product should be time delayed and temperature sensitive to implement properly.

BRIGHTWATER™ is a thermally active polymer (TAP), water-soluble submicron amphoteric polymeric particle system for modification of injection profile used for in-depth waterflood sweep efficiency improvement [47]. The particles are based on a highly crosslinked, sulfonate-containing Polyacrylamide microparticles that are crosslinked with two crosslinkers: a stable organic crosslinker that controls the size of the particle and prevent it from decomposition, and a temporary (labile) crosslinker that degrades as a function of time and temperature, allowing the particle to be exposed to the carrying water and hydrolyzed to expand in size [48,49]. The particle's size before activation varies in the range of 0.05–1 μm in diameter with the average size of 0.5 μm [50] and after activation

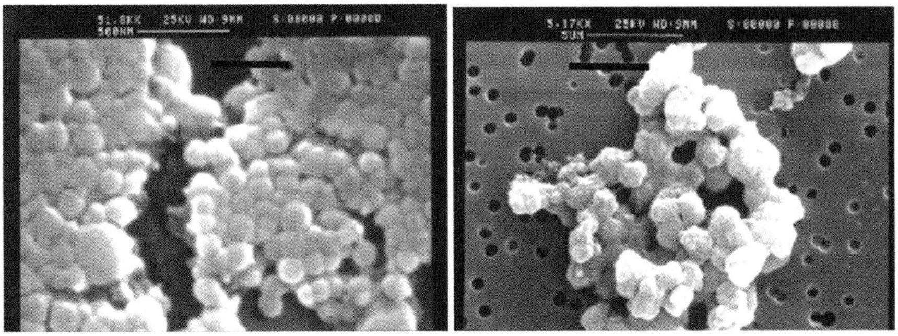

FIGURE 1.14

Particle size before and after activation. Left: BRIGHTWATER™ at submicron form before activation (500-nm scale). Right: BRIGHTWATER™ after activation at target reservoir temperature (5000-nm scale).

up to 10 μm. Fig. 1.14 shows the size of the particle before and after activation on 500 and 5000 nm, respectively.

1.7.1 Working mechanism

Slimtube tests are performed to evaluate the product performance in porous media. Resistance factor (RF) and residual resistance factors (RRF) are the two main parameters to be estimated from slimtube tests. RF reflects the water viscosity increase and retention due to polymer presence in the porous media and is calculated from pressure drop measurements during treatment injection compared to base brine flood stage [51]. A low RF value indicates that the product can be easily injected in the field without pressure increase concerns. RRF reflects the degree of resistance that BRIGHTWATER™ has created against the water phase after activation. Fig. 1.15 is an example of slim-tube results and RF and RRF estimations. In general, the lower the permeability the stronger resistance that can be obtained at a certain concentration (Fig. 1.16).

The expanded particles will block the pore throats to water phase in the thief zones and prevent further water flow through the high permeability rocks. As a result, the rest of the injected water will be diverted into unswept zones increasing oil recovery. Fig. 1.17 shows the concept of in-depth conformance in a field-scale.

1.7.2 Applications

This technology can be used to improve waterflood sweep efficiency far from the wellbore, in-depth conformance modification, under following conditions:

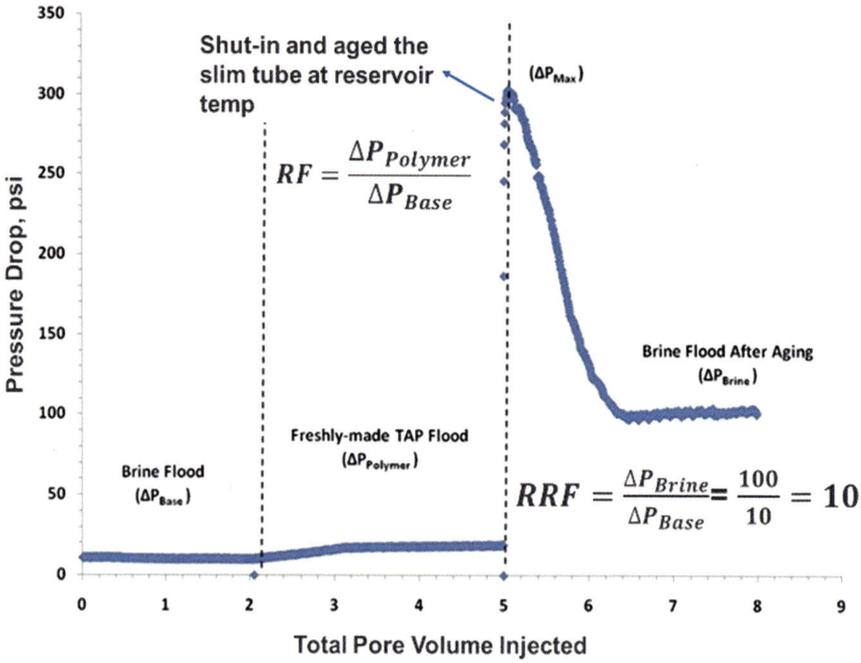

FIGURE 1.15

Example of slim-tube test and concept of BRIGHTWATER resistance and residual resistance factors.

- Reservoirs or well patterns with the unswept zones and existing movable remaining oil
- Reservoirs or well patterns suffering from early water breakthrough resulting in high immature water cut
- Reservoirs or well patterns with high degree of heterogeneity and permeability contrast
- BRIGHTWATER is an easy to operate technology and can be used in both onshore and offshore and remote locations.
- Unlike conventional conformance treatments, there is no need to shut-in the injector after treatment
- It can be used at various reservoir conditions, low to high water TDS and temperatures.

Some new applications are:

- Even though BRIGHTWATER was developed to address the conformance issues far from the injection point, it has recently been used to change the injection profile in tight formations (with avg. permeability <30 mD). This tight formation has been under waterflood for several decades in which

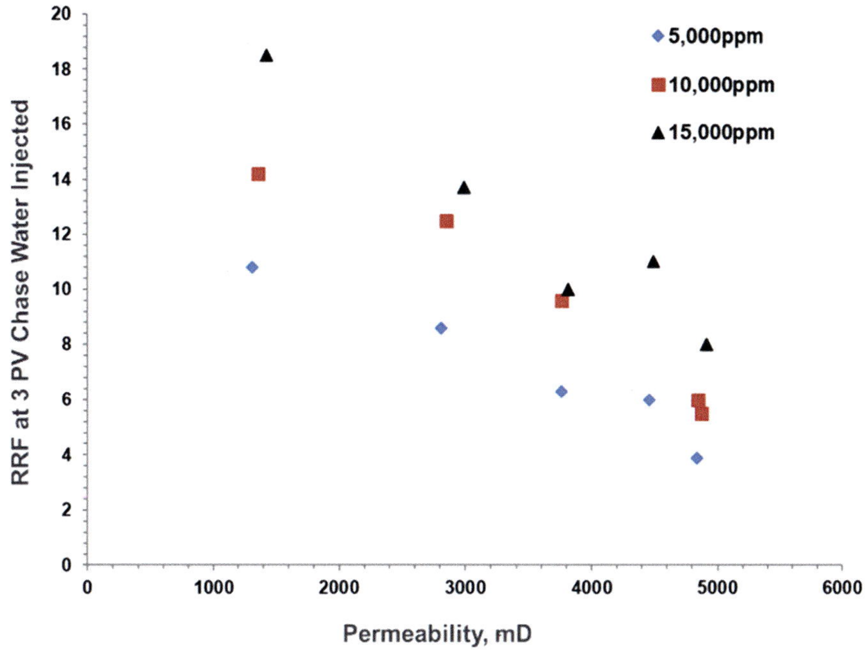

FIGURE 1.16

Thermally active polymer water mobility reduction versus permeability at various thermally active polymer concentrations.

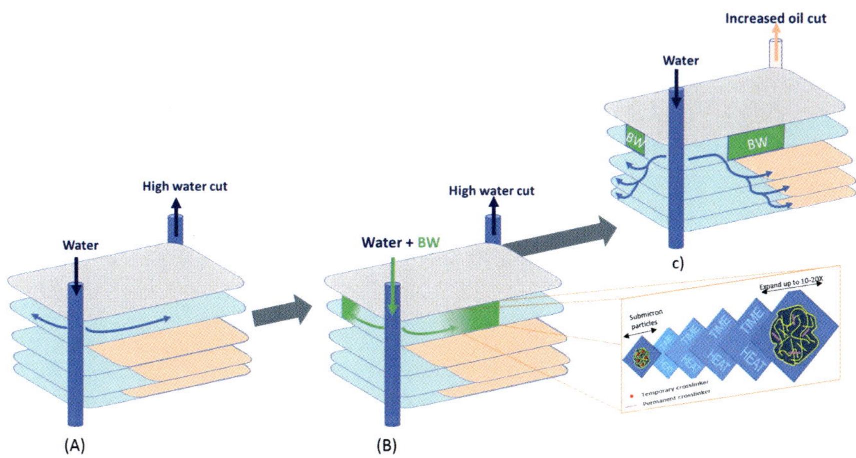

FIGURE 1.17

(A) Before BRIGHTWATER treatment. (B) BRIGHTWATER injection and its activation. (C) After BRIGHTWATER activation and in-depth flow divergence.

watered out layers have been established through time yielding with uneven injection profile. Thus based on the new experience, it could be used for injection profile modification at the wellbore where the common polymer bulk gel systems cannot be placed due to the concerns over possible face plugging.
- Some research projects are focusing on the use of BRIGHTWATER for WSO purposes on the production side by mixing it with compatible cross-linkers.

In the aforementioned cases, there is a need to shut-in the wellbore (injector or producer) for a while to let the product becomes activated near wellbore.

1.7.3 Basic screening criteria

Pritchett et al. [52] presents some screening criteria for evaluation of potential candidates for this technology; however, the list is not current due to new improvements in the technology and its design. Roussennac and Toschi [50] describe feasibility, screening, execution, simulation, and lab studies that led to performing a TAP trial in Salema Field, Brazil. Garmeh et al. [53] and Manrique et al. [54] provided a more detailed information on the screening criteria and pattern selection for this technology in Prudhoe Bay field, Alaska. Basic screening of BRIGHTWATER candidate is necessary to assess applicability of this technology. These criteria include followings:

- Presence of unswept zones and moveable oil.
- Early water breakthrough, resulting in high water cuts (high WOR) indicating water channeling (recovery factor against the reservoir/well- pattern life). And of course, the source of produced water should be from the injection side, which needs to be confirmed by tracer, injection and production profiles, pressure transient test interpretations.
- Moderate to high permeability contrast.
- Injection water pH ≥ 6.
- Reservoir temperatures below 150°C (300°F).
- Sandstone formations are preferred; however, it could be implemented in carbonates if the conditions met.
- Injection-water salinity less than 200,000 ppm (no restrictions on formation water salinity).
- Minimal natural fractures with low transmissibility.

1.7.4 Development history and field statistics

Approximately 130 BRIGHTWATER treatment projects have been carried out around the world. The first BRIGHTWATER treatment was injected in the Minas Field, Indonesia, in 2001. Since 2004–2015 a total of 85 treatments have been conducted only in North Slope, Alaska [55–58]. The cited publications deliver useful information about field types, injection water quality, injection and reservoir temperature, thief zone and channels permeability, heterogeneity variations, and the success rate of these projects. In addition, several publications describe

Table 1.2 Basic information of Cerro Dragon pilot area field study.

Lithology	Sandstone
Reservoir temperature, °C	85–100
Injection temp., °C	55
No. of injectors	7
No. of producers	15
Avg. permeability, mD	100
Thief zone permeability, mD	500–1000
TAP concentration, ppm	5000
TAP total mass, tones	360

TAP, *Thermally active polymer.*

the projects in Argentina [59–61], Tunisia and Brazil [50]. The first offshore project was in a UK North Sea field during the summer of 2002. This treatment proved that this technology could be injected offshore without any injectivity loss into 400-mD sandstone formation [57,58].

1.7.5 Field case study

Cerro Dragón is a giant field located in San Jorge Gulf basin in Argentina with multiple reservoir layers. Some basic information is listed in Table 1.2. The waterflood oil recovery had deteriorated due to the presence of highly conductive channels and unfavorable mobility ratios resulting in excessive water production from the field. TAP was recently injected at a segment scale in one of the blocks of Cerro Dragón field [58]. In this section, we will review the design and the deployment of TAP technology and its outcomes.

To have a proper pre- and posttreatment field performance comparison, the injection and production rates were kept unchanged. Fig. 1.18 presents the performance for both treated and untreated zones. In the treated zones, the WOR has been stabilized and the oil decline rate has been flattened, whereas untreated zone in which, the oil rate and WORs continued their pretreatment trends. As a conclusion of this trial, the TAP treatment not only significantly improved the waterflood performance and recover additional oil, but also reduced the produced water treatment costs.

1.8 Preformed particle gels

Preformed particle gel (PPG) is another type of crosslinked polyacrylamide particle gel that was developed by the University of Missouri Rolla and PetroChina Company specifically for mature Chinese fields. Most Chinese oilfields have high permeability features. It is believed that due to extended periods of waterflooding,

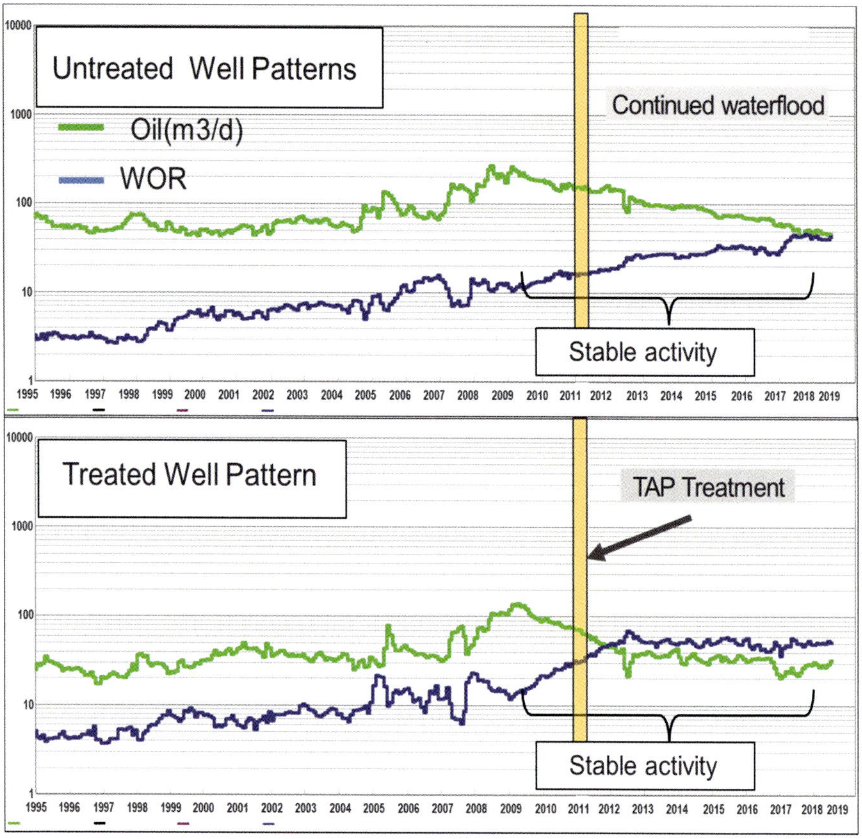

FIGURE 1.18

Field study results on thermally active polymer performance in Cerro Dragón.

the water erosion results in ultra-high-permeability features. Thus excess water production becomes a challenge later in life of these mature fields. PPG could be an alternative for the traditional bulk gel systems (polymer plus crosslinkers) to treat high permeability channels. Depending on the treatment volume, PPG can address the high heterogenous zones either near wellbore when designed in small volume or further into reservoir when the larger volumes are designed.

PPG is synthesized using an acrylamide monomer, a crosslinker, an initiative, and additives at room temperature in the surface facilities. Then, the synthetized product is crushed into particles. After that, the particles are dried at a higher temperature to form xerogel particles. Finally, the dried particles are sieved based on the target reservoir characteristics and will be injected with water. The particle size can be customized from hundreds of microns to few centimeters. In the reservoir, it will swell 20–200 times of its original size when mixing with water. Due to its large particle size, it is recommended to be applied in the reservoirs with the permeability >1D and fractured reservoirs; thus it is not suitable to be injected in the matrix [62,63].

1.8.1 Basic screening criteria

Some screening criteria for evaluation of potential candidates for the PPG technology are summarized as follows [64]:

- Presence of unswept zone and movable oil
- High permeability contrast
- Presence of very high permeability features >1D is required (not recommended for matrix injection)
- Reservoir temperatures below 126°C
- Sandstone formations are preferred; however, it could be implemented in fractured carbonates as well
- No restrictions on formation water salinity
- Injection water pH ≥6
- Candidate injector should have low pressure and high injection rate (good injectivity)
- In mature sandstone fields, the candidate pilot should have high water cut and also show a quick water transit time (a matter of few days or less) between injector and producer, indicating the presence of high permeability channels

1.8.2 Deployment history

Historically, PPG has been applied in numerous high permeability mature oil fields in China [63]. The first successful field application was in the Zhongyuan oilfield in 1999 [62]. PPG treatments have been injected in different types of reservoirs including unfractured sandstone, unconsolidated sandstone, fractured sandstone, and conglomerate ones. There have been 678 treatments reported on the injection side [64]. Almost 268 treatments were implemented in unfractured sandstones, 199 in unconsolidated sandstone, and 197 in fractured sandstone and the rest in conglomerates. Historical data shows concentrations from 1000 to 4000 ppm has been practiced for field application [64].

The Zhongyuan reservoir has been under waterflooding since 1979 [62]. This trial included two injection wells, as listed in Table 1.3, and their offset producers which were in direct connection with the two candidate injectors with high water cut (>85%). Two injectors had high water injectivity and enough pressure margin made them suitable to operate PPG. Injection profile and tracer tests showed that the area was suffering from extreme heterogeneity (conformance issue). Tracer breakthrough was in few days.

As a result of this treatment, the injection pressure was increased in both injectors which indicated the injection profile modification. The pressure

Table 1.3 Basic information of preformed particle gel (PPG) field study in Zhongyuan oil field.

Lithology	Sandstone
Reservoir temperature, °C	80–107
No. of injectors	2
No. of producers	3
Avg. permeability, mD	50 – 430
Thief zone permeability, mD	Multi Darcy
PPG concentration, ppm	3000
PPG total, m^3	6800

PPG, *Preformed particle gel.*

remained high for 2 years following the treatment, demonstrating that the PPG was stable under the reservoir condition. Water cut was decreased from 80% to 70% and a total incremental oil production of 3239 tons. Based on Bai et al. [62], this treatment resulted in producing 158 tons of oil per 1000 kg of PPG.

1.9 Foams

Hydrocarbon gases, air, CO_2, N_2, and steam are among the most widely used enhanced oil recovery methods in volatile, light, and condensate reservoirs. The presence of heterogeneity, as well as viscosity difference between gas and other reservoir fluids, causes early gas breakthrough in gas injection projects. As water flood, the handling of produced gas and recycling is very costly. As a solution, foams have been used for gas shut-off in production wells, and for mobility control on injection wells. Strong foams are suitable for producer gas shut-off, while weaker foams are suitable for the purpose of mobility control deeper in the reservoir without affecting field/ well injectivity. If the field/well injectivity is not a concern, then foam can be used near the wellbore on injection wells to improve the injection profile.

Foam is generated by mixing gas and surfactant solutions on the surface or in the reservoir. Foams can be generated at very low surfactant concentrations even sometimes below the critical micelle concentration. The surfactant is the most significant parameter in foam's generation and its stability. Surfactants will reduce the interfacial tension between the liquid and gas. If the surfactant concentration decreases, the foam weakens [65]. Based on Hirasaki [66], in foam, lamella is the thin films of fluid that surrounds the gas bubbles and reduces gas movement in the pore throat due to the surface tension and drag force which exists between the lamella and pore throat walls. Three mechanisms have been described by [67,68] in which lamella can form in porous media: leave behind, lamella division, or snap-off. In the leave behind mechanism, stabilized liquid films are formed in pores. In this mechanism, many lamellas may form. Division simply refers to the creation of two or more lamella from the original lamella. This happens when the initial mobile lamella crosses two or more smaller pore throats. Through snap-off, lamella will be generated in the gas occupied pore throats in which the capillary pressure is much less than the threshold entry capillary pressure. The foam quality and its stability under reservoir conditions are the key parameters in achieving effective mobility reduction. Foam will quickly disappear if is too dry (foam quality above 90%) or too wet (foam quality less than 45%) [69]. Foams can be generated on the surface (preformed) or can be generated in-situ by coinjecting gas and surfactant into the reservoir or through surfactant alternative gas (SAG) slugs.

1.9.1 Basic screening criteria

Foam stability and its effectiveness in mobility reduction depends on rock and fluid properties, such as rock permeability, foam quality, and slug size. Foam is reported to be not stabilized in the presence of oil at the edge of the foam front [70,71]. Short foam slug size might result in an unstable slug. Basic screening criteria for foam application for gas mobility improvement or GOR control are as following:

- Lithology: surfactant (foamer) selection considers, surfactant adsorption rock surface (low surfactant adsorption is more favorable).
- Temperature between 25 and 150°C (77 − 300°F): foam injection has been applied frequently for steam mobility control [72]. Selection is based on reservoir temperature and well spacing (thermal stability).
- Rock permeability: Higher permeability contrast is more desirable. Foam might form better in higher permeability zones showing better effectiveness [73]. Foams have been in vuggy carbonates with tight matrix and intense fracture networks.
- Oil viscosity: 0.1−30 cP; Foam stability in the presence of oil is part of the laboratory tests to screen surfactants.
- Reservoir pressure between 250 and 6000 psi.
- Reservoir net pay <1000 ft: There are field tests in thick reservoirs (e.g., Cantarell) but treatments have been conducted in fraction of the total net pay.
- Water salinity: field tests have been performed up to sea water salinity.
- Injection rate: Quality of foam depends on the injection rate [74,75].
- Injection and formation brines compositions which affect surfactant phase behavior and its adsorption.

1.9.2 Development history

There are numerous references in literature categorizing foam applications for either near wellbore injection profile modification or for the purpose of in-depth mobility control [76−79].

Table 1.4 Data related to Snorre field foam pilot.

Lithology	Sandstone
Reservoir temperature, °C	90
Avg. permeability, mD	300–3500
Thief zone permeability, mD	8000
Total surfactant, tons	32

Steam-foams in California are reported in many references [80–82]. CO_2-foam application projects in Texas have been documented by [83]. Also, there has been N_2-foam trials in the Cantarell reservoir complex in Mexico [84]. However, CO_2 foam has been more attractive than N_2 foam. The largest foam implementation documented is in Snorre field, Norway [85–87]. In this field, foam was first used in a trial as gas shut-off agent in a production well in 1996. In the other pilot, foam was injected to control the mobility in a WAG project. In other instances, the foamed gels, a combination of conventional polymer gels, and foam have been reported to address the conformance issue more effectively [88,89]. The gel's viscoelastic properties will make more stable and eventually a stronger agent to alter the injection profile [76].

1.9.3 Case study

Skauge [87] reported the foam injection for GOR control in production well P18 in Central Block of the Snorre field (Table 1.4). The well suffered from high GOR due to early gas breakthrough during the WAG injection.

Foam was considered to shutoff the excessive gas production. The foam injection was completed in mechanically isolated 7.2-m sand interval layer of Statfjord formation. The treatment was done through the injection of two cycles of 1–2 wt% SAG (16 tons of surfactant in total) and followed by coinjection of another 16 tons of surfactant and gas.

As a result of this treatment, pressure increased. However, analysis showed that foam generation during coinjection was more effective than surfactant gas alternative. Two months after treatment, GOR was reported a reduction up to 50%.

1.10 Summary

Both polymer- and surfactant-based conformance technologies have made a significant impact in improving injection and production profile modification in near wellbore and deep in the reservoir. These technologies have been widely used for shut-off treatments, well and zone abandonments, water coning, and leaking issues. In general, system selection depends on thermal and rheological characteristics under target reservoir/well conditions, mineralogy, pressure, temperature, salinity, and well communications. Depending on reservoir conditions, only one or a combination of multiple techniques could be applied. Table 1.5 summarizes a comparison between some of the selected conformance technologies and the main criteria that needs to be considered when selecting the technology. Of course, in selecting each technology, both technical and economic aspects of the project should always be considered.

Table 1.5 Conformance and water shut-off technologies in a nutshell.

Technology	Application	Pros	Cons	Criteria
Bulk gel	Water shut-off and conformance	Rig less, low sensitivity to high salinity	Specialized mixing equipment, shut in time required	Pressure margin, reasonable permeability (>300 mD)
RPMs	Water shut-off	Rig less, ability to reduce water production without impacting oil	Very thorough understanding of the reservoir condition required for success, low success rate historically, low resistance factor	No crossflow, different path of water and oil
Sodium silicate gels	Water shut-off and conformance	Flexibility in formulation, fit for high temperature applications	Not the best fit for high permeability formations, may not be able to withstand high differential pressure	N/A
CDG	Conformance and Mobility Control	Better injectivity compared to bulk gel In-depth conformance Self-adjusting properties based on shear rate	Difficult to characterize in the lab-scale, specialized mixing equipment	May not form in some brines with high TDS and high hardness level
BrightWater (TAP)	In Depth Conformance	No change in injection profile, shear resistive, no shut-in required, water-like viscosity	Not applicable for highly fractured with large aperture values	Fluid transit time more than 20 days is required

(Continued)

Table 1.5 Conformance and water shut-off technologies in a nutshell. *Continued*

Technology	Application	Pros	Cons	Criteria
Foam	Gas Shut-off and gas mobility improvement	More suitable tech for improving the performance of gas injection projects	The foam quality and stability under reservoir condition	Foam quality and slug size
PPG	Near wellbore conformance	Could be an alternative for bulk gels in high permeability reservoirs	Not applicable for low permeability reservoirs with $K < 1$ Darcy	Presence of high perm features >1D. Injector should have low pressure and high injection rate

CDG, *Colloidal dispersion gel;* PPG, *preformed particle gel;* RPMs, *relative permeability modifiers;* TAP, *thermally active polymer.*

Nomenclature

ATBS Acrylamide tertiary butyl sulfonic acid
bbls Barrels of liquid
bpd Barrels per day
CDG Colloidal dispersion gel
cP Centipoises
EOR Enhanced oil recovery
GOR Gas–oil ratio
HMP Hydrophobically modified polymer
HPAM Hydrolyzed polyacrylamide
KB Kelly bushing
mD Millidarcies
nm Nanometers
OOIP Original oil in place
PPG Preformed particle gel
ppm Parts per million
RF Resistance factor
RPM Relative permeability modifier
RRF Residual resistance factor
SAG Surfactant alternative gas
TAP Thermally active polymer
TBD To be determined
WAG Water alternating gas
WF Waterflooding
WOR Water–oil ratio
WSO Water shut-off

References

[1] C.R. Smith, Mechanics of Secondary Oil Recovery, Reinhold Publishing Corporation, New York, 1966.
[2] F.F. Craig, The Reservoir Engineering Aspect of Water Flooding, SPE Monograph, vol. 3, Society of Petroleum Engineers, Richardson, 1971.
[3] G.P. Willhite, Waterflooding, SPE Textbook Series, vol. 3, Society of Petroleum Engineers, Richardson, 1986.
[4] G.C. Thakur, A. Sattar, Integrated Waterflood Asset Management, PennWell Publishing Company, Tulsa, 1998.
[5] R.D. Sydansk, L. Romero-Zeron, Reservoir Conformance Improvement, Society of Petroleum Engineers (2011). Richardson.
[6] R.S. Seright, Gel propagation through fractures, Production and Facilities 16 (2001) 225–231.
[7] R.D. Sydansk. A new conformance-improvement-treatment chromium (III) technology, in: SPE-17329-MS, SPE Enhanced Oil Recovery Symposium, Tulsa, OK, USA, April 16–21, 1988.
[8] D.G. Perez, G.A. Pedersen, R.E. Lehu, et al., High resolution dynamic modelling to revitalize a mature oil field in a low oil price scenario, in: SPE-185535-MS, SPE Latin America and Caribbean Petroleum Engineering Conference, Buenos Aires, Argentina, May 18–19, 2017.
[9] D.G. Perez, M.D. Cocco, G.A. Pedersen, et al., Combining data analysis and simulation to design the largest polymer injection pilot in Argentina, in: SPE-199147-MS, SPE Latin America and Caribbean Petroleum Engineering Conference, Bogota, Colombia, March 17–19, 2020.
[10] D.G. Perez, G. Pedersen, R. Lehu, et al., Reservoir characterization to assess large field opportunity in Los Perales field, Golfo San Jorge Basin, Argentina, in: SPE-184905-MS, SPE Latin America and Caribbean Mature Fields Symposium, Salvador Bahai, Brazil, March 15–16, 2017.
[11] M. Unomah, S. Thach, R. Shong, et al., Performance of conformance gels under harsh conditions, in: SPE-190266-MS, SPE Improved Oil Recovery Conference, Tulsa, OK, USA, April 14–18, 2018.
[12] A. Mishra, S. Abbas, J. Braden, et al., Comprehensive review of fracture control for conformance improvement in the Kuparuk River Unit—Alaska, in: SPE-179649-MS, SPE Improved Oil Recovery Conference, Tulsa, OK, USA, April 11–13, 2016.
[13] A. Roostapour, M. Qaid, E. Tudor, et al., Optimizing tight oil assets on water flood utilizing polymer gel technology; a cost-effective approach with high rate of success, in: SPE-200401-MS, SPE Improved Oil Recovery Conference, Tulsa, OK, USA, August 31–September 4, 2020.
[14] B.B. Sandiford, Laboratory and field studies of water floods using polymer solutions to increase oil recovery, Journal of Petroleum Technology 16 (1964) 917–922.
[15] R.D. Sydansk, R.S. Seright, When and where relative permeability modification water-shutoff treatments can be successfully applied, in: SPE-99371-MS, SPE/DOE Symposium on Improved Oil Recovery, Tulsa, OK, USA, April 22–26, 2006.
[16] R.S. Seright, Impact of permeability and lithology on gel performance, SPE-24190-MS, SPE/DOE Symposium on Enhanced Oil Recovery, Tulsa, OK, USA, April 22–24, 1992.

[17] D.J. Ligthelm, Water shut off in gas wells: is there scope for a chemical treatment, in: SPE-68978-MS, European Formation Damage Conference, The Hague, The Netherlands, May 21–22, 2001.

[18] R.S. Seright, Optimizing disproportionate permeability reduction, in: SPE-9944-MS, SPE/DOE Symposium on Improved Oil Recovery, Tulsa, OK, USA, April 22–26, 2006.

[19] M.J. Pietrak, M.R. Stanley, B.J. Weber, et al., Relative permeability modifier treatments on Gulf of Mexico frac-packed and gravel-packed oil and gas wells, in: SPE-96945-MS, SPE Annual Technical Conference and Exhibition, Dallas, TX, USA, October 9–12, 2005.

[20] L.J. Kalfayan, J.C. Dawson, Successful implementation of resurgent relative permeability modifier (RPM) technology in well treatments requires realistic expectations, in: SPE-90430-MS, SPE Annual Technical Conference and Exhibition, Houston, TX, USA, September 26–29, 2004.

[21] C.B. Gludicellie, R.A. Truchetet, Prediction of the oil rate increment with water profile control by polymer injection, in: SPE-2660-MS, Annual Technical Conference and Exhibition, Houston, TX, USA, October 3–6, 1993.

[22] G.A. Farrera Romo, H. Hernández Leyva, R.B. Aguilar, et al., Advanced technology to reduce water cut: case studies from the Pemex southern region, SPE Production & Operations 25 (2010) 139–144.

[23] A.B. Ziad, D. Gromakovskii, A. Al-Sagr, et al., First successful application of temporary gel plug replacing calcium carbonate chips to isolate depleted reservoir, case study from Saudi Arabia gas field, in: SPE-178986-MS, SPE International Conference and Exhibition on Formation Damage Control, Lafayette, LA, USA, February 24–16, 2016.

[24] P.J. Boul, A. Ye, X. Pang, et al., Nanosilica-based conformance gels, in: SPE-86547-MS, SPE Eouropean Formation Damage Conference and Exhibition, Budapest, Hungary, June 3–5, 2015.

[25] H. Jia, X.Y. Yang, Environmental and strength-enhanced nanosilica-based composite gel for well temporary plugging in high-temperature reservoirs, Asia-Pacific Journal of Chemical Engineering 14 (2019) e2270.

[26] Y. Bai, X. Shang, Z. Wang, et al., Experimental study of low molecular weight polymer/nanoparticle dispersed gel for water plugging in factures, Colloids and Surfaces A: Physicochemical and Engineering Aspects 551 (2018) 95–107.

[27] A.A. Adewunmi, S. Ismail, A.S. Sultan, Study on strength and gelation time of polyacrylamide/ polyethyleneimine composite gels reinforced with coal fly ash for water shut-off treatment, Journal of Applied Polymer Science 132 (2014) 41392.

[28] D.G. Hatzingnatiou, N.H. Giske, Sodium silicate gelants for water management in naturally fractured hydrocarbon carbonate formations, Chemical Engineering Research and Design 132 (2018) 40–56.

[29] H.A. Nasr-El.Din, K.C. Taylor, Evaluation of sodium silicate/urea gels used for water shut-off treatments, Journal of Petroleum Science and Engineering 48 (2005) 141–160.

[30] I. Lakatos, J. Lakatos-Szabo, Reservoir conformance control in oilfield using silicates: State-of-the-arts perspectives, in: SPE-159640-MS, SPE Annual Technical Conference and Exhibition, San Antonio, TX, USA, August 8–10, 2012.

[31] L.T. Pham, D.G. Hatzignatiou, Rheology evaluation of sodium silicate gel system for water management in mature, naturally fractured oilfields, Journal of Petroleum Science and Engineering 138 (2016) 218–233.

[32] I. Lakatos, J. Lakatos-szabo, B. Kosztin, et al., Application of iron hydroxide-based well treatment technique at the Hungarian oil fields, in: SPE-59321-MS, SPE/DOE IOR Symposium, Tulsa, OK, USA, April 3–5, 2002.
[33] J.E. Smith, The transition pressure: a quick method for quantifying polyacrylamide gel strength, in: SPE-18739-MS, SPE International Symposium on Oilfield Chemistry, Houston, TX, USA, February 8–10, 1989.
[34] J.E. Smith, Performance of 18 polymers in aluminum citrate colloidal dispersion gels, in: SPE-28989-MS, SPE International Symposium on Oilfield Chemistry, San Antonio, TX, USA, February 14–17, 1995.
[35] E Manrique, S Reyes, J Romero, et al., Colloidal dispersion gels (CDG): field projects review, in: SPE-169705-MS, SPE EOR Conference at Oil and Gas West Asia, Muscat, Oman, March 31–April 2, 2014.
[36] R.S. Seright. Mechanism for gel propagation through fractures, in: SPE-55628-MS, Rocky Mountain Regional Meeting, Gillette, WY, USA, May 15–18, 1999.
[37] D. Wang, R.S. Seright, Examination of literature on colloidal dispersion gels for oil recovery, Petroleum Science 18 (2021) 1097–1114.
[38] J.E. Smith, J.C. Mack, A.B. Nicol, The Adon road—an in-depth gel case history, in: SPE-35352-MS, SPE/DOE 10th Symposium on Improved Oil Recovery, Tulsa, OK, USA, April 21–24, 1996.
[39] B.F. Sloat, Measuring engineered oil recovery, Journal of Petroleum Technology 43 (1991) 8–13.
[40] R.C. Fielding, D.H. Gibbons, F.P. LeGrand, In-depth drive fluid diversion using an evolution of colloidal dispersion gels and new bulk gels: an operational case history of North Rainbow Ranch Unit, in: SPE-27773-MS, SPE/DOE 9th Symposium on Improved Oil Recovery, Tulsa, OK, USA, April 17–20, 1994.
[41] D.R. Meyer, The water quality control station: a new monitoring tool for injection water, in: SPE-17524-MS, SPE Rocky Mountain Regional Meeting, Casper, WY, USA, May 11–13, 1988.
[42] J.E. Smith, J.C. Mack, Gels correct in-depth reservoir permeability variation, Oil & Gas Journal 95 (1997) 33–39.
[43] C.A. Norman, J.E. Smith, R.S. Thompson, Economics of in-depth gel processes, in: SPE-55632-MS, SPE Rocky Mountain Regional Meeting, Gillette, WY, USA, May 15–18, 1999.
[44] J.E. Smith, H. Liu, Z. D. Guo, Laboratory studies of in-depth colloidal dispersion gel technology for Daqing oil field, in: SPE-62610-MS, SPE/AAPG Western Regional Meeting, Long Beach, CA, USA, June 19–23, 2000.
[45] J.M. Leon, M. Izadi, A. Castillo, et al., Use of cross-linked polymer systems to improve volumetric sweep efficiency and alternative full field development strategy for a mature waterflooding optimization processes—Dina Cretaceous field case, in: SPE-190313-MS, SPE Improved Oil Recovery Conference, Tulsa, OK, USA, April 14–18, 2018.
[46] D. Broseta, O, Marquer, P. Baylocq, et. al., Gel treatment optimization by rheological measurements, in: SPE-50750-MS, SPE International Symposium on Oilfield Chemistry, Houston, TX, USA, February 16–19, 1999.
[47] K. Chang, H. Frampton, J.C. Morgan, Method of recovering hydrocarbon fluids from a subterranean reservoir, Patent US6454003 (2002).
[48] A. Fletcher, S. Flew, I.N. Forsdyke, et al., Deep diverting gels for very cost-effective waterflood control, Journal of Petroleum Science and Engineering 7 (1992) 33–43.

[49] M. Husband, D. Ohms, H. Frampton, et al. Results of a three-well waterflood sweep improvement trial in the Prudhoe Bay field using a thermally activated particle system, in: SPE-129967-MS, SPE Improved Oil Recovery Symposium, Tulsa, OK, USA, April 24–28, 2010.

[50] B. Roussennac, C. Toschi, BrightWater Trialin Salema field (Campos Basin, Brazil), in: SPE-131299-MS, SPE EUROPEC/EAGE Annual Conference and Exhibition, Barcelona, Spain, June 14–17, 2010.

[51] M. Salehi, M.C. Thomas, R. Kevwitch, et al. Performance evaluation of thermally activated polymers for conformance correction applications, in: SPE-154022-MS, 18th SPE IOR Symposium, Tulsa, OK, USA, April 14–18, 2012.

[52] J. Pritchett, H. Frampton, J. Brinkman, et al., Field application of a new in-depth waterflood conformance improvement tool, in: SPE-84897-MS, International Improved Oil Recovery, Kuala-Lumpur, Malaysia, October 20–21, 2003.

[53] R. Garmeh, M. Izadi, M. Salehi, et al., Thermally active polymer to improve sweep efficiency of water floods: Simulation and pilot design approaches, SPE Reservoir Evaluation & Engineering 15 (2012) 86–97.

[54] E. Manrique, R. Garmeh, M. Izadi, et al. In-depth sweep efficiency improvement: screening criteria and engineering approach for pattern evaluation and potential field implementation, in: SPE-160749-MS, SPE Russian Oil and Gas Exploration and Technical Conference, Moscow, Russia, October 16–18, 2012.

[55] D. Ohms, J. McLeod, C. Graff, et al. First commercial trial of a novel waterflood conformance control reagent, in: Proceedings of the Pacific Section AAPG-Western Region SPR Annual Meeting, Bakersfield, CA, USA, March 31–April 2, 2008.

[56] D. Ohms, J. McLeod, C. Graff, et al., Incremental oil success from waterflood sweep improvement in Alaska, SPE Production & Operations 25 (2010) 247–254.

[57] H. Frampton, P. Denyer, D. Ohms, et al., BrightWater sweep improvement from the lab to the field, in: Proceedings of the 15th European Symposium on Improved Oil Recovery, Paris, France, April 27–29, 2009.

[58] H. Frampton, J. Morgan; S. Cheung et al., Development of a novel waterflood conformance control system, in: SPE-89391-MS, SPE/DOE Symposium on Improved Oil Recovery, Tulsa, OK, USA, April 17–21, 2004.

[59] J.L. Mustoni, F. Cabrera, E. Manrique, et al., Application of a thermally activated particle system for improved sweep efficiency in a multilayer, geologically complex waterflood, in: SPE-153331-MS, SPE Improved Oil Recovery Symposium, Tulsa, OK, USA, April 14–18, 2012.

[60] A. Luliano, J. Gomez, C. Martinez, et al., Thermally activated particles injection for deep conformance control to improve oil recovery in an Argentina mature waterflooded reservoir: Cerro Dragon field. Design, field implementation and results, in: SPE-200464-MS, SPE IOR Symposium, Tulsa, OK, USA, April 14–18, 2020.

[61] J.M Gavilan, J.L. Mustoni, V. Alvarado, et al., Evaluation and implementation of BrightWater™ technology in a complex multilayered reservoir with wells producing under high water cut, in 16th European Symposium on Improved Oil Recovery, Cambridge, UK, April 12–14, 2011.

[62] B. Bai, L. Li, Y. Liu, et al., Preformed particle gel for conformance control: factors affecting its properties and applications, SPE Reservoir Evaluation & Engineering 10 (2007) 415–422.

[63] Y. Liu, B. Bai, P.J. Shuler, Application and development of chemical-based conformance control treatments in China oilfields, in: SPE-99641-MS, SPE/DOE Symposium on Improved Oil Recovery, Tulsa, OK, USA, April 22–26, 2006.

[64] Y. Qiu, M. Wei, B. Bai, Descriptive statistical analysis for the PPG field applications in China: screening guidelines, design considerations, and performances, Journal of Petroleum Science and Engineering 153 (2017) 1–11.

[65] O. Apaydin, A. Kovscek, Transient foam flow in homogeneous porous media: surfactant concentration and capillary end effects, in: SPE-59286-MS, SPE/DOE Improved Oil Recovery Symposium, Tulsa, OK, USA, April 3–5, 2000.

[66] G. Hirasaki, The steam-foam process, Journal of Petroleum Technology 41 (1989) 449–456.

[67] H. Chen, M. Ke, T. Chuang, et al., Experimental studies of capillary pressure effects of foams in porous media, in: SPE-20069-MS, SPE California Regional Meeting, Ventura, CA, USA, April 4–6, 1990.

[68] Q. Chen, M. Gerritsen, A. Kovscek, Improving steam-assisted gravity drainage using mobility control foams: foam assisted-SAGD (FA-SAGD), in: SPE-129847-MS, SPE Improved Oil Recovery Symposium, Tulsa, OK, USA, April 24–28, 2010.

[69] S. Chang, R. Grigg, Effects of foam quality and flow rate on CO_2-foam behavior at reservoir conditions, in: SPE-39679-MS, SPE/DOE Improved Oil Recovery Symposium, Tulsa, OK, USA, April 19–22, 1998.

[70] F. Friedmann, J. Jensen, Some parameters influencing the formation and propagation of foams in porous media, in: SPE-15087-MS, SPE California Regional Meeting, Oakland, CA, USA, April 2–4, 1986.

[71] A. Nikolov, D. Wasan, D. Huang, et al., The effect of oil on foam stability: Mechanisms and implications for oil displacement by foam in porous media, in: SPE-15443-MS, SPE Annual Technical Conference and Exhibition, New Orleans, LA, USA, October 5–8.

[72] S. Zhdanov, A. Amiyan, L. Surguchev, et al., Application of foam for gas and water shut-off: Review of field experience, in: SPE-36914-MS, European Petroleum Congress, Milan, Italy, October 22–24, 1996.

[73] J. Tsau, J. Heller, How can selective mobility reduction of CO2-foam assist in reservoir floods, in: SPE-35168-MS, The Permian Basin Oil and Gas Recovery Conference, Midland, Texas, USA, March 27–29, 1996.

[74] S. Chang, R. Grigg, Effects of foam quality and flow rate on CO_2-foam behavior at reservoir temperature and pressure, Reservoir Evaluation and Engineering Journal 2 (1999) 248–254.

[75] L. Schramm, Foam: Fundamentals and Applications in the Petroleum Industry, Advances in Chemistry, Series 242, American Chemical Society, Washington, DC, 1994.

[76] S. Farzaneh, M. Sohrabi, A review of the status of foam applications in enhanced oil recovery, in SPE-164917-MS, EAGE Annual Conference & Exhibition, London, UK, June 10–13, 2013.

[77] J. Hanssen, T. Holt, L. Surguchev, Foam processes: An assessment of their potential in North Sea reservoirs based on a critical evaluation of current field experience, in: SPE-27768-MS, SPE/DOE Improved Oil Recovery Symposium, Tulsa, Oklahoma, USA, April 17–20, 1994.

[78] D. Smith, Injectivity and surfactant-based mobility control: field tests, in: D. Smith (Ed.), Surfactant-Based Mobility Control-Progress in Miscible Flood Enhanced Oil Recovery, American Chemical Society, Washington, DC, 1998, pp. 429–438.
[79] A. Turta, A. Singhal, Field foam applications in enhanced oil recovery projects: Screening and design aspects, in: SPE-48895-MS, International Oil and Gas Conference and Exhibition, Beijing, China, November 2–6, 1998.
[80] N. Djabbarah, S. Weber, D. Freeman, et al., Laboratory design and field demonstration of steam diversion with foam, in: SPE-20067-MS, SPE California Regional Meeting, Ventura, CA, USA, April 4–6, 1990.
[81] A. Kovscek, C. Radke, Fundamentals of foam transport in porous media, in: L.L. Schramm (Ed.), Foams: Fundamentals and Applications in the Petroleum Industry, American Chemical Society, Washington, DC, 1994, pp. 115–163.
[82] J. Ploeg, J. Duerksen, Two successful steam/foam field tests, Sections 15A and 26C, Midway-Sunset field, in: SPE-13609-MS, SPE California Regional Meeting, Bakersfield, CA, USA, March 27–29, 1985.
[83] M. Hoefner, E. Evans, J. Buckles, T. Jones, CO_2 foam: results from four developmental field trials, SPE Reservoir Engineering 10 (1995) 273–281.
[84] F. Skoreyko, A. Pino, H. Prada, et al., Development of a new foam EOR model from laboratory and field data of the naturally fractured Cantarell field, in: SPE-145718-MS, SPE Reservoir Characterization and Simulation Conference and Exhibition, Abu Dhabi, UAE, October 9–11, 2011.
[85] T. Blaker, H. Celius, T. Lie, et al., Foam for gas mobility control in the Snorre field: the FAWAG project, in: SPE Annual Technical Conference and Exhibition, Houston, Texas, USA, October 3–6, 1999.
[86] T. Blaker, M. Aarra, A. Skauge, et al., Foam for gas mobility control in the Snorre field: the FAWAG project, Journal of Reservoir Evaluation & Engineering 5 (2002) 317–323.
[87] A. Skauge, M. Aarra, L. Surguchev, et al., Foam-assisted WAG: experience from the Snorre field, in: SPE-75157-MS, SPE/DOE Improved Oil Recovery Symposium, Tulsa, Oklahoma, USA, April 13–17, 2002.
[88] F. Friedmann, T. Hughes, M. Smith, Development and testing of a foam-gel technology to improve conformance of the Rangely CO_2 flood, SPE Reservoir Evaluation & Engineering 2 (1999) 4–13.
[89] A. Kantzas, D. Marrenete, K. Allsop, Utilization of polymer gels, polymer enhanced foams, and foamed gels for improving reservoir conformance, Journal of Canadian Petroleum Technology 38 (1999) 1–8.

CHAPTER

Low-salinity (enhanced) waterflooding in carbonate reservoirs

2

Hassan Mahani[1] and Geoffrey Thyne[2]

[1]Department of Chemical and Petroleum Engineering, Sharif University of Technology, Tehran, Iran

[2]ESal, Laramie, WY, United States

2.1 Introduction

The proportion of the world's "conventional" petroleum that is in place in carbonate rocks is commonly estimated at around 50%–60% [1], particularly in the middle-east. The primary recovery factor from carbonate reservoirs is commonly low, at around 10%–20%, compared with sandstone reservoirs [2–5].

Low-salinity waterflooding (LSWF) is a promising technology to increase recovery from carbonates [6]. LSWF is an enhanced waterflood aimed at increasing oil production from a reservoir through modification of the injection water composition (proportion and type of mono- and multivalent cations and anion), often leading to a lower salinity (LS) relative to the formation brine. LSWF has elements of both conventional waterflood and chemical enhanced oil recovery (EOR) (Fig. 2.1). It may be employed at any stage of hydrocarbon production and is very desirable because of the low cost of operations, low CO_2 footprint, lack of environmental impact, reduced risk of scaling and reservoir souring, and industry-wide experience with water injection during secondary recovery. The synergy of the technology with EOR methods has led to the creation of hybrid LS EOR methods, which take advantage of both technologies, and enhances oil recovery

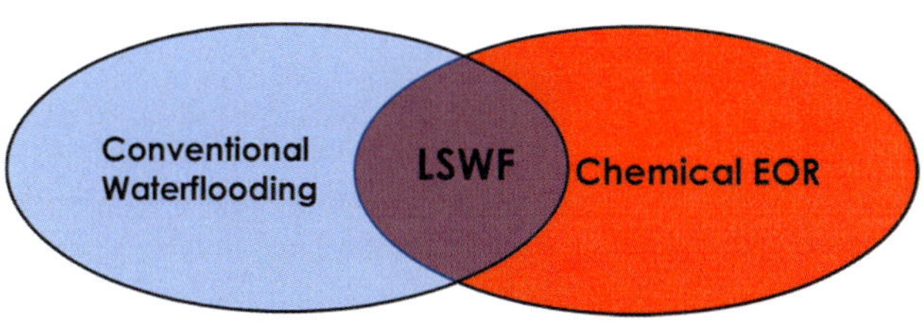

FIGURE 2.1

Classification of LSWF technology. *LSWF*, Low-salinity waterflooding.

Recovery Improvement. DOI: https://doi.org/10.1016/B978-0-12-823363-4.00007-8

efficiency. A few examples include LS polymer, LS surfactant, LS nanoparticle, LS foam, LS carbonate water, and LS CO_2 flood. LSWF as a preflush can facilitate the deployment of polymer and surfactant flooding due to the better efficiency of polymer and surfactant under LS conditions.

The earlier studies on LSWF focused primarily on sandstones (between 1960 and 1999). Initial laboratory-scale experiments on LSWF in carbonates started in early 2000 [7–9] where it was successfully demonstrated that the LSWF concept can be extended to carbonates as well. LSWF has become a favorite topic for both academic and industry researchers with some 360 scientific papers written about LSWF from 2015 to 2019, including 16 + review papers [10–23].

2.1.1 Different names of the technology and the misconceptions

The technology has been termed differently in different institutions and industries for marketing and branding purposes. However, the terms should not be taken literally as they all refer to one technology: for instance, LSWF, designer waterflooding, LoSal, RightWater®, smart waterflooding, advanced ion management, controlled-salinity waterflooding, engineered salinity waterflooding, and ion-tuned waterflooding.

2.1.2 The ultimate aim of low-salinity waterflooding

Some regard greater recovery or estimated ultimate recovery (EUR) as the key benefit but the point of view that increasing current rates of production (or production acceleration) is equally valid. It should be noted that reducing S_{or} (true residual oil saturation as defined by zero relative permeability of oil) is not the prime goal of LSWF as many target field candidates are either under waterflood or still in the primary production stage (decline phase) and are not even close to S_{or} in the majority of the reservoir. The main benefit expected for LSWF is the improvement of the sweep efficiency at oil saturation ranges well above S_{or}; leading to faster or accelerated oil recovery which can be of critical importance in improvement of project economics and cost recovery. Otherwise, targeting only S_{or} could make the process inefficient, because there are much better techniques to locally reduce S_{or}, for example, by surfactant flooding.

2.1.3 Target field candidates

LSWF can be deployed in both green and brown fields, offshore and onshore, albeit with different challenges. Experimental evidence suggests that oil-wet to mixed-wet reservoirs are favorable target candidates, since waterflooding is not very successful under this wettability condition and leaves a substantial amount of unswept oil. It is not necessary to wait and deploy the technology after conventional waterflood in tertiary mode, with the belief that it is better to recover the easier oil by waterflood before moving to a more advanced technology. It has been repeatedly observed in the experiments that early deployment of the technology can increase the incremental oil recovery (IOR) such that the incremental recovery in secondary mode is higher than that in tertiary mode. By delaying the water breakthrough and controlling the water production rate, produced water management can become less problematic. From a logistic point of view, it is also preferred to consider LSWF in the concept assessment and select phase of the opportunity realization process to avoid the additional cost of deployment at later stages of the field development. Fig. 2.2 shows a conceptual diagram of the two modes of employing LSWF.

Additional oil recovery gain from LSWF varies depending on the history of water injection in the field and reservoir complexities. Recent field evidence indicates that the estimated IOR of LSWF in carbonates can be as much as 6%−10% beyond conventional seawater injection in Saudi Arabian fields [24].

In the following sections, this chapter focuses on the state-of-the-art review of the fundamentals, candidate selection criteria and assessment workflow, field implementation, case histories, and numerical modeling and simulation methodologies. This chapter concludes by discussing the outstanding issues, new developments, and the future of LSWF technology

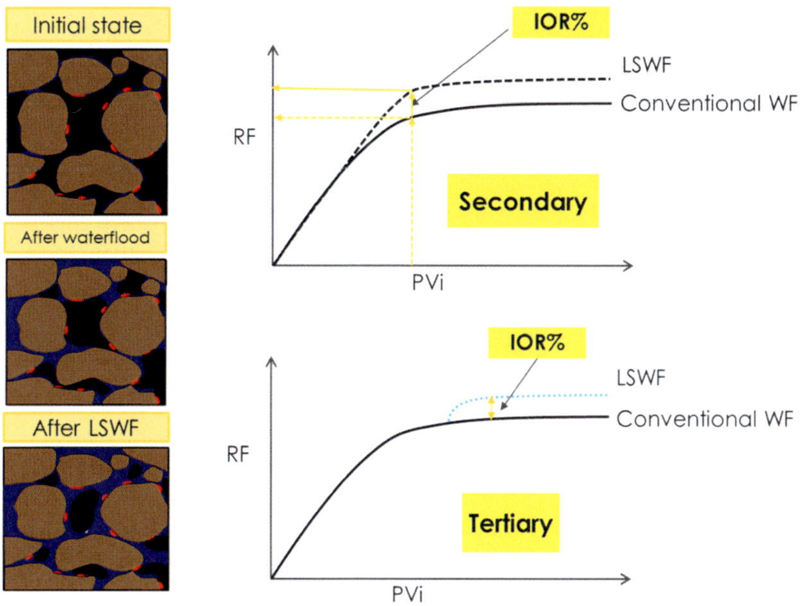

FIGURE 2.2

IOR by LSWF in secondary and tertiary injection modes. The secondary mode represents the recovery after the primary depletion, while the tertiary mode represents a case where the low-salinity is injected after a conventional high-salinity waterflood. *LSWF*, Low-salinity waterflooding; IOR, incremental oil recovery.

2.2 Fundamentals

2.2.1 Carbonate chemistry

The inorganic portion of the carbonate reservoir system consists of inorganic aqueous species (H_2CO_3, HCO_3^-, CO_3^{2-}), dissolved carbon dioxide, and carbonate minerals. There are other minor aqueous species, such as the neutral aqueous species ($NaHCO_3$ and $CaCO_3$), but these are usually ignored.

The relevant chemical reactions are:

$$CO_2(g) = CO_2(aq) \tag{2.1}$$

$$CO_2(aq) + H_2O = H_2CO_3 \tag{2.2}$$

$$H_2CO_3 = H^+ + HCO_3^- \tag{2.3}$$

$$HCO_3^- = H^+ + CO_{3^{2-}} \tag{2.4}$$

$$Ca^{2+} + CO_{3^{2-}} = CaCO_3(s) \tag{2.5}$$

Chemical equilibrium between gas phase and dissolved gas (CO_2 aq) is described by Henry's law. The values of the partitioning coefficient are dependent on temperature and pressure [25].

$$K_{CO_2} = \frac{CO_2(g)}{CO_2(aq)} \tag{2.6}$$

Thermodynamic models of carbonate chemistry often simplify the equations by directly relating gas-phase CO_2 and carbonic acid, the dissolved neutral species. In petroleum reservoirs, carbon dioxide gas can be present in the gas cap if the amount of dissolved carbon dioxide exceeds saturation (bubble point). In that case, the dissolved carbon dioxide is represented as the partial pressure (fugacity) of CO_2 in thermodynamic modeling. Carbonate reservoirs are initially at chemical and thermal equilibrium between the rock, brine, and oil until they are drilled. This chemical equilibrium is changed when drilling and completion fluids are introduced into the reservoir. These changes continue during primary production as the removal of oil and gas lowers the reservoir pressure. The changes continue as secondary recovery starts with the injection of water to displace oil and increase reservoir pressure.

The connate water chemistry of carbonate reservoirs varies over a wide range of TDS, but most of the variation is due to NaCl content. The dissolved calcium, magnesium, and bicarbonate (carbonate alkalinity) are dependent on chemical equilibrium with carbonate minerals, temperature, and pH. Sulfate content is dependent on chemical equilibrium with gypsum or anhydrite if present. Carbonate minerals have retrograde solubility, that is, they are less soluble at higher temperatures. However, the partial pressure of carbon dioxide increases with depth, and the pH declines meaning the equilibrium concentration of dissolved carbonate species may be quite high in deeper reservoirs. For instance,

Kaszuba et al. [26] found the pH was about 4.0 at 110°C and total dissolved solids (TDS) of 77,000 mg/L in the Madison limestone on the Moxa Arch of southwestern Wyoming, USA. Most of the TDS was dissolved carbonate and during production, the increasing pH caused precipitation of carbonate scale causing the TDS at the wellhead to be about 28,000 mg/L. This example shows the complexity of the carbonate system.

Carbonate reservoirs are composed of biogenic particles with small amounts of organic matter and clastic detrital particles (quartz, feldspars, and clays) with amounts depending on the depositional environment. The common carbonate minerals in carbonate reservoirs include calcite ($CaCO_3$), and dolomite ($CaMg(CO_3)_2$). Common accessory minerals are pyrite, anhydrite (gypsum), glauconite, phosphate, quartz, and clays. The common lithologies are based on the amount of carbonate material: chalk ($>95\%$), chalky marl (65%–95%), marl (35%–65%), marly shale (5%–35%), and shale ($<5\%$). The lithology is directly related to the depositional environment and the amount of noncarbonate material available for deposition. Nearshore environments can include significant amounts of terrigenous material producing the marl lithology. For instance, biogenic particles in lacustrine environments are derived from algal activity (phytoplankton) and aquatic macrophytes. Lacustrine carbonates also contain significant amounts of organic matter, biogenic silica (diatoms), and detrital silicates [27]. Many of these environments are associated with high salinity (HS) driven by seasonal changes in recharge.

Marine carbonates are composed of a variety of carbonate sources that strongly depend on depositional environments. Nearshore dunes, tidal flats, and beaches tend to have greater detrital input from terrestrial sources and are more influenced by seasonal variations. Carbonate sources include algae, gastropods, and stromatolites. Limestone reservoirs deposited in near-shore environments that include lakes, dunes, tidal flats (sabkha is the arid equivalent), beaches, shelves, and reefs and are composed of aragonite and high-Mg calcite, less chemically stable forms of calcite [27–31]. These calcite minerals are prone to chemical transformation during burial causing diagenetic alteration and changes in porosity and permeability. Near-shore carbonates may have larger grain sizes because they are derived from corals, shells, and algal colonies. These depositional environments will also have more clastic components leading to the marls and marly shales lithology depending on the particle size. Marine carbonates are prone to dolomitization, the transformation of calcite to dolomite in the presence of high dissolved magnesium that creates dolostones. This transformation can occur in the near-surface environment or during later burial. The smaller unit cell size of dolomite compared to calcite causes an overall increase in porosity by dolomitization. Few dolomites are completely transformed so most dolostones still contain some percentage of calcite in the rock.

Chalk reservoirs are the result of pelagic or deep-water deposition. The reservoirs tend to have uniform grain size and are composed of biogenic remains of coccolithophores and foraminifera with lesser amounts of pteropods, belemnites, crinoids, etc. The original mineralogy of chalks is usually low magnesium calcite, a very chemically stable form of calcite; thus chalks undergo little alteration with subaerial exposure or burial. Chalk reservoirs are prone to fractures which generate additional porosity and permeability during burial.

2.2.2 Carbonate surface charge

Carbonate surfaces have two types of reactive sites. The calcium and or magnesium sites (>Ca or >Mg), where > denotes the surface site, are positively charged while the carbonate sites (>CO_3) are negatively charged. The electrical charge measured at the edge of the diffuse layer is known as the zeta-potential. Fig. 2.3 shows the zeta-potential versus pH for some natural systems [32]. The data show the zeta-potential varies from an overall positive to an overall negative as pH shifts from acidic to basic. Al Mahrouqi et al. [33] argued that the observed pH dependency of zeta-potential is due to the moderation of equilibrium pCa and

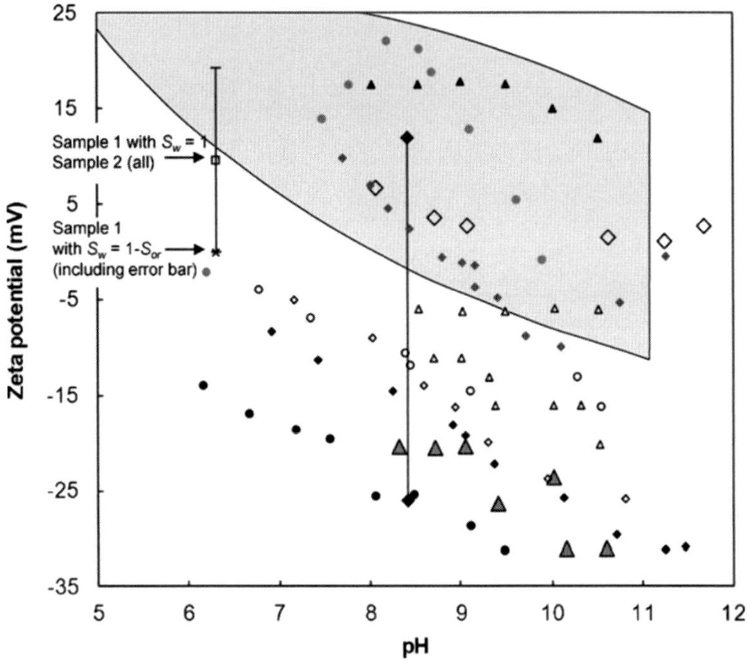

FIGURE 2.3

Zeta-potential of carbonate samples as a function of pH in various brine compositions including NaCl, $NaHCO_3$, and $CaCl_2$.

Reproduced with permission from M.D. Jackson, J. Vinogradov, Impact of wettability on laboratory measurements of streaming potential in carbonates, Colloids and Surfaces A: Physicochemical and Engineering Aspects 393 (2012) 86–95.

that if pCa is kept constant, there would not be a pH dependency. They reported that surface charge of calcite is a weak function of pH.

Surface charge is also sensitive to brine ionic composition. Therefore it is believed that by manipulation of brine composition the surface charge or polarity can be adjusted in such a way that the maximum electric double layer (EDL) repulsion between the carbonate/brine and oil/brine interface can be achieved. This condition would shift wettability toward more water-wet assuming that oil wetting is related to EDL attraction and repulsion between oil and mineral surfaces [34].

Both positively charged ions and negatively charged ions influence the carbonate surface charge, thus point of zero charge and isoelectric point. Depending on their charge density, they can get sufficiently close to the solid surface and be present in the Stern layer or in the diffuse layer. When present in the Stern layer, they can bind with calcium and carbonate sites on the calcite surface and form surface complexes with either neutral, negative, or positive charges (Fig. 2.4). These ions are known as potential determining ions (PDIs) and mainly include divalent ions, such as Ca^{2+}, Mg^{2+}, SO_4^{2-}, and CO_3^{2-}. Multivalent ions, such as PO_3^{3-} and BO_3^{3-}, have a high charge density thus a tendency to get adsorbed and turn the carbonate surface charge more negative [35]. However, these ions are not found in significant concentrations in reservoir brines and seawater.

Positively charged monovalent ions, such as Na^{+}, which are largely present in reservoir brine and seawater, are commonly known as indifferent (inactive) ions [8], even detrimental toward the low-salinity effect (LSE) and are believed to be present in a diffuse electrical layer or close to outer Helmholtz layer. However, there is recent evidence that at certain concentrations they can become present in the Stern layer, particularly in the inner Helmholtz layer, thus be more influential on carbonate surface charge, than earlier thought. Their effect on carbonate wettability and LSWF mechanism has been underestimated and requires further investigation [[36–38]].

2.2.3 Petrophysical characteristics of carbonate rocks

The petrophysical properties of carbonate rock are highly variable, but some generalizations are useful. The porosity can be quite high, as much as 40% even in deeply buried reservoirs. For instance, the Ekofisk and Dan chalks in the North Sea have porosity values as high as 40% and as low as 10% in the same reservoir interval. The range of porosity is not related to depth (1750–3200 m), but instead dependent on the depositional facies, grain size, and cementation [39]. While chalk reservoirs typically have a narrow range of grain and pore size, nearer shore depositional environments can produce a broad range of pore-size distributions. Reservoir porosity can include micropores, macropores, vugs, and fractures.

Unlike sandstones, the permeability of carbonate rocks is usually low in the range of 1–10 mD [39] and there is little change in permeability with burial depth. Many carbonate reservoirs have fractures that provide higher permeability features producing reservoirs with significant anisotropy under flow. In the case

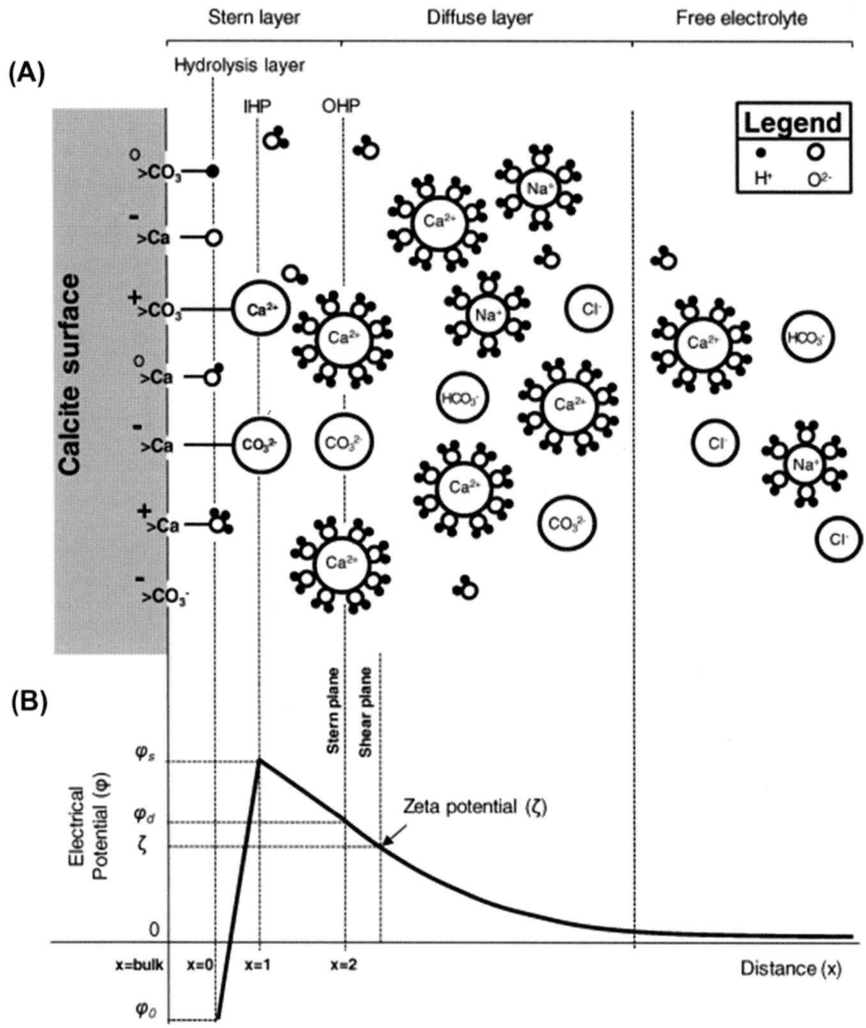

FIGURE 2.4

The electric double layer at calcite–brine interface and the surface complexes formed at calcite surface with the PDIs. Here, the calcite surface is negatively charged, but due to the adsorption of Ca^{2+} and CO_3^{2-} the zeta-potential is positive. *PDIs*, Potential determining ions.

Reproduced with permission from D. Al Mahrouqi, J. Vinogradov, M.D. Jackson, Zeta potential of artificial and natural calcite in aqueous solution, Advances in Colloid and Interface Science 240 (2017) 60–76.

of LSWF, the injected fluids move through the fractures and are imbibed into the matrix pores displacing oil. This mechanism can produce EUR values up to 50% under waterflood, similar to clastic reservoirs.

Mineral heterogeneity including variations in the amounts of calcite and dolomite is common in dolostones. Since dolomitization will increase porosity locally. This can lead to porosity variations related to mineralogy, and dolostones tend to have higher porosity than limestones. Anhydrite and pyrite cements are common in carbonates and may alter local petrophysical log responses however, the minor amounts of clay content mean gamma-ray response is low. Carbonate wettability is often measured as neutral-wet to oil-wet in limestone and dolomite. In chalks weakly oil-wet to weakly water-wet is possible, perhaps due to the generally finer grain size. Chalk also is known to be sensitive to seawater flooding and reacts by losing mechanical strength which can lead to subsidence [5]. Limestone and dolomite do not react negatively to seawater flooding.

2.2.4 Current opinions on low-salinity waterflooding mechanisms

Determining the LS mechanisms is complicated by the number of potentially important interactions between the mineral surface and the brine and oil, together with the brine–oil interactions. A schematic representation of various forces and interactions that can be involved between mineral surface, oil, and brine has been described by the following crude oil–brine–rock (COBR) ternary diagram (Fig. 2.5). The interactions include solid–liquid (oil–rock, rock–brine) and liquid–liquid (oil–brine) interactions. Introducing LS brine disturbs the

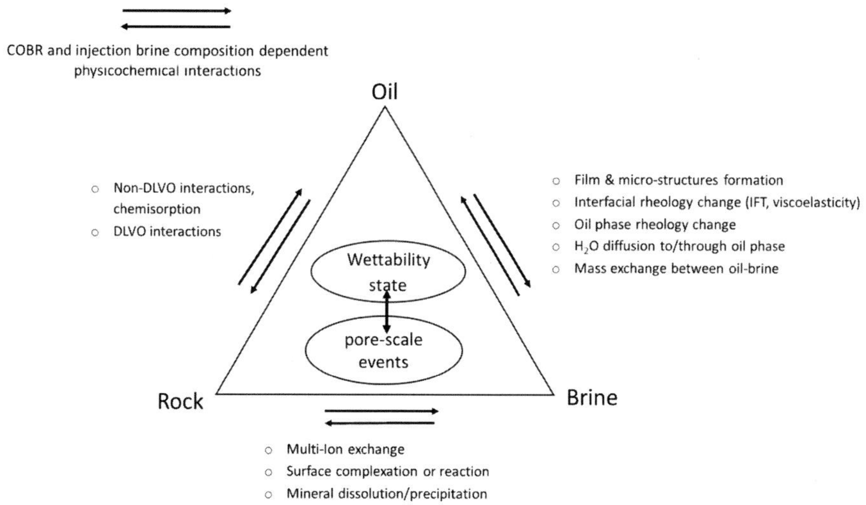

FIGURE 2.5

COBR ternary diagram highlighting different types of physicochemical interactions relevant to LSWF. These interactions are strongly brine composition dependent. LSWF efficiency is controlled by those interactions which are most influenced by LSW. *COBR*, Crude oil–brine–rock; *LSWF*, low-salinity waterflooding.

thermodynamic equilibrium of the system; thus different types of interactions are triggered. The type and magnitude of the interactions depend on brine chemistry and its interaction with rock and oil, which determines the wettability state of the reservoir system. In addition, when the LS water and the in-situ formation brine mix, fluid compatibility adds another element of complexity.

2.2.5 Low-salinity waterflooding "cause and effect" issue

The LSE is the observation that lowering the brine salinity or changing the brine chemistry during waterflooding can increase oil recovery. While there are many successes, there are also many failures. Statistical analyses of laboratory results have been used to try and identify important variables and develop predictive relationships with only limited success [40]. A large number of variables and a lack of systematic testing limits this approach since most experiments are designed only to assess response in specific reservoirs. Ultimately, we need to understand the basic mechanisms of the LSE to develop mechanistic predictive models. This requires designing and executing experiments that can separate *cause* from *effect* rather than relate changes in recovery to changes in water chemistry. In fact, there are few systematic investigations of the LSE in carbonates [41], rather most investigations have focused on the application of LSWF to specific systems of economic interest. These experiments cannot provide sufficient data to evaluate all the potential interactions in the experimental system. In part, this is due to a large number of interactions between oil, brines, and surfaces and the many methods to measure them as summarized in the COBR diagram above.

There have been several mechanisms proposed to explain the LSE. The early work on LS was focused on sandstones and based on observations made during formation damage testing related to clay mobility [42]. The groundbreaking work by Austad and his collaborators revealed the same effect can be found in carbonate rocks [24,43,44]. This finding has triggered research by several groups and fostered the range of mechanisms proposed. Table 2.1 lists the proposed mechanisms and includes both sandstone and carbonates for reference.

The mechanisms can be divided into two main groups: solid–liquid interactions and liquid–liquid interactions. The solid–liquid interactions occur at the contact point of the oil–brine–rock interface, while liquid–liquid interactions occur at the oil–brine interface. In recent years there has been a shift in better understanding the liquid–liquid interactions. The relative importance of the two interactions is dictated by the initial condition of the COBR system and the injected brine chemistry. Some of the mechanisms require flow conditions, such as emulsification, viscoelasticity changes and fines migration, while others can occur without flow, such as cation exchange, osmosis, microdisperion formation, EDL expansion, and IFT change. This means that during flow, additional interactions occur which contribute to the efficiency of LSWF and those cannot be accounted for by merely doing static tests.

Table 2.1 Proposed mechanisms of low-salinity waterflooding relevant to carbonates.

Group of mechanism	Mechanism	Key references
Solid–liquid	Surface reactions (adsorption–desorption of oil components)	[45–47]
	Surface charge change by complexation or MIE/electric double layer expansion	[34,48–51]
	Fines migration and flow diversion	[52–57]
	Mineral dissolution (e.g., calcite, anhydrite)	[58,59]
	pH change	[60,61]
	Asperities	[62]
Liquid–liquid	Interfacial tension change, emulsification	[60,63–65]
	Microdispersion formation	[66,67]
	Salt in effect	[68]
	Osmosis	[69,70]
	Changing oil–brine interface rheology	[71,72]
	Natural acid generation	[73]

MIE, *Multi-ion exchange.*

Review of the mechanisms proposed shows that most researchers identified wettability alteration or some form of surface reactions as the controlling process (or the main phenomena). Wettability has a significant effect on oil recovery [74]. Wettability alteration leads to an increase in the fraction of water-wet pores or an increase in oil-wet pores depending on the direction of shift. The use of the term wettability can be confusing. Strictly speaking, wettability is the tendency of a surface to prefer contact with one liquid versus another liquid. The common contact angle measurement best represents this definition but may have limited applicability to complex reservoir systems.

In petroleum systems wettability involves mineral surfaces with oil and water, that is, crude oil and reservoir brine. Wettability or wetting is usually measured by techniques that include contact angles, spontaneous or forced imbibition, capillary pressure, and nuclear magnetic resonance (NMR) response in cores [75], however, wettability is also the principal control of multiphase fluid flow and determines the permeability of crude oil in the presence of brine, a.k.a. relative permeability. Therefore directly relating a static contact angle measurement to flow properties must be made with some caution.

Reservoir wettability is the macroscale expression of multiple nanoscale interactions between rock surfaces, brine, and oil (solid–liquid interactions) that together with liquid–liquid interactions control the movement of oil and brine in the porous medium. There are multiple mechanisms by which oil wets the rock surface including direct contact without a water film, surface precipitation of organic molecules, such as asphaltenes, electrostatic interactions between polar portions of the oil, and the rock surface through a water film and ion binding between charged sites and higher valency ions [76,77]. EDL expansion and interfacial properties are sometimes identified as mechanisms [78,79], but the multiple surface reactions involved in wettability show how difficult it is to separate *cause* from *effect*. In the absence of such wettability alteration, the LSE has been also observed in some COBR systems which are based on favorable liquid–liquid interactions (Table 2.1) at oil–brine interface leading to increased oil phase connectivity and decreased break-up and trapping. Both can occur simultaneously with variable degrees depending on the brine composition and initial wettability of the COBR system.

Clearly, some proposed mechanisms, such as clay migration, are unique to sandstones and not applicable to carbonates [52,53]. However, other mechanisms may apply. Nor is it clear that a single mechanism is an answer. For instance, many experiments observe a shift in pH to more alkaline values and a shift in pH may trigger the generation of natural surfactants from oil. This situation leads to the "cause and effect" issue. The experimental design that can isolate and test a single variable is essential but often very difficult to achieve.

LSWF experiments show several common observations. First, in cases that exhibit the LSE, there is an increase in incremental recovery after salinity is lowered. Usually, a time lag (delay) is noted before the oil recovery increase can be detected in the core outlet which is related to oil bank buildup, the establishment of LS condition in the flooded zones and the electrokinetics involved in the process [48,80,81]. Most of the incremental recovery occurs after the first pore volumes (PV) following the change in water chemistry. Additional PV of injected water can yield small amounts of incremental oil. Recovery increases with salinity reduction until recovery reaches a plateau and further reduction in salinity does not produce additional oil [43]. This implies that there is an optimum salinity and further salinity reduction is not fruitful or negatively affects the wettability state [82,83].

2.3 Low-salinity waterflooding from laboratory to field

While the attraction of LSWF is great, an essential step in implementing any new technology in the industry is field studies. The basis for field studies is successful laboratory studies that generate confidence and predictability.

Fig. 2.6 shows a representative distribution of incremental recovery from several LS experiments. This data shows that the LSE can be significant to enhance recovery. The data show that while about one-half of the studies found between 0% and 5% incremental OOIP, many studies had 5%–15% and a few reached recovery as high as 25% OOIP in carbonate rocks for both imbibition and coreflood experiments. These levels of recovery are similar to or higher than CO_2

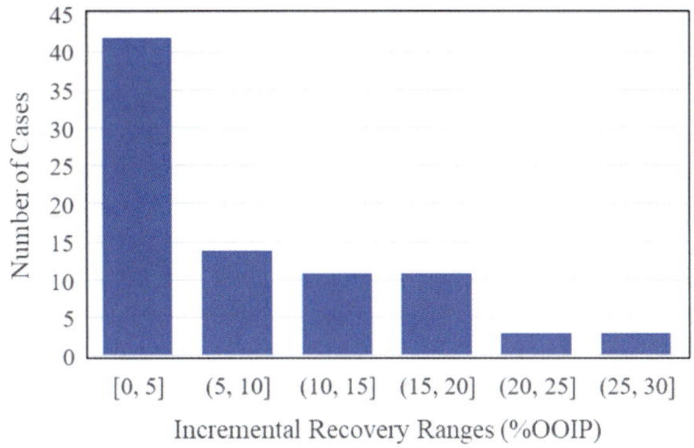

FIGURE 2.6

Histogram of ranges of incremental recovery for experiments using carbonate rocks. Recovery data include spontaneous imbibition and forced imbibition (coreflood).

From Refs. [41,43,45,85–87].

EOR [84]. Most of these experiments did not measure the initial or final wettability thus there is no way to correlate wettability shift and recovery. Nor were most experiments the result of systematic salinity change, but instead just employed various degrees of dilution.

Published studies can be divided into eight categories based on the techniques used: (1) core flooding experiments; (2) spontaneous imbibition (SI) experiments; (3) wettability assessment and contact angle measurements; (4) interfacial tension (IFT) and interfacial rheology measurements; (5) microscale experiments (micromodels and microcomputed tomography); (6) surface image and composion analysis using SEM-EDX, AFM, XPS, SAXS, etc.; (7) mathematical modeling; and (8) simulations. Many studies combine several of these techniques. Modeling and simulations include the use of modified reservoir simulators, molecular dynamic (MD) simulations, and geochemical surface complexation models and solute modeling—covered in the next sections.

The coreflooding and imbibition experiments are designed to measure increased recovery, which is the most practical aspect of LSWF, but few of these experiments measure related parameters, such as wettability, oil composition, and effluent chemistry. For instance, directly comparing coreflooding data, which is widely used in Middle Eastern studies to imbibition data from North Sea chalks, is difficult. Thus the reconnaissance data from coreflood and imbibition experiments often have limited value for the scientific questions regarding mechanisms.

The contact angle and IFT data, simple measurements that are proxies for wettability, can be useful but contact angle measurements do not involve natural surfaces and IFT changes have been shown to have little relationship to LSWF recovery. The contact angle data do show that lower salinity appears to shift wettability to more water-wet conditions in most cases, but wettability shift alone may not increase recovery.

Microscale experiments and surface analysis explore the LSE at the finer scale offering more insight into fundamental mechanisms but offer little immediate knowledge at the larger scale of application. Modeling studies test the various methods to incorporate predictive abilities into standard reservoir simulators. The creation of reliable simulators is a key component of moving LSWF to the field, but currently, there is no agreement on fundamental mechanisms, thereby limiting models to empirical data matching by changing the relative permeability curves.

2.3.1 Length- and time-scale consistency

LSWF is a multiscale process where several length scales are involved between the scale that the effect is initiated (atomistic-molecular scale) and the scale that increased oil recovery can be observed or measured (at and beyond pore network-scale, > a few mm; more conveniently core-scale). The order of change in scale between the molecular scale and core-scale is 10^7, which is very large. LSWF can be tested/evaluated at different length scales for different purposes (Fig. 2.7). For

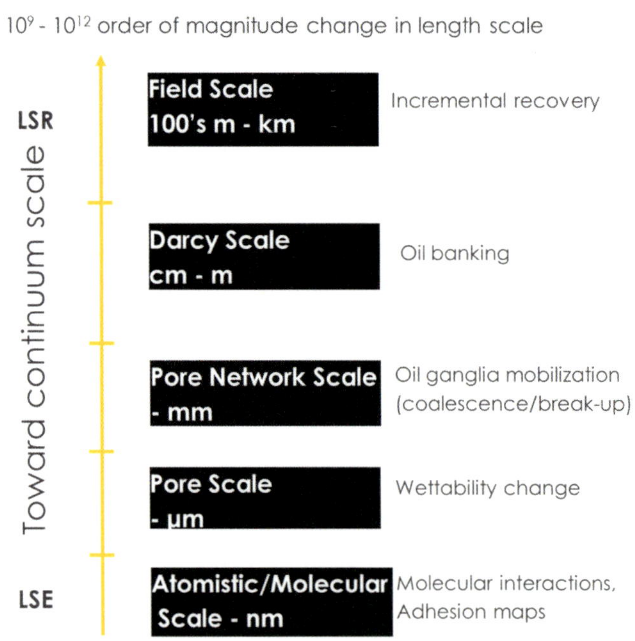

FIGURE 2.7

LSWF is a multilength-scale process, spanning from atomistic/molecular scale to field scale. *LSWF*, Low-salinity waterflooding.

instance, measurement of recovery is irrelevant if the system size under investigation is below the representative elementary volume (REV). Further, studies using AFM or even contact angle measurements are aimed at studying mechanisms not recovery. Studies at smaller scales can yield valuable information on the LSE, while studies at scales larger than REV provide information on low-salinity recovery (LSR) [80].

LSR is a manifestation of all the small-scale LSEs on a scale sufficient to characterize by volumetric measure. It has been also noted by Bartels et al. [80,88] that the small-scale effect or LSE cannot directly lead to a large-scale effect because the intermediate scales play an important role in transferring the LSE to LSR. For instance, if a contact angle change or wettability alteration (or the LSE) does not lead to oil detachment from the rock surface, no LSR can be observed. Or if the LSE causes local re-distribution of oil but does not lead to changes in local saturations and oil bank formation, no LSR is observed [89]. The multilength-scale dependency may explain the success or failure of some core-scale experiments. In this respect, any small-scale experiment in which the LSE is observed, should not be directly used for large-scale predictions without considering what happens at intermediate scales.

Another aspect is the time dependency of the LSE. When LSW is injected, slow and fast interfacial phenomena occur at various solid–liquid and liquid–liquid interfaces (see Bartels et al. [80] for further discussion). Wettability alteration and osmosis, as they require electrodiffusion through the water film, and water diffusion through the oil ganglion, respectively, can be slow in order of minutes to hours/days [90], but some processes occurring at oil–brine interface, such as IFT change and viscoelasticity, can be relatively faster. Therefore a range of slow and fast phenomena occur influencing the timing of observing the LSE/LSR. If most of the remaining oil is sticking to the solid surface (as in oil-wet and mixed-wet systems), then the time scale of wettability alteration would determine the response time. If the remaining oil is mainly in the center of pores (as in more water-wetting systems), then the time scale of IFT and viscoelasticity change would determine the response time of LSE. The wide range of time scales spanning seconds to hours/days should be accounted for in the design of laboratory experiments to avoid false-negative results and modeling/simulation of LSWF, and also in the field design.

2.3.2 Assessment of LSWF efficiency

The recommended workflow for laboratory assessment of LSWF in carbonates is outlined below. First, an initial screening or identification phase should be performed. This initial screening is based on the field conditions and uses the screening criteria described later to identify the prospective candidates and provide the basis for feasibility studies and technical, economic, commercial, organization, and political risk assessments [91–93]. Laboratory assessment is not useful for candidate fields that do not meet the initial screening criteria. Assuming the candidate reservoir screening provides positive indications, the next step is a laboratory assessment. Laboratory assessment can involve several levels of tests (Fig. 2.8). This stage can involve successive or simultaneous testing methodologies and scales. The goals are to establish if there is an LSE and, depending on methodology, quantify the additional oil production (IOR or LSR).

Subcore-scale experiments are used to establish indications of the LSE. This can include tests probing the importance of solid–liquid interactions, such as wettability/contact angle tests to assess formulations or test other water sources available to an asset. This may use rock thin sections or cleaved calcite crystal surface with reservoir crude and brines. Complementary tests (optional) such as zeta-potential measurement can supplement and provide insight into the contact angle results. Another evaluation or reconnaissance method is the modified flotation test to quickly measure the wettability alteration and screen candidate COBR systems [94].

Tests probing the importance of liquid–liquid interactions are available and use interfacial rheology (viscoelasticity) and IFT measurement. Assessment of microdispersion formation potential can be useful as well [67]. However, the predictive capability of these methods for increased recovery is not yet established for a wide range of COBR systems. Finally, microscale experimentation (optional) for visual observation to support the results of initial screening and confirm an increase in oil mobilization and recovery at the REV scale is available.

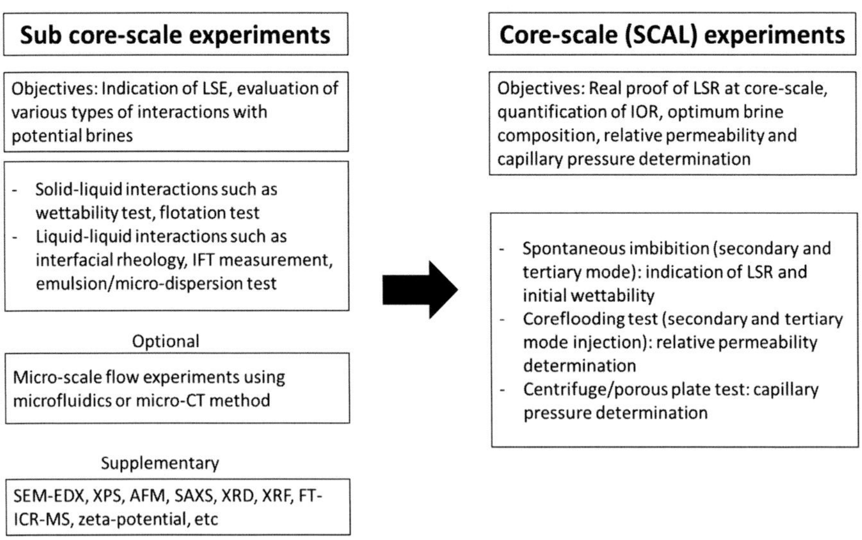

FIGURE 2.8

LSWF laboratory assessment workflow. *LSWF*, Low-salinity waterflooding.

Core-scale or special core analysis laboratory (SCAL) experimentation is currently the most reliable proof that LSWF works. There are several test methods used to determine the magnitude of the LSR and the suitable brine composition. One common method is spontaneous imbibition (SI) or the Amott test. Although it produces quantitative data, the test is often treated as "qualitative" or "indicative" and is used to establish if there is LSR at the core scale before more sophisticated tests are done, as well as test core initial wettability (as the "true" confirmation of wettability in a porous medium and at core scale) and its alteration by LSW. The basic procedure is straightforward. The core is initialized with oil and brine and is then placed in target brine and any displaced oil volume is measured over time. Imbibition of the brine is represented by the positive part of the imbibition capillary pressure (Pc) curve. This can be highly relevant in fractured reservoirs (where imbibition can be an important production mechanism) or where the oil has been trapped due to local heterogeneity.

SI is simpler to perform and less labor intensive than corefloods. There is less issue with core heterogeneity as the imbibition data are not used in the same way as coreflood data (dynamic flow data) in reservoir models. Typically, a larger incremental oil volume is produced in SI than in coreflood tests facilitating accurate measurements. The imbibition data obtained in both tertiary and secondary modes can also be used in the interpretation of the coreflooding experiments. However, it is not straightforward to use the imbibition results for the field scale unless it is upscaled properly. Moreover, reservoir simulators do not use SI data directly but instead require Pc curves derived from the data.

A baseline with HS/formation water is necessary to provide the basis to evaluate other fluids potential for wettability alteration. This means performing tests in tertiary mode as well as secondary mode. Typically, if SI does not show an effect, little or no effect is observed in coreflood, but this limited observation needs to be verified with a larger dataset. Based on the authors' experience, there is some correlation between SI recovery and coreflood recovery (unpublished) but not enough for rigorous prediction compared to core flood data.

Corefloods are the largest scale practical laboratory test. Data from coreflood tests are upscaled to field conditions in reservoir models. These models can provide economic evaluations and design pilot projects. Pilot project results will further modify the reservoir models and refine economic analyses. Forced imbibition includes unsteady-state (USS) and steady-state (SS) corefloods and centrifuge tests. Core flooding is under the visco-capillary regime and it is argued that this condition is more relevant to the field as we have forced displacement during water flood, particularly in nonfractured reservoirs. In coreflooding oil-wet and mixed-wet reservoirs, the negative part of the Pc curve is most relevant. Centrifugation (forced displacement) can be used to derive the Pc curves and the relative permeability of the displaced phase, particularly near the S_{or} which is difficult to measure with precision with coreflooding. Any risk of formation damage can be best identified during the testing, with the most common signature being the continuous increase of pressure drop over the core. However, the technique is more time-consuming and the results are sensitive to the protocol used to conduct the flooding. Sample homogeneity is particularly important for uniform initial saturation and displacement to facilitate data interpretation.

USS/SS data interpretation is straightforward. The USS/SS methods do not measure relative permeabilities or try to predict but rather provide estimates. For USS data interpretations, JBN [95], modified JBN [96], or Toth et al. [97] methods are available. These methods are analytical (i.e., the governing equations are solved directly), are rapid, and produce unique curves. However, they mostly ignore the Pc effect or assume sufficiently high flow rates during flooding that the capillary-end-effect is negligible. These methods do produce after-breakthrough relative permeability curves. Semianalytical methods include the Civan and Donaldson Method [98] and Udegbunam Method [99].

Another option is numerical interpretation with a (numerical) flow simulator: This approach is an inverse problem entailing history matching (typically) the pressure drop and oil recovery curves obtained during the two-phase displacement to find the relative permeability curves which match the transient or steady-state data. This approach is recommended compared to analytical methods; however, independently derived Pc curves are necessary. These can be (alternatively) obtained by measuring in-situ saturation profiles along core via X-ray scanning as a function of pore volume injected (PVi). The in situ saturation data should include the initial saturation and final saturation as well as intermediate saturation profiles along the core length. It should be noted that numerical simulation cannot be used to correct bad data, but rather to account for capillary pressure in the relative permeability estimation.

The general issues and challenges encountered in SCAL testing include the inherent heterogeneity and core contamination from mud invasion and additives during coring. This problem can pose serious issues in selecting suitable plugs, let alone identical (twin) plugs. The sample selection can be aided by using CT scanning, mercury intrusion capillary pressure, NMR (pore and throat size distribution), and porosity-permeability measurements to identify the rock types and the most homogeneous plugs suitable for testing. The lack of sufficient reservoir core for detailed experiments is a common problem. The use of model rocks, such as outcrops and 3D-printed rocks, may be of interest in understanding mechanisms, but the results cannot be related to fields as different recoveries can be obtained owing to the different surface chemistry of the reservoir and outcrops [74].

The current, specific challenges and issues in the laboratory assessment of LSWF are centered on the lack of standard workflows and protocols with uniform measures (core preparation, SCAL program design, and execution, data interpretation). This situation has produced inconsistent and incompatible results by different groups, core laboratories, and universities. The conventional workflow and assessment methods used in SCAL and reservoir engineering as outlined in Bartels et al. [80] are not adequate. The lack of standard workflow makes comparing the results, metadata analysis, and correlations difficult and inhibits arriving at conclusive statements about mechanisms. For instance, few studies measure both incremental production and the change in wettability, a critical shortcoming. Finally, to be representative, core-scale experiments should be performed under reservoir conditions. Proper assessment of LSWF requires an adequate amount of reservoir samples and is costly and time intensive. A more rigorous approach that is widely applied is necessary.

2.3.3 Considerations for the design of laboratory experiments

2.3.3.1 Core preparation steps

It is recommended to use "restored cores" than preserved or "native state" cores. This allows having more control over the core preparation procedure: cleaning, initialization to S_{wi}, aging, and flooding. Initialization by oil flooding is often not adequate to reach immobile connate water saturation. Here, thinning (and breaking up of) the water film is key to ensure achieving representative initial wettability after aging, unless sufficiently high viscous pressures are applied [100]. The porous-plate technique ensures reaching a uniform saturation in the core but is time intensive. The centrifugation method can give acceptable results in terms of saturation profile in the core and takes less time.

Determining the relative permeability as function of salinity/brine composition requires dedicated experiments at the target salinity and measurement of the core output. The coreflooding can be performed under SS and USS conditions. The USS is most common, while SS has the capability to produce the relative permeability under the full range of mobile oil saturation. In the industry, this has been only reported in a few studies [101].

A representative core initialization protocol is crucial. Whether LSW is going to be injected in secondary or tertiary mode, the core has to be initialized and aged with full composition formation water or HSW. In brine preparation, it is common to remove some ions, such as Sr^{2+} and Ba^{2+}, that can produce inorganic scales, but they can be essential for fully restoring core wettability [102]. Core initialized with LSW would result in unrepresentative initial wettability and oil recovery due to different initial oil distribution, displacement mechanisms, and oil trapping. It has to be recognized that injecting LSW triggers a dynamic wettability alteration process that is directly related to the initial conditions.

2.3.3.2 Coreflooding experimental protocols

The USS core flooding can be performed in both secondary and tertiary modes. It has been reported that the LSWF recovery depends on the timing of the LS injection. In the secondary mode, there is a higher oil saturation in the core which can be targeted by the LS. There would be also less issues with mixing between in situ HS and the injected LS brine. Therefore higher efficiency of the flood via the quicker establishment of LS in the core is observed. In the tertiary mode, the mobile oil saturation range is definitely smaller than that in the secondary mode and oil can be less connected and present as disconnected oil ganglia [103]. The job of LS is to access/invade new pores, establish a new flow path, and make the oil get re-connected, which would not be easy, requiring mobilization of oil ganglia, increasing their coalescence into larger ganglia and oil bank formation.

Unlike the sandstone reservoirs, LSWF in carbonates is not necessarily about reaching a threshold total salinity or range which is typically below 10,000 ppm TDS. It has been shown in many studies that diluted formation water, seawater, modified seawater, and diluted seawater can all be potential brines [23,43,104,105]. Therefore the experimental design should cover a wide range of brine salinity and PDI concentration, particularly testing those brinesthat are practically accessible in the field.

- Core size and orientation:

The core orientation in the flooding, particularly for long cores, is recommended to be vertical. A bottom-to-top water injection is recommended to ensure gravity stable flow and minimize gravity underrun and viscous fingering. Short cores (e.g., 5 cm long), as mentioned before, have issues related to 1D flow requirement and capillary-end-effects. The end effect is related to the capillary discontinuity at the core end face (outlet) where outside the core and at the core face the $Pc = 0$, while inside the core it is negative (for oil-wet cores); this capillary pressure gradient forces the oil accumulation near the outlet of plug.

Long cores (e.g., 30 cm or longer) are recommended if they are homogeneous. By using long cores, the pore volume increases which is also better for the observation of any produced oil bank and its quantification. However, long cores are more difficult to get. The use of a composite core, made of several short core plugs, can be problematic due to the issue of capillary contact between the plugs leading to capillary-end-effects due to near-zero Pc in the opening between the two adjacent cores, as well as differences between core properties, which make data interpretation difficult.

- Coreflood injection rate:

It is often argued that coreflooding should be performed at field relevant or lower rates, such as 0.03–2 m/day [106], to mimic the capillary number expected in the field. This poses two issues: (1) this is not generally the case in all field operations and very much depends on the reservoir settings, well completion, distance from well, and injection rate and (2) the capillary number corresponding to low flooding rates usually ranges from 10^{-8} to 10^{-6}, which designates a capillary-viscous regime dominated by capillary pressure. At low rates, it is not possible to properly quantify waterflooding efficiency in the coreflood, particularly in mixed- to oil-wet cores and in short cores (e.g., $\leq$5-cm long). Becasue at these conditions, low rate injections do not create enough viscous pressure to push the oil out of the core, which then results in capillary-end-effect artefact or oil accumulation near the core outlet. The capillary- end-effect zone can cover about a one-third to half-core length at low rates. This can often lead to overestimation of the benefits of LSWF and EOR in tertiary mode.

To overcome the end effect, the injection rate can be ramped up (known as bump rate) to increase the viscous forces and produce the accumulated oil at the core end. The test can also use long cores or put in oil-wet end pieces. The Rapoport–Leas [107] calculation can be used to estimate if the capillary end effect is an issue. If the core is CT scanned during the experiment, the in-situ saturation profile can be determined and monitored to identify any end effects.

As a bare minimum, it is necessary to collect oil production volume and pressure drop as a function of the PVi with sufficient frequency to properly determine the breakthrough time and postbreakthrough recovery. Experimental duration (or PVi) should be sufficient to ensure that production has ceased or reached equilibrium before the residual or remaining oil saturation can be calculated or injection of another brine. Pressure profiles should be recorded to signal oil bank formation or formation damage in the core.

The accuracy of oil volumetrics, the basis of material balance, is particularly important during LSWF. Because incremental recovery with LSW is often modest ($<10\%$) and can accumulate over many PVs, accurate measurement of oil volume as low as 0.05 cc may be necessary—see also the core analysis book [108] and the guideline in this paper [109].

Minimization of dead volumes (in the inlet and outlet sides) in the design of the coreflooding apparatus will reduce uncertainty in timing and oil recovery calculations. Use of techniques other than direct production, such as in-situ saturation monitoring—most commonly with X-ray CT scanner, and more recently T2-NMR [110]—along with the core and as a function of PVi are highly recommended. The data can be used in SCAL interpretation and help identify the capillary-end-effect.

The core effluent can be collected using fraction collectors and specifically analyzed to gain insight into the events/mechanisms that occurred during flooding; oil and water are usually separated before measurement and individual samples are collected for later chemical analysis. Produced water analyses can be used to determine changes in water chemistry, such as pH, conductivity, resistivity, and ion balances, and monitor fines production. Continuous measurement of conductivity or resistivity can detect the injected water breakthrough. Discrete oil samples can provide information on compositional changes as injection salinity is altered.

In some advanced core flooding equipment, it is possible to monitor the geomechanical behavior of rock as brine salinity/composition changes as well as core creep, elongation, or compaction/shrinkage, that is, geomechanical effects. The pore size distribution can also be remeasured to understand which pore sizes were affected most by the LSWF. Finally, the measurement of core permeability/porosity postcoreflood can help identify any rock dissolution or formation damage.

- Relative permeability and S_{or} change by LSWF

One of the most reported effects of LSW is a shift of relative permeability curves from a more-oil-wetting state to a more-water-wetting state. With this change, it is anticipated that oil connectivity is increased, and an oil bank is formed. This should lead to an increase in oil relative permeability as seen in Fig. 2.9. However, to observe a pronounced oil bank it is necessary to use a longer core than the standard 10–15 cm length. Otherwise a gradual rather than a

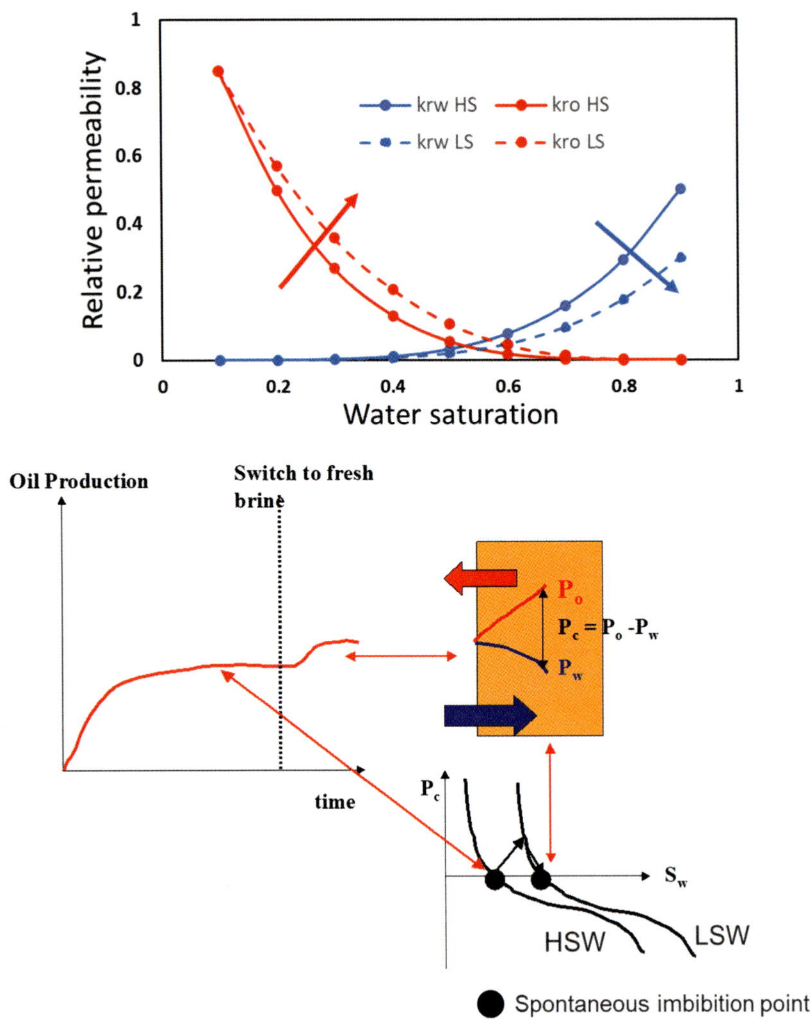

FIGURE 2.9

Oil–water relative permeability and (imbibition) capillary pressure curves under high- and low-salinity conditions.

sharp oil bank production is observed (see also the recent calcite micromodel study by Mohammadi and Mahani [111]).

Under more-water-wet conditions, it is expected that the water-end-point is decreased. If the water end point is less than 0.1 it can be considered strongly water-wet. A reduction in water end point has been reported in many studies; however, in some studies, the water end point and relative permeability did not change or were assumed to be unchanged [112,113]. Under a more-water-wet condition, oil relative permeability is increased concomitantly with either S_{or} decrease or no-S_{or} change. It should be noted that the starting point of oil relative permeability in both HS and LS is the same because it is the initial condition of the core with both brines.

While the main aim of LSWF is to accelerate the production compared to HSW, it is desirable that S_{or} is also decreased by LSW to increase the range of mobile oil saturation and recovery. But it is also likely that the S_{or} does not change with salinity or the change be negligible. This is quite possible because the capillary pressure reduction (either via IFT reduction or contact angle reduction) by LSW may not be enough to reduce the residual oil, compared to recovery by chemical EOR with surfactants where, typically, one to several orders of magnitude decrease in capillary number are required to reduce S_{or} significantly according to the capillary desaturation curve.

The increase of S_{or} by LSW is debatable. As discussed earlier, LSW can shift the wettability to a more-water-wetting state. It is reported repeatedly in the literature (observations both in corefloods and in pore-scale experiments) that a more water-wetting state causes further trapping of oil. However, this view is applicable to and relevant for two-phase flow under fixed wettability conditions. LSWF is a dynamic wettability alteration process, which divides porous medium into two regions: the region of altered wettability (more-water-wetting) and the region at original wettability (e.g., mixed-wet to oil-wet) as the flood advances. Due to the wettability alteration, oil blubs are released and bank-up along with the interface between HS brine and invading LSW. Thus rock becomes more wetting behind the front as the oil is mobilized. It is therefore not envisaged to be trapped behind the front [80,114]. Detailed pore-scale studies can shed further light on this.

- Capillary pressure changes with LSW

By the virtue of making rock more water-wet, the positive part of the Pc curve is extended, and the spontaneous water imbibition increases. Thus the point of SI at which Pc = 0 shifts to a higher water saturation (see Fig. 2.9). The negative Pc is controlled by the required entry Pc to invade the oil-wet pores and the smallest pores containing oil. The negative Pc also is expected to be reduced at saturations between SI and S_{or}.

In oil-wet fractured reservoirs, where there is no SI by formation water (i.e., Pc = 0 at S_{wi}), a positive change in capillary pressure by using an LSW can increase matrix drainage.

2.3.4 Field candidate screening criteria

The field screening criteria are not robust yet, and any screening should be confirmed via the workflow discussed earlier to validate the LSE and quantify the LSR. However, there are several key factors that should be considered in screening.

2.3.4.1 Initial reservoir wettability

Initial wettability is influenced by the formation of water salinity/composition, rock type, mineralogy, and oil composition, as well as reservoir temperature and pressure. The presence of pH buffering agents, such as dissolved CO_2 and carbonate/bicarbonate ions, and silicate minerals, such as clays/feldspar/plagioclase, can affect the initial rock wettability. It is often reported that the mixed-wettability and weak oil-wet states are the most suitable condition for wettability-alteration-induced LSE. Strongly oil-wet or strongly water-wet cases have been found to react less or the degree of alteration is limited [19,48,115,116]. In these cases, oil adheres very strongly to the rock surface and is difficult to detach by change of brine composition only.

The starting oil saturation is important. If oil is already present as a connected phase, e.g., in the secondary LSWF, it should be ensured that oil break up is reduced. If oil is a rather disconnected phase, it is important to increase coalescence of the oil ganglia. Thus different brine formulations would be necessary to make this work in the right direction. This is strongly a function of brine composition, which must be meticulously optimized to reduce the electrostatic repulsion between the droplets (something which is increased under LSW) while increasing the interface rigidity to help coalesce the droplets. These are two opposing mechanisms [117].

The basic assumption in many of the LSWF experiments is that increasing water-wet conditions are favorable for oil production and that, as mentioned earlier, waterflooding in oil-wet to mixed-wet reservoirs is often described as not very successful. However, simply shifting the wettability of the rock surface toward a more water-wetting state is not always sufficient to increase recovery. Recent findings from Chávez-Miyauchi et al. [71,118] show that in water-wet cores and even in the absence of wettability alteration, the liquid–liquid interactions that produce interfacial rheology changes toward more visco-elastic behavior can increase the oil phase connectivity, facilitating coalescence or reducing oil break-up [64,119]. Some of the increases in recovery may be wholly or in part due to oil rheology changes instead of rock wetting changes.

Many authors report that wettability shifts toward more water-wet with lower salinity and this wettability change is linked to increased recovery. However, these data are based on contact angle measurements that are relevant to oil-rock wettability on polished surfaces, not reservoir rocks. Many studies show the best recovery occurs at neutral wettability [74,120–126]. Therefore experiments that measure only surface-liquid wettability changes (i.e., contact angles) may not fully capture the LSE that leads to greater recovery as mentioned above.

Many studies show the best recovery occurs at neutral wettability [73,114–120]. However it is important to note that oil recovery expectations based on the initial wettability are applicable to waterflooding experiments where the whole period of displacement is at fixed or unaltered wettability conditions. During LSWF, the wetting condition can be altered dynamically therefore the relevance of those results should be carefully assessed (see also earlier discussion on S_{or} change). In addition, the high S_{or} data in the oil-wet conditions is still debatable, but most groups agree on the S_{or} trends under neutral and water-wet conditions. Examination of the laboratory data from oil-wet cases (which form the basis of this conclusion) suggests the influence of capillary-end-effect during these experiments that was not considered in the S_{or} estimation. Given that the relative permeability of oil is extremely low in oil-wet systems near S_{or}, many PV are required to establish true S_{or}, a condition not met in many experiments. Findings from Humphry et al. [127] suggest that S_{or} is likely lowest under oil wet conditions.

Considering the possibility of different mechanisms during LSWF, the degree and direction of the shift in wettability, and the timing of LSWF injection, it would be difficult to predict which initial wettability state is most favorable for LSWF application. Normally oil-wet to mixed-wet systems leave more oil unwept (before the water breakthrough time) than water-wet systems because of high water relative permeability; therefore this wettability state is (technically and economically) favorable. However, it has been observed that mechanisms relying on liquid–liquid interactions (such as oil-brine viscoelasticity) may have the possibility to dominate in water-wet systems, and may not be the primary mechanism in oil-wet to mixed-wet rocks [128], but are additive (synergistic) to solid–liquid interactions. Based on the insights gained from the published literature highlighting the dependence of solid–liquid and liquid–liquid interactions on the initial wettability, the expected IOR versus initial wettability index can be shown as in Fig. 2.10.

2.3.4.2 Formation brine composition and pH

Formation brine salinity, composition, and pH would impact the initial wettability as discussed earlier. Increasing brine salinity shifts wettability to more oil-wet [51,129]. Formation water salinity can be more than 200,000 ppm in some middle eastern reservoirs and about 90–100,000 ppm in the North Sea. The super-giant Ghawar oil field in Saudi Arabia has a salinity of 225,000 ppm. Ekofisk salinity is about 68,000 ppm. The HS condition is thought to be favorable for the oil-wet condition, but extensive measurements show that the Ghawar field has near-neutral wettability [130]. Formation brine pH may be an even greater controlling factor on the reservoir wettability, than salinity. The buffering capacity of the carbonate formation controls pH variation, resulting in brine pH typically in the range of 7–8 [7,8,19]. The pH range (6.5–7.5) at formation water conditions is

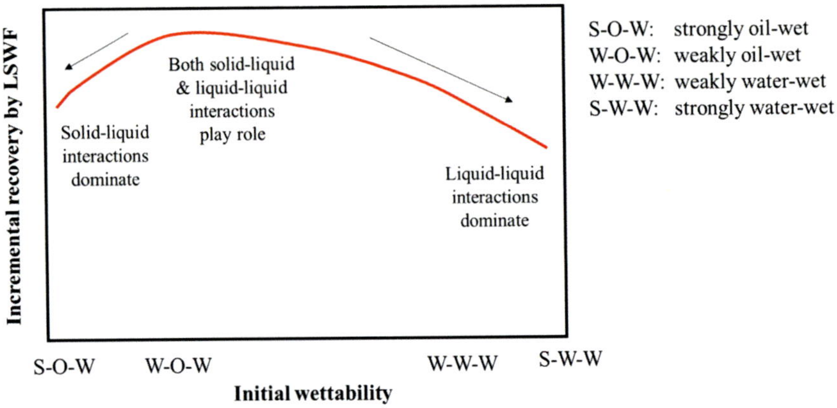

FIGURE 2.10

Expected incremental oil recovery by LSWF versus initial wettability. Under strongly oil-wet conditions, wettability alteration by LSWF is more difficult than in weakly oil-wet conditions due to the strong adhesion of oil to the surface. *LSWF*, Low-salinity waterflooding.

likely to generate oil wetness suitable for LSW application [131]. However, lower pH values <5 are likely to trigger water wetness and limit brine salinity influence.

2.3.4.3 Oil composition

The presence of polar oil components is key in LSE. Model laboratory oil free from any polar groups produces weak interactions with a rock when brine salinity is changed. With nonpolar model oils, the initial wettability is toward water-wet [103]. The polar fractions of oil are usually found in the resin and asphaltene fractions and are sometimes characterized by total acid number (TAN) and base number (TBN). Most acids will be negatively charged above pH of 6 (due to deprotonation of carboxylic functional groups), while the bases will be positively charged below pH of 5. The TAN and TBN measurements do not identify specific acids and bases in oils but instead quantify the total amounts. Experiments with specific acids and bases added to laboratory oil (e.g., decane) have shown that specific acids and bases may not have the same effect on wettability [132]. Most oils have more bases than acids although the ratio is variable. Fig. 2.11 shows a compilation of acid and base numbers for 177 oil samples from around the world. The data show there are three general groups, acid oils (TAN/TBN $\approx$ 5), neutral oils (TAN/TBN $\approx$ 1), and basic oils (TAN/TBN $\approx$ 0.2) but that about 75% are basic oils.

Although TAN/TBN of oil characterizes the acid and base content of oil, there is no strong correlation between TAN/TBN, asphaltene content, and LSR [133]. Both TAN/TBN can be important because of their charge difference and the in-situ pH condition which affects the carbonate rock charge. However, the presence

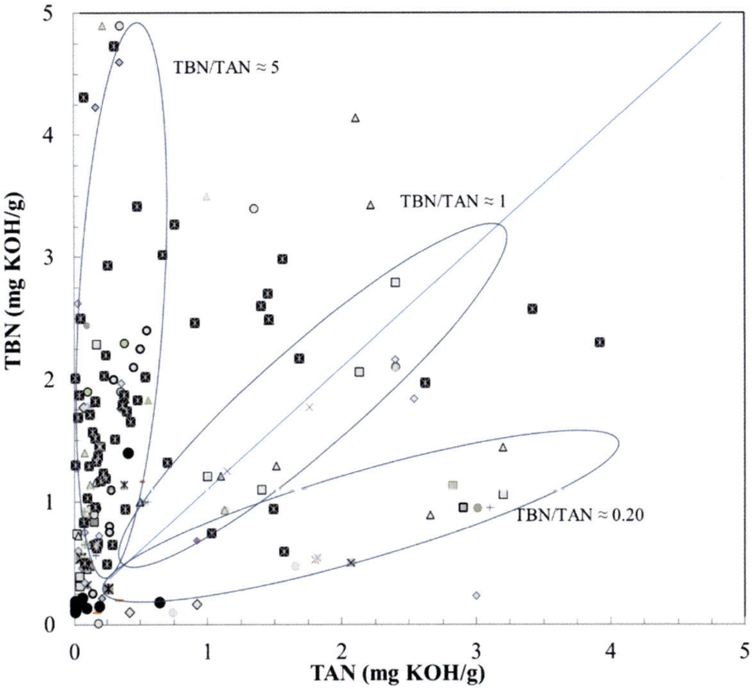

FIGURE 2.11

Total base number versus total acid number for 177 oils.

of asphaltenes (with sulfur and oxygen atoms), oxygen-bearing organic groups with carboxylic and long-chained acids, such as naphthenic acids ($C_7H_{10}O_2$), myristic acid ($CH_3(CH_2)_{12}COOH$), stearic acid, and oleic acid, have been found to influence the interfacial properties as well as making rock more oil-wet and increasing the response toward LSW [85,132,133].

TAN/TBN (as bulk oil properties) does not indicate the surface activity of polar groups or their number at the oil–brine interface. This is more important toward determining the LSE/LSR. The exact type/chemical structure of polar groups, which are responsible for the LSE, has yet to be determined. The polar groups can be divided into heavy polar fractions, insoluble in water but active as surfactants, and those water soluble that have been found to impact the LSE. Longer-chained water-soluble compounds in brine remain surface active but very short molecules, such as acetic acid, were found to be of less importance [132]. Fathi et al. [85] demonstrated that oil depleted in water-soluble carboxylic acids showed more wetness and less efficiency toward LSW. These observations indicate that in addition to the type of polar group, their organic structure and surface activity would influence the wettability alteration as well.

Aromatics attain slight polarity (weakly polar) in the oil due to $\pi-\pi$ and π–hydrogen interactions, which increase their surface activity as determined from recent AFM/CFM studies [134]. However, to date, the literature data in this context are lacking. AlMaskari et al. [135] showed that even for a nonpolar component, such as octadecane, in contact with calcite and in the presence of NaCl of 10,000 ppm salinity, the pH increases from 6.5 to 11 and leads to wettability alteration toward more-water-wetting (125 to 78 degrees), tracking the zeta-potential trends and AFM results. The conclusion is that the essential physiochemical process can even occur in reservoirs with low concentrations of polar components in oil. Microdispersion test has been proposed to probe the oil potency toward LSE[133]. It is a useful technique to find the potent oils (i.e., those containing polar, surface active components triggering LSE), although the correlation between oil recovery by LSW and microdispersion ratio needs to be supported by sufficient data from other groups.

2.3.4.4 Reservoir temperature

The temperature of carbonate formations has been reported to be as low as 50°C to as high as 195 – 210°C for 5000–6000-m-deep reservoirs in the Gulf of Mexico and China [136]. The temperature within 70 – 120°C is most common. Fundamentally there is no temperature limitation for the application of LSWF. Core experiments at elevated temperatures as high as 100 and 120°C showed the LSE/LSR. Zhang and Austad [137] reported a negligible effect of aging temperature on the wetting conditions; however, it is expected that reservoir temperature impacts the initial wettability, also the degree of potential wettability alteration by LSW.

Fig. 2.12 shows that incremental recovery in some cases increases with temperature. Overall, temperature above 70°C increases the likelihood of observing LSR. Increased recovery is usually interpreted as wettability shifting to more water-wet; however, these experiments lacked wettability data. It has been proposed by Austad's group that temperatures beyond 70°C–80°C are necessary to increase the LSE as the rock becomes more water-wet. Rezaei-Gomari and Hamouda [138] also observed an increase in the degree of water wetness using contact angle measurements. This is proposed as the link to the observations of increased activity of the PDIs particularly Mg^{2+} and SO_4^{2-} at elevated temperature. However, Sohal et al. [116] found chalk became more oil-wet at higher temperatures as did Zhang and Austad [137]. Mahani et al. [139] found that even at ambient temperature wettability alteration can occur for limestone samples. Among all cases of limestone and dolomite surface studied, increased temperature usually increased the likelihood of wettability alteration for the samples that did not show the effect at ambient temperature. However, there are cases where the increase in temperature apparently decreased the degree of wettability alteration based on a reduction in the magnitude of zeta-potential.

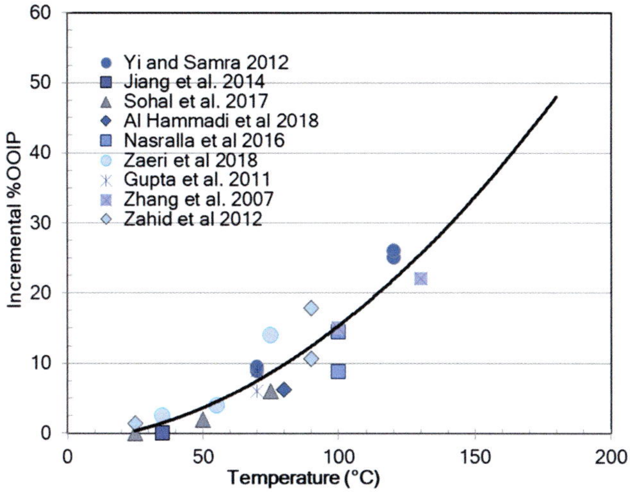

FIGURE 2.12

Relationship between temperature and incremental recovery for carbonate experiments.

Data from Refs. [34,40,104,140–145].

2.3.4.5 Rock composition and presence of dissolvable minerals, such as anhydrite

Chalks, limestones, and dolomite rocks have been found to react toward LSW with varying degrees of wettability alteration and oil recovery. Chalks were found to be more reactive than limestone and dolomite and have a more negative surface charge compared to limestone and dolomite even in HS conditions. The more crystalline the carbonate rock, the less the surface activity and free energy [139]. But for each rock-specific brine composition that can be tuned to generate the LSE effect, the LSR can greatly differ. Austad et al. [146] argued that the LSE (by brine dilution) exists only in rocks containing anhydrite. However, results from other groups have observed the LSE in absence of anhydrite even with brine depleted of PDIs. The brine ions can act in different layers of the EDL. Some ions in the Stern layer, such as Mg^{2+}, Ca^{2+}, and $SO_4{}^{2-}$, have a much stronger effect than monovalent ions, such as Na^+.

As most carbonates do not contain clay, clay mineral presence is not important to observe the LSE. Even in sandstones, clay presence is not a requirement, although the clay presence can increase the amount of adsorbed oil (or the remaining oil) and move the initial wettability toward more oil wetness [147]. In carbonate rock containing a substantial amount of clay (e.g., 5%–10% or more), the cation exchange capacity of clays can influence the local pH and the divalent concentrations, the propagation of divalents along the flow path, the depth of mineral dissolution and precipitation, and wettability alteration [88]. These simultaneous effects can influence the efficiency of LSWF (see also Shabani et al. [148]). Determining these complex interactions requires dedicated studies at the core-scale.

2.3.4.6 Initial oil saturation/water saturation

Initial fluid saturations have two effects. First, results from Puntervold et al. [148] and Strand et al. [8] show that initial water saturation of chalk cores influences the wettability of cores such that an increase in oil wetness was observed as initial water saturation was reduced from 50% to 30% to 0%. Similar results have been reported for sandstone reservoirs [149,150]. Conversely, high-water saturation limits oil wetting making rock more mixed-wet or water-wet. For instance, reservoirs in the capillary transition zone (e.g., some fields in Oman) with high S_{wi} (30%–50%) are likely to be mixed-wet to water-wet and thus have less scope for LSWF, because conventional waterflooding may work rather efficiently. Second, from the more practical perspective, reservoirs with low oil saturation (high water saturation) or that have been extensively waterflooded may not be economically viable candidates for LSWF.

2.3.4.7 Other important considerations

These include the injection water source, its availability, and its compatibility with the formation of brine and oil. The compatibility of LSW with oil is a less addressed issue. Upon contact with reservoir oil, specific ions, such as Ca^{2+} and Mg^{2+}, in the LSW may destabilize the polar groups, such as asphaltene leading to organic precipitation [151]. This has not been generally observed, but in some cases risks of associated formation plugging has been reported [152].

Compatibility between formation water and injection water containing Ca^{2+}, Mg^{2+}, $SO_4{}^{2-}$, and less abundant ions, such as Sr^{2+} and Ba^{2+}, should be evaluated to reduce the risk of inorganic scaling (mineral precipitation) and souring. Seawater is often regarded as a potential LSWF source in offshore reservoirs, but seawater contains a significant amount of $SO_4{}^{2-}$ (3000–5000 ppm). HS formation brines are often rich in Mg^{2+} (1–2000 ppm) and to a lesser extent Ca^{2+} (4–600 ppm). Mirzaalian et al. [153] studied the risk of $CaSO_4$ precipitation due to the incompatibility of seawater with formation water. They reported that though the magnesium ion can act as an inhibitor to prevent the formation of calcium sulfate, there exists an optimum concentration in seawater that reduces calcium sulfate precipitation by sixty percent. However, they found that calcium sulfate formation is complex and does not follow a regular trend, such that it may not increase at higher sulfate concentration for a specific calcium concentration.

Sulfate effects on souring can be mitigated by utilizing a sulfate removal unit (SRU) in the surface facilities. Sulfate removal also reduces the TDS [154]. However, since sulfate is an important (active) ion in wettability alteration, the negative effect of removal on LSWF efficiency should be evaluated and may be compensated by further dilution of the brine or supplementing with necessary ions. Seawater dilution and then enrichment with sulfate have been reported useful by several studies [153], nevertheless, the risk of reservoir souring can still be high. Other relevant issues, particularly with chalks, are the increased risk of rock softening as observed in the Valhall field (Ekofisk) due to the seawater injection, leading to increased subsidence of reservoir and offshore platforms [155].

2.3.5 Upscaling workflow and field assessment

The recovery at the field scale is affected by the design (well placement and completion), small- and large-scale heterogeneities, and inter-well connectivity. Lack of knowledge about mechanisms inhibits predictive capability but is not a showstopper for field deployment. It should be noted that the recovery data obtained in the laboratory cannot be directly used for field prediction because the field-scale volumetric sweep efficiency may be considerably lower than the near 100% sweep efficiency achieved in the core-scale experiments. Also, in the laboratory, the dozen or more PVi injected, which are necessary for measurement accuracy and plateau recovery, is not the case in the field. Numerical simulations [156,157] have shown that it is possible to recover the oil by LSWF with less than 0.5 PVi. As such the concept of slug LSWF injection instead of continuous injection may be more practical and economically more favorable.

To upscale the laboratory data to the field, the following workflow can be followed as depicted in Fig. 2.13:

- Laboratory assessment (focusing on core-scale tests) and data interpretation as described before to
  - determine IOR by LSWF
  - obtain salinity-dependent saturation functions for different rock types in the reservoir.

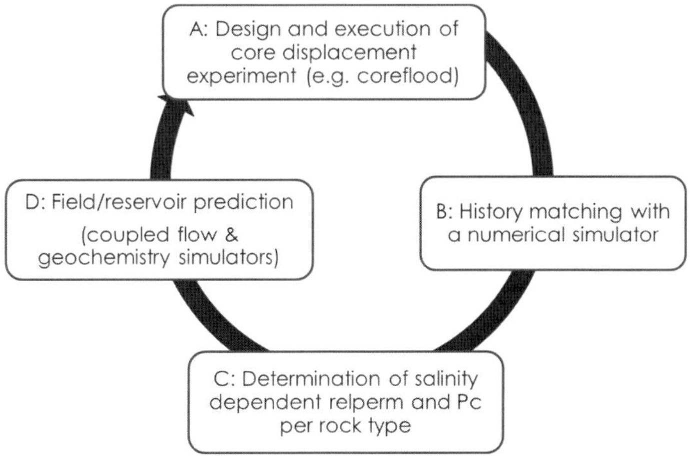

FIGURE 2.13

LSWF upscaling workflow from laboratory to field scale. *LSWF*, Low-salinity waterflooding.

- Piloting to better understand the subsurface uncertainties and test the technology readiness, and to de-risk prior to full-field implementation.
- Full-field scale design and evaluation:
  - Building (dynamic) reservoir model
  - Field-scale production forecasting using a fit-for-purpose reservoir simulator
  - Field design (surface and subsurface facilities), project economics [capital expenditure (CAPEX), operational expense (OPEX)], and profitability indicators (e.g., net present value, unit technical cost, value investment ratio)
  - Amendment to the design to improve the economics
  - Investment decision (final investment decision)
- Implementation in the field, and monitoring/evaluation of the performance
  - Phased deployment or full scale

The technology has promise for offshore and onshore applications. In the offshore environment, seawater is an accessible source of water, which can be treated and desalinated to the specifications required for the target field [158]. The risk of reservoir souring and (inorganic and organic) scaling should be evaluated when designing the injection brine composition.

2.4 Field case histories

As mentioned previously, there is an increasing number of papers published by different groups in academia and industry, which clearly help to advance the science and engineering related to this field. These publications provide more data for analysis to identify the commonalities and differences and develop predictive capabilities. Field data show that successful waterfloods make successful EOR (and likely successful LSWF), because of good dynamic connectivity between the injection and production wells. An essential aspect of the future of LSWF technology is its successful implementation in the field. "Learning by doing" as a guide is important for successful deployment. So far, this has not happened for LSWF as there are limited field cases. This may be an over expectation for a relatively young technology, while other more mature water-based technologies such as CEOR (polymer, SP, or alkali-surfactant-polymer (ASP) flooding) still have only a handful of full-field cases.

The main challenge faced by the technology is that most EOR methods are known to work if certain conditions are present. As such, preliminary screening methods have been developed The principle of EOR is to increase oil connectivity, which can be done by increasing capillary number (Ca) and improving the mobility ratio (*M*) toward favorable conditions. In polymer flooding, stabilizing the front in a viscous oil displacement can be done by adding polymers. The questions in applying the technique are the best type of polymer, the amount of polymer and the economic evaluation of cost and benefit. In miscible gas injection, attaining the miscibility condition is key and can be answered by PVT or phase behaviour studies. In surfactant or emulsion EOR, reducing IFT via using the right type and concentration of surfactant is based on laboratory testing coupled with field experience.

In LSW, the brine composition is optimized to get the best recovery factor. This would mainly be through the improvement of M and shift of fractional flow curve due to changes in oil and water relative permeabilities. However, how brine composition affects oil recovery is not clear and depends on the specific oil-brine-rock system, making it difficult to design the optimal salinity. The brine composition can be tuned by systematic changes in composition, but in the reservoir setting, the accessible water sources are critical factors to assess the balance between cost and reward.

To the best of our knowledge, the only study published is the field trials in carbonate reservoirs in Saudi Arabia. Yousef et al. [159] reported the first field test of LSWF in a carbonate reservoir based on the previously reported experimental results [43,44]. They reported two field trials with two single-well chemical tracer tests using various dilutions of seawater. For well A, a slug of seawater and twice-diluted seawater was sequentially injected, while for well B, a slug of seawater, twice-diluted and 10 times diluted seawater were sequentially injected, and after each injection cycle, three different tracers were injected to estimate the residual oil saturation (S_{or}). They observed that diluted seawater resulted in 7% and 3% S_{or} reduction at well A and B respectively, beyond conventional seawater injection. A further 3% reduction beyond twice-diluted seawater was achieved by ten times diluted seawater injection. They concluded that the experimental and field IOR were similar.

Additional considerations in designing optimization experiments include knowledge of the flow regime/force balance. The prevailing force regime at the field scale can influence the recovery factor by LSWF. This has not received much attention in the published studies. Mahani et al. [160] showed the gravity force to viscous force ratio is important. In fields where the operation is gravity dominated, oil–water segregation may cause the remaining oil to be quite low. Thus the scope for LSWF is limited in these circumstances. Well design is also a factor to consider. The WF well placement is typically for pressure maintenance and is peripheral, with large well distances and injection into underlying aquifers. This is not suitable for efficient displacement, particularly for LSWF. The optimum well placement and completion are necessary to ensure the highest contact of the LSWF with the target oil to increase the volumetric sweep efficiency. The injection should be to the oil leg/column to avoid injecting into the aquifer which increases reservoir mixing. The optimum design/well completion can be found with the aid of reservoir simulation.

Finally, for deep, offshore reservoirs injected seawater is usually cooler (e.g. around 4°C at the top hole) than the reservoir temperature. Although the injection water gets warmed up as it reaches the bottom-hole perforations, it may be still colder than the reservoir. This can introduce thermal-induced fracturing and thermo-poro-elasticity-induced geomechanical effects, which can adversely influence the volumetric sweep and wellbore stability [161]. Temperature gradients in the reservoir can affect the geochemical interactions and LSWF performance too. Paying attention to these aspects is important in the design of experiments to make sure the laboratory data can be effectively applied to the field trials.

2.5 Modeling and numerical simulation of low-salinity waterflooding

2.5.1 Length scales of simulation

The objective and output of modeling and simulation depend on the scale of the simulation. The scales of simulation are atomic and molecular, pore and pore network, core, and reservoir. The smaller scales (atomic and molecular, and pore and pore network) aim to mimic observations usually based on fundamental physical and chemical processes. The core-scale simulations aim at matching experimental data and validating the models which are based on macroscopic mechanisms (e.g., capillary pressure). The reservoir-scale models try to predict field-scale behavior and assess development and deployment scenarios for economic evaluations and investment decisions.

2.5.2 Current approaches to simulation at core and reservoir scale

All LSWF simulation methodologies existing in standard/commercial software for field- and/or core-scale rely on changing relative permeability and capillary pressure based on the local concentration of a tracer [162]. The brine tracking option using a tracer approach must be activated to enable LSWF simulations. Implicit or explicit coupling with geochemistry and geomechanics is necessary, becasue LSWF is a 'mild' reactive trasnport process. Geochemical modeling allows for multiple ion tracking in addition to total salinity tracking. This is typically based on solving the advection-dispersion-reaction/sorption equation for the aqueous phase. Eclipse, CMG, and UTCHEM/UTCOMP are popular commercial simulators for this purpose. Some companies and research institutes have their own proprietary simulators. The Jerauld et al. [156] model has been implemented in ECLIPSE and CMG-STARS and the ion exchange model proposed by Dang et al. [11] in CMG-GEM. The LSWF simulation philosophy and workflow are shown in Fig. 2.14.

To properly simulate LSWF at the core and field scales, several types of inputs are necessary. The first is correctly representing the reservoir rock and fluids and heterogeneity. The necessary inputs include rock properties, such as spatial distribution of porosity and permeability, and fluid properties, such as

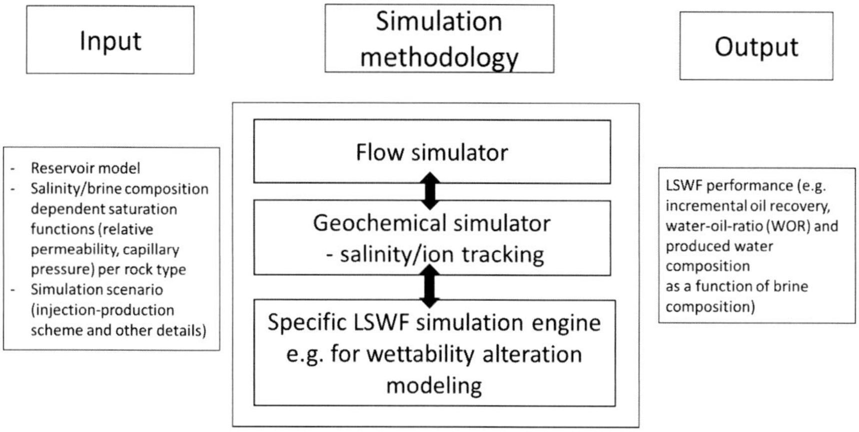

FIGURE 2.14

LSWF simulation workflow at core and reservoir scales. *LSWF*, Low-salinity waterflooding.

appropriate PVT for brine and oil, viscosity, and phase behavior, particularly for live fluids. Although oil properties and compostion can be specified in detail (SARA, TAN, TBN, etc.), simulators mainly use bulk properties. For instance, the polar fraction of oil is not explicitly defined in common simulators. Some geochemical simulators do accept chemical concentrations as inputs as well as TDS and more complex simulators can even model water–rock reactions. An additional consideration is modeling the distribution of CO_2 between oil and brine phases since the carbonate system can act as a pH buffer and change the viscosity of the oil. CO_2 presence affects the geochemical reactions, thus wettability alteration by LSW.

To simulate dynamic fluid movement, we need inputs for saturation functions that include relative permeabilities and capillary pressure curves as a function of brine composition, per rock type. Models need a wettability alteration (interpolation) criterion or weighting factor to control the relative permeability functions used for HSW and LSW. Inputs for physical dispersion of salts/ion species are also required. Dispersion coefficients vary with the scale of the system. Dispersion coefficients incorporate the effects of rock heterogeneity and molecular diffusion that serve to spread out the salinity interface as the injected fluid moves through the reservoir. Generally, dispersion coefficients incorporate diffusion, but some models required both parameters explicitly. Evaluation of the relative magnitudes of dispersion and diffusion is made by calculating the *Peclet* number. When the *Peclet* number is less than 60, physical dispersion becomes important [156]. The result of dispersion is the development of an in-situ mixing zone instead of a sharp salinity front as reservoir brine is displaced by injected water. This mixing phenomenon reduces the efficiency of microscopic mechanisms responsible for the LSE. To compensate for the mixing effect, additional PV of LSW would be required [157,163].

Interfacial properties, such as IFT and liquid–liquid interactions, and fluid rheology also change with brine composition and phase behavior, but this is usually not explicitly accounted for in LSWF simulations. The change of IFT affects wettability and relative permeability by changing the capillary number. However, the change in IFT with brine salinity is relatively minor in most reported cases, although there are exceptions showing a sizeable change on the order of 10 mN/m. Finally, the physical changes to the porous medium associated with rock/mineral dissolution/precipitation, fines migration, and formation damage that impact porosity, permeability, and pore connectivity should be captured in the simulation.

2.5.3 Wettability alteration criterion

Wettability alteration is the prime LSWF effect implemented in simulators. In phenomenological approaches, bulk salinity or concentration of a particular ion is used as a criterion for change of the saturation functions using interpolation schemes. The interpolation is usually linear for a given range of salinity which covers the range of salinity from HS to injection water LS. Alternatively, the function is interpolated over a narrower range of salinity, i.e. an effective salinity range, wherein the wettability is shifted. Jerauld et al. [156] proposed the first modeling approach with linear interpolation between the HS and LS relative permeabilities and capillary pressure:

$$k_{rw} = \theta k_{rw}^{HS}(S^*) + (1-\theta)k_{rw}^{LS}(S^*), \tag{2.7}$$

$$k_{row} = \theta k_{row}^{HS}(S^*) + (1-\theta)k_{row}^{LS}(S^*), \tag{2.8}$$

$$P_{cow} = \theta P_{cow}^{HS}(S^*) + (1-\theta)P_{cow}^{LS}(S^*), \tag{2.9}$$

$$\theta = (S_{orw} - S_{orw}^{LS})/(S_{orw}^{HS} - S_{orw}^{LS}), \tag{2.10}$$

$$S^* = (S_o - S_{orw})/(1 - S_{wr} - S_{orw}), \tag{2.11}$$

where k_{rw} and k_{row} are water and oil relative permeability respectively. P_{cow} designates capillary pressure. S_{orw}, S_{wr} and S_o represent residual oil saturation, irreducible water saturation and oil saturation respectively. The superscripts HS and LS indicate the HS and LS conditions respectively. This model is widely used in LSWF simulations.

Al-Ibadi et al. [164] investigated nonlinear interpolation functions as well giving more flexibility toward LSWF simulation. However, these approaches can result in different predictions of oil recovery and need to be further explored and verified with experimental data.

Mahani et al. [160] proposed a threshold salinity or mixing factor (*f*) at which wettability is changed. The approach was incorporated in MoReS, the Shell proprietary reservoir simulator. This approach is based on the idea that during the in-situ mixing, the salinity profile is smeared out (becoming "S" shape) and a portion of the LSWF becomes ineffective. For instance, at a mixing ratio of 50%, the salinity profile is sharp (representing the case where dispersion is very small), and at mixing ratios above 50%, particularly above 90% (a case where the mixing zone length is significant compared to the well spacing), wettability alteration can only occur if the LS fraction of the mixture is 90% or above. This condition is satisfied only in a small portion of the mixing zone.

Other variations to this general approach include the adsorption of ion species on the surface of the rock in addition to advection diffusion [165]. Improved oil recovery was incorporated by salinity-dependent relative permeability, capillary pressure, and residual oil saturations. Al-Shalabi et al. [113] developed an empirical model that used contact angles and the modified CDC approach (linking residual oil by LS to the trapping number) for LSWF simulation. The main consideration/assumption was that the residual oil is often lowered by LSW. This model was implemented in UTCHEM. Furthermore, it was assumed that water relative permeability is not affected by the salinity, while oil permeability (mainly the end point) is a function of trapping number. Aljaberi and Sohrabi [166] focused more on the fluid–fluid interactions driving the wettability alteration than geochemical reactions and developed a correlation between the microdispersion formation and relative permeability. This correlation was developed based on their experimental data whereby the LS relative permeability could be predicted. Based on their correlation, a microdispersion ratio from 1 to 25 by a suitable LSW leads to IOR as high as 6%–7% in carbonates.

In (more) mechanistic approaches, sometimes referred to as thermodynamic or geochemical-based modeling, sophisticated simulations coupling flow simulators with a geochemistry package, such as IPHREEQC, or with embedded geochemical simulation capabilities have been attempted. In these cases, the nonequilibrium effect imposed by the change of brine composition is captured. Instead of using bulk concentrations of ion or total salinity, the surface concentration of specific surface species is used to define the wettability alteration. In other words, the consequence of bulk (or pore water) salinity change on rock–fluid interactions are modeled. This is a more appropriate method, because the physicochemical interaction of ions with the mineral and oil surfaces determines the LSE.

The coupled simulations can account for rock–brine interactions, such as dissolution, precipitation, ion exchange, and complexation (chemical reactions) [162,167]. The models implemented in most cases closely follow the mechanism proposed by Austad and collaborators highlighting the role of SO_4^{2-} as a catalyst and that the ion is present in the injection brine or produced in-situ by anhydrite dissolution. All the models have been employed to interpret published experimental data from chalks and limestone and are reported to be in reasonable agreement with the data. However, the universal validity of these models cannot be determined since there are several matching parameters that are varied to fit the data.

For carbonates, often the concentration of a surface species, such as $>CaSO_4^-$ (where > represents the surface site), or sites that bond with carboxylic groups of oil, such as $>CaOH_2^+$ (COOH), is used to define the interpolation criterion between the HS and LS relative permeabilities [168]. Though a step forward, this is an oversimplification of a complex process wherein other PDIs and their simultaneous effect influence the wettability. Also, when the injection brine does not contain sulfate there would be no wettability alteration in their formulation, which is not consistent with some experimental data. Another mathematical model was proposed by Hiorth et al. [58] for SI experiments in chalk core plugs. The results showed that seawater shifted the wettability toward a more water-wet condition. The wettability alteration is modeled by interpolating the relative permeability and capillary pressure as a function of the dissolution of calcite ($CaCO_3$). The wettability shifted toward more water-wet conditions when the dissolution of calcite occurred.

Awolayo et al. [18,169] developed a coupled flow and reaction model where they considered the carbonate surface as an ion exchanger (similar to clay in sandstones) and used the adsorbed concentration Ca^{2+} and Mg^{2+} attached to the carbonate surface (Ca-X, Mg-X), for interpolation, rather than directly using the concentration of adsorbed SO_4^{2-}. The contribution of dissolution and precipitation of different minerals as they alter the distribution of PDIs available for surface sorption were captured and simulated. The fraction of the free surface sites that could adhere to oil was observed to reduce as the brine salinity reduced and sulfate concentration in injected water increased, which was then used to transit between two sets of flow functions.

Korrani et al. [170] recently reported coupling flow and geochemistry using UTCOMP-IPHREQC. Their interpolation was performed either on total ionic strength or the sum of adsorbed ions on the exchanger. The coupled simulator was applied to simulating sandstones, though. Yogarajah et al. [171] developed an integrated geochemical model to predict solute transport and oil recovery in carbonate reservoirs. Surface complexation modeling (SCM) of both oil–brine and calcite–brine interfaces were predicted with PHREEQC to account for ion interactions with oil and rock surface with a triple-layer model. The flow simulation was not handled with a two-phase simulator. Oil was modeled as colloidal particles, which can be transported. While the good agreement was obtained between the predicted and measured values of pH and relative ionic profiles, there was a large discrepancy between the predicted and measured oil recovery. This is mainly due to the inadequacies of PHREEQC to simulate two-phase flow and the lack of a specific mechanism for LSE.

Finally, Sanaei et al. [172] developed a new approach that allows for a more mechanistic simulation where the wettability alteration is steered by the DLVO force calculations at any salinity/brine composition—the DLVO force calculation is linked to SCM of both oil/brine and rock/brine interfaces. This is a notable step forward that captures the role of surface-charge-change and electrical-double-layer forces in the simulation, in addition to capturing the adsorption/desorption of oil components.

2.5.4 Pore-scale simulation of LSWF

Pore-scale modeling has evolved over the past 10 years and has become a powerful (complementary) tool in the area of core analysis. The pore network and subpore-scale simulations aim at reliable prediction of flow functions, such as relative permeability and capillary pressure curves, based on the full physics and chemistry involved in the flow and transport [173]. This is particularly important to upscale the micro-scale results to the macro scale of the reservoir, one of the long-standing objectives of digital core analysis or rock physics that has not been fully achieved. These simulations hope to nail down and substantiate the pore-scale mechanisms leading to the LSE and perform sensitivity analysis on the effect of flow regimes (viscous and capillary), initial wettability and saturation, pore-size distribution, surface roughness, mineralogy, heterogeneity, ionic transport mechanisms, and interfacial properties. This could lead to identifying conditions that produce the greatest LSWF efficiency. These simulations could, in theory, provide mechanistic insights into the results at a higher scale, e.g., core-scale and propose the suitable conditions to observe LSR and obtain average (macroscopic) flow properties, which can be used for upscaling to a higher scale. Pore-scale simulations can be performed even below the REV scale.

The physics/chemistry in the pore-scale model/simulator can be confirmed by quantitative comparison and matching of the experimental results at the same scale but for now, these outcomes are still lacking. There have been major efforts recently in this direction. The simulators still have issues to properly represent the pore-scale mineralogy and actual pore geometries. Wettability at the pore-scale is nonhomogeneous due to changes of rock mineralogy and surfaces and is spatially correlated (not random) [174]. The physics of flow related to mixed wettability and nonuniform wettability distributions need to be better understood to properly handle the corner flow (in noncircular throats) (Fig. 2.15), oil trapping, film flow/swelling, and other important pore-scale physics critical to simulate LSWF.

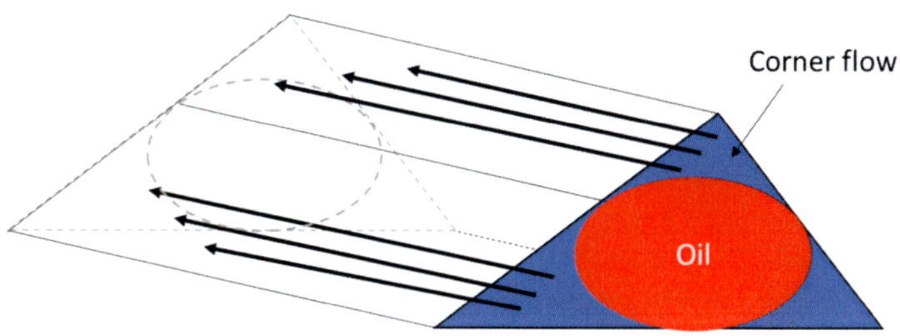

FIGURE 2.15

Corner flow at pore-scale relevant to low-salinity waterflooding [175].

Inspired from M. Tuller, D. Or, Hydraulic conductivity of variably saturated porous media: film and corner flow in angular pore space, Water Resources Research 37 (2001) 1257–1276.

Platforms for pore-scale simulation of the flow of two or more phases have been developed. There is a variety of mesh-based and mess-less methods [176,177]. In the mesh-based methods, the Navier–Stokes (N–S) equations or the governing flow equations are directly solved on a sufficiently fine grid model of the porous medium. In the mesh-less methods, the N–S is solved without tracking the two-phase interface or indirectly on a simplified pore-network model. The pore-scale simulation approaches can be divided into (1) particle/lattice-based methods, such as Lattice–Boltzmann and smoothed particle hydrodynamics methods, (2) upscaled continuum methods, such as phase-field models, level-set and volume-of-fluid (VOF), and (3) topological methods, such as pore-network models.

Sorbi and Collins [178] were the first to explore the effect of wettability alteration by LSWF on the oil recovery using pore-network modeling simulation. Watson et al. [179] developed a steady-state 3D pore-network model in which LS brine displaces oil from an HS-bearing network. The model had uniform wettability and no initial water saturation was assumed. A threshold salinity concept was used to alter the contact angle. The local wettability alteration model has been implemented following the philosophy proposed by Sorbi and Collins [178] where film flow is neglected. Both studies report a lack of improved LS recovery in strongly water-wet networks and demonstrate that, in general, only small alterations to wettability may be necessary to elicit significant improvements in oil recovery. They recognize two principal effects of dynamic contact angle modification at the pore-scale: a "pore filling sequence" effect characterized by an alteration to the standard HS pore filling sequence. This can impact the overall amount of oil produced, and a "sweep efficiency" effect is demonstrated by a change in the overall fraction of pores displaced by LS compared with HS. The study indicates that any LS effect will depend on the relative (scenario-dependent) influence of each mechanism, where factors. such as the initial wettability state of the system and the pore-size distribution of the underlying network, are found to play crucial roles. In addition, the important role played by end point capillary pressure in determining LS efficiency.

Boujelben et al. [180], building upon Watson et al. [179], used a dynamic pore network model to simulate the LSWF under USS. The authors investigated the impact of several key parameters on secondary and tertiary LS injection including injection rate, viscosity ratio, and salinity-induced wettability. Greater variation in LS outcome was observed when LS brine modified pores to close to neutral-wet conditions. But the displacement regime (capillary number) impacts the magnitude of the recovery. Due to the large parameter space, when compared against HS data, both positive and negative LS effects have been observed depending upon the prevailing system properties and the timing of LS initiation.

Maes and Geiger [181] used OpenFOAM coupled with PHREEQC to implement a surface complexation model and capture geochemical reactions at calcite–brine interface. They used the VOF method in OpenFOAM. The contact angle was a linear function of the rock surface zeta-potential, which in turn depends on the brine chemistry. The oil–brine interface was not considered in the simulations. They reported S_{or} change and improved sweep efficiency with LSWF. Aziz et al. [182] used the VOF method in OpenFOAM to simulate LSWF. In their model, contact angle and salinity are assumed to be linearly related. The main focus was on understanding the effect of mixing between LS and HS that occurs at the displacement front. They found that in tertiary mode flooding the efficiency is less than in secondary flood due to the existence of stagnant regions where the LS displacement is by diffusion only. This would delay wettability alteration and proposed the "push–pull" mechanism.

Akai et al. [183] used the Lattice–Boltzmann method (DNS-N-S) to simulate LSWF in 3D (Fig. 2.16). To simulate the process at the pore-scale, they proposed that the alteration in surface wettability - mediated by thin water films which are below the resolution of simulation grid blocks as observed in experiments by Bartels et al. [88] - must be considered. The novelty of the study is in incorporating the adsorption of ions onto the rock surface to model kinetics of wettability alteration as proposed by Mahani et al. [48]. They also echo the finding from Aziz et al. [182] on the higher efficiency of secondary LSWF than tertiary.

Martínez-Mendoza et al. [184] used the PNM approach in OpenPNM to estimate capillary pressure and relative permeability for LSWF using a pore-network representation of rock. This has not been done in previous work. The model was applied to sandstone and carbonates under drainage conditions, although forced imbibition is more relevant to LSWF. Similar to Aziz et al. [182], a linear relation between contact angle and salinity was assumed. Nevertheless, the reliability of saturation functions is still questionable due to inability of the existing approaches

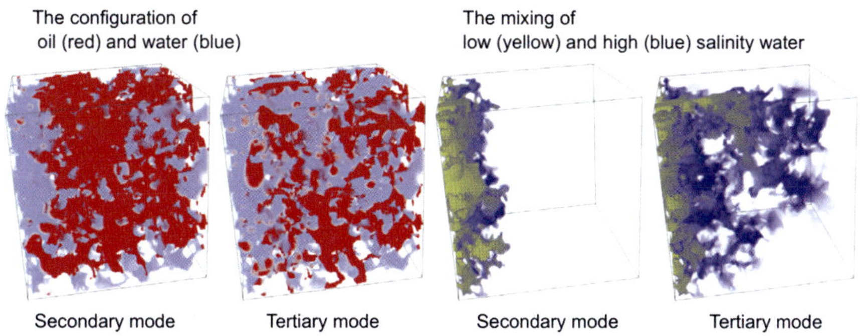

FIGURE 2.16

The pore-scale configuration of oil and water, and brine concentration during secondary and tertiary LSWF simulation. The simulation was performed by incorporating wettability alteration into a Lattice–Boltzmann pore-scale simulator. *LSWF*, Low-salinity waterflooding.

Reproduced with permission from T. Akai, M. Blunt, B. Bijeljic, Pore-scale numerical simulation of low salinity water flooding using the Lattice Boltzmann method, Journal of Colloid and Interface Science 566 (2020) 444–453.

in capturing the key physics/chemistry of flow as well as the model/image resolution used.

Abu-Al-Saud et al. [185], building upon their previous publications [186], simulated the movement of an oil ganglion/slug displacement by LS in a single pore with a multiscale level-set method with a capillary number (Ca) of 4×10^{-5} to produce film-stability maps. The novelty is accounting for the double-layer forces and surface roughness effects. Surface roughness increases film collapse during HS. They show that the double-layer expansion in LSW decreases the pressure drop available to move the oil drop. Wilmott et al. [187] used a dynamic pore network model to describe the macroscopic oil recovery following the microscale mechanisms. They assume a capillary-dominated flow (Ca $\sim$ 10^{-7}). Film flow/expansion was modeled, but not the corner flow which can occur in noncircular throats. Building on their previous work [188], focused on the channel scale and incorporated a multicomponent ionic exchange (MIE) mechanism to account for the salinity dependency of the thin water film thickness separating the oil phase from the solid surface. This results in a salinity-dependent slip condition on the effective oil–solid interface. This is an important addition compared to previous studies. Their results show that about 5% increased oil recovery can be achieved as the salinity of the injection brine decreases. The recovery scales linearly with the viscosity ratio. Though the work focuses on clayey sandstone with exchange sites, in principle it can be applied to carbonates as well as there is adsorption–desorption of PDIs occurs on the carbonate sites as well.

There is an issue around the limitation of capillary numbers to handle LSWF. Currently, with direct numerical methods, the visco-capillary regime can be practically handled at moderate to high capillary numbers (i.e., $\mathrm{Ca} > 10^{-4}$) not representative of actual flow in rocks. The issue is that the computational time significantly increases as the capillary number decreases below 10^{-4}–10^{-5}. However, at high Ca, capillary desaturation can occur. The majority of the pore-scale network models developed earlier could handle quasi-static conditions where the capillary flow is dominant ($\mathrm{Ca} \sim 10^{-8}$ or 10^{-7}). Dynamic pore network modeling can be applied at $\mathrm{Ca} > 10^{-5}$. VOF in OpenFOAM [182], Lattice–Boltzmann [183], PNM in OpenPNM, or other propriety packages have been used to simulate LSWF at the pore-scale. To enable LS simulation, as in the core-scale simulation, the salt transport model (advection–dispersion-adsorption) or brine tracking must be implemented to track salinity or individual ions in the whole pore structure. This is often performed explicitly after solving the pressure and saturation equations. Then based on the spatial salinity/ion profiles and defining a wettability alteration model or criterion flow functions are switched from HS to LS and the oil configuration and saturation are updated. Ionic transport in water film (linked to kinetics of wettability alteration) and cornerflow has to be captured too [189]. Finally, almost all studies focus on rock–fluid interactions, such as wettability alteration and the liquid–liquid interactions, which have been largely neglected.

2.5.5 Molecular dynamic simulation

We can examine atomic and molecular scale interactions using MD simulation, density functional theory (DFT), and SCM. MD simulation is a well-known approach in chemistry, physics, solid-state materials, medicine, bioscience, and chemical engineering fields; however, its application toward understanding EOR at the molecular scale, rock–fluid interaction (e.g., wettability), and other relevant phenomena affecting oil recovery and porous media, such as organic and inorganic precipitation and scaling, has been so far limited. The number of such studies in LSWF has been increased recently. Such studies can help substantially in the atomistic and molecular level understanding of the LSWF. This is particularly important to substantiate the role of interactions between individual ions in the brine and (acidic and basic) functional groups in oil and rock surface species, which helps discern potential mechanisms.

The molecular simulations can be used to estimate properties and constants difficult to measure directly. For instance, diffusion coefficients, dielectric constants, and water viscosities in thin films, as well as surface charge density and slip coefficient [190]. The major limitation in this area is related to the simplicity and number of atoms/molecules used in the models, and short simulation times due to computational challenges faced. It should be noted that a positive LSE identified at this scale should be verified at larger scales to evaluate the benefits of additional oil recovery. Open sources software packages, such as LAMMPS or GROMACS, can be used.

Unlike the simulations at other scales, the simulations ought to show if there is an LSE, without making a presumption as "a priori." The main considerations are to properly represent the intermolecular and interspecies forces (e.g., between individual oil components and brine ions), and those with the solid surface (e.g., adsorption/desorption and surface speciation) include a representative number of solid and liquid phases/species to represent the active ions and interfacially active oil groups. One observation in the existing MD models is that the net charge of the calcite surface is zero. While, they bear nonzero charge due to calcite dissolution, defects in the crystalline structure, or substitution of atoms and diagenesis. Inclusion of surface and mineral heterogeneity and nonequilibrium effects (when salinity is altered) is preferred. Model calibration with scale-appropriate tools, such as AFM or CFM, is required to increase the reliability of the results.

Important insights have been obtained in these studies that can provide fundamental insights on rock–brine–oil interactions to advance the models. Chang et al. [191] performed simulations to classify calcite among water-wet minerals. Based on the inherent water wetness of calcite, the tuning of ions in LSW is expected to impact the mineral wettability. Santos et al. [192] reported the formation of an electrical double layer (EDL) within a calcite/aqueous electrolyte interface at the atomistic level. Zhao et al. [177] verified that the EDL is composed of oppositely charged layers of Na^+ and Cl^- ions and attributed it to the DLE mechanism.

Kirch et al. [193] reported the structure of water and ions at a calcite/aqueous solution interface. They detected a strong ordering of water molecules in contact with the solid surface. Ricci et al. [194] showed that highly ordered water layers at the calcite surface prevent the hydrated ions from directly interacting with calcite due to the energy penalty incurred by the necessary restructuring of the ion solvation shells. Therefore the Na^+ and Cl^- electrolyte ions occupied the available sites near the solid/brine interface, resulting in a screening effect. The study by Brekke-Svaland and Bresme [195] provided molecular insight into the solvation structure of calcite. They reported well-ordered three hydration layers at a calcite/water interface up to 1 nm above the surface.

Kim et al. [196] showed that polar oil components locate at the water/oil interface and the nonpolar ones are repelled from the water phase. They reported that the polar components are responsible for creating the link between the nonpolar fraction of the oil and the polar calcite surface through an adsorbed layer of water. Prabhakar and Melnik [197] showed that the first charged layer close to calcite is composed of sodium cations. They proposed that the formation of Na–carboxylate complexes over the calcite surface makes the mineral more oil-wet. The addition of Mg^{2+} and SO_4^{2-} ions alters the relative wettability of calcite to a less oil-wet state.

Koleini et al. [36] addressed the LSWF case more specifically. Their main finding was the formation/development of ion aggregates/clustering, mainly comprising Na^+ and Cl^-, within the HS thin brine film. These aggregates act as pinning entities to localize polar oil components within the oil/brine interface and connect the hydrocarbon phase to the calcite surface. Their simulation results suggest that the aggregates do not form in LSW.

Fang et al. [190] used MD simulation to study oil recovery from nanopores by reducing salinity. This is an effort toward the multiscale simulation of LSWF by linking molecular scale to nanoscale and continuum scale. They highlight that when ion depletion from thin brine films is suppressed [48,198], the osmosis of water can play a fundamental role in the expansion of the brine films; thus the EOR. The slow osmosis of water through thin brine films and thus the slow displacement of oil from the pore may help explain the anomalously slow oil recovery reported in micromodeling experiments of LSW. Chai et al. [199] studied the effect of polarity of four oil components (acetic acid, pyridine, toluene, and hexane) on their interaction with calcite surfaces. Electrostatic interaction is dominant with more polar oils, while in less polar oil the VDW force is dominant. An acetic acid–pyridine–water–toluene–hexane adsorption sequence was observed on the water-wet calcite surface. Xin et al. [200] showed that under the dispersion force and electrostatic interactions, the oil molecules (dodecane) migrated to the calcite surface and array in two layers and form an oil-wet surface. However, the oil adsorption was performed in the absence of the water film, creating direct contact between the oil and calcite surface, which is not the case in the reservoir.

Looking at the LS-surfactant and nanoparticle-aided LS concept, Aminian and ZareNezhad [201] showed that a mixture of titania nanoparticles, ionic surfactant, and LS brine leads to the maximum synergism leads to the lowest contact angle and maximum oil detachment from the calcium carbonate surface. The arrangement of nanoparticles near the oil–brine interface and interaction with surfactant at the interface generates a surface pressure that in combination with MIE and cation bridging leads to the oil detachment from the calcite solid surface.

A less addressed, but equally important issues are the side effects of injecting LSW and related effects on the brine–brine and oil–brine interactions. One such effect is the dependency of asphaltene instability and aggregation on brine salinity and composition. Tirjoo et al. [151] reported that the presence of salty water containing (Na^+, Ca^{2+}, and Mg^{2+}) ions can intensify the asphaltene sedimentation process. Electronegative sites in the asphaltene structure can absorb more ions and increase asphaltene sedimentation rates. There also exists the risk of inorganic scaling due to brine incompatibility. Mirzaalian et al. [153] studied the role of Na^+ and Mg^{2+} ions on $CaSO_4$ precipitation via MD simulation and showed that the increase of NaCl concentration can increase the solubility of calcium sulfate in the solution. It was shown that the increase of NaCl concentration can increase the solubility of calcium sulfate in the solution. Magnesium at optimum concentration decreases the calcium sulfate precipitation which inhibits the scale deposition to some extent.

2.5.6 Density functional theory (DFT)

DFT is a computational quantum mechanical modeling method used in physics, chemistry, and materials science to investigate the electronic structure of many-body systems, in particular atoms, molecules, and condensed phases. DFT is among the most popular and versatile methods available in condensed-matter physics, computational physics, and computational chemistry. This method has made its way into the study of LSWF and can be of added value when used together with MD simulations. Andersson et al. [202] used DFT to predict how solid–fluid interfacial energy is changed, when divalent cations substitute into a calcite surface. The effect on wettability can be dramatic. As a result of divalent cation substitution, calcite–fluid interfacial energy is significantly altered, enough to change the macroscopic contact angle by tens of degrees. Substitution of Sr^{2+}, Ba^{2+}, and Pb^{2+} makes the carbonate surfaces more hydrophobic. In contrast, the substitution of Mg and the transition metals causes calcite to become more hydrophilic weakening organic compound adsorption.

Prabhakar and Melnik [197] used acetic acid as a model compound to represent crude oil molecules and found that the binding energy of acetic acid is larger than water molecules on pure calcite surface, which indicates that the calcite surface is oil-wet. In this case, the Na^+ ions precipitated on the calcite surface and formed Na acetate. The binding energy of Na acetate is larger than the original oil molecule (acetic acid), which reduces oil recovery. On the other hand, Mg^{2+} and SO_4^{2-} ions can also reach the calcite surface and modify the surface face. The binding energy of oil molecule on the modified calcite surface is smaller than on the pure calcite surface, which can enhance oil recovery. Alghamdi et al. [203] used Gibbs free energy calculations and found the most stable surface complexes formed between the calcite surface and the major ions in seawater are: $>CO_3H$, $>CaO^-$, $>CO_3Ca^+$, $>CO_3Mg^+$, $>CaCO_3$, $>CaHCO_3$, $>CaH_2O^+$, and $>CaSO_4$, where $>$ denotes the surface site. These complexes form spontaneously without any activation energy, particularly $>CaSO_4$. Their calculation also demonstrated a thermodynamic tendency for these complexes to give up their hydration shell except for $>CO_3H$. Using this approach, they determined the ionic species that alter the wettability of calcite surface to a water-wet condition are: Ca^{2+}, Mg^{2+}, SO_4^{2-}, $MgOH^+$ (as an ion complex), OH^-, and NaCl (as a neutral complex).

2.5.7 Surface complexation modeling (SCM)

SCM is the modeling of sorption based on surface reaction equilibrium that includes aqueous complexation and fast surface complexation reactions. The prime objective of SCM in LSWF is a better representation of rock-oil-brine interactions, the electrokinetics, predictions of surface species formed at the oil–brine and rock–brine interfaces, and how they relate to salinity and ionic composition. This information is then used to describe the surface charges, wetting tendency, and interfacial behavior. The SCM is used in the mechanistic simulation of LSWF as mentioned previously. There are three commonly used SCMs, the constant capacitance model (CCM), the diffuse layer model (DLM), and the triple-layer model (TLM). In many cases both DLM and TLM work well, however as the brine composition becomes more complex, TLM is expected to become superior.

The surface species are formed on the reactive sites at the carbonate surface, usually $>Ca$ and $>CO_3$. The very first layer is formed by the hydration of these sites forming "master species" which then form complexes with the ions present in the brine. These complexes are charged, or uncharged, which controls the surface charge and potential of the carbonate surface. The interaction between the complexes on oil–brine and carbonate–brine moieties represent adhesion between oil and rock which underlies the electrostatic aspects of wettability. SCM has been shown to be a powerful tool to study the wettability state and its alteration by change of brine salinity/composition. This approach helps us understand the surface reactions and the type of functional groups controlling the charge of carbonate–brine and oil–brine interfaces and its trend with pH, brine salinity/composition, mineralogy (surface chemistry), temperature and partial pressure of CO_2. SCM has been used to predict the zeta-potential of oil–brine and carbonate–brine interfaces. The surface complexes formed at the oil-brine and the carbonate-brine interface can be found in various publications [34, 139, 204, 205].

Brady et al. [34] and Brady and Krumhansl [204] have used this type of modeling to examine the proposed wettability-alteration theories and to give insight into the primary oil—(polar functional groups) mineral-coordination reactions leading to oil adhesion on clay and calcite. Mahani et al. [139] developed a new SCM to explain the upward trend of zeta-potential of carbonate−brine interface with pH obtained with seawater and brines containing sulfate. They made changes to the reaction of SO_4^{2-} with the calcite sites. They found that increasing pH turns the negatively charged $>CaSO_4^-$ surface sites into negatively charged $>CaCO_3^-$ and neutral $>CaOH^\circ$ sites. This substitution reduces the negative charge of the surface as reflected in the zeta-potential. Brines without sulfate ions show a downward trend with pH as sometimes reported in the literature. Brine dilution was observed to lead to a more negative surface charge due to the increase in the concentration of negatively charged species, decrease in positively charged species, and formation of neutral species. Song et al. [206] incorporated four impurities in the form of organic acids, that is, humic acid, in their complexation model to better match limestone brine zeta-potential. This is of particular importance as outcrop material behaves differently than reservoir rocks in terms of surface reactivity. This is partly related to contaminants deposited on their surface over geological times.

The equilibrium constants of the SCM have some uncertainty due to the type and nature of rock surface (surface chemistry). In several studies, experimental coreflooding data were used to better fit the preliminary values. Awolayo et al. [18] obtained temperature-dependent equilibrium constants for the reactions by fitting with the single-phase breakthrough curves of different ions and temperatures. The model proposes that Ca^{2+} is strongly adsorbed at temperatures below 100°C in the presence of sulfate. They also found that the same thermodynamic parameters can be used in modeling brine-dependent recovery in both chalk and limestone. In a similar type of study, Eftekhari et al. [205] developed an SCM reactive transport model to determine the reaction equilibrium constant for natural carbonates by using a nonlinear optimization technique to fit the model with zeta-potential data and single-phase breakthrough curves on intact chalk cores.

On the oil side, similar complexes are formed between the two major functional groups of oil: R-COOH (carboxylic) and R-NH (amine) groups. Acid groups are believed to be more polar than amine groups. Amine groups have been overlooked as agents in most cases. The success of these SCM depends largely on the definition of active 'sites' on oil. Takeya et al. [207] developed a triple-layer surface complexation model for the crude oil/brine interface. The carboxyl groups (−COOH) were attributed to the surface charge and electrical triple-layer development of the crude oil in LSWF—see Fig. 2.17. They found that the ions' affinity toward the crude oil follows the order: $Ca^{2+} < Mg^{2+} < OH^-$. The model was used to predict the zeta-potential of the oil−brine interface. Bonto et al. [208] developed an SCM to account for the interaction of Na and acid sites and defined weak acid sites to account for the negative surface charge observed for nonpolar hydrocarbons in water. The model was validated with experimental

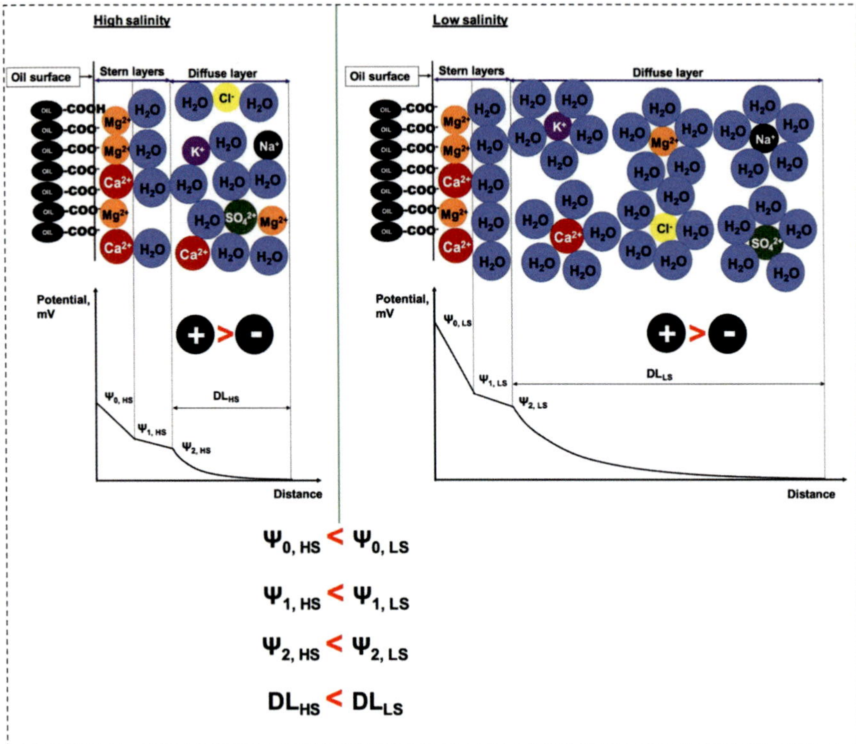

FIGURE 2.17

Surface complexes formed at oil–brine interface between the carboxylic groups of oil and potential determining ions. The distribution of species in the electric double layer at high salinity and low salinity are compared. The inner Helmholtz layer is primarily occupied by the positively charged potential determining ions, while the outer Helmholtz layer is occupied by water molecules. The diffuse layer in low salinity has expanded and the ions are strongly hydrated. In low salinity, the surface is more negatively charged than in high salinity.

Reproduced with permission from M. Takeya, M. Shimokawara, Y. Elakneswaran, et al., Predicting the electrokinetic properties of the crude oil/brine interface for enhanced oil recovery in low salinity water flooding, Fuel 235 (2019) 822–831.

zeta-potential data; however, the performance of the SCM was best at low ionic strength, due to the inherent assumption of the Poisson–Boltzmann equation at HS condition.

Integrating the SCM for both the oil–brine and calcite–brine interface, Brady and Thyne [209] developed a surface complexation model to characterize the wettability of oil–brine–calcite using a "bond-product-sum" concept to represent the potential to bind organics to the surface, which is equal to [$-NH^+$] [$>CO_3$] + [$-NH^+$][$>CaCO_3$] + [$-COOCa^+$][$>CO_3$] + [$-COOCa^+$][$>CaCO_3$] + [$-COO$][$>CaOH_2{}^+$] + [$-COO$] [$>CO_3Ca^+$]. They assume that the potential adhesion between the polar oil and carbonate surface increases with increasing bond product (more oil-wet system) and decreases with decreasing bond product (less oil-wet system). The strongest oil–calcite attraction bridge at reservoir pH is [$-COO$] [$>CaOH_2{}^+$], which can be reduced by increasing Ca^{2+} and/or Mg^{2+} (which could reverse the charge of [COO −] species), and/or increasing $SO_4{}^{2-}$, which can coordinate with [$>CaOH_2{}^+$] eliminating the positive charge to produce a negative surface. Similarly, Eftekhari et al. [205] suggested a correlation between the remaining oil in several imbibition tests on chalk and [$>CaOH_2{}^+$] [$-OOC$], which was estimated using equilibrium constants analogous to aqueous acid–base reactions.

2.6 Concluding remarks, remaining challenges, and future of low-salinity waterflooding technology

This chapter has presented information concerning LSWF in carbonate reservoirs. There have been numerous studies over the past 15 years showing increased recovery. While approximately half the studies found between 0% and 5% incremental OOIP, almost as many studies had 5%–15% additional recovery while a few reached recovery as high as 25% OOIP. Clearly, there are challenges to bring this technology to fruition, but the most pressing the lack of definitive mechanisms, the lack of consistent and standardized experimental protocols and procedures, and the lack of field-scale projects and practical knowledge.

2.6.1 Mechanisms

Investigation of fundamental mechanisms requires an experimental design that can separate causation from correlation. The lack of understanding of the mechanism(s) is seen in the review papers published over the last 5 years that summarize the LSWF studies performed (Table 2.1). These papers discuss proposed mechanisms of LSWF in carbonates based on investigations at three different scales: the core-to reservoir-scale, the porenetwork-scale; and the subpore-scale [80]. Substantiating mechanisms by direct observations and focusing more on micro- and submicron-scale experimentation is required to determine the fundamental mechanisms. There are likely multiple mechanisms at work in LSWF, depending on COBR settings. It is also essential to focus further on understanding interfacial and surface phenomena between rock–fluid and fluid–fluid in addition to measuring bulk properties, as they play a prominent role in the LSWF mechanisms and its efficiency. Understanding the mechanisms would allow process-based predictive models, but experimental designs often fail to separate correlation from causation as discussed.

Sufficient systematic data could be used to construct empirically based predictive models if experimental protocols and procedures were the same. This would allow the development of empirical relationships between factors that control increased recovery, but this approach requires systematic experimental design. Examples of such studies include Romanuka et al. [41]. This investigation used imbibition with one chalk, two limestones and dolostones, three crude oils, and systematically lowered reservoir salinity and altered divalent ion composition at different temperatures. The data even included the effluent water chemistry. Using increased recovery as the primary metric the authors found some combinations of rock and oil were sensitive to salinity while other combinations were sensitive to brine composition, specifically sulfate content.

2.6.2 Improved evaluation of LSWF—experiments

Despite the volume of experiments conducted in industry and academia to assess LSWF, we still do not have standard laboratory experiments with uniform measures. Many studies do not perform essential measurements for the acid and base content of oil, brine pH, pH change, effluent chemistry, in-situ saturations and pressure drop data. There is a wide range of experimental protocols and measurement techniques. For instance, core flooding is widely used in Middle Eastern carbonate studies, but North Sea chalk studies generally use imbibition [14].

Most experiments are performed at the macroscale to assess the potential for increased recovery rather than to isolate and investigate the micro- and nanoscale where changes in wettability occur. Nor do most experiments separate the contributions of altering rock–oil wettability from oil–water interactions, both of which contribute to oil mobility. Finally, most experiments did not usually have reservoir-specific LS fluid design based on extensive testing [210] but used guidelines from previous experiments or general recommendations [211].

Among the challenges that remain are systematic experimental designs with common experimental techniques so data could be compatible between studies [12,212]. The core flooding and imbibition experiments are designed to measure increased recovery, which is the fundamental practical aspect of LSWF, but few of these experiments measure related parameters, such as wettability, oil composition, and effluent chemistry. The contact angle and IFT data, simple measurements that are proxies for wettability, can be useful but contact angle measurements do not involve natural surfaces and IFT changes have been shown to have little relationship to LSWF recovery. Spontaneous and forced imbibition and drainage data can be used for the wettability alteration measurements in the natural condition of the reservoir cores and thus are more reliable than the contact angle measurements.

2.6.3 Improved simulation of LSWF—next-generation, multiscale, multiphysics simulation

In almost all the previously discussed simulation methods, one tries to link a phenomenon occurring at submicroscale or microscale, such as wettability alteration, to its macroscale manifestation, to flow function changes such as relative permeability change. This is a two-scale approach, missing out on other intermediate length scales. Practical multiscale approaches must be developed which link all the involved length scales (e.g., from the nanoscale to cm-scale for core-scale simulation, with intermediate upscaling step) as discussed in the earlier sections. This is very challenging and computationally expernsive, though, and still requires major research efforts.

Other mechanisms which should be accounted for are brine–brine (i.e., injected brine–reservoir brine) interaction and their compatibility as they come in contact, mineral dissolution (which is of importance in the injection of diluted brines), precipitation (more important with seawater and incompatible brines), and any related fines migration. This would require full geochemical simulation and linking the microscopic geochemical interactions to macroscopic Darcy-scale changes in reservoir petrophysical properties. In addition, accounting for the mass transfer between phases, kinetics of reactions and electrokinetic phenomena is necessary. The existing simulators assume that wettability alteration is instantaneous, while this is not the case as per the micro-scale (e.g., micromodel) and droplet scale experiments [48,80, 147].

The results of LSWF simulations (either histroy matching or forward simulation) are driven by the type of model implemented in the simulator. Fully mechanistic models have yet to be developed. The uncertainty in the model by itself makes the derived relative permeabilities and consequently production forecast uncertain. This would particularly affect the transient production period, though. The relative permeability at end points and ultimate production can be still accurately predicted, provided a sound experimental design is employed as covered earlier. Hysteresis affecting oil sweep efficiency by LSWF, as in many other recovery techniques, should be considered too, This may explain success or lack of response in some cases due to the simultaneous effect of imbibition and drainage processes.

2.6.4 Field projects

There are still few field-scale projects because the technique is perceived as experimental rather than operational. Field projects and the practical information they provide are essential to the normal development of new IOR techniques. The cited barriers to field projects include inconsistent results, lack of complete understating of mechanisms, and lack of information to evaluate project economics. While it is true that results vary from good to little or no increased recovery, these differences should be expected given variations in rock, oil, and water compositions, temperatures, and essential factors, such as degree of dilution or brine compositions. No IOR/EOR technique is always effective and both successes and failures are essential in formulating screening criteria, as well as gaining knowledge to constrain and improve economic evaluations. Consider the example of CO_2 flooding which has been widely used for decades without full knowledge of mechanisms [84]. Industry experience shows that complete understanding is not required to implement new processes that increase recovery.

2.6.5 Candidate screening

From the industry point of view, it is important to have an identified series of steps, or workflow, for the implementation of LSWF. The standard industry methodology for deploying new recovery technology usually includes preliminary assessments (screening), economic studies (scoping), laboratory testing, reservoir modeling and simulations, and field pilots that may include single-well tracer tests and/or multiwell pilots.

IOR/EOR project screening traditionally consists of applying a series of rules to evaluate the likelihood that specific techniques will work on a candidate reservoir. In contrast, current industry workflows for LSWF include extensive laboratory testing and reservoir modeling. The availability of a more traditional screening tool is a critical step to transition LSWF academic studies to production technology [80]. In traditional screening, the rules can be qualitative or quantitative and are based on experience where specific criteria are related to the historic success or failure of a technique. Examples of criteria include flow response, oil-in-place, temperature, salinity, depth, oil properties (API gravity, viscosity) rock properties (porosity, permeability, mineralogy, clay content), pay thickness, and heterogeneity [213–215].

One example of LSWF screening criteria is from the Burgan reservoir in Kuwait [216]. The authors used macroscale parametric screening that incorporated estimated steam–oil ratio, permeability, oil properties (API, SARA, viscosity), temperature, water characteristics (TDS, divalent), and drives mechanism. Thyne [217] used empirically based criteria to evaluate the suitability of LSWF of Alaskan reservoirs. He considered laboratory and field examples to establish correlations between increased recovery and water salinity, rock, and oil compositions to formulate the screening algorithm. Recently, Fjelde et al. [218] suggested a fast screening methodology that consists of three steps: rock and fluids studies; flotation experiments to measure wettability; and geochemical simulations with PHREEQC.

Instead of using the relatively simple approach of traditional screening, current projects have used expensive and time-consuming approaches. Researchers at Eni proposed an LSWF workflow that consists of six main stages [92,219]. They proposed screening, 3-D preliminary modeling, followed by laboratory analysis, single-well tracer testing, interwell pilot tests, and full-field implementation.

Suijkerbuijk et al. [102,154] included surface chemistry evaluations, pore-scale physics, pore network models, SCAL experiments, and field tests before reservoir-scale deployment. This approach was implemented for the West Salym field by upscaling the laboratory and field data to evaluate the potential of LSWF implementation by reservoir modeling. The reservoir modeling shifted the relative permeability curves to emulate the LSWF process. They concluded that project economics based on laboratory testing were promising for the offshore West Salym field, but reservoir modeling predicted lower recovery than the experiments [220].

2.6.6 Economics

The economics of LSWF is the ultimate determining factor for a project but few papers address the profitability of LSWF. The basic considerations are the benefit of the additional recovery versus the costs, both OPEX (operational costs) and CAPEX (capital costs). Preliminary estimates of recovery can be made based on empirical relationships between recovery and other known parameters, such as dimensionless curves based on established performance in other analogous conditions, or the results of reservoir models. Dimensionless curves are based on documented performance and standard practice but LSWF lacks sufficient field histories to generate dimensionless curves. Reservoir modeling is based on extensive laboratory and reservoir data typically gathered by larger companies but often unavailable to smaller operators. This means preliminary calculation of benefits must be based on laboratory estimates of increased recovery coupled with reservoir modeling.

Factors, such as initial reservoir salinity, the required dilution factor, and the slug size, are key in estimating OPEX. Al-Shalabi et al. [221] performed a sensitivity analysis for LSWF project economics. They found that slug size was the most influential factor regarding project economics followed by reservoir heterogeneity and salinity. Fani et al. [222] found that smaller slug sizes were as effective as larger slugs in altering wettability. Sadeed et al. [93] found the main parameters when optimizing LSWF economics in a Middle Eastern carbonate reservoir were the number and duration of slugs, degree of dilution, and injection and production rates.

The salinity of produced water can be lowered before injection by membrane filtration (reverse osmosis). Water treatment is an important OPEX component in this option and the evaluation of desalination costs suggests that water treatment was between \$5 and \$10 per barrel for lowering salinity to between 500 and 57,000 ppm TDS, respectively [93]. This process generates HS waste streams (25% volume) that must be disposed of, an additional cost. Onshore fields can use sources, such as lakes, streams, groundwater, or graywater, to mix with produced water to lower injection salinity. For example, in the Permian Basin typical water cost is between \$0.50 and \$2.0 per barrel of water for brackish water supplies.

Accurate prediction of oil breakthrough time during LSWF should be considered as well. Taheriotaghsara et al. [81] investigated if breakthrough time could be predicted as a function of PDIs, such as Ca^{2+}, Mg^{2+}, and SO_4^{2+}. Results showed that the delay can be predicted by a tuned surface complexation model and that Ca^{2+} was the best PDIs for predicting breakthrough time. Aljuboori et al. [223] reported simulations for a field-scale fractured reservoir model showed sustained oil productions were up to 32,000 and 25,000 bpd for LS and HS water injection, respectively. Muriel et al. [224] presented a cost–benefit analysis with a focus on offshore fields that compared LS water and different chemical EOR methods, such as surfactant flooding, alkaline–surfactant, polymer flooding, alkaline–surfactant–polymer flooding, and nanoparticle injection. Two different scenarios were considered: injection from the first day of production and injection after secondary production. LSWF resulted in the lowest CAPEX/OPEX.

Finally, optimization and increased economic robustness for LSWF should be tackled to take this technology to the field faster. LSWF can also be an enabler for the successful application of chemical EOR, either as a preflush or as a hybrid. LSWF should be considered a useful low-cost technique to improve economics producing both short-term increases in production and long-term asset values (reserves). Successful, full-field implementation of LSWF paves the way for further consideration of the technique by companies and securing funding for further research and development.

Nomenclature

AFM	Atomic force microscopy
CAPEX	Capital expenditure
COBR	Crude oil–brine–rock
CDC	Capillary desaturation curve
CFM	Chemical force microscopy
DFT	Density functional theory
DLVO	Derjaguin, Landau, Verwey and Overbeek
EDL	Electric double layer
EDX	Energy-dispersive X-ray
EOR	Enhanced oil recovery
EUR	Estimated ultimate recovery
FPSO	Floating production storage and offloading
FT-ICR-Ms	Fourier-transform ion cyclotron resonance-mass spectrometer
IOR	Improved oil recovery/Incremental oil recovery
LSE	Low-salinity effect
LSR	Low-salinity recovery
LSW	Low-salinity water
LSWF	Low-salinity waterflooding
MICP	Mercury intrusion capillary pressure

MIE	Multi-ion exchange
MD	Molecular dynamics
NMR	Nuclear magnetic resonance
OPEX	Operational expense
PV	Pore volume
PVi	Pore volume injected
REV	Representative elementary volume
ROS	Remaining oil saturation
S_{or}	Residual oil saturation
S_{wi}	Initial water saturation
SAXS	Small-angle X-ray scattering
SCAL	Special core analysis laboratory
SCM	Surface complexation modeling
SI	Spontaneous imbibition
SP	Surfactant polymer
SS	Steady state
TAN	Total acid number
TBN	Total base number
TDS	Total dissolved solids
USS	Unsteady state
VOF	Volume of fluid
VDW	van der Waals
XRD	X-ray diffraction
XRF	X-ray fluorescence
XPS	X-ray photoelectron spectroscopy

References

[1] T.P. Burchette, Carbonate rocks and petroleum reservoirs: a geological perspective from the industry, Geological Society, London, Special Publications 370 (2012) 17–37.

[2] R. Sloan, Quantification of uncertainty in recovery efficiency predictions: Lessons learned from 250 mature carbonate fields, in: SPE-84459-MS, SPE Annual Technical Conference and Exhibition, Denver, CO, USA, 5-8 October, 2003.

[3] X. Xie, W.W. Weiss, Z.J. Tong, et al., Improved oil recovery from carbonate reservoirs by chemical stimulation, SPE Journal 10 (2005) 276–285.

[4] E.J. Manrique, V.E. Muci, M.E. Gurfinkel, EOR field experiences in carbonate reservoirs in the United States, SPE Reservoir Evaluation & Engineering 10 (2007) 667–686.

[5] T. Austad, S. Strand, M.V. Madland, et al., Seawater in chalk: an EOR and compaction fluid, SPE Reservoir Evaluation & Engineering 11 (2008) 648–654.

[6] S. Rassenfoss, Scaling up smart water, Journal of Petroleum Technology 68 (2016) 39–41.

[7] T. Austad, S. Strand, E. J. Hognesen, et al., Seawater as IOR fluid in fractured chalk, in: SPE-93000-MS, SPE International Symposium on Oilfield Chemistry, The Woodlands, TX, USA, 2–4 February, 2005.

[8] S. Strand, E.J. Høgnesen, T. Austad, Wettability alteration of carbonates—effects of potential determining ions (Ca^{2+} and SO_4^{2-}) and temperature, Colloids and Surfaces A: Physicochemical and Engineering Aspects 275 (2006) 1–10.

[9] A. Lager, K.J. Webb, C. Black, et al., Low salinity oil recovery—an experimental investigation, Petrophysics 49 (2008) 28–35.

[10] J. Hao, S. Mohammadkhani, H. Shahverdi, et al., Mechanisms of smart waterflooding in carbonate oil reservoirs—a review, Journal of Petroleum Science and Engineering 179 (2019) 276–291.

[11] C. Dang, L. Nghiem, Z. Chen, et al., Modeling low salinity waterflooding: ion exchange, geochemistry and wettability alteration, in: SPE-166447-MS, SPE Annual Technical Conference and Exhibition, New Orleans, LA, USA, 30 September–2 October, 2013.

[12] M. Jackson, J. Vinogradov, G. Hamon, et al., Evidence, mechanisms and improved understanding of controlled salinity waterflooding, part 1: sandstones, Fuel 185 (2016) 772–793.

[13] S. Strand, T. Puntervold, T. Austad, Water based EOR from clastic oil reservoirs by wettability alteration: a review of chemical aspects, Journal of Petroleum Science and Engineering 146 (2016) 1079–1091.

[14] M.A. Sohal, G. Thyne, E.G. Søgaard, Review of recovery mechanisms of ionically modified waterflood in carbonate reservoirs, Energy & Fuels 30 (2016) 1904–1914.

[15] D.A. Afekare, M. Radonjic, From mineral surfaces and coreflood experiments to reservoir implementations: comprehensive review of low-salinity water flooding (LSWF), Energy & Fuels 31 (2017) 13043–13062.

[16] A. Kilybay, B. Ghosh, N.C. Thomas, A review on the progress of ion-engineered water flooding, Journal of Petroleum Engineering 2017 (2017) 1–9. 2017.

[17] P. Purswani, M.S. Tawfik, Z.T. Karpyn, Factors and mechanisms governing wettability alteration by chemically tuned waterflooding: a review, Energy & Fuels 31 (2017) 7734–7745.

[18] A. Awolayo, H. Sarma, L. Nghiem, Brine-dependent recovery processes in carbonate and sandstone petroleum reservoirs: review of laboratory-field studies, interfacial mechanisms and modeling attempts, Energies 11 (2018) 3020.

[19] M. Derkani, A. Fletcher, W. Abdallah, et al., Low salinity waterflooding in carbonate reservoirs: review of interfacial mechanisms, Colloids and Interfaces 2 (2018) 20.

[20] A. Katende, F. Sagala, A critical review of low salinity water flooding: mechanism, laboratory and field application, Journal of Molecular Liquids 278 (2019) 627–649.

[21] M. Chavan, A. Dandekar, S. Patil, et al., Low-salinity-based enhanced oil recovery literature review and associated screening criteria, Petroleum Science 16 (2019) 1344–1360.

[22] H. Ding, S. Rahman, Experimental and theoretical study of wettability alteration during low salinity water flooding-an state of the art review, Colloids and Surfaces A: Physicochemical and Engineering Aspects 520 (2017) 622–639.

[23] E.W. Al-Shalabi, K. Sepehrnoori, A comprehensive review of low salinity/engineered water injections and their applications in sandstone and carbonate rocks, Journal of Petroleum Science and Engineering 139 (2016) 137–161.

[24] A.A. Yousef, S. Al-Saleh, M.S. Al-Jawfi, Improved/enhanced oil recovery from carbonate reservoirs by tuning injection water salinity and ionic content, in: SPE-154076-MS, SPE Improved Oil Recovery Symposium, Tulsa, OK, USA, 14–18 April, 2012.

[25] Z. Duan, R. Sun, C. Zhu, et al., An improved model for the calculation of CO_2 solubility in aqueous solutions containing Na^+, K^+, Ca^{2+}, Mg^{2+}, Cl^-, and SO_4^{2-}, Marine Chemistry 98 (2006) 131–139.

[26] J.P. Kaszuba, A. Navarre-Sitchler, G. Thyne, et al., Supercritical carbon dioxide and sulfur in the Madison Limestone: a natural analog in southwest Wyoming for geologic carbon–sulfur co-sequestration, Earth and Planetary Science Letters 309 (2011) 131–140.

[27] W. Dean, T. Fouch, Lacustrine environment, in: P.A. Scholle, D.G. Bebout, C.H. Moore (Eds.), Carbonate Depositional Environments, American Assoication of Petroleum Geologists, Tusla, 1983, pp. 97–130.

[28] E. Shinn, Tidal flats, in: P.A. Scholle, D.G. Bebout, C.H. Moore (Eds.), Carbonate Depositional Environments, American Assoication of Petroleum Geologists, Tulsa, 1983, pp. 171–210.

[29] R. Inden, C. Moore, Beach, in: P.A. Scholle, D.G. Bebout, C.H. Moore (Eds.), Carbonate Depositional Environments, American Assoication of Petroleum Geologists, Tulsa, 1983, pp. 211–266.

[30] P. Enos, Shelf, in: P.A. Scholle, D.G. Bebout, C.H. Moore (Eds.), Carbonate Depositional Environments, American Assoication of Petroleum Geologists, Tulsa, 1983, pp. 267–296.

[31] P. Scholle, M. Arthur, A. Ekdale, Pelagic, in: P.A. Scholle, D.G. Bebout, C.H. Moore (Eds.), Carbonate Depositional Environments, American Assoication of Petroleum Geologists, Tulsa, 1983, pp. 619–692.

[32] M.D. Jackson, J. Vinogradov, Impact of wettability on laboratory measurements of streaming potential in carbonates, Colloids and Surfaces A: Physicochemical and Engineering Aspects, 393, 2012, pp. 86–95.

[33] D. Al Mahrouqi, J. Vinogradov, M.D. Jackson, Zeta potential of artificial and natural calcite in aqueous solution, Advances in Colloid and Interface Science 240 (2017) 60–76.

[34] P.V. Brady, J.L. Krumhansl, P.E. Mariner, Surface complexation modeling for improved oil recovery, in: SPE-153744-MS, SPE Improved Oil Recovery Symposium, Tulsa, OK, USA, 14–18 April, 2012.

[35] R. Gupta, G.G. Smith, L. Hu, et al., Enhanced waterflood for middle east carbonate cores—Impact of injection water composition, in: SPE-142668-MS, SPE Middle East Oil and Gas Show and Conference, Manama, Bahrain, 25–28 September, 2011.

[36] A. Sanaei, K. Sepehrnoori, Implication of oil/brine/rock surface interactions in modeling modified salinity waterflooding in carbonate and sandstone reservoirs, in: SPE-191639-MS, SPE Annual Technical Conference and Exhibition, Dallas, TX, USA, 24–26 September, 2018.

[37] M.M. Koleini, M.H. Badizad, R. Hartkamp, et al., The impact of salinity on the interfacial structuring of an aromatic acid at the calcite/brine interface: an atomistic view on low salinity effect, Journal of Physical Chemistry B 124 (2020) 224–233.

[38] F. Bordeaux-Rego, M. Mehrabi, A. Sanaei, K. Sepehrnoori, Improvements on modelling wettability alteration by Engineered water injection: Surface complexation at the oil/brine/rock contact, Fuel 284 (2021) 118991.

[39] I.L. Fabricius, B. Røgen, L. Gommesen, How depositional texture and diagenesis control petrophysical and elastic properties of samples from five North Sea chalk fields, Petroleum Geoscience 13 (2007) 81–95.

[40] M.A. Sohal, S. Kucheryavskiy, G. Thyne, et al., Study of ionically modified water performance in the carbonate reservoir system by multivariate data analysis, Energy & Fuels 31 (2017) 2414–2429.

[41] J. Romanuka, J. Hofman, D.J. Ligthelm, et al., Low salinity EOR in carbonates, in: SPE-153869-MS, SPE Improved Oil Recovery Symposium, Tulsa, OK, USA, April 14–18, 2012.

[42] G. Bernard, Effect of floodwater salinity on recovery of oil from cores containing clays, in: SPE-1725-MS, SPE California Regional Meeting, Los Angeles, CA, USA, October 26–27, 1967.

[43] A.A. Yousef, S.H. Al-Salehsalah, M.S. Al-Jawfi, New recovery method for carbonate reservoirs through tuning the injection water salinity: smart waterflooding, in: SPE-143550-MS, SPE EUROPEC/EAGE Annual Conference and Exhibition, Vienna, Austria, May 23–26, 2011.

[44] A.A. Yousef, S. Al-Saleh, M.S. Al-Jawfi, The impact of the injection water chemistry on oil recovery from carbonate reservoirs, in: SPE-154077-MS, SPE EOR Conference at Oil and Gas West Asia, Muscat, Oman, April 16–18, 2012.

[45] T. Austad, S.F. Shariatpanahi, S. Strand, et al., Conditions for a low-salinity enhanced oil recovery (EOR) effect in carbonate oil reservoirs, Energy & Fuels 6 (2012) 569–575.

[46] A. RezaeiDoust, T. Puntervold, T. Austad, Chemical verification of the EOR mechanism by using low saline/smart water in sandstone, Energy & Fuels 25 (2011) 2151–2162.

[47] T. Austad, A. Rezaeidoust, T. Puntervold, Chemical mechanism of low salinity water flooding in sandstone reservoirs, in: SPE-129767-MS, SPE Improved Oil Recovery Symposium, Tulsa, OK, USA, April 24–28, 2010.

[48] H. Mahani, A.L. Keya, S. Berg, et al., Insights into the mechanism of wettability alteration by low-salinity flooding (LSF) in carbonates, Energy & Fuels 29 (2015) 1352–1367.

[49] D.J. Ligthelm, J. Gronsveld, J. Hofman, et al., Novel waterflooding strategy by manipulation of injection brine composition, in: SPE-119835-MS, EUROPEC/EAGE Conference and Exhibition, Amsterdam, The Netherlands, June 8–11, 2009.

[50] Y. Lee, H. Park, J. Lee, et al., Enhanced oil recovery efficiency of low-salinity water flooding in oil reservoirs including Fe^{2+} ions, Energy Exploration & Exploitation 37 (2019) 355–374.

[51] B. Suijkerbuijk, J. Hofman, D.J. Ligthelm, et al., Fundamental investigations into wettability and low salinity flooding by parameter isolation, in: SPE-154204-MS, SPE Improved Oil Recovery Symposium, Tulsa, OK, USA, 14–18 April, 2012.

[52] F. Hussain, A. Zeinijahromi, P. Bedrikovetsky, et al., An experimental study of improved oil recovery through fines-assisted waterflooding, Journal of Petroleum Science and Engineering 109 (2013) 187–197.

[53] A. Zeinijahromi, R. Farajzadeh, J. Bruining, et al., Effect of fines migration on oil–water relative permeability during two-phase flow in porous media, Fuel 176 (2016) 222–236.

[54] G. Tang, N.R. Morrow, Influence of brine composition and fines migration on crude oil/brine/rock interactions and oil recovery, Journal of Petroleum Science and Engineering 24 (1999) 99–111.

[55] H. Pu, X. Xie, P. Yin, et al., Application of coalbed methane water to oil recovery by low salinity waterflooding, in: SPE-113410-MS, SPE Symposium on Improved Oil Recovery, Tulsa, OK, USA, 20–23 April, 2008.

[56] M. Kumar, A. Fogden, N.R. Morrow, et al., Mechanisms of Improved oil recovery from sandstone by low salinity flooding, in: SCA2010-25, International Symposium of the Society of Core Analysts, Halifax, Canada, 4 – 7 October, 2010.
[57] A.A. Hamouda, O.M. Valderhaug, Investigating enhanced oil recovery from sandstone by low-salinity water and fluid/rock interaction, Energy & Fuels 28 (2014) 898–908.
[58] A. Hiorth, L.M. Cathles, M.V. Madland, The Impact of pore water chemistry on carbonate surface charge and oil wettability, Transport in Porous Media 85 (2010) 1–21.
[59] H. Pu, X. Xie, P. Yin, et al., Low salinity waterflooding and mineral dissolution, in: SPE-134042-MS, SPE Annual Technical Conference and Exhibition, Florence, Italy, 19–22 September, 2010.
[60] P.L. McGuire, J.R. Chatham, F K. Paskvan, et al., Low salinity oil recovery: An exciting new EOR opportunity for Alaska's North Slope, in: SPE-93903-MS, SPE Western Regional Meeting, Irvine, CA, USA, 30 March–1 April, 2005.
[61] I.D. Piñerez T., T. Austad, S. Strand, et al., Linking low salinity eor effects in sandstone to pH, mineral properties and water composition, in: SPE-179625-MS, SPE Improved Oil Recovery Conference, Tulsa, OK, USA, 2016.
[62] P.V. Brady, N.R. Morrow, A. Fogden, et al., Electrostatics and the low salinity effect in sandstone reservoirs, Energy & Fuels 29 (2015) 666–677.
[63] M.B. Alotaibi, H.A. Nasr-El-Din, Effect of brine salinity on reservoir fluids interfacial tension, in: SPE-121569-MS, EUROPEC/EAGE Conference and Exhibition, Amsterdam, The Netherlands, 8–11 June, 2009.
[64] V. Alvarado, G. Garcia-Olvera, P. Hoyer, et al., Impact of polar components on crude oil-water interfacial film formation: a mechanisms for low-salinity waterflooding, in: SPE-170807-MS, SPE Annual Technical Conference and Exhibition, Amsterdam, The Netherlands, 27–29 October, 2014.
[65] F. Moeini, A. Hemmati-Sarapardeh, M.H. Ghazanfari, et al., Toward mechanistic understanding of heavy crude oil/brine interfacial tension: the roles of salinity, temperature and pressure, Fluid Phase Equilibria 375 (2014) 191–200.
[66] A. Emadi, M. Sohrabi, Visual investigation of oil recovery by low salinity water injection: Formation of water micro-dispersions and wettability alteration, in: SPE-166435-MS, SPE Annual Technical Conference and Exhibition, New Orleans, LA, USA, 30 September–2 October, 2013.
[67] M. Sohrabi, P. Mahzari, S.A. Farzaneh, et al., Novel insights into mechanisms of oil recovery by use of low-salinity-water injection, SPE Journal 22 (2017) 407–416.
[68] A. RezaeiDoust, T. Puntervold, S. Strand, et al., Smart water as wettability modifier in carbonate and sandstone: a discussion of similarities/differences in the chemical mechanisms, Energy & Fuels 23 (2009) 4479–4485.
[69] S.B. Fredriksen, A.U. Rognmo, M.A. Fernø, Pore-scale mechanisms during low salinity waterflooding: oil mobilization by diffusion and osmosis, Journal of Petroleum Science and Engineering 163 (2018) 650–660.
[70] K. Sandengen, A. Kristoffersen, K. Melhuus, et al., Osmosis as mechanism for low-salinity enhanced oil recovery, SPE Journal 21 (2016) 1227–1235.
[71] T.E. Chávez-Miyauch, Y. Lu, A. Firoozabadi, Low salinity water injection in Berea sandstone: effect of wettability, interface elasticity, and acid and base functionalities, Fuel 263 (2020) 116572.
[72] S.C. Ayirala, S. H. Saleh, A.A. Yousef, Microscopic scale study of individual water ion interactions at complex crude oil-water interface: a new smartwater flood

recovery mechanism, in: SPE-179590-MS, SPE Improved Oil Recovery Conference, Tulsa, OK, USA, 11–13 April, 2016.

[73] H. Gachuz-Muro, M. Sohrabi, M.D. Benavente, Natural generation of acidic water as a cause of dissolution of the rock during smart water injection in heavy oil carbonate reservoirs, in: SPE-181167-MS, SPE Latin America and Caribbean Heavy and Extra Heavy Oil Conference, Lima, Peru, 19–20 October, 2016.

[74] N.R. Morrow, H.T. Lim, J.S. Ward, Effect of crude-oil-induced wettability changes on oil recovery, SPE Formation Evaluation 1 (1986) 89–103.

[75] W. Abdallah, J.S. Buckley, Fundamentals of wettability, Oilfield Review, Summer Issue (2007) 44–61.

[76] J.S. Buckley, Y. Liu, Some mechanisms of crude oil/brine/solid interactions, Journal of Petroleum Science and Engineering 20 (1998) 155–160.

[77] J.S. Buckley, Y. Liu, S. Monsterleet, Mechanisms of wetting alteration by crude oils, SPE Journal 3 (1998) 54–61.

[78] E.W. Al Shalabi, K. Sepehrnoori, M. Delshad, Does the double layer expansion mechanism contribute to the LSWI effect on hydrocarbon recovery from carbonate rocks? in: SPE-165974-MS, SPE Reservoir Characterization and Simulation Conference and Exhibition, Abu Dhabi, UAE, 16–18 September, 2013.

[79] R.A. Nasralla, H.A. Nasr-El-Din, Double-layer expansion: is it a primary mechanism of improved oil recovery by low-salinity waterflooding?, in: SPE-154334-MS, SPE Improved Oil Recovery Symposium, 1 Tulsa, OK, USA, 4–18 April, 2012.

[80] W.B. Bartels, H. Mahani, S. Berg, et al., Literature review of low salinity waterflooding from a length and time scale perspective, Fuel 236 (2019) 338–353.

[81] M. Taheriotaghsara, M. Bonto, A.A. Eftekhari, et al., Prediction of oil breakthrough time in modified salinity water flooding in carbonate cores, Fuel 274 (2020) 117806.

[82] S. Aslan, N. Fathi Najafabadi, A. Firoozabadi, Non-monotonicity of the Contact Angle from NaCl and $MgCl_2$ Concentrations in Two Petroleum Fluids on Atomistically Smooth Surfaces, Energy & Fuels 30 (2016) 2858–2864. Available from: https://doi.org/10.1021/acs.energyfuels.6b00175.

[83] N.K. Karadimitriou, H. Mahani, H. Steeb, V. Niasar, Nonmonotonic effects of salinity on wettability alteration and two-phase flow dynamics in PDMS micromodels, Water Resources Research 55 (2019) 9826–9837. Available from: https://doi.org/10.1029/2018WR024252.

[84] P.M. Jarrell, C.E. Fox, M.H. Stein, et al., Practical Aspects of CO_2 Flooding, SPE Monograph Series 22, Society of Petroleum Engineers, Richardson, 2002.

[85] S.J. Fathi, T. Austad, S. Strand, "Smart water" as a wettability modifier in chalk: the effect of salinity and ionic composition, Energy and Fuels 24 (2010) 2514–2519.

[86] S.J. Fathi, T. Austad, S. Strand, Water-based enhanced oil recovery (EOR) by "smart water": optimal ionic composition for EOR in carbonates, Energy & Fuels 25 (2011) 5173–5179.

[87] S.F. Shariatpanahi, S. Strand, T. Austad, Evaluation of water-based enhanced oil recovery (EOR) by wettability alteration in a low-permeable fractured limestone oil reservoir, Energy & Fuels 24 (2010) 5997–6008.

[88] W.-B. Bartels, M. Ruecker, S. Berg, et al., Fast X-ray micro-CT study of the impact of brine salinity on the pore-scale fluid distribution during waterflooding, Petrophysics 58 (2017) 36–48.

[89] A. Namaee-Ghasemi, S. Ayatollahi, H. Mahani, Pore-Scale Simulation of the Interplay between Wettability, Capillary Number, and Salt Dispersion on the Efficiency of Oil Mobilization by Low-Salinity Waterflooding,The, Society of Petroleum Engineers Journal 26 (2021) 4000−4021. Available from: https://doi.org/10.2118/206728-PA.

[90] M. Mohammadi, H. Nikbin-Fashkacheh, H. Mahani, Pore network-scale visualization of the effect of brine composition on sweep efficiency and speed of oil recovery from carbonates using a photolithography-based calcite microfluidic model, in: Journal of Petroleum Science and Engineering, 208, Part D, 2022, p. 109641. Available from: https://doi.org/10.1016/j.petrol.2021.109641.

[91] C. Reddick, T. Buikema, D. Williams, Managing risk in the deployment of new technology—getting LoSal EOR into the business, in: SPE-153933-MS, SPE Improved Oil Recovery Symposium, Tulsa, OK, USA, 14−18 April, 2012.

[92] C. Callegaro, M. Bartosek, M. Nobili, et al., Design and implementation of low salinity waterflood in a North African brown field, in: SPE-177590-MS, Abu Dhabi International Petroleum Exhibition and Conference, Abu Dhabi, UAE, 9−12 November, 2015.

[93] A. Sadeed, Z. Tariq, A.N. Janjua, et al., Smart water flooding: an economic evaluation and optimization, in: SPE-192330-MS, SPE Annual Technical Symposium and Exhibition, Dammam, Saudi Arabia, 23−26 April, 2018.

[94] P. Mwangi, G. Thyne, D. Rao, Extensive experimental wettability study in sandstone and carbonate-oil-brine systems: part 1—screening tool development, in: SCA2013-084, International Symposium of the Society of Core Analysts, Napa Valley, CA, USA, 16−19 September, 2013.

[95] E.F. Johnson, D.P. Bossler, V. Bossler, Calculation of relative permeability from displacement experiments, Trans. AIME 216 (1959) 370−372.

[96] J. Kalbus, R. Christiansen, New data reduction developments for relative permeability determination, in: SPE-30799-MS, SPE Annual Technical Conference and Exhibition, Dallas, TX, USA, 22−25 October, 1995.

[97] J. Toth, T. Bodi, P. Szucs, et al., Convenient formulae for determination of relative permeability from unsteady-state fluid displacements in core plugs, Journal of Petroleum Science and Engineering 36 (2002) 33−44.

[98] F. Civan, E.C. Donaldson, Relative permeability from unsteady-state displacements with capillary pressure included, SPE Formation Evaluation 4 (1989) 189−193.

[99] E. Udegbunam, A FORTRAN program for interpretation of relative permeability from unsteady-state displacements with capillary pressure included, Comput. Geosci. 17 (1991) 1351−1357.

[100] M. Rücker, W.-B. Bartels, G. Garfi, et al., Relationship between wetting and capillary pressure in a crude oil/brine/rock system: from nano-scale to core-scale, Journal of Colloid and Interface Science 562 (2020) 159−169.

[101] T.G. Sorop, S.K. Masalmeh, B. Suijkerbuijk, et al., Relative permeability measurements to quantify the low salinity flooding effect at field scale, in: SPE-177856-MS, Abu Dhabi International Petroleum Exhibition and Conference, Abu Dhabi, UAE, 9−12 November, 2015.

[102] B. Suijkerbuijk, H. Kuipers, C. Van Kruijsdijk, et al., The development of a workflow to improve predictive capability of low salinity response, in: IPTC-17157-MS, International Petroleum Technology Conference, Beijing, China, 26−28 March, 2013.

[103] M. Golmohammadi, S. Mohammadi, H. Mahani, S. Ayatollahi, The non-linear effect of oil polarity on the efficiency of low salinity waterflooding: A pore-level investigation, Journal of Molecular Liquids 346 (2022) 117069. Available from: https://doi.org/10.1016/j.molliq.2021.117069.

[104] A. Zahid, E.H. Stenby, A.A. Shapiro, Smart waterflooding (high Sal/low Sal) in carbonate reservoirs, in: SPE-154508-MS, SPE EUROPEC/EAGE Annual Conference, Copenhagen, Denmark, 4–7 June, 2012.

[105] A. Al Harrasi, R.S. Al-maamari, S.K. Masalmeh, Laboratory investigation of low salinity waterflooding for carbonate reservoirs, in: SPE-161468-MS, Abu Dhabi International Petroleum Conference and Exhibition, Abu Dhabi, UAE, 11–14 November, 2012.

[106] L.W. Lake, R.R. Johns, B. Rossen, et al., Fundamentals of Enhanced Oil Recovery, Society of Petroleum Engineers, Richardson, 2014.

[107] L.A. Rapoport, W.J. Leas, Properties of linear waterfloods, Journal of Petroleum Technology 5 (1953) 139–148.

[108] C. McPhee, J. Reed, I. Zubizarreta, Core Analysis: A Best Practice Guide (Developments in Petroleum Science), Vol. 64, Elsevier, Amsterdam, 2015.

[109] A. Al-Sarihi, et al., Admissible parameters for two-phase coreflood and Welge-JBN method, Transport in Porous Media 131 (2020) 831–871.

[110] B. Medina-Rodriguez, T.M. Reilly, H. Wang, et al., Time-domain nuclear magnetic resonance determination of wettability alteration: analysis for low-salinity water, Applied Sciences 10 (2020) 1017.

[111] M. Mohammadi, H. Mahani, Direct insights into the pore-scale mechanism of low-salinity waterflooding in carbonates using a novel calcite microfluidic chip, Fuel 260 (2020) 116374.

[112] A. Aladasani, B. Bai, Y. Wu, et al., Studying low-salinity waterflooding recovery effects in sandstone reservoirs, Journal of Petroleum Science and Engineering 120 (2014) 39–51.

[113] E.W. Al-Shalabi, K. Sepehrnoori, M. Delshad, et al., A Novel method to model low-salinity-water injection in carbonate oil reservoirs, SPE Journal 20 (2015) 1154–1166.

[114] S.K. Masalmeh, T. Sorop, B. Suijkerbuijk, et al., Low salinity flooding: experimental evaluation and numerical interpretation, in: IPTC-17558-MS, International Petroleum Technology Conference, Doha, Qatar, 19–22 January, 2014.

[115] P. Mwangi, The Role of Surface Active Compounds in Crude Oil on Reservoir Wettability (Ph.D. dissertation), Louisiana State University, 2017.

[116] M.A. Sohal, G. Thyne, E.G. Søgaard, Effect of the temperature on wettability and optimum wetting conditions for maximum oil recovery in a carbonate reservoir system, Energy & Fuels 31 (2017) 3557–3566.

[117] S.C. Ayirala, Z. Li, S.H. Saleh, et al., Effects of salinity and individual ions on crude-oil/water interface physicochemical interactions at elevated temperature, SPE Reservoir Evaluation & Engineering 22 (2019) 897–910.

[118] T.E. Chávez-Miyauchi, A. Firoozabadi, G.G. Fuller, Nonmonotonic elasticity of the crude oil–brine interface in relation to improved oil recovery, Langmuir 32 (2016) 2192–2198.

[119] S.C. Ayirala, Z. Li, S.H. Saleh, et al., Effects of salinity and individual water ions on crude oil-water interface physicochemical interactions at elevated temperature,

in: SPE-190387-MS, SPE EOR Conference at Oil and Gas West Asia, Muscat, Oman, 26–28 March, 2018.

[120] E. Amott, Observations relating to the wettability of porous rock, Trans. AIME 216 (1959) 156–162.

[121] W.G. Anderson, Wettability literature survey—part 6: the effects of wettability on waterflooding, Journal of Petroleum Technology 39 (1987) 1605–1622.

[122] E.C. Donaldson, R.D. Thomas, P.B. Lorenz, Wettability determination and its effect on recovery efficiency, SPE Journal 9 (1969) 13–20.

[123] P.P. Jadhunandan, N.R. Morrow, Effect of wettability on waterflood recovery for crude-oil/brine/rock systems, SPE Reservoir Engineering 10 (1995) 40–46.

[124] N.R. Morrow, J. Buckley, Improved oil recovery by low-salinity waterflooding, Journal of Petroleum Technology 63 (2011) 106–112.

[125] C.C. Agbalaka, A.Y. Dandekar, S.L. Patil, et al., The effect of wettability on oil recovery: a review, in: SPE-114496-MS, SPE Asia Pacific Oil and Gas Conference and Exhibition, Perth, Australia, 20–22 October, 2008.

[126] N.J. Hadia, A. Ashraf, M.T. Tweheyo, et al., Laboratory investigation on effects of initial wettabilities on performance of low salinity waterflooding, Journal of Petroleum Science and Engineering 105 (2013) 18–25.

[127] K.J. Humphry, B. Suijkerbuijk, H.A. van der Linde, et al., Impact of wettability on residual oil saturation and capillary desaturation curves, Petrophysics 55 (2013) 313–318.

[128] T. Amirian, M. Haghighi, P. Mostaghimi, Pore scale visualization of low salinity water flooding as an enhanced oil recovery method, Energy & Fuels 31 (2017) 13133–13143.

[129] U. Karabakal, S. Bagci, Determination of wettability and its effect on waterflood performance in limestone medium, Energy & Fuels 18 (2004) 438–449.

[130] T.M. Okasha, J.J. Funk, H.N. Al-Rashidi, Fifty years of wettability measurements in the Arab-D carbonate reservoir, in: SPE-105114-MS, SPE Middle East Oil and Gas Show and Conference, Manama, Bahrain, 11–14 March, 2007.

[131] Q. Xie, A. Sari, W. Pu, et al., pH effect on wettability of oil/brine/carbonate system: implications for low salinity water flooding, Journal of Petroleum Science and Engineering 168 (2018) 419–425.

[132] P. Mwangi, P.V. Brady, M. Radonjic, et al., The effect of organic acids on wettability of sandstone and carbonate rocks, Journal of Petroleum Science and Engineering 165 (2018) 428–435.

[133] J. Song, S. Rezaee, W. Guo, et al., Evaluating physicochemical properties of crude oil as indicators of low-salinity–induced wettability alteration in carbonate minerals, Scientific Report 10 (2020) 3762.

[134] J. Wu, F. Liu, Y. Hui, et al., Effect of specific functional groups on oil adhesion from mica substrate: implications for low salinity effect, Journal of Industrial and Engineering Chemistry 56 (2017) 342–349.

[135] N. Maskari, A. Sari, M. Hossain, et al., Response of non-polar oil component on low salinity effect in carbonate reservoirs: adhesion force measurement using atomic force microscopy, Energies 13 (2019) 77.

[136] C. Xu, W. Zou, Y. Yang, et al., Status and prospects of deep oil and gas resources exploration and development onshore China, Journal of Natural Gas Geoscience 3 (2018) 11–24.

[137] P. Zhang, T. Austad, Wettability and oil recovery from carbonates: effects of temperature and potential determining ions, Colloids and Surfaces A: Physicochemical and Engineering Aspects 279 (2006) 179–187.
[138] S.R. Gomari, A.A. Hamouda, Effect of fatty acids, water composition and pH on the wettability alteration of calcite surface, Journal of Petroleum Science and Engineering 50 (2006) 140–150.
[139] H. Mahani, R. Menezes, S. Berg, et al., Insights into the impact of temperature on the wettability alteration by low salinity in carbonate rocks, Energy & Fuels 31 (2017) 7839–7853.
[140] Z. Yi, H. Sarma, Improving waterflood recovery efficiency in carbonate reservoirs through salinity variations and ionic exchanges: a promising low-cost "smart-waterflood" approach, in: SPE-161631-MS, Abu Dhabi International Petroleum Conference and Exhibition, Abu Dhabi, UAE, 11–14 November, 2012.
[141] H. Jiang, C.G. Chopping, C. Forsman, et al., Lab observation of low salinity waterflooding for a phosphoria reservoir rock, in: SPE-169546-MS, SPE Western North American and Rocky Mountain Joint Meeting, Denver, CO, USA, 17–18 April, 2014.
[142] M. AlHammadi, P. Mahzari, M. Sohrabi, Fundamental investigation of underlying mechanisms behind improved oil recovery by low salinity water injection in carbonate rocks, Fuel 220 (2018) 345–357.
[143] R.A. Nasralla, E. Sergienko, S.K. Masalmeh, et al., Potential of low-salinity waterflood to improve oil recovery in carbonates: demonstrating the effect by qualitative coreflood, SPE Journal 21 (2016) 1643–1654.
[144] M.R. Zaeri, R. Hashemi, H. Shahverdi, et al., Enhanced oil recovery from carbonate reservoirs by spontaneous imbibition of low salinity water, Petroleum Science 15 (2018) 564–576.
[145] P. Zhang, M.T. Tweheyo, T. Austad, Wettability alteration and improved oil recovery by spontaneous imbibition of seawater into chalk: impact of the potential determining ions Ca^{2+}, Mg2 + , and $SO_4$2 − , Colloids and Surfaces A: Physicochemical and Engineering Aspects 301 (2007) 199–208.
[146] T. Austad, S.F. Shariatpanahi, S. Strand, et al., Low salinity EOR effects in limestone reservoir cores containing anhydrite: a discussion of the chemical mechanism, Energy & Fuels 29 (2015) 6903–6911.
[147] W.-B. Bartels, H. Mahani, S. Berg, et al., Oil configuration under high-salinity and low-salinity conditions at pore scale: a parametric investigation by use of a single-channel micromodel, SPE Journal 22 (2017) 1362–1373.
[148] T. Puntervold, S. Strand, T. Austad, New method to prepare outcrop chalk cores for wettability and oil recovery studies at low initial water saturation, Energy & Fuels 21 (2007) 3425–3430.
[149] S.M. Ma, X. Zhang, N.R. Morrow, et al., Characterization of wettability from spontaneous imbibition measurements, Journal of Canadian Petroleum Technology 38 (1999) 1–8.
[150] X. Xie, N.R. Morrow, Oil recovery by spontaneous imbibition from weakly water-wet rocks, Petrophysics 42 (2001) 313–322.
[151] A. Tirjoo, B. Bayati, H. Rezaei, et al., Molecular dynamics simulation of the effect of ions in water on the asphaltene aggregation, Journal of Molecular Liquids 277 (2019) 40–48.

[152] S. Ayatollahi, M.M. Zerafat, Nanotechnology-assisted EOR techniques: new solutions to old challenges, in: SPE-157094-MS, SPE International Oilfield Nanotechnology Conference and Exhibition, Noordwijk, The Netherlands, 12–14 June, 2012.

[153] A.M. Dastjerdi, Z. Kargozarfard, B. Najafi, et al., Microscopic Insight into kinetics of inorganic scale deposition during smart water injection using dynamic quartz crystal microbalance and molecular dynamics simulation, Industrial & Engineering Chemistry Research 59 (2020) 609–619.

[154] T.G. Sorop, B. Suijkerbuijk, S.K Masalmeh, et al., Integrated approach in deploying low salinity waterflooding, in: SPE-165277-MS, SPE Enhanced Oil Recovery Conference, Kuala Lumpur, Malaysia, 2 – 4 July, 2013.

[155] J.E. Sylte, L.K. Thomas, D.W. Rhett, et al., Water induced compaction in the Ekofisk field, in: SPE-56426-MS, SPE Annual Technical Conference and Exhibition, Houston, TX, USA, 3–6 October, 1999.

[156] G.R. Jerauld, K.J. Webb, C.Y. Lin, et al., Modeling low-salinity waterflooding, in: SPE-102239-MS, SPE Annual Technical Conference and Exhibition, San Antonio, TX, USA, 24–27 September, 2006.

[157] A. Attar, A. Muggeridge, Low salinity water injection: impact of physical diffusion, an aquifer and geological heterogeneity on slug size, Journal of Petroleum Science and Engineering 166 (2018) 1055–1070. Available from: https://doi.org/10.1016/j.petrol.2018.03.023.

[158] Global Water Intelligence, Water for onshore oil and gas: Water desalination report, 50, 2014, 18.

[159] A.A. Yousef, J. Liu, G. Blanchard, et al., SmartWater flooding: industry's first field test in carbonate reservoirs, in: SPE-159526-MS, SPE Annual Technical Conference and Exhibition, San Antonio, TX, USA, 8–10 October, 2012.

[160] H. Mahani, T. Sorop, D. Ligthelm, et al., Analysis of field responses to low-salinity water flooding in secondary and tertiary mode in Syria, in: SPE-142960-MS, SPE EUROPEC/EAGE Annual Conference and Exhibition, Vienna, Austria, 23–26 May, 2011.

[161] P.J. van der Hoek, B. Hustedt, M. Sobera, et al., Dynamic induced fractures in waterflooding and EOR, in: SPE-115204-RU, SPE Russian Oil and Gas Technical Conference and Exhibition, Moscow, Russia, 28–30 October, 2008.

[162] D. Kuznetsov, S. Cotterill, M.A. Giddins, et al., Low-salinity waterflood simulation: Mechanistic and phenomenological models, in: SPE-174615-MS, SPE Asia Pacific Enhanced Oil Recovery Conference, Kuala Lumpur, Malaysia, 11–13 August, 2015.

[163] A. Darvish Sarvestani, B. Rostami, H. Mahani, Polymer-enhanced low-salinity brine to control In situ mixing and salt dispersion in low-salinity waterflooding. Energy & Fuels (2021). Available from: https://doi.org/10.1021/acs.energyfuels.1c00871.

[164] H. Al-Ibadi, K.D. Stephen, E.J. Mackay, Extended fractional-flow model of low-salinity waterflooding accounting for dispersion and effective salinity range, SPE Journal 24 (2019) 2874–2888.

[165] Y. Wu, B. Bai, Efficient simulation for low salinity waterflooding in porous and fractured reservoirs, in: SPE-118830-MS, SPE Reservoir Simulation Symposium, The Woodlands, TX, USA, 2–4 February, 2009.

[166] A. Aljaberi, M. Sohrabi, A new approach to simulate low salinity water flooding in carbonate reservoir, in: SPE-195081-MS, SPE Middle East Oil and Gas Show and Conference, Manama, Bahrain, 18–21 March, 2019.

[167] I. Fjelde, S.M. Asen, A.V. Omekeh, Low salinity water flooding experiments and interpretation by simulations, in: SPE-154142-MS, SPE Improved Oil Recovery Symposium, Tulsa, OK, USA, 14–18 April, 2012.

[168] C. Qiao, R. Johns, L. Li, Modeling low-salinity waterflooding in chalk and limestone reservoirs, Energy & Fuels 30 (2016) 884–895.

[169] A.N. Awolayo, H.K. Sarma, L.X. Nghiem, A Comprehensive geochemical-based approach at modeling and interpreting brine dilution in carbonate reservoirs, in: SPE-182626-MS, SPE Reservoir Simulation Conference, Montgomery, TX, USA, 20–22 February, 2017.

[170] A. Korrani, G. Jerauld, K. Sepehrnoori, Mechanistic modeling of low-salinity waterflooding through coupling a geochemical package with a compositional reservoir simulator, SPE Reservoir Evaluation & Engineering 19 (2016) 142–162.

[171] Y. Elakneswaran, M. Takeya, A. Ubaidah, et al., Integrated geochemical modelling of low salinity waterflooding for enhanced oil recovery in carbonate reservoir, in: IPTC-20250-MS, International Petroleum Technology Conference, Dhahran, Saudi Arabia, 13–15 January, 2020.

[172] A. Sanaei, A. Varavei, K. Sepehrnoori, Mechanistic modeling of carbonated waterflooding, in: SPE-190319-MS, SPE Improved Oil Recovery Conference, Tulsa, OK, USA, 14–18 April, 2018.

[173] M.J. Blunt, B. Bijeljic, H. Dong, et al., Pore-scale imaging and modelling, Advances in Water Resources 51 (2013) 197–216. Available from: https://doi.org/10.1016/j.advwatres.2012.03.003.

[174] G. Garfi, C.M. John, M. Rücker, et al., Determination of the spatial distribution of wetting in the pore networks of rocks, Journal of Colloid and Interface Science 613 (2022) 786–795. Available from: https://doi.org/10.1016/j.jcis.2021.12.183.

[175] M. Tuller, D. Or, Hydraulic conductivity of variably saturated porous media: film and corner flow in angular pore space, Water Resources Research 37 (2001) 1257–1276.

[176] J. Su, L. Wang, Z. Gu, et al., Advances in pore-scale simulation of oil reservoirs, Energies 11 (2018) 1132.

[177] B. Zhao, C.W. MacMinn, B.K. Primkulov, et al., Comprehensive comparison of pore-scale models for multiphase flow in porous media, Proceedings of the National Academy of Sciences of the United States of America 116 (2019) 13799–13806.

[178] K.S. Sorbie, I.R. Collins, A proposed pore-scale mechanism for how low salinity waterflooding works, in: SPE-129833-MS, SPE Improved Oil Recovery Symposium, Tulsa, OK, USA, 24–28 April, 2010.

[179] M.G. Watson, I. Bondino, G. Hamon, et al., A pore-scale investigation of low-salinity waterflooding in porous media: uniformly wetted systems, Transport in Porous Media 118 (2017) 201–223.

[180] A. Boujelben, S. McDougall, M. Watson, et al., Pore network modelling of low salinity water injection under unsteady-state flow conditions, Journal of Petroleum Science and Engineering 165 (2018) 462–476.

[181] J. Maes, S. Geiger, Direct pore-scale reactive transport modelling of dynamic wettability changes induced by surface complexation, Advances in Water Resources 111 (2018) 6–19.

[182] R. Aziz, V. Joekar-Niasar, P.J. Martínez-Ferrer, et al., Novel insights into pore-scale dynamics of wettability alteration during low salinity waterflooding, Scientific Reports 9 (2019) 9257.

[183] T. Akai, M. Blunt, B. Bijeljic, Pore-scale numerical simulation of low salinity water flooding using the lattice Boltzmann method, Journal of Colloid and Interface Science 566 (2020) 444–453.

[184] E.G. Martínez-Mendoza, M.A. Díaz-Viera, M. Coronado, et al., Capillary pressure and relative permeability estimation for low salinity waterflooding processes using pore network models, Journal of Petroleum Science and Engineering 182 (2019) 106253.

[185] M. Abu-Al-Saud, A. Gmira, S. Al-Enezi, et al., Pore-scale simulation of fluid flow in carbonates using micro-CT scan images, in: IPTC-19832-MS, International Petroleum Technology Conference, Dhahran, Saudi Arabia, 13–15 January, 2020.

[186] M.O. Abu-Al-Saud, A. Riaz, H.A. Tchelepi, Multiscale level-set method for accurate modeling of immiscible two-phase flow with deposited thin films on solid surfaces, J. Comput. Phys. 333 (2017) 297–320.

[187] Z.M. Wilmott, C. Breward, S. Chapman, A dynamic network model for the action of low salinity on two-phase flow, Advances in Water Resources 137 (2020) 103520.

[188] Z.M. Wilmott, C. Breward, S. Chapman, Modelling low-salinity oil recovery mechanisms using an ion dissociation model, Transport in Porous Media 127 (2019) 685–709.

[189] A. Pourakaberian, H. Mahani, V. Niasar, The impact of the electrical behavior of oil-brine-rock interfaces on the ionic transport rate in a thin film, hydrodynamic pressure, and low salinity waterflooding effect, Colloids and Surfaces A: Physicochemical and Engineering Aspects 620 (2021) 126543. Available from: https://doi.org/10.1016/j.colsurfa.2021.126543.

[190] C. Fang, Y. Yang, S. Sun, et al., Low salinity effect on the recovery of oil trapped by nanopores: a molecular dynamics study, Fuel 261 (2020) 116443.

[191] X. Chang, Q. Xue, X. Li, et al., Inherent wettability of different rock surfaces at nanoscale: a theoretical study, Applied Surface Science 434 (2018) 73–81.

[192] M.S. Santos, M. Castier, I.G. Economou, Molecular dynamics simulation of electrolyte solutions confined by calcite mesopores, Fluid Phase Equilibria 487 (2019) 24–32.

[193] A. Kirch, S.M. Mutisya, V.M. Sánchez, et al., Fresh molecular look at calcite–brine nanoconfined interfaces, Journal of Physical Chemistry C 122 (2018) 6117–6127.

[194] M. Ricci, P. Spijker, K. Voïtchovsky, Water-induced correlation between single ions imaged at the solid–liquid interface, Nature Communications 5 (2014) 4400.

[195] G. Brekke-Svaland, F. Bresme, Interactions between hydrated calcium carbonate surfaces at nanoconfinement conditions, Journal of Physical Chemistry C 122 (2018) 7321–7330.

[196] S. Kim, M.C. Marcano, U. Becker, Mechanistic study of wettability changes on calcite by molecules containing a polar hydroxyl functional group and nonpolar benzene rings, Langmuir 35 (2019) 2527–2537.

[197] S. Prabhakar, R. Melnik, Influence of Mg^{2+}, SO_4^{2-} and Na^+ ions of sea water in crude oil recovery: DFT and ab initio molecular dynamics simulations, Colloids and Surfaces A: Physicochemical and Engineering Aspects 539 (2018) 53–58.

[198] V. Joekar-Niasar, H. Mahani, Nonmonotonic pressure field induced by ionic diffusion in charged thin films, Industrial & Engineering Chemistry Research 55 (2016) 6227–6235.

[199] R. Chai, Y. Liu, Y. Zhang, et al., Wettability alteration mechanism of calcite surface: molecular dynamics simulation, 81st EAGE Conference and Exhibition, 2019 (2019) 1–5. Available from: https://doi.org/10.3997/2214-4609.201900958.

[200] J. Xin, C. Li, R. Chai, Effect of sulfate ions on oil detachment from calcite surface: experiments and molecular dynamics simulations, Arabian Journal of Geosciences 12 (2019) 305.

[201] A. Aminian, B. ZareNezhad, Oil-detachment from the calcium carbonate surfaces via the actions of surfactant, nanoparticle and low salinity brine: an insight from molecular dynamic simulation, Chemical Engineering Science 202 (2019) 373–382.

[202] M.P. Andersson, K. Dideriksen, H. Sakuma, et al., Modelling how incorporation of divalent cations affects calcite wettability—implications for biomineralisation and oil recovery, Scientific Reports 6 (2016) 28854.

[203] A.O. Alghamdi, M.B. Alotaibi, A.A. Yousef, Atomistic simulation of calcite interaction with ionic species and oil components in water-flooding, Colloids and Surfaces A: Physicochemical and Engineering Aspects 529 (2017) 760–764.

[204] P.V. Brady, J.L. Krumhansl, A surface complexation model of oil–brine–sandstone interfaces at 100°C: low salinity waterflooding, Journal of Petroleum Science and Engineering 81 (2012) 171–176.

[205] A.A. Eftekhari, K. Thomsen, E.H. Stenby, et al., Thermodynamic analysis of chalk–brine–oil interactions, Energy & Fuels 31 (2017) 11773–11782.

[206] J. Song, Y. Zeng, L. Wang, et al., Surface complexation modeling of calcite zeta potential measurements in brines with mixed potential determining ions (Ca^{2+}, CO_3^{2-}, Mg^{2+}, SO_4^{2-}) for characterizing carbonate wettability, Journal of Colloid and Interface Science 506 (2017) 169–179.

[207] M. Takeya, M. Shimokawara, Y. Elakneswaran, et al., Predicting the electrokinetic properties of the crude oil/brine interface for enhanced oil recovery in low salinity water flooding, Fuel 235 (2019) 822–831.

[208] M. Bonto, A.A. Eftekhari, H.M. Nick, An overview of the oil-brine interfacial behavior and a new surface complexation model, Scientific Reports 9 (2019) 6072.

[209] P.V. Brady, G. Thyne, Functional wettability in carbonate reservoirs, Energy & Fuels 30 (2016) 9217–9225.

[210] C. Polat, M. Parlaktuna, A study on the effect of the chemical composition of brine to improve oil recovery from carbonates, Energy Sources, Part A: Recovery, Utilization, and Environmental Effects 39 (2017) 2151–2156.

[211] S.C. Ayirala, A.A. Yousef, Injection water chemistry requirement guidelines for IOR/EOR, in: SPE-169048-MS, SPE Improved Oil Recovery Symposium, Tulsa, OK, USA, 12–16 April, 2014.

[212] S. Patil, A.Y. Dandekar, S.L. Patil, et al., Low salinity brine injection for EOR on Alaska North Slope (ANS), in: IPTC-12004-MS, International Petroleum Technology Conference, Kuala Lumpur, Malaysia, 3–5 December, 2008.

[213] R. Henson, A. Todd, P. Corbett, Geologically Based screening criteria for improved oil recovery projects, in: SPE-75148-MS, SPE/DOE Improved Oil Recovery Symposium, Tulsa, OK, USA, 13–17 April, 2002.

[214] J.J. Taber, F.D. Martin, R.S. Seright, EOR screening criteria revisited—part 1: introduction to screening criteria and enhanced recovery field projects, SPE Reservoir Evaluation & Engineering 12 (1997) 189–198.

[215] J.J. Taber, F.D. Martin, R.S. Seright, EOR screening criteria revisited—part 2: applications and impact of oil prices, SPE Reservoir Evaluation & Engineering 12 (1997) 199–206.

[216] M.T. Al-Murayri, H.E. Al-Mayyan, A.A. Faraj, et al., Evaluation of enhanced oil recovery technologies for the Sabriyah Lower Burgan reservoir Kuwait, in: SPE-186026-MS, SPE Reservoir Characterization and Simulation Conference and Exhibition, Abu Dhabi, UAE, 8–10 May, 2017.

[217] G. Thyne, Wettability alteration in reservoirs: how it applies to Alaskan oil production, in: SPE-180370-MS, SPE Western Regional Meeting, Anchorage, AK, USA, 23–26 May, 2016.

[218] I. Fjelde, A.V. Omekeh, P.E. Haugen, Screening of the potential for different injection water compositions to alter wettability to more water-wet, in: SPE-184918-MS, SPE Latin America and Caribbean Mature Fields Symposium, Salvador, Brazil, 15–16 March, 2017.

[219] M. Rotondi, C. Callegaro, F. Masserano, et al., Low salinity water injection: eni's experience, in: SPE-171794-MS, Abu Dhabi International Petroleum Exhibition and Conference, Abu Dhabi, UAE, 10–13 November, 2014.

[220] S.I. Erke, Y.E. Volokitin, I.Y. Edelman, et al., Low salinity flooding trial at West Salym field, in: SPE-179629-MS, SPE Improved Oil Recovery Conference, Tulsa, OK, USA, 11–13 April, 2016.

[221] E. Al-Shalabi, K. Sepehrnoori, G. Pope, An improved method for estimating volumetric sweep efficiency of low salinity water injection, in: SPE-172279-MS, SPE Annual Caspian Technical Conference and Exhibition, Astana, Kazakhstan, 12–14 November, 2014.

[222] M. Fani, H. Al-Hadrami, P. Pourafshary, et al., Optimization of smart water flooding in carbonate reservoir, in: SPE-193014-MS, Abu Dhabi International Petroleum Exhibition & Conference, Abu Dhabi, UAE, 12–15 November, 2018.

[223] F.A. Aljuboori, J.H. Lee, K.A. Elraies, et al., The effectiveness of low salinity waterflooding in naturally fractured reservoirs, Journal of Petroleum Science and Engineering 191 (2020) 107167.

[224] H. Muriel, S. Ma, S. Sofla, et al., Technical and economical screening of chemical EOR methods for the offshore, in: OTC-30740-MS, Offshore Technology Conference, Houston, TX, USA, 4–7 May, 2020.

CHAPTER

Enhanced oil recovery by Smart Water injection in sandstone reservoirs

3

Tina Puntervold, Skule Strand, Aleksandr Mamonov and Iván Darío Torrijos Piñerez
The National IOR Centre of Norway, University of Stavanger, Stavanger, Norway

3.1 Introduction

Fossil fuel is being challenged as the future world energy source. In the process of developing efficient methods to produce energy from renewable sources, there is still a need for oil and gas and their related products. According to the International Energy Agency's (IEA) World Energy Outlook, the global oil demand will become stable around the 2030s, but a rapid decline is not expected in near future [1]. The focus in many countries worldwide is, however, on sustainable recovery and to reduce emissions into the air. Several countries have set net-zero emission goals or visions, which should be reached within the mid-century, around 2050. To reach such ambitious environmental goals, the oil and gas industry is reliant on energy-efficient and environmentally friendly recovery processes. Current oil production is mainly being produced from mature fields [2], and therefore there is a need for improved oil recovery (IOR) from the existing fields. Most recently, IEA published a roadmap for the global energy sector in which it was emphasized that exploration for new oil and gas fields should seize, to be able to reach the ambitious emission goals of becoming net-zero by 2050 [3]. As such, it is even more important to energy-efficiently sweep the current fields already under or planned for production. Waterflooding of oil reservoirs is not only an environmentally friendly recovery method but also a very efficient method for aiding the production of crude oil from a reservoir. Other field operations aiming at maintaining reservoir pressure and oil production performance, such as gas and chemicals injection, are typically more intensive in terms of energy and raw material consumption. Waterflooding of reservoirs has been ongoing for more than a century; however, the choice of water source, for other reasons than availability, has only been in focus for the last three decades. In the mid-1990s systematic research on the composition of formation water (FW) and injection water composition commenced, and since then the amount of research and written publications within the topic has skyrocketed.

Recovery Improvement. DOI: https://doi.org/10.1016/B978-0-12-823363-4.00002-9

Oil recovery is often divided into three stages: primary, secondary, and tertiary oil recovery. During primary recovery, oil is produced by the natural energy stored in the reservoir, and drive mechanisms are mainly depletion drive, gas cap drive, water drive, rock and liquid expansion drive, gravity drainage, and combination drive [4]. When the remaining reservoir energy is not sufficient to produce the oil, or the pressure of the reservoir reaches the bubble point of the oil, extra energy must be supplied to the reservoir. The most common methods in this secondary oil recovery stage are waterflooding or gas injection. Water or gas is injected into the reservoir to maintain pressure above the bubble point pressure and at the same time to displace the oil [4]. The secondary oil recovery stage is efficient and, in high-quality reservoirs, oil recovery can reach up to 70% of oil originally in place (OOIP) [5]. Worldwide, the average secondary oil recovery is in the range of 20%–40% of OOIP [6]. Because of reservoir heterogeneity in the form of mineralogy, wettability and large permeability contrasts between rock matrix and fractures, or high permeability reservoir layers, the volumetric reservoir sweep efficiency is reduced. The injected fluid is not able to contact the entire reservoir area and height, thus the macroscopic sweep efficiency is decreased. In addition, on the microscopic scale, the injected fluid is not able to displace the oil from all pores in the reservoir area contacted by the injected fluid; thus, the microscopic sweep efficiency is decreased. The residual oil saturation (S_{or}) after the secondary recovery stage will therefore vary from reservoir to reservoir. Traditionally, the tertiary oil recovery stage was initiated to target and reduce the S_{or} by deploying various alternative methods, such as chemical, thermal, microbial, and other methods, to increase the reservoir sweep efficiency. However, in some cases, it is beneficial to produce an oil reservoir in nonchronological order, for example, deploy early water injection to maintain high reservoir pressure or deploy thermal methods to heavy oil reservoirs at an early stage. Therefore the term "enhanced oil recovery" (EOR) is nowadays widely used instead of "tertiary recovery" for this stage.

EOR, in most cases, involves the injection of a fluid that interacts with the crude oil–brine–rock (COBR) system in the reservoir, by, for example, reducing the interfacial tension (IFT), swelling the oil, improving oil mobility by reducing oil viscosity or increasing water/gas viscosity, or modifying the reservoir wettability [4,5]. Waterflooding has conventionally been considered a secondary recovery method. However, since the mid-1990s injection of modified water has been more in focus and is by many now considered an EOR method due to the injection water being different from the FW in composition and is thus not a fluid already present in the reservoir. Depending on the properties of the reservoir and its mineralogy, injection water composition can be optimized, resulting in significant, additional oil production from sandstones [7,8] and carbonates [9,10].

With regards to water injection, the terms "low-salinity (LS) water injection" and "Smart Water injection" are widely used these days to signify brines containing low salt content or a modified ionic composition, respectively, however, both with the aim to improve oil recovery beyond standard waterflooding. These terms may be used intermittently in the literature, still one should be aware of the differences between them. In LS waterflooding, the salinity of the injection water is the most important characteristic. Generally, salinities below 5000 ppm have been considered low [11]; however, the composition of the LS brine may vary significantly, being composed, for example, of only NaCl or containing divalent ions as in the case of a diluted seawater (SW) brine. Smart Water, on the other hand, does not emphasize the salinity in the same sense but rather focuses on the actual composition of monovalent and divalent anions and cations, due to their differences in properties and reactivities in water and toward surfaces/interfaces, such as the mineral surface, oil phase, and an additional water phase. Research in both sandstone and carbonate systems has shown that it is not only the salinity that is influencing the oil recovery but also the chemical composition of the injection brine [10,12–15]. Therefore the term "Smart Water" is sometimes used, instead of the term "LS water." Both terms refer to an injection brine, different from the FW, which can disturb the initial wettability established between COBR in the reservoir and alter the wettability toward more water-wet conditions. Increased water wetness induces capillary forces, promoting water imbibition into the oil-filled pores of water-wet reservoirs, thereby improving sweep efficiency, and increasing oil recovery. The reader should be aware that both terms are being used intermittently and this is to some extent a cause of some of the confusion and inconsistencies existing in the published literature today.

In Fig. 3.1, results from a typical oil recovery experiment demonstrating a tertiary Smart Water EOR effect in a sandstone core plug are shown [16]. Both produced oil (%OOIP) and pressure drop (ΔP) at a constant injection rate of 4 pore

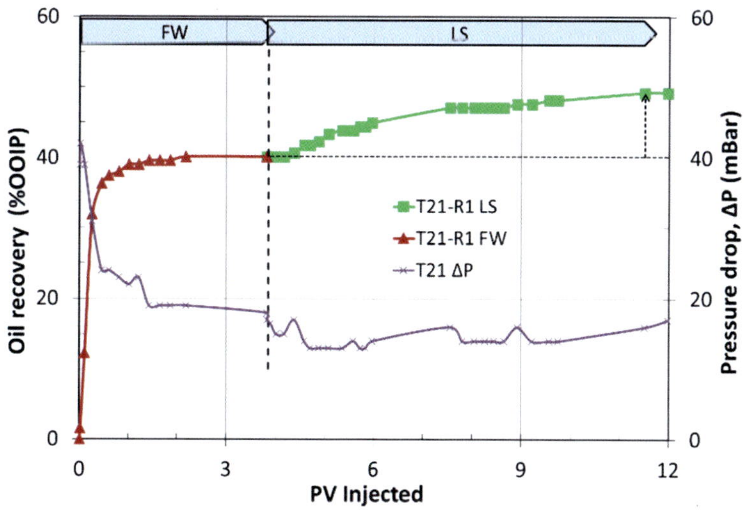

FIGURE 3.1

Oil recovery test from a sandstone core containing $S_{wi} = 20\%$ and crude oil. Formation water–low salinity was flooded through the core at a rate of 4 pore volume/day at 60°C. Oil recovery (%OOIP) and pressure drop (ΔP) were monitored during brine flooding.

volume (PV)/day are monitored when first flooding FW followed by Smart Water, which in this case is a LS 1000 ppm NaCl brine. A recovery plateau by FW injection was reached at 40%OOIP, while a significant increase in oil recovery of 25% was observed during LS injection, ending with an ultimate recovery plateau of 50%OOIP. There was no increased pressure drop accompanying the extra oil recovered during LS injection.

Over the past three decades, an increasing amount of published articles has been written about Smart Water or LS water injection, and this number is continuing to grow since both methods are environmentally friendly, cost-efficient and have a large potential to improve oil recovery. Some of that work is summarized in the previously published review articles [17–21] and in books [22,23]. According to Alvarado and Manrique [2], sandstone reservoirs have been the most targeted for EOR projects and in 2010 about 78% of all international EOR projects were devoted to sandstone reservoirs. In 2018 Smalley et al. [24] published an EOR-screening study in which they screened the EOR potential from a number of large sandstone oil fields on the Norwegian Continental Shelf (NCS). In that study, several EOR methods were evaluated, and it was found that the technical potential amounted to approximately 592 MSm^3 ($\sim$3.7 billion barrels), and that injection of LS water was one of the more promising methods. In 2020 the same researchers published a follow-up study where they added economic and environmental considerations into the study [25]. They also included Smart Water injection into carbonate fields in the new study. Accounting for both technical, economic, and environmental considerations, it was found that LS water injection in sandstone reservoirs was a very promising recovery method having large EOR potential on the NCS [24,25].

This chapter is limited to a chemical view on important aspects of LS and Smart Water flooding in sandstone reservoirs, and it only considers wettability alteration as a cause of the extra oil recovery observed. It is important to highlight that sandstone and carbonate reservoirs are chemically very different and should be treated separately. This has not always been done in the past in previously published literature and this is also a cause of misconceptions in the research community and industry. To read more about carbonates, see Chapter 1.

In this chapter, first, the initial wettability of sandstone reservoirs will be introduced, with a focus on the phases and factors that influence the wettability. Then, Smart Water EOR effects in outcrop T sandstone core material, which has shown very good reproducibility in parallel experiments, will be discussed. Lastly, opportunities for applying Smart Water injection in reservoir systems and effects of injection strategies will be discussed.

3.2 Initial wettability of sandstone reservoirs

LS water injection or Smart Water injection are water-based EOR methods aiming at increasing oil recovery by wettability alteration to more water-wet conditions. To observe any extra oil recovery by wettability alteration to more water wet it is required that the reservoir be not completely water wet. Otherwise, there is no EOR potential. The wettability of reservoirs, which dictates the location of fluids and their flow in porous media, has been established over millions of years. Sandstone reservoirs consist of the three main phases that are hydrocarbons, initial FW, and rock. The rock phase, contributing to the pore surface minerals, was deposited in an aqueous environment and is initially assumed water wet before the hydrocarbon phase invasion took place by migration from the matured source rock, accumulating in the reservoir trap. The chemical interactions between the phases, largely influenced by temperature and less by pressure and time, dictate the initial wettability established in the reservoir. Examples of phase interactions and factors influencing wettability are illustrated in Fig. 3.2.

Treiber et al. [26] performed a large study on the wettability of carbonate and sandstone reservoirs and concluded that most of the carbonates tested were in the oil-wet range and that a larger percentage of the sandstones tested were on the water-wet side. Chilingar and Yen [27] observed that a majority of carbonates worldwide were on the oil-wet side. Salathiel [28] suggested that the reservoir might not be all water wet and introduced the term "mixed wettability," which describes a situation where oil-wet pores form a continuous path through the porous medium. Spotted, dalmatian, or fractional wettability have also been suggested, describing a situation where the rock surfaces are partly oil wet and partly water wet [29].

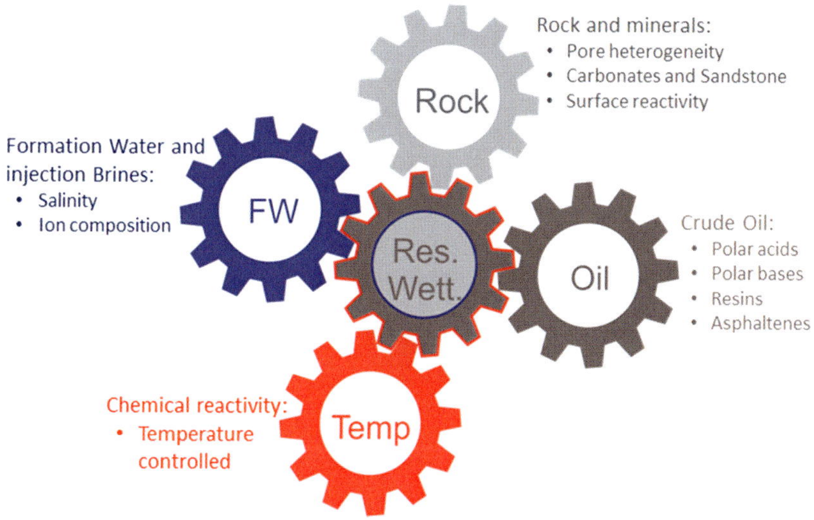

FIGURE 3.2

Main parameters influencing reservoir wettability.

One question to be asked is why the concern about the initial wettability of the core/reservoir? In brief, for any wettability alteration to more water-wet conditions to be successful, it is inherent that there is an initially mixed/intermediate/less water-wet state in the core/reservoir. This implies that the more water wet a core/reservoir is, the lower is the EOR potential by wettability alteration during Smart Water or LS water injection.

Morrow and colleagues published experimental work on sandstone cores showing that initially neutral to mixed-wet conditions lead to the highest incremental oil recovery [11,30,31]. Particularly, it was reported from a waterflooding study of 50 restored Berea sandstone cores with different initial wettability, that the highest oil recovery and the lowest S_{or} were observed at slightly water-wet conditions [30], as seen in Fig. 3.3.

Buckley et al. [32] and Aghaeifar et al. [33] have later questioned whether these experimental observations are limited to waterflooding of Berea outcrop material of rather high permeability. With carbonate, Johannesen and Graue [34] performed a systematic study of S_{or} after waterflooding in low-permeable chalk. They observed that at very low flow rates (with $\Delta P < 0.05$ bar/cm) and as the

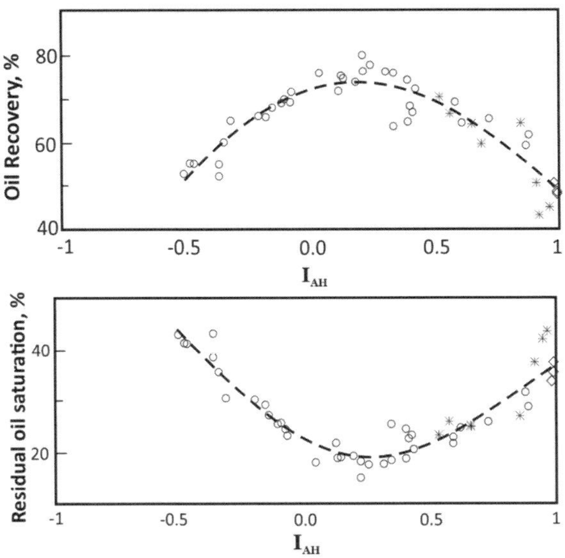

FIGURE 3.3

Ultimate oil recovery and residual oil saturation from Berea sandstone cores as a function of wettability, where I_{AH} refers to the difference between the Amott wettability indices to water, I_w, and oil, I_o. I_{AH} of 1 represents a very water-wet core, −1 a very oil-wet core, and 0 represents a neutral-wet core [30].

Reproduced with modifications from P.P. Jadhunandan, N.R. Morrow, Effect of wettability on waterflood recovery for crude-oil/brine/rock systems, SPE Reserv. Eng. (1995) 40–46.

water wetness of the carbonate cores increased, capillary forces increased and more oil was displaced by water imbibition, thus reducing the S_{or}. The trend observed in Fig. 3.3 cannot be a universal trend valid for all COBR systems. Bartels et al. [20] even referred to the results in Fig. 3.3 as a paradox, because it is frequently reported in the literature that additional recovery is observed when moving to more water-wet conditions in LS waterflooding, but according to the results in Fig. 3.3, the S_{or} increases for water-wet systems.

Thus even after decades of research, wettability remains a topic for discussion, therefore it is important to study and understand all phases present: crude oil, rock minerals, and brine composition and saturation. The different phases and related phenomena that influence wettability will now be looked at in more detail.

3.2.1 Mineralogy in porous sandstones

Sandstone pore surfaces are covered with different minerals as seen in the scanning electron microscope (SEM) picture in Fig. 3.4.

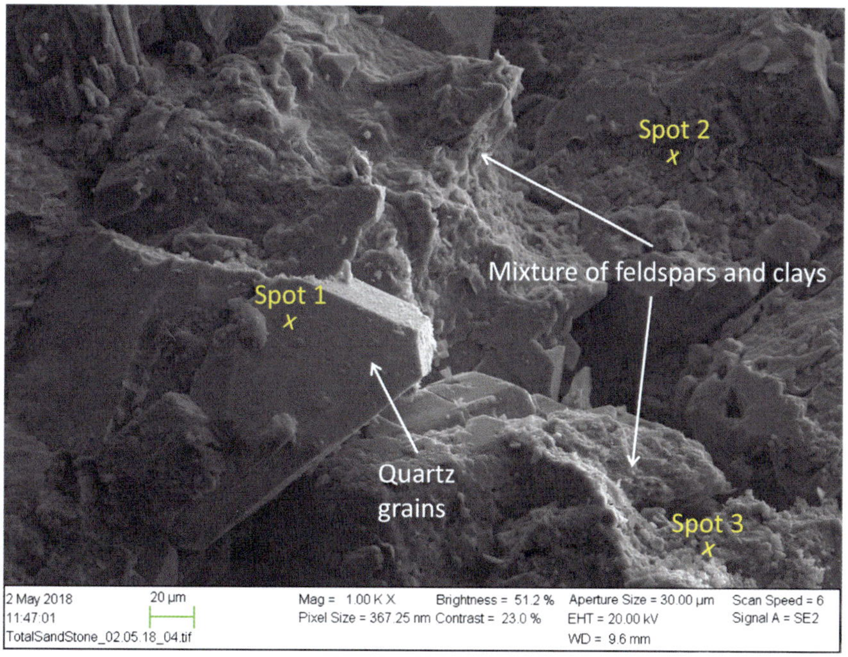

FIGURE 3.4

Scanning electron microscope picture (left) at 1000× magnification of outcrop T sandstone rock sample.

Both quartz, feldspars and clays, can be qualitatively detected based on atomic content from SEM analysis combined with energy-dispersive X-ray spectroscopy analysis [35], as listed in Table 3.1.

The framework mineral, quartz, is often the dominant mineral, but feldspar and clay minerals are also important and need to be accounted for if the pore wall wettability should be discussed and understood. Quartz often accounts for more than 50% of the bulk minerals, while feldspars typically are in the range of 5%–15% but have also been observed at concentrations above 30% in specific cases [36]. Clays are commonly found at concentrations in the range of 5%–15%, and at higher concentrations, the permeability in reservoir layers can be significantly reduced [37]. The Brent group is an important reservoir rock in the North Sea basin where the sandstones are dominated by arkoses and subarkoses. The area has been extensively drilled in the major fields of the East Shetland basin and the northern Viking Graben. Bjørlykke et al. [37] have studied the diagenesis of these sandstones at burial depts of <2 to >4 km. The results confirm the presence of significant amounts of feldspar minerals, both albite and K-feldspars, and significant amounts of kaolinite and illite clay minerals at all investigated depths.

X-ray diffraction (XRD) analyses have been performed on outcrop T sandstone samples presented in this chapter. The most abundant mineral present was quartz, as expected in sandstone cores. Approximately one-third of the minerals was plagioclase feldspar (albite), and about 10% of clay minerals were present. Note that most of the clay minerals present were illite and that no kaolinite was detected. Core mineralogical composition in weight percent (wt.%) is reported in Table 3.2.

The pore size (pore throat radius) distributions and permeabilities of sandstone rocks are controlled by the grain sizes and the amount of the individual matrix minerals. Mercury injection capillary pressure (MICP) tests of rock samples give a clear indication of the pore heterogeneity of porous sandstone rocks and provide pore size distributions. In Fig. 3.5, it can be observed that the tested outcrop T sandstone material consisted of a substantial amount of macropores (>2 μm), in

Table 3.1 Energy-dispersive X-ray spectroscopy analysis results of the outcrop T sandstone rock sample in Fig. 3.4.

Element	Atomic %			
	Spot 1	Spot 2	Spot 3	Avg.
Si	97	66.9	32.1	75.4
Al	1.8	18.2	13.2	13.8
Na	<1	11.7	10.4	6.5
Ca	<1	<1	41.1	<1
Fe	<1	<1	<1	<1
Mg	<1	<1	2.2	<1
K	<1	2.7	<1	2.0

Table 3.2 Mineralogical composition (in wt.%) of outcrop T sandstone obtained from X-ray diffraction analyses.

Mineral	Outcrop T sandstone
Quartz (wt.%)	56–57
Albite (wt.%)	30–33
Chlorite (wt.%)	1.7–2.2
Illite (wt.%)	8.1–9.4
Kaolinite (wt.%)	0.0
Calcite (wt.%)	0.3
Others (wt.%)	0.8–0.9
Total (wt.%)	100

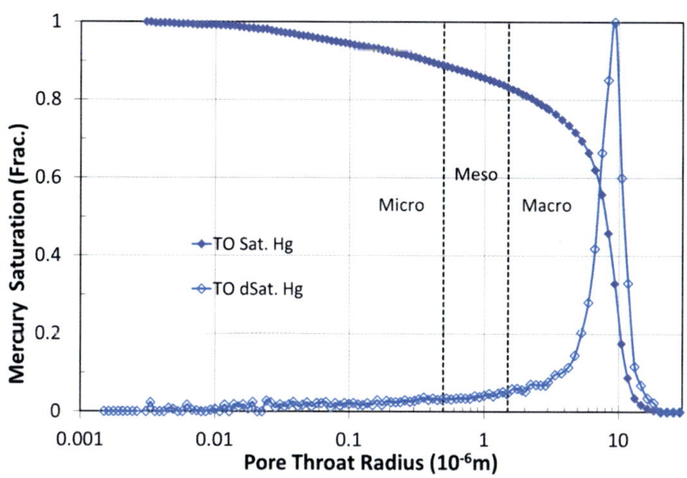

FIGURE 3.5

Pore throat radius distribution of porous outcrop T sandstone measured by mercury injection capillary pressure test.

addition to some meso (>0.5 to <2 μm) and micro (<0.5 μm) pores, the two latter accounting for 17% of the total PV.

The pore throat radii vary by three orders of magnitude, from 10×10^{-9} m (nm) to 20×10^{-6} m (μm) confirming heterogeneous pore size distribution, which needs to be accounted for and which could affect the displacement efficiency during waterflooding of reservoirs or core experiments in the laboratory. Large pores are more easily displaced than smaller pores, and stronger capillary forces or displacement forces are needed to displace oil from the smallest pores.

The petrophysical properties of the outcrop T sandstone cores are presented in Table 3.3.

Both porosity (~20%) and Brunauer–Emmett–Teller (BET)-specific surface area (~1.8 m^2/g) were reproducible between the tested samples, while the measured permeabilities of the individual cores varied significantly (20–140 mD). This confirms a heterogeneous pore distribution as observed from the MICP data (Fig. 3.5).

3.2.1.1 Quartz

Quartz (SiO_2) is the most abundant sandstone mineral, and sandstone depositions typically consist of 60%–70% (or more) quartz. It has a three-dimensional crystalline structure classified as tectosilicates or framework silicates. Quartz contains silicate tetrahedrons, each of which consists of a central silicon atom surrounded by four oxygen atoms at the corners of the tetrahedron with unit composition SiO_4^{2-}. The tetrahedrons are arranged in a three-dimensional lattice where each oxygen atom of a given tetrahedron is shared with another tetrahedron through covalent bonds, as seen in Fig. 3.6. Each tetrahedron is therefore linked to four

Table 3.3 Outcrop T sandstone core properties.

Core #	Pore volume (mL)	Porosity (φ), %	Permeability k^a (mD)	BET (m^2/g)
T15	15.6	19.6	–	1.81 ± 0.02
T21	23.3	20.8	48.4	
T22	23.0	20.5	135.1	
T24	22.5	19.9	22.3	
T25	23.6	21.0	50.2	
T26	22.5	19.9	34.6	
T32	15.9	20.1	3.4	

BET, Brunauer–Emmett–Teller.
[a]Permeability measured at 100% brine saturation.

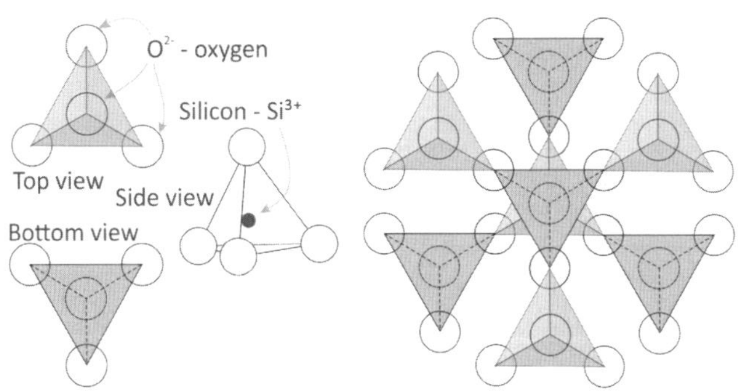

FIGURE 3.6

Crystalline structure of quartz. (Left) the building block is SiO_4 tetrahedra sharing all corner oxygens with other tetrahedra. (Right) the tectosilicate crystalline structure of quartz [39].

Reproduced with modifications from C. Klein, C.S. Hurlbut, J.D. Dana, Manual of Mineralogy (After James D. Dana) (sixth ed.), Wiley, New York, 1999.

others and can build an infinite tetrahedral network. The chemical formula of quartz is SiO_2. The SiO_2 structure is electrically neutral and requires no additional cation or anion species, which gives quartz high chemical stability and low ion exchange capacity [38].

Since the Si–O bonds are strong covalent bonds and since the structure is interlocking, the tectosilicate minerals tend to have high hardness and resist disintegration and decomposition in weathering processes. Quartz is a relatively stable mineral with a solubility in water of 5 ppm at surface temperature and pH 7–8. At higher pH, in basic/alkaline environment, the solubility increases [40].

3.2.1.2 Feldspar minerals

The feldspar group minerals are the most common silicates in the earth's crust and occur mainly in igneous rocks. The crystalline structure of feldspar is similar to the SiO_2 polymorphs, consisting of an infinite network of interconnected tetrahedra, bridged by oxygen atoms. In contrast to the SiO_2 group, the tetrahedra may be AlO_4^{5-} as well as SiO_4^{4-} Minerals are rendered electrically neutral as a result of being "stuffed" with alkali (Na^+, K^+) or alkali-earth (Ca^{2+}) element cations in available voids within the crystalline structure, giving the mineral increased chemical reactivity. The crystalline structure of potassium feldspar is described in Fig. 3.7. Like quartz, silicon and oxygen are the base constituents, but the substitution of Si^{2+} by Al^{3+} within some of the tetrahedrons requires additional electrons to satisfy the valency.

Important feldspars arc K-feldspar (microcline/orthoclase), with chemical formula $K[AlSi_3O_8]$, Na-feldspar (albite), with chemical formula $Na[AlSi_3O_8]$, and Ca-feldspar (anorthite), with thc chcmical formula $Ca[Al_2Si_2O_8]$. Naturally occurring feldspar minerals are chemically classified using a ternary phase diagram

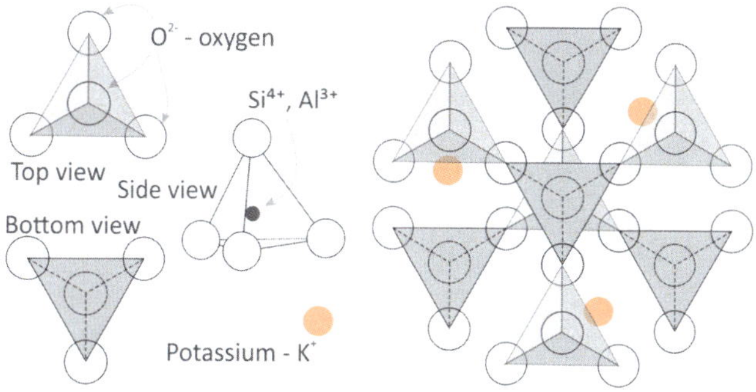

FIGURE 3.7

Crystalline structure of K-feldspar. (Left) the building block is $(Si,Al)O_4$ tetrahedra sharing all corner oxygens with other tetrahedra. (Right) a hypothetical tectosilicate crystalline structure of K-feldspar with exchangeable K^+ ions for electrical neutrality.

Reproduced with modifications from C. Klein, C.S. Hurlbut, J.D. Dana, Manual of Mineralogy (After James D. Dana) (sixth ed.), Wiley, New York, 1999.

(Fig. 3.8), where orthoclase (K-feldspar), anorthite (Ca-feldspar), and albite (Na-feldspar) represent the three apexes. Feldspars with compositions lying between Na-feldspar and K-feldspar are grouped as alkali feldspars, while feldspars composed of a mixture of Na-feldspar and Ca-feldspar are grouped as plagioclase feldspars. Alkali feldspars are not homogeneous but contain separate K-rich and Na-rich phases unless they have crystallized rapidly from solid solutions at high temperatures (> 660°C) [41].

The feldspars typically make up approximately 20% of the earth's sandstones and are more easily weathered than quartz [43]. High feldspar content in sandstone formations implies that the chemical weathering of the deposits has not been extensive. Their grains have a size distribution comparable to that of quartz. Large, rounded particles will not significantly contribute to mineral surfaces, giving a low specific surface area, in m^2/g rock, quantified by BET analysis. By zeta-potential measurements in NaCl brine, feldspars have been found to be negatively charged in the pH range 1–10 [44]. A point of zero charge (PZC) at pH 2–2.4 was identified by Stumm and Morgan [45]. In the pH range 4–9 normally found in reservoirs, feldspars possess a negative surface charge. The presence of nonframework cations Ca^{2+}, Na^+, and/or K^+ in void spaces in the crystalline lattice is a main characteristic of the feldspars, providing charge balance. Feldspars have low but fast cation-exchange capacity (CEC) [38,46] and can contribute to the initial FW pH in the reservoir [47]. The cations can be exchanged with H^+ from water (H_2O) by hydrolysis [41,48]. The exchange has been found to be

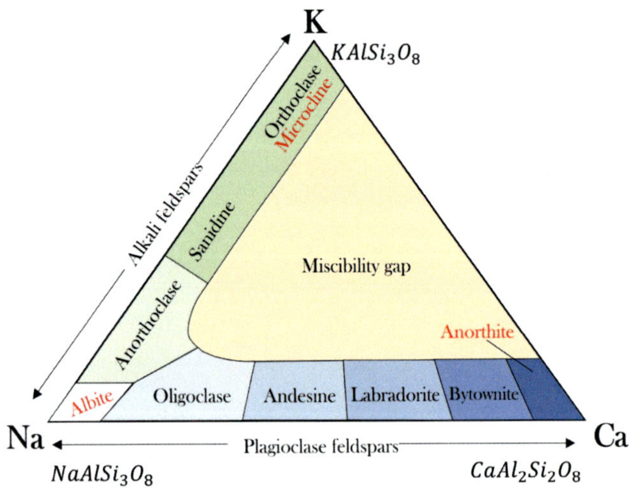

FIGURE 3.8

Ternary phase diagram of feldspars (at 900°C) [42].

Reproduced with modifications from L. Van Der Plas, The Identification of Detrital Feldspars. Developments in Sedimentology, vol. 6, first ed., Elsevier Science, 1966.

instantaneous and reversible and is accompanied by an increase in pH [48,49]. Eqs. (3.1)–(3.3) show the chemical cation-exchange reactions in the feldspar minerals leading to an increased pH:

$$NaAlSi_3O_8 + H_2O \rightleftarrows HAlSi_3O_8 + OH^- + Na^+ \quad (3.1)$$

$$KAlSi_3O_8 + H_2O \rightleftarrows HAlSi_3O_8 + OH^- + K^+ \quad (3.2)$$

$$CaAl_2Si_2O_8 + H_2O \rightleftarrows (HAl_2Si_2O_8)^- + OH^- + Ca^{2+} \quad (3.3)$$

3.2.1.3 Clay minerals

Clay minerals occur as metamorphic and eruptive minerals, as clastic minerals in clay sediments and they are phyllosilicates containing sheet-like lattice structures consisting of oxygen, silicon, aluminum, magnesium, iron, and water. Clay minerals can form in weathering reactions between minerals and pore water during the breakdown of feldspars and micas, and clay minerals that are formed in sediments after deposition are called authigenic clay minerals [40].

The structure of sheet silicates contains sheets of SiO_4^{2-} tetrahedra and octahedra alternating with cation layers of Al^{3+}, Fe^{2+}, and Mg^{2+} [50,51]. In the tetrahedra layers, the cation is surrounded by six oxygen or hydroxyl ions. The octahedral layer can be filled with both di- and trivalent ions. If the sheet silicates contain trivalent ions, only ⅔ of the positions are occupied, but if they are divalent ions, all positions are filled (Fig. 3.9).

Common reservoir clay minerals are illite, montmorillonite, and kaolinite, and their building block structures are seen in Fig. 3.10. Depending on the structure, the different clay minerals exhibit different properties [40]. Illite is the most abundant clay mineral with its stable 2:1 structure, consisting of two tetrahedra layers and one octahedra layer bonded together by potassium ions preventing the sheets from

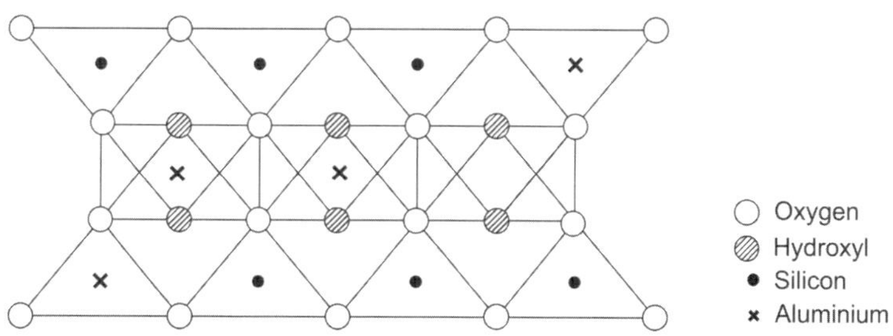

FIGURE 3.9

Clay structure of two tetrahedra layers and one octahedra layer filled with trivalent Al^{3+} ions.

Reproduced with modifications from C. Klein, C.S. Hurlbut, J.D. Dana, Manual of Mineralogy (After James D. Dana) (sixth ed.), Wiley, New York, 1999.

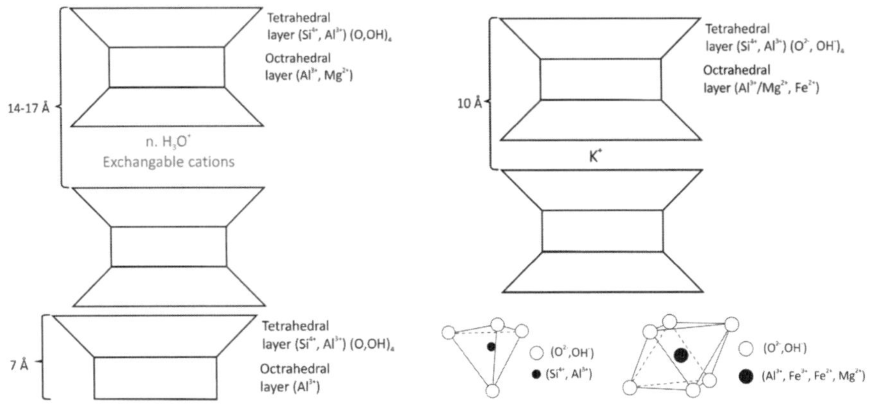

FIGURE 3.10

Clay mineral structures: montmorillonite (upper left), illite (upper right), and kaolinite (lower left). The building blocks (lower right) in the tetrahedral and octahedral layers in clays: $(SiO_4)^{4-}$ tetrahedron and $(Al(OH)_6)^{3-}$ octahedron.

Reproduced with modifications from K. Bjørlykke, Sedimentology and Petroleum Geology, Springer, Berlin, 1989.

expanding easily [40]. Montmorillonite has a similar 2:1 structure, but most of the potassium is replaced by hydrogen and water, other cations, or organic compounds. Montmorillonite has a net negative charge that is balanced by Na^+, K^+, and Ca^{2+} cations. Montmorillonite is known for its swelling property in contact with water and it also has a very high ion exchange capacity. Montmorillonite is converted to illite at an increasing temperature [40].

The kaolinite structure consists of one tetrahedral layer and one octahedral layer. CEC is lower for kaolinite than montmorillonite due to the 1:1 structure. This structure contains no room for water or large cations between the layers, which makes them chemically and mineralogically simple compared to other clays. Kaolinite sheets are neutral and held together by hydrogen bonding. Kaolinite attracts external cations only by negative charges on the edges of the sheets. Because of this, grain size affects the CEC [52]. Kaolinite is stable at low temperatures but becomes unstable and will be altered to illite at 120°C−130°C if potassium ions are available, but if potassium is not available kaolinite will remain stable even at high temperatures [40,51]. Some basic clay mineral properties are contained in Table 3.4 [53].

Clay minerals have some properties that distinguish them from other minerals. Because of the sheet-like structure, clays have building blocks of low thickness (7−17 Å) and large surface areas (Fig. 3.10). Due to isomorphous substitutions of Si^{4+} by Al^{3+} in the tetrahedral layer, and/or Al^{3+} by Fe^{2+}/Mg^{2+} in the octahedral layer, the surfaces require an additional electron to satisfy the valency. Ca^{2+}, Na^+, K^+, or H^+ from water (H_2O) by hydrolysis are nonframework cations that

Table 3.4 Clay mineral properties.

Property	Kaolinite	Illite	Chlorite	Smectite
Sheet structure	1:1	2:1	2:1:1	2:1
Particle size (μm)	0.5–5	Large sheets up to 0.5	0.1–5	0.1–2
Cation-exchange capacity (meq/100 g)	3–15	10–40	10–40	80–150
Specific surface area BET (m^2/g)	15–25	50–110	140	30–80

BET, Brunauer–Emmett–Teller.

can provide charge balance. This gives the clays their unique CEC property [52]. Clay minerals in general have high CEC, because of crystal lattice charge deficiencies giving the clay surfaces a permanent negative charge, which needs to be charge balanced by cations, and because of broken bonds at the edges of the clay platelets. In the reservoir pH range, 4–9, the surfaces of most clay minerals are permanently negatively charged, and cation-exchange reactions in clay minerals can be simply described by the following equation:

$$Na^+ \cdots Clay + H^+ \rightleftarrows H^+ \cdots Clay + Na^+ \tag{3.4}$$

It has been found that the clays in general have the following relative affinity to cations [54]:

$$Li^+ < Na^+ < K^+ < Rb^+ < Cs^+ < Mg^{2+} < Ca^{2+} < Sr^{2+} < Ba^{2+} \ll H^+ \tag{3.5}$$

The affinity to a proton, H^+, is by far the largest, followed by the divalent cations and the monovalent cations. However, a high concentration of a monovalent ion (e.g., Na^+) can compete with divalent ions (e.g., Ca^{2+}) for charge balancing the clay surface. The large surface area in combination with high CEC leads to the clay surfaces also acting as adsorbents for polar organic components (POC), surface-active material in the crude oil which is regarded as the main wetting material in sandstone reservoirs.

3.2.1.4 Evaporites

In addition to the siliciclastic minerals, other minerals can be generated in situ by chemical processes during deposition, as evaporites, or be formed during diagenesis with increasing temperature and increasing salinity of FW. Evaporites are sedimentary rocks formed from brine crystallization and contain soluble ions. Calcite, aragonite, and magnesite are common carbonate evaporites. Gypsum and anhydrite are common sulfates while halides exist in the form of, for example, halite, sylvite, or carnallite. In sandstone reservoirs, anhydrite ($CaSO_4$), gypsum ($CaSO_4 \cdot 2H_2O$), and calcite ($CaCO_3$) can be present as cementing material in the pore space and can affect water chemistry, chemical reactions, and wettability alteration processes taking place during Smart Water or LS water injection [55,56]. Note that sulfate minerals at low concentrations might not be detected by standard XRD analysis [57]. Therefore laboratory core experiments involving waterflooding and subsequent effluent analysis should be used to properly detect any presence of these minerals, as even small concentrations significantly could affect brine chemistry and wettability alteration processes [58].

3.2.2 Formation water

The initial water saturation (S_{wi}) of sandstone reservoirs varies from one reservoir to another, and also within a reservoir oil zone, with higher S_{wi} in the transition zone close to the oil–water contact. But values in the range of 10%–25% are commonly seen.

The ionic compositions, salinities, and ionic strengths of four FW from different North Sea sandstone oil reservoirs are given in Table 3.5. The FW acts as an electrolyte and influences the equilibrated chemistry established between mineral surfaces, FW, and crude oil in sandstone reservoirs. FW salinity varies significantly from less than 20,000 ppm (ppm is equivalent to mg/l) to more than 250,000 ppm. As seen in Table 3.5, all ion concentrations (in millimolar—mM) increase with increasing total dissolved solids (TDS). Na^+ and Cl^- constitute the majority of the ions, and regarding the divalent cations, Ca^{2+} and Mg^{2+}, the concentration of Ca^{2+} is 4–8 times higher than that of Mg^{2+} in the given FW compositions.

The divalent ions are important for the equilibrium pH in FW, as they form complexes $[CaOH]^+_{(aq)}$ and $[MgOH]^+_{(aq)}$, with hydroxyls (OH^-), in the water phase, thus affecting pH by reducing the activity of hydroxyls in solution, especially at high temperatures.

Table 3.5 Ionic compositions of North Sea sandstone formation water.

	FW 1 (mM)	FW 2 (mM)	FW 3 (mM)	FW 4 (mM)
Na^+	355.2	939.5	2563.2	2842.2
K^+	1.6	17.8	58.8	144.3
Mg^{2+}	3.4	7.0	18.3	95.6
Ca^{2+}	12.2	44.2	123.8	753.1
Ba^{2+}[a]	0.0	5.2	0.6	3.5
Sr^{2+}[a]	0.0	3.0	0.9	12.4
HCO_3^-	0.0	7.7	3.4	1.3
Cl^-	387.9	1068.5	2905.7	4714.5
SO_4^{2-}	0.0	0.0	0.0	0.0
TDS, mg/l	22,551	63,567	170,010	272,282
Ionic Strength (M)	0.403	1.136	3.053	5.580

FW, Formation water; TDS, total dissolved solids.
[a]Absolute concentrations of trace ions are uncertain.

Sour gases, CO_2 and H_2S, present in the hydrocarbon phase will equilibrate in the brine and contribute to a slightly acidic FW pH in sandstone reservoirs, $pH < 7$ (measured at standard conditions).

FW chemical composition, salinity (=TDS), and pH are key factors influencing the reservoir wettability [59]. The individual ions present in the FW affect mineral surface charges and the charges of oil–water interfaces [60–63]. The FW ionic composition has also shown an effect on oil recovery from sandstone cores, where cores saturated with FW with high divalent cation (Ca^{2+}, Mg^{2+}) content showed higher oil recovery than cores saturated with FW containing only monovalent (Na^+) cations [64]. The effect of brine chemistry and rock composition on the initial wetting has been validated experimentally and slightly acidic conditions in the reservoir should promote adsorption of POC from the hydrocarbon phase, which induces mixed wettability in the reservoir [11,65–68]. Salathiel [28] suggested that the wetting state was also dependent on S_{wi}, and this was later supported by other researchers claiming that water wetness increased as the S_{wi} increased [30,69,70]. If this is the case, then reproducible FW S_{wi} should be obtained in core preparation prior to wettability studies.

3.2.3 Brine–rock interactions

One very important feature of solid surfaces is their interaction with water. Solid surfaces can be dissolved in water and leach its constituent ions to the water phase. Another property, which is less destructive, but an important interaction between the water phase and solid surface, is ion exchange. The ion exchanges are determined by the structure and chemical composition of the mineral and the chemical elements in the solution in contact with the mineral. Interactions between the minerals and FW may lead to changes in water phase composition, pore surface mineralogy, and pH. The rock–brine interactions control the conditions dictating the initial reservoir wettability established when crude oil is introduced, and the wettability alteration process during Smart Water or LS water flooding of sandstone reservoirs.

3.2.3.1 Surface area and ion exchange

The amount of water present in the pores of an example sandstone formation similar to the outcrop T sandstone with a porosity of 20% and a specific surface area (BET) of 1.8 m^2/g rock is illustrated in Fig. 3.11.

There is a huge surface area available for fluid–rock interactions, but the PV in 1 g of outcrop T sandstone is close to only 0.1 mL. At 100% water saturation, only two droplets of water (of 0.05 mL each) fit in the pore space and interact with 1.8 m^2 mineral surfaces. With oil present in the pore system at $S_{wi} = 0.2$ and initially strongly water-wet conditions, only 0.4 droplet of FW is available for mineral surface interactions in 1 g of this example rock.

Initially, the dissolution of carbonate minerals, for example, calcite $CaCO_3$ or dolomite $CaMg(CO_3)_2$ establishes a carbonate buffer system resulting in a slightly

FIGURE 3.11

Illustration of the amount of water present in 1 g of sandstone T rock. The droplet volume of water is 0.05 mL, and at $S_{wi} = 1$, only two droplets are interacting with 1.8 m^2 of mineral pore surfaces. At $S_{wi} = 0.2$, the formation water volume only equals 0.4 water droplet.

alkaline carbonate reservoir pH. The pH of sandstone reservoirs is normally more acidic due to the presence of silicate minerals. With high concentrations of divalent cations in FW, carbonate surface charge is positive [45], while it is negative for sandstones because of the presence of clays and other silicates with PZC in the range of pH 2.2–2.8 [44,45,71].

Ion exchange reactions take place when a solution containing cations and anions contacts a mineral surface. The reactions are determined by the structure and chemical composition of the mineral and the chemical elements in the solution in contact with the mineral. An ion held by a charge near a mineral surface can exchange with another ion present in the solution in contact with the mineral. This phenomenon is called ion exchange and it is a reversible process governed by physicochemical laws. Ion exchange occurs in substances, such as organic matter, minerals, finely crushed rocks, and amorphous material. Both cations and anions can exchange from surfaces, and their exchange is simply referred to as cation exchange and anion exchange, respectively. Surfaces, depending on their properties, may have CEC and/or anion-exchange capacity (AEC). Ion exchange is affected by several factors, such as the type of mineral and associated ions, the nature and concentrations of the replacing ions, the pH of the solution regulating the amounts of protons, H^+, and hydroxyls, OH^-, in solution. According to Carroll [52], there are especially two structural causes for the cation ion exchange: (1) Unsatisfied valences produced by broken bonds at surfaces and edges of particles. Broken bonds are the most important cause of cation exchange in clays and in fine particles of minerals. (2) Unbalanced charges caused by the isomorphous substitution of cations, for example, Al^{3+} substituted for Si^{4+}.

Clay minerals in general have high CEC, because of crystal lattice charge deficiencies giving the clay surfaces a permanent negative charge, which is charge balanced by cations, and because of broken bonds at the edges of the clay platelets. Minerals show differences in exchange capacity due to the availability of exchange sites on basal planes and due to irregularities in the lattice structure and on the edges due to broken bonds. Particle size affects the exchange capacity due to larger surface area, and more broken bonds at the edges.

- CEC

CEC is a measure of how many cations can be retained on soil and rock particle surfaces. Negatively charged surfaces bind positively charged atoms or molecules (cations), allowing these to exchange with other positively charged particles in the surrounding water. The CEC depends on the quantity of clay and silt fractions [52]. The clay fraction has a higher CEC than the silt fraction. And it depends on the type of clay mineral—illite and montmorillonite have greater exchange capacity than kaolinite [38]. CEC affects many aspects of the behavior of mineral surfaces, soils, and porous media. CEC is used as a measure of soil fertility, as it indicates the capacity of the soil to retain several nutrients (e.g., K^+, NH_4^+, Ca^{2+}) in plant-available form. It also indicates the capacity to retain pollutant cations (e.g., Pb^{2+}). CEC is used to explain the reactivity of clay minerals present in porous rocks which are involved in the adsorption of POC and reservoir wettability in sandstone reservoirs. The CEC can be measured by displacing all the bound cations with a concentrated solution of another cation and then measuring either the displaced cations or the amount of added cation that is retained. The measurement is normally done at constant pH. The CEC is expressed in cmol/kg (centimol positive charge per kg of soil) or meq/100 g (milli-equivalents per 100 grams of soil). Both expressions are numerically identical (10 cmol/kg = 10 meq/100 g).

The order of cation exchange toward clays is described in Eq. (3.5), and it shows that the divalent cations have a high affinity for the clay mineral surface and will exchange cations most readily, replacing other ions associated with the clay surface. The divalent cation affinity is higher than the monovalent affinity, and monovalent ions are thus less prone to exchange with clay-associated ions. However, if the concentration of monovalent ions increases in solution, their cation replacement affinity increases and becomes more important. Therefore in high saline solutions containing mostly monovalent Na^+, its replacement order might become higher if the divalent cation concentration is low. The common-ion effect and Le Chatelier's principle, which states that a system will respond to counteract any disturbances imposed on the system equilibrium, are here central terms, which could become increasingly important if ions are present in the water phase. Note, however, that it is the proton, H^+, which has the absolute highest affinity for the clay surface, and which is most prone to exchange with a cation. The amount of free H^+ in solution is dependent on the pH of the water phase; consequently, the system pH will significantly influence, or dominate, the cation exchange taking place between the solid surface and ions in solution.

Allard et al. [38] measured and reported surface area (both BET and measured by ethylene glycol), CEC, and AEC for a variety of mineral surfaces, among them quartz, feldspars, and clay minerals, as shown in Table 3.6. For most minerals, CEC was reported higher than AEC.

It is important to keep in mind that cation exchange between rock minerals and the water (brine) phase impacts the pH of the reservoir brine before the crude oil migration into the reservoir. Minerals and FW, in addition to the oil phase itself, are therefore related to the generation of the initial wettability of the reservoir.

- AEC

Adsorption of anions and AEC are induced by positive charges on mineral surfaces originated either from the rupture of structural unit planes and resulting edge charges or from iron and aluminum oxides that cover crystalline clays or occupy interlayer positions in lattice layers [72]. Methods for measuring AEC are based on the same principles as the methods for measuring CEC, as described above, and are subject to the same constraints. Nonetheless, the valence signs have to be reversed and AEC is expressed in cmol/kg soil or meq/100 g. Before 1975 AEC was rarely studied, mainly because the interactions of anions with the surface are weaker than those of cations, which are linked to negatively charged surfaces. This is particularly true for 2:1 clays, for which PZC is difficult to define. High CEC to AEC ratios are obtained as the permanent negative charge is too high to be balanced by the less significant positive charges, which appear at the edge of the rupture zones of the structural units [73]. As the CEC is relatively low in 1:1 clays, the effects of the edge charges give low CEC to AEC ratios, and the influence of AEC is consequently more significant, especially if the particles are small and display a significant level of disorder [73]. The surfaces of kaolinite and iron and aluminum oxides often carry positive charges, which has been reported for highly weathered soils [74].

Table 3.6 Surface area, cation-exchange capacity, and anion-exchange capacity of relevant minerals measured at pH 5.

Mineral	Surface area (m^2/g)		CEC at pH 5 (meq/kg)	AEC at pH 5 (meq/kg)
	BET	Ethylene glycol		
Quartz	0.3	2.8	0.1	0.1
Kaolinite	11	27	9.0	0.8
Montmorillonite	–	800	700	
Albite	–	2.9	1.8	0.2
Microcline	1.1	4.2	1.8	0.2
Anorthite	–	–	2.2	3

AEC, Anion-exchange capacity; BET, Brunauer–Emmett–Teller; CEC, cation-exchange capacity.

3.2.3.2 Surface reactivity of sandstone cores

Interactions between pore surface minerals and injection brines, such as FW, LS, or other potential injection brines, can be studied in a waterflooding process in absence of crude oil. An outcrop T sandstone core T21 was used in a surface reactivity test by flooding the FW-saturated core successively with FW–LS–FW at 60°C, while measuring effluent pH. Effluent pH evolution results are presented in Fig. 3.12.

During FW injection, the effluent pH was slightly acidic and stabilized close to pH 6. pH increased rapidly to approximately 9 when the injection brine was switched from FW to LS brine, which is more than three pH units higher than the LS bulk pH. Ion exchange reactions between protons, H^+, from the brine and easily accessible cations at the mineral surfaces, in this case, illite clay or plagioclase feldspar present in outcrop T sandstone cores, cause the rapid pH changes. The following period of slower pH increase is due to cation exchange of the less accessible cations in, for example, illite interlayers, or from minerals present in smaller, low-permeable pores [54]. The surface reactivity test confirms that the outcrop T sandstone possesses both low initial pH in the presence of FW, favorable for generating mixed wettability, and the ability to induce a significant increase in pH during LS flooding, which is favorable for observing wettability alteration. It is however important

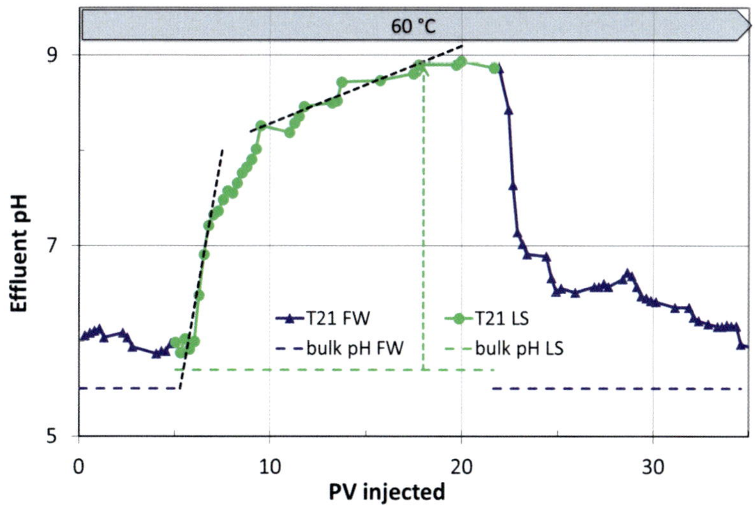

FIGURE 3.12

Surface reactivity test on outcrop T sandstone core T21 at 60°C. The core contained 100% formation water before it was flooded with formation water–low salinity–formation water at a rate of 4 pore volume/day. Formation water bulk pH was 5.5 and low-salinity brine bulk pH was 5.7.

to note that the test was performed without crude oil present. The presented core flood results on outcrop T sandstone indicate that for transporting alkalinity through a sandstone core, the injected brine has to play along with the rock minerals [75].

3.2.4 Crude oil and adsorption of its components onto sandstone surfaces

Sandstone rocks are initially assumed water wet before the crude oil migrates into and accumulates in the reservoir trap, and the surface charge of silicate minerals is negative in the typical reservoir pH range from 4 to 9 [44,45]. The crude oil is the main wetting phase in sandstone reservoirs and contains the components responsible for wetting the sandstone surfaces. Reservoir crude oils are complex mixtures composed of organic components from methane to semisolid resins and asphaltenes. Organic nitrogen–sulfur–oxygen (NSO) components contribute to acidic and basic POC which are present both in the liquid phase and in the resin and asphaltene fractions of the crude oil. The content of crude oil acids and bases can be determined by measuring the acid number (AN) and base number (BN), both given in mgKOH needed to neutralize the acids or bases (equivalent to the mgKOH neutralizing the acids needed to neutralize the bases) in 1 g of crude oil. The charged POC interact with charged mineral surfaces and their reactivity is dependent on the chemical conditions in the reservoir. The crude oil also contains some inorganic sour gases, such as CO_2 and H_2S, which affect reservoir chemistry.

There have been several interpretations of wettability in sandstone reservoirs. The asphaltene and resin fractions contain the majority of the NSO components in crude oils and these fractions have been considered important wetting parameters [76]. But light crude oil with low content of asphaltenes are also influencing surface wettability, and significant EOR potential has been observed by wettability alteration toward more water-wet conditions after the cores were restored with light crude oil containing POC [77].

Mineral surfaces have also been highlighted as important for surface wettability. In sandstone systems, various silicate minerals with different chemical and physical properties are present, which will influence crude oil interactions [47,62,78,79]. With the small amount of electrolytic pore fluid compared to available mineral pore surfaces, significant changes in the mineral surfaces by dissolution or precipitation within the production time of a reservoir are not likely to happen affecting wettability.

With clay and feldspar minerals contributing to reactive mineral surfaces and CEC, surface charges and brine pH can affect the reservoir chemistry during reservoir operations. In the next section, the reactivity of POC present in crude oils toward surface minerals will be discussed with respect to wettability and wettability alteration processes in sandstone systems.

3.2.4.1 Polar organic components

It has been observed that pH has an effect on initial wettability and the wettability alteration process in sandstones by Smart Water/LS waterflooding [7,47,80–83]. The fractions of protonated and deprotonated POC species, and thus, their reactivity toward charged silicate mineral surfaces, are pH dependent as seen in Fig. 3.13. Note that when the fraction of protonated species increases, the corresponding fraction of deprotonated species decreases, and the two fractions summarize unity.

In the acidic region, pH < 7, the fraction of protonated basic material, R_3NH^+, increases with reduced pH, and the positively charged molecules can be adsorbed onto the negatively charged surfaces by electrostatic interaction. The fraction of protonated acidic material, R-COOH, also increases with reduced pH, and these molecules can interact with negatively charged surfaces by hydrogen bonding [65,66]. Acidic and basic POC have a similar pH-dependent affinity toward negatively charged mineral surfaces (Fig. 3.14).

Clay minerals influence reservoir wettability due to their large surface area and their permanent localized negative charges, which give them the unique CEC. Since the clays contribute to most of the available adsorption sites for crude oil components, they are regarded as the main wetting mineral in sandstone reservoirs.

The negative charges on the clay surfaces are charge balanced by organic components from the oil phase and inorganic cations from the brine phase. Ca^{2+} has a high affinity for the clay surface, its concentration is relatively high in FW, and it will compete with POC for adsorption sites on the clay surface establishing initial wettability, as illustrated in Fig. 3.15. Increased temperature will reduce the hydration of divalent ions, such as Ca^{2+} or Mg^{2+}, and increase their reactivity [84], giving increased adsorption of Ca^{2+} and reduced adsorption of positive charged POC on the clay surfaces [85]. Reduced water wetness in reservoirs can be linked to the affinity of POC toward charged mineral surfaces controlled by reservoir temperature, pH, and composition/salinity of FW, facilitating a mixed wetting [7,58,67,68,78,85]. When crude oil initially invaded the water-wet and water-filled reservoir rock, to obtain mixed-wet conditions some of the active cations already present on the mineral surfaces have been exchanged by surface-active POC.

According to the model illustrated in Fig. 3.15, the following can increase the water wetness of a sandstone reservoir:

- Increased Ca^{2+} concentration in the FW,
- Reduced amount of POC in the crude oil,
- Increased pH of the equilibrated FW,
- Increased reservoir temperature (T_{res}).

Note that while positively charged carbonate surfaces at slightly alkaline pH have a high affinity for acidic POC, negatively charged sandstone surfaces at acidic pH favor adsorption of positively charged basic POC [77,86]. This

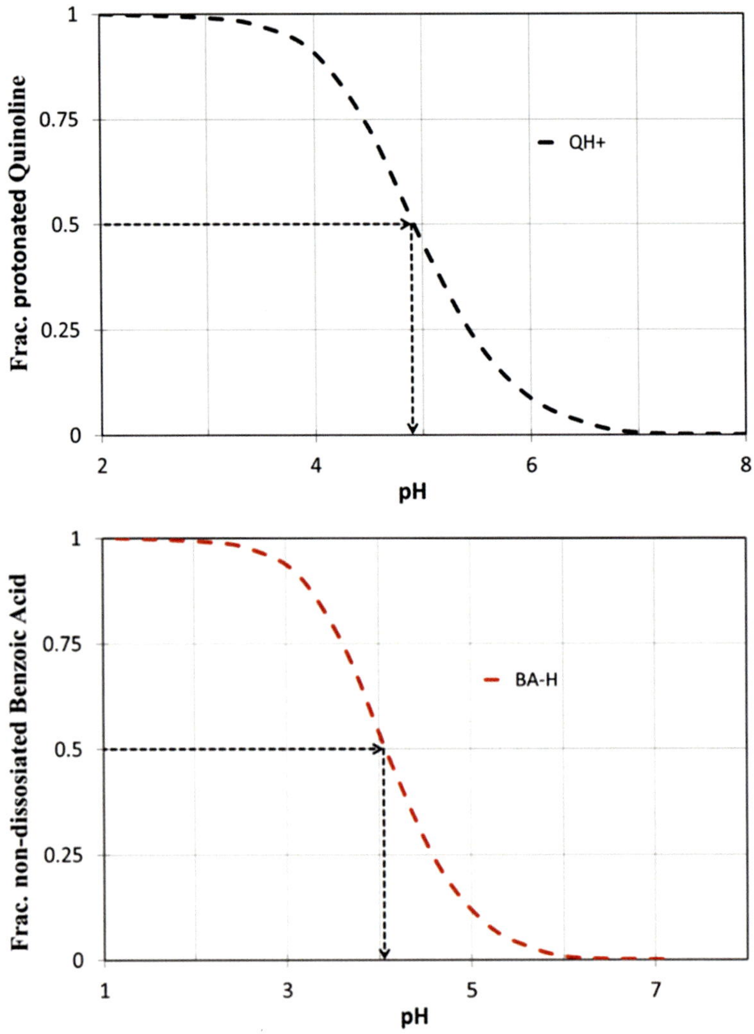

FIGURE 3.13

Top: fraction of protonated quinoline against pH. Bottom: fraction of non-dissociated benzoic acid against pH.

Top: Reproduced with modifications from W.D. Burgos, N. Pisutpaisal, M.C. Mazzarese, et al., Adsorption of quinoline to kaolinite and montmorillonite, Environmental Engineering Science. 19 (2002) 59–68. Bottom: Drawn after calculation with Marvin Sketch.

observation questions the multicomponent ion exchange (MIE) mechanism proposed by BP [87] in which divalent cations are bridging ions for the adsorption of negatively charged acidic POC onto negatively charged mineral surfaces.

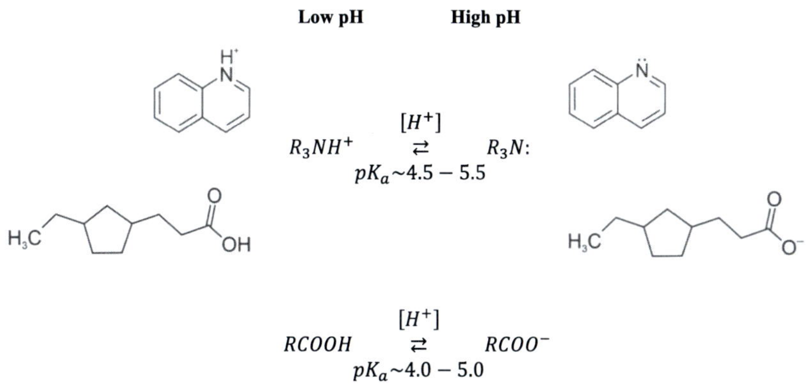

FIGURE 3.14

Basic (upper) and acidic (lower) POC in crude oils and their pH-dependent charges.

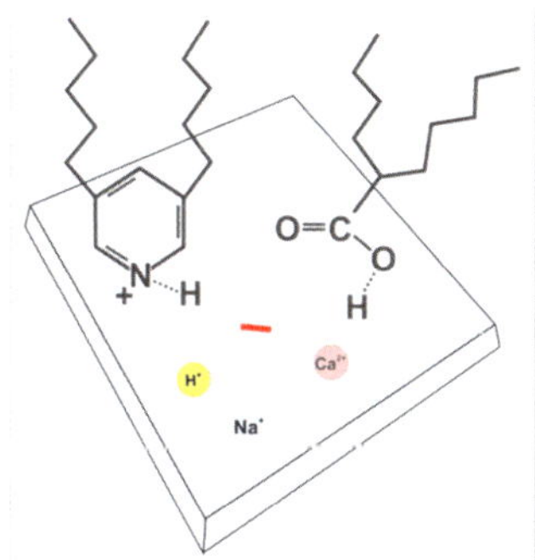

FIGURE 3.15

Competitive adsorption of ions and POC onto sandstone surfaces.

3.2.4.2 pH-dependent polar organic component adsorption on clay mineral surfaces

Since crude oil consists of numerous large and complex compounds complicating parametric laboratory experiments, tests can instead be performed on smaller model components. Burgos et al. [66] studied adsorption of quinoline, a small base molecule present in crude oil, onto kaolinite clay at conditions of varying pH and Ca^{2+} concentration at ambient conditions. It was observed that at constant pH, adsorption of quinoline was the highest at low Ca^{2+} concentration, and the adsorption of quinoline was reported to decrease as pH increased at both high and low Ca^{2+} concentrations (Fig. 3.16).

Several published studies have confirmed similar pH dependence of adsorption of POC on clays and silica surfaces [7,67,76,80,88]. From adhesion studies of 22 crude oils, Buckley and Morrow noticed that adhesion of crude oil onto silica surfaces was

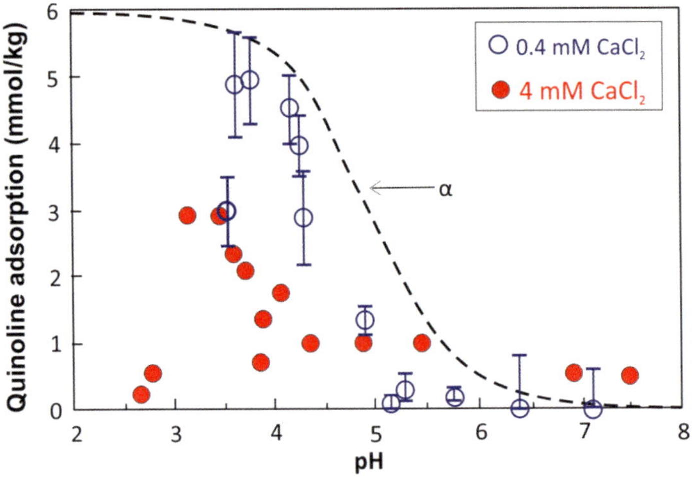

FIGURE 3.16

Quinoline adsorption onto kaolinite clay as a function of pH and Ca^{2+} concentration at ambient conditions.

Reproduced with modifications from W.D. Burgos, N. Pisutpaisal, M.C. Mazzarese, et al., Adsorption of quinoline to kaolinite and montmorillonite, Environmental Engineering Science. 19 (2002) 59–68.

dependent on pH, and that above pH 6–7 crude oil did not adhere to the silica surfaces at any brine salinity tested. From that work, they concluded that pH was a key driver in the wetting process [76]. No adhesion of de-asphalted crude oil was noticed at low pH, which led to the conclusion that the pH effect on adhesion was linked to surface-active components in the crude oil. Didier et al. [80] observed that crude oil adhesion onto two types of sand was more pronounced in presence of LS brine. Adsorption of quinoline onto kaolinite clay mineral at varying pH and brine salinity was investigated by Rezaeidoust et al. [67]. It was found that adsorption was reversible in the pH range of 5–9 in both LS and high-salinity (HS) brines (Fig. 3.17).

They also observed that quinoline adsorption was lowered at pH 2.5 even though quinoline is fully protonated at pH 3 and below. At low pH, the increased activity of the H^+ ion toward clay surfaces overcomes the increased concentration of protonated quinoline, and thus quinoline adsorption was lowered.

Experimental data on adsorption of quinoline onto illite clays confirm the same pH and salinity trends [58]. The highest adsorption of quinoline onto illite clay minerals was observed close to the pK_a value of quinoline in a LS brine. Reduced maximum adsorption was observed when the brine salinity was increased in a HS brine (Fig. 3.18).

To significantly reduce adsorption in both HS and LS brines, the pH had to be increased, such that the amount of protonated quinoline dropped to zero. Alternatively, pH could be lowered, thereby increasing H^+ concentration, which also decreased quinoline adsorption.

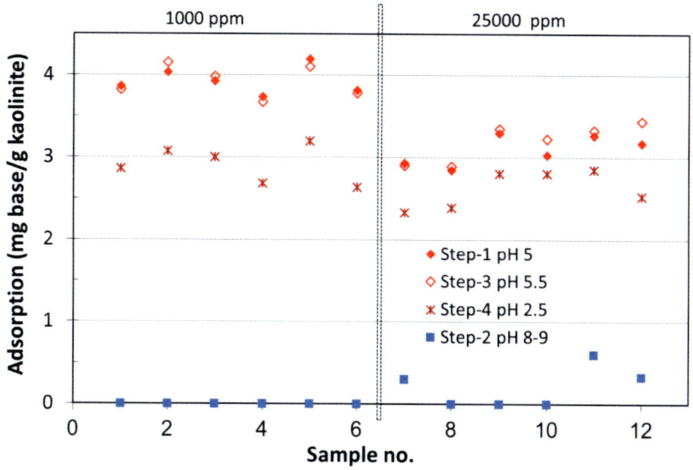

FIGURE 3.17

Adsorption reversibility of quinoline onto kaolinite surfaces in low-salinity brine (1000 ppm) and high-salinity brine (25,000 ppm) at pH 5, pH 9, and then back to pH 5.5 and pH 2.5.

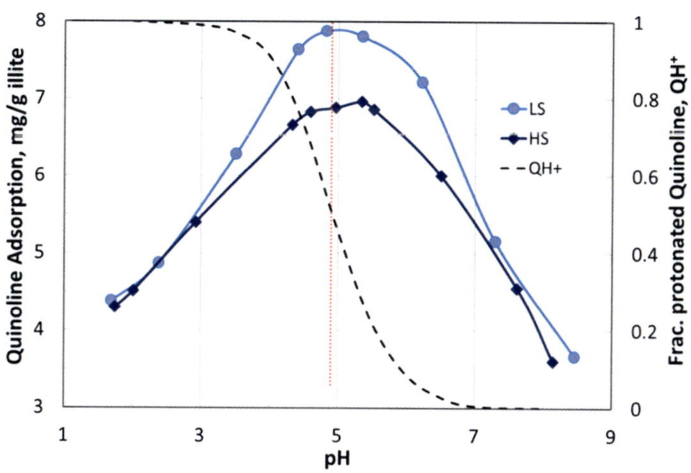

FIGURE 3.18

Quinoline adsorption toward illite clay as a function of pH.

Adsorption of the small carboxylic acid molecule, benzoic acid, onto kaolinite clay mineral was studied by Madsen and Lind [65] who found that adsorption decreased when brine pH increased from 5 to 9. Thus for both basic and acidic POC, adsorption decreased when pH increased, and adsorption was highest at slightly acidic pH.

These above-presented results question using models, which do not include pH considerations, such as the electrical double layer (EDL) expansion model, Guoy-Chapman, and Derjaguin–Landau–Verwey–Overbeek theories, to explain LS EOR effects.

3.2.4.3 Adsorption of polar crude oil components in sandstone cores

Adsorption of model compounds onto pure clay minerals presented in the previous section may indicate the adsorption behavior of crude oil components in consolidated sandstone cores. Both crude oils and sandstone rocks are multicomponent systems consisting of a variety of POC and rock minerals, respectively. The POC must be transported through the brine phase for adsorption onto the pore surface minerals. How polar organic acids and bases in crude oil adsorb and affect the wettability of an initially water-wet outcrop sandstone was tested by crude oil flooding tests. Outcrop T sandstone core T15 containing typical sandstone minerals, properties given in Table 3.2, was used and $S_{wi} = 0.21$ was established with a 50,000 ppm FW brine. Modified crude oil (Oil M) containing similar amounts of naturally occurring acidic and basic components ($AN \approx BN \approx 0.2$ mgKOH/g) was prepared in the laboratory and flooded through the core. The amount of adsorbed POC was determined by analyzing the effluent oil AN and BN and comparing them against influent AN and BN in Oil M (bulk values) [77], as shown in Fig. 3.19.

The crude oil flooding results showed adsorption of POC immediately upon contact with the mineral pore surfaces, and especially effluent BN decreased, Fig. 3.19 (bottom). From approximately 2 PV of crude oil injected and onwards, the measured effluent BN was $\sim$0.08 mg KOH/g lower than the initial crude oil bulk BN value, which suggested continuous adsorption of bases. Basic component adsorption equilibrium had not been reached after 10 PV of Oil M was injected, meaning that core T15 still was able to adsorb basic POC when the experiment was completed. On the other hand, only a very small decrease in AN relative to the Oil M bulk AN value was noticed, Fig. 3.19 (top). Already at $\sim$5PV Oil M injected, adsorption equilibrium was reached as effluent and influent ANs were the same. The percentages of adsorption of acidic and basic POC were quantified versus PV injected and are illustrated in Fig. 3.20. Note that only parts of the crude oil POC adsorbed, with $\sim$30% of the basic POC adsorbed, and only $\sim$10% of the acidic POC adsorbed. The results show that only a fraction of the POC in Oil M interacted with the mineral pore surfaces. The more pronounced decrease in effluent BN compared to that of AN is likely due to the higher affinity of positively charged basic molecules for negatively charged silicate mineral surfaces.

These adsorption results agree with previously published work investigating the adsorption of POC on carbonate minerals, where it was shown that negatively charged dissociated carboxylic acids ($R\text{-}COO^-$) had a high affinity for positively charged chalk mineral surfaces, by adsorbing and affecting rock wettability [89,90]. Adsorption of crude oil acidic components onto chalk has been found to be more pronounced than that of basic components [91], and Madsen and Lind

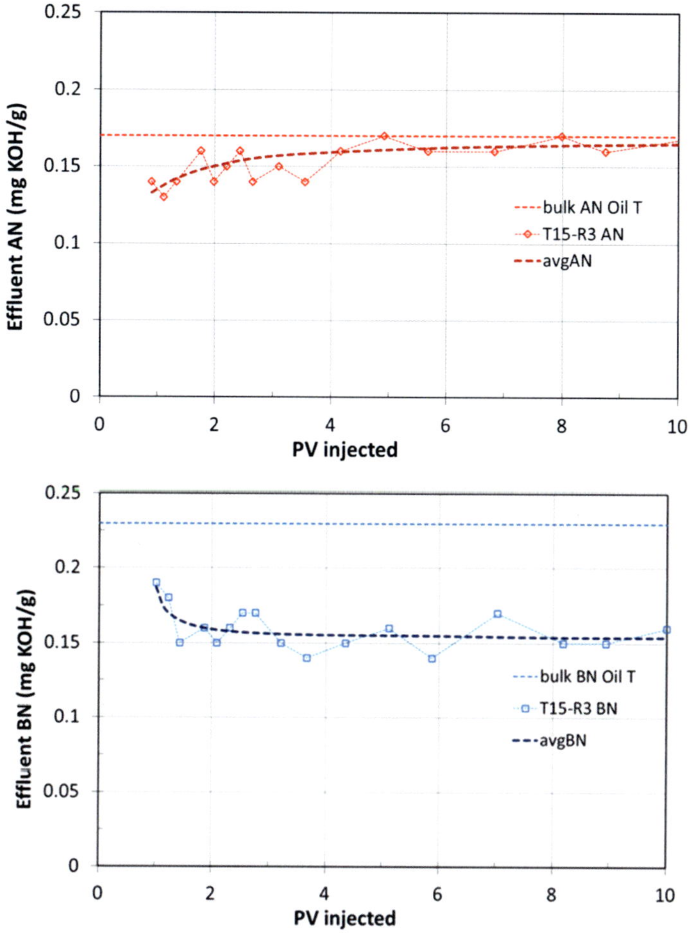

FIGURE 3.19

Effluent oil acid number (top) and base number (bottom) measured, reflecting adsorption of acidic and basic polar organic components during Oil M flooding of Core T15 with S_{wi} of 21% of formation water. Oil M, containing base number = 0.23 and acid number = 0.17 mg KOH/g, was flooded at 0.1 mL/min through the core at 50°C [86].

Reproduced with permission from T. Puntervold, A. Mamonov, I.D. Piñerez Torrijos, et al., Adsorption of crude oil components onto carbonate and sandstone outcrop rocks and its effect on wettability. Energy & Fuels 35 (2021) 5738–5747.

[65] showed that long-chain carboxylic acids had a high affinity for and adsorbing onto calcite mineral surfaces. Thus literature suggests that acidic crude oils readily adsorb onto carbonate mineral surfaces while basic crude oil components are more prone to adsorb onto silicate minerals, affecting wettability [86].

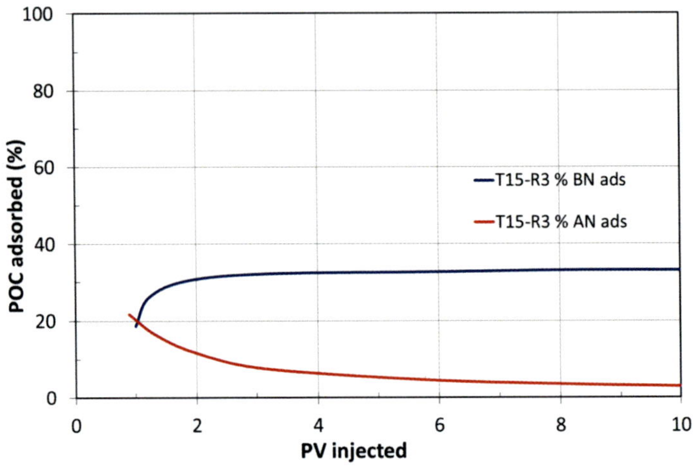

FIGURE 3.20

Percentage of adsorption of basic and acidic polar organic components onto core T15 during Oil M flooding.

Reproduced with permission from T. Puntervold, A. Mamonov, I.D. Piñerez Torrijos, et al., Adsorption of crude oil components onto carbonate and sandstone outcrop rocks and its effect on wettability. Energy Fuels 35 (2021) 5738–5747.

After the 10 PV Oil M was injected into core T15 in Fig. 3.19, spontaneous imbibition (SI) at 60°C was performed to evaluate the influence of the adsorbed POC on core wettability and capillary forces. By using FW as imbibing fluid the chemical equilibrium initially established between FW and mineral surfaces is not disturbed while introducing LS Smart Water could induce wettability alteration. The oil recovery by SI is shown in Fig. 3.21.

Since no wettability alteration is expected by SI of FW, the initial core wettability is reflected by the oil recovery profile and amount of oil recovered. After 12 days of FW imbibition, the oil recovery plateau of 12%OOIP was achieved. In this case, the initial wetting state of the core can be described as slightly water wet, which means that parts of the pore surfaces are water wet and will imbibe FW, while others are oil wet. Initial rock wettability determines the potential for wettability alteration by Smart Water injection, and experiments have shown that too water wet initial wettability significantly reduced the Smart Water EOR potential due to the limited possibility of changing rock wettability to more water-wet conditions [36,85]. At initial oil-wet conditions, the EOR potential can be reduced due to the decrease in available water-wet surfaces and limited access to pore mineral surfaces, which is necessary to initiate the wettability alteration by Smart Water. Therefore the mixed, slightly water-wet initial state should be optimal wettability conditions for observing Smart Water EOR effects in sandstone rock [55].

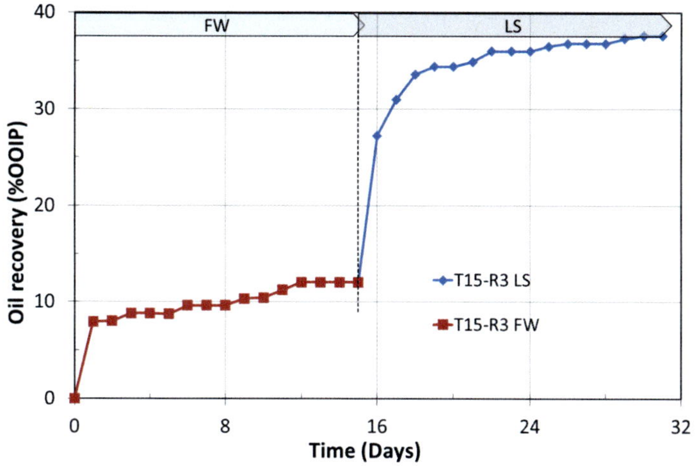

FIGURE 3.21

Oil recovery by SI at 60°C on core T15-R3 containing $S_{wi} = 20\%$ and having been exposed to 10 PV of Oil M without additional aging. The SI test was conducted with FW followed by LS Smart Water brine as imbibition fluids. Note that R3 signifies that this is the third restoration of the core.

When the recovery plateau by SI of FW had been reached, FW was replaced by LS and SI continued with LS as imbibing brine (Fig. 3.21). An immediate increase in water imbibition and oil production was observed. The LS brine behaved as Smart Water mobilizing extra oil from the heterogeneous pore system. The oil recovery increased from 12% to 27%OOIP in less than 1 day, confirming altered wettability, generated positive capillary forces, and high speed of imbibition. The mobilization of oil continued and reached the ultimate recovery plateau of 38%OOIP after a total of 30 days of imbibition.

3.2.5 Core restoration

Core wettability affects the mobilization of oil during water injection from heterogeneous pore systems, both in viscous flooding (VF) (Fig. 3.1) and SI (Fig. 3.21) experiments. When an initial water-wet sandstone core is exposed to crude oil during fluid restoration, POC interact with pore surface minerals, establishing a COBR equilibrium in the core at reduced water wetness. The impact of wettability on oil mobilization should be kept in mind when special core analysis (SCAL) is performed in laboratories. Relative permeability and capillary pressure curves developed for reservoir systems are wettability dependent, and the results dictate both the oil mobility at increasing water saturations and the final S_{or}, which are the key parameters in the reservoir simulator. Therefore replicating reservoir wettability in fluid restoration in preserved reservoir cores prior to the SCAL is of utmost importance. Restoring the native wettability of reservoir cores in front of reliable SCAL is a major challenge [92]. Even under ideal circumstances, the cores have been exposed to several handling and treatment processes from reservoir coring operations, transport to surface, core cutting, and preservation before it is received at the laboratory. During these operations, the cores are exposed to drilling mud, mechanical forces, temperature, and pressure reductions.

In front of SCAL in the laboratory, the brine and oil phases need to be restored to realistic water saturation and wettability representative of the reservoir. Laboratory core restoration consists of two main operations: core cleaning and fluid saturation and aging. The core cleaning process aims to remove crude oil, brine, and possibly drilling mud components from the preserved reservoir core samples, but without removing components belonging to the solid rock phase. The core cleaning process is challenging, because the chosen solvent can significantly change the rock wettability [93]. Common core cleaning involves cleaning the core to a completely water-wet stage, representing the core before crude oil invasion into the reservoir. Various cleaning solvents are used for this purpose; organic solvents and water are used to displace the hydrocarbon and connate water phases and precipitated salts, respectively.

During fluid saturation, FW and crude oil need to be introduced to the cleaned core at correct volume ratios to establish a core wettability representative of reservoir conditions, and normally an aging period follows. Often S_{wi} is established by saturating and flooding a core to 100% FW saturation, followed by crude oil flooding to lower the water saturation, a process which is mimicking crude oil invasion into a water-filled reservoir [59]. Aging the core is thereafter normally performed to assure evenly distributed crude oil component adsorption onto the rock surface, generating a reservoir representative initial core wettability. However, the length of a sufficient aging period has been varying in the literature from weeks to days, and for a long time, it has been presumed that aging is required for creating mixed wettability. Studies on chalk have confirmed that instantaneous adsorption of polar components occurred when the crude oil entered the core rendering the core mixed-wet without additional aging [90].

Core restoration is not a main topic here, but it must be kept in mind when discussing the efficiency of water injection in oil production. There have been significant improvements in the understanding of initial wettability and core restoration methods, which is of high importance for realistic SCAL and IOR/EOR studies. As an example of the complexity involved in the restoration of core wettability, the reservoir oil contributes with POC that significantly affect the wettability of silicate minerals. The amount of crude oil injected during core restoration has been shown to affect chalk wettability [94,95], which questions the correctness of using an arbitrary amount of crude oil to reduce S_{wi} during core restoration. Instead, in an attempt to preserve reservoir wettability in reservoir cores, it has been suggested to mildly clean the cores with less strong solvents, such as kerosene and *n*-heptane, aiming at preserving initially adsorbed POC on the rock surface [96]. In a recently published paper, chalk core wettability was reproduced using a mild core cleaning procedure and exposing the core to only $(1 - S_{wi})$PV of crude oil [95]. It was also shown that when the crude oil exposure increased, the water wetness of the restored core decreased. A similar study has been performed on reservoir sandstone cores, and again core wettability was reproduced when performing mild cleaning with kerosene and *n*-heptane and exposing the cores to a minimum amount of crude oil of $(1 - S_{wi})$PV [97].

3.2.6 Wettability measurements

Wettability can be difficult to determine in multicomponent and multiphase systems. Several methods exist for measuring wettability on smooth surfaces and in porous media, but common methods are contact angle measurements on smooth surfaces and SI in consolidated porous media, such as core plugs.

3.2.6.1 Contact angle measurements on smooth surfaces

Contact angle measurements are often used to evaluate the wetting state of a rock surface, and the wettability of pure fluids on plain surfaces [98,99], such as crystals representing calcite or limestone surfaces, and mica or glass representing different types of silicate minerals linked to sandstone systems.

Contact angle measurements have also been suggested used for obtaining a qualitative indication of reservoir wettability [26]. The equilibrated and static oil–water–solid system is illustrated in Fig. 3.22 and defined by Young's Eq. (3.7):

$$\sigma_{OS} = \sigma_{WS} + \sigma_{OW}\cos\theta_c \tag{3.7}$$

where θ_c = contact angle measured through the denser phase, σ_{os} = oil–solid IFT, σ_{ow} = oil–water IFT, and σ_{ws} = water–solid IFT.

Wettability is described by the contact angle, θ_c, measured through the denser phase, which here is the water phase, and classified according to the value of the contact angle as described in Table 3.7.

The main challenge with contact angle measurements is that it represents homogeneous wettability on a smooth surface. The surface is treated prior to analysis, which could affect the original surface wettability. Contact angle measurements cannot be performed in typical pore spaces of porous rocks because of the large droplet size (μL) that cannot fit into pores with diameters of nano- to

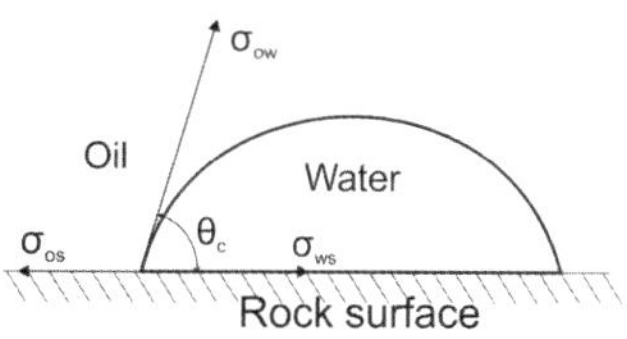

FIGURE 3.22

Static contact angle (θ_c) measurements.

Table 3.7 Wettability classification as a function of the measured contact angle.

Contact angle (°)	Wettability
0–30	Strongly water wet
30–90	Water wet
90	Neutral wet
90–150	Preferentially oil wet
150–180	Strongly oil wet

microscale. Nor is it possible to picture fluid droplets in representative micron-sized pores, even using micro-CT scanning with typical voxel resolution down to 3 μm.

3.2.6.2 Wettability in porous media

Reliable estimations of reservoir wettability are important for any reservoir process relying on wettability alteration, such as LS or Smart Water flooding. As contact angle measurements ideally should be performed on smooth and homogeneous mineral surfaces, they are not necessarily giving representative results measured on inhomogeneous porous media. For measuring wettability in consolidated porous media several experimental methods have been proposed and the most common are Amott, Amott–Harvey, and the United States Bureau of Mines method. Some of these methods are destructive, time-consuming, and involve numerous experimental operations reducing reproducibility and the value of the measurements. It is also difficult to perform most of these experiments at elevated and realistic temperatures for reservoir systems.

SI can be used to evaluate wettability and quantify the degree of water wetness established in restored cores containing initial FW and an oil phase [59]. A SI experiment is performed by placing the restored core in an Amott cell surrounded by imbibition fluid as illustrated in Fig. 3.23. In water imbibition studies the core is immersed in water, for example, FW, and if the core is water wet, FW will imbibe into the core by positive capillary forces, displacing oil from the core. The displaced oil is measured during the experiment, and both rate of imbibition and the ultimate recovery are monitored since the rate is proportional to the imbibition capillary pressure [98]. To interpret the results, it is important to compare with reference results that are close to the initial core wettability [78,100].

3.3 Smart Water enhanced oil recovery in sandstones

Waterflooding of oil reservoirs has been performed for decades for pressure support and improving the flow of oil towards the producers. The preferred injection water has been the most easily available water source, such as SW in offshore

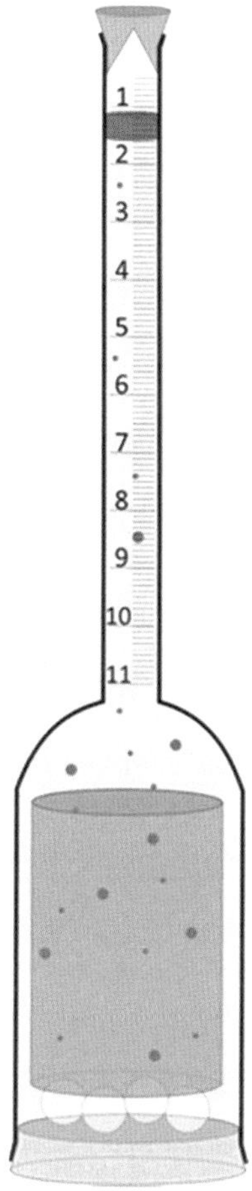

FIGURE 3.23

Spontaneous imbibition cell.

operations, and surface, aquifer, or produced water (PW) onshore. No thought was given to the composition of the water phase injected.

Generally, FW injection is not expected to disturb the established chemical equilibrium between COBR in the reservoir, since the composition of the injected FW is the same and the water residing in the pores in the reservoir. An oil recovery experiment with FW injection in the laboratory can then be regarded as the baseline recovery for the reservoir system tested. If the composition of the injected water is deliberately modified by changing its chemical constituents, then this modified waterflooding can be considered an EOR method. Smart Water EOR implies an increase in waterflooding efficiency caused by wettability alteration of the reservoir toward more water-wet conditions by the modification of the ionic composition of the injected brine. At more water-wet conditions, capillary forces increase, and IOR is a result of phase interactions between reservoir fluids, such as crude oil and brine, and the pore surface minerals.

In this section, the principles of Smart Water EOR in sandstones will be discussed from a chemical viewpoint, and special attention is given to COBR interactions.

3.3.1 Mechanistic studies and proposed mechanisms

Injection of brines with modified ionic composition to improve oil recovery has been given increased attention in the last 25–30 years. Although an effect of floodwater salinity in oil recovery from clay-containing cores was observed in the 1960s [101], more systematic research began in the 1990s when Tang and Morrow [14] observed and reported that oil recovery was improved if LS brine instead of HS brine was used as both imbibition and flooding brine in Berea sandstone cores. The reason for the large interest in modified water injection is first the reported significant incremental oil recovery potential of approximately 10%–30%OOIP [81], second its environmental friendliness because it is mostly based on fresh water or SW, third its simplicity of implementation since oil producing companies already know water injection, and fourth the relatively low cost of water treatment compared to, for example, chemical EOR methods involving polymers and surfactants.

Due to the complex nature of the multicomponent phases in the reservoir, the crude oil, FW and rock, comparative studies of high statistical validity can be difficult to perform. Reservoir cores have heterogeneous pore structures and mineralogical distributions and are most often not readily available for research purposes. To replace reservoir cores, it is standard to use outcrops; however, they too have heterogeneous pore structures and mineralogical distributions, but maybe to a lower degree. Because of the phase complexities, in addition to the different laboratory procedures, there is no unanimous consensus in the industry and academia of the cause of the extra oil that can be produced by modified waterflooding, such as Smart Water or LS water injection.

Several mechanisms explaining the incremental oil recovery have been proposed in the literature [35,102], most of which can be divided into two main groups: (1) solid–liquid mechanisms describing surface reactions between reservoir fluids and rock minerals and (2) liquid–liquid mechanisms describing reactions at the interface between crude oil and brine in porous media. Some commonly known proposed mechanisms are classified in Fig. 3.24.

Understanding the underlying mechanisms for the extra oil recovered by modified water injection is crucial for knowing when to apply the EOR method, and when to choose another method. The mechanistic understanding is also important if the EOR process is to be correctly modeled and upscaled to reservoir conditions. Liquid–liquid interface reactions at the crude oil–brine interface, such as change in IFT [103], osmotic processes [104], and microdispersions [105], have

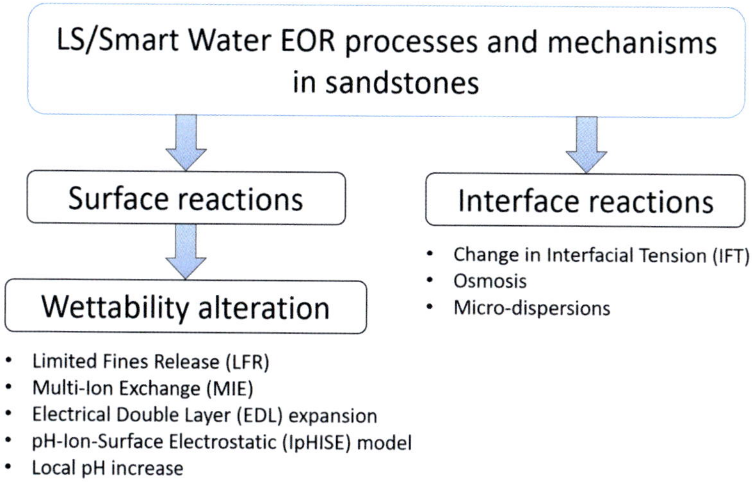

FIGURE 3.24

Low salinity/Smart Water enhanced oil recovery mechanisms proposed for sandstones.

been suggested as important interactions during LS/Smart Water flooding explaining the extra oil recovered. Yet, the above-mentioned processes are likely influenced by fluid properties at the test conditions and are thus not able to explain all surface interactions observed during LS or Smart Water flooding. Brooks and Purcell [106] found that the BET surface area of most sandstones ranged from 0.5 to 4 m^2/g rock, and the outcrop T sandstone used here has a measured surface area of 1.8 m^2/g. According to the calculations related to Fig. 3.11 and upscaling to the reservoir scale, a huge number of m^2 of mineral surfaces are available for fluid–rock interactions in the reservoir. However, the in situ formation fluid volumes are relatively low. Thus solid–liquid interactions should play a major role in reservoir processes. One very important surface interaction phenomenon is wettability. Wettability alteration has been proposed as the main cause of the extra oil recovered by modified waterflooding, and wettability alteration mechanisms are the focus of this chapter. Several mechanisms leading to wettability alteration have been proposed in the literature:

3.3.1.1 Migration of fines/limited fines release

The fines migration mechanism, first introduced by Tang and Morrow [31], describes wettability alteration caused by detachment of partly oil-coated fines (mostly kaolinite clay particles) during LS water injection. The fines are detached because of increased electrostatic interactions between particles in LS brine environment leading to expansion of the EDL. The observed increase in oil recovery was explained by shifting rock wettability toward a more water-wet state by the removing mixed-wet fines from the surface of the pore minerals [31].

3.3.1.2 In situ surfactant generation

McGuire et al. [107] suggested that the LS EOR effect was similar to the EOR effect observed during alkaline and surfactant flooding; that in situ generation of surfactants from the residual oil at elevated pH would promote additional oil recovery by altering wettability, and acting as emulsifying agents for the dispersion of oil into water. This would cause a reduction in IFT and improve miscibility between the oil phase and the injected LS brine, with increased oil recovery as a result.

3.3.1.3 Multicomponent ion exchange

In the mid-2000s Lager et al. [87,108] introduced MIE as the cause of the extra oil recovered by LS waterflooding. They suggested that multivalent cations present in the FW could bridge the negatively charged organic components from the crude oil and the negatively charged clay surface, promoting an oil-wet surface. The adsorption mechanisms suggested by Sposito [109] were the basis for the discussion of crude oil adhesion onto rock surfaces where Ca^{2+} bridged negatively charged clay surfaces and polar crude oil components. They also suggested that ion exchange could remove adsorbed organic material from the surfaces [108,109]. During LS injection the multivalent cation bridge would break, removing POC and organo-metallic complexes from the surface and exchanging them with monovalent cations [87,108]. Their claims were based on experimental observations in which LSE was not detected when the FW used during fluid restoration was stripped of divalent ions [108].

3.3.1.4 Electrical double layer expansion

In 2009 Ligthelm et al. [110] proposed that expansion of the EDL surrounding the rock surface during LS waterflooding was responsible for the removal of POC from mineral surfaces and recovering additional oil. An EDL consisting of a surface charge layer and a diffuse layer, is formed around a charged surface due to attraction of ions present in an electrolyte solution [111]. The EDL thickness can be estimated from of zeta (ζ)-potential measurements, and some researchers have reported EDL expansion during LS injection from measuring ζ-potentials [112,113].

3.3.1.5 Local pH increase—pH-induced wettability alteration

In several laboratory studies a pH increase has been observed when switching from an HS or FW brine to LS waterflooding of sandstone cores [67,107,114–116]. In 2010 Austad et al. [7] proposed a chemical mechanism, in which the pH increase played a central role, explaining the extra oil recovered by LS waterflooding. They suggested that a local pH increase caused by cation exchange at the sandstone mineral surfaces would change the charges of adsorbed POC, causing them to desorb from the rock surface and thus altering the rock surface wettability toward more water wet. The pH-induced wettability alteration mechanism will be discussed in more detail in Section 3.3.2.

3.3.1.6 Integrated pH–ion–surface electrostatics model

In 2015 Brady et al. [81] published an integrated pH–ion–surface electrostatics model (IpHISE), which considered effects of ion exchange and pH on oil and mineral surface electrostatics. The model contains a methodology for calculating speciation, that is, in which form the various compounds exist, and complexation of ions at oil interfaces and mineral (here kaolinite) surfaces. These calculations can then be used to indicate under which conditions of salinity, pH, and oil surface parameters LS EOR effects can be observed.

Several mechanisms have been proposed as the main driver for the extra oil recovered by LS or Smart Water injection in sandstone systems. The reason for lack of consensus regarding the dominating mechanism explaining the wettability alteration and IOR by Smart Water or LS waterflooding lies in the complexity of the COBR system. Both crude oil, rock and brine are multicomponent systems having differing chemical reactivity. For every mechanism proposed there are contradicting results reported in the literature. Over the past decades, produced fines have been observed and the fines migration mechanism has been confirmed by some researchers [31,117], but in other LS/Smart Water core flooding studies fines production has not been observed [67,108,118]. These contradictory observations and discrepancy between results can be ascribed to the use of different core material, outcrop or reservoir samples, of various origin, and probably also to different laboratory protocols. Several reported cases of fines migration were recorded in experiments using high permeability Berea outcrop sandstone cores containing kaolinite clay minerals [31,117], while no fines migration has been observed using lower permeable illite-containing outcrops [7,119]. As the detachment and migration of very small particles can be difficult to detect visually in the effluent, it has been suggested that additional analyses should be performed [120].

It was argued by Lager et al. [108] that in situ surfactant generation at increased pH proposed by McGuire et al. [106] could not be the main driver behind the extra oil recovery by LS waterflooding. The reason for their argument was that the highest incremental oil recovery was observed in a system containing a crude oil with very low acidic content, AN $<$ 0.05 mgKOH/g oil. Previous work on alkaline flooding has suggested that AN $>$0.2 mgKOH/g oil is needed for wettability alteration and emulsion formation to take place [121]. Therefore although a pH increase has been observed, it seems doubtful that in situ surfactant generation is the main mechanism for the extra oil produced by LS flooding. In the chemical wettability alteration mechanism proposed by Austad et al. [7], a local pH increase close to the surface caused by ion exchange between the water phase and ions adsorbed at the surface is sufficient for altering rock wettability toward more water-wet conditions. In tests performed at reservoir conditions no pH increase was observed accompanying the extra oil recovered by LS waterflooding [122]. However, presence of sour gases, such as H_2S and CO_2, in reservoirs or at laboratory reservoir conditions will prevent the pH increase because of buffering effects [67]. Static adhesion tests have shown that the wettability of silica surfaces was dependent on solution pH [76,80]. The theory of expansion of the EDL in LS brine does not agree with the reported increased POC adsorption in presence of LS brine compared to that in presence of HS brine at constant pH [66,67,82]. Zeta (ζ)-potential measurements performed on sandstone mineral solutions of various brine compositions became more negative as the brine salinity decreased [64,123]. This observation suggests that the EOR effect caused by LS injection in sandstone rock is not induced by zeta-potential changes, since silicate surfaces are negatively charged in the pH range 5–9 normally present at reservoir and laboratory conditions. Instead, this observation supports the increased adsorption of POC observed in presence of LS brine, shown in Figs. 3.16 and 3.18.

The results presented by Lager et al. [122] discussing the MIE mechanism showed a drop in the Mg^{2+} concentration during the LS waterflood. The concentration drop was larger than that of Ca^{2+}, and it was interpreted to be due to an exchange mechanism with the surface. However, according to Austad et al. [7] these observations could be due to complexation with hydroxyl ions (OH^-) and/or precipitation of divalent ions in the form of hydroxides, $Mg[OH]_2$ and $Ca[OH]_2$, in the pores at increased local pH occurring due to ion exchange.

In the next chapter, the proposed pH-induced chemical wettability alteration mechanism will be looked at in more detail [7,67]. All experimental results presented later in this chapter are discussed related to this mechanism.

3.3.2 pH-induced wettability alteration mechanism in sandstones

Many scientists agree with the increased oil recovery observed by LS or Smart Water injection being caused by wettability alteration, although the mechanism for the wettability alteration is still under discussion. Chemically, the change in wettability takes place due to the chemical imbalance occurring when FW is replaced by LS/Smart Water brine in the porous network. Physically, the change in wettability induces capillary forces, which increase oil recovery due to improved sweep efficiency as described in Fig. 3.25.

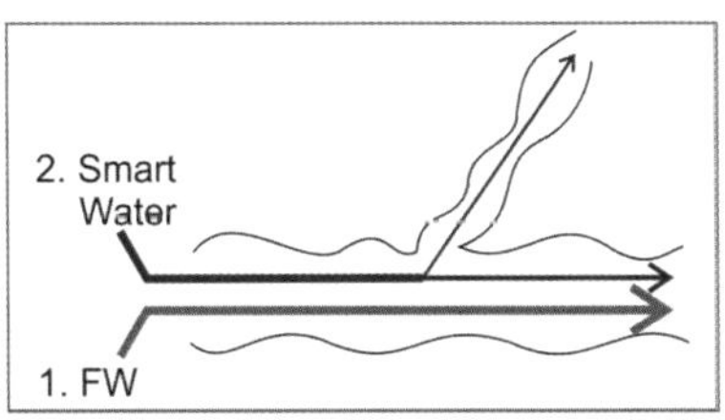

FIGURE 3.25

Sketch showing how wettability alteration with Smart Water increases sweep efficiency due to increased capillary forces.

First, flooding with FW does not induce wettability alteration since FW is already in equilibrium with the in situ crude oil and minerals. FW displaces oil from the larger pores and the oil recovery is a result of the initial wettability in the system. Second, flooding with Smart Water or LS brine in sandstone can alter wettability of the system toward more water wet, thereby inducing positive capillary forces and the injected brine can displace crude oil from the smaller pores, thus improving the sweep efficiency.

The wettability alteration process during LS/Smart Water injection is dependent on both initial wettability in the core or reservoir system, as well as on the wettability alteration conditions. Initial wettability refers to the wettability state prior to waterflooding, when the crude oil, FW and rock minerals present are in chemical equilibrium. POC from the crude oil, represented by basic nitrogen-containing aromatic species of the pyridine type having a reactive electron pair, R_3N:, and acidic species containing carboxylic groups, R-COOH, present at the oil–water interface can participate in acid–base reactions by accepting or releasing protons, H^+ as described in Eqs. (3.8) and (3.9).

$$\text{Basic material} \quad R_3N: + H_2O \rightleftarrows R_3NH^+ + OH^- \tag{3.8}$$

$$\text{Acidic material} \quad R-COOH + H_2O \rightleftarrows RCOO^- + H_3O^+ \tag{3.9}$$

Since the acid–base reactions in the above equations are dependent on pH, the POC charges are dependent on the pH of the brine in contact with the crude oil. Low pH and acidic environment (high $[H^+]$) promote an increased fraction of protonated species, R-COOH and R_3NH^+. Higher pH (low $[H^+]$ and high $[OH^-]$) and alkaline environment promote increased fraction of the deprotonated species R-COO^- and R_3N:. Note that pK_a values are quite similar and in the range of about 4.5–5 for both acidic and basic material. It means that at pH 4.5–5 equal amounts of charged and uncharged species are present for both acidic and basic material, and that the most reactive species toward a negatively charged sandstone surface, R-COOH and R_3NH^+, are present at higher fractions at low pH compared to high pH [55].

Reservoir pH is commonly in the range of 5–9 because of the presence of sour gases, which can lower pH, and minerals, which can exchange ions or partly dissolve, thus increasing pH to alkaline conditions. The PZC for silicate minerals is at ~ pH 2–3, which means that silicates are negatively charged above this pH level [71]. The initial wettability is therefore a result of COBR interactions involving adsorption of both inorganic cations from the FW and protonated POC (both neutral and positively charged) from the crude oil onto the negatively charged silicate surface. The initial wettability depends on the amount and nature of adsorbed POC, inorganic ion type and concentration, solution pH and temperature in the competition illustrated in Fig. 3.15.

When LS or Smart Water is injected into the reservoir, the existing COBR equilibrium is disturbed due to the injected brine composition being different from the FW composition, and wettability alteration can be induced. The wettability alteration process is initiated by cation exchange between initially adsorbed cations on clay mineral surfaces and cations or protons present in the injected brine, as shown by Eq. (3.4). In addition, cation exchange can take place between injected brine and feldspar minerals, such as albite, anorthite and microcline, often present in sandstone rocks, as described in the following equations:

$$\text{Fast reaction } X - \text{Na} + \text{H}^+ \leftrightarrow \text{X} - \text{H} + \text{Na}^+ \tag{3.10}$$

$$\text{Fast reaction } \text{X}_2 - \text{Ca} + 2\text{H}^+ \leftrightarrow 2(\text{X} - \text{H}) + \text{Ca}^{2+} \tag{3.11}$$

X represents a cation-exchange site on the feldspar mineral surface

Cation-exchange reactions in feldspars are fast [46]; hence, even though CEC in feldspars have been reported to be much lower than that of clay minerals [38], feldspars can influence FW pH due to the favorable kinetics. Induced local pH increase close to the mineral surfaces can affect the charges of the adsorbed POC, according to Eqs. (3.8) and (3.9), increasing the fractions of POC with lower affinity for the negatively charged silicate surfaces, and desorption of POC takes place. Austad et al. [7] proposed a chemical mechanism for the wettability alteration in sandstone by injection of Smart Water. The mechanism is illustrated in Fig. 3.26, including the initial mixed-wet conditions, introduction of Smart Water and the resulting water-wet conditions.

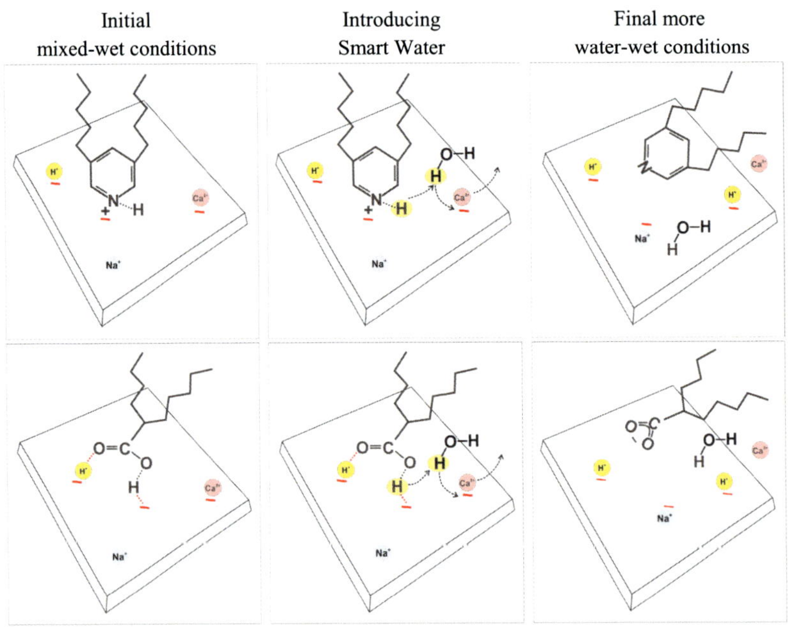

FIGURE 3.26

pH-induced chemical mechanism of wettability alteration during Smart Water injection in sandstones. An example of desorption of initially adsorbed protonated basic crude oil material (upper) and protonated acidic crude oil material (lower) from clay mineral surfaces.

Since sandstone material is not homogeneous but consisting of varying amounts of a collection of minerals, each sandstone may react individually toward reservoir fluids, generating varying initial pH and difference in POC adsorption and responding differently to Smart Water injection. Therefore initial wettability and the wettability alteration process are dependent on the properties of all phases involved, the rock minerals, crude oil, FW, and injection water. Reservoir temperature also has influence since most chemical reactions and processes are temperature dependent. In the following chapters, experimental observations from tests performed on individual minerals as well as on real outcrop and reservoir systems will be highlighted and discussed.

3.3.3 Smart Water enhanced oil recovery on a model sandstone system

To study the LS EOR effect reported in sandstones, a COBR system that shows good reproducibility in core experiments should be chosen. With optimized core restoration and experimental procedures, it should be possible to reproduce experiments in the laboratory using the same core. Reproducible results during FW flooding were obtained by Rezaeidoust et al. [67,124] using the outcrop T sandstone core material (properties given in Table 3.2), FW brine (FW_T) (properties given in Table 3.8) and crude oil (Oil T) (properties given in Table 3.9)

Table 3.8 Brine compositions used in experiments on outcrop T sandstone cores.

Ions	FW_T (mM)	LS (mM)	SW (mM)
Na^+	1540	17.1	450
K^+	–	–	10
Ca^{2+}	90	–	13
Mg^{2+}	–	–	45
Cl^-	1720	17.1	525
SO_4^{2-}	–	–	24
TDS, (mg/l)	100 000	1000	33 390
pH	5.5	5.7	7.8
μ @25°C, (cp)	1.18	1.01	1.10
μ @60°C, (cp)	0.63	0.60	0.62

FW, Formation water; LS, low salinity; SW, seawater; TDS, total dissolved solids.

Table 3.9 Properties of Oil T used in experiments on outcrop T sandstone cores.

Crude oil	AN (mg KOH/g)	BN (mg KOH/g)	Density (g/cm^3)	Viscosity (cP) @20°C
Oil T	0.10	1.80	0.846	17.6

AN, Acid number; BN, base number.

system provided by Total E&P. Recall that the pore size distribution in outcrop T sandstone is heterogeneous as shown in Fig. 3.5, with a three orders of magnitude span in pore radius, a peak around 10 μm, and with a significant contribution to the PV from pores down to 0.01 μm radii. Note that this COBR system including outcrop T sandstone, oil T and the brines listed in Table 3.8 was used to generate the results included in this chapter.

Rezeidoust et al. [67] performed oil recovery tests on four outcrop T sandstone cores, all with similar properties as those presented in Table 3.3. Four different compositions of LS brine were used, containing either NaCl, KCl, $MgCl_2$, or $CaCl_2$, but all had the same ionic strength and a salinity of ~1000 ppm. In Fig. 3.27 the results from the oil recovery core flooding experiments are presented.

The equally restored cores gave an ultimate recovery close to 40%OOIP after FW_T injection, confirming very good reproducibility between core experiments. During LS injection all cores responded similarly with increased oil recovery of 6%–7%OOIP, confirming a tertiary LS EOR effect despite the difference in ionic composition of the LS brine injected. This COBR system was therefore deemed suitable for further parametric and mechanistic studies on LS Smart Water EOR.

3.3.3.1 Tertiary Smart Water enhanced oil recovery effect

The tertiary oil recovery effect from core T21 (core properties given in Table 3.3) obtained by flooding LS brine at 60°C is shown in Fig. 3.28. The oil recovery curve is the same as that shown in Fig. 3.1, however now the belonging effluent

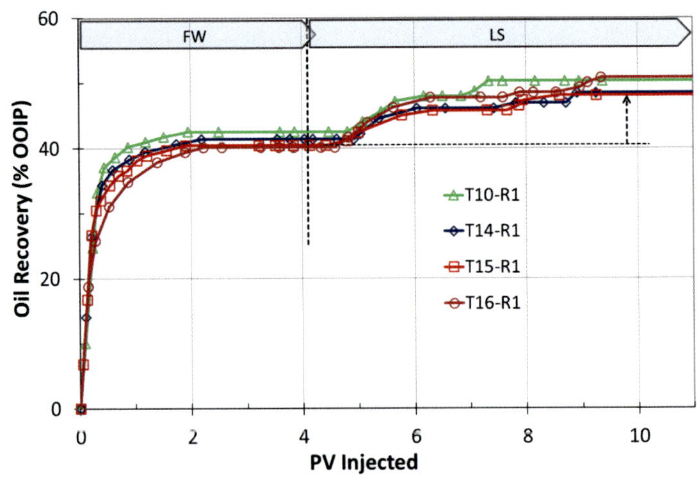

FIGURE 3.27

Oil recovery tests performed on four different outcrop T sandstone cores equally restored with S_{wi} = 20% FW_T, exposed to and aged in Oil T. The cores were flooded at 40°C at a rate of 4 pore volume/day with FW_T, followed by four different low-salinity brines of varying composition but all with a salinity close to 1000 ppm.

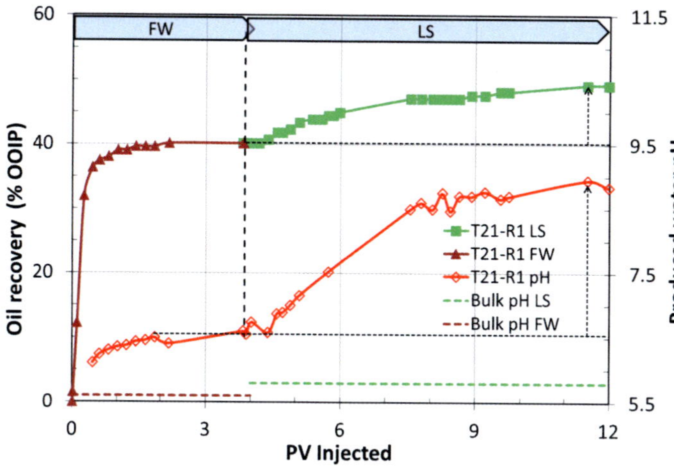

FIGURE 3.28

Oil recovery test by flooding FW_T—low salinity at a rate of 4 pore volume/day in core T21 at 60°C. The core was restored with $S_{wi} = 0.20$ FW_T, exposed to 5 pore volumes of Oil T and aged. Bulk pH values are 5.5 for FW_T and 5.7 for low salinity brine.

pH curve is also plotted. The core was restored with $S_{wi} = 0.20$ FW_T, saturated, flooded and aged with Oil T at the test temperature, and then successively flooded with FW_T–LS brine at a rate of 4 PV/day. The choice of that specific rate was based on the typical reservoir injection rate of 1 ft/day. Oil recovery was logged against PV injected, and the PW pH was monitored. During FW_T injection an oil recovery plateau of 40%OOIP was reached after 2 PV injected, and the first PW pH detected was 6.0, indicating slightly acidic initial conditions that are beneficial for generating initial mixed wettability. The flooding fluid was changed from FW_T to LS after 4 PV brine injected. During LS injection PW pH gradually increased and a simultaneous tertiary LS EOR effect was observed. The oil recovery increased to a plateau of 49%OOIP after 12 PV brine injected accompanied by a pH increase of 2.5 pH units. Note that measured pH in the effluent is higher than the brine bulk pH because of cation-exchange interactions with the sandstone surfaces. Some dissolution of carbonate cement is also contributing to the increase in pH.

The large EOR effect of 9% OOIP or 23% extra oil could not be described by changes in viscous forces or improved mobility ratio (M) when FW_T is substituted by LS brine. The mobility ratio is described by Eq. (3.12):

$$M = \frac{\lambda_D}{\lambda_d} = \frac{\lambda_w}{\lambda_o} = \frac{\left(\frac{k_{rw}}{\mu_w}\right)_{S_{or}}}{\left(\frac{k_{ro}}{\mu_o}\right)_{S_{wi}}} \tag{3.12}$$

where λ is the mobility of D—displacing fluid, d—displaced fluid, w—water, and o—oil, k_{rw} and k_{ro} are relative permeability to water and oil, μ_w and μ_o are

Table 3.10 Viscosity and viscosity ratios for the outcrop T sandstone system at 60°C.

Viscosity Oil T at 60°C (cP)	Viscosity FW_T at 60°C (cP)	Viscosity LS at 60°C (cP)	$\frac{\mu_o}{\mu_{wFW}}$	$\frac{\mu_o}{\mu_{wLS}}$
4.5	0.63	0.60	7.29	7.65

FW, Formation water; LS, low salinity.

viscosities of water and oil, S_{or} is the residual oil saturation, and S_{wi} is the initial water saturation. When M is ≤ 1 the sweep efficiency is higher than that when M is 1, and thus higher oil recovery is expected.

The LS brine has similar, but slightly lower viscosity than the FW_T at 60°C as seen in Table 3.10, which gives a slight unfavorable change in the viscosity ratio and the mobility ratio if the relative permeabilities are assumed unchanged when the injection brine is changed from FW_T to LS brine.

Previously, Rezaeidoust et al. [67] demonstrated a similar LS EOR effect after bumping the rate four times during FW_T injection, thereby excluding the possibility that the extra oil from this COBR system was recovered because of capillary end effects when working with rather short cores in the laboratory. The tertiary Smart Water EOR effect observed in the outcrop T sandstone system could also not be explained by any increased pressure drop, as the pressure drop observations in Fig. 3.1 did not show significant pressure increase when switching to LS brine injection. It seems that increased pressure drop during the waterflood experiments is not a requirement for observing EOR by LS or Smart Water brines. As such, explanations for the extra oil recovered linked to varying mobility ratio, clay swelling, fines migration, capillary end effects or diverted flow can be excluded being universal.

3.3.3.2 Wettability alteration and capillary forces

Capillary forces are one of the main oil recovery mechanisms in porous media. The importance of capillary forces on oil mobilization increases with increasing heterogeneity of the pore system. MICP data from the outcrop T sandstone samples confirmed a three orders of magnitude span in the pore throat distribution. The wettability of the rock systems controls the contribution of the capillary forces in the oil recovery process. When the wettability is water wet, positive capillary forces will contribute to the mobilization of oil by water imbibition, especially in small pores. It is well known that capillary forces and imbibition of water into the rock matrix is needed to obtain a high recovery from fractured, mixed-wet carbonate rock. Otherwise only the oil residing in the fractures will be produced.

To study capillary forces in the outcrop T sandstone system a SI test was performed. An outcrop T sandstone core was restored and spontaneously imbibed with FW_T at 60°C to evaluate its initial wettability after restoration with oil T. Afterwards, wettability alteration potential by LS brine was investigated by

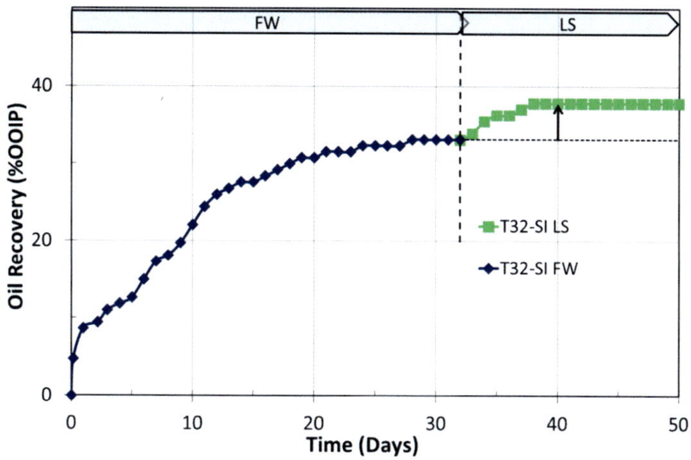

FIGURE 3.29

SI test on core T32 at 60°C. The core was restored with $S_{wi} = 0.20$ FW_T, exposed to 5 pore volumes of Oil T and aged. The core was initially imbibed with FW_T until recovery plateau was reached, followed by imbibition of low-salinity brine.

switching to LS as imbibing brine in tertiary mode. The oil recovery results by SI of FW_T and LS are presented in Fig. 3.29.

During SI of FW_T brine in core T32, a steady production of oil was observed, reaching an ultimate oil recovery plateau of 33%OOIP after 28 days. Recall that about 40%OOIP was recovered by FW_T injection in Fig. 3.28. These results indicate that the core is quite water wet since capillary forces can mobilize more than 80% of the oil produced by VF of FW_T. The imbibing fluid was changed from FW_T to LS after 32 days of imbibition, and a significant increase in oil recovery to a new plateau of 38%OOIP was observed (Fig. 3.29). This means that the LS brine induces wettability alteration toward more water-wet conditions, generating increased capillary forces and mobilizing 15% extra oil compared to that by FW_T.

The parameters affecting capillary forces in porous media can be derived from the equation describing capillary pressure (P_c) for a capillary tube model:

$$P_c = \frac{2 \cdot \sigma \cdot \cos\theta}{r} \tag{3.13}$$

where σ is the IFT (sometimes also denoted γ) between the oil and water phases, θ is the contact angle between the tube wall and the fluid meniscus, measured through the water phase, and r is the pore radius. Both wettability and the IFT affect capillary forces, and in a heterogeneous porous media the capillary forces will also be influenced by the pore size distribution controlling the rock permeability. The above core flooding and SI results show that capillary forces need to be accounted for when fluid flow in heterogeneous porous media should be described in core experiments and when fluid flow in porous media should be properly described in fluid flow models and upscaled in reservoir simulation.

3.3.3.3 Effect of oil–brine interfacial tension

IFT between the oil and water phases is one of the parameters affecting capillary forces, Eq. (3.13). For the outcrop T sandstone system used, IFT has been measured between Oil T, FW_T, and LS to evaluate if changing brine system would significantly affect the oil–water IFT, thereby explaining the significant increase in oil recovery observed when changing injection brine from FW_T to LS. Pendant drop IFT measurements were performed by analyzing the shape of an oil droplet positioned in a continuous brine phase at constant pressure and temperature. IFT results from the equilibrated Oil T–FW_T/LS brine system confirmed that there was no significant change in IFT observed when the brine was changed from FW_T to LS (Table 3.11).

Only a slight increase in IFT was observed when the brine salinity was decreased, which is in line with the general observation for IFT measurements that there is a slight IFT reduction when the water phase is polluted with salts. The small increase in IFT of maximum 5%–10% when FW_T is changed by LS brine cannot explain the huge EOR effect of 23% extra oil observed during tertiary LS injection (Fig. 3.28). Studies on clay-coated micromodels also concluded that IFT change was not the main driver for the observed LS EOR effect [125].

3.3.3.4 Effect of temperature on surface reactivity and Smart Water enhanced oil recovery potential

During tertiary LS injection into core T21 (Fig. 3.28), the increase in oil production was accompanied by a large increase in PW pH, from slightly acidic conditions with FW_T, to alkaline conditions with LS brine. The reactivity of POC present in crude oils toward negatively charged silicate surfaces is pH dependent with significant desorption of basic POC at alkaline conditions from both kaolinite (Fig. 3.16) and illite clay minerals (Fig. 3.18). Hence, a pH increase during water injection, caused by ion exchange reactions between brine and available pore surface area, could facilitate wettability alteration toward more water-wet conditions. Note that the bulk pH of the LS brine was acidic.

Table 3.11 Oil–water interfacial tension between Oil T and FW_T or low-salinity brine as a function of temperature at 10 bar.

Temperature (°C)	IFT Oil T/FW_T (mN/m)	IFT Oil T/LS (mN/m)
40	19	21
80	19	21
100	19	20
130	20	21

IFT, Interfacial tension; LS, low salinity.

To evaluate the brine–rock reactivity between the injection brines FW_T–LS and outcrop T sandstone and its dependence on temperature, a surface reactivity test was performed at 60 (Figs. 3.12), 90°C and 120°C by successively flooding FW_T–LS–FW_T brines, collecting the effluent samples and analyzing their pH values (Fig. 3.30).

During FW_T flooding acidic pH was measured at all tested temperatures. Acidic conditions enable protonated POC to interact with negatively charged silicate surfaces promoting mixed-wet initial conditions. When the LS brine was introduced, a significant increase in pH was observed at all temperatures, with a pH gradient of $\Delta pH \approx 3$. With a bulk pH of 5.7 for the LS brine, the three orders of magnitude decrease in proton concentration is caused by ion exchange at pore surface minerals, increasing pH. Alkaline conditions are favorable for desorption of POC, which is needed for inducing wettability alteration toward more water-wet conditions and an EOR effect. Both feldspars, that are considered basic minerals, and clay minerals having permanent CEC can contribute as cation exchangers.

Recall that acidic initial conditions in presence of FW must occur to promote adsorption of protonated POC onto negatively charged silicate minerals, which are primarily clays with large surface areas and permanently negatively charged sites. If an initial alkaline environment is observed with FW, the amount of protonated POC which can adsorb will be significantly reduced, see Fig. 3.13. Too water-wet initial conditions reduce the potential for observing Smart Water EOR

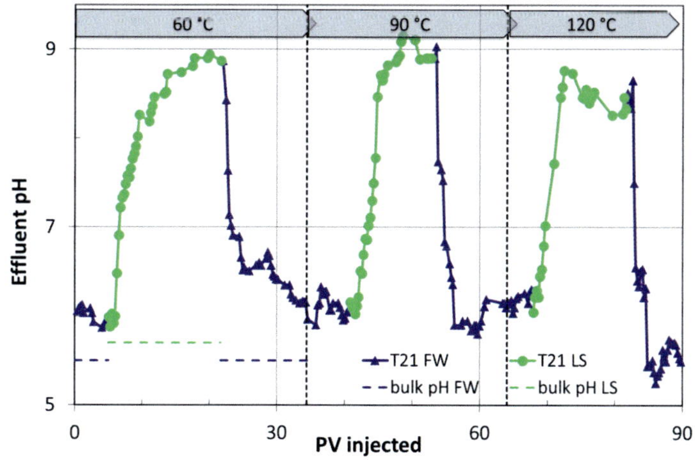

FIGURE 3.30

Surface reactivity test on core T21 performed at 60°C, 90°C, and 120°C. A 100% brine-saturated core was flooded with FW_T–low salinity–FW_T at each temperature at a rate of 4 pore volume/day. Effluent pH was monitored and compared with bulk pH of 5.5 for FW_T and 5.7 for low-salinity brine.

effects, since the wettability alteration potential to more water-wet conditions is very small [85]. Too water-wet initial conditions can possibly explain the observations made in the Snorre field pilot on the NCS [36,126]. The reservoir cores, which contained a high content of plagioclase (feldspar) behaved alkaline during injection of the rather LS FW (35,000 ppm), which are unfavorable conditions for promoting mixed-wet conditions. Thus there was limited potential to alter wettability by LS Smart Water injection.

Working with reservoir cores, the surface reactivity test could be performed at reservoir temperature to evaluate the interactions between the genuine reservoir minerals present and relevant formation and injection brines. The pH development during FW injection gives important information about conditions for reservoir wettability, and the change in pH during Smart Water injection gives important information regarding the potential for inducing the pH increase needed for wettability alteration.

To investigate the effect of temperature on the Smart Water EOR potential, an oil recovery test at 120°C was performed on an equally restored outcrop T sandstone core, T22. Oil recovery and PW pH are presented in Fig. 3.31.

The oil recovery plateau of 35%OOIP reached after 3 days of FW_T injection, was slightly lower compared to that reached at 60°C (Fig. 3.28). This difference could be due to slightly varying initial wettability conditions and differences in the mobility ratios at 60°C and 120°C. After 5 PV injected, the LS brine was introduced. The new oil recovery plateau at 44%OOIP was reached after a total of 12 PV brine injected. The LS EOR effect of 9%OOIP extra oil was the same

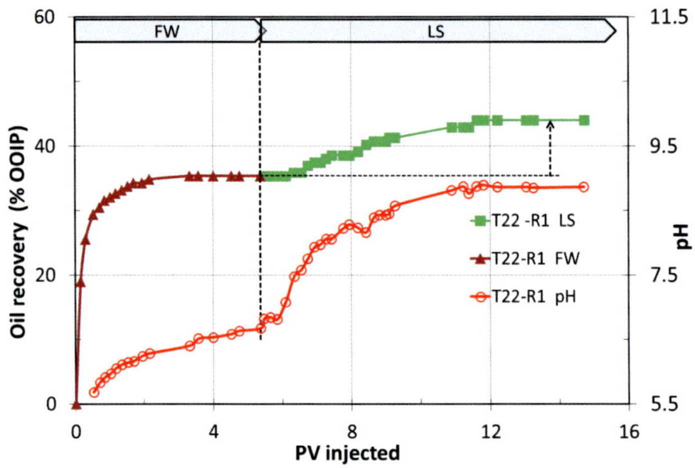

FIGURE 3.31

Oil recovery test by flooding FW_T—low salinity at a rate of 4 pore volume/day at 120°C in core T22. The core was restored with $S_{wi} = 0.20$ FW_T, exposed to 5 pore volumes of Oil T and aged at 120°C. Bulk pH values are 5.5 for FW_T and 5.7 for low salinity brine.

as that observed at 60°C (Fig. 3.28). The PW pH increased accompanied by a mobilization of more oil. Note that the increase in oil recovery is synchronized with the increase in pH at both 60°C and 120°C, indicating that there is a link between the observed pH gradient and the LS EOR effect. Vaidya and Fogler [114] also pointed to a strong link between salinity change and pH increment during waterflooding of Berea sandstone cores.

The pH development during LS injection was similar at both temperatures but note that the PW pH was lower during the oil recovery test compared to the values obtained in the surface reactivity test in Fig. 3.30. The difference in pH could be explained by the crude oil. Oil T contains POC that can buffer pH changes, and in the presence of crude oil, there is reduced amount of water-wet mineral sites, which can participate in rock–brine interactions, compared to the case at strongly water-wet conditions and 100% S_{wi}. The results from the pH development and tertiary EOR effects are summarized in Table 3.12. The oil recovery results agree with the rock–brine interactions observed during the surface reactivity screening test. Neither pH increase nor the LS EOR effect were significantly affected by the temperature change in this system.

A fluid property, which is dependent on temperature, is the viscosity. Viscosities of the various fluids at different temperatures are given in Table 3.13. The viscosity ratio between Oil T and FW_T is slightly more favorable at 60°C

Table 3.12 Influence of temperature on the low salinity enhanced oil recovery effect and pH gradients from surface reactivity and oil recovery tests.

Test temperature (°C)	Surface reactivity test FW_T–LS injection (ΔpH)	Oil recovery test FW_T–LS injection	
		LS EOR effect (%OOIP)	Produced water (ΔpH)
60	3.0	9.2	2.3
90	3.0	–	–
120	2.7	8.7	2.2

EOR, Enhanced oil recovery; LS, low salinity.

Table 3.13 Viscosity and viscosity ratios of crude oil and brines at varying temperature.

H	Oil T (cp)	FW_T (cp)	LS (cp)	Viscosity ratio ($\mu_o/\mu_{w_{FW_T}}$)	Viscosity ratio ($\mu_o/\mu_{w_{LS}}$)
25	17.60	1.18	1.01	14.92	17.43
60	4.59	0.63	0.60	7.29	7.65
120	2.02	0.29	0.23	6.97	8.78

FW, Formation water; LS, low salinity.

compared to that at 120°C, and the viscosity ratio between Oil T and FW_T is more favorable than that between Oil T and LS brine at 60°C and 120°C. These viscosity data confirm that the lower recovery with FW_T at 120°C and the extra oil recovered during LS brine injection are not results of a favorable difference in viscous forces.

3.3.3.5 Salinity effects on Smart Water enhanced oil recovery in sandstones

To observe a LS EOR effect there is a general acceptance in the literature that the typical LS brine has a salinity below 5000 ppm [11]. However, from a scientific perspective relevant information could be overlooked if salinity is the only parameter considered. Salinity is just a measure of salt content, not differentiating between the nature of the ions present. Only considering the salinity as an important parameter is excluding any effects caused by the individual ions present in the brines, pore surface mineralogy, or initial wettability.

To evaluate if injection brine salinity was the main factor dictating the Smart Water EOR effect in sandstone, an oil recovery test using 25,000 ppm NaCl as injection brine was performed on the outcrop T sandstone system. Core T25 was restored similarly as core T21 with $S_{wi} = 0.20$ FW_T and Oil T, and an oil recovery test was performed at 60°C by flooding FW_T followed by 25,000 ppm NaCl brine ($NaCl_{25'}$). Note that the latter is not considered a LS brine due to its salinity being >5000 ppm. Oil recovery results and PW pH are presented in Fig. 3.32.

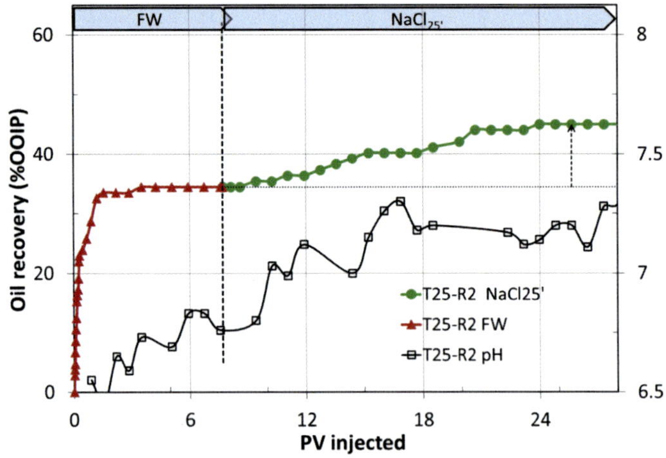

FIGURE 3.32

Oil recovery by flooding FW_T (100,000 ppm)–$NaCl_{25}$ (25,000 ppm NaCl) at 4 pore volume/day in core T25 at 60°C. The core was restored with $S_{wi} = 0.20$ FW_T, exposed to 5 pore volumes of Oil T and aged at 60°C.

During secondary FW_T injection an oil recovery plateau of 35%OOIP was reached. The pH of the first PW was 6.5 and it increased slightly during FW_T flooding. After 8 PV injected, the injection brine was changed to $NaCl_{25}$. As the PW pH slowly increased to 7.3, a gradual production of oil was observed reaching a new recovery plateau at 45%OOIP after 16 PV of 25,000 ppm NaCl injection. These results show that the oil production rate was significantly lower during 25,000 ppm NaCl injection compared to that during 1000 ppm NaCl (LS) injection into core T21 in Fig. 3.28. The pH change was also slower at 25,000 ppm NaCl compared to that at 1000 ppm NaCl injection. The smaller pH gradient could be explained by the higher NaCl concentration in the $NaCl_{25}$ brine. A high Na^+ concentration will reduce the H^+-ion exchange with the reactive feldspar minerals giving reduced alkalinity as described in Eqs. (3.1)–(3.3) [127]. This could also explain the reduced rate of wettability alteration compared to that in core T21 in Fig. 3.28.

Even though the salinity of the Smart Water brine used in core T25 was 25,000 ppm, which is five times higher than recommended salinity of 5000 ppm for an LS brine, 10%OOIP extra oil was mobilized during the core flooding experiments. Previously published results on the same outcrop T sandstone system, but with a different FW composition, showed that 4%OOIP extra oil was produced using a 40,000 ppm NaCl brine [124]. These results clearly indicate that the ionic composition of the injection brine is more important than the salinity for the EOR effect, and that the term "LS EOR" could be misleading for the wettability alteration process observed in sandstone systems. Increased attention should instead be given to the ionic composition of the injection brine, and thus the term "Smart Water" could be a better notation for brines designed to induce wettability alteration.

3.3.4 Smart Water injection strategies for sandstones

Tertiary Smart Water EOR effects have been demonstrated, however the wettability alteration process and therefore the increased oil recovery are time demanding as observed in Figs. 3.1, 3.28, 3.31, and 3.32. Thus there is much to gain by investigating Smart Water injection strategies, such as slug injection, secondary injection, or choice of injection brine, to reduce the EOR response time and improve the project economics.

3.3.4.1 LS slug injection in sandstones

To improve project economics and reduce the cost of LS water injection, slug instead of continuous LS injection has been suggested. According to Seccombe et al. [128] and their work at BP, a 40% PV LS slug was determined to be fully effective in corefloods, maintaining slug salinity throughout the core plug and recovering approximately 88% of the obtainable recovery by continuous LS injection. These results were stated to be consistent with single-well tracer tests performed at the Endicott field in Alaska, USA.

A close relationship between the LS EOR effect and pH increase was observed in oil recovery tests by continuous LS injection. Piñerez et al. [115] investigated slug injection in the outcrop T sandstone system, and the effect slug injection would have on the pH increase initiating the wettability alteration process. Based on the oil recovery results from core T21 in Fig. 3.28, about half of the extra oil was recovered within 2 PV of LS brine injection. This volume, 2 PV, was therefore proposed as an appropriate slug volume for testing Smart Water slug injection in the outcrop T sandstone system. Core T24 was restored similarly as in the previously reported experiments. The oil recovery plateau at 60°C with FW_T was reached at 39%OOIP (Fig. 3.33), which is similar to that from core T21. The first PW pH from core T24 was 6.8.

After 8 PV of FW_T injected, the 2 PV LS slug was introduced, and the PW pH rapidly increased by 1.6 pH units to approximately 8.6, accompanied by a 5% OOIP increase to a total oil recovery of 44%OOIP. Following the LS slug, FW_T was again injected, and there was an immediate cessation of the oil mobilization. No extra oil was produced, and pH fell rapidly back down to 6.8. After a total of 16 PV brine had been injected, continuous LS brine injection was initiated. The pH rapidly increased to 8.3 and continued to a maximum of 9.3, however at a slightly lower rate. The oil mobilization rate was significantly lowered as seen by the two production slopes in Fig. 3.33, indicated by the dotted lines. The final ultimate recovery plateau of 49%OOIP was achieved after a total of 22 PV brine

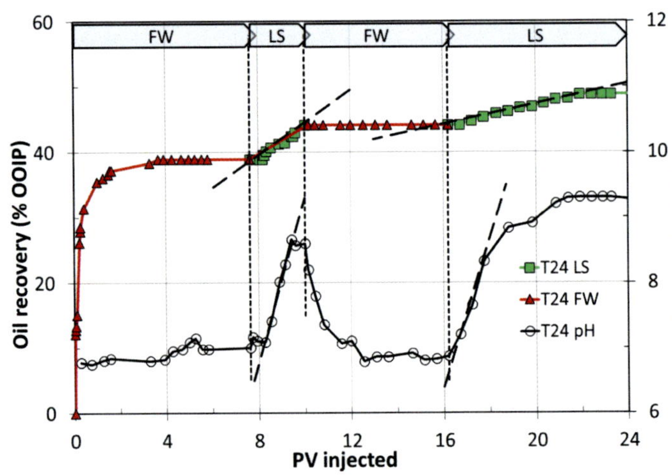

FIGURE 3.33

Oil recovery test by flooding FW_T—low salinity (2 pore volume)—FW_T—low salinity at a rate of 4 pore volume/day in core T24 at 60°C. The core was restored with S_{wi} = 0.20 FW_T, exposed to 5 pore volumes of Oil T and aged at 60°C. Bulk pH values were 5.5 for FW_T and 5.7 for low salinity brine.

injected, which is a significantly longer time than that observed in tertiary continuous LS injection in Fig. 3.28.

How can these observations be explained? It is plausible that this behavior is linked to the core material's heterogeneous pore distribution, seen in Fig. 3.5. At initial mixed- and quite water-wet conditions, the first FW_{T1} flood mobilizes oil by both viscous and capillary forces from large and easily accessible pores, reaching a recovery plateau of 39%OOIP and a residual oil saturation $(S_{orw})_{FW_{T1}}$. Injection of a LS slug (LS_1), having lower viscosity than FW_T, results in a less favorable mobility ratio during oil displacement compared to that of FW_T (Table 3.8), and the injected LS_1 follows the same water path established through the core by FW_T. By wettability alteration, the main recovery mechanisms are mobilization of wettability-trapped oil from mineral surfaces in large pore bodies and mobilization of trapped oil in small and less accessible pores by increased positive capillary forces. Thus oil mobilization and the new recovery plateau at 44%OOIP and $(S_{orw})_{LS_1}$ have not been achieved by increased viscous forces, although the brine flux contributes to the transportation of oil through the core. The second FW_T injection, FW_{T2}, follows the path of the pores with the highest water saturation and the lowest restrictions, that is, the larger pores. Simultaneously, pH decreases and halts the wettability alteration process according to the mechanism proposed by Austad and colleagues [7,67]. No extra oil was produced by flooding FW_T, and the residual oil saturation $(S_{orw})_{FW_{T2}}$ remained the same as $(S_{orw})_{LS_1}$. The second LS injection period, LS_2, resulted in a rapid pH increase mobilizing more oil from less accessible and bypassed pores by wettability alteration and positive capillary forces, reaching the ultimate recovery plateau of 49%OOIP and a residual oil saturation $(S_{orw})_{LS_2}$. This capillary-influenced Smart Water recovery mechanism can be illustrated as in Fig. 3.34.

In the oil recovery test by slug and continuous injection of LS shown in Fig. 3.33, the residual oil saturations decreased in the following order: $(S_{orw})_{FW_{T1}} > (S_{orw})_{LS_1} = (S_{orw})_{FW_{T2}} > (S_{orw})_{LS_2}$. Only half of the tertiary Smart Water EOR effect was produced after the 2 PV LS slug injected. An additional 6 PV of LS brine was needed to reach the ultimate EOR plateau of 49%OOIP, in this outcrop T sandstone possessing a heterogeneous pore size distribution. A fast tertiary Smart Water EOR response can only be obtained if the LS brine is able to mobilize oil from pores preflooded with FW_T brine. A significant pH increase confirms that wettability alteration during LS injection could contribute to POC desorption and oil mobilization. Even though the conditions for a tertiary Smart Water EOR effect are present, secondary injection of a brine with unfavorable ion composition, here FW_T, will increase the water saturation and contaminate the porous network with unfavorable ions, which could reduce or slow down the wettability alteration process. To evaluate this effect, experiments should be performed to compare the tertiary Smart Water EOR potential with Smart Water flooding in secondary mode.

3.3.4.2 Secondary Smart Water enhanced oil recovery effect

Tertiary Smart Water injection has shown a slightly delayed EOR response, where the extra oil was produced some time after the start of Smart Water injection.

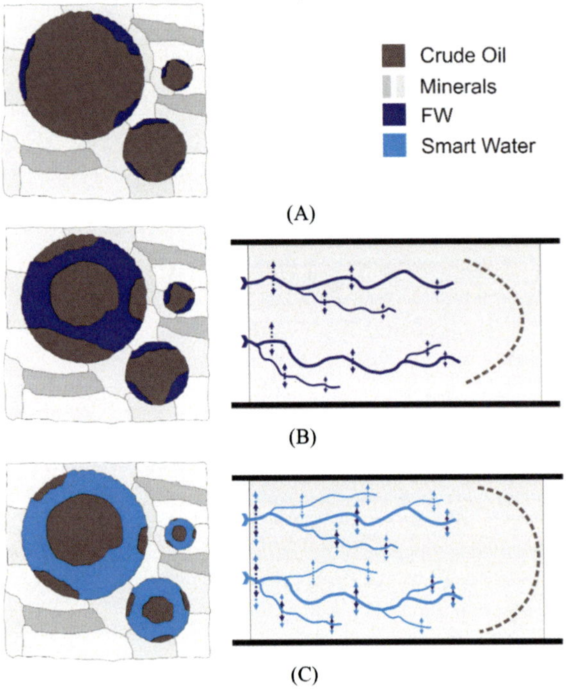

FIGURE 3.34

Fluid distributions and displacement efficiency during formation water and Smart Water injection in a heterogeneous porous network containing large, medium and small pores [33]. (A) Initial fluid distribution at fractional, slightly water-wet conditions, (B) S_{or} after formation water injection where the oil displacement is controlled by viscous and capillary forces, and (C) S_{or} after wettability alteration with Smart Water where the oil displacement is controlled by viscous and stronger capillary forces.

Reproduced with permission from Z. Aghaeifar, S. Strand, T. Puntervold, Significance of capillary forces during low-rate waterflooding. Energy & Fuels 33 (2019) 3989–3997.

Secondary injection of FW or other types of easily available injection brines, such as SW, aquifer water, PW, or surface water, increase the water saturation in the reservoir and contaminate it with ions that are unfavorable for wettability alteration with Smart Water. What would be the effect of injecting the Smart Water in secondary mode, when water saturation is low and of initial FW composition?

To compare secondary and tertiary Smart Water EOR effects, the outcrop T sandstone core T26, prepared similarly as the previous cores to establish comparable initial wettability, underwent an oil recovery test by flooding LS in secondary mode at 60°C. The results from the test are shown in Fig. 3.35, where oil recovery and PW pH are plotted against PV LS injected.

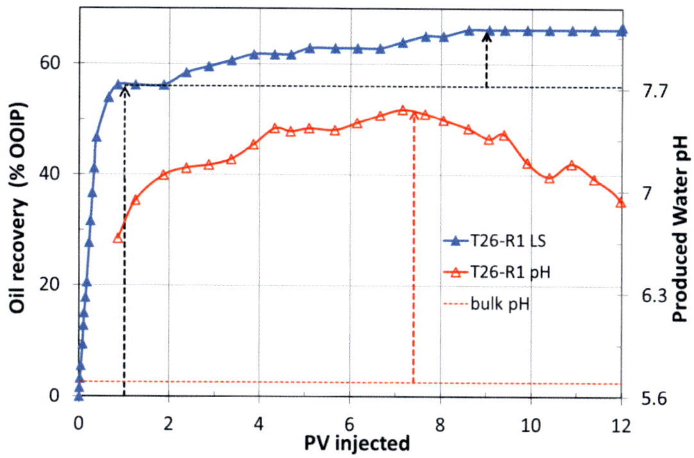

FIGURE 3.35

Oil recovery test by flooding low salinity brine in secondary mode at a rate of 4 pore volume/day at 60°C in core T26-R1. The core was restored with $S_{wi} = 0.20$ FW_T, exposed to 5 pore volumes of Oil T and aged at 60°C. Bulk pH value for low salinity brine is 5.7.

When LS injection commenced, the oil recovery increased rapidly, and the first PW with pH 6.7 was not observed until 46%OOIP had been produced. The observed slightly acidic conditions are favorable for generating mixed wettability in the core, and a steady PW pH increase to more alkaline conditions during LS injection is favorable for inducing wettability alteration. The mobilization of oil continued and reached 56%OOIP after 1 PV of LS injection, and ultimate recovery plateau of 67%OOIP after 9 PV of LS injection. Secondary and tertiary Smart Water EOR effects are compared in Fig. 3.36.

Secondary Smart Water injection is extremely efficient, and a delayed water breakthrough was experienced. After 1 PV brine injected a recovery difference of 17%OOIP was observed when injecting LS instead of FW_T in secondary mode. The oil recovery by 1 PV LS injected was even 7%OOIP higher than the tertiary Smart Water ultimate oil recovery plateau reached in T21 after 7 PV of LS brine injection. The ultimate secondary Smart Water recovery plateau of 67%OOIP is 68% higher compared to the secondary FW_T ultimate recovery plateau of 40% OOIP.

The PW pH during secondary Smart Water injection became slightly alkaline and reached a maximum value of 7.5 after 4 PV injected, which is significantly lower compared to the pH observed during tertiary LS injection. This moderate increase is explained by pH-buffering effects from the POC in the crude oil phase being more pronounced at the higher oil saturation present in a secondary flood, and due to a more efficient wettability alteration process desorbing significant amounts of POC from mineral surfaces.

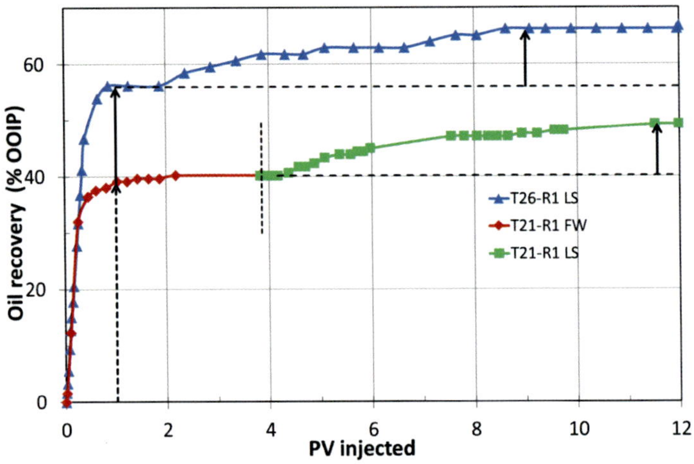

FIGURE 3.36

Comparison of Smart Water enhanced oil recovery effects in secondary and tertiary mode in equally restored outcrop T sandstone cores. At 60°C, low salinity brine was injected in secondary mode into core T26, while core T21 was exposed to low salinity brine after secondary FW_T injection.

Hamon [18] summarized results from more than 500 studies, and reported that secondary LS injection results in faster and higher oil recoveries compared to injection of HS brine. It was also reported that the incremental oil was observed early; at water breakthrough or briefly afterwards, as also observed here in Fig. 3.36. The huge difference in secondary oil recovery between FW_T and Smart Water cannot be explained by mobility ratios and viscous forces. However, wettability alteration toward more water-wet conditions have been confirmed in SI experiments, such as that observed in Fig. 3.29. FW_T is already in equilibrium with the oil and rock minerals and will not disturb the established chemical wettability equilibrium during FW_T injection. LS brine on the other hand, has a completely different ion composition and is able to disturb the established chemical COBR equilibrium. This is confirmed by the significant increase in PW pH during LS flooding to alkaline conditions, which promotes desorption of POC in crude oils from silicate mineral surfaces as described in Section 3.2.4.

The presented experimental results point in the direction of the Smart Water EOR effect observed in heterogeneous sandstones, such as the outcrop T sandstone system, being a wettability alteration process toward more water-wet conditions with a significant reduction in S_{or}. There are two main processes responsible for the extra oil produced: (1) desorption of oil components from mineral surfaces and (2) increased positive capillary forces and improved sweep efficiency in smaller and less accessible pores. Viscous forces seem not to contribute significantly to the production of the extra oil, but the flux of fluid through high permeable pores with low flow restrictions participate in the transportation of mobilized oil to the outlet of the core. In heterogeneous pore systems, more time is needed to access and alter wettability in the smaller and more challenging pores, delaying the mobilization of extra oil, which could explain the production profile in Fig. 3.36. In addition, it cannot be excluded that capillary forces may play a significant part in the oil recovery during secondary FW_T injection, in Fig. 3.28. 33%OOIP was mobilized by capillary forces in the SI with FW_T, seen in Fig. 3.29, while VF with FW_T (Fig. 3.28), reached a recovery plateau of 40% OOIP, which means that only 7%OOIP of the extra oil can be ascribed directly to the applied viscous forces.

Briefly summarized, not only is production significantly increased by LS Smart Water injection, it should also be highlighted again that a significant delay in the water breakthrough has been observed during LS Smart Water compared to that during FW_T injection. The first PW was detected after 32%OOIP produced during FW_T injection, while it was delayed to 47%OOIP produced during Smart Water injection. This delay in water breakthrough could have a tremendous effect on the production strategies for reservoirs. With increased environmental focus in the industry and by authorities, both water production and the need of wastewater treatment could be significantly reduced though optimized water injection strategies.

3.3.4.3 Seawater as Smart Water in sandstone

Because of its availability, SW is the standard injection brine for offshore reservoirs. SW is a Smart Water EOR fluid in chalk due to its favorable chemical properties [9]. With numerous reservoirs on SW flooding and new discoveries being developed and planned, the effect of SW injection on the oil recovery from mixed-wet sandstone core systems should be of great interest for oil companies. SW salinity is dependent on its location, as a salinity of 33,000 ppm is common in the North Sea area while 58,000 ppm is common in parts of the Persian/Arabic Gulf. SW ionic composition is distinctly different from FW compositions, with typical Ca^{2+}/Mg^{2+} molar ratios of approximately 1:4 in SW, but 5:1 in FW.

A restored outcrop T sandstone core T26 was flooded with SW in secondary mode to compare the oil recovery against the results with secondary FW_T and secondary LS flooding previously presented in Fig. 3.36. Oil recovery results and PW pH are presented in Fig. 3.37.

Secondary SW injection reached an ultimate oil recovery plateau of 44%OOIP and water breakthrough and the first PW were detected after 36%OOIP had been produced. Even though SW bulk pH was 7.8, the PW pH decreased by more than 1 pH unit and remained acidic during the test. Acidic conditions are unfavorable for wettability alteration and induced EOR effects according to the wettability alteration mechanism (Fig. 3.26), proposed by Austad and coworkers [7,67]. Alkaline conditions promote significant reduction in protonated POC, and several studies have shown that pH changes influence the affinity of POC to negatively charged silicate surfaces, dictating the surface wettability [65,66,68,76].

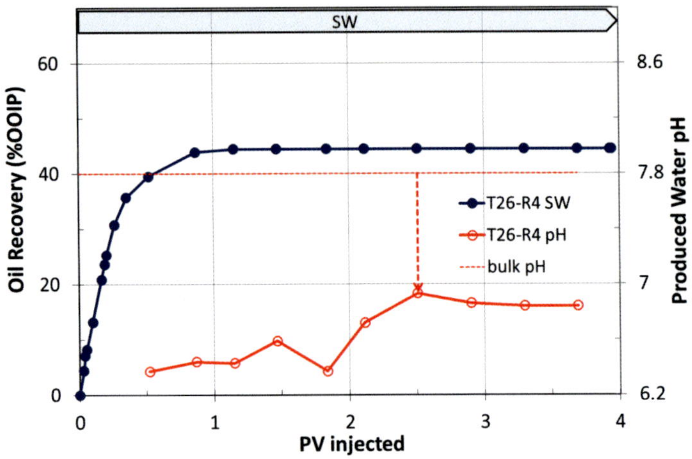

FIGURE 3.37

Oil recovery test by secondary flooding of seawater in core T26-R4 at 60°C. The core was restored with $S_{wi} = 0.20$ FW_T, exposed to 5 pore volumes of Oil T and aged at 60°C. Seawater with a bulk pH of 7.8 was injected at 4 pore volume/day and produced water pH was monitored.

Oil recovery and PW pH profiles during secondary SW, FW_T, and LS injection are compared in Fig. 3.38. The secondary SW recovery plateau is slightly higher than the FW_T recovery plateau, and both plateaus are significantly lower than that observed during secondary LS flooding (Fig. 3.38, top). It is also interesting to note that the pH of the three PWs follow the same trend (Fig. 3.38, bottom). Significantly higher PW pH was observed during LS flooding compared to that by SW and FW_T flooding.

The results show that SW injection into a restored mixed to slightly water-wet outcrop T sandstone core has both pH development and oil recovery plateau similar to FW injection. SW is thus not behaving like a Smart Water by altering wettability to increase recovery.

3.3.5 Smart Water enhanced oil recovery observations in reservoir systems

Large Smart Water EOR effects have been demonstrated in the outcrop T sandstone system, and it has been shown in laboratory core floods that SW injection is not an optimal strategy when it comes to ultimate oil recovery plateau. However, outcrops are sampled in surface quarries and have not been in contact with crude oil at high temperatures and pressures. Thus reservoir cores might not behave in the same way as outcrops in laboratory studies. It should also be mentioned that different outcrops are mineralogically very diverse and may also behave

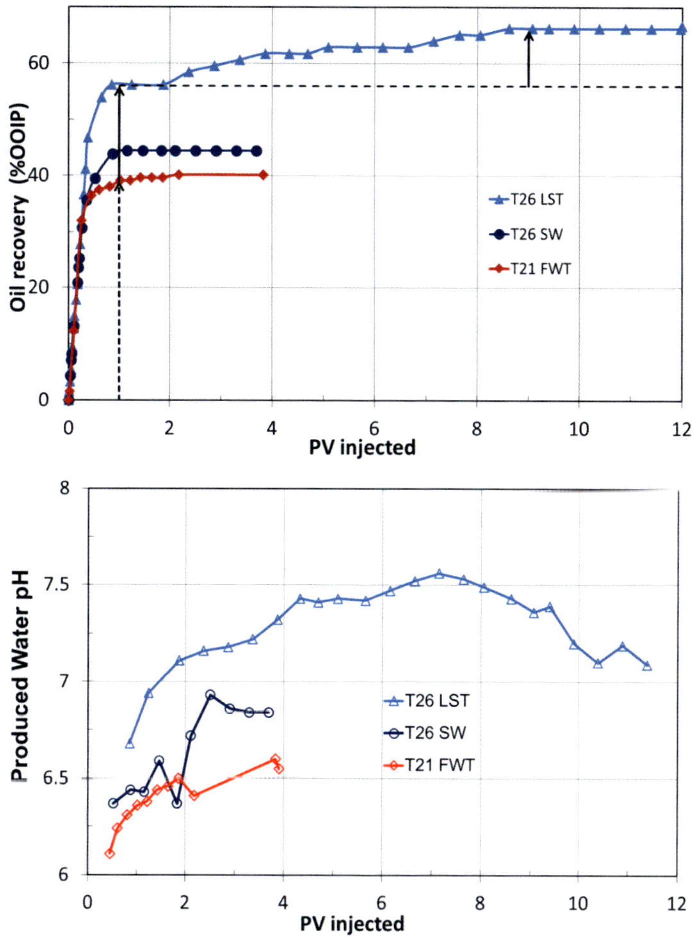

FIGURE 3.38

Oil recovery tests by secondary FW_T, seawater, and low-salinity brine injection in the cores T21 and T26. The tests were performed at 60°C at a rate of 4 pore volume/day. Oil recovery (%OOIP) (top) and produced water pH (bottom) are plotted against brine injected.

differently regarding wettability and wettability alteration in Smart Water/LS EOR studies [83].

In the next sections, Smart Water EOR observations in reservoir systems are presented and discussed. Note that contrary to outcrop cores in parametric studies, well-preserved reservoir cores are a rarity and not easily available. Thus reservoir studies are often done with limited core material available therefore reproducible multiple restorations are of high importance. Multiple restorations on the same core have in the past shown increasing EOR effects in LS EOR studies [129]. Therefore to avoid doubt about the efficiency of the Smart Water/LS injection, it is recommended to vary the injection/restoration order when working with, for example, only two reservoir cores [130].

The reservoir cores used in this chapter are sampled from a high temperature, $T_{res} > 130°C$, North Sea sandstone reservoir, named Reservoir C, having FW salinity of 63,000 ppm (FW_C) [33,131]. The reservoir rock consists predominantly of quartz, but with total clay and feldspar content in the range of 14–20 and 4–5 wt.%, respectively. Stabilized reservoir crude oil (Oil C) containing AN and BN of 0.16 and 0.76 mgKOH/g, respectively, was used in core restorations prior to the oil recovery tests.

3.3.5.1 Enhanced oil recovery effects by viscous flooding

The preserved reservoir core C5 was used in several oil recovery tests, between which the core was restored using the same optimized core cleaning and core restoration process [132]. Four distinctly different brines were injected in secondary mode; FW_C, SW, sulfate-reduced modified SW (mSW) containing very low concentration of SO_4^{2-} and reduced concentrations of Ca^{2+} and Mg^{2+}, and LS_C, which was 20 times diluted mSW. This nanofiltrated mSW is typically used in high temperature reservoirs to reduce the risk of sulfate salt scaling during water injection. The oil recovery profiles and PW pH from the four oil recovery tests are compared in Fig. 3.39.

Oil recovery with FW_C reached a production plateau of 44%OOIP after approximately 2 PV brine injected (Fig. 3.39, top). FW_C is already in equilibrium with the COBR system, and the recovery results represent the recovery baseline with no wettability alteration. After all tests were completed clearly LS_C injection was the most efficient by reaching an oil recovery plateau of 58%OOIP after 1.3 PV injected. LS_C injection also resulted in the most delayed water breakthrough. It was observed that injection of mSW and SW both resulted in ultimate oil recoveries of 39%OOIP, which is not only very much lower than the oil recovery obtained by LS_C injection, but even significantly lower compared to FW_C injection. SW was the least beneficial injection brine in terms of the time taken to reach the production plateau. During the oil recovery tests the PW pH was monitored and stable pH trends were observed after 2 PV brine injected, except for some fluctuations during SW injection, as shown in Fig. 3.39 (bottom). Out of the four brines tested, the highest PW pH level reached was that during LS_C injection, which started at pH 5.5 and increased to slightly above pH 7. The lowest pH level reached was that during SW injection, corresponding with the lowest ultimate oil recovery obtained. Again, a correlation between oil recovery plateau and PW pH was noticed, confirming the results from the outcrop T sandstone system in Fig. 3.38. Interestingly, these results showed that SW and mSW are not optimal injection brines in this reservoir system, in agreement with the observations made for the outcrop T sandstone system (Fig. 3.38). Thus contrary to carbonates,

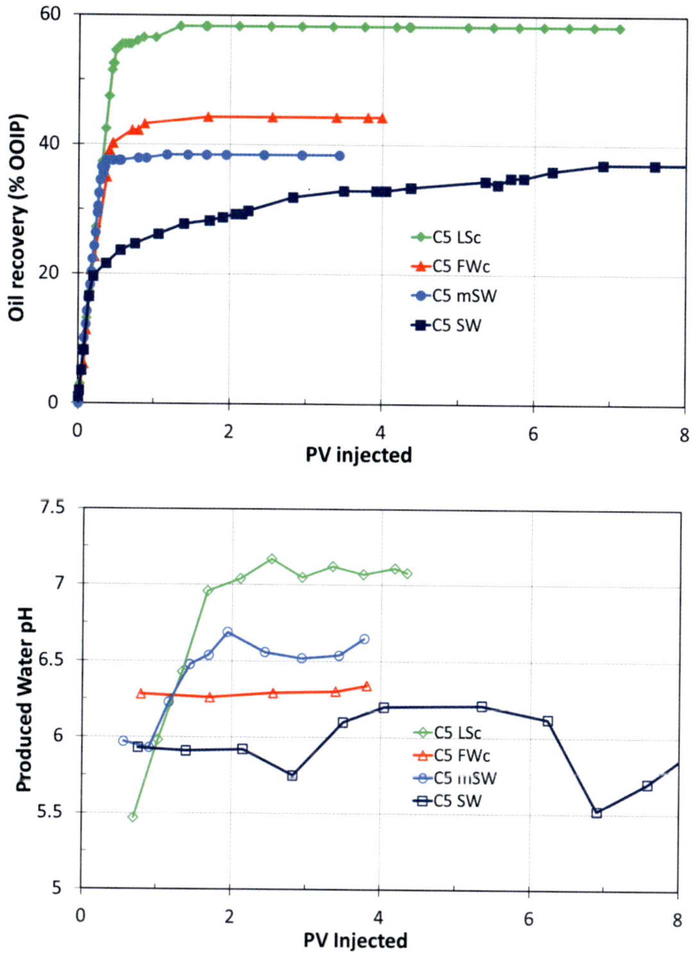

FIGURE 3.39

Oil recovery tests by secondary flooding of LS_C, modified seawater, seawater, or FW_C at rate 4 pore volume/day in core C5 at $T_{res} > 130°C$. The core was mildly cleaned before fluid restoration with $S_{wi} = 0.15$ FW_C and Oil C exposure. Oil recovery (%OOIP) (top) and produced water pH (bottom) are plotted against pore volume injected.

SW-based brines cannot be regarded as Smart Water with potential for significantly altering wettability of sandstone reservoirs and improving oil recovery.

3.3.5.2 Enhanced oil recovery effects by spontaneous imbibition

Wettability alteration involves several COBR interactions, such as cation exchange and acid–base reactions between reservoir fluids and minerals and is therefore time dependent. In core flooding experiments low flow rates might be needed to allow wettability alteration to take place to be able to observe a LS EOR effect. Low flow rates in the middle of the reservoir are also expected in the field, hence, mobilization of oil from matrix blocks is then perhaps more dependent on capillary forces than on viscous forces. Clay-coated micromodel studies have shown that wettability alteration can take time [133]. Another core flood study concluded that desorption of oil components from the rock surface was a slow process, and it was suggested to allow sufficient time for interactions between the injected LS brine and the established COBR system [134]. Results from pore-scale simulations suggest that to correctly simulate the LS EOR effect core flood experiments should be performed at lower capillary numbers, that is, at lower injection rate and consequently lower viscous-to-capillary-force ratio [135].

In the above-discussed VF experiments low flow rates have been chosen, 4 PV/day, which allows for time-dependent chemical wettability alteration reactions to take place. With core lengths of 7–8 cm, a flow rate of 4 PV/day corresponds approximately to the industry standard rate of 1 ft/day. Capillary forces are then allowed to contribute to the oil recovery processes during VF of heterogeneous reservoir cores.

The efficiency of LS brine injection has been demonstrated in several VF tests presented in the previous sections. Now the idea of capillary-driven Smart Water EOR effects will be discussed. VF and SI tests at reservoir temperature ($T_{res} > 130°C$) were performed on a sister core C3 from Reservoir C. The core was equally restored before every oil recovery experiment, following the same procedure as that used for core C5 in the previous section; mild cleaning before fluid restoration with $S_{wi} = 0.15$ FW_C and Oil C exposure. Prior to performing SI tests, the core went through two VF experiments. In the first test (C3-R2), secondary LS_C injection was performed and in the second test (C3-R3), the core was successively flooded with mSW–LS_C brines. The results from the VF experiments are summarized in Table 3.14.

The oil recovery results are in line with the VF experiments reported for the sister core C5 in Fig. 3.39. Significantly higher recovery by LS_C injection was observed, reaching a recovery plateau of 63%OOIP, while a plateau at 51%OOIP was reached by secondary mSW injection. However, by switching from mSW to LS_C flooding a significant Smart Water EOR effect of 9%OOIP was obtained.

Table 3.14 Summary of the oil recovery test results by viscous flooding in core C3 at $T_{res} > 130°C$.

Test no.	Test type	Brines	Secondary recovery plateau (%OOIP)	Tertiary oil recovery plateau (%OOIP)
C3-R2	VF	LS_C	63	–
C3-R3	VF	mSW–LS_C	51	60

LS, Low salinity; mSW, modified seawater; VF, viscous flooding.

To verify the potential of FW_C, mSW, and LS_C to alter wettability and generate positive capillary forces, secondary and tertiary SI tests were performed on the restored core C3. The oil recovery results are presented in Fig. 3.40. The baseline recovery by SI of FW_C was 42%OOIP, which was reached after 5 days of imbibition. Since water did imbibe to such an extent, the core was regarded as being quite water wet, possessing significant positive capillary forces. Wettability alteration was confirmed when changing imbibition fluid to LS_C and observing that 6%OOIP extra oil was gradually produced over 5 days of imbibition. In the next test, the core C3-R5 was imbibed with LS_C in secondary mode, and in agreement with the results discussed above, oil recovery and capillary forces increased when imbibing LS_C compared to FW_C in secondary mode. Both imbibition rate and oil mobilization increased significantly due to wettability alteration toward more water-wet conditions, reaching an ultimate recovery plateau of 67%OOIP after 6 days of imbibition. Even though the core behaved quite water wet during FW_C imbibition by mobilizing and producing 42%OOIP (baseline recovery), LS_C imbibition produced an additional 60% extra oil. By comparing recoveries in the same time frame, the LS_C imbibition rate was higher, which is very beneficial for optimized recovery processes.

Tertiary LS_C imbibition efficiency after secondary mSW imbibition was tested after the sixth core restoration (C3-R6). The results from this test are also displayed in Fig. 3.40 and show that the ultimate oil recovery by mSW imbibition was 38%OOIP, comparable with FW_C, but produced at a lower rate of imbibition. Again, it was confirmed that SW—in this case modified (mSW)—was not able to induce positive capillary forces by wettability alteration and is therefore not a Smart Water in sandstone systems. Yet, by switching to LS_C, a substantial amount

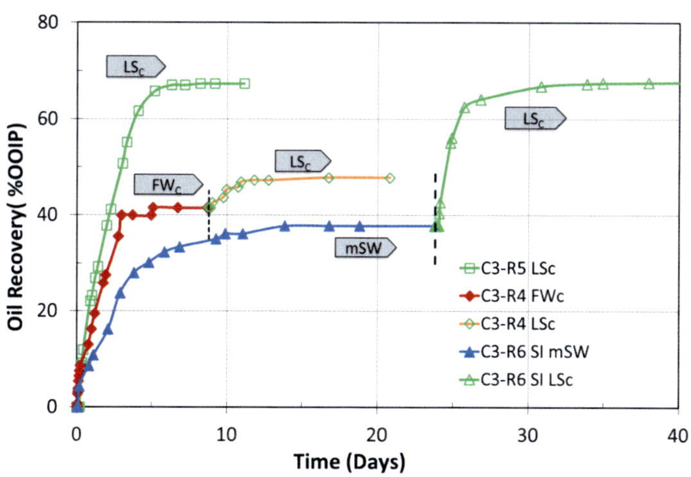

FIGURE 3.40

Oil recovery tests by spontaneous imbibition on restored core C3 at $T_{res} > 130°C$. Imbibing fluids were: C3-R4: FW_C–LS_C, C3-R5: LS_C, and C3-R6: modified seawater–LS_C.

Table 3.15 Summary of the oil recovery test results by spontaneous imbibition in core C3 at $T_{res} > 130°C$.

Test no.	Test type	Brines	Secondary recovery plateau (%OOIP)	Tertiary LS oil recovery plateau (%OOIP)
C3-R4	SI	FW_C–LS_C	42	48
C3-R5		LS_C	67	–
C3-R6		mSW–LS_C	38	68

LS, Low salinity; mSW, modified seawater; SI, spontaneous imbibition.

of extra oil was recovered after 6 more days of imbibition reaching an ultimate oil recovery plateau of 68%OOIP. The results from all three SI tests performed on core C3 are summarized in Table 3.15.

If the ultimate recoveries obtained by SI and VF of LS_C brine in secondary mode are compared, the results show that higher oil recovery was in fact observed by SI over VF. 67%OOIP was produced by SI, while only 63%OOIP was produced by VF. In a series of waterflooding experiments in chalk, Johannesen and Graue [34] showed that SI and VF recovery curves reached almost the same production plateau (similar S_{or}) at the lowest flooding rate. Thus results from both sandstones and carbonates indicate that in the main part of the reservoir, where the pressure drop is the lowest, and in heterogeneous porous networks, SI is the main driving force in Smart Water flooding due to positive capillary forces. Performing brine injection experiments at low rates is therefore essential for observing capillary force effects in the laboratory.

The experimental work presented has shown that LS_C promoted the most water-wet system and behaved as Smart Water resulting in high sweep efficiency and delayed water breakthrough in the VF tests. In addition, results from SI tests showed that the highest oil recovery was achieved in the most water-wet system. These results suggest that the results in Fig. 3.3 and statement by Jadhunandan and Morrow [30], that the highest oil recovery will be achieved at neutral to slightly water-wet conditions, is not universal for all sandstone COBR systems.

3.4 Summary

Reservoir wettability is established over geological time because of chemical interactions between FW, pore surface minerals, and surface-active POC present in the hydrocarbon phase. Mobilization and production of crude oil from the reservoir are controlled by gravity, viscous, and capillary forces. SI experiments are used to determine wettability, the degree of water wetness and presence of capillary forces in the COBR system. Strongly water-wet conditions in cores are confirmed by a high rate of water imbibition, expelling a large amount of the oil initially present in the core within a short time period. These experiments confirm the importance of positive capillary forces in mobilization of oil from heterogenous pore systems. Significant secondary and tertiary Smart Water EOR effects have been demonstrated both in an outcrop sandstone COBR system and in a reservoir sandstone COBR system. The extra oil recovered by LS Smart Water was ascribed to wettability alteration toward more water-wet conditions. Experimental results question favorable mobility ratios, fines migration, diverted flow, and EDL expansion as likely explanations for the EOR effect observed. During tertiary Smart Water injection, a significant increase in PW pH from slightly acidic to alkaline conditions was observed. An increase in pH at pore surfaces transforms protonated POC to uncharged basic and negatively charged acidic components, thus promoting wettability alteration. It is important to highlight that the wettability alteration is not controlled by change in the surface charge of the silicate minerals, which is negative in the pH range 5–9 observed in the experiments. It has been seen that the tertiary Smart Water EOR effect was slightly delayed, and that LS slug injection seemed to be less efficient than continuous tertiary LS injection in displacing the oil from an outcrop core. However, pronounced, accelerated and higher oil production and delayed water breakthrough were demonstrated by secondary Smart Water injection, emphasizing the importance of a timely injection strategy. The high recovery efficiency observed in secondary LS injection is a combined effect of the wettability alteration shock toward a more water-wet state and the high oil connectivity through the porous media at low initial water saturation.

It should be highlighted that SW-based injection brines gave significantly lower oil recovery compared to LS injection in both the outcrop T sandstone system at 60 °C and in the reservoir C system at high temperature >130°C. The reason is that the ion compositions in these SW-based brines are unfavorable for inducing wettability alteration. PW pH confirmed significantly higher pH during LS injection compared to that during FW or SW injection.

It is interesting and important to note that Smart Water/LS injection in secondary mode, in addition to the extra oil produced, significantly delays water breakthrough reaching oil production plateau at earlier time and thus the total amount of PW should dramatically decrease. This should reduce production costs linked to the lower volumes of injection water needed, reduce costs for PW handling and treatment before disposal, and could also reduce the need for production well maintenance and scale treatment when the Smart Water is a LS brine. In consequence, due to the more energy- and cost-efficient drainage of the reservoir by Smart Water/LS water injection, the oil could be recovered more environmentally friendly at lower emissions.

Nomenclature

AEC anion-exchange capacity
AN acid number, mg KOH/g oil
BET Brunauer–Emmett–Teller, Specific surface area, m^2/g rock
BN base number, mg KOH/g oil
CEC cation-exchange capacity

COBR	crude oil–brine–rock
EDL	electrical double layer
EOR	enhanced oil recovery
ESEM	environmental scanning electron microscope
FW	formation water
HS	high salinity
$\mathbf{I_{AH}}$	Amott–Harvey wettability index
IEA	International Energy Agency
IFT	interfacial tension
$\mathbf{I_o}$	Amott oil index
IOR	improved oil recovery
IpHISE	pH–ion–surface electrostatic model
$\mathbf{I_w}$	Amott water index
k	permeability
$\mathbf{k_r}$	relative permeability
LS	low salinity
LSE	low salinity effect
M	mobility ratio
MIE	multicomponent ion exchange
MICP	mercury injection capillary pressure
mSW	sulfate-reduced SW (modified SW)
NCS	Norwegian continental shelf
NSO	nitrogen–sulfur–oxygen
OOIP	original oil in place
P_c	capillary pressure
pK_a	negative log of the dissociation constant
POC	polar organic components
PV	pore volume
PW	produced water
PZC	point of zero charge
r	radius
R	restoration
SCAL	special core analysis
SEM	scanning electron microscope
SI	spontaneous imbibition
$\mathbf{S_{or}}$	residual oil saturation
SW	seawater
$\mathbf{S_{wi}}$	initial water saturation
$\mathbf{T_{res}}$	reservoir temperature
TDS	total dissolved solids
VF	viscous flooding
XRD	X-ray diffraction
ΔP	pressure drop, differential pressure over the core
φ	porosity
θ	contact angle
λ	mobility
μ	viscosity
σ	interfacial tension

References

[1] International Energy Agency (IEA), World Energy Outlook 2020, Paris, 2020.

[2] V. Alvarado, E. Manrique, Enhanced oil recovery: an update review, Energies 3 (2010) 1529–1575.

[3] International Energy Agency (IEA), Net Zero by 2050—a roadmap for the global energy sector, Paris, 2021.

[4] D.W. Green, G.P. Willhite, Enhanced Oil Recovery, second ed., Society of Petroleum Engineers, Richardson, 2018.

[5] L.W. Lake, Enhanced Oil Recovery, Prentice Hall Inc, Hoboken, 1989.

[6] A. Muggeridge, A. Cockin, K. Webb, et al., Recovery rates, enhanced oil recovery and technological limits, Philosophical Transactions of the Royal Society A: Mathematical, Physical and Engineering Sciences 2014 (2006) 372.

[7] T. Austad, A. RezaeiDoust, T. Puntervold, Chemical mechanism of low salinity water flooding in sandstone reservoirs, in: SPE-129767-MS, SPE Improved Oil Recovery Symposium, April 24–28, 2010, Tulsa, Oklahoma, USA.

[8] N.R. Morrow, G.-Q. Tang, M. Valat, et al., Prospects of improved oil recovery related to wettability and brine composition, Journal of Petroleum Science and Engineering 20 (1998) 267–276.

[9] T. Austad, S. Strand, M.V. Madland, et al., Seawater in chalk: an EOR and compaction fluid, SPE Reservoir Evaluation & Engineering 11 (2008) 648–654.

[10] S.J. Fathi, T. Austad, S. Strand, Water-based enhanced oil recovery (EOR) by "Smart Water": optimal ionic composition for EOR in carbonates, Energy & Fuels 25 (2011) 5173–5179.

[11] N.R. Morrow, J. Buckley, Improved oil recovery by low-salinity waterflooding, Journal of Petroleum Technology 63 (2011) 106–112.

[12] I.D. Piñerez Torrijos, T. Puntervold, S. Strand, et al., Optimizing the low salinity water for EOR effects in sandstone reservoirs—composition vs salinity, in: Proceedings of the 78th EAGE Conference & Exhibition, May 30–June 2, 2016, Vienna, Austria.

[13] S.J. Fathi, T. Austad, S. Strand, "Smart Water" as wettability modifier in chalk: the effect of salinity and ionic composition, Energy & Fuels 24 (2010) 2514–2519.

[14] G.Q. Tang, N.R. Morrow, Salinity, temperature, oil composition, and oil recovery by waterflooding, SPE Reservoir Engineering 12 (1997) 269–276.

[15] G.Q Tang, N.R. Morrow, Oil recovery by waterflooding and imbibition—invading brine cation valency and salinity, in: SCA9911, International Symposium of the Society of Core Analysts, August 1–4, 1999, Golden, CO, USA.

[16] I.D. Piñerez Torrijos, Enhanced oil recovery from sandstones and carbonates with "Smart Water", Ph.D. thesis, University of Stavanger, 2017.

[17] E.W. Al-Shalabi, K. Sepehrnoori, A comprehensive review of low salinity/engineered water injections and their applications in sandstone and carbonate rocks, Journal of Petroleum Science and Engineering 139 (2016) 137–161.

[18] G. Hamon, Low-salinity waterflooding: facts, inconsistencies and the way forward, Petrophysics 57 (2016) 41–50.

[19] J.J. Sheng, Critical review of low-salinity waterflooding, Journal of Petroleum Science and Engineering 120 (2014) 216–224.

[20] W.B. Bartels, H. Mahani, S. Berg, et al., Literature review of low salinity waterflooding from a length and time scale perspective, Fuel 236 (2019) 338–353.

[21] A.N. Awolayo, H.K. Sarma, L.X. Nghiem, Brine-dependent recovery processes in carbonate and sandstone petroleum reservoirs: review of laboratory-field studies, interfacial mechanisms and modeling attempts, Energies 11 (2018) 3020.

[22] E.W. Al-Shalabi, K. Sepehrnoori, Low Salinity and Engineered Water Injection for Sandstone and Carbonate Reservoirs, Elsevier Science, 2017.

[23] K.S. Lee, J.H. Lee, Hybrid Enhanced Oil Recovery Using Smart Waterflooding, Elsevier Science, 2019.

[24] P.C. Smalley, A.H. Muggeridge, M. Dalland, et al., Screening for EOR and estimating potential incremental oil recovery on the Norwegian continental shelf, in: SPE-190230-MS, SPE Improved Oil Recovery Conference, April 14–18, 2018, Tulsa, Oklahoma, USA.

[25] P.C. Smalley, A.H. Muggeridge, S.S. Amundrud, et al., EOR screening including technical, operational, environmental and economic factors reveals practical EOR Potential offshore on the Norwegian continental shelf, in: SPE-200376-MS, SPE Improved Oil Recovery Conference, August 31–September 4, 2020, Tulsa, Oklahoma, USA.

[26] L.E. Treiber, D.L. Archer, W.W. Owens, A laboratory evaluation of the wettability of fifty oil producing reservoirs, SPE Journal 12 (1972) 531–540.

[27] G.V. Chilingar, T.F. Yen, Some notes on wettability and relative permeabilities of carbonate rocks, II, Energy Sources 7 (1983) 67–75.

[28] R.A. Salathiel, Oil recovery by surface film drainage in mixed-wettability rocks, Journal of Petroleum Technology 25 (1973) 1216–1224.

[29] R. Nolen-Hoeksema, Wettability, Oilfield Review, The Defining Series, Schlumberger, Houston, 2016.

[30] P.P. Jadhunandan, N.R. Morrow, Effect of wettability on waterflood recovery for crude-oil/brine/rock systems, SPE Reservoir Engineering (1995) 40–46.

[31] G.Q. Tang, N.R. Morrow, Influence of brine composition and fines migration on crude oil/brine/rock interactions and oil recovery, Journal of Petroleum Science and Engineering 24 (1999) 99–111.

[32] J.S. Buckley, C. Bousseau, Y. Liu, Wetting alteration by brine and crude oil: from contact angles to cores, SPE Journal 1 (1996) 341–350.

[33] Z. Aghaeifar, S. Strand, T. Puntervold, Significance of capillary forces during low-rate waterflooding, Energy & Fuels 33 (2019) 3989–3997.

[34] E.B. Johannesen, A. Graue, Systematic investigation of waterflood reducing residual oil saturations by increasing differential pressures at various wettabilities, in: SPE-108593-MS, SPE Offshore Europe Oil and Gas Conference and Exhibition, September 4–7, 2007, Aberdeen, Scotland, UK.

[35] A. Mamonov, EOR by Smart Water flooding in sandstone reservoirs—effect of mineralogy on rock wetting and wettability alteration, Ph.D. thesis, University of Stavanger: Stavanger, 2019.

[36] A.J. Reinholdtsen, A. RezaeiDoust, S. Strand, et al., Why such a small low salinity EOR - potential from the Snorre formation?, in: Proceedings of the 16th European Symposium on Improved Oil Recovery, April 12–14, 2011, Cambridge, UK.

[37] K. Bjørlykke, T. Nedkvitne, M. Ramm, et al., Diagenetic processes in the Brent Group (Middle Jurassic) reservoirs of the North Sea: an overview, Geology of the Brent Group 61 (1992) 263–287.

[38] B. Allard, M. Karlsson, E.-L. Tullborg, et al., Ion exchange capacities and surface areas of some major components and common fracture filling materials of igneous rocks, KBS Technical Reports, Göteborg, 1983.

[39] C. Klein, C.S. Hurlbut, J.D. Dana, Manual of Mineralogy (After James D. Dana), sixth ed., Wiley, New York, 1999.
[40] K. Bjørlykke, Sedimentology and Petroleum Geology, Springer, Berlin, 1989.
[41] N.N. Greenwood, A. Earnshaw, Chemistry of the Elements, Butterworth-Heinemann, Oxford, 1997.
[42] L. Van Der Plas, Volume 6 The Identification of Detrital Feldspars. Developments in Sedimentology, first ed., Elsevier Science, 1966.
[43] F.J. Pettijohn, Sedimentary Rocks, third ed., Harper & Row, New York, 1975.
[44] E.H. Oelkers, S.V. Golubev, C. Chairat, et al., The surface chemistry of multi-oxide silicates, Geochimica et Cosmochimica Acta 73 (2009) 4617–4634.
[45] W. Stumm, J.J. Morgan, Aquatic Chemistry—Chemical Equilibria and Rates in Natural Waters, third ed., John Wiley & Sons, New York, 1996.
[46] L. Chou, R. Wollast, Steady-state kinetics and dissolution mechanisms of albite, American Journal of Science 285 (1985) 963–993.
[47] A. Mamonov, T. Puntervold, S. Strand, et al., Contribution of feldspar minerals to pH during smart water EOR processes in sandstones, Energy & Fuels 34 (2020) 55–64.
[48] I. Parsons, Feldspars and Their Reactions (NATO ASI Series), Vol. 421, Springer, Dordecht, 1994.
[49] J. Grotzinger, T.H. Jordan, Understanding Earth, seventh ed., W.H.Freeman & Co Ltd., New York, 2014.
[50] R.H. Worden, S. Morad, Clay minerals in sandstones: controls on formation, distribution and evolution, in: R.H. Worden, S. Mora (Eds.), Clay Mineral Cements in Sandstones, Blackwell Publishing, Hoboken, 2003, pp. 3–41.
[51] D.R. Prothero, Sedimentary Geology: An Introduction to Sedimentary Rocks and Stratigraphy, second ed., Freeman, New York, 2003.
[52] D. Carroll, Ion exchange in clays and other minerals, Bulletin of the Geological Society of America 70 (1959) 749–779.
[53] International Drilling Fluids (IDF), Clay Chemistry—Technical Manual, International Drilling Fluids Limited, Aberdeen, 1982.
[54] B. Velde, Origin and Mineralogy of Clays: Clays and the Environment, Springer, Berlin, 1995.
[55] S. Strand, T. Puntervold, T. Austad, Water based EOR from clastic oil reservoirs by wettability alteration: a review of chemical aspects, Journal of Petroleum Science and Engineering 146 (2016) 1079–1091.
[56] I.D. Piñerez Torrijos, M. Risanger, T. Puntervold, et al., Impact of anhydrite on the low salinity EOR effect in sandstone material with high clay content, in: Proceedings of the 19th European Symposium on Improved Oil Recovery, April 24–27, 2017, Stavanger, Norway.
[57] J.L. Lopez-Salinas, G.J. Hirasaki, C.A. Miller, Determination of anhydrite in reservoirs for EOR, in: SPE-141420-MS, SPE International Symposium on Oilfield Chemistry, April 11–13, 2011, The Woodlands, TX, USA.
[58] H. Aksulu, D. Håmsø, S. Strand, et al., Evaluation of low salinity EOR-effects in sandstone: effects of temperature and pH gradient, Energy & Fuels 26 (2012) 3497–3503.
[59] W.G. Anderson, Wettability literature survey—part 1: rock/oil/brine interactions and the effects of core handling on wettability, Journal of Petroleum Technology 38 (1986) 1125–1144.
[60] J.S. Buckley, Chemistry of the crude oil/brine interface, in: Proceedings of the 3rd International Symposium on Evaluation of Reservoir Wettability and Its Effect on Oil Recovery, September 21–23, 1994. Laramie, WY, USA.

[61] J.S. Buckley, T. Fan, Crude oil/brine interfacial tensions, Petrophysics 48 (2007) 175–185.
[62] J.S. Buckley, Y. Liu, Some mechanisms of crude oil/brine/solid interactions, Journal of Petroleum Science and Engineering 20 (1998) 155–160.
[63] J.S. Buckley, Y. Liu, S. Monsterleet, Mechanisms of wetting alteration by crude oils, SPE Journal 3 (1998) 54–61.
[64] A.M. Shehata, H.A. Nasr-El-Din, Laboratory investigations to determine the effect of connate-water composition on low-salinity waterflooding in sandstone reservoirs, SPE Reservoir Evaluation & Engineering 20 (2017) 59–76.
[65] L. Madsen, I. Lind, Adsorption of carboxylic acids on reservoir minerals from organic and aqueous phase, SPE Reservoir Evaluation & Engineering 1 (1998) 47–51.
[66] W.D. Burgos, N. Pisutpaisal, M.C. Mazzarese, et al., Adsorption of quinoline to kaolinite and montmorillonite, Environmental Engineering Science 19 (2002) 59–68.
[67] A. RezaeiDoust, T. Puntervold, T. Austad, Chemical verification of the EOR mechanism by using low saline/smart water in sandstone, Energy & Fuels 25 (2011) 2151–2162.
[68] A. Fogden, Removal of crude oil from kaolinite by water flushing at varying salinity and pH, Colloids and Surfaces A: Physicochem. Eng. Aspects 402 (2012) 13–23.
[69] J.S. Buckley, Asphaltene precipitation and crude oil wetting—crude oils can alter wettability with or without precipitation of asphaltenes, SPE Advanced Technology Series 3 (1995) 53–59.
[70] X. Zhou, N.R. Morrow, S. Ma, Interrelationship of wettability, initial water saturation, aging time, and oil recovery by spontaneous imbibition and waterflooding, SPE Journal 5 (2000) 199–207.
[71] R.K. Iler, The Chemistry of Silica: Solubility, Polymerization, Colloid and Surface Properties, and Biochemistry, Wiley, New York, 1979.
[72] L.W. Zelazny, L. He, A. Vanwormhoudt, Charge analysis of soils and anion exchange, in: D.L. Sparks, A.L. Page, P.A. Helmke, et al. (Eds.), Methods of Soil Analysis: Part 3. Chemical Methods, American Society of Agronomy, Madison, 1996, pp. 1231–1253.
[73] F.T. Bingham, J.R. Sims, A.L. Page, Retention of acetate by montmorillonite, Soil Science Society of America Journal 29 (1965) 670–672.
[74] N.C. Brady, R.R. Weil, Nature and Properties of Soils, fourteenth ed., Pearson, Upper Saddle River, 2008.
[75] A. Mamonov, M.A.I. Khan, T. Puntervold, et al., Optimized alkalinity for EOR purposes in sandstone reservoirs, in: SPE-192015-MS, SPE Asia Pacific Oil and Gas Conference and Exhibition, October 23–25, 2018, Brisbane, Australia.
[76] J.S. Buckley, N.R. Morrow, Characterization of crude oil wetting behavior by adhesion tests, in: SPE-20263-MS, SPE/DOE Enhanced Oil Recovery Symposium, April 22–25, 1990, Tulsa, Oklahoma, USA.
[77] A. Mamonov, O.A. Kvandal, S. Strand, et al., Adsorption of polar organic components onto sandstone rock minerals and its effect on wettability and enhanced oil recovery potential by Smart Water, Energy & Fuels 33 (2019) 5954–5960.
[78] M.O. Denekas, C.C. Mattax, G.T. Davis, Effects of crude oil components on rock wettability, Petroleum Transactions AIME 216 (1959) 330–333.
[79] T. Puntervold, A. Mamonov, Z. Aghaeifar, et al., Role of kaolinite clay minerals in enhanced oil recovery by low salinity water injection, Energy & Fuels 32 (2018) 7374–7382.

[80] M. Didier, A. Chaumont, J. Thibaut, et al., Contradictory trends for smart water injection method: role of pH and salinity from sand/oil/brine adhesion maps, in: SCA2015-005, International Symposium of the Society of Core Analysts, August 16–21, 2015, St. John's Newfoundland and Labrador, Canada.

[81] P.V. Brady, N.R. Morrow, A. Fogden, et al., Electrostatics and the low salinity effect in sandstone reservoirs, Energy & Fuels 29 (2015) 666–677.

[82] L. Shi, M.H.M. Olsson, T. Hassenkam, et al., A pH resolved view of the low salinity effect in sandstone reservoirs, Energy & Fuels 30 (2016) 5346–5354.

[83] I.D. Piñerez Torrijos, T. Austad, S. Strand, et al., Linking low salinity EOR effects in sandstone to pH, mineral properties and water composition, in: SPE-179625-MS, SPE Improved Oil Recovery Conference, April 11–13, 2016, Tulsa, OK, USA.

[84] J. Burgess, Metal Ions, In Solution, John Wiley & Sons, New York, 1978.

[85] Z. Aghaeifar, S. Strand, T. Austad, et al., The influence of formation water salinity/composition on the low salinity EOR effect in high temperature sandstone reservoirs, Energy & Fuels 29 (2015) 4747–4754.

[86] T. Puntervold, A. Mamonov, I.D. Piñerez Torrijos, et al., Adsorption of crude oil components onto carbonate and sandstone outcrop rocks and its effect on wettability, Energy & Fuels 35 (2021) 5738–5747.

[87] A. Lager, K.J. Webb, C.J.J. Black, Impact of brine chemistry on oil recovery, in: Proceedings of the 14th European Symposium on Improved Oil Recovery, April 22–24, 2007, Cairo, Egypt.

[88] J. Chorover, M.K. Amistadi, W.D. Burgos, et al., Quinoline sorption on kaolinite-humic acid complexes, Soil Science Society of America Journal 63 (1999) 850–857.

[89] P. Hopkins, S. Strand, T. Puntervold, et al., The adsorption of polar components onto carbonate surfaces and the effect on wetting, Journal of Petroleum Science and Engineering 147 (2016) 381–387.

[90] P. Hopkins, K. Walrond, S. Strand, et al., Adsorption of acidic crude oil components onto outcrop chalk at different wetting conditions during both dynamic adsorption and aging processes, Energy & Fuels 30 (2016) 7229–7235.

[91] J.E.S. Mjøs, S. Strand, T. Puntervold, et al., Effect of initial wetting on Smart Water potential in carbonates, in: SPE-190414-MS, SPE EOR Conference at Oil and Gas West Asia, March 26–28, 2018, Muscat, Oman.

[92] C. McPhee, J. Reed, I. Zubizarreta, (Vol. 64) Core Analysis—A Best Practice Guide. Developments in Petroleum Science, Elsevier, 2015.

[93] S.F. Shariatpanahi, S. Strand, T. Austad, et al., Wettability restoration of limestone cores using core material from the aqueous zone, Petroleum Science and Technology 30 (2012) 1082–1090.

[94] P.A. Hopkins, I. Omland, F. Layti, et al., Crude oil quantity and its effect on chalk surface wetting, Energy & Fuels 31 (2017) 4663–4669.

[95] I. Piñerez, T. Puntervold, S. Strand, et al., Core wettability reproduction: a new solvent cleaning and core restoration strategy for chalk cores, Journal of Petroleum Science and Engineering 195 (2020) 107654.

[96] P. Hopkins, T. Puntervold, S. Strand, Preserving initial core wettability during core restoration of carbonate cores, in: SCA2015-017, International Symposium of the Society of Core Analysts, August 16–21, 2015, St. John's Newfoundland and Labrador, Canada.

[97] P. Aslanidis, S. Strand, I.D. Piñerez Torrijos, et al., Reproducing wettability in sandstone reservoir core material in laboratory core restorations, Journal of Petroleum Science and Engineering 208 (2022) 109531. Available from: http://doi.org/10.1016/j.petrol.2021.109531.

[98] N.R. Morrow, Wettability and its effect on oil recovery, Journal of Petroleum Technology 42 (1990) 1476–1484.

[99] W.G. Anderson, Wettability literature survey—part 2: wettability measurement, Journal of Petroleum Technology 38 (1986) 1246–1262.

[100] S.M. Ma, N.R. Morrow, X. Zhang, et al., Characterization of wettability from spontaneous imbibition measurements, Journal of Canadian Petroleum Technology 38 (1999) 1–8.

[101] G.G. Bernard, Effect of floodwater salinity on recovery of oil from cores containing clays, in: SPE-1725-MS, 38th Annual California Regional Meeting of the Society of Petroleum Engineers of AIME, October 26–27, 1967, Los Angeles, California, USA.

[102] G. Thyne, Wettability alteration in reservoirs: how it applies to Alaskan oil production, in: SPE-180370-MS, SPE Western Regional Meeting, May 23–26, 2016, Anchorage, Alaska, USA.

[103] V. Alvarado, G. Garcia-Olvera, P. Hoyer, et al. Impact of polar components on crude oil-water interfacial film formation: a mechanism for low-salinity waterflooding, in: SPE-170807-MS, SPE Annual Technical Conference and Exhibition, October 27–29, 2014, Amsterdam, The Netherlands.

[104] K. Sandengen, M.T. Tweheyo, M. Raphaug, et al., Experimental evidence of low salinity water flooding yielding a more oil-wet behaviour, in: SCA2011-16, International Symposium of the Society of Core Analysts, September 18–21, 2011, Austin, TX, USA.

[105] P. Mahzari, M. Sohrabi, Crude oil/brine interactions and spontaneous formation of micro-dispersions in low salinity water injection, in: SPE-169081-MS, SPE Improved Oil Recovery Symposium, April 12–16, 2014, Tulsa, OK, USA.

[106] C.S. Brooks, W.R. Purcell, Surface area measurements on sedimentary rocks, in: SPE-222-G, Fall Meeting of the Petroleum Branch of AIME, October 1–3, 1952, Houston, TX, USA.

[107] P.L. McGuire, J.R. Chatham, F.K. Paskvan, et al., Low salinity oil recovery: an exciting new EOR opportunity for Alaska's North Slope, in: SPE-93903-MS, SPE Western Regional Meeting, March 30–April 1, 2005, Irvine, CA, USA.

[108] A. Lager, K.J. Webb, C.J.J. Black, et al., Low salinity oil recovery—an experimental investigation, Petrophysics 49 (2008) 28–35.

[109] G. Sposito, The Chemistry of Soils, Oxford University Press, New York, 2008.

[110] D.J. Ligthelm, J. Gronsveld, J.P. Hofman, et al., Novel waterflooding strategy by manipulation of injection brine composition, in: SPE-119835-MS, SPE EUROPEC/EAGE Annual conference and exhibition, June 8–11, 2009: Amsterdam, The Netherlands.

[111] K. Holmberg, (Vol. 1-2) Handbook of Applied Surface and Colloid Chemistry, John Wiley & Sons, Hoboken, 2002.

[112] R.A. Nasralla, H.A. Nasr-El-Din, Double-layer expansion: is it a primary mechanism of improved oil recovery by low-salinity waterflooding? SPE Reservoir Evaluation & Engineering 17 (2014) 49–59.

[113] Q. Xie, Y. Liu, J. Wu, et al., Ions tuning water flooding experiments and interpretation by thermodynamics of wettability, Journal of Petroleum Science and Engineering 124 (2014) 350–358.

[114] R.N. Vaidya, H.S. Fogler, Fines migration and formation damage: influence of pH and ion exchange, SPE Production Engineering 7 (1992) 325–330.

[115] I.D. Piñerez Torrijos, T. Puntervold, S. Strand, et al., Experimental study of the response time of the low-salinity enhanced oil recovery effect during secondary and tertiary low-salinity waterflooding, Energy & Fuels 30 (2016) 4733–4739.

[116] G. Tang, N.R. Morrow, Injection Of Dilute Brine And Crude Oil/Brine/Rock Interactions, American Geophysical Union Geophysical Monograph Series, Washington DC, 2002.

[117] P. Bedrikovetsky, F.D. Siqueira, C.A. Furtado, et al., Modified particle detachment model for colloidal transport in porous media, Transport in Porous Media 86 (2011) 353–383.

[118] S. Boussour, M. Cissokho, P. Cordier, et al., Oil recovery by low salinity brine injection: laboratory results on outcrop and reservoir cores, in: SPE-124277-MS, SPE Annual Technical Conference and Exhibition, October 4–7, 2009, New Orleans, Louisiana, USA.

[119] I.D. Piñerez Torrijos, T. Puntervold, S. Strand, et al., Impact of temperature on the low salinity EOR effect for sandstone cores containing reactive plagioclase, Journal of Petroleum Science and Engineering 156 (2017) 102–109.

[120] A. Fogden, M. Kumar, N.R. Morrow, et al., Mobilization of fine particles during flooding of sandstones and possible relations to enhanced oil recovery, Energy Fuels 25 (2011) 1605–1616.

[121] R. Ehrlich, R.J. Wygal Jr., Interrelation of crude oil and rock properties with the recovery of oil by caustic waterflooding, Society of Petroleum Engineers Journal 17 (1977) 263–270.

[122] A. Lager, K.J. Webb, I.R. Collins, et al., LoSalTM enhanced oil recovery: evidence of enhanced oil recovery at the reservoir scale, in: SPE-113976-MS, SPE/DOE Improved Oil Recovery Symposium. April 20–23, 2008, Tulsa, OK, USA.

[123] A.M. Shehata, H.A. Nasr-El-Din, Zeta potential measurements: Impact of salinity on sandstone minerals, in: SPE-173763-MS, SPE International Symposium on Oilfield Chemistry, April 13–15, 2015, The Woodlands, TX, USA.

[124] A. RezaeiDoust, Low salinity water flooding in sandstone reservoirs—a chemical wettability alteration mechanism (Ph.D. thesis), University of Stavanger, 2011.

[125] W.-B.-B. Bartels, H. Mahani, S. Berg, et al., Oil configuration under high-salinity and low-salinity conditions at pore scale: a parametric investigation by use of a single-channel micromodel, SPE Journal 22 (2017) 1362–1373.

[126] K. Skrettingland, T. Holt, M.T. Tweheyo, et al., Snorre low-salinity-water injection—coreflooding experiments and single-well field pilot, SPE Reservoir Evaluation & Engineering 14 (2011) 182–192.

[127] S. Strand, T. Austad, T. Puntervold, et al., The impact of plagioclase on the low salinity EOR-effect in sandstone, Energy & Fuels 28 (2014) 2378–2383.

[128] J.C. Seccombe, A. Lager, K.J. Webb, et al., Improving waterflood recovery: LoSalTM EOR field evaluation, in: SPE-113480-MS, SPE/DOE Improved Oil Recovery Symposium, April 20–23, 2008, Tulsa, OK, USA.

[129] N. Loahardjo, X. Xie, N.R. Morrow, Oil recovery by cyclic waterflooding of mixed-wet sandstone and limestone, in: Proceedings of the 10th International Symposium on Reservoir Wettability, October 27–28, 2008, Abu Dhabi, UAE.

[130] Z. Aghaeifar, T. Puntervold, S. Strand, et al., Low salinity EOR effects after seawater flooding in a high temperature and high salinity offshore sandstone reservoir, in: SPE-191334-MS, SPE Norway One Day Seminar, April 18, 2018, Bergen, Norway.

[131] Z. Aghaeifar, S. Strand, T. Puntervold, et al., Smart Water injection strategies for optimized EOR in a high temperature offshore oil reservoir, Journal of Petroleum Science and Engineering 165 (2018) 743–751.

[132] Z. Aghaeifar, EOR by seawater and "Smart Water" flooding in high temperature sandstone reservoirs (Ph.D. thesis), University of Stavanger: Department of Energy Resources, 2019.

[133] H. Mahani, S. Berg, D. Ilic, et al., Kinetics of low-salinity-flooding effect, SPE Journal 20 (2014) 8–20.

[134] M.D. McMillan, H. Rahnema, J. Romiluy, et al., Effect of exposure time and crude oil composition on low-salinity water flooding, Fuel 185 (2016) 263–272.

[135] A. Namaee-Ghasemi, S. Ayatollahi, H. Mahani, Pore-scale simulation of the interplay between wettability, capillary number, and salt dispersion on the efficiency of oil mobilization by low-salinity waterflooding, SPE Journal (2021) 1–22.

CHAPTER

4 Chemical enhanced oil recovery

Xindi Sun[1] and Baojun Bai[2]

[1]*Department of Physics and Engineering, Slippery Rock University of Pennsylvania, Slippery Rock, PA, United States*

[2]*Geosciences and Geological and Petroleum Engineering Department, Missouri University of Science and Technology, Rolla, MO, United States*

4.1 Introduction

Oil and gas are the world's major energy resources even with the booming of renewable resources [1]. After the primary and secondary recoveries, a considerable amount of remaining oil is left in reservoirs. On average, 65% of the original oil in place remains in reservoirs. The remaining in the fields can range from 40% to 95% [2,3]. This is either caused by unfavorable sweep efficiency or the trapping of oil drops in rock pores due to capillary forces [4]. To unlock these considerable reserves, enhanced oil recovery (EOR) methods are employed. Among all the EOR methods, chemical EOR is one of the most trending EOR topics of interest because of its high efficiency and technical and economic feasibilities.

Chemical EOR is the process where one or more chemicals are mixed with water and injected into reservoirs. This chapter will focus on chemical EOR processes, including polymer flooding (PF), surfactant flooding (SF), alkaline flooding (AF), and any combinations of these processes. The fundamental recovery mechanism of PF is to provide better mobility control by adding polymers into the displacing fluids to increase viscosity. However, some limitations can significantly reduce the PF efficiency, such as high temperature, high formation water salinity and divalent concentration. The key EOR mechanism of SF is to reduce the interfacial tension (IFT) between the oil and the displacing fluids. In addition, rock wettability in the contact area may also be altered. In AF, in-situ surfactants can be generated due to the interaction between alkali and naphthenic acids in the oil. More importantly, alkali can reduce surfactant and polymer adsorption if injected alternatively.

In this chapter, the principles, EOR mechanisms, synergy, remaining challenges, field applications, and future developments of various chemical EOR technologies are presented.

Recovery Improvement. DOI: https://doi.org/10.1016/B978-0-12-823363-4.00003-0

4.2 Polymer flooding

The study of PF can be traced back to the 1950s. The United States initiated the laboratory evaluation in the late 1950s and early 1960s and conducted the first field application in 1964. Since 1970 Russia, Canada, England, France, Romania, Germany, etc., tested PF performance in different reservoirs. As of now, PF has been widely applied in more than 200 fields or units worldwide [5]. As shown in Fig. 4.1, 34% of PF pilot tests were conducted in China, 30% in the United States, 14% in Germany, and 4% in Brazil. After analyzing more than 400 PF cases, a summary can be drawn that more than 77% of the projects were conducted in sandstone reservoirs, around 20% of the projects were carried out in carbonate reservoirs, and only 3% of the projects were carried out in reservoirs with other lithologies.

4.2.1 Principles

PF is a relatively mature EOR technique using water that is viscosified with soluble polymers. By improving displacing fluid viscosity, the mobility ratio of the system is reduced, which can minimize fingering problems and delay water breakthrough, thereby increasing sweep efficiency [6]. Fig. 4.2A illustrates a typical waterflooding process. Since water has lower mobility than oil, in this scenario, water fingering occurs and leaves a considerable amount of oil in the unswept area. In contrast, Fig. 4.2B shows an improved sweep efficiency by PF. The

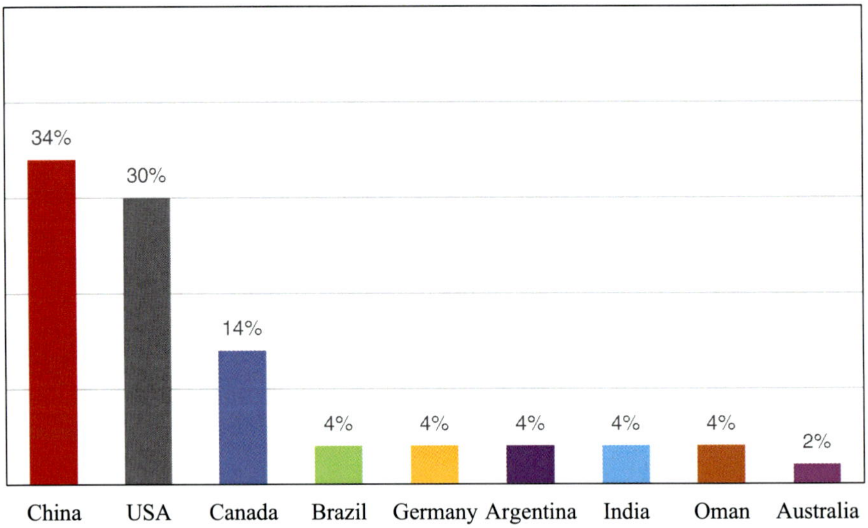

FIGURE 4.1

World polymer pilot projects in percentage.

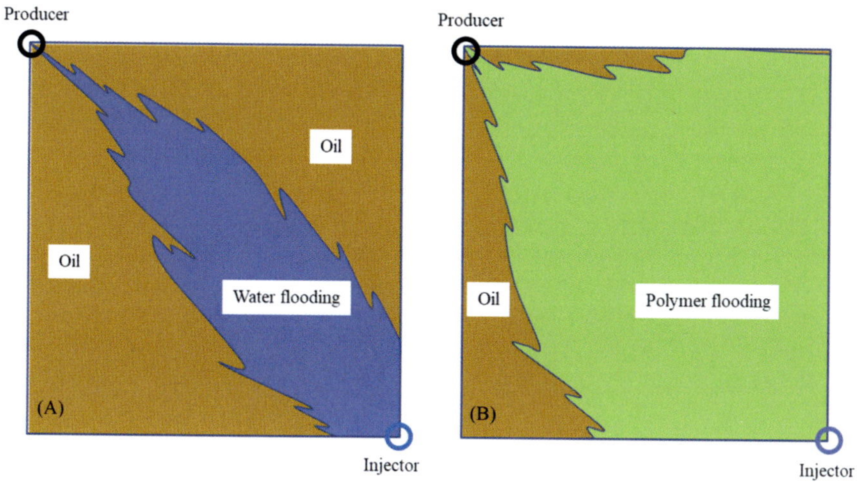

FIGURE 4.2

(A) Unfavorable mobility ratio by waterflooding and (B) favorable mobility ratio by polymer flooding.

polymer can improve displacing phase viscosity significantly, which will reduce mobility ratio and improve sweep efficiency.

The mobility ratio (M) is defined as the ratio of displacing phase mobility to displaced phases. Typically, in a PF oil field, the polymer is defined as displacing phase while the oil is defined as the displaced phase, as shown in the equation below.

$$M = \frac{\lambda_D}{\lambda_d} \tag{4.1}$$

where λ_D is the displacing phase mobility and λ_d is the displaced phase mobility.

Conventionally, the favorable mobility ratio is equal to or less than 1 while the unfavorable mobility ratio is more than 1. The higher the mobility ratio is, the more severe the fingering problem is. Consequently, a lower sweep efficiency will be caused. The PF case with the lowest mobility ratio yields the best sweep efficiency and the case with the highest mobility ratio shows the most severe fingering problem. The sweep efficiency decreases with the increase of the mobility ratio.

4.2.2 Characterization and evaluation of commonly used polymer agents

Polymers are used as fluid viscosifier to control mobility and are also applied in drilling fluids, fracturing fluids, and completion fluids [7,8]. The polymers most widely used in oil field recovery projects include hydrolyzed polyacrylamide (HPAM), polysaccharide (e.g., xanthan gum) and hydrophobically associating polymers (HASP).

4.2.2.1 Hydrolyzed polyacrylamide

HPAM is the most widely used polymer in chemical EOR to improve displacing fluid viscosity [9]. The key factor in increasing chemical EOR efficiency is to reduce chemical adsorption. Therefore polyacrylamide (PAM) is partially hydrolyzed to minimize polymer adsorption onto mineral surfaces of a porous medium. During the hydrolysis process, the PAM molecules react with a base (i.e., sodium hydroxide, sodium carbonate, and potassium hydroxide), which converts some of the amide groups ($-CONH_2$) to carboxyl groups ($-COO-$) (Fig. 4.3) [6,10,11]. Since both the carboxyl groups and the mineral surfaces are negatively charged, the electrical repulsion between negative charges will reduce adsorption on the rock surface [6,10,12]. In addition, it can increase the viscosity of water more than unhydrolyzed PAM. The repulsion between negative charges of carboxylate groups on the polymer chain makes the chains of the polymer extend. This extension will enlarge the fluid mechanical volume of the polymer and increase the viscosity [6,10–12].

The monomeric unit is the acrylamide molecule. The typical molecular weight of HPAM is within the range of $2-20 \times 10^6$ g/mole. The degree of hydrolysis is defined as the ratio of the number of carboxylate groups to the number of acrylamide and carboxylate groups [13]. Typical degrees of hydrolysis are 15%–35%. The degree of hydrolysis can impact polymer viscosity, water solubility, and retention. The polymers with low hydrolysis degrees are difficult to dissolve, even not soluble. Contrarily, polymers with high hydrolysis degrees are more sensitive to salinity and hardness [14]. HPAM is available in both powder and liquid (emulsion) forms, which is convenient for transportation and application in different situations. It is inexpensive and relatively resistant to bacterial attacks.

4.2.2.2 Polysaccharide

Polysaccharide polymers are biomacromolecules derived from plants directly or through a fermentation process. This type of polymer has multiple functions, including thickening, crosslinking, adsorption, etc. Typically, the polysaccharide is susceptible to bacterial attach but insensitive to brine salinity and hardness, however, it degrades at elevated temperatures. Polysaccharide increases the water viscosity by adding a more rigid structure to the solution. They can be in a raw or chemical modified form and have been applied in a wide range of industries, such as food, chemical, construction, and petroleum [15]. In PF, thickening is the major function of polymers, as a consequence, only a limited number of polymers are suitable for PF. As listed in Table 4.1, xanthan gum, scleroglucan, guar gum,

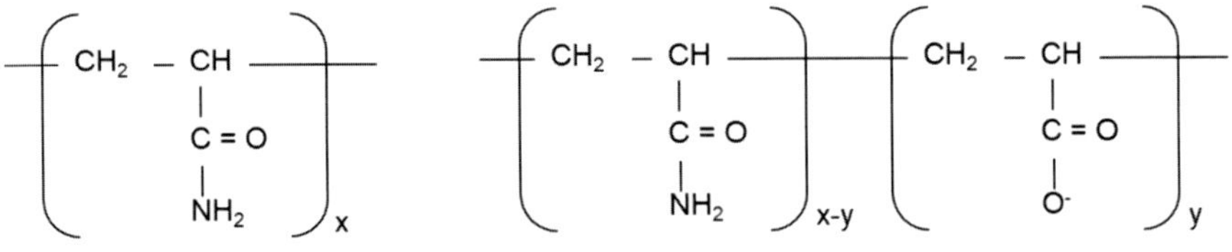

FIGURE 4.3

Chemical structures of polyacrylamide (left) and hydrolyzed polyacrylamide (right).

Table 4.1 Characteristics of biopolymers used in polymer flooding.

Biopolymer	Source	Monomers	Molecular weight	Properties	Modification
Xanthan gum	Fermentation product of *Xanthomonas campestris*	D-Mannose, D-glucose, pyruvic acid, D-glucuronic acid	2×10^6 to 20×10^6 Da	Thickening Crosslinking	Carbonate modified Formaldehyde modified Propylene oxide modified
Scleroglucan	Fermentation product of *Sclerotium rolfsii*	D-glucose	1.3×10^5 to 6×10^6 Da	Thickening Crosslinking	Hydrophobic modified
Guar gum	Endosperm componnet of *Cyamopsis tetragonolobus*	D-Mannose, D-galactose	10^6 to 2×10^6 Da	Thickening Crosslinking	Hydroxypropyl modified Carboxymethyl modified Carboxymethyl hydroxypropyl modified
Cellulose	Lignocellulose of plantsFermentation product of Acetobacter xylinam	D-Glucose	2×10^6 Da	Thickening Filtration Adsorption	Hydroxyethyl modified Carboxymethyl modified Amphoteric modified

Source: *Reproduced with modification from S. Xia, L. Zhang, A. Davletshin, et al., Application of polysaccharide biopolymer in petroleum recovery, Polymers, 12 (2020) 1860.*

and cellulose have outstanding thickening properties and have been studied in the oil and gas industry as viscosifiers to increase water phase viscosity.

4.2.2.2.1 Xanthan gum

Xanthan gum is the most commonly used polysaccharide biopolymer, which can be supplied as a dry powder or as a broth [10]. Xanthan gum is produced from a microorganism fermentation process by the naturally occurring bacterium *Xanthomonas campestris*. Xanthan gum has high solubility, reliable thermostability, and high viscosity yield, which makes it a good viscosifier in PF [15].

Xanthan gum has a complex structure (Fig. 4.4), which can protect the backbone of the polymer molecules from enzymatic attach and backbone cleavage. The molecular weight is around 2×10^6 to 20×10^6 Da. The viscosity of 1000 ppm xanthan gum solution is around 13–35 cp. Xanthan gum shows a high

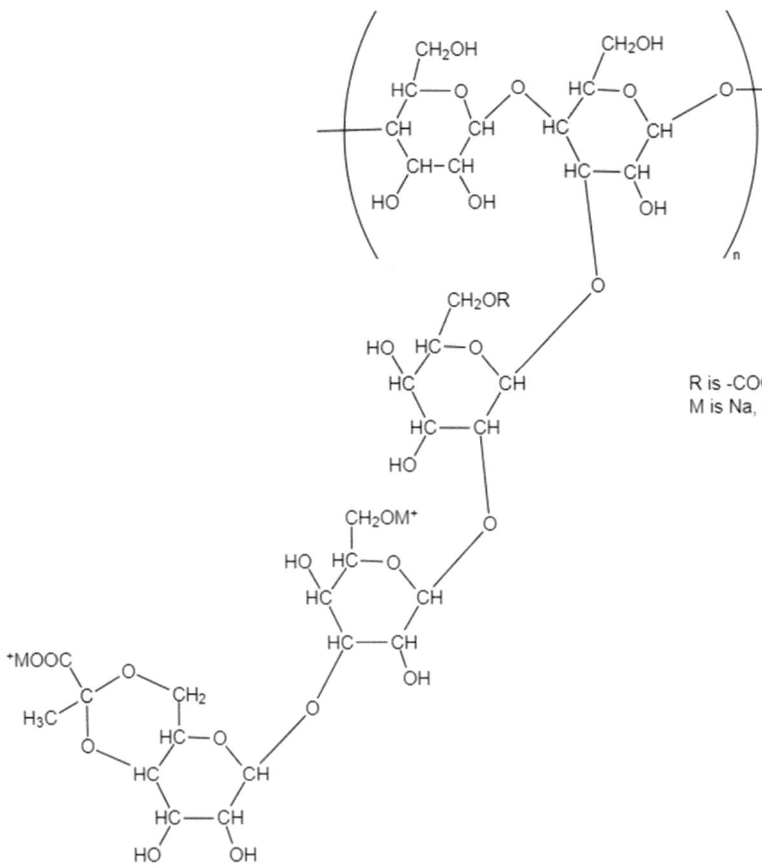

FIGURE 4.4

Molecular structure of xanthan gum.

viscosity in high-salinity brine and it has good resistance to shear degradation. Because of the complex molecular structure, xanthan gum shows good resistance to acid (pH ≥ 3), salinity (up to 30,000 ppm), and temperature (up to 80°C) [16]. To meet the PF requirements, xanthan gum is modified to improve the thickening effect, solubility, and thermostability. The widely used modification methods are carbonate-modified, formaldehyde-modified, and propylene oxide-modified approaches. Carbonate modification and propylene oxide modification can significantly increase xanthan gum molecular weight, which leads to increased viscosity. Formaldehyde modification can shorten polymer dissolution time to achieve a quick injection.

4.2.2.2.2 Scleroglucan

Scleroglucan is the extracellular polysaccharide of the fungus of the genus sclerotium. It has been reported that scleroglucan has outstanding stability under high temperature and low pH conditions. The molecular structure of scleroglucan is shown in Fig. 4.5. The molecular weight of scleroglucan ranges from 1.3×10^5 to 6×10^6 Da, which depends on different production strains and fermentation conditions [17]. 35 ppm of scheroglucan can increase water viscosity to 10 cp. It can also be crosslinked by bentonite to form a gel with 4000-cp viscosity. Compared to xanthan, scleroglucan has better stability under harsh environments, however, the recommended application temperature should not exceed 135°C to avoid significant viscosity loss. Besides the good thermostability, scleroglucan also has a high tolerance to alkaline conditions. Due to the outstanding thermostability, rheology performance, and low cost, scleroglucan is the second most widely used polysaccharide polymer in PF [15].

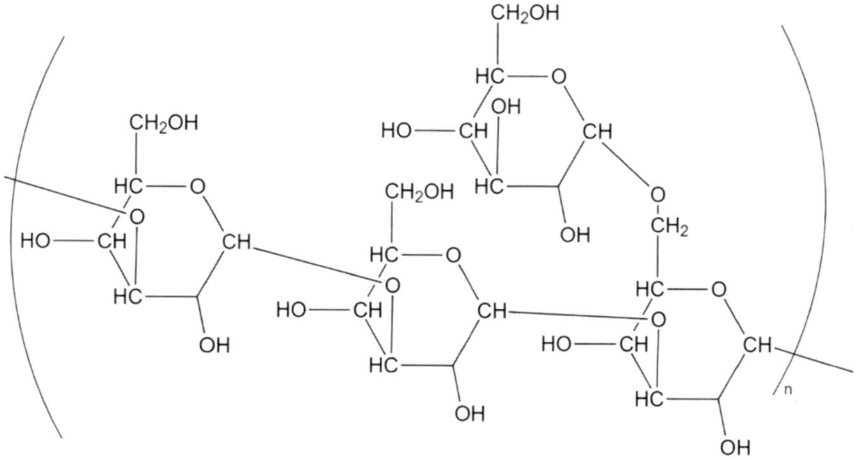

FIGURE 4.5

Molecular structure of scleroglucan.

4.2.2.2.3 Guar gum

Guar gum is a natural nonionic polymer, which is extracted from the *Cyamopsis tetragonolobus* plant. Guar gum was commonly used in the food industry as an emulsifier, viscosifier, and stabilizer. Guar gum is obtained directly from the plant without the fermentation process; therefore, the price of guar gum is lower than other polysaccharide biopolymers. Due to the low price and good ability to increase viscosity, guar gum has drawn more attention in the petroleum industry. Nowadays, more than 40% of the world's guar gum is used in hydraulic fracturing fluids.

The guar gum molecule has linear backbone chains consisting of (1–4)-β-D-mannopyranosyl units and the attached α-D-galactopyranosyl units (Fig. 4.6). The average molecular weight of guar gum is from 10^6 to 2×10^6 Da. As a good viscosifier, 1% of guar gum can increase water viscosity to 10,000 cp. Similar to most of the polymers, guar gum has shear thinning behavior and the viscosity is highly impacted by the pH and temperature. Typically, guar gum can achieve its highest viscosity at pH 6–9 and the lowest viscosity at pH 3.5. However, guar gum is not suitable for treatments with a temperature higher than 83°C due to its poor thermostability.

4.2.2.2.4 Cellulose

Cellulose is one of the most abundant polymers that exist in natural materials. For example, 40%–50% in wood is cellulose, and 90%–95% in cotton is cellulose. Besides obtaining cellulose from plants, cellulose can also be fermented

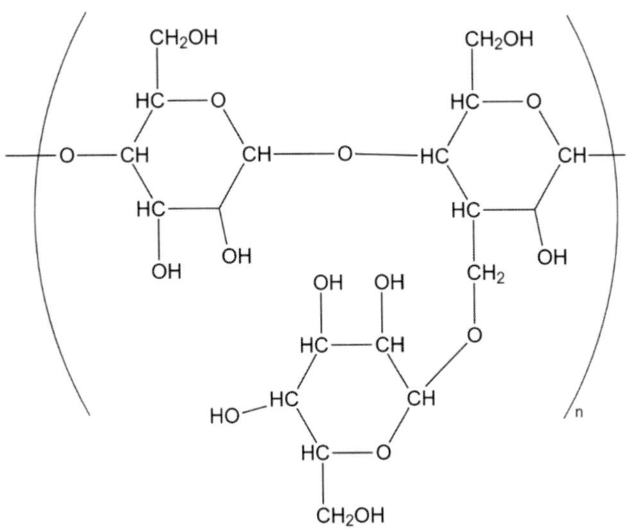

FIGURE 4.6

Molecular structure of guar gum.

using acetobacter xylinam [18]. The molecular weight of native cellulose is around 2×10^6 Da when the polymerization degree is around 100,000 [19]. Carboxymethyl cellulose is the most commonly used cellulose derivatives in PF. It is a cost-friendly, nontoxic, highly biocompatible, and biodegradable material, which can be used as a drilling fluid loss control agent and PF viscosifier. Compared to guar gum, carboxymethyl cellulose can maintain viscosity at temperatures up to 160°C [20]. The molecular structure of cellulose is shown in Fig. 4.7.

4.2.2.3 Hydrophobically associating polymers

Hydrophobically associating polymer is similar to conventional polymers used for PF except that it contains a small fraction of hydrophobic monomers (<1%) attached to the polymer backbone [21], as shown in Fig. 4.8. The advantage of hydrophobic groups is that they can interact together to form an associativity polymer in an aqueous solution, which has superior viscosity at a similar level of molecular weight (Fig. 4.9).

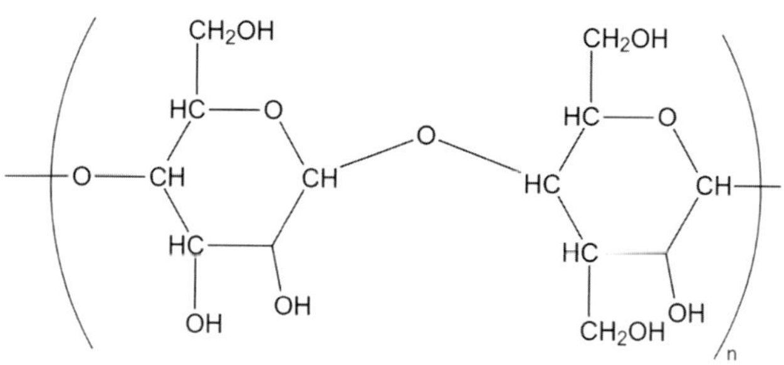

FIGURE 4.7

The molecular structure of cellulose.

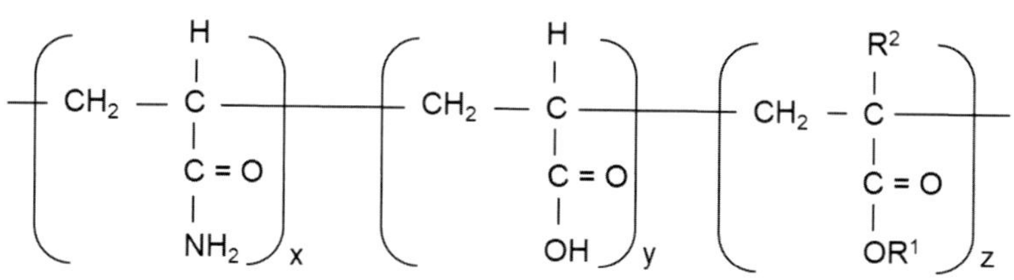

R^1 = hydrophobic group (ie. $C_8 - C_{20}$)
R^2 = H or CH_3

FIGURE 4.8

Structure of hydrophobically associating polymers.

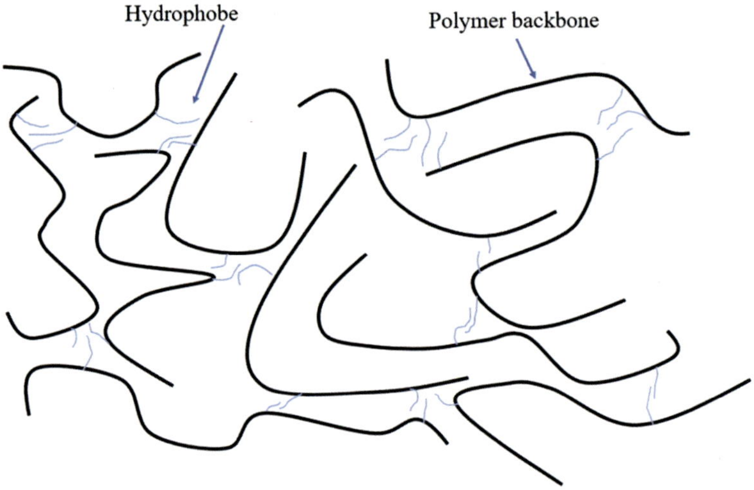

FIGURE 4.9

Hydrophobic groups can make polymer associate together to increase viscosity.

4.2.3 Impacting factors on treatment efficiency

The primary EOR mechanism of PF is to reduce the water–oil mobility ratio by increasing displacing phase (water/polymer) viscosity. The PF efficiency is directly related to the polymer viscosity. In this case, the parameter of the greatest importance to improve sweep efficiency is viscosity. Several factors can impact polymer viscosity, such as molecular weight, polymer concentration, salt concentration, temperature, and shear.

4.2.3.1 Molecular weight

Molecular weight is the main factor that affects the flow behavior of polymers. In general, the polymer with a higher molecular weight has a larger molecular size. The polymer solutions with higher molecular weight have a higher absorption degree, higher resistance, greater viscosity, and higher oil recovery factor. Fig. 4.10 compares two polymers with different molecular weights. As shown in the figure, the polymer with a higher molecular weight yields a high polymer viscosity. The viscosity can be increased by increasing polymer concentration. The high-molecular-weight polymer has a larger viscosity improvement if increase the polymer concentration by the same amount [22].

4.2.3.2 Polymer concentration

Conventionally, increasing polymer concentration is one of the practical approaches to increase viscosity. Generally, the apparent viscosity has a proportional relationship with polymer concentration where an increase in the polymer concentration results in an increase in the apparent viscosity. As shown in Fig. 4.10 and Fig. 4.11,

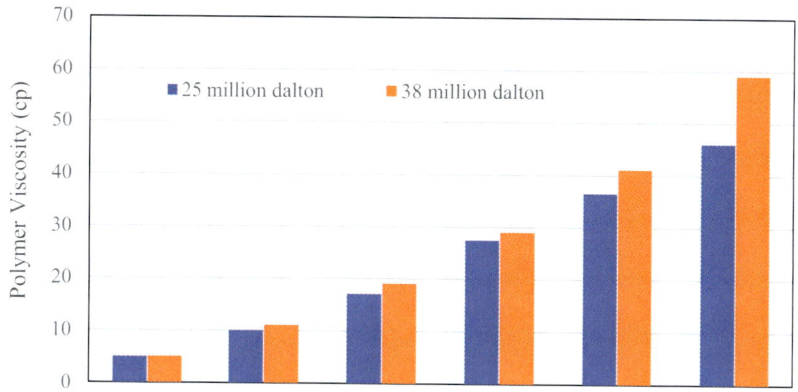

FIGURE 4.10

Impact of molecular weight on polymer viscosity.

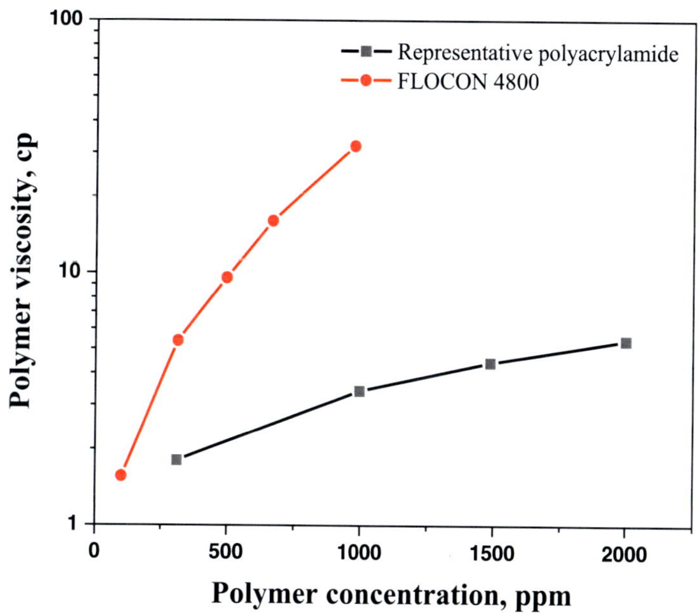

FIGURE 4.11

Impact of polymer concentration on hydrolyzed polyacrylamide and xanthan (FLOCON 4800) viscosities.

a higher polymer concentration yields a high polymer viscosity. As illustrated in Fig. 4.11, with the same polymer concentration, xanthan polymer has a higher viscosity than HPAM [21]. Fig. 4.12 shows the comparison between HASP and HPAM. The HASP has a higher viscosity than HPAM once the polymer concentration is higher than 1000 ppm [23].

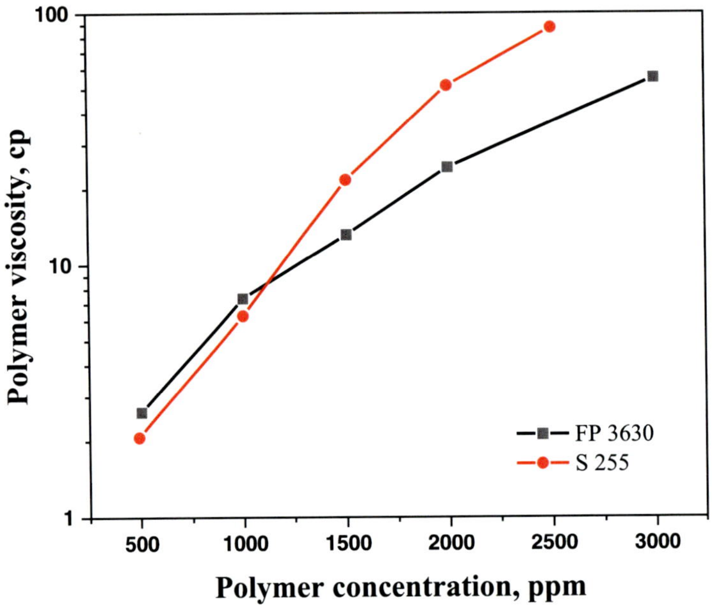

FIGURE 4.12

Viscosity versus polymer concentration for hydrophobically associating polymer (S 255) and hydrolyzed polyacrylamide (FP 3630) at a shear rate of 10/s.

4.2.3.3 Salt concentration

The salt concentration is another parameter that impacts polymer viscosity. The presence of salt decreases the polymer solution viscosity because the salt ions weaken the electrostatic resistance among the chains, which leads to the coiling of the polymers (Fig. 4.13). This coiling effect can cause a reduction in fluid mechanical volume first and then reduce the viscosity of the polymer solution.

The degree of hydrolysis is a factor that dominates the sensitivity of polymer to salt, which is measured by the ratio of the number of carboxylate groups to the number of acrylamide and carboxylate groups [24]. With a sufficient hydrolysis degree, polymer adsorption can be reduced. If the degree of hydrolysis is too low, the polymer solubility in water will decrease. If the hydrolysis is excessive, the polymer viscosity will decrease due to the sensitivity to salt ions [10,25]. In this case, optimizing the degree of hydrolysis is crucial. Typically, the optimum degree of hydrolysis is between 15% and 33% [10]. An HPAM with a greater degree of hydrolysis will cause greater viscosity reduction in the presence of divalent ions. This reduction is caused by the interaction of a larger number of anionic and cationic species [26]. The viscosity decreases as the salinity increases, as shown in Fig. 4.14.

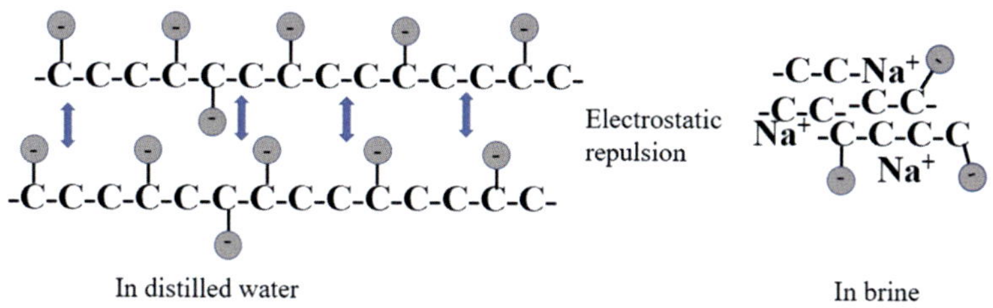

FIGURE 4.13

Illustration of salt concentration effect.

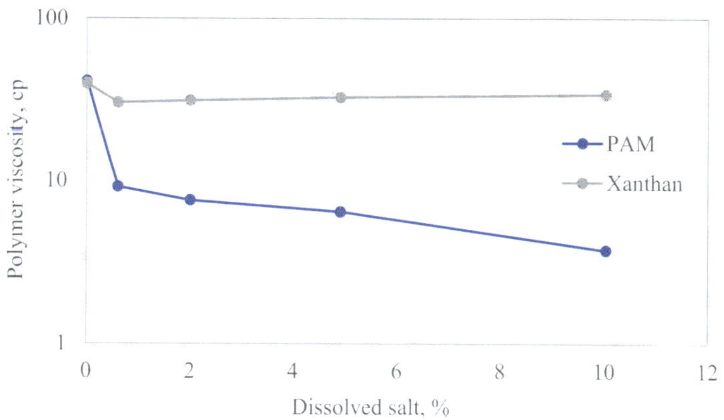

FIGURE 4.14

Salinity effect on polymer viscosity.

Some of the polymers are sensitive to salt but some of them are not. Typically, PAM is sensitive to salt, which means that the polymer solution viscosity reduces significantly in high-salinity water. In contrast, biopolymers, such as xanthan are not that sensitive. Fig. 4.14 illustrates the viscosity of PAM and xanthan in different salinity brines. The major viscosity reduction occurs while the brine salinity increases to 1% [21]. The viscosity of xanthan increases slightly with brine salinity above 1% but the viscosity of PAM solution keeps reducing up to 10% TDS, where the PAM begins to precipitate. The hardness ions appear to have no effect on xanthan solution viscosity beyond their contribution to salinity.

Fig. 4.15 shows the salinity effect on the viscosities of hydrophobically associating polymer water solution and HPAM solution [27]. HPAM viscosity decreases when a monovalent salt, such as NaCl, is added because the added salt neutralizes the charge in HPAM side chains. When the salt is added, the $-COO^-$ groups in

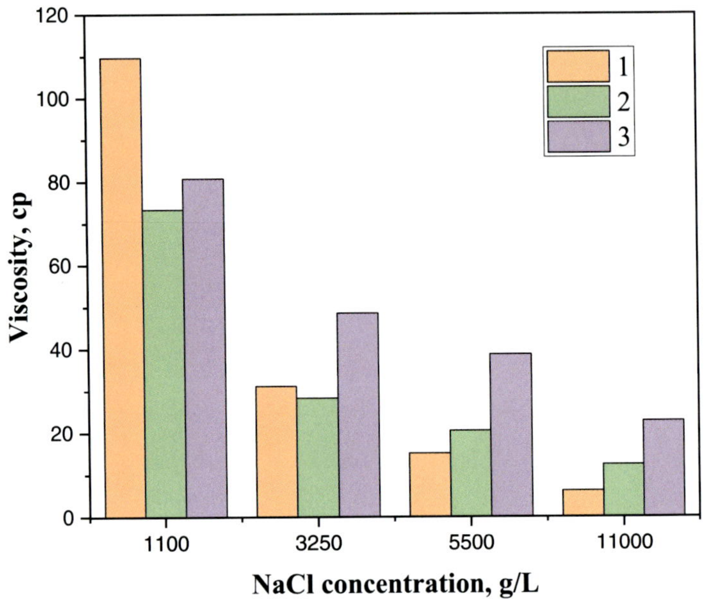

FIGURE 4.15

Polymer viscosity of the hydrolyzed polyacrylamide (1) and hydrophobically associating polymers (2 and 3) as a function of NaCl concentration.

the high molecular chains are surrounded by some Na^+, which shields the charge. This phenomenon is called the shielding effect, which reduces electrostatic repulsion among groups on polymer chains. As a result, polymer chains tend to coil up, thus the polymer solution viscosity decreases with the increase of the salt concentration. The viscosity decrease of hydrophobically associating polymer can be partially retarded the electrolyte because the addition of salt improves the polarizability of the solution and results in the increase of the associated effect of the hydrophobic groups. As a consequence, the viscosity of the hydrophobically associating polymer solution decreases much slower than the viscosity of HPAM solution with increased salt concentration.

4.2.3.4 Temperature

Fig. 4.16 shows the temperature effect on polymer solution viscosity. As temperature increases, the viscosity of HPAM water solution decreases steadily but the hydrophobically associating polymer (HASP) water solution behaves differently [27]. If the temperature is lower than 45°C, the viscosity of HASP with lower hydrophobic monomer content decreases with the increase in temperature, however, the viscosity of HASP with higher hydrophobic monomer content increases slightly and reaches the maximum at around 45°C. The reason is that endothermal association is favored by higher temperature in the hydrophobic entropy-driving process, which yields an increase in the polymer viscosity. In addition, the

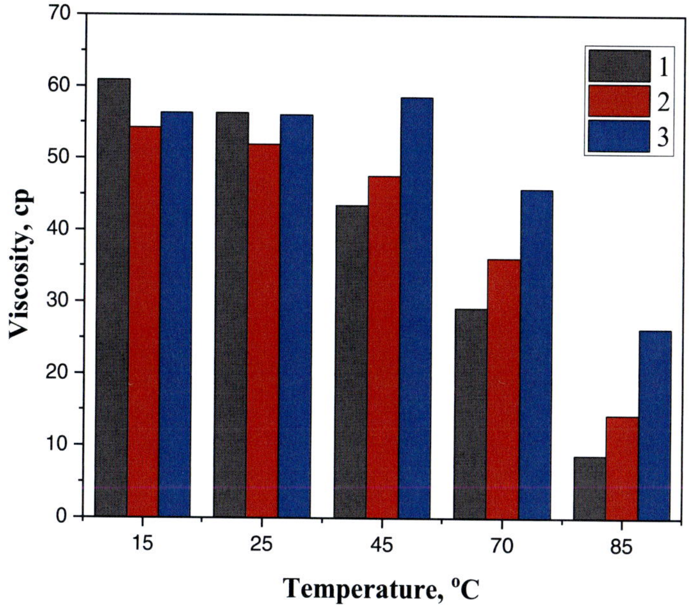

FIGURE 4.16

Temperature effect on hydrolyzed polyacrylamide (1), higher hydrophobic monomer content hydrophobically associating polymers (2), and lower hydrophobic monomer content hydrophobically associating polymers (3).

stronger thermal vibration of water molecules retards the association effect at a higher temperature. As demonstrated by Fig. 4.16, the viscosity of the HASP solution is always higher than that of high-molecular-weight HPAM solution at relatively high temperatures, which clearly reveals that the HASP has a better temperature resistance than HPAM.

4.2.3.5 Shear rate

Generally, a polymer solution exhibits pseudoplastic fluid behavior. Polymer viscosity decreases with the increase in shear rate, which can be described by the power–law model below [24]:

$$\mu_p = K\dot{\gamma}^{(n-1)} \tag{4.2}$$

where K is the flow consistency index and n is the flow behavior index. Typically, n is between 0.4 and 0.7 for pseudoplastic fluids. For a Newtonian fluid, $n = 1$ and K is constant. Table 4.2 compares the n and K values for four different types of polymers. At the same polymer concentration, the xanthan gum has the highest K value while the carboxymethyl cellulose has the lowest K value. For most of the polymers, the increased concentration yields a lower n and a higher K value. However, for xanthan gum, both n and K values increase with concentration. As shown in Fig. 4.17, with the increase in shear rate, the viscosity

Table 4.2 Comparison between different polymers.

Polymer type	Concentration (%)	*n*	*K* (mPa•s^n)
HPAM	1	0.28	1080
	2	0.26	2050
	5	0.25	5770
Xanthan gum	0.5	0.58	1190
	1	0.65	3163
	2	0.71	6526
Schleroglucan	0.5	0.49	55
	1	0.31	272
	2	0.2	1073
Carboxymethyl cellulose	1	0.95	50
	2	0.85	450
	4	0.61	830

HPAM, *Hydrolyzed polyacrylamide.*
Source*: Reproduced with permission from S. Xia, L. Zhang, A. Davletshin, et al., Application of polysaccharide biopolymer in petroleum recovery, Polymers, 12 (2020) 1860.*

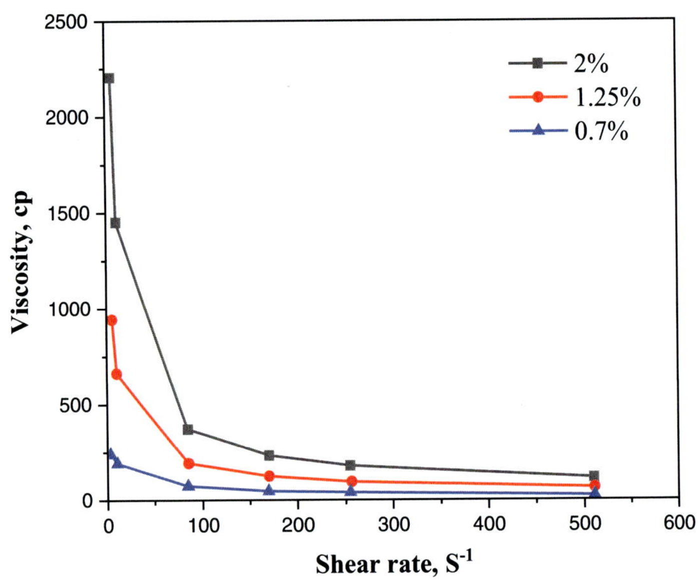

FIGURE 4.17

Shear rate effect on polymer viscosity [28].

Reproduced with permission from W. Zheng, X. Wu, Y. Huang, Impact of polymer addition, electrolyte, clay and antioxidant on rheological properties of polymer fluid at high temperature and high pressure, Journal of Petroleum Exploration and Production Technology 10 (2020) 663–671.

of polymer (Driscal-D, Chevron Phillips Chemical Co., Ltd, United States) reduces, which is also referred to as shear thinning behavior.

4.2.3.6 Polymer stability

As aforementioned, the temperature can reduce polymer viscosity and even degrade polymer at high temperatures. In this case, the study of polymer thermal stability is crucial. Typically, PF projects will last several months or even years; therefore the longer the polymer can maintain its viscosity, the longer the treatment will remain effective.

4.2.3.6.1 Chemical stability

Chemical degradation refers to polymer chain breakdown, which can be caused by a short-term attack by oxygen and irons or a long-term attack by hydrolysis processes. The application of polymers as mobility control agents in low temperature and low-salinity reservoirs is very common. Many commercial polymer products are available in the market that can be used for low-temperature and low-salinity purposes.

Oxygen level is another factor that impacts polymer thermal stability. To eliminate oxidation, the oxygen level should be very low, otherwise, the polymer will interact with oxygen under high temperature and degradation will occur. At a low temperature, the presence of oxygen in HPAM solution does not have a significant effect on polymer viscosity and the polymer can stay stable for a long time. However, the elevated temperature will cause oxidative degradation severely even if only a small amount of oxygen is present, which can cause a fast viscosity reduction in a short time. As the oxygen level increases, the polymer viscosity reduces even faster.

Under high temperature and high-salinity conditions, divalent cations (e.g., Ca^{2+} and Mg^{2+}) and oxygen impact the stability of HPAM polymers significantly. Due to the thermal degradation, long polymer chains (larger molecular weight) are broken down to shorter chains (smaller molecular weight), which will reduce the polymer solution viscosity. In addition, the amine groups in the polymer molecules change to carboxyl due to hydrolysis, which will react with ions in brine (e.g., Ca^{2+} and Mg^{2+}) and form precipitates and lose viscosity.

4.2.4 Screening criteria and status of the application

PF has been applied worldwide since the 1940s. Manning et al. [29] conducted a statistical study on field PF projects. The statistical analysis proves that polymer flooding is one of the efficient methods to improve oil recovery.

4.2.4.1 Screening criteria

Currently, most of the polymer flooding cases were carried out in sandstone reservoirs, but some of the projects were conducted in carbonate reservoirs [30]. Applying polymer flooding in heavy oil reservoirs is a trending topic. According

Table 4.3 Screening criteria for polymer flooding and Daqing oilfield average properties.

Parameters	Suggested in literature	Daqing
Dykstra–Parsons coefficient of permeability variation (V_K or V_{DP})	0.7 ± 0.1	0.5–0.8
Average permeability, mD	>10	100–600
Reservoir temperature, °C	<100	45
Formation water salinity, mg/L	<10,000	3000–7000
Water cut, %	Low water cut	90–96
Oil viscosity, cp	10–5400	6–9
Recovery, %	Low	35–45
Well spacing, m/well pattern	200–300/5-spot	150–300/5-spot

Source: *Reproduced with modification from D. Wang, Polymer flooding practice in Daqing, in: J.J. Sheng (Ed.), Enhanced Oil Recovery Field Case Studies, Elsevier, 2013, pp. 83–116.*

to the survey prepared by Standnes et al. [31], HPAM is more often used than biopolymers. To achieve the best performance, many reservoir properties need to be considered when selecting the reservoir candidate, such as reservoir heterogeneity, temperature, permeability, formation water salinity, and oil viscosity. The screening criteria for polymer flooding are listed in Table 4.3.

4.2.4.2 Application status

PF has been widely applied around the world. Table 4.4 summarizes some of the projects conducted in different countries. In this section, several typical field application cases will be discussed, including Daqing oilfield in China (sandstone), North Oregon Basin Field and Byron Field in the United States (carbonate), East Bodo reservoir in Canada (heavy oil), and SZ36-1 field in Bohai Bay (offshore).

4.2.4.2.1 Daqing oilfield, China

As one of the earliest oilfields that applied PF to improve oil recovery, the Daqing oilfield is the most representative field. Daqing oilfield started laboratory research in the 1960s and started to conduct PF pilot studies in the 1970s [49]. PF has gained remarkable success in the Daqing oilfield, where 10%–12% oil recovery improvement was achieved between 1996 and 2010. As listed in Table 4.3, the Daqing oilfield has many favorable properties, which makes it one of the best candidates for PF.

A set of PF pilot tests was conducted in the WN block, WM block, and EN1 block in the Daqing oilfield in 2001 [50]. The average reservoir permeabilities of these three blocks range from 625 to 912 mD. The formation thickness

Table 4.4 Some polymer flooding projects in different fields.

Field	Polymer	Reservoir temperature, °C	Formation water salinity (mg/L)
Daqing (China)	HPAM	45	9000
Pelican (Canada)	HPAM	23	6800
Gudong (China)	HPAM	68	3022
Tambaredjo (Suriname)	HPAM	36	N/A
Bohai Bay (China)	HPAM	65	6070
Henan oil field (China)	HPAM	75	5060
XingLongTai, Liaohe oil field (China)	HPAM	56.6	3112
West Khiel (United States)	HPAM	57	46,480
Viraj (India)	HPAM	81	13,250
Matzen (Austria)	HPAM	50	20,000
Tanner (United States)	PAM	80	66,800
Cambridge Minnelusa (United States)	PAM	55.6	N/A
Milne Point field (United States)	HPAM	21.7	27,500
Sanand (India)	PAM	85	N/A
David Pool (Canada)	PAM	31	6660
Eddesse-Nord sandstone reservoir (Germany)	Xanthan	22	120,000
Vorhop-Knesebeck (Germany)	Xanthan	56	210,000
Romashkino field in Tatarstan (Russia)	HEC	36	250,000

HPAM, *Hydrolyzed polyacrylamide;* PAM, *polyacrylamide.*
Source: *Data from [32–48].*

is 10.1–17.9 m. The well spacing is around 245 m. These three blocks were initially flooded using 16 million daltons molecular weight (MW) polymer with a concentration of 1000 mg/L. Afterward, the polymer with a high MW of 25 million daltons was injected at a higher concentration (2000 mg/L). The high MW and high concentration PF successfully reduced the water cut by 25%–30%.

The success of these pilot tests encouraged the large-scale application of high MW and high-concentration PF in the EWN1 block. This block had 160,688 million tons of OOIP. The effective thickness was around 12.2 m. The well spacing was about 125 m. The formation permeability was 652 mD. Starting from January 2009, 0.61 PV of 2500 million daltons polymer solution with 2050 mg/L concentration was injected, which reduced the water cut significantly and improved oil recovery by more than 10%.

4.2.4.2.2 North Oregon Basin Field and Byron Field, United States

Most of the PFs are conducted in sandstone reservoirs. According to the treatment results, sandstone reservoirs are better candidates for PF. However, PF can still improve oil recovery from some of the carbonate reservoirs.

A PF treatment was carried out in two large fields in Wyoming's mature Big Horn Basin oil production area [51]. The two fields are the North Oregen Basin field and Byron field. The Embar formation in the Big Horn Basin was defined as fossiliferous limestone, sandy limestone, dolomitic limestone, and limestone. The major production intervals were porous dolomitic limestone with barren carbonates. Pinpoints, vagulars, and vertical fractures commonly existed in this formation. The Tensleep Formation was characterized as porous sandstone layers with dolomite and anhydrite zones. The net pay thickness of Embar and Tensleep members in Byron field was 22 and 97 ft, respectively. The net pay thickness of Embar and Tensleep in North Iregon Basin was 26 and 83 ft, respectively. The average porosity in the Byron field was 13.8% and the average porosity in North Oregon Basin was 17.5%. The permeability of the Embar reservoir in Byron was 4.5 mD and the permeability of Tensleep in Byron was 78 mD. The permeability of Embar in North Oregon was 68 mD and the Tensleep in North Oregon Basin was 793 mD. The oil viscosity in Byron was 17 cp and the viscosity in Tensleep was 9.1 cp. Partially hydrolyzed polyacrylamide (PHPA) was used in both fields. The PF in the North Oregon Basin field started with 0.3 PV of 500 ppm PHPA-11, followed by 0.1 PV of 330 ppm and then 0.1 PV of 165 ppm PHPA polymer solution.

The incremental oil production was observed in both the North Oregon Basin field and Byron field. However, the tertiary oil response in the Byron field is lower than in the North Oregon Basin, which might be caused by the existence of extensive natural fractures and the use of the recycled polymer. Overall, the oil production response from both fields was around 3500 bbls/day. A decreased water–oil ratio was observed as well.

4.2.4.2.3 Milne Point Field, Alaska's North Slope, United States

In this field, PF is combined with low-salinity waterflooding to improve oil recovery from a heavy oil reservoir. The sandstone consisted of 1.5% illite, 1.5% chlorite, 1% dolomite, approximately 10% albite, and around 86% quartz. The oil viscosity at reservoir temperature (71°F) was 202 cp and the gravity was 19.0°API. The pattern includes two horizontal injection wells and two horizontal production wells.

A series of laboratory studies were carried out before the pilot test. The low-salinity brine used in the experiments had a salinity of 2498 ppm and the ionic strength was 0.046. The polymer used in this study was Flopaam 3630S (Pfizer Inc., New York, United States). As revealed by laboratory results, the combination of polymer and low-salinity water could improve oil recovery by approximately 10% of OOIP.

The pilot test was initiated on August 28, 2018 [52]. Based on the availability, cost, and laboratory studies, the selected polymer for this pilot test was Flopaam

3630S. As of May 2020, the water cut reduced from 70% to less than 15%, and no polymer breakthrough was observed at that time [41].

4.2.4.2.4 East Bodo reservoir, Canada

The East Bodo reservoir locates in Alberta, Canada, which had a porosity of 27%–30% and permeability of 1000 mD. The oil viscosity varies from 600 to 2000 cp and the oil gravity was 14°API. The formation water salinity was between 25,000 and 29,000 ppm. The average thickness of this reservoir was 3.2 m.

After conducting a series of studies using coreflooding tests and numerical simulation, a pilot test was carried out in an area with the highest injectivity, which had been waterflooded. The project started in May 2006. As designed, 1500 ppm polymer solution was prepared using produced water but the polymer viscosity was only 10 cp, which was much less than the desired 25 cp. Therefore, fresher water with a salinity of 3700 ppm was used. As a consequence, the viscosity of 1500 ppm polymer solution reached 60 cp at the surface. The collected produced polymer from adjacent wells showed a viscosity of 100 cp. The polymer was injected using a 200 m^3/day flowrate, which increased the injection pressure to around 870 psi. Without polymer, a 250 m^3/day injection rate was required to achieve a similar injection pressure.

4.2.4.2.5 SZ36-1 field in Bohai Bay, China

A pilot offshore PF test was conducted in the SZ26-1 field in Bohai Bay, China from September 2003 to May 2005. The formation was unconsolidated and the wells were completed using gravel packing. The porosity was 196%. Due to the poorly cemented rock, the permeability of this formation was up to 8541 mD. However, the permeability of some areas was only around 10 mD due to reservoir heterogeneity. The Dykstra–Parsons coefficient of permeability variation factor in this area was 0.76. The formation depth was 1300–1500 m. The target formation area was 0.396 km^2 and the OOIP was around 737,000 tons. The formation temperature was 65°C. The average oil viscosity at reservoir temperature was 70 cp. The average thickness of this formation was 61.5 m. The average well spacing was 370 m.

Since this is an offshore field, seawater (TDS = 32,423 mg/L) was injected to displace oil initially. Due to the 8-year seawater injection history, the produced water salinity and divalent concentration were higher than formation water. In this case, the produced water was mixed with Guantao formation water, which lowered the TDS to 9048 mg/L and Ca^{2+} and Mg^{2+} concentration to 800 mg/L. The polymer used in this project was hydrophobically associating polymer AP-P4. The polymer injection rate was 500 m^3/day. The polymer concentration of the first slug was 3000 mg/L and the second slug was 1750 mg/L. The viscosity of the second slug was 78 cp at the surface. As a consequence, a decreased water cut and increased oil rate were observed from producers after 10 months of polymer injection.

4.2.5 Future development of polymer flooding

Some of the polymers are sensitive to temperature and salinity. Therefore the application of polymers in high-salinity and high-temperature reservoirs is still challenging [14]. The development of salinity-resistant polymers with robust thermostability is one of the new development directions for PF.

As stated earlier in this section, PF has been applied in heavy oil reservoirs, especially in thin reservoirs, where the thermal recovery is not economic due to heat loss. Adding polymers can improve the viscosity of the displacing fluids. However, in most cases, the injected polymer viscosity is lower than the heavy oil viscosity due to the injectivity limitation. Recently, the use of horizontal wells and control injection rates higher than the fracturing pressure encourage the application of PF in heavy oil reservoirs [53].

Some of the field application results indicate that the combination of PF with AF, SF, or both can improve oil recovery more than singular PF. Therefore applying PF with the assistance of other chemical or mechanical processes is one of the future development directions.

4.2.6 Remaining challenges

The PF technique is relatively mature, however, some problems still exist. After the polymer is injected into a reservoir, thermal degradation and further hydrolysis will occur under high-temperature conditions, which will impact polymer stability and reduce the displacement efficiency significantly. Typically, the low salinity of the formation water and the injected water can increase the viscosity of polymer solutions. However, in some of the reservoirs, the high-salinity water causes the reduction of polymer viscosity, which leads to a higher polymer consumption and higher project cost. Therefore, more efforts are being spent on the development of polymers with robust thermostability and salinity-resistant ability.

Nowadays, some studies are conducted to develop novel cost-effective and high-quality polymers, such as hydrophobic associates and modified PAM. For example, the Shengli oilfield in China is conducting field tests of new polymers suitable for high temperature, high salinity, and high divalent concentration formations.

4.3 Surfactant flooding and wettability alternation

SF is an EOR process in which a small amount of surfactant is mixed with the injected water to reduce the IFT between the oil and water phases and also change the wettability of a reservoir rock to improve oil recovery.

4.3.1 Characterization and evaluation of commonly used surfactant agents

Surfactant is a chemical structure that consists of two different molecular components, known as hydrophilic and hydrophobic groups. A hydrophilic group is a water-soluble component and a hydrophobic group is an oil-soluble component. In the standard surfactant terminology, the water-soluble component, or hydrophilic group, is called the "head," and the hydrophobic group is called the "tail." As demonstrated in Fig. 4.18, the head and tail surfactants attack the interface between two immiscible surfaces, thus decreasing the interfacial forces between the two surfaces.

4.3.1.1 Types of surfactants

Based on the ionic nature of the headgroup, surfactants can be classified into four types: anionic, nonionic, cationic, and zwitterionic. As shown in Fig. 4.19, the anionic surfactant has a negative headgroup; the cationic surfactant has a positively charged head group, and the nonionic surfactant has a neutral head group.

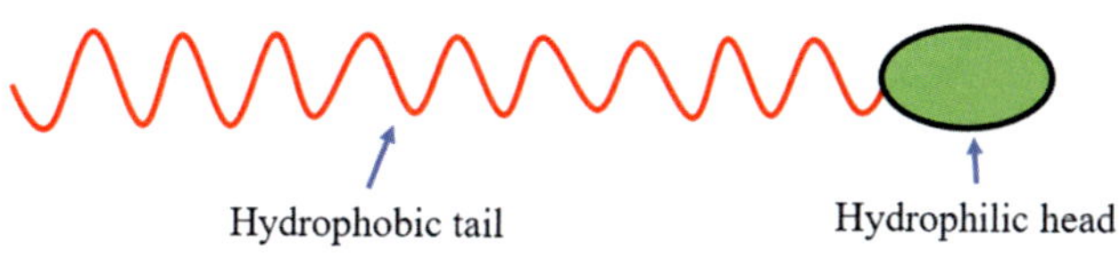

FIGURE 4.18

Structure of a surfactant molecule.

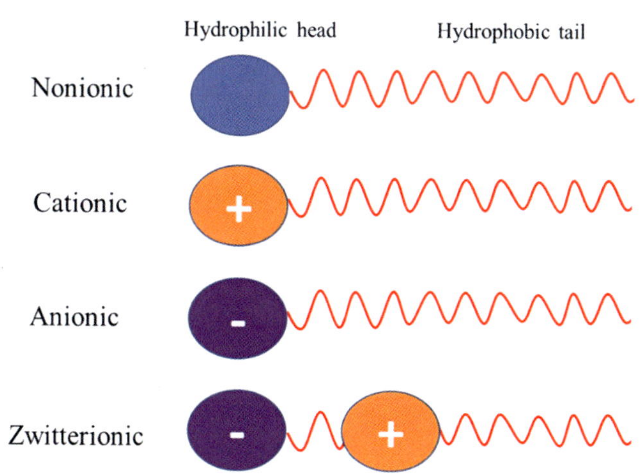

FIGURE 4.19

Illustrations of surfactant types.

The zwitterionic surfactants have one negative as well as a positively charged head group. Since sandstone reservoir rock surface is negatively charged, the anionic surfactants are most commonly used in the EOR process due to the low surfactant adsorption at neutral to high pH on both sandstones and carbonates. Anionic surfactants can be tailored to a wide range of conditions and are widely available at a low cost. Nonionic surfactants have higher tolerance of salinity; however, nonionic surfactants have less ability to reduce IFT than anionic surfactants. Therefore nonionic surfactants have not performed as well for EOR as anionic surfactants. However, in some cases, anionic and nonionic surfactants are used as a mixture to achieve a higher salinity tolerance and better IFT reduction ability. In contrast, cationic surfactants are not used in sandstone rocks because of strong adsorption, but they can be applied to carbonate reservoirs to change the wettability from oil-wet to water-wet. Zwitterionic surfactants can be nonionic–anionic, nonionic–cationic, or anionic–cationic. These surfactants show a superior tolerance to salinity and temperature; however, the major drawback for zwitterionic surfactants is the high cost.

4.3.1.2 Surfactant characterization

Some commonly used parameters to characterize surfactants are introduced in this section, including hydrophile–lipophile balance (HLB), critical micelle concentration, and Krafft temperature.

4.3.1.2.1 Hydrophile–lipophile balance

The most common parameter to characterize surfactant is the HLB. This number can indicate if the water-in-oil (W/O) or oil-in-water (O/W) emulsion will be formed once adding surfactant into an oil/water system. A low HLB number reveals that the surfactants tend to be more soluble in oil than water, which can form W/O emulsions. Typically, for the formations that have low salinity, the surfactant with a low HLB number should be selected since this type of surfactant can form microemulsion at low salinity. If the formation salinity is high, a high HLB surfactant should be selected. A completely hydrophobic molecule has 0 HLB while a completely hydrophilic molecule has 20 HLB. If the HLB value is between 0 and 3, it represents an antifoaming agent; 4–6 HLB indicates a W/O emulsifier; 7–9 represents a wetting agent; 8–18 indicates an O/W emulsifier.

Griffin's equation can be used to calculate the HLB number for nonionic surfactants:

$$\mathrm{HLB} = \frac{20\mathrm{MW_b}}{\mathrm{MW}} \tag{4.3}$$

where $\mathrm{MW_b}$ is the molecular weight of the hydrophilic part of the molecule and MW is the molecular weight of the whole molecule.

4.3.1.2.2 Micelles and critical micelle concentration

In a water/oil system, a suitable concentration of surfactant can solubilize a significant amount of oil. The concentration of surfactant plays a significant role in determining surfactant characteristics. The surfactant molecules disperse randomly in solutions at low surfactant concentrations. The surfactant molecules in this condition are defined as monomers. As the monomer concentration increases, they tend to aggregate to form micelles, which can solubilize the insoluble phase. As illustrated in Fig. 4.20, the aggregated surfactant molecules form micelles, which have a spherical shape and enfold droplets of liquid substance. In the left figure, the hydrophilic heads of the surfactant molecules enfold a droplet of oil and the tails extend to the water phase, which is called a W/O micelle. In the right figure, the hydrophilic heads joint together, and the tails are incorporated with the oil droplet in the middle of the sphere, which is called O/W micelle. These micelles can later form microemulsions.

Critical micellar concentration (CMC) of a surfactant is defined as the surfactant concentration above which the micelles will be formed. The CMC of a surfactant is impacted by surfactant molecular structure, pressure conditions, solution salinity, ionic composition, pH, temperature, etc. [54–57].

As demonstrated in Fig. 4.21, the micelles only start forming once the surfactant concentration passes the CMC. Once reaching the CMC, surfactant monomers will no longer disperse in the solution but aggregate to form micelles. This phenomenon is important for SF design because the later injected surfactant will aggregate into micelles and no longer contribute to IFT reduction significantly. In the water–oil system, the micellization occurs at a narrow concentration range. For the commonly used surfactants from the perspective of EOR, this concentration range is about

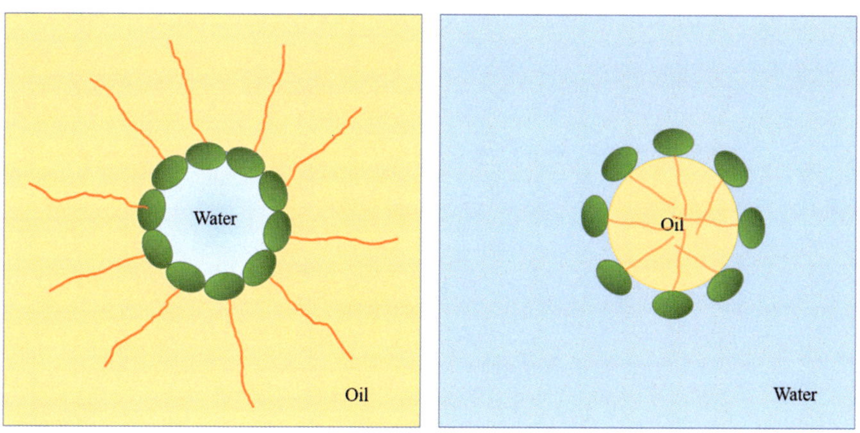

FIGURE 4.20

Micelles of water in oil (left) and oil in water (right).

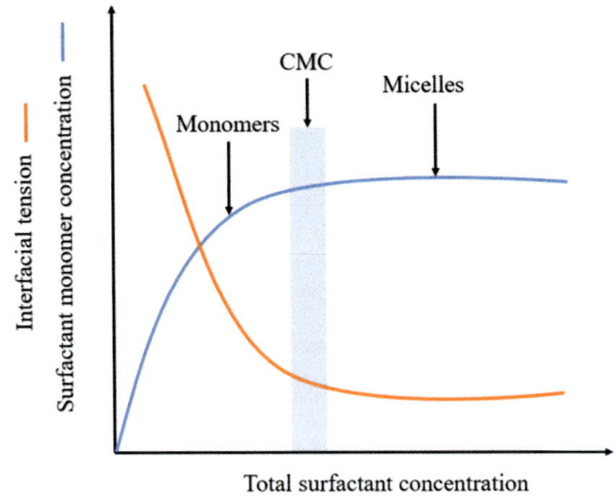

FIGURE 4.21

Relationship between critical micellar concentration and micelles.

10^{-5} to 10^{-4} mol/L [10]. Therefore the CMC for a given system is in the range of a few ppm to tens of ppm.

Many methods have been developed to measure the CMC, such as the surface tension method, the dye adsorption method, and the solubilization method [58]. The measurement methods can be classified into direct and indirect methods. The direct methods detect the changes in some properties, such as solution viscosity, electrical conductivity, refractive index, and osmotic pressure. The indirect methods measure the CMC by monitoring the changes of a certain property of a probe as a function of the change of surfactant concentration. The indirect methods include the spectrometric method and voltammetric method [59].

4.3.1.2.3 Krafft temperature

One parameter related to CMC is Krafft temperature or critical micelle temperature. This is the minimum temperature at which surfactants form micelles. Below the Krafft temperature, the surfactant tends to be insoluble and the micelles cannot form. At the Krafft temperature, the solubility of the surfactant increases dramatically. The micelles can form if the temperature is above the Krafft temperature and the CMC is reached. Krafft temperature can be measured by detecting the electrical conductivity of a surfactant solution.

4.3.1.3 Macroemulsion and microemulsion

Macroemulsion and microemulsion need to be distinguished. Macroemulsion can be defined as a suspension of liquid droplets (dispersed phase) of a certain size within a second immiscible liquid (continuous phase). Macroemulsion typically do not form spontaneously. The formation of macroemulsion requires energy input, such as shaking, stirring, and spraying. The macroemulsion solutions can

Table 4.5 Comparison between microemulsion and macroemulsion.

Microemulsion	Macroemulsion
Thermodynamically stable	Kinetically stable
Clear and translucent	Cloudy
Small size (10–200 μm)	Large size (> 1000 μm)
O/W, W/O, bicontinuous types	W/O or O/W

O/W, Oil-in-water; W/O, water-in-oil.

be opaque. The size of the macroemulsion can be larger than 1000 μm. The microemulsion is a stable dispersion of immiscible liquids. It can be formed when there is a large amount of solubilized materials, either W/O or O/W. The droplet of dispersed phase size is approximately 100 nanometers or less. The microemulsion solutions are typically clear and transparent. Sometimes microemulsion may not look clear to naked eyes due to the existence of dark heavy oil. A detailed comparison of macro- and microemulsion is shown in Table 4.5. The fundamental difference between macroemulsion and microemulsion is thermodynamic stability. Macroemulsion is kinetically stable but not thermodynamically stable. The microemulsion is thermodynamically stable [60], which means microemulsion can change with temperature but can form again once the temperature is changed back.

4.3.1.4 Types of microemulsions

Three types of microemulsions may be formed once surfactants are injected into reservoirs: Winsor type I, II, and III (Fig. 4.22). Winsor's R ratio is helpful to understand the phase behavior in a three-phase system that contains water, oil, and surfactant. The R ratio is defined as below [61]:

$$R = \frac{A_{\mathrm{co}}}{A_{\mathrm{cw}}} \tag{4.4}$$

where A_{co} and A_{cw} represent the interaction between the surfactant and oil and water, respectively. As shown in Table 4.6, different R values indicate different types of microemulsions.

Winsor type I is an O/W microemulsion, in which the surfactants stay in the aqueous phase. It is also called lower phase microemulsion or type II (−) microemulsion. The surfactant-rich water phase coexists with the oil phase. In this type of microemulsions, it is difficult to achieve ultralow IFT. Winsor type II is a W/O microemulsion, which is also called upper phase microemulsion or type II (+) microemulsion. In this system, the surfactant is mainly in the oil phase, which is observed as surfactant retention. Due to a large amount of surfactant retention, type II microemulsion should be avoided in surfactant EOR. Winsor type III is a three-phase system where a separate microemulsion phase coexists with oil and water phases. It is also called middle phase microemulsion. The formation of the Winsor type III system indicates the equilibrium status between the microemulsion,

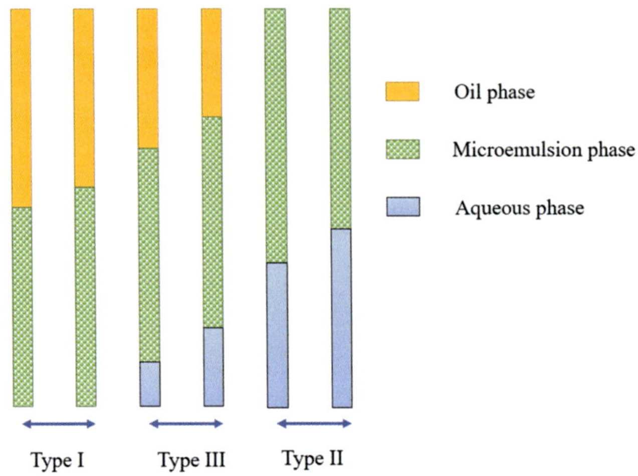

FIGURE 4.22

Illustration of Winsor types.

Table 4.6 *R* ratios and the associated microemulsion types.

R ratio	Interaction	Types of microemulsion
$R<1$	Water–surfactant interaction is stronger than oil–surfactant interaction	Winsor type I
$R>1$	Oil–surfactant interaction is stronger than water–surfactant interaction	Winsor type II
$R=1$	Water–surfactant and oil–surfactant interactions reach equilibrium	Winsor type III

water, and oil. It is the most effective surfactant EOR system since it reduces the IFT to the ultralow level (~0.001 dynes/cm).

Surfactant phase behavior is strongly affected by the brine salinity. Generally, the solubility of the anionic surfactant in the brine increases with the decrease of brine salinity, which is caused by the increased electrolyte concentration. The oil/water solubilization ratio (SR) is defined as the volume of oil/water solubilized by a unit volume of surfactant in the microemulsion phase [62,63]. The SR can be expressed as follows [64]:

$$SR_{oil} = \frac{Volume_{oil}}{volume_{surfactant}} \tag{4.5}$$

$$SR_{water} = \frac{Volume_{water}}{volume_{surfactant}} \tag{4.6}$$

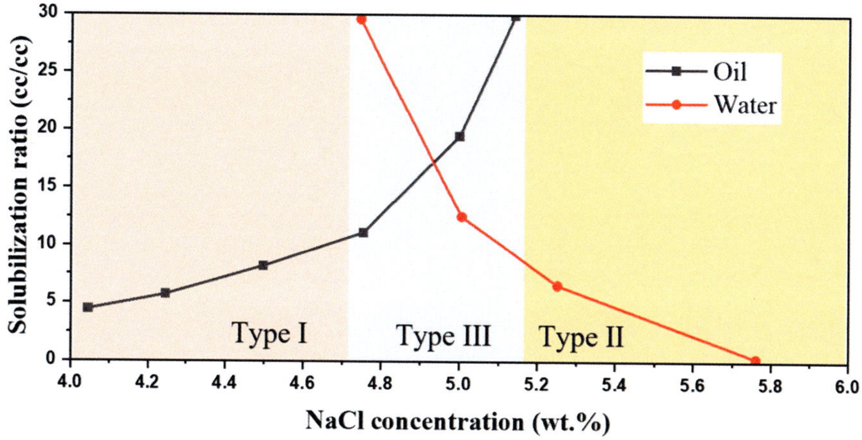

FIGURE 4.23

Determination of optimum salinity for an oil–water–surfactant system using the solubilization ratio curves.

The SR curves for water and oil are plotted in Fig. 4.23. As indicated by the figure, the oil SR increases with salinity, while the water SR decreases with the increase of salinity. To achieve equilibrium between the water, oil, and microemulsion, an optimum salinity should be determined. As illustrated in Fig. 4.23, the intersection point represents the optimum solubilization occurring at the optimum salinity [62,65]. Studying the optimum solubilization and salinity is crucial since the lowest IFT in the water–oil system is usually achieved at the optimum salinity.

4.3.1.5 Surfactant retention

Surfactant retention in rocks can be caused by adsorption, precipitation, or phase trapping. Typically, more than half of the expenditure is the surfactant cost for an SF project [66]. In this case, the study of surfactant retention is presented in terms of cost optimization and treatment effectiveness in a surfactant EOR project.

4.3.1.5.1 Adsorption

Surfactant adsorption depends on the surfactant type, concentration, equivalent weight, surfactant molecular weight, solution pH, salinity, temperature, flow rate, reservoir properties (i.e., rock minerals, clay content, porous media surface area), and so on.

Adsorption of surfactants on rock surfaces (especially carbonate reservoirs) is one of the crucial factors that impact the success of SF projects. A great amount of the injected surfactant is lost due to the large amount of adsorption, which can cause an increase in project expenditure and also reduce the treatment performance. As illustrated in Fig. 4.24, surfactant adsorption is caused by the attraction

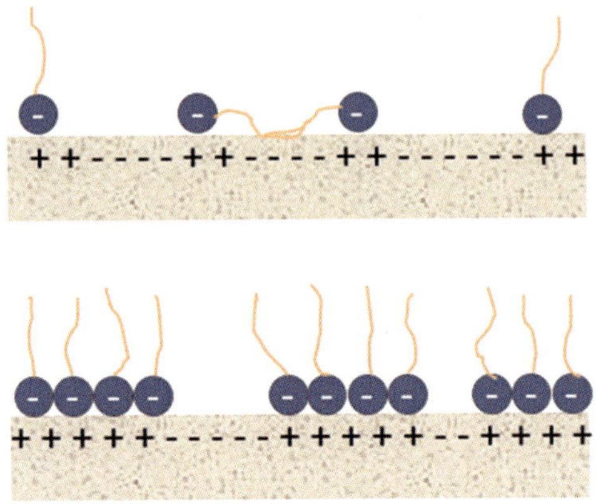

FIGURE 4.24

Surfactant adsorption on the rock surface.

between the charged solid surface (i.e., rock surface) and the charged hydrophilic head of the surfactant [67]. In general, surfactant adsorbs on rock surfaces as single monomers rather than micelles. The adsorption of nonionic and anionic surfactants is studied using both sandstone and carbonate rocks [68]. Typically, nonionic surfactants have much higher adsorption in sandstone rock than anionic surfactants. Additionally, the adsorption of nonionic surfactant on calcite is much less than that of anionic surfactant with no Na_2CO_3. In this case, the application of nonionic surfactants in carbonate reservoirs might have less adsorption.

Regarding the surfactant type and rock type, nonionic surfactants have much higher adsorption on a sandstone surface than anionic surfactants [68]. However, Liu's initial experiments indicated that the adsorption of nonionic surfactant on calcite was much lower than that of anionic surfactant without the presence of Na_2CO_3 and was of the same order of magnitude as that of anionic surfactant with the presence of Na_2CO_3. Thus nonionic surfactants might be candidates for use in carbonate formations from the adsorption point of view. The performance of both nonionic and anionic surfactants is not highly impacted by salinity. However, nonionic surfactants are more sensitive to temperature than anionic surfactants [69].

The surface area of the porous media has a significant impact on the adsorption of surfactants. Liu [68] studied the surfactant adsorption in three carbonate rock samples with different surface areas (Fig. 4.25). The surfactant is a 1:1 mixture by weight of dodecyl 3 ethoxylated sulfate and iso-tridecyl 4 propoxylated sulfate from Stepan. The results revealed that the adsorption in the three rocks is similar to each other in terms of adsorption per surface area (Fig. 4.25, top). However, if the adsorption is calculated per weight of porous media, the results are very different even though the mineral composition of the three samples is similar (Fig. 4.25, bottom). This experiment indicates that the surfactant adsorption should be calculated using surface area rather than weight in terms of comparison.

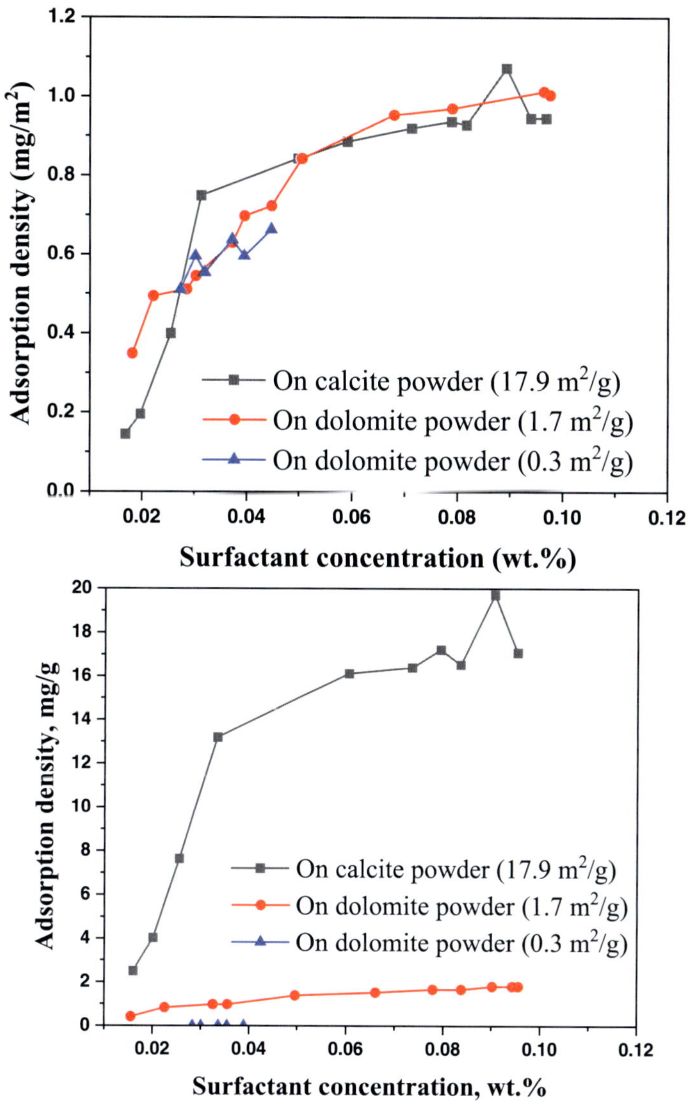

FIGURE 4.25

Adsorption of the 1:1 mixture of iso-tridecyl 4 propoxylate sulfate and dodecyl 3 ethoxylated sulfate on different samples in terms of adsorption per surface area (top) and expressed in mg/g on the same three samples (bottom).

Reproduced with permission from S. Liu, Alkaline Surfactant Polymer Enhanced Oil Recovery Process (Ph.D. dissertation), Rice University, Houston, 2008.

In addition to the abovementioned factors, the pH of the surfactant plays a major role in surfactant adsorption. The concentration of divalent ions (Ca^{2+}, Mg^{2+}) is also an important parameter controlling the amount of surfactant adsorption. Diffusion of surfactants on pores of a rock (phase trapping) can cause surfactant retention.

4.3.1.5.2 Precipitation

Besides adsorption, precipitation is also a factor that causes increased surfactant retention. Generally, the change in temperature dissolute the surfactant into salts, which can cause precipitation [70]. In terms of reducing surfactant precipitation to increase surfactant treatment efficiency, a surfactant with robust stability is needed, which can resist temperature change, lower dissolution to salt, and reduce precipitation.

4.3.1.5.3 Phase trapping

Phase trapping can be mechanical trapping, phase partitioning, or hydrodynamic trapping [25]. Mechanical trapping mechanisms are complex. The amount of surfactant retention caused by mechanical trapping depends on the condition of multiple-phase flow. Even though the mechanisms of phase trapping are complex, the types of microemulsions are believed to be a major factor that is closely related to phase trapping. In a Winsor II microemulsion system, which is a W/O emulsion, the emulsion viscosity is higher than the chase water and the IFT near the chase water than emulsion is high; therefore the chase water can bypass the microemulsion bank easily, which causes phase trapping. In a Winsor I microemulsion system, water is the continuous phase. In this case, the chase water is miscible with emulsion; therefore, the microemulsion can be displaced by the chase water without phase trapping.

4.3.2 Surfactant enhanced oil recovery mechanisms

Surfactants reduce the water/oil IFT by emulsifying two immiscible phases into a single solution. In addition, surfactants can also change reservoir wettability from oil-wet to a more favorable water-wet in fractured carbonate reservoirs. Reduction in IFT improves microscopic displacement efficiency, which is a pore-scale oil displacement efficiency enhancement. In some of the cases, polymers are introduced after SF as a mobility control agent to improve macroscope sweep efficiency (volumetric sweep efficiency). Another surfactant EOR mechanism is emulsion formation.

4.3.2.1 Interfacial tension reduction

IFT exists when two immiscible phases are present. It is the contractile tendency between two immiscible phases and is normally measured in dynes/cm. IFT is a function of pressure, temperature, and the composition of each phase. The oil and brine IFT is usually between 20–30 dynes/cm in reservoir rocks. The IFT between oil and water molecules increases the capillary force, which can cause trapping in porous media. High IFT leads to a high residual oil saturation in the reservoirs. Some surface-active agents, such as surfactants, can increase surface area at the immiscible interface to reduce IFT, hence improving microscopic displacement efficiency to reduce residual oil saturation. Surfactant solutions can also mobilize and displace oil until the surfactant is diluted or lost due to adsorption.

The capillary number is defined as the ratio of viscous forces to capillary forces [71], which can be expressed by the following equation. This equation is the simplest form, but it is also the most widely used one.

$$N_c = \frac{F_v}{F_c} = \frac{v\mu}{\sigma\cos\theta} \tag{4.7}$$

where N_c is a capillary number, v is the effective velocity (m/s), μ is the viscosity of the displacing fluid (cp), and σ is the IFT between displacing and displaced fluids (dyne/cm).

The capillary desaturation curve (CDC) is used to calculate the capillary number in a typical waterflooding case. A typical CDC is shown in Fig. 4.26. CDC can relate residual oil saturation to the physical properties, such as capillary forces and IFT, on a microscopic scale (pore scale) [72]. Taber [73] indicated that the residual oil saturation decreased with the increase of capillary number. The reduction of IFT can reduce the capillary number, hence reducing the residual oil saturation. The inversely proportional relationship between IFT and capillary number can be a method to reduce residual oil saturation.

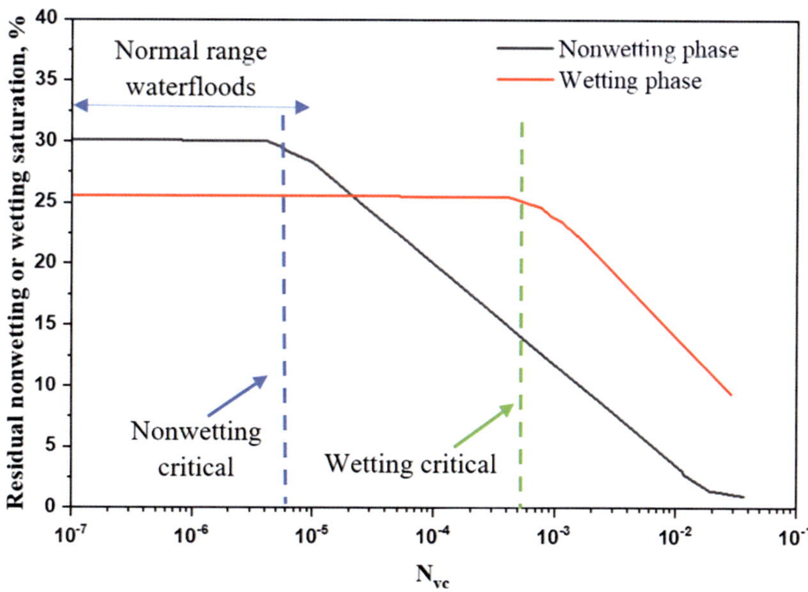

FIGURE 4.26

Capillary desaturation curve.

In a typical waterflooding project, the water injection velocity is 1 ft/day, the water viscosity is assumed to be 1 cp, and the IFT is 30 dynes/m, the corresponding capillary number is calculated to be around 10^{-7}. Taber suggested that the capillary number after waterflooding can be further increased by adding surfactants into displacing fluids to decrease the IFT of the system.

The viscous force used to calculate capillary number is a function of reservoir permeability, applied pressure drop, and the viscosity of the displacing fluids. The capillary force is inversely proportional to capillary numbers and it is a function of IFT between displacing and displaced fluids, reservoir wettability, and the pore structures [74]. As aforementioned, the capillary number can be increased by either increasing the viscous forces or decreasing the capillary forces. However, increasing the viscous forces may not always be practical. Theoretically, the viscous forces can be increased by increasing the applied pressure drop or increasing the displacing fluid viscosity. There is a high risk that the injection pressure will exceed the formation fracturing pressure if using a high-pressure drop or injecting water with too high viscosity. Therefore, adding surfactants to reduce the capillary forces or using gas to reach miscible flooding is a safer and more practical method to reduce residual oil saturation.

4.3.2.2 Wettability alteration of the rock

Rock wettability is one of the major factors that impact ultimate oil recovery because it can impact the degree of immiscible displacement, electrical properties, capillary pressure, relative permeability, saturation profiles, and distribution of fluids in a reservoir.

Reservoirs can be water-, oil-, or intermediate-wet. Compared to sandstone reservoirs, most of the carbonate reservoirs are oil- or intermediate-wet with natural fractures. Typically, water is injected into the reservoirs to improve oil recovery. The efficiency of waterflooding in fractured reservoirs highly depends on the spontaneous imbibition of water to displace the oil in the matrix. This spontaneous process is more effective in water-wet reservoirs since the capillary driving forces are strong. However, the capillary driving in fractured oil- or intermediate-wet reservoirs are weak, which yields a lower oil recovery during waterflooding [75]. Furthermore, in nonfractured oil-wet rock, oil is coated on the matrix rock surface due to the lipophilic surface properties, which causes high residual oil in the reservoirs. In the case of water-wet or intermediate-wet reservoirs, since the rock surface is hydrophilic, the injected water has a better displacement efficiency than that in the oil-wet reservoirs. Most of the carbonate reservoirs are oil wet and have natural fractures, which act as high-permeability conduits for both water and the displaced oil. The oil retains on the fracture surface and matrix in the oil-wet carbonate reservoirs, the injected water flows through high-permeability fractures and cannot efficiently displace the oil to production wells. If the wettability changes from oil-wet to water-wet, the injected fluid will displace the residual oil on the fracture surface and the remaining oil in the matrix, which is dominated by capillary force. Therefore, the wettability alteration can improve oil recovery. Using surfactants to change rock wettability is one of the techniques that have been investigated by many researchers to improve oil recovery from fractured oil-wet carbonate and tight sandstone reservoirs. As shown in Fig. 4.27, the rock wettability can be altered from oil-wet to a favorable water-wet by using surfactants. Wettability alteration is impacted by the aging time, surfactant type, concentration, formation temperature, crude oil composition, reservoir rock properties, etc. The elevated temperature changes the solubility of the adsorbed carboxylate groups and desorbs them from the rock surface. Accordingly, the contact angle decreases with increasing temperature, which indicates a more water-wet rock.

The most intuitive method to measure wettability is to measure the contact angle between the rock surface and the oil droplet. The contact angle is always measured through the denser phase. The contact angle can also be measured using the static sessile drop method, the pendant drop method, the dynamic sessile drop method, the dynamic Wilhelmy method, Washburn's equation capillary rise method, etc.

4.3.2.2.1 Wettability alteration by cationic surfactants

The organic components in crude oil contain negative-charged groups, which are normally adsorbed on negatively charged carbonate rocks. Using cationic surfactant monomers interact with the anionic materials (mostly carboxylate) in the crude oil, which is "coated" on the rock surface (Fig. 4.28). The formation of ion

FIGURE 4.27

The wettability of a water-wet solid surface in contact with an oil droplet increases when surfactant molecules adsorb on the surface.

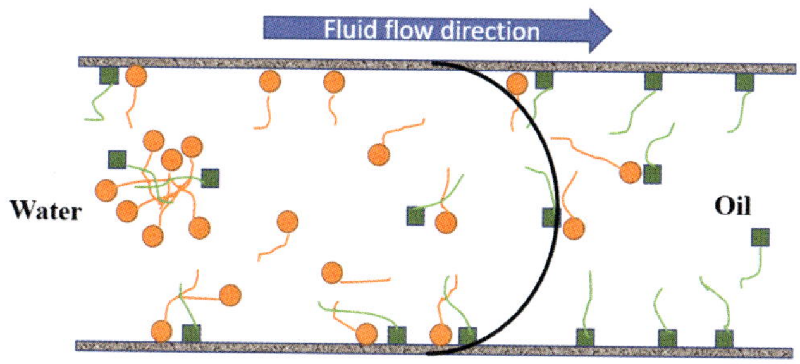

FIGURE 4.28

Mechanism of wettability alteration for cationic surfactants. Circles represent the anionic surfactant, and squares represent the carboxylic substance.

Reproduced with permission from M.S. Kamal, I.A. Hussein, A.S. Sultan, Review on surfactant flooding: phase behavior, retention, IFT, and field applications, Energy & Fuels 31 (2017) 7701–7720.

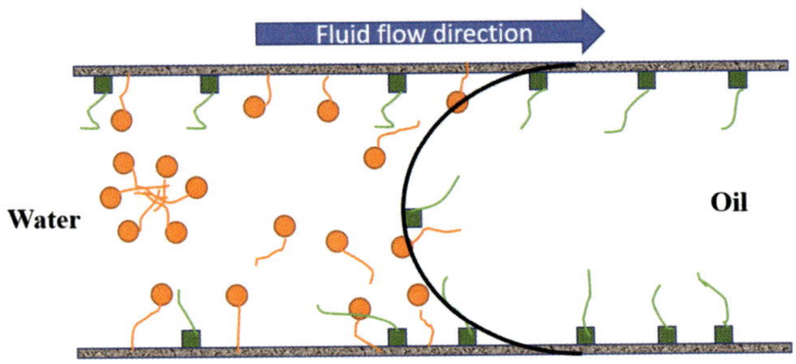

FIGURE 4.29

Mechanism of the formation of the bilayer with ethoxylated sulfonate. Circles represent the anionic surfactant, and squares represent the carboxylic substance.

Reproduced with permission from M.S. Kamal, I.A. Hussein, A.S. Sultan, Review on surfactant flooding: phase behavior, retention, IFT, and field applications, Energy & Fuels 31 (2017) 7701–7720.

pairs between the cationic monomer and anionic groups in crude oil causes the desorption of the adsorbed material at the water, oil, and rock interface. The formed ion pairs are not soluble in the aqueous phase, such as water, but soluble in the oleic phase, such as oil. Therefore the water will invade into the rock minerals and oil will be displaced from the rock surface. When the adsorbed materials are desorbed from the surface, the rock becomes more water wet, and the residual oil that is adsorbed on the rock can be displaced by water.

4.3.2.2.2 Wettability alteration by anionic surfactants

Unlike cationic surfactants, anionic surfactants cannot desorb carboxylate groups from the rock. Generally, the tails of the anionic surfactants interact with the negatively charged adsorbed groups through hydrophobic interaction to weaken capillary forces. A water-wet bilayer between the carbonate rock surface and oil will be formed during the minor displacement by ethoxylated sulfonates from carbonate rock. The formation mechanism of the water-wet bilayer is illustrated in Fig. 4.29.

4.3.2.3 Formation of emulsions

Surfactant solutions can improve oil recovery by forming emulsions. The two main mechanisms are emulsification and entrainment [76–78]. In the case of emulsification and entrainment, oil is emulsified into small droplets in the water phase. Typically, the sizes of the formed oil droplets are equal to or smaller than the pore sizes, which allows the transportation of crude oil in the form of small droplets. In the case of emulsification and entrapment, since the IFT is not low enough to reduce the average size of the emulsion droplets, the formed droplets are larger than the pore throats, which can plug the pore throats and water-flowing paths. As a consequence, the surfactant solution is diverted to an unswept area, which improves the macroscopic sweep efficiency [79–81].

4.3.3 Characterization and evaluation of commonly used surfactant agents

The effect of salinity, temperature, and pH on surfactant performance is discussed in this section. As listed in Table 4.7, the performance of some surfactants has been tested under various salinity conditions. It is observed that the adsorption of cationic, zwitterionic, and anionic surfactants increases with salinity.

Table 4.8 shows some laboratory studies of temperature effect on surfactant adsorption. For both anionic and nonionic surfactants, the adsorption decreases with increased temperature.

Table 4.7 Summary of salinity effect on surfactant adsorption.

Surfactant name and type	Rock type	Salinity (ppm)	Main conclusions
Dodecyltrimethylammonium chloride, tetradecyltrimethylammonium chloride, and octadecyltrimethylammonium chloride (cationic)	Carbonate	13,987.7, 27,975.4, 55,950.7, 111,901.4, and 223,802.8	The adsorption of surfactant on limestone increases with salinity.
Hexadecyl-*N*,*N*-dimethyl-2-ammonio-1-ethanecarbonate (Zwitterionic)	Carbonate and sandstone	0, 15,000, 30,000, and 45,000	• Adsorption in carbonate rock is higher than in sandstone rock. • At CMC, adsorption increases with salinity because the increased salinity reduces the surfactant solubility and reduces the repulsion between the adsorbed surfactant molecules
Soap-nut surfactant (anionic)	Sandstone, carbonate, and bentonite	0, 10,000, 20,000, and 30,000	• Adsorption increases with salinity • Surfactant adsorption on bentonite is the highest and on sandstone is the lowest • Rock minerology and morphology affected the surfactant adsorption

(*Continued*)

Table 4.7 Summary of salinity effect on surfactant adsorption. *Continued*

Surfactant name and type	Rock type	Salinity (ppm)	Main conclusions
Triton X-100 (nonionic) and cetyltrimethylammonium bromide (cationic)	Shale	0, 10,000, and 30,000	• For both nonionic and cationic surfactants, the adsorption on shale increases with salinity • The high quartz content in the shale contributes to a high adsorption of cationic surfactant
Aerosol-OT (anionic)	Quartz sand, illite, kaolinite, and montmorillonite	0, 20,000, 35,000, and 60,000	• The value of CMC increases with salinity with the presence of minerals • The adsorption rate in quartz is the highest and the rate in kaolinite is the lowest

CMC, *Critical micellar concentration.*
Source*: Data from [82–86].*

Table 4.8 Summary of temperature effect on surfactant adsorption.

Surfactant name and type	Rock type	Temperature (°C)	Main findings
Synthesized surfactant (C18 carbon chain and a sulfonate headgroup) (anionic)	Berea sandstone	25 and 70	Surfactant adsorption was 0.9604 mg/g at 25°C and it decreases to 0.7491 mg/g at 70°C
Alkyl polyglycoside (APG) (nonionic)	Sandstone	20, 50, 80, and 115	The adsorption of 0.6 wt.% APG decreases from 24.2 to 21.6 mg/g when the temperature changes from 20°C to 115°C
Saponin (nonionic)	Shale-sandstone surface	28, 40, 60, and 75	Surfactant adsorption decreases with the increased temperature at different surfactant concentrations
Sodium dodecyl sulfates (anionic)	Kaolinite	25.85, 49.85, and 79.85	Surfactant adsorption decreases with increased temperature

Source*: Data from [82,87–91].*

Table 4.9 Summary of pH effect on surfactant adsorption.

Surfactant name and type	Rock type	pH values	Main findings
Sodium dodecyl sulfates (SDS) (anionic)	Porous SiO_2 powder gel	3–7	SDS adsorption on SiO_2 surface decreases with increased pH, which is caused by the increasing of electrostatic repulsion
Internal olefin sulfonates (IOS) 2024 (anionic)	Sandstone	8.24–9.57	The adsorption of anionic surfactant (3000 ppm) decreases from 0.76 to 0.161 meq/100 gram of rock when the pH changes from 8.24 to 9.57
Soap-nut surfactant (anionic)	Sandstone Carbonate Bentonite	Alkali addition = 0, 1%, 2%, and 3% (Na_2CO_3)	The presence of Na_2CO_3 increases the pH and yields more electrostatic repulsion of anionic surfactants. Surfactant adsorption decreases with increased Na_2CO_3 concentration
IOS (anionic) Tridecyl alcohol propoxy sulfate TDA (anionic) Guerbet alkoxylate carboxylate (anionic)	Indiana limestone	~7.6 to ~12	The adsorption of two surfactant mixtures decreases nearly linearly once the pH exceeds 9
Cetyltrimethylammonium bromide (CAPB) (zwitterionic)	Dolomite	7.3–11.2	CAPB adsorption reduces from 0.795 to 0.32 mg/g rock when the pH increases from 7.3 to 11.2
Alcohol alkoxy sulfate (anionic)	Clay surface	6–11	With the presence of Na + , the elevated pH value reduces alcohol alkoxy sulfate adsorption. The opposite was observed in the presence of Ca^{2+} ions

Source: *Data from [82,85,92–96].*

Table 4.9 shows the pH effect on surfactant adsorption, which is commonly discussed in surfactant–AF. As shown in the table, for anionic and zwitterionic surfactants, the adsorption on different types of surfaces is disproportional to solution pH.

4.3.4 Surfactant screening criteria

Surfactant screening aims at identifying the most suitable surfactant formulation for specific reservoir conditions. Surfactant screening can also help with the prediction of technical problems before it happens during the field application.

Therefore, researchers can find solutions to these problems and optimize EOR treatment design to prevent any damages.

Generally, a good surfactant candidate for SF should have good tolerance to reservoir salinity, exhibit good thermal stability under reservoir temperatures, be capable to reduce the IFT under reservoir conditions, and have low adsorption on reservoir rock. Due to the economic considerations, the selected surfactant should be effective at a low concentration (0.1%–0.3%) to reduce the required chemical amount, hence lowering the cost. In addition, the surfactant should be commercially available at a reasonable cost. Typically, surfactants are used in combination with other chemicals, such as polymers, alkalis, and nanoparticles; therefore, the selected surfactant candidates should not interact with these chemicals [97–100].

The procedures for surfactant screening can be described as follows:

- Test the compatibility of the candidate surfactants with chemicals (polymer, alkalis, etc.) that involve in the same project under reservoir conditions;
- Examine phase behavior of candidate surfactant. The samples showing Winsor type III microemulsion are preferred;
- Evaluate the ability of candidate surfactants to reduce IFT between crude oil and formation brine;
- Measure surfactant adsorption on reservoir rocks;
- Test long-term stability under reservoir temperature and salinity;
- Evaluate the ability of candidate surfactants to improve oil recovery using coreflooding experiments;
- Select the most suitable surfactant for pilot tests or field applications by combining all the test results and considering the cost.

4.3.5 Application limitation and the status of the application

As aforementioned, surfactants can improve microscopic sweep efficiency and also change rock wettability. However, SF is not applicable in any reservoirs, some restrictions are applied. Generally, surfactants are applied in reservoirs with high permeability and porosity, the oil saturation of higher than 25%, and the oil viscosity of less than 50 cp. However, a recent trend is to apply surfactants viscous oils (up to 200 cp) or even higher viscosity. To achieve a sufficient treatment, the reservoirs should have good geological continuity and the reservoir temperature is less than 300°F. If the polymer is combined with surfactant for mobility control, the temperature should be less than 220°F. In addition, a good source of high-quality water can improve the performance of surfactants.

Many studies have proven that surfactants can improve oil recovery; however, it also has some limitations and challenges that require further research. Due to the cost limitation and surfactant adsorption, the application of singular surfactants is not common. Typically, surfactants are combined with polymers, alkalis, or both, which is called surfactant–polymer (SP) flooding, surfactant–alkali flooding, and alkaline–surfactant–polymer (ASP) flooding, respectively. Table 4.10 summarizes some field applications of chemical EOR which involve surfactants in the recent

Table 4.10 Summary of chemical enhanced oil recovery field application cases with surfactants.

Field	Location	Formation type	Formation properties	Oil properties	Surfactants	Other chemicals	Performance
Daqing oilfield (B-1-D Block)	China	Sandstone	$k = 670$ mD $P = 858.6$ psi $T = 42.4°C$ Salinity = 5611 ppm	Density: 0.864 g/cm^3 Viscosity: 8.2–9.3 cp	ABS (0.1%–0.3%)	HPAM polymer (1300–2000 mg/L)NaOH (0.8%–1.2%)	Alkaline–surfactant–polymer (ASP) EOR increased oil recovery by 30%
Daqing oilfield (Chao 522 Block)	China	Sandstone	$k = 18.7$ mD	N/A	Nonionic alkanol acid amide surfactant (1.0%)	Auxiliary agent	In a single well, the oil recovery increased by 30%
Daqing oilfield (northern part)	China	Sandstone	$k = 577$ mD $T = 45°C$ Salinity = 4456 ppm	Density: 0.86 g/cm^3 Viscosity: 9–10 cp	Primary surfactant: ABS (0.1%–0.15%) Sacrificial agent: RH biosurfactant (0.2%)	HPAM polymer (800–1800 mg/L) NaOH (1.2%)	The ASP pilot test yielded a total of 16.6% of OOIP. Meanwhile, the project cost was reduced significantly
Dagang oilfield (Guan-109 Reservoir)	China	Sandstone	$k = 163.1$ mD $T = 78°C$ Salinity = 26,983 ppm	Viscosity: 67 cp	Petroleum sulfonate surfactant PS-985 (0.3 wt.%)	Associate polymer AP-P7 (2000 mg/L)	The combination of surfactant and associate polymer flooding increased the oil production rate from 27.7 to 51.7 m^3/day
Gudong oilfield (the tertiary Guantao member)	China	Sandstone	$k = 3.8$ mD $P = 1842$ psi $T = 68°C$ Salinity = 3025 ppm	Viscosity: 41.3 cp	OP-10 (0.2%) CY1 (0.2%)	Na_2CO_3 (1.5%) 3530S polymer (0.1%)	A total of ASP injection of 0.55 PV was designed. The oil recovery from ASP flooding increased by 13.4% OOIP
Karamay EZ district	China	Sandstone	Salinity = 6693 ppm	Density: 0.8659 g/cm^3 Viscosity: 41 cp	Petroleum sulfonate (0.35%)	Na_2CO_3 (1.2%) 3530S polymer (0.13%) $Na_5P_3O_{10}$ (0.1%)	The reservoir was preflushed using NaCl brine, then followed by ASP flooding and polymer flooding. The 24% OOIP incremental oil recovery was achieved in the whole study area

(*Continued*)

Table 4.10 Summary of chemical enhanced oil recovery field application cases with surfactants. *Continued*

Field	Location	Formation type	Formation properties	Oil properties	Surfactants	Other chemicals	Performance
Cottonwood creek field	Wyoming (United States)	Carbonate	$k = 0.2-130$ mD $T = 60°C$ Salinity = 30,755 ppm	Oil gravity: 30°API Viscosity: 24.3 cp	Polyoxyethylene alcohol (1500 ppm)	N/A	23 wells were treated using the surfactant huff-n-puff method. Nearly 70% of the treatments were failed
Cambridge Minnelusa Reservoir	Wyoming (United States)	Eolian sandstone with dolomite and anhydrite cement	$k = 845$ mD $P = 1792$ psi $T = 55.6°C$	Oil gravity: 20°API Viscosity: 31 cp	Petrostep B-100 surfactant (0.1 wt.%)	Na_2CO_3 (1.25 wt.%) Alcoflood 1275A polymer (1450 mg/L)	ASP flooding increased oil recovery by 26.8%
Tanner Minnelusa Field (Campbell County)	Wyoming (United States)	Carbonate	$k = 200$ md $T = 79.4°C$ Salinity = 66,800 ppm	Oil gravity: 21°API Viscosity: 11 cp	ORS-41 (0.1 wt.%)	NaOH (1.0 wt.%) Alcoflood 1275A polymer (1000 mg/L)	ASP flooding increased oil recovery by 17% OOIP. ASP treatment yielded a 23.5% peak oil cut
East Seminole Field (San Andres Formation)	Texas (United States)	Carbonate	$k = 3$ mD $P = 3400$ psi $T = 40°C$ Salinity = 70,000 ppm	Oil gravity: 31°API	Linear ethoxylated alcohol (0.5 wt.%)	CO_2	Compared to baseline CO_2 injection, pilot CO_2 injectivity was reduced by 70% after each surfactant slug
Reservoir SF	Oklahoma (United States)	Sandstone	$k_{avg} = 74$ mD $T = 42°C$ Salinity = 165,000 ppm	Oil gravity: 40°API Viscosity: 4 cp	Primary surfactant: sodium sulfosuccinate (0.084%) Cosurfactant: sodium alkyl ethoxylate sulfate (0.105%) and branched sodium diphenyl oxide disulfonate (0.042%)	CPXG biopolymer (1800 ppm)	In a single-well test, oil recovery increased by 19%

Viraj oilfield (Ahmedabad-Mehsana tectonic block)	India	Sandstone	$k = 4500-9900$ mD $P = 1920$ psi $T = 81°C$	Oil gravity: 18.9°API Viscosity: 50 cp	Petroleum sulfonate (0.2 wt. %, 100% active)	Na_2CO_3 (1.5 wt.%) Anionic polymer (PHPA) (800 ppm)	ASP injection increased the oil rate from 24.4 to 98.23 m^3/day
Tanjung Field (Zone A)	Indonesia	Sandstone	$k = 10-100$ mD $T = 60-63°C$ Salinity = 3800 ppm	Oil gravity: 40°API Viscosity: 1.14 cp	Anionic surfactant A127 (2.05%)	N/A	A significant oil recovery increment was observed
Semoga Field (Baturaja Formation)	Indonesia	Carbonate	$k = 3.79-14.9$ mD $P = 845-895$ psi	Viscosity: 0.98–1.28 cp	Nonionic surfactants	N/A	Water cut reduced by 8% and the oil recovery increased by 5800 bbls in three months
West Salym Field	West Siberia (Russia)	Sandstone	$k = 10-250$ mD $T = 83°C$ Salinity = 14,000–16,000 ppm	Density: 0.87 g/cm^3 Viscosity: 2 cp	IOS 24–28 (0.5 wt.%) IOS 15–18 (0.2 wt.%)	FLOPAAM 3230 polymer (0.25 wt.%) Na_2CO_3 (0.8–2.0 wt.%) C4H10O (2.0 wt.%)	ASP pilot test yielded a 16% oil recovery increment
N/A	Colombia	Sandstone	$P = 1350.3$ psi $T = 50°C$	Oil gravity: 19°API	Anionic surfactant (350–1000 mg/L)	SiO_2 nanoparticles (100 mg/L)	The production from two pilot wells increased by 30,035 bbls in 5 months after treatment

ABS, *Alkylbenzene sulfonate;* ASP, *alkaline–surfactant–polymer;* EOR, *enhanced oil recovery;* HPAM, *hydrolyzed polyacrylamide.*
Source: *Data from [40,48,102–117].*

20 years. The data indicate that surfactants have been used in many oil fields around the world and most of these reservoirs are sandstone reservoirs. As listed in Table 4.10, the most widely used surfactant is sulfonate surfactants. Most laboratory experimental results reveal that surfactants can improve oil recovery by 20%–30% [101]. However, the incremental oil recovery is only in the range of 12%–30% in the fields due to reservoir complexity [47].

4.3.6 New development of surfactant flooding

Table 4.11 summarizes several new research directions proposed in the most recent published papers. The new development of SF includes the elimination of alkali utilization, combination of surfactants and nanoparticles, controlled surfactant release, development of natural/biosurfactants, the combination of various types of surfactants, and surfactant optimization.

Table 4.11 Summary of the new development of surfactant flooding.

Categories	Surfactant names	Surfactant type	Remarks
Elimination of alkali utilization	POAPMB	Zwitterionic	A novel surfactant–polymer system is developed to eliminate the utilization of alkali to alleviate the formation of scale. This SP system increased oil recovery by about 15% under high-salinity conditions.
	AOS APS ABS CAHS	Anionic Anionic Anionic zwitterionic	A novel surfactant–polymer system is developed to eliminate the utilization of alkali. This SP system increased oil recovery by about 24.5%–34.8% under high-salinity conditions.
Combination of surfactants and nanoparticles	Commercial anionic surfactants	Anionic	Nanoparticles are added to surfactant solutions to improve EOR efficiency. The results indicate that the silica nanoparticles (100 mg/L) increase oil recovery by 5% than surfactant flooding.
	SDS Aerosol-OT LAPE Betaine-based surfactant	Anionic Anionic Nonionic Zwitterionic	Nanoparticles are implanted in surfactant treatments. 1 wt.% of silica nanoparticles are added to the methane-surfactant foam. This combination improves oil recovery by 30%.
	CTAB TR-880	Cationic Zwitterionic	A combination of nanofluid (500 ppm) prepared from Fe_3O_4/eggshell and CTAB solution (1000 ppm) is developed to

(Continued)

Table 4.11 Summary of the new development of surfactant flooding. *Continued*

Categories	Surfactant names	Surfactant type	Remarks
			improve oil recovery. The experimental results demonstrate that this combination can improve oil recovery by 8.16%.
	SDS	Anionic	Surfactant solutions are used to stabilize nanoparticles. Appropriate SDS concentrations can prevent nanoparticles from aggregation. The SDS concentration depends on solution salinity.
Controlled surfactant release	CTAB	Cationic	Nanocapsules are developed to control surfactant releasing. Experimental results indicate that after exposure to 56,000 mg/L brine for 12 days, only 46% of CTAB is released.
	Alkyl betaine surfactant	Zwitterionic	Encapsulation technique is utilized to achieve controlled surfactant release. Surfactant encapsulation increased oil recovery by 17%–26%.
	SDS	Anionic	Encapsulation technique can reduce dynamic surfactant adsorption on the sand-kaolin surface by 43.3%.
Development of natural or biosurfactants	Saponin	Nonionic	Saponin is a natural surfactant. Results indicate that 5 g/L saponin solution can improve oil recovery by 6.2%–8.4%.
	Saponin	Nonionic	Saponin is a natural surfactant. Results indicate that 4 wt.% saponin solution can recover 19.2% of OOIP using spontaneous imbibition.
	APG	Nonionic	APG is a biosurfactant, which can be used to enhance oil recovery. 2% APG solution can recover 52.1% of OOIP.
	Vitagnus plant extract	N/A	Vitagus plant is a natural surfactant that was developed for EOR purposes. The vitagnus surfactant with a 3000 ppm concentration can improve oil recovery by 10.6%.

(*Continued*)

Table 4.11 Summary of the new development of surfactant flooding. *Continued*

Categories	Surfactant names	Surfactant type	Remarks
Combination of anionic and zwitterionic surfactants	APS ABS AAS AES ADAO CAHS	Anionic Anionic Anionic Anionic Zwitterionic Zwitterionic	Mixtures of anionic and zwitterionic surfactants are tested for chemical EOR purposes. A proper anionic–zwitterionic surfactant solution can reduce the IFT below 0.001 dynes/cm, which outperforms the singular surfactants.
Comparison between different surfactants' ability on wettability alteration	ADAC ATAC APS IOS AES CAPB	Cationic Cationic Anionic Anionic Nonionic Zwitterionic	Several surfactants are tested to study the ability of wettability alteration. The results reveal that the nonionic surfactants have better wettability alteration ability depending on the tail length and EO number.

ABS, *Alkylbenzene sulfonate;* CTAB, *cetrimonium bromide;* EOR, *enhanced oil recovery;* IFT, *Interfacial tension.*
Source: *Data from [82,118–131].*

Alkalis are used to reduce surfactant adsorption on rock surfaces in low-salinity reservoirs. In high-salinity reservoirs, conventional alkalis may precipitate and damage the formation. Recent research suggests that sodium metaborate can be applied in high-salinity reservoirs. However, adding alkalis may cause precipitation of calcium carbonate. Using chelating agents can prevent precipitation, but it is questionable from an economic perspective. In this case, some studies are proposing new surfactants to reduce the usage of alkalis to avoid these problems.

Another trending study is combining surfactants with nanoparticles. As reported in the literature, nanoparticles, such as polysilicon, iron oxide, and silica particles, can change rock wettability, stabilize emulsions, stabilize CO_2 foam, and reduce IFT. The combination of surfactant and nanoparticles can reduce the IFT more than a singular surfactant and yield a higher oil recovery. In addition, the presence of nanoparticles can reduce surfactant adsorption.

4.3.7 Remaining challenges

Despite the abovementioned new developments of SF, the applications of SF in high-temperature and high-salinity reservoirs are still challenging. Compared to other surfactants, sulfonate surfactants have better performance under high-temperature and low-salinity conditions [47]. However, in high-temperature and high-salinity reservoirs, surfactant precipitation, high adsorption, and chemical degradation may occur. Another challenge for SF is to reduce adsorption or eliminate the reactions with minerals, especially clays. The loss of surfactants increases with clay content. To reduce the precipitation and adsorption, a sacrificial agent may be injected as preflush. However, this method may not be always practical for the consideration of the project budget.

Typically, laboratory evaluation of surfactants yields higher oil recovery than field applications. In this case, the prediction of field application results using laboratory experiments is also a challenge. Laboratory results should be viewed with caution because the laboratory-scale experimental models cannot reflect reservoir heterogeneity and complexity.

4.4 Alkaline flooding

AF, also known as caustic flooding, is an EOR technique in which the injected alkaline chemicals interact with certain types of oils to form in-situ surfactants in the reservoirs. The formed in-situ surfactant can reduce IFT between oil and water and increase oil production. Common alkaline chemicals include sodium hydroxide, sodium orthosilicate, and sodium carbonate. However, alkalis are not recommended for carbonate reservoirs because the abundant calcium ions in the reservoirs can interact with alkalis to produce hydroxide precipitation, which can damage the formation.

AF is a technique that injects a water-soluble high-pH agent into reservoirs. The injected agents can react with components (naphthenic acids) of the crude oil to produce inexpensive in-situ surfactants. AF can be applied singularly or in combination with surfactants, polymers, or both. The combination of alkalis with surfactants or polymers improves the performance by reducing the adsorption onto the formation rocks. It can also increase the activity of polymers and surfactants by reducing divalent ion concentrations (e.g., Mg^{2+} and Ca^{2+}). Carbonate reservoirs are typically positively charged at neutral pH, in which the anionic surfactants are adsorbed easily. The injected alkalis can alter the rock surface to be negatively charged, hence the adsorption decreases significantly.

4.4.1 Alkaline enhanced oil recovery mechanisms

The EOR mechanisms of AF can be summarized as follow:

- IFT Reduction by in situ-generated surfactants
- Wettability alteration
- Emulsification and entrapment to improve mobility control
- Solubilization of rigid oil films at oil–water interfaces.

4.4.1.1 Interfacial tension reduction

The interaction between naphthenic acid in crude oil and alkalis can generate in-situ surfactants, which can reduce the IFT. Castor et al. [80] tested the capillary numbers of AF and waterflooding. The results indicated that the AF capillary number was 100 times higher than the waterflooding. Their results also demonstrated that the oil recovery efficiency was closely related to the stability of emulsions and wettability alteration rather than IFT. The IFT during AF is around 0.1 mN/m, which does not reach ultralow IFT conditions. In this case, the improved oil recovery may not significantly correlate with the IFT reduction. However, the low IFT created by AF is favorable for the emulsification of crude oil, which is another proposed EOR mechanism of AF.

4.4.1.2 Emulsification

In a mixture with a water–oil ratio greater than 1, alkalis interact with reservoir fluids and form emulsifiers. Fig. 4.30 shows the structure of an emulsifier and the formed emulsion. The hydrophobic tails of the formed emulsifiers insert into the oil drops and the hydrophilic heads face the water phase. In this case, the oil drops (internal phase) are suspended in water (continuous phase) to form an O/W emulsion. In the scenario that the water–oil ratio is smaller than 1, the oil serves as the continuous phase and the water drops are suspended to form W/O emulsion.

4.4.1.2.1 Emulsification and entrainment

Emulsification and entrainment occur when the crude oil is emulsified by the generated in-situ surfactant. The IFT is reduced by the surfactant, which contributes to the formation of the O/W emulsion system. Typically, emulsification and entrainment happen in the O/W system, the formed oil drops can flow through the pore throats, which can be entrained by the alkaline solution towards the production wells. In this mechanism, the viscosity of displacing fluid is improved by

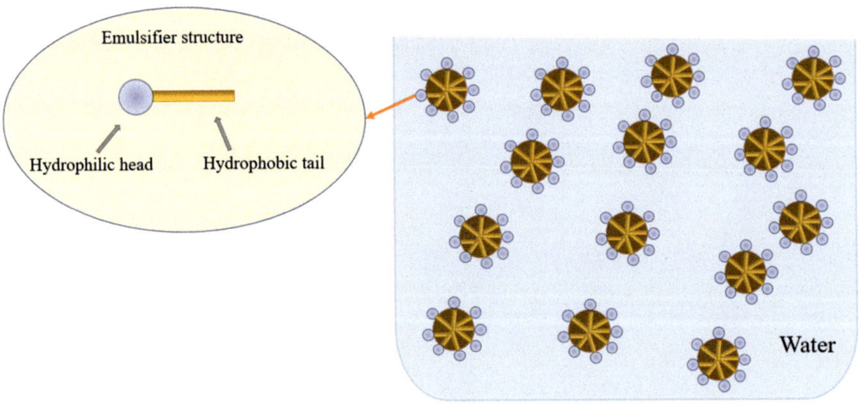

FIGURE 4.30

Explanation of emulsification.

emulsions, hence the viscosity difference between displacing phase and displaced phase is reduced. This is a process that benefits oil recovery.

4.4.1.2.2 Emulsification and coalescence

Emulsification and coalescence are related to spontaneously formed unstable W/O emulsion [80] or mixed emulsion. The injected alkaline solution can emulsify the isolated oil drops. These emulsified droplets coalesce with each other to form larger droplets. This coalescence occurs because the films of W/O emulsion are not strong and can be collapsed and ruptured easily. The coalesced larger droplets may be greater than pore throats. Some of the oil drops coalesce into a continuous oil bank, which can be displaced to production wells easily and improve the microscopic displacement efficiency.

Castor et al. conducted experiments to study the AF performance in acidic oils with hydroxides of certain divalent cations and oils with hydroxide of univalent ions with or without high electrolyte concentration. The experimental results indicate that the existence of divalents helps the formation of W/O emulsions and yields a higher oil recovery. Therefore, the dominant condition for emulsification and coalescence is that the reservoir oil contains hydroxides of certain divalent cations.

4.4.1.2.3 Emulsification and entrapment

Emulsification and entrapment occur when the oil drops are larger than the pore throats. The formed droplets plug the pore throats during the transportation and divert the following injected fluids to the unswept area. This mechanism can improve waterflooding sweep efficiency but has little to no contribution to residual oil reduction. The dominant conditions for this mechanism are high pH, moderate acid number, low salinity, and O/W emulsion size greater than pore throat diameter.

4.4.1.3 Wettability reversal (oil wet to water wet)

The wettability of a real reservoir depends on the polar compounds in the crude oil. The adsorption of these compounds can make a hydrophilic surface to be less hydrophilic or even hydrophobic. Sandstone wettability is affected by the compounds with higher molecular weight fractions in crude oil. Carbonate wettability is affected by basic nitrogen compounds.

Generally, more water-wet reservoirs can produce more oil. A lower contact angle indicates water-wet rocks, which have higher oil recovery than oil-wet rocks. The injected water tends to cling to the rock surface and push the oil passing through the pore throats. Changing the wettability from oil-wet to water-wet can reduce the negative capillary pressure effect in oil-wet conditions, which leads to favorable oil production. If the reservoir wettability is water-wet originally, the residual oil in the rock is discontinuous and immobile. In contrast, the residual oil in an originally oil-wet reservoir is continuous. Therefore, changing wettability from oil-wet to water-wet can improve oil recovery from oil-wet reservoirs [132]. The dominant condition for this mechanism is that the reservoir candidate should be an oil-wet reservoir and this process is dependent on temperature [133].

4.4.1.4 Wettability reversal (water wet to oil wet)

This mechanism has not been widely discussed since it is contrary to the common perception that water-wet rock has higher oil recovery. Cooke et al. [134] studied this mechanism in detail and concluded that the wettability of porous media could be altered from oil-wet to water-wet and generate W/O emulsion under proper conditions of pH, salinity, and temperature during NaOH flooding. What reaches a consensus is that the discontinuous nonwetting residual oil is converted to a continuous wetting phase and W/O emulsion can be generated during the AF. As shown in Fig. 4.31, a W/O emulsion is formed during AF. The adsorption of surface-active agents has changed the sand grain surface to oil wet. During the transportation of W/O emulsion through a pore throat, oil films appear on both ends of the droplet. These films close the pore throat efficiency and yield a high-pressure gradient. In this case, the formed W/O emulsion blocks the pore throats. One explanation of this mechanism is that the blocking of pore throats yields a high-pressure gradient, which diverts the injected displacing fluid to an unswept area. Another explanation is that the yield of high pressure by W/O can overcome the capillary pressure hence reducing the residual oil saturation. The continuous oil phase provides a flow path for these residual oil droplets.

Kowalewski et al. [135] conducted experiments using Berea sandstone, NaCl brine, *n*-decane, and hexadecylamine. The results indicate that the adsorption of hexadecylamine on rock surfaces changes the wettability from water-wet to neutral. Gong et al. [136] proposed that the incremental oil recovery during AF was caused by the wettability reversal mechanism induced improved sweep efficiency. They used micromodels and sandpack models to study the oil recovery due to the wettability reversal mechanism. The sandpack flooding experiments reveal that the oil recovery improved by 10% of the initial oil in place by NaOH-only flooding or NaOH and Na_2CO_3 flooding. The experimental results show that the wettability of micromodels is changed from water-wet to oil-wet by AF. The transparent micromodels show clearly that the oil is converted to a continuous

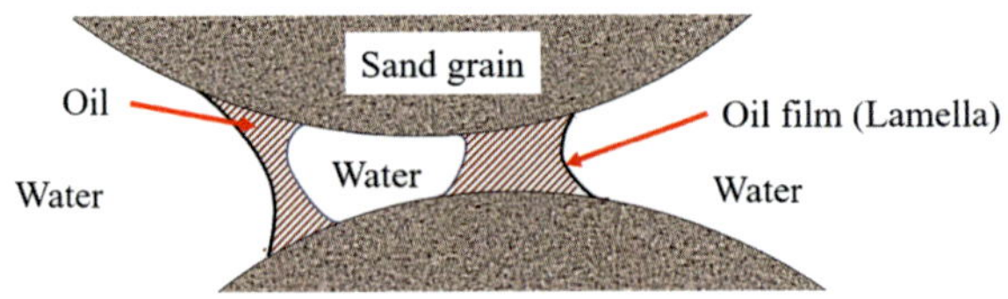

FIGURE 4.31

Oil and water distribution near displacement front.

phase, which adsorbs on the rock surface and plugs the water channels, thereby improving the sweep efficiency.

4.4.2 Characterization and evaluation of commonly used alkaline agents

Commonly used alkalis as EOR agents include sodium hydroxide, sodium carbonate, sodium orthosilicate, sodium tripolyphosphate, sodium metaborate, ammonium hydroxide, and ammonium carbonate. In the past, sodium hydroxide and sodium carbonate are the two most widely used alkalis in the oil fields. However, the field application results in China reveal that sodium hydroxide may cause scaling and corrosion problems. Therefore, sodium hydroxide is not commonly used in oil fields recently.

4.4.2.1 pH values of commonly used alkalis

Alkalis have pH values greater than 7, which is caused by the generation of hydroxide during alkali dissociation. Hydroxide generation can be direct or indirect. For example, hydroxide can be generated directly by sodium hydroxide dissociation. The dissociation of sodium carbonate can generate hydroxide indirectly. The first reaction of sodium carbonate dissociation generates Na^+ and CO_3^{2-}. The second reaction hydrolysis of the CO_3^{2-} that was generated during the first reaction to HCO_3^- and OH^-. The chemical reactions are shown below:

$$\text{Direct generation:}\quad NaOH \rightarrow Na^+ + OH^- \tag{4.8}$$

$$\text{Indirect generation:}\quad Na_2CO_3 \rightarrow Na^+ + CO_3^{2-} \tag{4.9}$$

$$H_2O + CO_3^{2-} \rightarrow HCO_3^- + OH^- \tag{4.10}$$

Alkalis have different pH values due to different chemical formulas, solution concentration, and solution salinities. The pH values of abovementioned commonly used alkalis are plotted in Fig. 4.32. As illustrated in the figure, pH values of sodium hydroxide and sodium orthosilicate increase significantly with alkaline concentrations. Sodium carbonate pH increases slightly with solution concentration. In contrast, the pH of sodium tripolyphosphate decreases with the increase of alkaline concentration. With the same concentration, the pH values of the compared alkalis show the following trend:

Sodium hydroxide > sodium orthosilicate > sodium carbonate > sodium tripolyphosphate

Besides chemical formula and alkaline concentration, solution salinity is also a factor that impacts pH. Generally, the pH of caustic solutions decreases with increased salinity. However, not all alkalis are strongly affected by salinity. For example, the pH of sodium carbonate solution is less impacted by solution salinity [137].

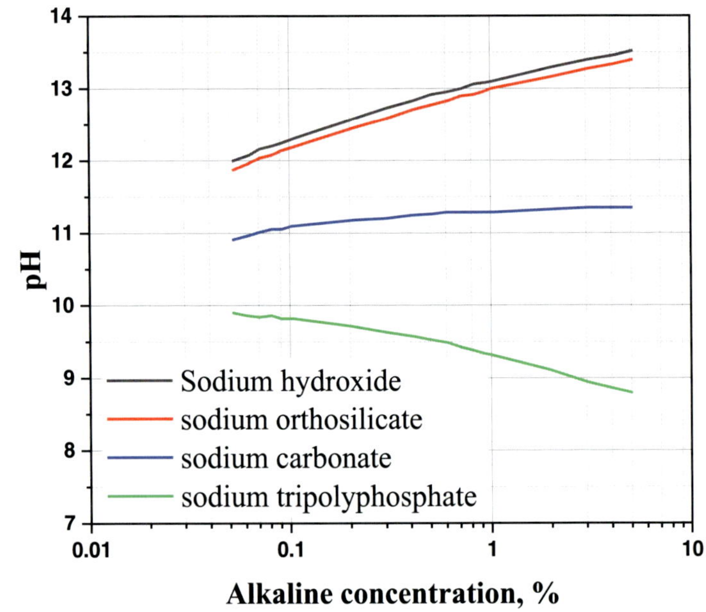

FIGURE 4.32

pH values of commonly used alkalis at different concentrations at 25°C.

Reproduced with permission from J.J. Sheng, Modern Chemical Enhanced Oil Recovery: Theory and Practice, Gulf Professional Publishing, 2010.

4.4.2.2 Properties of commonly used alkalis

The ability of alkalis to reduce IFT is slightly different from each other. In general, the minimum IFT (0.01 mN/m) occurs in a relatively narrow range of alkaline concentrations (0.05–0.1 wt.%) [10]. The properties of commonly used alkalis are listed in Table 4.12. In terms of generating in-situ surfactants and reducing IFT, sodium hydroxide, sodium carbonate, and sodium orthosilicate are good candidates. Sodium tripolyphosphate and ammonium hydroxide have little to no ability to reduce IFT. However, sodium tripolyphosphate has an outstanding ability to minimize divalent precipitation [138]. Therefore, it is more commonly used as a sequester of Ca^{2+} and Mg^{2+} rather than an IFT reducer. As shown in Table 4.12, sodium hydroxide, sodium carbonate, and sodium orthosilicate can precipitate with the presence of divalent. In some cases, sodium tripolyphosphate is used together with these alkalis when divalent is a problem to reduce precipitation [25]. Except for ammonium hydroxide, the other four alkalis can be good emulsifiers to mobilize residual oil. In addition, the abovementioned five different alkalis show an ability to alter wettability and reduce surfactant adsorption.

Table 4.12 Properties of commonly used alkalis as enhanced oil recovery agents.

Alkali formula	Sodium hydroxide NaOH	Sodium carbonate Na_2CO_3	Sodium orthosilicate Na_4SiO_4	Sodium tripolyphosphate $Na_5P_3O_{10}$	Ammonium hydroxide NH_4OH
Reduces IFT	Yes	Yes	Yes	No	No
Precipitates Ca^{2+}	Yes	Yes	Yes	–	–
Precipitates Mg^{2+}	Easier than Ca^{2+}	No or much more difficult than Ca^{2+}	Yes	–	–
Sequesters Ca^{2+} and Mg^{2+}	–	–	–	Yes	–
Emulsifier	Good	Yes	Good	Good	–
Wettability alteration	Yes	Yes	Yes	Yes	Good
Reduces surfactant adsorption	Yes	Yes	Yes	Yes	Yes

IFT, *Interfacial tension.*
Source: *Reproduced with permission from J.J. Sheng, Modern Chemical Enhanced Oil Recovery: Theory and Practice, Gulf Professional Publishing, 2010.*

4.4.3 Impacting factors on treatment efficiency

AF performance is closely related to crude oil acid numbers, alkaline types, and alkaline concentrations.

4.4.3.1 Acid number

The acid number of oil is defined as the mass of potassium hydroxide (KOH) that can neutralize one gram of crude oil. The reaction between alkalis and crude oil to generate in-situ surfactant is one of the main mechanisms for AF. As shown in Fig. 4.33, most people believe that more surfactants will be generated if the crude oil has a higher acid number, thereby more oil will be recovered [139]. In addition, a minimum acid number is required to achieve effective AF. Cooke et al. [134] conducted laboratory experiments to study the oil recovery of different crude oils and synthetic oils using AF. The results indicated that a minimum of 1.5 acid number was required to successfully improve oil recovery after conventional waterflooding. Chang [140] reported that the minimum acid number for an effective AF was 0.3 mg KOH/g oil. Ehrlich and Wygal [141] concluded that the acid number should be greater than 0.1–0.2 mg KOH/g oil to maximize AF efficiency.

4.4.3.2 Alkaline concentration and type

Several studies were conducted to study the impact of alkaline type and concentration on AF efficiency. Campbell et al. [142] conducted experiments using Berea sandstone cores to compare the displacement efficiency of sodium silicate

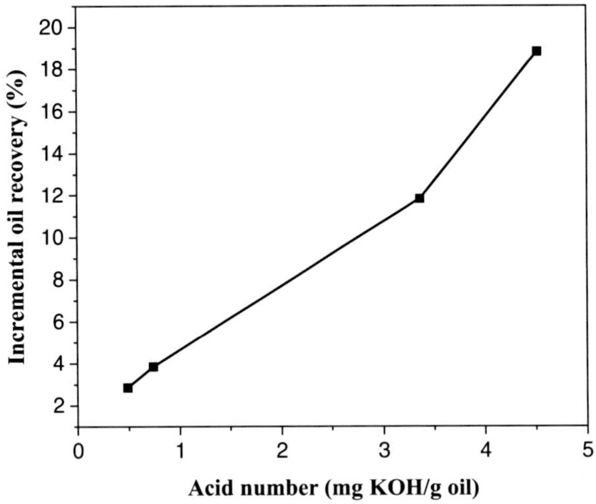

FIGURE 4.33

Influence of the oil acid number on the displacement efficiency.

Reproduced with permission from J. Ge, A. Feng, G. Zhang, et al., Study of the factors influencing alkaline flooding in heavy-oil reservoirs, Energy & Fuels 26 (2012) 2875–2882.

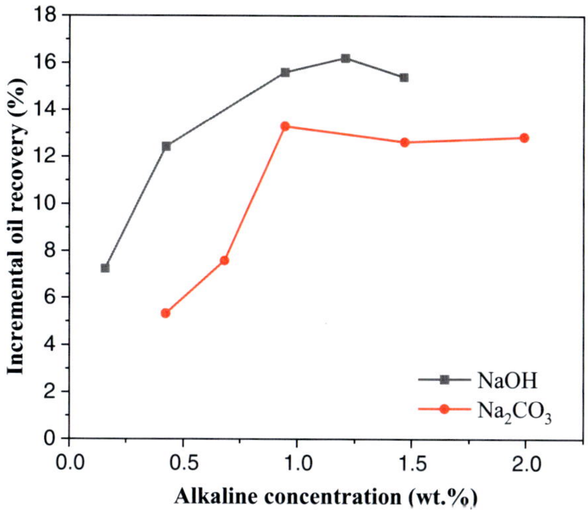

FIGURE 4.34

Alkaline concentration and type effect on oil recovery.

and sodium hydroxide solutions. The results indicated that sodium orthosilicate solution could reduce residual oil saturation much more than sodium hydroxide solution. Dong et al. [143] studied the displacement efficiency of alkaline solutions with different concentrations in Brintnell heavy oil and observed that the oil recovery was not proportionate to the alkaline concentration. In contrast, the maximum oil recovery was achieved at an optimum alkaline concentration. Arhuoma et al. also observed the same phenomenon for Alberta heavy oil. Ge et al. [139] compared the performance of NaOH and Na_2CO_3 using sand-packs. As demonstrated in Fig. 4.34, NaOH shows a higher oil recovery than Na_2CO_3 under the same concentration. At lower concentrations, the oil recovery increased with alkaline concertation for both alkaline solutions. Once the optimum concentration was reached, the oil recovery during NaOH flooding decreased with increased alkaline concentration, while the recovery during Na_2CO_3 flooding was almost constant, which is consistent with the observation of Dong et al. and Arhuoma et al.

4.4.4 Application limitation and the status of the application

Similar to other EOR methods, AF has limitations. This section introduces the application limitations of AF and also presents examples of field applications.

4.4.4.1 Screening criteria

The screening criteria for AF are discussed, such as lithology, formation water salinity, divalent content, oil acid number, and oil viscosity.

4.4.4.1.1 Oil properties

One of the main EOR mechanisms for AF is that the injected alkalis react with crude oil to form in-situ surfactants. The degree of the reaction is related to the acid number in the crude oil. Typically, a higher acid number is preferred in terms of the oil–alkaline reaction. The minimum acid number is still controversial, however, the median of the acid number reported in the literature is 0.82. Typically, heavy oil has a high acid number. Based on the EOR mechanism of AF, some research is focusing on using AF to improve heavy oil recovery. Generally, the oil viscosity of AF candidates is less than 200 cp and the median oil viscosity is 17 cp [144].

4.4.4.1.2 Reservoir properties

According to the survey conducted by Sheng [144], most of the AF projects are carried out in sandstone reservoirs and only a few are conducted in carbonate reservoirs. One main reason is that the existence of anhydrite in carbonates can cause precipitation during AF and yield a high alkaline consumption. In addition, clay content can also cause high alkaline consumption. Therefore one screening criterion for AF is that the clay content should not exceed 15%–25%.

As aforementioned, one of the mechanisms of AF is to generate in-situ surfactant via a reaction between alkaline solution and crude oil. To generate a sufficient amount of surfactant, the minimum oil saturation in the target formation should be higher than waterflooding residual oil saturation. The minimum oil saturation should be greater than 0.35 and the median oil saturation of the actual projects is 0.52 [144]. To achieve a better AF performance, the injection wells should be located in the oil zone.

The alkaline solution can react with the multivalent in the reservoirs to form precipitation. To reduce precipitation and lower the risk of AF, the multivalent content in the reservoirs should be low. Sheng [144] reported that the average multivalent cation concentration of the actual AF projects is 145 ppm.

The reservoir temperature is preferred to be less than 200°F. In the actual projects, the median reservoir temperature is 113.4°F. However, the reservoir temperature is not a strict requirement for AF. Accordingly, the reservoir depth is preferred to be less than 9000 ft, depending on the reservoir temperature.

To assure good injectivity, the formation permeability of candidates of AF should be close to waterflooding. Therefore the formation permeability should be greater than 20 mD. The median permeability is 240 mD [144].

4.4.4.2 The status of the application

Currently, AF is commonly combined with SF and/or PF in the fields. Only a few projects used the singular AF technique to improve oil recovery. In this section, some typical field application cases will be discussed.

4.4.4.2.1 North Gujarat oilfield, India

Goyal et al. [145] reported a caustic flooding project in North Gujarat oilfield in India. The formation consisted of fine sandstone, silts, and shale intercalations with the occasional presence of carbonaceous shales. The reservoir was heterogeneous in both vertical and horizontal directions. The formation temperature was 84 °C. The average horizontal permeability was 800 mD and the oil viscosity in place was 10 cp. The porosity of the reservoir was 20%–25%. The original oil saturation was around 70%. The acid number of crude oil was 1.5 mg KOH/g oil. The formation water contained 1280 mg/L chloride, 120 mg/L carbonate, 671 mg/L bicarbonate, 235 mg/L sulfate, 24 mg/L calcium, and 98 mg/L magnesium.

Before the pilot test, a laboratory evaluation was conducted to study the acid number and IFT reduction and optimize the alkaline concentration. The laboratory results revealed that the acid number of the crude oil was 1.5 mg KOH/g oil and the optimum concentration that could yield the minimum IFT was 0.25%. The well pattern was inverted 5-spot with 200-m well spacing. A total of 0.2 PV caustic slug was injected at 100 m^3/day. After caustic flooding, chasing waterflooding was conducted. The treatment increased the injection pressure of the order of 284.5 psi but the pressure increase was short-lived, which was caused by the in-situ emulsification and demulsification.

4.4.4.2.2 The North Ward-Estes field, TX, United States

The pilot test of caustic flooding was conducted in the Queen sand, East Flat area of the North Ward-Estes field [146]. The Queen sand was a channel-type deposit, which had strongly directional permeability. The highest permeability existed in the direction of Northwest–Southeast. The East Flat Queen sand was a fine-grained argillaceous formation and the off-center portion was anhydritic and dolomitic. The production sand had 80% quartz and 10% feldspar and varying amounts of illite and montmorillonite clays. The reservoir depth was 3140 ft and the pore volume was 283,000 res bbls. The average sand thickness was 35.75 ft. The permeability was 25 mD and the average porosity was 20.6%. The average oil saturation was 64%. The oil viscosity was 2.28 cp. Oil gravity was 34-degree API. The oil acid number was 0.22 mg KOH/g oil.

The laboratory evaluation was conducted before the pilot test to study the applicability of AF. The experimental results indicated that the high-porosity and low-clay-content cores had higher oil recovery. In addition, the NaOH flooding recovered 8% of OOIP after waterflooding. The pilot test used a 5% NaOH solution, which was prepared using freshwater. Before the AF, 12,750 bbls of freshwater were injected. A total of 33,000 bbls of 4.85% caustic solution was injected into a 5-spot well pattern. The oil production during freshwater injection was 9,000 bbls (3.2% OOIP in the 5-acre test area) and the oil production during caustic flooding was 45,000 bbls (15.9% OOIP in the 5-acre test area). The oil production rate of Well 874 increased from 220 bbls/day to 550 bbls/day one month after the treatment. The production rate increased to a maximum of about

600 bbls/day in two months. However, the production rate dropped rapidly to 350 bbls/day. Six months after treatment, the well plugged up completely with calcium sulfate precipitate for a few days. The formation of scale forced a reperforation over the entire sand.

4.4.4.2.3 The Court Bakken heavy oil reservoir in Saskatchewan

The Court Bakken heavy oil reservoir was located in west central Saskatchewan [147]. The Middle Bakken sand was fine to very fine-grained and the majority of the mineral composition was quartz with very minor amounts of feldspar and clay. The reservoir is unconsolidated sand with minor interbedded shale layers. The reservoir temperature was 31°C. The original reservoir pressure was 960 psi. The average porosity was 29% and the average permeability was 2100 mD. The initial water saturation was about 23% and the average thickness of the production sand was about 7.6 m. The oil viscosity at reservoir condition was 155 cp and the dead oil viscosity was 860 cp. The oil gravity was 17-degree API and the acid number of Court heavy oil was 0.77 mg KOH/g oil.

Some laboratory tests were conducted to study the interaction between alkalis, water, and oil. The original oil/water IFT was about 25 dynes/cm. A much lower IFT (0.05 dynes/cm) was achieved by adding 500–2500 ppm NaOH. The experimental results revealed that NaOH was more effective than Na_2CO_3 in reducing IFT at low concentrations.

A single-well test was carried out in the Court main pool in 2004. The treatment consisted of a slug of water followed by a slug of 5000 ppm NaOH solution and then finished with a slug of push water. After the first slug of water injection, the residual oil saturation in the area affected by the injected caustic solution was 0.28. The NaOH flooding reduced the residual oil saturation to 0.17. Apparently, the NaOH flooding successfully displaced around 39% of the residual oil after waterflooding.

The promising performance of NaOH in the single-well test encouraged the implementation of AF on a larger scale in 2006. 0.44 PV of 5000 ppm NaOH solution was injected during a pilot test in Court. The pilot pattern was located at the north side of the Court pool and close to the erosional edge of the Bakken reservoir. The pilot area was 7 acres and had half of a 5-spot pattern. The well spacing ranged from 150 to 170 m. The average net pay thickness in the pilot area was about 6 m. The injection rate of the NaOH solution with 5000 ppm concentration was maintained at 300 m^3/day. The AF was successfully implemented in the pilot area. However, the low reservoir pressure and water channeling were the major challenges for this pilot test.

4.4.5 Future development of alkaline flooding

Currently, AF has been applied to only a limited number of field projects, which is caused by a low incremental oil recovery from the previous field applications and other problems, such as corrosion and emulsion. Compared to conventional oils, heavy oils have a high content of acid components, which is favorable for AF in theory. In addition, thermal EOR is one of the most commonly used methods to develop heavy oil reservoirs. However, due to the limitation of heat transfer, this method may not yield an efficient recovery in thin heavy oil reservoirs. In this circumstance, AF has been investigated to improve oil recovery from heavy oil reservoirs. Surfactants and polymers may be added to enhance the performance of AF.

4.4.6 Remaining challenges

The precipitation and the scaling problems are of particular concern. Injecting alkalis into reservoirs can increase the concentrations of OH^-, CO_3^{2-}, and SiO_3^{2-}. These multivalent hardness ions react with the divalent in the reservoirs to form inorganic scales and precipitates. The formed scales can damage the formation and also plug production wells. Scales can be classified into carbonates, hydroxides, and silicates. The removal of carbonate scales has a mature procedure in the field. The first choice of removing the carbonate scales is using chemical techniques, especially when the scales are not easily accessible. For example, carbonate minerals are highly reactive with hydrochloric acid and can be dissolved easily. The conventional mechanical methods can also remove the scales but with a limited range of applicability. Mechanical methods need to be selected carefully based on the well and the scale deposit conditions. Hydroxide scales are dissolvable in low pH conditions. However, the removal of silicates using acid treatment needs to be conducted with caution.

Another challenge of AF is how to break the formed emulsion in the produced fluids to separate oil and water. The formed O/W or W/O emulsion in the produced fluids is too stable to be broken. The main deemulsification mechanism is partially replacing deemulsifier molecules with emulsifying molecules and lowering the film intensity.

When designing an AF project, the scaling and deemulsification problems need to be fully considered. In addition, alkalis can cause corrosion to the production and injection wells. To achieve a higher oil recovery, a high caustic solution consumption is required. In this case, the cost of alkalis should be taken into consideration.

4.5 Surfactant–polymer flooding

SF and PF are two of the commonly used EOR methods. In some fields, these two EOR methods are combined to achieve a higher oil recovery. Depending on the ultimate purposes of the chemical EOR treatments, either surfactant or polymer can be injected first during the SP flooding. If the main purpose is to reduce surfactant adsorption, polymers are injected first as sacrificial agents [148]. If the main purpose is to modify the mobility ratio, polymers are injected after surfactant slugs. Typically, SP flooding achieves higher oil recovery than individual surfactant or PF owing to the synergy.

4.5.1 Synergy between surfactant and polymer

Many researchers have studied the interaction between surfactant and polymer. The SP interaction and compatibility are summarized below [149].

- Many people believe that polymer has little to no impact on IFT reduction.
- The viscosity of HPAM is not significantly impacted by surfactants.
- Competitive adsorption. If a polymer slug is injected first, the pore surface available for surfactant adsorption is reduced. If a surfactant slug is injected ahead of the polymer, some of the adsorption sites are occupied by surfactant molecules; therefore the adsorption of polymer can be reduced. As mentioned above, the order of surfactant and polymer injection is determined on the aim of the projects.
- Some experimental results indicate that the performance of SF can be improved by a polymer preflush [150–152].

4.5.2 Surfactant–polymer flooding enhanced oil recovery mechanisms

The detailed mechanisms of surfactant and PF have been discussed in previous sections. The main mechanism of SF is to improve microscopic displacement efficiency by reducing IFT. The ultralow IFT induced a high capillary number, which leads to low residual oil saturation. In addition, the low IFT condition created by surfactants contributes to the formation of emulsions, which makes the residual oil movable. The main mechanism of PF is to improve macroscopic displacement efficiency. As a consequence, the EOR mechanism of SP flooding is the synergy of the mechanisms of these two EOR technologies.

4.5.3 Field applications

Unlike PF and ASP flooding, only a few SP flooding field cases were reported. Most of the field cases discussed in this section were conducted in fields in China, such as Daqing, Shengli, and Liaohe. Several cases in other countries are also introduced.

4.5.3.1 Status of surfactant–polymer flooding in China

Ten field cases of SP flooding conducted in China were summarized. Table 4.13 listed reservoir rock, formation water, and oil properties.

Table 4.14 summarized SP flooding designs of different SP projects. In these 10 cases, 9 fields used HPAM during polymer slug and only 1 field used SP9, which is an associate polymer. As shown in Table 4.14, various surfactants were selected based on different reservoir conditions.

Table 4.13 Typical surfactant–polymer flooding field applications in China [153].

Oilfield	Liaohe	Xinjiang	Jilin	Changqing	Dagang	He'nan	Shengli	Shengli	Daqing	CNOOC
Start time	05/2011	08/2011	01/2010	09/2014	03/2012	06/2009	09/2003	01/2006	12/2008	10/2010
No. injectors	24	18	9	16	11	15	10	37	6	8
No. producers	35	26	16	25	26	20	16	6	12	23
Well pattern	5-spot	5-spot	5-spot	5-spot	N/A	Irregular	Line	Inverted 9-spot	5-spot	Inverted 9-spot
Well spacing (m)	150	150	141	150	150–200	202–210	300 × 150	212 × 106	250	400
Area (km^2)	1.28	1.21	0.68	1.12	0.45	0.93	0.94	2.8	1.08	N/A
Depth (m)	1255–1460	1146	1202	1670	1300	1415	1261–1294	1350	1040–1050	N/A
Pressure (psi)	1798	2330	1110	1955	1325–1863	2001	1798	1939	1493	1943–2146
Temperature (°C)	55	40	55	50	53	67	68	67.5	45	57
Porosity (%)	29.1	17	21	15	31	17.5	34	33.9	22–26	22–36
Permeability (mD)	3442	94	115	110	936	569	1320	1079	706	10–50,000
Dykstra–Parsons coefficient	0.76–1.78	1–2	0.7	0.9	2.6	0.72	0.58	0.6–0.8	0.72	N/A
In situ oil viscosity (cp)	14.3	6	12.9	2.3	19–37.5	7.6	45	66.9	12.8	17.1
Acid number (mg KOH/g oil)	1.2	0.7	0.14	0.9	2.6	N/A	3	3	0.02	N/A
OOIP (10^4 ton)	298	120.8	93	105.8	205.4	296	277	645	262.41	4874
Initial oil saturation (%)	60.6	70	64.5	65	N/A	65–70	72	65.8	N/A	N/A
Formation brine TDS (ppm)	2467	14,250	15,168	19,000	14,350	2765	8207	8436	4264	6401–9182
Injected water TDS (ppm)	2748.9	3681.1	1200	5000	6726	5012	8246	7749	4012	2902
Oil recovery (% OOIP)	16.3	13.7	N/A	N/A	N/A	2	16.8	6.19	2.4	~3

TDS, *Total dissolved solids.*

Source: *Reproduced with modification from C. Sun, H. Guo, Y. Li, et al., Recent advances of surfactant-polymer (SP) flooding enhanced oil recovery field tests in China, Geofluids 2020 (2020) 8286706.*

Table 4.14 Summary of surfactant–polymer flooding designs.

		Liaohe	Xinjiang	Jilin	Changqing	Dagang	He'nan	Shengli	Shengli	Daqing	CNOOC
Preslug	Slug PV	0.04	0.06	0.045	0.06	0.1/0.15	0.05	0.05	0.24	0.0375	0.23
	Polymer con. (ppm)	2000	1800	2000	2000	2000/ 1500	2000	1500	2266	2500	2000
	Polymer MW (10^4 Da)	3000	2500	2500	2000 + 800	2500	2200	2000	2000	2500	2200
Main slug	Slug PV	0.35	0.5	0.3	0.225	0.08	0.35	0.3	0.3	0.45	0.3
	Polymer con. (ppm)	1600	1600	2000	1500	2000	1700/ 1500	1700	1800	1500	1200–1500
	Polymer MW (10^4 Da)	3000	2500	2500	2000 + 800	2500	2200	2000	2000	2500	2200
	Viscosity (cp)	120	24–48	80	N/A	40–80	37.5	30	30	50	34.1
	S con. (%)	0.25	0.25	0.2	0.12	0.25	0.4	0.6	0.4	0.2/0.1	0.2–0.24
	Surfactant type[a]	Amphoteric	KPS + Co-S	PS	Amphoteric + anioinc–nonionic	PS Co-S	PS Co-S	0.45% SPS + 0.15% Co-S	0.27% SPS + 0.13% Co-S	Amphoteric	PS
Vice slug	Slug PV	0.2	–	0.21	0.315	0.3	–	–	–	–	–
	Polymer con. (ppm)	1600	–	2000	1500	1500	–	–	–	–	–
	Polymer MW (10^4 Da)	3000	–	2500	2000 + 800	2500	–	–	–	–	–
	Surfactant con. (%)	0.15	–	0.1	0.1	0.2	–	–	–	–	–
Postslug	Slug PV	0.1	0.1	0.15	0.12	0.02	0.1	0.05	0.05	–	0.035
	Polymer con. (ppm)	1400	1400	1000	1000	700	1500	1500	1500	–	1200
	Polymer MW (10^4 Da)	3000	2500	2500	2000 + 800	2500	2200	2000	2000	–	2200
Total slug PV		0.69	0.66	0.75	0.65	0.65	0.5	0.4	0.59	0.4875	0.565

[a]*Surfactant type:* KPS, *Karamay petroleum sulfonate;* Co-S, *cosolvent;* PS, *petroleum sulfonate;* SPS, *Shengli petroleum sulfonate.*

Source: *Reproduced with modification from C. Sun, H. Guo, Y. Li, et al., Recent advances of surfactant-polymer (SP) flooding enhanced oil recovery field tests in China, Geofluids 2020 (2020) 8286706.*

Shengli Gudong oilfield is the first field that conducted SP flooding on a large scale and it is also the only field in China that carried out more than two SP flooding pilot tests. The oil production from the Shengli oilfield using SP flooding was more than 2 million tons in 2016, which was one of the most successful SP flooding projects in the world. In Daqing oilfield, the experience of implementing PF is over 20 years and the ASP flooding also has been extensively applied. However, only one SP flooding was conducted in Daqing.

4.5.3.1.1 Shengli Gudong oilfield, China

An SP flooding pilot test was conducted in Shengli Gudong oilfield in September 2003. The pilot area was 0.9 km^2. The project was carried out using 10 injectors and 16 producers in a linear well pattern. The average permeability was 1320 mD. The Dykstra–Parsons coefficient was 0.58 in the pilot area. The average porosity was 34%. The initial oil saturation was 72%. The estimated OOIP was 277×10^4 tons. The reservoir temperature was 68°C. The oil viscosity at reservoir temperature was 45 cp. The formation brine TDS was 8207 ppm, and the injected water TDS was 8246 ppm.

The surfactant used in this project was Shengli petroleum sulfonate (SPS) and Co-S. The polymer was HPAM. The project injection design is:

- Preslug: 0.05 PV of 1500 mg/L polymer
- Main slug: 0.35 PV of 1700 mg/L polymer, 0.45% SPS, and 0.15% Co-S. The solution viscosity was 30 cp
- Postslug: 0.05 PV of 1500 mg/L polymer.

The reservoir response was observed after 0.04 PV of the main slug injection. The peak oil rate increased from 10.4 to 127.5 tons/day and the water cut reduced from 98.3% to 60.4%. As of December 2011, the incremental oil recovery was 16.8%.

4.5.3.1.2 CNOOC JZ 9–3 oilfield, Bohai Bay, China

The SP flooding test in Bohai Bay is different from any other cases summarized in Table 4.13 because Bohai Bay is an offshore field. The pilot test was started in October 2010. The test was conducted using 8 injectors and 23 producers in an inverted 9-spot well pattern [154]. The well spacing was about 400 m. The reservoir pressure was 1943–2146 psi. The reservoir temperature was 57°C. The reservoir had a severe heterogeneity problem. The permeability ranged from 10 to 50,000 mD and the porosity ranged from 22% to 36%. The in-situ oil viscosity was 17.1 cp. The estimated OOIP was 4874×10^4 tons. The formation water TDS varied from 6401 to 9182 ppm and the injected water TDS was 2902 ppm.

Conventional waterflooding was conducted in this oilfield in June 2000. By January 2007, the water cut increased rapidly to 75.7% but the recovery was only 13.6%. This was probably caused by the water channeling through the high permeability features in the reservoir. The SP flooding design is shown below:

- Preslug: 0.23 PV of 2000 ppm 3640 C polymer solution

- Main slug: 0.3 PV of 1200–1500 ppm polymer and 0.2%–0.24% petroleum sulfonate (HDS series from HOPEC company). The solution viscosity was 34.1 cp
- Postslug: 0.035 PV of 1200 ppm 3640 C polymer solution.

In January 2009 the injection of preslug yielded 860 m^3/day of oil production and 80% water cut with a 23% oil recovery rate. In October 2010 the main slug was injected. By October 2014 the full field oil production increased by 40,000 tons and the oil recovery was 3% OOIP. In the central well W7–4, the water cut was decreased by 42% and the oil production was doubled.

4.5.3.2 Status of surfactant–polymer flooding in other countries

In the recent decade most chemical EOR projects are polymer or ASP flooding. Most of the reported applications of SP flooding were conducted in the 1990s or early. In this section, two field applications in the United States are presented.

4.5.3.2.1 Loma Novia Field, Texas, United States

In the mid-1960s an SP flooding test was conducted in the Loma Novia field, Duval County, Texas, United States. The project had four injectors, one producer, and two observation wells in a 5-acre pattern. The surfactant used in this pilot test was a petroleum sulfonate blend with an equivalent weight of about 415 Da. The laboratory tests indicated that the lowest IFT occurred at 0.5% sodium chloride and 0.15% petroleum sulfonate. Before the SP flooding, a slug of salt water was injected due to the high clay content. Then 0.12 PV surfactant slug with 1.1% sodium chloride, 1.9% petroleum sulfonate, and a small amount of sodium carbonate and sodium tripolyphosphate was injected. The postslug was 0.1 PV of 0.33% sodium chloride and a small amount of the sacrificial chemicals. The peak oil cut, which was 20% was observed 84 days after the surfactant injection. Four cores were drilled to study the vertical sweep efficiency during the SP flooding. The core analysis showed that the conducted SP flooding did not yield a satisfying vertical sweep efficiency, especially in this reservoir with only 10–12 ft net pay thickness.

4.5.3.2.2 Delaware-Childers, Oklahoma, United States

Unlike other water-wet sandstone reservoirs, the production sand in the Delaware-Childers field was the oil-wet Bartlesville sandstone. The minerals in the production zone included 50% quartz, 12% carbonate, 10% clay, 5% feldspars, 3% anhydrite and gypsum, and 20% others. The reservoir depth was 620–700 ft and the average thickness was 52 ft. The average permeability was 100 mD and the average porosity was 21%. The estimated oil saturation was 32%–36%. The reservoir temperature was 86°F. The oil viscosity in place was 9.6 cp. The formation water TDS was 110,000 ppm at the beginning of the preflush in 1975 and the TDS decreased to 7000 ppm at the end of the project in 1980. This TDS reduction was caused by waterflooding. The TDS in the preflush was 8800 ppm, in the micellar slug was 8200 ppm, and in the polymer slug was 2900 ppm [155].

The project was conducted in a 2.5-acre area using inverted 5-spot well patterns. The 0.1 PV of the micellar slug was injected in April 1976 followed by 0.4 PV of the polymer slug in August 1979. The micellar was petroleum sulfonate blend with ethoxylated alcohol (Cosurfactant 121). The polymer used was Nal-flo B with $8-10\times10^6$ MW.

In this project, only a small amount of incremental oil was produced. The oil saturation near the observation wells was redistributed but not significantly reduced [149]. Technically, this project was not a success.

4.6 Alkaline–surfactant flooding

Alkalis can react with the naphthenic acids in crude oil to generate in-situ surfactant, which can reduce IFT and improve oil recovery. To promote EOR performance, alkalis can be coinjected with surfactants. Some people believe that adding alkalis to SF can reduce the dose of synthetic surfactants since the alkalis can generate in-situ surfactants. However, generating in-situ surfactant is only one of the alkaline EOR mechanisms. Introducing alkalis to SF increases the pH of the solution, which leads to a reduction in surfactant adsorption. In addition, the synergy between the in-situ-generated surfactant and synthetic surfactant can widen the salinity range that yields ultralow IFT.

4.6.1 Interaction between surfactant and alkali

In AF, ultralow IFT is preferred to achieve a better EOR performance. However, as reported by many researchers, the ultralow IFT typically occurs at a very low concentration of alkali [156]. If a low alkaline concentration is used, the high alkaline consumption may limit the swept area by AF, which means the oil in the in-depth area may not be touched by the injected alkaline solution. The salinity that achieves the lowest IFT is called optimum salinity. If the salinity is above the optimum salinity, the system is over-optimum, which will bring up the IFT. Adding a synthetic surfactant can increase the range of salinity that yields the lowest IFT owing to the synergy of the in-situ surfactant and the injected synthetic surfactant. Liu [68] studied the synergy of in-situ surfactants and synthetic surfactants. The results indicated that the IFT values were reduced by adding Na_2CO_3 into surfactants and the width of the low IFT region is much wider with the addition of alkali. Other than synergy, the alkaline solution can reduce surfactant adsorption. By adding alkalis, the solution pH will be increased and the mineral will gradually become negatively charged, which may retard the adsorption of anionic surfactant. In addition, alkalis can reduce multivalent ions, which can contribute to the reduction of surfactant retention.

4.6.2 Alkaline–surfactant flooding enhanced oil recovery mechanisms

The EOR mechanisms of AS flooding can be summarized below:

- Reduction of surfactant adsorption
- Generation of in situ surfactant by alkaline solution
- The synergy between in-situ-generated surfactants and injected synthetic surfactants
- Wettability alteration

4.6.3 Field application examples

Currently, some laboratory studies are conducted on AS flooding applications in heavy oil reservoirs, carbonate reservoirs, and tight sand reservoirs. However, the field cases are not as many as PF or ASP flooding. One reason may be that the oil price was low, and another possible reason may be the lack of mobility control in AS flooding. In this section, an AS flooding carried out in the Big Sinking field (East Kentuck, United States) is presented.

Big Sinking field was a tight oil reservoir with a high water cut. Because the water injection rate was restricted, the fluid production rate, especially the oil rate was limited. The main purpose of AS flooding in this field was to reduce residual oil saturation and improve water effective permeability [157]. If the water effective permeability could be increased, the water injection rate could be lifted and the oil production rate could be increased.

The pilot area was in the EL Rogers tract. The effective thickness in the pilot area was 25 ft and the average permeability was 45 mD. The alkali and surfactant used in this test were 1.5% sodium hydroxide and 0.1% ORS-62HF, respectively. The injection design in a single-well test is shown below:

- Preslug: 1500 bbls of fresh water
- Main slug: 1500 bbls of AS solution
- Postslug: fresh water.

The main objective of this project was to improve injectivity. After the treatment, the results showed that the injectivity was significantly improved, which was increased by more than twofold.

4.7 Alkaline–polymer flooding

As introduced in Section 4.4, AF can improve oil recovery in some cases, but some field applications indicated that the incremental oil recovery during AF was not significantly higher than conventional waterflooding. One possible reason for this phenomenon is the lack of mobility control in AF [158]. In this case, adding polymers as mobility control agents is one option to promote alkaline treatment efficiency.

4.7.1 Interaction between alkali and polymer

Typically, binary chemical EOR systems can improve oil recovery more than individual chemical systems. This section introduces the interaction, synergy, and impact between the alkalis and polymers.

4.7.1.1 Alkaline effect on polymer viscosity

The reaction between alkalis and polymers hydrolyzes polymers, which causes alkali consumption. As a consequence, the alkaline concentration and solution pH value decreases. Polymers or buffer alkalis may be injected as sacrificial agents to eliminate pH reduction and precipitate divalents to protect polymers. Adding alkalis into polymer solutions prepared by distilled water causes polymer viscosity reduction owing to the salt effect; then the viscosity increases gradually over time due to the hydrolysis process; finally, the viscosity reaches to be stable. In brines, hydrolysis increases at the beginning, which leads to increased viscosity. As the salt concentration increases with hydrolysis, the solution viscosity decreases.

The polymer viscosity is mainly impacted by alkalis from two aspects: increased salt effect and increased hydrolysis effect [159]. The increased salt effect decreases polymer viscosity while the increased hydrolysis effect increases viscosity. The final viscosity is determined by the degrees of these two effects. Generally, the salt effect has more impact than the hydrolysis impact. As shown in Fig. 4.35,

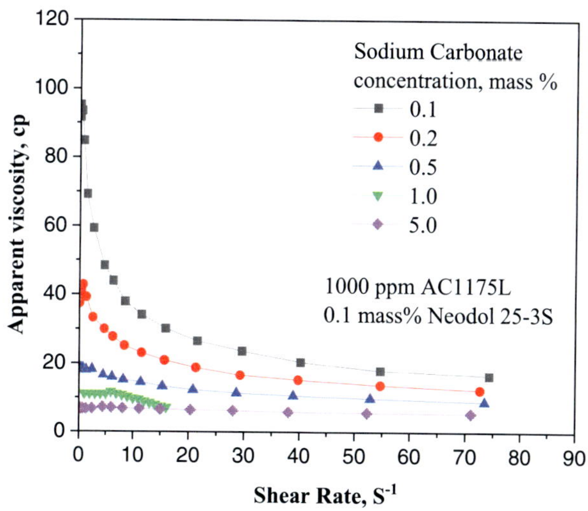

FIGURE 4.35

Alkaline concentration effect on polymer viscosity [160].

Reproduced with permission from K.C. Taylor, H.A. Nasr-El-Din, The effect of synthetic surfactants on the interfacial behaviour of crude oil/alkali/polymer systems. Colloids and Surfaces A: Physicochemical and Engineering Aspects 108 (1996) 49–72.

HPAM viscosity was tested with the presence of NaOH. The figure shows that the HPAM viscosity decreases with the increase of NaOH concentration. Overall, polymer viscosity decreases with increased alkaline concentration.

4.7.1.2 Alkaline consumption in alkaline–polymer flooding

In some laboratory tests, it was observed that the alkaline consumption in alkaline–PF is lower than in singular AF. One possible reason is that some pore surfaces are covered by polymer, which reduces the contact area between alkali and rock. During alkaline–PF, alkali competes with polymer for positively charged sites. In this case, the reduction of polymer adsorption is caused by the more negatively charged rock surface [161].

4.7.2 Synergy between alkali and polymer

The synergy between alkali and polymer in AP flooding is summarized below:

- Polymers can reduce alkaline consumption and alkali can reduce polymer adsorption.
- Polymer improves the viscosity of the AP system, which yields better mobility control and higher sweep efficiency.
- Due to a higher solution viscosity, alkaline solutions can sweep more area to reduce residual oil saturation, hence improving macroscopic displacement efficiency.
- The mixing of alkali and polymer decreases the viscosity of the polymer. In most cases, this is not a favorable effect. However, in tight sand reservoirs, the reduction of polymer viscosity caused by alkali may help with near wellbore injectivity problems.

4.7.3 Field applications

The reported field applications of AP flooding are very limited. Only eight field cases were found. Four out of eight were conducted in fields in Canada, two projects in the United States, and two cases in China. All of these eight fields that implemented AP floodings are onshore sandstone reservoirs. Na_2CO_3 is the alkaline agent that had been applied in most of the field cases. The reservoir rock, formation water, and oil properties are summarized in Table 4.15. In this section, some examples of AP flooding projects are discussed.

4.7.3.1 Cessford Basal Colorado A, Alberta, Canada

The Cessfor Basal Colorado A was a heavy oil sandstone reservoir. The reservoir depth was 920 m. The reservoir temperature was 29°C. The average porosity was 24% and the average permeability was 350 mD. The oil gravity and viscosity are 23-degree API and 24 cp, respectively. The oil acid number was 0.5 mg KOH/g oil. The average water saturation before AP flooding was 30% [162].

Table 4.15 Summary of alkaline–polymer flooding field application cases.

Field	Cessford	David Pool	Horsefly Lake	Wrentham	Xing Long Tai	Yang San Mu	Isenhour Unit	Moorcroft West
Country	Canada	Canada	Canada	Canada	China	China	United States	United States
Start date (years)	1981	1987	1984	2014	1995	1999	1980	1989
No. injectors	7	7	4	N/A	3	4	3	1
No. producers	19	18	5	N/A	N/A	18	6	1
Permeability (mD)	350	1400	260	N/A	2063	841	21	32
Temperature (°C)	29	30.56	34.4	N/A	56.6	62	36.1	48.9
Formation water TDS (ppm)	N/A	N/A	28,700	N/A	3112	N/A	N/A	42,370
Oil viscosity (cp)	24	34	13	N/A	6.3	120	2.8	42
Oil saturation before EOR (%)	N/A	N/A	51	N/A	N/A	52.8	38.1	68.6
Water cut before EOR (%)	N/A	98.5	N/A	N/A	96.1	96.4	N/A	N/A

EOR, *Enhanced oil recovery;* TDS, *total dissolved solids.*
Source: *Data from [36,43,158,162–172].*

Waterflooding was conducted in the pilot area from 1981. A preflush was conducted using softened salt water to prevent alkaline solution from contacting hard water. The AP flooding was conducted in May 1985 in the North and Central patterns and in January 1985 in the South area.

The treatment in this area was not successful. No incremental oil production was observed due to poor injectivity at the chemical injection wells. The poor injectivity might be caused by the precipitation formed during the alkaline injection. It might also be caused by the adsorption of polymer and the reaction between alkaline solution and clay minerals in the formation. Due to the poor performance, the AP flooding project was terminated in the North and South area in October 1990 and stopped in the Central area in 1991.

4.7.3.2 Moorcroft West, Wyoming, United States

The Moorcroft West Minnelusa Sand Unit was in the Powder River Basin, Wyoming, United States. The average porosity and water saturation were 15.2% and 28.1%, respectively. The reservoir permeability ranged from 6 to 150 mD and the average permeability was 114 mD. The permeability variation coefficient was 0.68. The average effective thickness was 8.4 ft. The estimated OOIP was 686,000 bbls. The reservoir temperature was 48.9°C. The oil viscosity at reservoir temperature was 42 cp.

The AP flooding was initiated in September 1991. The main objective was to reduce the IFT, hence reducing residual oil saturation. The AP flooding injected 0.2 PV of 0.2 wt. % potassium hydroxide and 225 mg/L anionic polymers. The injection design was tested in the laboratory and results indicated that adding KOH could reduce IFT by more than 100-fold. To eliminate the microbial scale, biocide was injected once a month at 127 mg/L concentration.

4.7.3.3 Xing Long Tai, Liaohe, China

An AP project was carried out in the Xing Long Tai field, which was a water-wet sandstone reservoir [36]. The block area was 2.05 km^2 and the reservoir thickness was 3 m. The estimated OOIP was 0.96 million tons. The average porosity and the average permeability were 27.6% and 2063 mD, respectively. The live oil at reservoir temperature was 6.3 cp. The formation water TDS was 3112 mg/L with Ca^{2+} and Mg^{2+} of 14 mg/L.

The selected polymer for this project was 1175 A (HPAM). The alkali used in this AP project was Na_2CO_3. The final injection concentrations are 2% of Na_2CO_3 and 1000 mg/L of 1175 A. The AP flooding was carried out from January 1995 to August 1998. Consequently, the AP flooding increased the oil production by 1.98% OOIP for the whole pilot area and 19.5% for the central area.

4.8 Alkaline–surfactant–polymer flooding

AF, SF, and PF can be applied individually or as combinations to further improve oil recovery. Two-component combinations, alkaline–surfactant (AS),

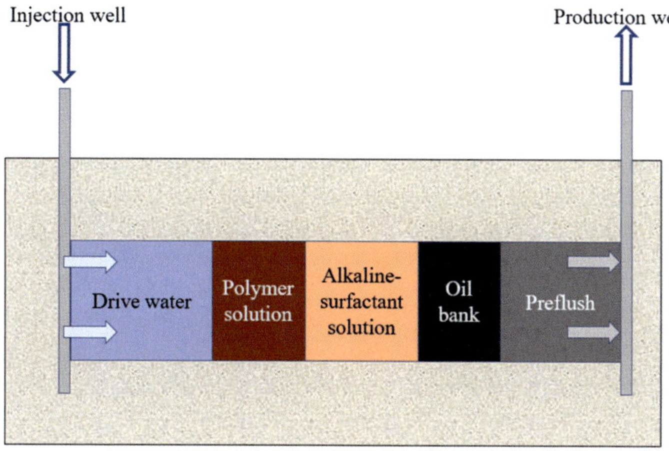

FIGURE 4.36

Illustration of alkaline–surfactant–polymer flooding process.

alkaline–polymer (AP), and surfactant-polymer (SP), have been introduced in the previous sections. This section discusses the technical characteristics, status of application, problems and challenges, and future development of ASP flooding.

The ASP flooding is an EOR technology that involves three components: alkalis, surfactants, and polymers. The development of ASP flooding started in the 1970s. The integrated synergy of the individual component of the injected chemical agents yields higher oil recovery than one or two-component chemical flooding, which makes the ASP flooding to be the most promising chemical EOR method. ASP flooding was developed based on single-component flooding (AF, SF, or PF) and two-component flooding (SP flooding, AP flooding, or AS flooding). The injection of three chemical agents with proper sequence and slug size can fully synergize the advantages of each component [173]. In general, polymers contribute to a higher sweep efficiency; surfactants can reduce IFT; alkalis generate in-situ surfactants. In the meantime, the combination of alkalis, polymers, and surfactants can overcome some disadvantages of each component, such as poor mobility control, high alkali consumption, and high surfactant adsorption.

The ASP flooding involves several slugs of chemical agent injection. The process is illustrated in Fig. 4.36. A preflush (low-salinity water or fresh water) is injected first to prepare the reservoir to a more favorable condition for later chemical flooding. After preflush, the first slug consists of alkali and surfactant, which can improve microscopic sweep efficiency. Subsequently, a slug of polymer is injected to reduce the mobility ratio and improve macroscopic sweep efficiency. Finally, a slug of freshwater and drive water is injected to displace oil in production wells.

4.8.1 Alkaline–surfactant–polymer flooding Enhanced oil recovery mechanisms

In ASP flooding, emulsification occurs due to the mixing of surfactant with alkali, crude oil, and formation water. One of the mechanisms is the generation of in-situ surfactant due to alkali's reaction with crude oil and emulsification. The injected surfactant enforces the stability of generated emulsions due to IFT reduction. It can also increase the interface strength and generate charge at the interface. Because of the low to ultralow IFT, the capillary number is further increased during ASP flooding, which can reduce residual oil saturation. In addition, adding polymers increases the displacing phase viscosity, which can alleviate unfavorable fingering problem, hence improving sweep efficiency. The viscosity contributed by polymers can reduce the diffusion of droplets, which reduces the possibility of coalescence and improves the emulsion stability [25]. As aforementioned, EOR mechanisms of AF include emulsification, entrainment, and entrapment of oil droplets. As one of the components in ASP flooding, alkalis also contribute to the oil recovery owing to these mechanisms.

4.8.2 Synergistic mechanisms of alkaline–surfactant–polymer flooding

ASP systems are complex due to the involvement of three components. The interaction among three chemical agents can synergize the improvement of oil recovery. Also, APS systems are sensitive to some factors that may impact EOR performance. In this section, the effects of alkalis, surfactants, and polymers in ASP systems are discussed.

4.8.2.1 Alkaline and surfactant effects in alkaline–surfactant–polymer systems

The existence of alkalis can reduce surfactant adsorption. However, this statement is only correct if the alkaline concentration is in a favorable range. If the alkaline concentration is too high, the static electric repulsion between the rock surface and surfactant molecules becomes weaker, which leads to more adsorption of surfactants. Another important mechanism is the synergy between the in-situ-generated surfactant and synthetic surfactant. The optimum salinity for the generated surfactant is very low and the optimum salinity for the synthetic surfactant is high. However, the combination of these two surfactants can widen the range of the salinity that the IFT reaches the lowest value (Fig. 4.37) [156,174]. The study conducted by Chen and Chen [175] revealed that the APS system viscosity reduces with the increase of anionic surfactant concentration. This phenomenon may be caused by the polymer chain aggregation owing to the electric shield effect.

Typically, the HPAM hydrolysis degree increases with the alkaline concentration and the COO− is increased. Due to the increase of electrostatic repulsion with anionic surfactants, the solubility of surfactants decreases. In addition, IFT

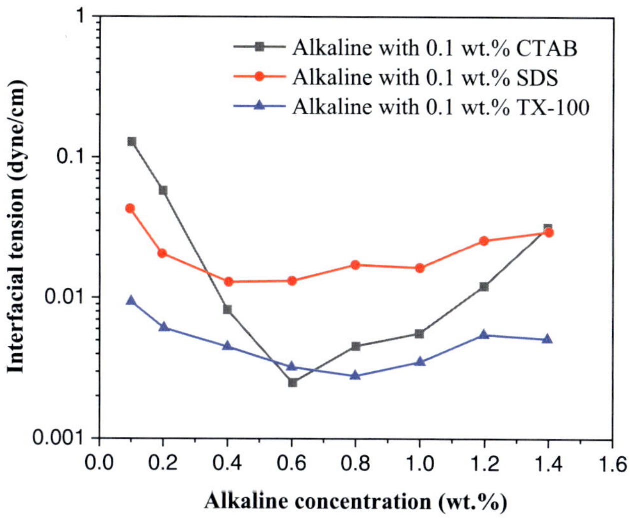

FIGURE 4.37

Effect of alkaline concentration on the interfacial tension of different surfactants and oils. *CTAB*, cetrimonium bromide; *SDS*, sodium dodecyl sulfate; *TX-100*, octyl phenol ethoxylate.

Reproduced with permission from E. Hosseini, F. Hajivand, R. Tahmasebi, The effect of alkaline–surfactant on the wettability, relative permeability and oil recovery of carbonate reservoir rock: experimental investigation, Journal of Petroleum Exploration and Production Technology 9 (2019) 2877–2891.

becomes higher with increased alkaline concentration, which reduces the fluid compatibility in the ASP system.

4.8.2.2 Polymer effect in alkaline–surfactant–polymer systems

As mobility control agents, polymers can improve sweep efficiency and also interact with surfactants and alkalis to improve oil recovery. Polymers slightly change the value of electrophoretic mobility and make the surfactant system emulsification better. Generally, polymers cannot reduce IFT, but it shortens the phase separation time of emulsion. Large polymer molecules may protect some electrically positive sites from occupation by anionic surfactants. Adding polymers in an ASP system reduces surfactant adsorption but does not significantly affect alkaline consumption. However, polymers may reduce the interaction between oil and alkalis when alkaline concentration is high.

4.8.3 Challenges associated with alkaline–surfactant–polymer

Many field applications have confirmed that the ASP technique can improve oil recovery and reduce water cut. However, there are some challenges that limit the wide application of ASP flooding.

4.8.3.1 Scaling problems

The reaction between alkalis and divalents can cause severe scaling problems in wellbores and facilities. During ASP flooding, the reaction between alkali and reservoir rock or formation water increases the ion concentration, which results in carbonate or silicate scales. Although different fields have different formation water compositions, the main compositions are NaCl and Na_2CO_3. The divalents are mainly Ca^{2+} and Mg^{2+} with a range of 20–2000 mg/L [25]. Typically, Mg^{2+} is lower than Ca^{2+} in formation water but higher in seawater. When alkalis are injected into a formation, the concentration of OH^-, ${CO_3}^{2-}$, and ${SiO_3}^{2-}$ are increased. These anions react with cations to form scales. Some inorganic scales can be formed as follows:

$$\text{Carbonate scales: } Ca^{2+} + CO_3^{2-} = CaCO_3\downarrow \tag{4.11}$$

$$Mg^{2+} + CO_3^{2-} = MgCO_3\downarrow \tag{4.12}$$

$$\text{Silicate scales: } Ca^{2+} + SiO_3^{2-} = CaSiO_3\downarrow \tag{4.13}$$

$$Mg^{2+} + SiO_3^{2-} = MgSiO_3\downarrow \tag{4.14}$$

$$\text{Sulfate scales: } Ca^{2+} + SO_4^{2-} = CaSO_4\downarrow \tag{4.15}$$

$$Mg + SO_4^{2-} = MgSO_4\downarrow \tag{4.16}$$

$$\text{Hydroxyl scales: } Ca^{2+} + 2OH^- = Ca(OH)_2\downarrow \tag{4.17}$$

$$Mg^{2+} + 2OH^- = Mg(OH)_2\downarrow \tag{4.18}$$

4.8.3.2 Formation damage

Formation erosion is one of the problems that occur during the alkaline injection. Reservoir permeability may be increased due to erosion in some cases. However, in most cases, the permeability is reduced, which is caused by grain and clay migration. In addition, the injected alkaline solution and surfactants may react with the formation of rock and water to form precipitates. The formed precipitates can cause permeability reduction as well, although the permeability may be restored to some degree during post waterflooding.

4.8.3.3 Surfactant precipitation

In reservoirs with high salinity and divalent concentration, the injection surfactant may precipitate and cause surfactant retention and formation damage. Factors that dominate surfactant precipitation include cation valence, salinity, surfactant concentration, alcohol concentration, temperature, etc. [10]. At lower salinity and divalent concentration, the multivalent cations react with the anionic surfactant to form a monovalent cation that can exchange with the cations that originally bond to the reservoir clays [176].

4.8.3.4 Emulsion treatment issues

Some emulsions are generated during ASP flooding owing to the alkaline and surfactant properties. Although the formed emulsions can improve oil mobility and increase sweep efficiency, they can also cause problems in transportation and oil/water separation. For example, the produced emulsion from the Shengli oilfield could stay stable under room temperature for more than 129 days, which made it difficult to separate the produced water and oil.

4.8.4 Status of application

ASP flooding has been extensively reported in the literature during the last two decades, especially in China. Daqing oilfield in China is one of the fields that massively use ASP flooding to improve oil recovery. ASP flooding has been studied and implemented in Daqing for more than 20 years. Besides the Daqing oilfield, Shengli, Karamay, Liaohe, and He'nan oilfields are some oilfields in China that conducted ASP flooding on various scales. Other than China, many ASP field tests were conducted in the United States, Canada, South America, and India. This section discusses the application status of ASP flooding in China and some examples in other oilfields, such as West Kiehl, Tanner, and Cambridge Minnelusa in the United States and Taber South and Taber Glauconitic fields in Canada.

4.8.4.1 Status of alkaline–surfactant–polymer flooding in China

ASP flooding has been widely applied in China, especially in the Daqing oilfield. Currently, 30 field cases were reported and 23 of them are from Daqing [177–190]. The field cases conducted in China are summarized in Table 4.16. In this section, some field application examples are discussed.

4.8.4.1.1 Status of alkaline–surfactant–polymer flooding in Daqing oilfield

- S-Z-X block in Daqing oilfield

The implementation of ASP flooding in Daqing oilfield was as early as 1994. The first ASP flooding pilot test was conducted in the S-Z-X block in Daqing.

In the S-Z-X block, the production layer was composed of braided channel sand [25]. The formation depth was 814 m and the temperature was 45°C. The production sand thickness was 8.6 m. The average permeability was 509 mD and the porosity was 26%. The estimated total pore volume was 203,300 m^3. OOIP was calculated as 117,300 tons. The in-situ oil viscosity was 11.5 cp and the acid number was 0.01 mg KOH/g oil. The formation water TDS was 6800 mg/L and freshwater salinity was 600 mg/L. The pilot test area was 0.09 km^2, which consisted of four 5-spot patterns. The test involved 4 injectors and 9 producers and the well spacing was 106 m. The detailed formation information and selected chemical agents are listed in Table 4.16.

Table 4.16 Properties of alkaline–surfactant–polymer oil fields.

Oilfield	Block	Date	Injector/ producer	Well spacing (m)	*k* (mD)	ϕ (%)	*T* (°C)	μ_o (cp)	Surfactant typea	Alkali type	Total ASP slug size	Incremental oil recovery (%)
Daqing	S-Z-X	1994.9–1996.5	4/9	106	509	26	45	11.5	B-100	Na_2CO_3	0.603	21.4
	X-5	1995.1–1997.4	1/4	141	589	26	45	6.1	ORS-41	NaOH	0.678	25
	X-2-X	1996.9–1999.1	4/9	200	675	26	45	7.1	ORS-41	NaOH	0.694	19.16
	B-1-DX	1997.9–1999.1	6/12	250	512	26	45	9.8	ORS-41	NaOH	0.653	21.9
	XJJ-B	1997.12–1998.12	3/4	75	567	26	45	9.8	ORS-41 + RH	NaOH	0.741	23.24
	XJJ-N	2004.1–2005.1	3/4	75	467	26	45	9.8	PS	Na_2CO_3	0.749	24.66
	N-4-D	1999.9–2001.11	4/9	170	N/A	26	45	9.8	Carboxylate	Na_2CO_3	0.55	15.4
	X-2-Z	2000.4–2007.1	17/27	250	404	25	45	9.8	HABS	NaOH	0.848	18.1
	B-3-X	2002.8–2008.12	13/14	250	657	26	45	9.8	HABS	Na_2CO_3	0.688	18.2
	B-2-X	2008.1–2013.12	35/44	125	533	26	45	9.8	PS (DPS)	Na_2CO_3	0.9788	29.4
	B-1-D	2006.7–2012.12	49/63	125	670	26	45	9.8	HABS	NaOH	0.923	30
	N-5	2006.7–2012	29/39	175	501	26	45	9.8	ABS	NaOH	0.934	19.8
	L-B-D	2008.1–2013.12	44/62	120	676	26	45	10.3	ABS	NaOH	0.864	20.3
	X-1–2-D	2007.12	112/143	150	517	26	45	9–11	N/A	NaOH	0.6875	18.5
	N-6-D	2008.11	93/94	175	520	26	45	9–12	N/A	NaOH	N/A	N/A
	X-6-D-I	2009.6	102/112	141	575	26	45	9–13	N/A	NaOH	N/A	18.5
	X-6-D-II	2009.11	109/105	141	526	26	45	9–14	N/A	NaOH	N/A	19
	B-1-DX-X	2012.6	90/118	125	466	26	45	9–15	N/A	NaOH	N/A	N/A
	DQ-N	2009.1	4/9	N/A	450	26	45	9–16	Carboxylate (PBH-808E)	NaOH	0.562	9.7
	B-3-D-X	2013.8	96/96	125	358	N/A	45	9–17	N/A	Na_2CO_3	0.84	21.59
	B-2-X-D	2015.5	172/178	150	330	26	45	9–18	N/A	Na_2CO_3	N/A	20.1
	N-4-D	2015.8	9/16	N/A	553	26	45	9–19	N/A	Na_2CO_3	N/A	10
	XQ-2	2015.7	99/131	125	546	26	45	9–20	N/A	Na_2CO_3	N/A	N/A
	ZY-HZJ	2000.1–2001.3	4/5	N/A	129–277	19.6	86	N/A		$NaHCO_3Na_2CO_3$	N/A	0.61

									Carboxylates alkyl sulfobetaines			
Gudong (Shengli)	SL-GD	1992–1994	4/9	50	2560	35	68	41.3	OP-10 + CY1	Na_2CO_3	0.592	16.8
Gudao-Xi (Shengli)	SL-GDX	1997–2002	6/13	212	1520	32	69	70	Anionic (BES) lignosulfonate PS	Na_2CO_3	0.497	15.5
Karamary (Xinjiang)	XJ-Kara	1995.8–1997	4/9	50	181	18.1	23	8.82	PS (KPS-1)	Na_2CO_3	0.3708	25
Laojunmiao (Yulin)	YM-LJM	1994.3–1995.8	1/4	70	374	24	28	7.14	PS (YPS-3A)	Na_2CO_3KCl	0.86	1.82
Honggang (Jilin)	JL-HG	1997.9–2000.6	3/22	200	163	21	55	12.9	PS	Na_2CO_3 NaOH	<0.27	20.4
Huanghe (He'nan)	HN-SH	2010.12	38/43	170–260	543	19	81	3.3	Mixed (SH6)	Na_2CO_3	0.6	6.9

ASP, *Alkaline–surfactant–polymer.*

[a]*Surfatant type:* HABS, *heavy alkylbenzene sulfonate;* ABS, *alkylbenzene sulfonate;* PS, *petroleum sulfonate;* DPS, *Daqing petroleum sulfonate.*

The designed ASP chemical compositions and concentrations for this block were 1.25% Na_2CO_3, 0.3% B-100 surfactant, and 1200 ppm 1275 A polymer. The IFT values were tested in the laboratory before conducting the pilot test. The results indicate that the ultralow IFT (10^{-3} mN/m) can be achieved in a wider concentration range, which is close to the designed injection concentrations [191].

As the first pilot test, the ASP flooding project in this block only consisted of the main slug and postslug. Before ASP flooding, water was injected into the pilot area to establish residual oil saturation as a reference for ASP performance evaluation. The waterflooding was conducted from June 1993 to September 1994. After waterflooding, 0.327 PV ASP slug, 0.273 PV polymer slug, and postwater were injected sequentially. The response to ASP flooding was observed after 0.0693 PV injection. The average water cut in the pilot area decreased from 82.7% to 59.7%. The peak oil production rate increased from 37 to 91.5 m^3/day. After the ASP flooding, a 21.4% incremental oil recovery over waterflooding was observed.

- X-2-Z block in Daqing oilfield

The ASP flooding in the X-2-Z block in Daqing oilfield was from April 2000 to January 2007. This project involved a much larger area (2.03 km^2) than other smaller pilot tests. The formation depth was 947–1029 m. The effective thickness was 7.2 m. The average permeability was 404 mD and the average porosity was 25%. The total PV was estimated as 3,654,000 m^3 and the OOIP was 2,017,000 tons. The reservoir temperature was 45 °C. The formation water TDS was 4144 mg/L.

This project involved 17 injectors and 27 producers that were drilled in regular 5-spot patterns. The average well spacing was 250 m. The surfactant used in this project was hard alkyl benzene sulfonate (HABS). The treatment included preslug, main slug, vice slug, postslug, and water drive. The preslug was 0.128 PV of 1400 mg/L polymer injection. The main slug consisted of 0.354 PV of mixed 1% NaOH, 0.2% HABS, and 1650 mg/L polymer. The vice slug was 0.1 PV of 1% NaOH, 0.1% HABS, and 1000 mg/L polymer. The postslug injection started with 0.1 PV of 1000 mg/L polymer and was followed by 0.1 PV of 630 mg/L polymer solution. The detailed injection design is listed in Table 4.17. After the ASP flooding, water was injected until the water cut in the central area reached 98%.

During the preslug stage, the injection pressure increased and the injection rate decreased. In the following ASP flooding, the injection rate and pressure increased slightly and gradually stabilized. The oil rate increased and the water cut decreased. The polymer breakthrough was observed from adjacent wells after about 0.164 PV injection; the surfactant breakthrough was observed after 0.17 PV injection on average; the alkaline breakthrough was detected after about 0.187 PV injection.

The ASP treatment received promising results, however, the breakdown of the produced emulsion was a problem in this field application. The collected emulsion sample from Well Xing2-D2-P3 was a W/O emulsion with a viscosity of 185 cp. The emulsion can stay stable under room temperature for more than 129 days without phase separation.

Table 4.17 Alkaline–surfactant–polymer flooding in the X-2-Z block in Daqing oilfield.

	Surfactant concentration (%)	Alkaline concentration (%)	Polymer concentration (mg/L)	Viscosity (cp)	Injection volume (PV)
Preslug	–	–	1400	40	0.13
Main slug	0.2	1	1650	40	0.35
Vice slug	0.1	1	1000	35	0.1
Postslug	–	–	1000	30	0.1
			630	15	0.1

4.8.4.1.2 Status of alkaline–surfactant–polymer flooding in other oilfields in China

In addition to the Daqing oilfield, other oilfields in China, such as Shengli, Karamay, Henan, Jilin, Zhongyuan, and Yulin oil fields, also implemented the ASP flooding EOR method.

- Shengli Gudong oilfield

Shengli oilfield started to investigate ASP flooding in laboratories in the early 1980s and applied this technology in the Gudong oilfield in 1992 [192,193].

Gudong oilfield is an unconsolidated sandstone reservoir with severe heterogeneity problems and high fluid viscosity. The tested area was 0.031 km^2. The formation depth was 1272 m. The reservoir average permeability was 2560 mD and the average porosity was 35%. The permeability variation coefficient (Dykstra–Parsons coefficient) was 0.33. Reservoir effective thickness was 11 m. The calculated total pore volume was 119,227 m^3 and the estimated OOIP was 77,952 tons. The reservoir temperature was 68°C and the oil viscosity at reservoir temperature was 41.3 cp. The oil acid number was 3.11 mg KOH/mg oil. The formation water TDS was 3259 mg/L and the formation water divalent concentration was 19–92 mg/L. The injected water salinity was 737 mg/L. This test included 4 injectors and 9 producers in 5-spot patterns. The average well spacing was 50 m.

Before the ASP pilot test, the field was waterflooded for 5 years. The water cut was 98.4% and the oil recovery was 54.4% after waterflooding. The ASP flooding pilot test was conducted from August 1992 to 1994. A total of 0.592 PV of ASP was injected. The ASP flooding injection design was summarized in Table 4.18. The preslug used 0.05 PV of 1000 mg/L 3530 S polymer. The flowing injection was AS flooding, which used 0.05 PV of a mixture of 0.2% CY1 and 0.2% OP-10 surfactants and 1.5% Na_2CO_3. The chemical agents used in the main ASP slug included 0.2% CY1 and 0.2% OP-10 surfactants, 1.5% Na_2CO_3, and 1000 mg/L 3530 S polymer. The volume of the main slug was 0.35 PV.

Table 4.18 Alkaline–surfactant–polymer flooding design in Shengli Gudong oilfield.

	Preslug	AS slug	Main ASP slug	Postslug
Surfactant concentration (%)	–	0.4	0.4	–
Alkaline concentration (%)	–	1.5	1.5	–
Polymer concentration (mg/L)	1000	–	1000	500
Viscosity (cp)	16.5	0.47	7.86	6.16
Injection volume (PV)	0.05	0.05	0.35	0.1

Table 4.19 Alkaline–surfactant–polymer flooding design in Huzhuangji oilfield.

	Preslug	Main slug	Postslug
Surfactant concentration (%)	–	1.58	–
Alkaline concentration (%)	–	0.5	–
Polymer concentration (mg/L)	1500	800	1000
Injection volume (m^3)	2920	49,640	2920

The postslug used 0.1 PV of 500 mg/L 3530 S polymer solution. Water drive was reestablished in 1994. The water cut was decreased to a minimum of 74.2% after 0.49 PV injection, however, it returned to the level before ASP treatment by July 1994. The incremental oil recovery was 16.8% over waterflooding.

- Zhongyuan Huzhuangji oilfield

The Huzhuangji oilfield is a high-temperature and high-salinity field operated by Zhongyuan oilfield. The ASP pilot area was 0.00041 km^2 [194]. The formation depth was 2000 m. The estimated OOIP was 737,000 tons. Reservoir porosity was 19.6% and the reservoir permeability ranged from 129–277 mD. The reservoir temperature was 86°C. The formation water TDS was 170,000 mg/L. Considering the high-salinity and high-temperature reservoir conditions, the polymer used in this pilot test was xanthan gum. The injection scheme was summarized in Table 4.19. The reservoir was firstly flushed by 2920 m^3 of 1% KCl to reduce formation salinity. The preslug was 2029 m^3 of 1500 mg/L xanthan gum. In the main slug, 49,640 m^3 ASP was injected. The alkaline solution was 0.5% mixed Na_2CO_3 and $NaHCO_3$ with a 1:1 ratio. Due to the high-salinity condition, two surfactants were mixed to increase the salinity tolerance. The surfactants were 1.5% natural-mixed carboxylates and 0.08% alkyl sulfobetaines (DSB). The combination of these two surfactants can resist up to 5000 mg/L divalent. The polymer was 800 mg/L xanthan gum. The postslug was 2920 m^3 of 1000 mg/L xanthan.

The test was initiated in January 2000 and ended in March 2001. Although the liquid production rate decreased from 377.5 to 219.8 tons/day, the oil rate increased from 10.5 to 20.1 tons/day and the water cut decreased from 97.2% to 90.9%. This project yielded a 0.16% incremental oil recovery but the large amount of chemical consumption brought in economic concerns.

- Xinjiang Karamay oilfield

An ASP pilot test was carried out in the Karamay oilfield from August 1995 to 1997. The reservoir was a heterogeneous conglomerate reservoir [107,108]. The pilot test involved 4 injectors and 9 producers. The average well spacing was 50 m. The average reservoir permeability was 181 mD and the average porosity was 18.1%. The reservoir temperature was 23°C. The oil viscosity in place was 8.82 cp.

The reservoir was preflushed using 0.4 PV 1.5% KCl. The main slug used 1.5% Na_2CO_3, 0.3% PS (KPS-1), and 1330 polymer. The injection volume of the main slug was 0.337 PV. Finally, a 0.15 PV postslug with 1000 mg/L polymer was injected.

The oil recovery after conventional waterflooding was around 50%. The reservoir response was observed after 0.04PV ASP injection and the peak oil recovery appeared when about 0.2 PV ASP slug was injected. The water cut reduced from 99% to 79%. The incremental oil recovery in the central well was 25% OOIP. Similar to the X-2-Z block in the Daqing oilfield, the produced emulsion was stable and difficult to break down in this oilfield.

4.8.4.2 Status of alkaline–surfactant–polymer in other countries

Other than China, many other ASP field cases were conducted in the United States, Canada, South America, and India. In this section, examples of cases in the United States and Canada are discussed.

4.8.4.2.1 West Kiehl, Wyoming, United States

West Kiehl field was one of the earliest fields in the United States that conducted ASP pilot tests [195]. The ASP project started in September 1987. The ASP slug consisted of 0.8% Na_2CO_3, 0.1% Petrostep B-100, and 0.105% Pusher 700. The total ASP injection volume was 0.37 PV. The incremental oil recovery reached 26% OOIP in 2.5 years after ASP flooding. The performance indicated that the ASP flooding successfully increased oil recovery in the West Kiehl field at a low cost of $2.00 per incremental barrel.

4.8.4.2.2 Tanner field, United States

Tanner field was discovered in 1991. The depth was 8915 ft and the average thickness was 25 ft. The reservoir average permeability was 200 mD and the average porosity was 20%. The reservoir temperature was 175°F. The oil gravity was 21-degree API and the viscosity in place was 11 cp. The calculated OOIP was 1280 Mbbls.

The ASP project was carried out from May 2000 to January 2005. The injected ASP slug included 1 wt.% sodium hydroxide, 0.1 wt.% active ORS-41HF, and

1000 mg/L Alcoflood 1275 A [39]. After the ASP injection, a polymer drive with tapered concentrations was conducted. The incremental oil recovery over the next one year reached 10% OOIP. The cost of chemicals, plant, and design was $5.85 per incremental barrel of oil.

4.8.4.2.3 Cambridge Minnelusa, United States

The Cambridge Minnelusa field was located in Crook County, Wyoming [40]. The reservoir depth was 2139 m and the reservoir temperature was 55.6°C. The average reservoir thickness was 8.75 m. The average permeability and porosity are 845 mD and 18%, respectively. In this pilot test, a total of 0.252 PV of ASP slug was injected. The ASP flooding increased oil recovery by 28% OOIP and the cost was $2.42 per incremental barrel of oil.

4.8.4.2.4 Taber South, Alberta, Canada

The ASP project started in the Taber South field in June 2006 [196,197]. The liquid production rate before ASP flooding was 300 bbl/day. The ASP flooding increased the production rate to 1502 bbl/day as of December 2007. The water cut decreased from 98.3% to 92.7%. In this pilot test, severe scaling problems and injectivity loss were encountered.

4.8.4.2.5 Taber Glauconitic, Canada

The first field-scale application of ASP flooding by Husky Oil Operations Limited was conducted in the Taber Glauconitic field in January 2008 [196,197]. The surfactants used in this project were sodium lignosulfonate (lignin) and alkyl polyglycosides (APG). Before the ASP flooding, the oil production rate was 26 m^3/day and the water cut was 99.1%. The implementation of ASP flooding increased the oil production rate to 127 m^3/day and reduced the water cut to 94.7%. As of April 2010, the total incremental oil recovery above the base decline was 0.4% OOIP.

4.8.5 Future developments

ASP has been studied and applied for more than 2 decades, but some problems are still needed to be addressed in the future. Since ASP flooding is a combination of surfactant, alkaline, and PF, the limitations and future developments for these chemical EOR methods, which are covered in previous sections are also applicable to ASP flooding. The advantage of ASP over individual chemical EOR is the synergy. To maximize this synergy, chromatographic separation should be avoided when mixing alkali, surfactant, and polymer. Therefore, more attention should be put on increasing chemical compatibility and optimizing the synergy of chemical components.

4.9 Summary

This chapter discussed some commonly used chemical EOR methods, including PF, SF, AF, and their combinations. Currently, PF is a relatively mature technology that can improve mobility ratio, which can increase oil displacement efficiency. However, the long-term thermostability and salinity sensitivity are two major challenges for PF. SF can alter reservoir wettability and reduce IFT. However, due to the surfactant adsorption, surfactants are commonly applied together with alkalis or alkalis and polymers to reduce adsorption and improve displacement efficiency. AF is not extensively used due to the risks of corrosion and scaling problems. Typically, it is used together with surfactants or polymers as combinations. ASP flooding has been applied in many fields or areas. The synergy among the three chemicals makes ASP flooding one of the most efficient chemical EOR methods to improve oil recovery. As summarized in this chapter, chemical EOR is facing some challenges, which will be the future development direction.

Nomenclature

A alkali
ABS alkylbenzene sulfonate
AF alkaline flooding
AP alkaline–polymer
AS alkaline–surfactant
ASP alkaline–surfactant–polymer
CDC capillary desaturation curve
CMC critical micellar concentration
Co-S cosolvent
Da Dalton
DPS Daqing petroleum sulfonate
EOR enhanced oil recovery
HABS heavy alkylbenzene Sulfonate
HASP hydrophobically associating polymers
HLB hyrophile–lipophile balance
HPAM hydrolyzed polyacrylamide
IFT interfacial tension
KPS Karamay petroleum sulfonate
M mobility ratio
mD millidarcy
MW molecular weight
P polymer
PAM polyacrylamide
PF polymer flooding
PS petroleum sulfonate

SF	surfactant flooding
SP	surfactant−polymer
SPS	Shengli sulfonate
SR	solubilization ratio
TDS	total dissolved solids

References

[1] F. Afolabi, T. Ojo, S.G. Udeagbara, et al., Bitumen extraction from tar sands using solvent techniques, International Journal of Scientific and Engineering Research 8 (2017) 783−790.

[2] A. Agi, R. Junin, M. Faddli, et al., Intermittent and short duration ultrasound in a simulated porous medium, Petroleum 5 (2019) 42−51.

[3] A.O. Gbadamosi, J. Kiwalabye, R. Junin, et al., A review of gas enhanced oil recovery schemes used in the North Sea, Journal of Petroleum Exploration and Production Technology 8 (2018) 1373−1387.

[4] A.O. Gbadamosi, R. Junin, M.A. Manan, et al., An overview of chemical enhanced oil recovery: recent advances and prospects, International Nano Letters 9 (2019) 171−202.

[5] L.D. Saleh, M. Wei, Y. Zhang, et al., Data analysis for polymer flooding that is based on a comprehensive database, SPE Reservoir Evaluation & Engineering 20 (2016) 876−893.

[6] H. Aluhwal, O. Kalifa, Simulation Study of Improving Oil Recovery by Polymer Flooding in a Malaysian Reservoir (MEng thesis), University Technology Malaysia, Johor, 2008.

[7] J. Chatterji, J. Borchardt, Applications of water-soluble polymers in the oil field, Journal of Petroleum Technology 33 (1981) 2042−2056.

[8] K. Taylor, H. Nasr-El-Din, Water-soluble hydrophobically associating polymers for improved oil recovery: a literature review, Journal of Petroleum Science and Engineering 19 (1998) 265−280.

[9] E. Manrique, M. Ahmadi, S. Samani, Historical and recent observations in polymer floods: an update review, CT&F-Ciencia, Tecnología y Futuro 6 (2017) 17−48.

[10] D.W. Green, G.P. Willhite, Enhanced Oil Recovery, 6, Henry L. Doherty Memorial Fund of AIME, Society of Petroleum Engineers, 1998.

[11] F. Martin, The effect of hydrolysis of polyacrylamide on solution viscosity, polymer retention and flow resistance properties, in: SPE-5339-MS, SPE Rocky Mountain Regional Meeting, April 7−9, 1975, Denver, Colorado, USA.

[12] G. Hirasaki, G. Pope, Analysis of factors influencing mobility and adsorption in the flow of polymer solution through porous media, Society of Petroleum Engineers Journal 14 (1974) 337−346.

[13] C. Gao, Viscosity of partially hydrolyzed polyacrylamide under shearing and heat, Journal of Petroleum Exploration and Production Technology 3 (2013) 203−206.

[14] J.J. Sheng, B. Leonhardt, N. Azri, Status of polymer-flooding technology, Journal of Canadian petroleum technology 54 (2015) 116−126.

[15] S. Xia, L. Zhang, A. Davletshin, et al., Application of polysaccharide biopolymer in petroleum recovery, Polymers 12 (2020) 1860.

[16] F. Garcıa-Ochoa, V.E. Santosa, J.A. Casasb, et al., Xanthan gum: production, recovery, and properties, Biotechnology Advances 18 (2000) 549–579.
[17] J. Fariña, F. Siñeriz, O.E. Molina, et al., Isolation and physicochemical characterization of soluble scleroglucan from Sclerotium rolfsii. Rheological properties, molecular weight and conformational characteristics, Carbohydrate Polymers 44 (2001) 41–50.
[18] S. Masaoka, T. Ohe, N. Sakota, Production of cellulose from glucose by Acetobacter xylinum, Journal of Fermentation and Bioengineering 75 (1993) 18–22.
[19] S. Ummartyotin, H. Manuspiya, A critical review on cellulose: from fundamental to an approach on sensor technology, Renewable and Sustainable Energy Reviews 41 (2015) 402–412.
[20] H. Fan, Z. Gong, Z. Wei, et al., Understanding the temperature-resistance performance of a borate cross-linked hydroxypropyl guar gum fracturing fluid based on a facile evaluation method, RSC Advances 84 (2017) 53290–53300.
[21] M. Auerbach, Prediction of viscosity for xanthan solutions in brines, in: SPE-13591-MS, SPE Oilfield and Geothermal Chemistry Symposium, April 9–11, 1985, Phoenix, Arizona, USA.
[22] H. Dong, S. Fang, D. Wang, et al. Review of practical experience & management by polymer flooding at Daqing, in: SPE-114342-MS, SPE Symposium on Improved Oil Recovery, April 20–23, 2008, Tulsa, Oklahoma, USA.
[23] M. Buchgraber, T. Clemens, L.M. Castanier, et al., A microvisual study of the displacement of viscous oil by polymer solutions, SPE Reservoir Evaluation & Engineering 14 (2011) 269–280.
[24] X. Gao, R. Jones, Nitrogen-15 labeled deoxynucleosides. Synthesis of 15N6- and 15N1-deoxyadenosines from deoxyadenosine, Journal of the American Chemical Society 109 (1987) 1275–1278.
[25] J.J. Sheng, Modern Chemical Enhanced Oil Recovery: Theory and Practice, Gulf Professional Publishing, 2010.
[26] F. Martin, The effect of hyrolysis of polyacrylamide on solution viscosity, polymer retention and flow resistance properties, in: SPE-5339-MS, SPE Rocky Mountain Regional Meeting, April 7–9, 1975, Denver, Colorado, USA.
[27] Y. Niu, O. Jian, Z. Zhu, et al. Research on hydrophobically associating water-soluble polymer used for EOR, in: SPE-65378-MS, SPE International Symposium on Oilfield Chemistry, February 13–16, 2001, Houston, Texas, USA.
[28] W. Zheng, X. Wu, Y. Huang, Impact of polymer addition, electrolyte, clay and antioxidant on rheological properties of polymer fluid at high temperature and high pressure, Journal of Petroleum Exploration and Production Technology 10 (2020) 663–671.
[29] R.K. Manning, A Technical Survey of Polymer Flooding Projects (Ph.D. dissertation), University of Texas at Austin, Austin, 1983.
[30] E.J. Manrique, V.E. Muci, M.E. Gurfinkel, EOR field experiences in carbonate reservoirs in the United States, SPE Reservoir Evaluation & Engineering 10 (2007) 667–686.
[31] D.C. Standnes, I. Skjevrak, Literature review of implemented polymer field projects, J. Pet. Sci. Eng. 122 (2014) 761–775.
[32] H. Pu, Q. Xu, An update and perspective on field-scale chemical floods in Daqing oilfield, China, in: SPE-118746-MS, SPE Middle East Oil and Gas Show and Conference, March 15–18, 2009, Manama, Bahrain.
[33] J.L. Mogollón, T. Lokhandwala. Rejuvenating viscous oil reservoirs by polymer injection: Lessons learned in the field, in: SPE-165275-MS, SPE Enhanced Oil Recovery Conference, July 2–4, 2013, Kuala Lumpur, Malaysia.

[34] Z. Qu, Y. Zhang, X. Zhang, et al. A successful ASP flooding pilot in Gudong oil field, in: SPE-39613-MS, SPE/DOE Improved Oil Recovery Symposium, April 19–22, 1998, Tulsa, Oklahoma.

[35] T. Chen, Z. Song, Y. Fan, et al., A pilot test of polymer flooding in an elevated-temperature reservoir, SPE Reservoir Evaluation & Engineering 1 (1998) 24–29.

[36] J. Zhang, K. Wang, F. He, et al. Ultimate evaluation of the alkali/polymer combination flooding pilot test in XingLongTai oil field, in: SPE-57291-MS, SPE Asia Pacific Improved Oil Recovery Conference, October 25–26, 1999, Kuala Lumpur, Malaysia.

[37] J. Meyers, M.J. Pitts, K. Wyatt, Alkaline-surfactant-polymer flood of the West Kiehl, Minnelusa Unit, in: SPE-24144-MS, SPE/DOE Enhanced Oil Recovery Symposium, April 22–24, 1992, Tulsa, Oklahoma, USA.

[38] T. Clemens, M. Deckers, M. Kornberger, et al. Polymer solution injection—near wellbore dynamics and displacement efficiency, pilot test results, Matzen Field, Austria, in: SPE-164904-MS, EAGE Annual Conference & Exhibition incorporating SPE Europec, June 10–13, 2013, London, UK.

[39] M.J. Pitts, P. Dowling, K. Wyatt, et al. Alkaline-surfactant-polymer flood of the Tanner field, in: SPE-100004-MS, SPE/DOE Symposium on Improved Oil Recovery, April 22–26, 2006, Tulsa, Oklahoma, USA.

[40] J. Vargo, J. Turner, B. Vergnani, et al., Alkaline-surfactant-polymer flooding of the Cambridge Minnelusa field, SPE Reservoir Evaluation & Engineering 3 (2000) 552–558.

[41] Y. Zhao, S. Yin, R.S. Seright, et al., Enhancing heavy-oil-recovery efficiency by combining low-salinity-water and polymer flooding, SPE Journal 26 (2021) 1535–1551.

[42] D. Tiwari; R.V. Marathe, N.K. Patel, et al. Performance of polymer flood in Sanand field, India—a case study, in: SPE-114878-MS, SPE Asia Pacific Oil and Gas Conference and Exhibition, October 20–22, 2008, Perth, Australia.

[43] M. Pitts, K. Wyatt, H. Surkalo, Alkaline-polymer flooding of the David Pool, Lloydminster Alberta, in: SPE-89386-MS, SPE/DOE Symposium on Improved Oil Recovery, April 17–21, 2004, Tulsa, Oklahoma, USA.

[44] S. Abbas, A.W. Sanders, J.C. Donovan, Applicability of hydroxyethylcellulose polymers for chemical EOR, in: SPE-165311-MS, SPE Enhanced Oil Recovery Conference, July 2–4, 2013, Kuala Lumpur, Malaysia.

[45] A. Loppinet, S. Iakovlev, P. Glenat. Five years of injection of hydroxyethylcellulose: An ecological pure product for enhanced oil recovery in the field of Romashkino, in: SPE-38464-MS, SPE Offshore Europe, September 9–12, 1997, Aberdeen, UK.

[46] M.S. Kamal, A.S. Sultan, U.A. Al-Mubaiyedh, et al., Review on polymer flooding: Rheology, adsorption, stability, and field applications of various polymer systems, Polymer Reviews 55 (2015) 491–530.

[47] M.S. Kamal, I.A. Hussein, A.S. Sultan, Review on surfactant flooding: phase behavior, retention, IFT, and field applications, Energy & Fuels 31 (2017) (2017) 7701–7720.

[48] M. Pratap, M. Gauma. Field implementation of alkaline-surfactant-polymer (ASP) flooding: a maiden effort in India, in: SPE-88455-MS, SPE Asia Pacific Oil and Gas Conference and Exhibition, October 18–20, 2004, Perth, Australia.

[49] D. Wang, Polymer flooding practice in Daqing, in: J.J. Sheng (Ed.), Enhanced Oil Recovery Field Case Studies, Elsevier, 2013, pp. 83–116.

[50] X. Zhang, Application of polymer flooding with high molecular weight and concentration in heterogeneous reservoirs, in: SPE-144251-MS, SPE Enhanced Oil Recovery Conference, July 19–20, 2011, Kuala Lumpur, Malaysia.

[51] T.G. DeHekker, J.L. Bowzer, R.V. Coleman, et al. A progress report on polymer-augmented waterflooding in Wyoming's North Oregon Basin and Byron Fields, in: SPE-14953-MS, SPE Enhanced Oil Recovery Symposium, April 20–23, 1986, Tulsa, Oklahoma, USA.

[52] A. Dandekar, B. Bai, J. Barnes, et al. First ever polymer flood field pilot to enhance the recovery of heavy oils on Alaska's North Slope pushing ahead one year later, in: SPE-200814-MS, SPE Western Regional Meeting, April 20–22, 2021, Virtual.

[53] R.S. Seright, Potential for polymer flooding reservoirs with viscous oils, SPE Reservoir Evaluation & Engineering 13 (2010) 730–740.

[54] A.R. Glennie, M.M. Mohareb, R.M. Palepu, Thermodynamic and related properties of alkyl cationic surfactants based on dimethyl and diethyl ethanol amines, Journal of Dispersion Science and Technology 27 (2006) 731–738.

[55] S. Thiruvengadam, M. Murphy, K. Miller, A generalised methodology using conformal geometric algebra for mathematical chemistry, Journal of Mathematical Chemistry 58 (2020) 1737–1783.

[56] A. Bratovcic, S. Nazdrajic, Viscoelastic behavior of synthesized liquid soaps and surface activity properties of surfactants, Journal of Surfactants and Detergents 23 (2020) 1135–1143.

[57] L.R. Harutyunyan, R.S. Harutyunyan, Effect of amino acids on micellization and micellar parameters of anionic surfactant alpha olefin sulfonate C14–C16 in aqueous solutions: Surface tension, conductometric, volumetric, and fluorescence studies, Journal of Chemical & Engineering Data 64 (2019) 640–650.

[58] H. Suzuki, Determination of critical micelle concentration of surfactant by ultraviolet absorption spectra, Journal of the American Oil Chemists Society 47 (1970) 273–277.

[59] K. Nesměrák, I. Němcová, Determination of critical micelle concentration by electrochemical means, Analytical Letters 39 (2006) 1023–1040.

[60] M. Bourrel, R.S. Schechter, Microemulsions and Related Systems: Formulation, Solvency, and Physical Properties, Editions Technip, Paris, 2010.

[61] P. Winsor, Hydrotropy, solubilisation and related emulsification processes, Transactions of the Faraday Society 44 (1948) 376–398.

[62] A. Das, N. Nguyen, Q.P. Nguyen, Low tension gas flooding for secondary oil recovery in low-permeability, high-salinity reservoirs, Fuel 264 (2020) 116601.

[63] S. Sarmah, S.B. Gogoi, F. Xianfeng, et al., Characterization and identification of the most appropriate nonionic surfactant for enhanced oil recovery, Journal of Petroleum Exploration and Production Technology 10 (2020) 115–123.

[64] N. Abalkhail, P.J. Liyanage, K.A.N. Upamali, et al., Alkaline-surfactant-polymer formulation development for a HTHS carbonate reservoir, Journal of Petroleum Science and Engineering 191 (2020) 107236.

[65] H. Hamidi, E. Mohammadian, R. Rafati, et al., The effect of ultrasonic waves on the phase behavior of a surfactant-brine-oil system, Colloids and Surfaces A: Physicochemical and Engineering Aspects 482 (2015) 27–33.

[66] K.O. Meyers, S.J. Salter, The effect of oil/brine ratio on surfactant adsorption from microemulsion, Society of Petroleum Engineers Journal 21 (1981) 500–512.

[67] C.T.Q. Dang, Z.J. Chen, N.T.B. Nguyen, et al., Lessons learned and experiences gained in developing the waterflooding concept of a fractured basement-granite reservoir: a 20-year case study, Journal of Canadian Petroleum Technology 50 (2011) 10–23.
[68] S. Liu, Alkaline Surfactant Polymer Enhanced Oil Recovery Process (Ph.D. dissertation), Rice University, Houston, 2008.
[69] J. Salager, J.C. Morgan, R.S. Schechter, et al., Optimum formulation of surfactant/water/oil systems for minimum interfacial tension or phase behavior, Society of Petroleum Engineers Journal 19 (1979) 107–115.
[70] S.M. Skjæveland, J. Kleppe, SPOR Monograph, Recent advances in improved oil recovery methods for North Sea sandstone reservoirs, Norwegian Petroleum Directorate, Norway, 1992.
[71] T. Moore, R. Slobod, Displacement of oil by water-effect of wettability, rate, and viscosity on recovery, in: SPE-502-G, Fall Meeting of the Petroleum Branch of AIME, October 2–5, 1955, New Orleans, Louisiana, USA.
[72] A. Bashiri, N. Kasiri, Properly use effect of capillary number on residual oil saturation, in: SPE-150800-MS, Nigeria Annual International Conference and Exhibition, July 30–August 3, 2011, Abuja, Nigeria.
[73] J.J. Taber, Research on enhanced oil recovery: past, present and future, Pure and Applied Chemistry 52 (1980) 1323–1347.
[74] M. Delshad, D. Bhuyan, G.A. Pope, et al. Effect of capillary number on the residual saturation of a three-phase micellar solution, in: SPE-14911-MS, SPE Enhanced Oil Recovery Symposium, April 20–23, 1986, Tulsa, Oklahoma, USA.
[75] A.S. Abushaikha, O.R. Gosselin, Matrix-fracture transfer function in dual-media flow simulation: review, comparison and validation, in: SPE-113890-MS, Europec/EAGE Conference and Exhibition, June 9–12, 2008, Rome, Italy.
[76] A. Mehranfar, M.H. Ghazanfari, Investigation of the microscopic displacement mechanisms and macroscopic behavior of alkaline flooding at different wettability conditions in shaly glass micromodels, Journal of Petroleum Science and Engineering 122 (2014) 595–615.
[77] S. Shi, F. Wang, L. Wang, et al., Potential of spontaneous emulsification flooding for enhancing oil recovery in high-temperature and high-salinity oil reservoir, Journal of Dispersion Science and Technology 36 (2015) 660–669.
[78] U. Türksoy, S. Bağci, Improved oil recovery using alkaline solutions in limestone medium, Journal of Petroleum Science and Engineering 26 (2000) 105–119.
[79] J. Bryan, A. Kantzas, Potential for alkali-surfactant flooding in heavy oil reservoirs through oil-in-water emulsification, Journal of Canadian Petroleum Technology 48 (2009) 37–46.
[80] T.P. Castor, W.H. Somerton, J.F. Kelly, Recovery mechanisms of alkaline flooding, in: D.O. Shah (Ed.), Surface Phenomena in Enhanced Oil Recovery, Springer US, Boston, 1981, pp. 249–291.
[81] C.E. Johnson, Status of caustic and emulsion methods, Journal of Petroleum Technology 28 (1976) 85–92.
[82] O. Massarweh, A.S. Abushaikha, The use of surfactants in enhanced oil recovery: a review of recent advances, Energy Reports 6 (2020) 3150–3178.
[83] L. Chen, G. Zhang, L. Wang, et al., Zeta potential of limestone in a large range of salinity, Colloids and Surfaces A: Physicochemical and Engineering Aspects 450 (2014) 1–8.

[84] A. Kumar, A. Mandal, Critical investigation of zwitterionic surfactant for enhanced oil recovery from both sandstone and carbonate reservoirs: adsorption, wettability alteration and imbibition studies, Chemical Engineering Science 209 (2019) 115222.

[85] N. Saxena, A. Kumar, A. Mandal, Adsorption analysis of natural anionic surfactant for enhanced oil recovery: the role of mineralogy, salinity, alkalinity and nanoparticles, Journal of Petroleum Science and Engineering 173 (2019) 1264–1283.

[86] N. Yekeen, E. Padmanabhan, A.K. Idris, et al., Surfactant adsorption behaviors onto shale from Malaysian formations: influence of silicon dioxide nanoparticles, surfactant type, temperature, salinity and shale lithology, Journal of Petroleum Science and Engineering 179 (2019) 841–854.

[87] M.R. Azam, I.M. Tan, L. Ismail, et al., Kinetics and equilibria of synthesized anionic surfactant onto berea sandstone, Journal of Dispersion Science and Technology 35 (2014) 223–230.

[88] P. Wei, J. Li, Y. Xie, et al., Alkyl polyglucosides for potential application in oil recovery process: adsorption behavior in sandstones under high temperature and salinity, Journal of Petroleum Science and Engineering 189 (2020) 107057.

[89] M.A. Ahmadi, S.R. Shadizadeh, Z. Chen, Thermodynamic analysis of adsorption of a naturally derived surfactant onto shale sandstone reservoirs, The European Physical Journal Plus 133 (2018) 1–11.

[90] N. Yekeen, M.A. Manan, A.K. Idris, et al., Influence of surfactant and electrolyte concentrations on surfactant adsorption and foaming characteristics, Journal of Petroleum Science and Engineering 149 (2017) 612–622.

[91] A.H. Abbas, A. Moslemizadeh, W.R.W. Sulaiman, et al., An insight into a di-chain surfactant adsorption onto sandstone minerals under different salinity-temperature conditions: chemical EOR applications, Chemical Engineering Research and Design 153 (2020) 657–665.

[92] P. Li, M. Ishiguro, Adsorption of anionic surfactant (sodium dodecyl sulfate) on silica, Soil Science and Plant Nutrition 62 (2016) 223–229.

[93] J.G. Southwick, E. van den Pol, C.H. van Rijn, et al., Ammonia as alkali for alkaline/surfactant/polymer floods, SPE Journal 21 (2016) 10–21.

[94] M. Tagavifar, S.H. Jang, H. Sharma, et al., Effect of pH on adsorption of anionic surfactants on limestone: experimental study and surface complexation modeling, Colloids and Surfaces A: Physicochemical and Engineering Aspects 538 (2018) 549–558.

[95] A. Rezaei, M. Riazi, M. Escrochi, et al., Integrating surfactant, alkali and nano-fluid flooding for enhanced oil recovery: a mechanistic experimental study of novel chemical combinations, Journal of Molecular Liquids 308 (2020) 113106.

[96] Z. Liu, M.K. Ghatkesar, E.J.R. Sudhölter, et al., Understanding the cation-dependent surfactant adsorption on clay minerals in oil recovery, Energy & Fuels 33 (2019) 12319–12329.

[97] A.O. Al-Amodi, U.A. Al-Mubaiyedh, A.S. Sultan, et al., Novel fluorinated surfactants for enhanced oil recovery in carbonate reservoirs, The Canadian Journal of Chemical Engineering 4 (2016) 454–460.

[98] Y.Y. Foo, R.D. Tewari, K.C. Kok, et al., Laboratory evaluation of chemical EOR Process for viscous and high EACN oil in East African oilfields, in: IPTC-18601-MS, International Petroleum Technology Conference, Bangkok, Thailand, November 14–16, 2016, Bangkok, Thailand.

[99] M. Han, A. AlSofi, A. Fuseni, et al. Development of chemical EOR formulations for a high temperature and high salinity carbonate reservoir, in: IPTC-17084-MS, International Petroleum Technology Conference, March 26–28, 2013, Beijing, China.

[100] S.S. Khalilinezhad, S. Mobaraki, M. Zakavi, et al., Mechanistic modeling of nanoparticles-assisted surfactant flood, Arabian Journal for Science and Engineering 43 (2018) 6609–6625.

[101] M.S. Kamal, A.S. Sultan, U.A. Al-Mubaiyedh, et al., Rheological properties of thermoviscosifying polymers in high-temperature and high-salinity environments, The, Canadian Journal of Chemical Engineering 93 (2015) 1194–1200.

[102] C. Sun, H. Guo, Y. Li, et al., Alkali effect on alkali-surfactant-polymer (ASP) flooding enhanced oil recovery performance: two large-scale field tests' evidence, Journal of Chemistry 2020 (2020) 1–22.

[103] D. Yin, H. Pu. A numerical simulation study on surfactant flooding and its field application in Daqing oilfield, in: SPE-112424-MS, Europec/EAGE Conference and Exhibition, June 9–12, 2008, Rome, Italy.

[104] D. Li, S. Lu, Y. Liu, et al., The effect of biosurfactant on the interfacial tension and adsorption loss of surfactant in ASP flooding, Colloids and Surfaces A: Physicochemical and Engineering Aspects 244 (2004) 53–60.

[105] Y. Guo, J. Zhang, X. Zhang, et al. Investigation and application of an associative polymer-surfactant binary system for a successful flooding pilot in a high-temperature, high-salinity, ordinary heavy oil reservoir, in: SPE-190411-M, SPE EOR Conference at Oil and Gas West Asia, March 26–28, 2018, Muscat, Oman.

[106] Z. Qu, Y. Zhang, X. Zhang, et al. A successful ASP flooding pilot in Gudong oil field, in: SPE-39613-MS, SPE/DOE Improved Oil Recovery Symposium, April 19–22, 1998, Tulsa, Oklahoma, USA.

[107] Q. Qi, H. Gu, D. Li, et al., The pilot test of ASP combination flooding in Karamay oil field, in: SPE-64726-MS, International Oil and Gas Conference and Exhibition in China, November 7–10, 2000, Beijing, China.

[108] H. Gu, R. Yang, S. Guo, et al., Study on reservoir engineering: ASP flooding pilot test in Karamay oilfield, in: SPE-50918-MS, SPE International Oil and Gas Conference and Exhibition in China, November 2–6, 1998, Beijing, China.

[109] W.W. Weiss, X. Xie, J. Weiss, et al., Artificial intelligence used to evaluate 23 single-well surfactant-soak treatments, SPE Reservoir Evaluation & Engineering 9 (2006) 209–216.

[110] X. Xie, W.W. Weiss, Z. Tong, et al., Improved oil recovery from carbonate reservoirs by chemical stimulation, SPE Journal 10 (2005) 276–285.

[111] M.J. Pitts, P. Dowling, K. Wyatt, et al. Alkaline-surfactant-polymer flood of the Tanner field, in: SPE-100004-MS, SPE/DOE Symposium on Improved Oil Recovery, April 22–26, 2006, Tulsa, Oklahoma, USA.

[112] Z.P. Alcorn, T. Føyen, L. Zhang, et al. CO_2 foam field pilot design and initial results, in: SPE-200450-MS, SPE Improved Oil Recovery Conference, August 31–September 4, 2020, Virtual.

[113] T.-P. Hsu, L. Prapas, B.L. Roberts, et al. Improved oil recovery by chemical flood from high salinity reservoirs-single-well surfactant injection test, in: SPE-154838-M, SPE EOR Conference at Oil and Gas West Asia, April 16–18, 2012, Muscat, Oman.

[114] B.M. Aslam, D. Ulitha, B. Swadesi, et al., History match to support interpretation of surfactant flooding pilot test in Tanjung field, in: SPE-186325-MS, SPE/IATMI Asia Pacific Oil & Gas Conference and Exhibition, October 17–19, 2017, Jakarta, Indonesia.

[115] N.A. Rilian, M. Sumestry, Wahyuningsih, Surfactant stimulation to increase reserves in carbonate reservoir: a case study in Semoga field, in: SPE-130060-MS, SPE EUROPEC/EAGE Annual Conference and Exhibition, June 14–17, 2010, Barcelona, Spain.

[116] A. Syaifudin, Reservoir simulation to increase recovery in Kaji Semoga field, in: Proceedings of the 28th Annual Convention, Indonesian Petroleum Association, 2002.

[117] Y. Volokitin, M. Shuster, V. Karpan, et al., Results of alkaline-surfactant-polymer flooding pilot at West Salym field, in: SPE-190382-MS, SPE EOR Conference at Oil and Gas West Asia, March 26–28, 2018, Muscat, Oman.

[118] M. Ge, S. Miao, J. Liu, et al., Laboratory studies on a novel salt-tolerant and alkali-free flooding system composed of a biopolymer and a bio-based surfactant for oil recovery, Journal of Petroleum Science and Engineering 196 (2021) 107736.

[119] X. Han, Z. Chen, G. Zhang, et al., Surfactant-polymer flooding formulated with commercial surfactants and enhanced by negative salinity gradient, Fuel 274 (2020) 117874.

[120] C.A. Franco, L.J. Giraldo, C.H. Candela, et al., Design and tuning of nanofluids applied to chemical enhanced oil recovery based on the surfactant–nanoparticle–brine interaction: from laboratory experiments to oil field application, Nanomaterials 10 (2020) 1579.

[121] Z. Xu, B. Li, H. Zhao, et al., Investigation of the effect of nanoparticle-stabilized foam on EOR: nitrogen foam and methane foam, ACS Omega 5 (2020) 19092–19103.

[122] A. Omidi, A.K. Manshad, S. Moradi, et al., Smart- and nano-hybrid chemical EOR flooding using Fe_3O_4/eggshell nanocomposites, Journal of Molecular Liquids 316 (2020) 113880.

[123] K.R. Chaturvedi, T. Sharma, Rheological analysis and EOR potential of surfactant treated single-step silica nanofluid at high temperature and salinity, Journal of Petroleum Science and Engineering 196 (2021) 107704.

[124] A.W. Alsmaeil, A. Enotiadis, M.A. Hammami, et al., Slow release of surfactant using silica nanosized capsules, SPE Journal 25 (2020) 3472–3480.

[125] O.F. Ojo, A. Farinmade, V. John, et al., A nanocomposite of halloysite/surfactant/wax to inhibit surfactant adsorption onto reservoir rock surfaces for improved oil recovery, Energy & Fuels 34 (2020) 8074–8084.

[126] L. Romero-Zerón, D. Wadhwani, R. Sethiya, Formulation of an encapsulated surfactant system, ESS, via β-CD host-guest interactions to inhibit surfactant adsorption onto solid surfaces, Industrial & Engineering Chemistry Research 59 (2020) 15542–15555.

[127] H. Khayati, A. Moslemizadeh, K. Shahbazi, et al., An experimental investigation on the use of saponin as a non-ionic surfactant for chemical enhanced oil recovery (EOR) in sandstone and carbonate oil reservoirs: IFT, wettability alteration, and oil recovery, Chemical Engineering Research and Design 160 (2020) 417–425.

[128] M. Eslahati, P. Mehrabianfar, A.A. Isari, et al., Experimental investigation of Alfalfa natural surfactant and synergistic effects of Ca^{2+}, Mg^{2+}, and SO_4^{2-} ions for

EOR applications: interfacial tension optimization, wettability alteration and imbibition studies, Journal of Molecular Liquids 310 (2020) 113123.
[129] T.O. do Vale, R.S. de Magalhães, P.F. de Almeida, et al., The impact of alkyl polyglycoside surfactant on oil yields and its potential effect on the biogenic souring during enhanced oil recovery (EOR), Fuel 280 (2020) 118512.
[130] S.R.M. Dashtaki, J.A. Ali, A.K. Manshad, et al., Experimental investigation of the effect of Vitagnus plant extract on enhanced oil recovery process using interfacial tension (IFT) reduction and wettability alteration mechanisms, Journal of Petroleum Exploration and Production Technology 10 (2020) 2895–2905.
[131] W. Chen, D.S. Schechter, Surfactant selection for enhanced oil recovery based on surfactant molecular structure in unconventional liquid reservoirs, Journal of Petroleum Science and Engineering 196 (2021) 107702.
[132] O. Wagner, R. Leach, Improving oil displacement efficiency by wettability adjustment, Transactions of the AIME 216 (1959) 65–72.
[133] N. Mungan, Interfacial effects in immiscible liquid-liquid displacement in porous media, Society of Petroleum Engineers Journal 6 (1966) 247–253.
[134] C. Cooke, R. Williams, P. Kolodzie, Oil recovery by alkaline waterflooding, Journal of petroleum technology 26 (1974) 1365–1374.
[135] E. Kowalewski, T. Holt, O. Torsaeter, Wettability alterations due to an oil soluble additive, Journal of Petroleum Science and Engineering 33 (2002) 19–28.
[136] H. Gong, Y. Li, M. Dong, et al., Effect of wettability alteration on enhanced heavy oil recovery by alkaline flooding, Colloids and Surfaces A: Physicochemical and Engineering Aspects 488 (2016) 28–35.
[137] J. Labrid, Modeling of high pH sandstone dissolution, Journal of Canadian Petroleum Technology 30 (1991). PETSOC-91-06-08.
[138] D.K. Olsen, M.D. Hicks, B.G. Hurd, et al. Design of a novel flooding system for an oil-wet Central Texas carbonate reservoir, in: SPE-20224-MS, SPE/DOE Enhanced Oil Recovery Symposium, April 22–25, 1990, Tulsa, Oklahoma, USA.
[139] J. Ge, A. Feng, G. Zhang, et al., Study of the factors influencing alkaline flooding in heavy-oil reservoirs, Energy & Fuels 26 (2012) 2875–2882.
[140] H.L. Chang, Oil recovery by alkaline-sulfonate waterflooding, Patent US3977470A, 1978.
[141] R. Ehrlich, R.J. Wygal, Interrelation of crude oil and rock properties with the recovery of oil by caustic waterflooding, Society of Petroleum Engineers Journal 17 (1977) 263–270.
[142] T.C. Campbell, P.H. Krumrine, Laboratory studies on alkaline waterflooding, in: SPE-8328-MS, SPE Annual Technical Conference and Exhibition, September 23–26, 1979, Las Vegas, Nevada, USA.
[143] M. Dong, S. Ma, Q. Liu, Enhanced heavy oil recovery through interfacial instability: a study of chemical flooding for Brintnell heavy oil, Fuel 88 (2009) 1049–1056.
[144] J.J. Sheng, Status of alkaline flooding technology, Journal of Petroleum Engineering & Technology 5 (2015) 44–50.
[145] K.L. Goyal, R.K. Gupta, S.K. Nanda, et al., Performance evaluation of caustic pilot project of North Gujarat oil field, India, in: SPE-21410-MS, Middle East Oil Show, November 16–19, 1991, Manama, Bahrain.
[146] P. Raimondi, B.J. Gallagher, R. Ehrlich, et al., Alkaline waterflooding: design and implementation of a field pilot, Journal of Petroleum Technology 29 (1977) 1359–1368.

[147] J. Xie, BC Chung, L. Leung, Design and implementation of a caustic flooding EOR pilot at Court Bakken heavy oil reservoir, in: SPE-117221-MS, International Thermal Operations and Heavy Oil Symposium, October 20–23, 2008, Calgary, Alberta, Canada.

[148] A. Samanta, K. Ojha, A. Sarkar, et al., Surfactant and surfactant-polymer flooding for enhanced oil recovery, Advances in Petroleum Exploration and Development 2 (2011) 13–18.

[149] J.J. Sheng, Surfactant-polymer flooding, in: J.J. Sheng (Ed.), Enhanced Oil Recovery Field Case Studies, Elsevier, 2013, pp. 117–142.

[150] T. Chen, C. Pu, Study on ultralow interfacial tension chemical flooding in low permeability reservoirs, Journal of Xi'an Shiyou University (Naturnal Science Edition) 3 (2006) 30–33.

[151] W. Gogarty, Enhanced oil recovery through the use of chemicals—part I, Journal of Petroleum Technology 35 (1983) 1581–1590.

[152] H. Murtada, C. Marx, Evaluation of the low tension flood process for high-salinity reservoirs-laboratory investigation under reservoir conditions, Society of Petroleum Engineers Journal 22 (1982) 831–846.

[153] C. Sun, H. Guo, Y. Li, et al., Recent advances of surfactant-polymer (SP) flooding enhanced oil recovery field tests in China, Geofluids 2020 (2020) 8286706.

[154] Q. Lu, Y. Ning, J. Wang, et al. Full field offshore surfactant-polymer flooding in Bohai Bay China, in: SPE-174591-MS, SPE Asia Pacific Enhanced Oil Recovery Conference, August 11–13, 2015, Kuala Lumpur, Malaysia.

[155] R.D. Thomas, K.L. Spence, F.W. Burtch, et al., Performance of DOE's micellar-polymer project in northwest Oklahoma, in: SPE-10724-MS, SPE Enhanced Oil Recovery Symposium, April 4–7, 1982, Tulsa, Oklahoma, USA.

[156] A.H. Falls, D.R. Thigpen, R.C. Nelson, et al., Cosurfactant-enhanced alkaline flooding, SPE Reservoir Engineering 9 (1994) 217–223.

[157] J.J. Sheng, Alkaline-surfactant flooding, in: J.J. Sheng (Ed.), Enhanced Oil Recovery Field Case Studies, Elsevier, 2013, pp. 179–188.

[158] J.J. Sheng, Critical review of alkaline-polymer flooding, Journal of Petroleum Exploration and Production Technology 7 (2017) 147–153.

[159] J.J. Sheng, Alkaline-polymer flooding, in: J.J. Sheng (Ed.), Enhanced Oil Recovery Field Case Studies, Elsevier, 2013, pp. 169–178.

[160] K.C. Taylor, H.A. Nasr-El-Din, The effect of synthetic surfactants on the interfacial behaviour of crude oil/alkali/polymer systems, Colloids and Surfaces A: Physicochemical and Engineering Aspects 108 (1996) 49–72.

[161] P. Krumrine, J. Falcone, Beyond alkaline flooding: Design of complete chemical systems, in: SPE-16280-MS, SPE International Symposium on Oilfield Chemistry, February 4–6, 1987, San Antonio, Texas, USA.

[162] K. Edinga, F. McCaffery, I. Wytrychowski, Cessford basal Colorado a reservoir—caustic flood evaluation, Journal of Petroleum Technology 32 (1980) 2103–2110.

[163] K. Manji, B. Stasiuk, Design considerations for dome's david alkali/polymer flood, Journal of Canadian Petroleum Technology 27 (1988). PETSOC-88-03-04.

[164] Housefly, Horsefly Lake Caustic-Polymer Pilot Project Lower Mannville Pool: Progress Report No. 6, PanCanadian Petroleum Limited, Calgary, 1987.

[165] E. Delamaide, B. Bazin, D. Rousseau, et al., Chemical EOR for heavy oil: The Canadian experience, in: SPE-169715-MS, SPE EOR Conference at Oil and Gas West Asia, March 31–April 2, 2014, Muscat, Oman.

[166] Y. Qiu, Z. Chen, S. Xu, Waterflooding of channel sandstone reservoirs, in: SPE-10559-MS, International Petroleum Exhibition and Technical Symposium, March 17–24, 1982, Beijing, China.
[167] D. Yang, J. Wang, L. Jing, et al., Case study of alkali-polymer flooding with treated produced water, in: SPE-129554-MS, SPE EOR Conference at Oil & Gas West Asia, April 11–13, 2010, Muscat, Oman.
[168] B. Sloat, The Isenhour Unit—a unique polymer-augmented alkaline flood, in: SPE-10719-MS, SPE Enhanced Oil Recovery Symposium, April 4–7, 1982, Tulsa, Oklahoma, USA.
[169] E.H. Mayer, R.L. Berg, J.D. Carmichael, et al., Alkaline injection for enhanced oil recovery—a status report, Journal of Petroleum Technology 35 (1983) 209–221.
[170] T.E. Doll, Performance data through 1987 of the Isenhour Unit, Sublette County, Wyoming, polymer-augmented alkaline flood, in: SPE-17801-MS, SPE Rocky Mountain Regional Meeting, May 11–13, 1988, Casper, Wyoming, USA.
[171] T.E. Doll, An update of the polymer-augmented alkaline flood at the Isenhour Unit, Sublette County, Wyoming, SPE Reservoir Engineering 3 (1988) 604–608.
[172] G.A. Bala, M.L. Duvall, J.D. Jackson, et al., A flexible low-cost approach to improving oil recovery from a (very) small Minnelusa sand reservoir in Crook County, Wyoming, in: SPE-24122-MS, SPE/DOE Enhanced Oil Recovery Symposium, April 22–24, 1992, Tulsa, Oklahoma, USA.
[173] P. Yang, Z. Li, B. Xia, et al., Comprehensive review of alkaline-surfactant-polymer (ASP)-enhanced oil recovery (EOR), in: Z. Qu, J. Lin (Eds.), Proceedings of the International Field Exploration and Development Conference 2017, Springer, Singapore, 2019, pp. 858–872.
[174] E. Hosseini, F. Hajivand, R. Tahmasebi, The effect of alkaline–surfactant on the wettability, relative permeability and oil recovery of carbonate reservoir rock: experimental investigation, Journal of Petroleum Exploration and Production Technology 9 (2019) 2877–2891.
[175] T. Chen, Z. Chen, Multiphase physicochemical flow in porous media and measurement of relative permeability curves, in: L. Yu, W. Song, W. Li, et al. (Eds.), Fundamentals and Advances in Combined Chemical Flooding, China Petrochemical Press, Beijing, 2002.
[176] H. Hill, Cation exchange in chemical flooding: part 3—experimental, SPE Journal 18 (1978) 445–456.
[177] H. Guo, Y. Li, D. Kong, et al., Lessons learned from alkali/surfactant/polymer-flooding field tests in China, SPE Reservoir Evaluation & Engineering 22 (2018) 78–99.
[178] D. Wang, Z. Zhang, J. Cheng Wang, et al., Pilot test of alkaline surfactant polymer flooding in Daqing oil field, SPE Reservoir Engineering 12 (1997) 229–233.
[179] D. Wang, J. Cheng, J. Wu, et al., An alkaline/surfactant/polymer field test in a reservoir with a long-term 100% water cut, in: SPE-49018-MS, SPE Annual Technical Conference and Exhibition, September 27–30, 1998, New Orleans, Louisiana, USA.
[180] D. Wang, J. Cheng, J. Wu, et al., Summary of ASP pilots in Daqing oil field, in: SPE-57288-MS, SPE Asia Pacific Improved Oil Recovery Conference, October 25–26, Kuala Lumpur, Malaysia, 1999.
[181] X. Wan, M. Huang, H. Yu, et al., Pilot test of weak alkaline system ASP flooding in secondary layer with small well spacing, in: SPE-104416-MS, International Oil & Gas Conference and Exhibition in China, December 5–7, 2006, Beijing, China.

[182] Y. Yan, Study the Formulation of Alkaline/Surfactant/Polymer Flooding System for the Gaotaizi Second Reservoir, Daqing Petroleum Institute, Daqing, China, 2006.

[183] Y. Li, F. Wang, J. Wu, et al., Current status and prospects of ASP flooding in Daqing oil fields, in: SPE-114343-MS, SPE Symposium on Improved Oil Recovery, April 20–23, 2008, Tulsa, Oklahoma, USA.

[184] H. Li, G. Liao, P. Han, et al. Alkaline/surfactant/polymer (ASP) commercial flooding test in Central Xing2 area of Daqing oilfield, in: SPE-84896-MS SPE International Improved Oil Recovery Conference in Asia Pacific, October 20–21, 2003, Kuala Lumpur, Malaysia.

[185] H.L. Chang, Z.Q. Zhang, Q.M. Wang, et al., Advances in polymer flooding and alkaline/surfactant/polymer processes as developed and applied in the People's Republic of China, Journal of Petroleum Technology 58 (2006) 84–89.

[186] Y. Zhu, Q. Hou, W. Liu, et al. Recent progress and effects analysis of ASP flooding field tests, in: SPE-151285-MS, SPE Improved Oil Recovery Symposium, April 14–18, 2012, Tulsa, Oklahoma, USA.

[187] J. Cheng, D. Xu, B. Wenguang, Commercial test of ASP flooding in Daqing oil field, in: SPE-117824-MS, International Petroleum Exhibition and Conference, November 3–6, 2008, Abu Dhabi, UAE.

[188] Z. Wang, R. Pang, X. Le, et al., Survey on injection-production status and optimized surface process of ASP flooding in industrial pilot area, Journal of Petroleum Science and Engineering 111 (2013) 178–183.

[189] L. Wang, Z. Yang, F. Zhou, Study and evaluation of ASP flooding in Daqing oilfield, Journal of Petrochemical University 29 (2016) 28–32.

[190] H. Chang, ASP Flooding Test in Henan Oil Field Succeed, China Petrochemical News, 2016.

[191] H. Wang, G. Liao, J. Song, Combined chemical flooding technologies, in: P.P. Shen (Ed.), Technological Developments in Enhanced Oil Recovery, Petroleum Industry Press, Beijing, 2006, pp. 126–188.

[192] C. Wang, B. Wang, X. Cao, et al. Application and design of alkaline-surfactant-polymer system to close well spacing pilot Gudong oilfield, in: SPE-38321-MS, SPE Western Regional Meeting, June 25–27, 1997, Long Beach, California, USA.

[193] Z. Qu, Y. Zhang, X. Zhang, et al., A successful ASP flooding pilot in Gudong oil field, in: SPE-39613-MS, SPE/DOE Improved Oil Recovery Symposium, April 19–22, 1998, Tulsa, Oklahoma, USA.

[194] J. Jiang, D. Guo, X. Li, et al., Pilot field trial of natural mixed carboxylates/xanthan flood at Well H5-15 block in Huzhuangli oil field, Oilfield Chemistry 20 (2003) 58–60.

[195] J.J. Meyers, M.J. Pitts, K. Wyatt, Alkaline-surfactant-polymer flood of the West Kiehl, Minnelusa Unit, in: SPE-24144-MS, SPE/DOE Enhanced Oil Recovery Symposium, April 22–24, 1992, Tulsa, Oklahoma, USA.

[196] A.A. Olajire, Review of ASP EOR (alkaline surfactant polymer enhanced oil recovery) technology in the petroleum industry: prospects and challenges, Energy 77 (2014) 963–982.

[197] M. Charest, Alkaline-surfactant-polymer (ASP) flooding in Alberta: small amounts of the right chemicals can make a big difference, Canadian Discovery Digest 1 (2013) 20–52.

CHAPTER

Molecular designs of enhanced oil recovery chemicals

5

Qisheng Ma[1,2] and Yongchun Tang[1]

[1]*Power Environmental Energy Research Institute, Covina, CA, United States*

[2]*ChemEOR Inc., Covina, CA, United States*

5.1 Introduction

Primary oil recovery from a hydrocarbon-bearing reservoir relies upon the use of natural energy present in the reservoir to displace the oil to the production wells. This process usually recovers only a small portion of the original oil-in-place (OOIP). Secondary recovery techniques extend a filed productive life generally by injection of water to displace oil and drive it to the producing wells, resulting in the additional recovery of 15%–35% of OOIP. However, much of easy-to-produce oils have already been recovered, oil producers have attempted several tertiary, or enhanced oil recovery (EOR), techniques that can offer the prospect of ultimately producing 30%–60% or more of the OOIP.

Chemical EOR processes involve the injection of a fluid with some types of chemicals into a reservoir to assist more oil flowing to the production wells. Two factors, volumetric sweep efficiency and microscopic oil displacement efficiency, are essential to the viability of a successful chemical EOR process. Water-soluble polymers and surfactants are the most common chemical additives aiming at improving these two efficiencies. Polymers, specifically the water-soluble, ultra-high-molecular-weight polyacrylamide (PAM) and its derivatives, can be used either as fluid viscosification agents (FVA) or fluid-diverting agents (FDA), aiming at improving the volumetric sweep efficiency (VSE) of waterflooding. The mobility control process is based on maintaining a favorable mobility ratio of injected fluid and residue oil. In essence, FVA polymers added to the aqueous solution increase its viscosity to achieve more stable oil sweeping, while FDA polymers are used to direct injection fluid to more oil-bearing zones.

Surfactant is another type of chemical used in the EOR processes. The most important function of a surfactant is to lower the interfacial tension (IFT) between the aqueous solution and the residue oil; which has been well recognized and documented almost one hundred years ago: "A definite relationship exists between interfacial tension of the oil against the flooding medium and the percentage recovery obtained by flooding" [1]. While the main purpose of the

Recovery Improvement. DOI: https://doi.org/10.1016/B978-0-12-823363-4.00006-6

addition of polymer is to modify the rheological properties of the injected fluid, surfactants could have multiple purposes, including lowering the oil–water IFT (liquid–liquid interaction), alert the surface wettability (liquid–solid interaction), and change the forming tendency between oil and gas (air–liquid interaction).

Combining uses of various chemical additives results in different EOR schemes, such as polymer–EOR, surfactant Huff-n-Puff, surfactant–polymer, and alkaline–surfactant–polymer processes. The improved oil recovery is affected by a number of factors. Among these, chemical selection and design become one of the most critical issues to achieve optimal operations and higher economic payoff. Computational modeling is expected to provide important insights to assist and guide the chemical optimization process by (1) providing theoretical background on the relationships between the molecular structure and the critical parameters related to their performances in the EOR operations; (2) basic understanding of the interactions between additive chemicals with water, oil, and reservoirs, especially the possible transformation of chemicals, both physically and chemically, during the EOR processes of chemicals (3) reliable predictive models and descriptions that can guide synthesis and production of special chemicals.

5.2 Computational modeling of enhanced oil recovery polymers

Polymeric materials are ubiquitous in modern society and they play an instrumental role in almost all industries, undoubtedly including energy production sectors. The success of chemical EOR operation necessitates the design and selection of advanced polymers with enhanced properties suitable for particular reservoir environments. Unfortunately, the structural and behavioral complexity of polymers renders such designs challenging and impeded. By far, intensive laboratory tests, such as the comprehensive and costly core-flooding experiments, along with a number of standardized lab testing, predominates the EOR polymer screening with limited information and guidance to the new discovery of novel polymeric materials.

To address this problem, scientists are developing various computational modeling techniques and leveraging their powers to depict the relationship between structural characteristics of polymers and their properties (such as stability, solubility, and rheological behaviors) and use such predictions to guide the design and synthesis of special polymers with enhanced chemical EOR performances. Unfortunately, predicting the relationships between polymer structure and composition with their properties and performances via atomistic modeling is still a major challenge because of the extremely complex molecular compositions of the polymeric materials and the extended time and length scales involved. For example, studying dynamic shear viscosity and linear viscoelasticity using molecular models requires capabilities that have been elusive, including the representation of large molecular weight chains with an effective internal scale capable of describing entanglement, a shear rate that is in the 1/s scale with accurate quantitative stresses, and chemically realistic combinations of both homogeneous and heterogeneous systems.

Nevertheless, significant progresses have been made toward fundamental understandings of the critical properties of polymers, for example, PAM and hydrolyzed polyacrylamide (HPAM) and their derivatives, and their performances in chemical EOR. A number of computational modeling studies have been conducted to relate the critical physical and chemical properties of PAM/HPAM polymers and the potential impacts on EOR. Without attempting to provide a comprehensive list of endless efforts on these issues, this section focuses on some special cases as examples of successes and challenges when computational modeling techniques are applied to the polymer predictions.

5.2.1 Stability of polyacrylamide/hydrolyzed polyacrylamide

In a typical polymer-based EOR process, the polymer solution resides in reservoirs for many months. Maintaining the polymer stability is essential for the success of EOR operation. The degradation of the polymer results in the loss of fluid viscosity as one of the main reasons for early water breakthroughs. The stability of polymer in solution can be affected by a number of factors. Thermal stability is the one that significantly limits the applicability of PAM/HPAM-based EOR processes, in particular under high-temperature/high-salinity (HT/HS) reservoir conditions.

PAM is composed of repeating units of acrylamide (AM) (Fig. 5.1A). At elevated temperature, some of the amide groups undergo the hydrolysis reaction to become acidic acid (AA) (Fig. 5.1B). The degree of hydrolysis τ of HPAM (Fig. 5.1C) is defined as:

$$\tau = \frac{Y}{X+Y}; \quad 0 \le \tau \le 1 \tag{5.1}$$

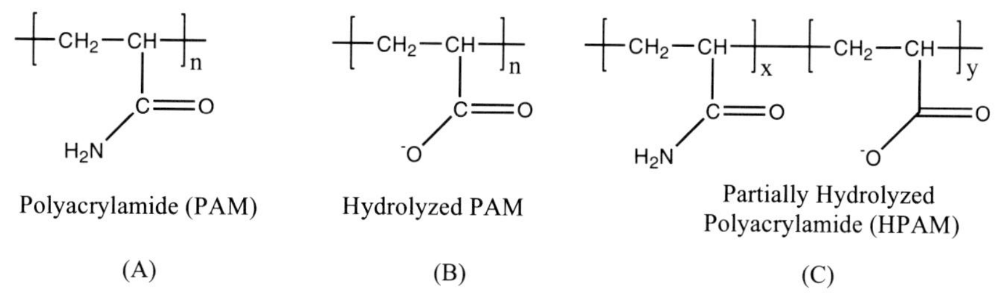

FIGURE 5.1

Molecular structures of (A) polyacrylamide, (B) hydrolyzed polyacrylamide, and (C) partially hydrolyzed polyacrylamide.

where X is the percentage of AM group and Y is for the AA group on an HPAM polymer.

Our first studies mainly focused on the hydrolysis reaction of the amide group. We modeled the PAM with a dangling $CONH_2$ group attaching to the center carbon of the C_5 molecule, that is, we neglect the possible neighbor group interactions.

Both AM and AA are soluble in water, so the effect of the change of the degree of hydrolysis of HPAM on the solubility will be small. However, their chemical reactivity will be significantly affected, especially under high-temperature (~200°F) and high-salinity (100,000–200,000 ppm) reservoir conditions. This is particularly important when there is a substantial amount of divalent (Ca^{2+}, Mg^{2+}, etc.) in brines. It has been well documented that cationic species strongly interact with HPAM, resulting in the rapid phase separation through the formation of gel or precipitate [2–4], where cations are assumed to be chelated by HPAM chains to form a stable cross-linking CaO_4 complex [5]. Laboratory studies conducted by Seright et al. [6] indicated that in the absence of dissolved oxygen and divalent cations, the PAM polymer can remain thermally stable so that the HPAM solutions were projected to maintain at least half of their original viscosity for over 8 years at 100°C and about 2 years at 120°C, which is much more stable than the cases where certain concentrations of the Ca^{2+} cation are present. Moradi-Araghi and Doe [7] have shown that the thermal stability limits for PAM varied significantly under brines containing different divalent cations (<96°C for 270 mg/L; <88°C for 500 mg/L; and <75°C for 2,000 mg/L), suggesting the significant adverse effects of the divalent cations on the stability of PAM/HPAM polymers.

The fundamental chemistry regarding the amide hydrolysis is very rich and has been considered an important model reaction for the enzymatic cleavage of peptide bonds [8–10]. The mechanism and kinetics of the amide hydrolysis under different conditions have been extensively studied from both experimental and theoretical aspects. Theoretical studies of the base-catalyzed amide hydrolysis reaction predicted the 20 ~ 30 kcal/mol transition barrier is in good agreement with experimental measurements [11,12]. Under the base-catalyzed condition, a two-step mechanism involving the nucleophilic attack of hydroxide ion on α-carbon to form a tetrahedral intermediate as the first step followed by the release of ammonia molecule has been established [13–17]. Acid-promoted amide hydrolysis reaction has also been studied [9,18–21]. Both O- and N-protonation pathways were found to compete under the acidic condition. Theoretically predicted the ~20 kcal/mol transition barriers are consistent with the experimental 21.5 kcal/mol value [22,23].

A systematic computational study has been conducted to explore and compare different reaction pathways of the PAM hydrolysis under different brine conditions, that is, acidic, neutral, and basic. All calculations used the same computational level, that is, DFT/B3LYP/6-31G(d,p), such that the computed energy results are comparable [24]. The water solvation effect is also considered using the continuum-solvation approach by solving the Poisson–Boltzmann equation with a default dielectric constant of 80.37 and a probe radius of 1.40 Å for water. Specially, all stable, *meta*-stable, and transit state energies of 11 different paths were determined. These included:

1. Basic condition
   - A two-stepwise reaction path: this is the well-documented text-book reaction path. Under the basic condition, the nucleophilic addition of OH^- group to α-Carbon with a transition barrier of 22.32 kcal/mol, followed by the ammonia elimination with a transition barrier of 31.51 kcal/mol are determined from our calculations.
2. Acidic condition
   - O-protonation: Under the acidic condition, proton (H^+ or $OH_3{}^+$) can first ionize the O-atom on the C = O bond, forming the O-protonated *meta*-stable state. The protonation process is energetically favorable, with the formation energy of −8.7 kcal/mol. The transition barrier from this O-protonated configuration to the *meta*-stable is 45.2 kcal/mol, followed by the second transition state with a much lower transition barrier (10.8 kcal/mol).
   - N-protonation: Instead of attacking the O-atom, the proton might also attach to the dangling NH^2 group, forming a less stable N-protonated configuration when compared to the O-protonation (formation energy = −7.9 kcal/mol). However, the transition barrier from the protonated state to the *meta*-stable state is relatively lower (−41 kcal/mol), such that the N-protonation path has a lower overall transition barrier compared to the O-protonation.
   - Water-assisted: Noting that all above studied transition states involved 4-member, 6-membered ring configuration in the presence of one water molecule has also been proposed. Our calculations indeed confirmed a much lower transition barrier when taking the water molecule into account to form the 6-membered configuration. With the assistance of the H_2O, the transition barrier of the O-protonation has been significantly lowered from 45.2 kcal/mol to only 37.4 kcal/mol.
3. Neutral condition
   - Stepwise: Under neutral conditions without other catalyzed reactions, the PAM group is much more stable. The stepwise reaction path rendered the transition barriers of 53.6 can 54.4 kcal/mol, significantly higher than the acidic and basic conditions.
   - Concerted: Similarly, the one-step concerted reaction path under the neutral condition yielded the transition barrier of 57.8 kcal/mol. This is good news for maintaining the highly stable PAM for the EOR applications, suggesting that the PAM polymers could be quite stable without being explored for the catalyzed reactions.
   - Water assisted: Similar to the acidic condition wherein the water-assisted pathway has lowered transition barrier than the 4-membered transition state, water will also lower the transition barrier under the neutral

condition. For the water-assisted concerted reaction pathway, the transition barrier has been lowered to 42.4 kcal/mol.

- Ammonium assisted (stepwise): In additional water, ammonium is produced as the direct product of the hydrolysis reaction. And like the proton, the ammonium could also catalyze the reaction. For the stepwise path, the transition barriers have been lowered to 35.6 and 37.8 kcal/mol.
- Ammonium assisted (concerted): For the concerted path, the transition barrier of ammonium-assisted reaction is lowered to 46.8 kcal/mol compared to the uncatalyzed 57.8 kcal/mol. Because the ammonium is continuously generated, one will consider the reaction is auto-accelerated, that once the reaction starts even under the neutral condition, the reaction could be accelerated by the products.
- Water assisted (ammonium catalyzed): The ammonium ion could also attack the O-atom on the C = O bond, while the 6-membered transition occurs on the dangling NH_2 group. The protonation of the ammonium ion on the O-atom yields a -7.6 kcal/mol more stable configuration. However, the transition barrier from this stable state is much lower than others (27.2 kcal/mol), rendering the least stable situation. Note that ammonia (NH_3) has a pK value of 9.25. Even under the neutral condition, nearly 99% of ammonia exists in the protonated state, for example, NH_4^+. Therefore, this auto-accelerated water-assisted pathway can occur under the neutral condition.
- Water assisted (Ca^{2+} catalyzed): These calculations suggested that the initial attack of the positive species on the O-atom could be responsible to lower transition barriers. By analyzing the Millikan charge distribution, we found that the charge on the α-Carbon substantially increased from $+0.56$ to $+0.61$ when the O-atom on the C = O bond is protonated by H_3O^+. This is even increased to $+0.71$ when a Ca^{2+} is attached to the O-atom. Consequently, the transition barrier is further reduced to as low as 17.8 kcal/mol (Fig. 5.2). The increase of the positive charge on the α-carbon makes the ammonium elimination step easier and is considered one of the reasons for this lowered transition barrier.

The thermal stability of PAM/HPAM under HT/HS environments is one of the most important factors preventing widespread oilfield applications of PAM chemicals in harsh reservoir conditions. A fundamental understanding of the basic chemistry related to the PAM hydrolysis mechanism is of great importance to the molecular design and syntheses of modified PAM and their oilfield applications, especially in HT/HS reservoir conditions. Computational modeling offers a great potential for a systematic understanding of the reaction mechanism under different environments. Some highlights of studies are:

- The 6-membered ring transition configuration has a lower transition barrier compared to the 4-membered transition. Consequently, the water-assisted reaction paths are more likely to occur.

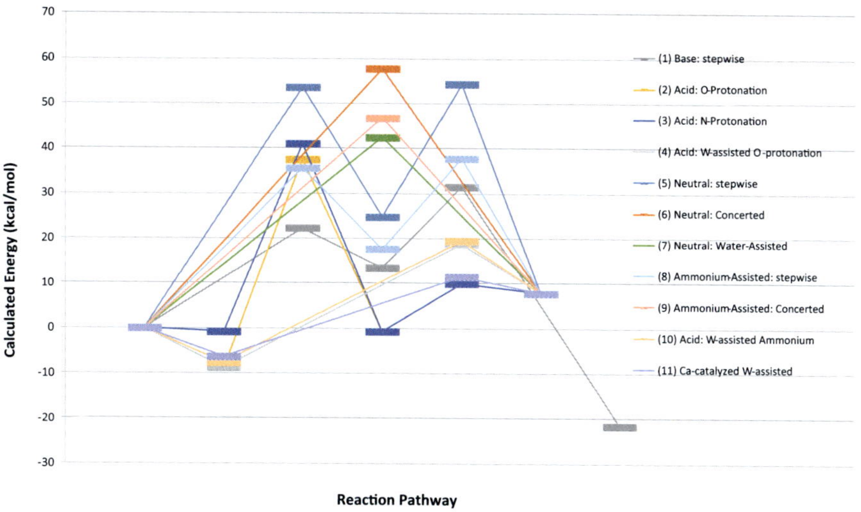

FIGURE 5.2

Calculated energies of various reaction pathways.

- Microscopically, we found the positive charge on the α-Carbon is responsive for the easier ammonium elimination step. This can be caused by the attacking of the positively charged species, such as proton, ammonium ion, and cations (Ca^{2+}) on the O-atom in the $C = O$ bond.
- Ammonium ion (NH_4^+) as a product of the reaction will also catalyze the reaction. Therefore, the reaction becomes auto-accelerated.
- The presence of the divalent cations, for example, Ca^{2+}, could significantly lower the transition barrier. In addition, these divalent cations are assumed to be chelated by the AA^- groups, which is another product of the reaction, to form a stable cross-linked complex, resulting in the reduction of the solubility or even precipitation.

5.2.2 Molecular shape of hydrolyzed polyacrylamide—stability and solubility

Mechanical or shear degradation is another main factor that affects the stability of HPAM polymers in the aqueous solution when injected underground. Unlike the polymers used in the fracture fluid as the friction reducer requiring achieving maximum effect within minutes, EOR polymers need to be maintained in the solution for several months. High-pressure pumping is used in hydraulic fracturing such that the HPAM polymers are typically under extremely high shear stress causing severe shear degradation. On the contract, the EOR injection typically involves relatively low-pressure injections, in which the ability to keep polymers in the solution is one of the key factors to maintain viscosity. The molecular shape and tropology of polymers in solution are important.

Fig. 5.3 illustrates eight different molecular structures of PAM/HPAM with various degrees of hydrolysis, τ. The polymer structures of totally 100 AM or AA units are randomly created, and the structure is optimized using the DREIDING force fields (FFs). Without the addition of the AA group, the PAM polymer ($\tau = 0$) exhibits a perfect coiling structure that could be more resistant to the shear stress. When fully hydrated in solution and under longitude direction, the PAM can maintain a linear structure to bypass low permeable zones. The molecular structures of HPAM become more irregular with the increase of the τ value. This could increase the possibility of interaction of the HPAM with the reservoir and become problematic when applied to low-perm or heterogeneous reservoir conditions.

The solubility of polymers is an important parameter in EOR applications. Classical theoretical predictive models have been built based on the modified Flory–Huggins model and the molecular modeling calculations to determine the compatibility of the binary mixtures:

$$\frac{\Delta G}{RT} = \frac{\phi_b}{n_b}\ln\phi_b + \frac{\phi_s}{n_s}\ln\phi_s + \chi\phi_b\phi_s \tag{5.2}$$

where ΔG is the Gibb's free energy change when a soluble (s) in the solution (b), ϕ is the volume fraction, n is the degree of polymerization of components s and b, R is the gas constant, T is the absolute temperature, and χ is the so-called interaction parameter, which is defined as:

$$\chi = \frac{E_{\text{mix}}}{RT} = \frac{1}{2}\left[Z_{bs}\langle E_{bs}\rangle_{\text{avg}} + Z_{sb}\langle E_{sb}\rangle_{\text{avg}} - Z_{bb}\langle E_{bb}\rangle_{\text{avg}} - Z_{ss}\langle E_{ss}\rangle_{\text{avg}}\right] \tag{5.3}$$

The mixed energy E_{mix} can be determined by calculating the average energies of b/s, s/b, b/b, and s/s interactions and Z is the coordinate number of the mixture. The average energies can be calculated computationally using the Monte Carlo (MC) simulations with defined FFs, for example, DREIDING, COMPASS, and Universal.

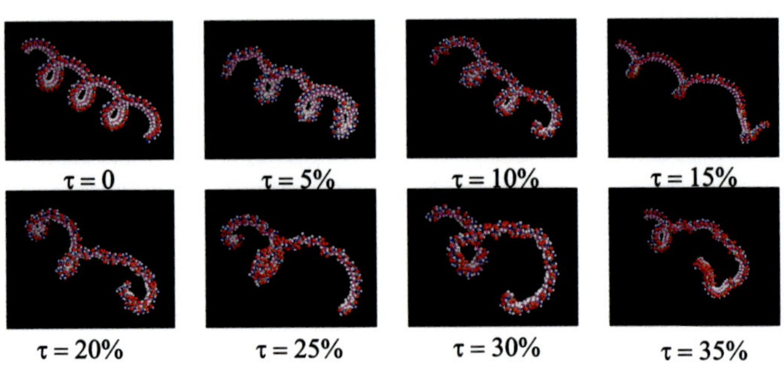

FIGURE 5.3

Molecular structures of HPAM with the change in the degree of hydrolysis (totally 100 acrylamide and acidic acid units).

As an attempt to test the efficiency and accuracy of the model, we conducted the preliminary calculations of the interaction parameters of HAPM with different τ values in water (Fig. 5.4). However, significant errors were obtained, because the experimental data of the interaction parameter of polymers are in the range of <1.0. When the polymers interact strongly with the solvent molecule, the mixing is nonideal then the interaction parameters are <0.5; when the polymer–polymer interaction is more favorable than the polymer/solvent interaction, the interaction parameters are >0.5. From analyzing the energy distributions of each energy term (Fig. 5.5), we can see that the large discrepancy is mainly caused by the computational errors from the polymer–polymer interaction energy ("green"). This could be caused by the complicated polymer–polymer interactions, which will require an extremely large number of configurations in the MC simulation, or more accurate FF to better describe the polymer interactions.

5.2.3 Improvements of polyacrylamide/hydrolyzed polyacrylamide stability—neighbor effects

Without involving free radical and high salinity, PAM/HPAM could be generally thermal stable under very little physical and chemical changes when the temperature is below 200°C. However, as temperature increases, HPAM polymers begin to undergo irreversible structural changes. A number of intra- and intermolecular reaction pathways have been proposed [25,26]. It has been found that the

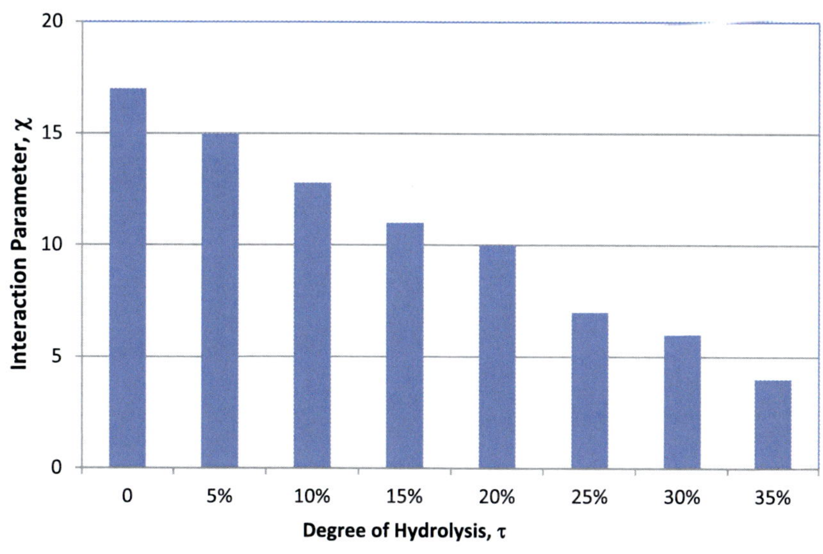

FIGURE 5.4

Calculation of interaction parameter χ as a function of the degree of hydrolysis τ of the hydrolyzed polyacrylamide polymers.

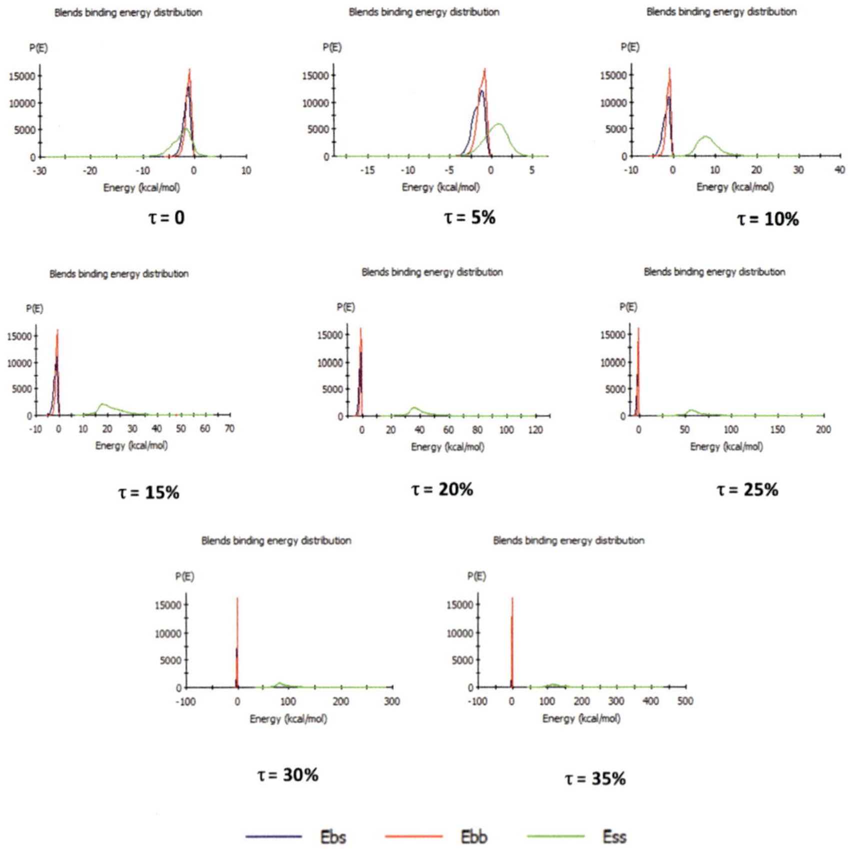

FIGURE 5.5

Energy distribution of binding energy of hydrolyzed polyacrylamide with different degrees of hydrolysis in water as base.

imidization reaction is the dominant reaction, with H_2O, NH_3, and CO_2 as the main products of the reactions [27,28]. It has also been noted that within this temperature range, there is relatively little participation of the polymeric backbone in these degradation reactions, and there is no evidence for the formation of the AM monomer. Based on these observations, we calculated the energetic relationships of a number of reaction pathways mainly involving the interactions between the neighboring groups in PAM/HAPM polymers [29].

1. AM−AM neighbors
   - The first reaction pathway is associated with the imidization reaction of two neighboring AM groups to form the 3-ethyl-5-methyl-2,6-piperidinedione as the intermediate. The transition state of this reaction path is a 4-membered ring structure with one H transferring from the $-NH_2$ group to another $-NH_2$

group to form the NH_3 byproduct, accompanied by the formation of the C–N bond. Because of the 4-member ring TS structure, the activation barrier is substantially high, falling in the range of ~ 43 kcal/mol.

- The second possible imidization reaction path undergoes a 6-member ring transition state with the formation of the same intermediate. Instead of the formation of the C–N in the above reaction path, a C–O bond is formed. Although the 6-membered ring TS usually requires much less activation energy than the 4-membered ring, the reaction products, however, are less stable than the products from the above reaction. As a result of this difference, the overall activation energy for this reaction path is ~ 44 kcal/mol, which is very similar to the first reaction.

2. AM–AA neighbors
   - Imidization is still the dominant reaction when one AM and one AA become neighbors, with H_2O as the byproduct. Here again, a 4-member ring TS pathway is possible. In the reaction pathway, the H transfers from the $-NH_2$ group to the –OH group to facilitate the formation of the H_2O byproduct, accompanied by the formation of the C–N bond. However, the calculated activation energy of ~ 50 kcal/mol is substantially higher than that of the AM–AM interaction.
   - The activation energy of the AM and AA neighbors undergoing the 6-membered ring is also higher than that of the AM–AM interaction, attributing to the less stable intermediate. These results suggest that by including a small portion of AA monomer in HPAM polymers could even increase the thermal stability of a pure PAM polymer.
3. AA–AA neighbors
   - If we keep increasing the degree of hydrolysis of the HPAM, the possibility of having adjacent AA groups also increases. We also examined the imidization reaction of two AA neighbors. For the 4-membered ring transition state, H transfers from one –OH group to another –OH group, accompanied by the formation of the C–O bond to form the intermediate product of 3,5-dimethyldihydro-2*H*-pyran-2,6(3*H*)-dione. The activation energy is substantially high (~42 kcal/mol) due to the 4-membered ring TS structure.
   - However, the activation energy (~38 kcal/mol) of the 6-membered ring TS pathway of two AA groups is much lower than all other pathways studied. The formed dione product is a more stable intermediate such that the 6-membered TS activation energy could be lowered.

   Table 5.1 lists the lowest activation energy corresponding to each potential neighbor interaction. From these results, we can see that AA–AA becomes the weakest link between the HPAM polymers with the lowest activation energy of 38 kcal/mol.
4. Effects of Special Monomers

   One practical method to improve the stability of HPAM is to include special monomers such as 2-acrylamido-2-methylpropane sulfonic acid (AMPS) or *n*-vinyl pyrrolidone (NVP) to improve the stability of HPAM.

Table 5.1 Calculated transition barriers of the neighbor interaction between polyacrylamide-based polymers.

Neighbors	Transition barrier (kcal/mol)
AM–AM	43
AM–AA	50
AA–AA	38
AA–AMPS	51
AM–AMPS	55

AA, Acidic acid; AM, acrylamide; AMPS, 2-acrylamido-2-methylpropane sulfonic acid.

One of the possible reactions involves the H-transfer from the –NH– group to the –OH, similar to those in the reaction. However, the activation barrier for this reaction is significantly higher than that of the AA–AA reaction. Therefore, by including AMPS in the copolymerization, one can effectively reduce the possibility of AA–AA interaction and further stabilize the HPAM polymers.

The addition of the AMPS monomer into the HPAM polymer is a well-known approach to improve the thermal stability and salt tolerance of the formed polymers. From our studies (Fig. 5.6), the presence of the AMPS monomer in HPAM could effectively prevent the neighbor–neighbor intermolecular interaction in polymer, thus rendering extra stability. In all examined reaction pathways, the hydrogen transfer from one dangling group, either in AM, AA, or AMPS, to the neighboring unit is a required step to facilitate the reaction. This implies that the imidization interactions between an amide whose active hydrogen is fully substituted by alkyl groups(such as *N*-vinyl pyrrolidone, NVP and dimethylacrylamide, DMAM) and neighboring units do not exist, because, in NVP, the N-atom is bonded inside the heterogeneous ring structure; in DMAM, the two amide hydrogens are replaced by the methyl group. This could be the reason that the monomers, such as AMPS, NVP, and DMAM, can render extra thermal stability when included in the associated PAM-based polymers.

5.2.4 Cross-linked polymers

Linear PAM/HPAM polymers can be cross-linked by chemical bonding (covalent bonding and ionic bonding) or physical interaction (van der Waals force and hydrogen bonding). Special cross-linking monomers are also included during the polymerization synthetic process. When copolymers are formed, these cross-linking monomers form covalent bonds with AM monomers giving rise to the preformed polymeric gels. In addition to the classical Zr- or Cr-based permanent cross-linked polymer gels, special monomers with certain relative weak chemical bonds are subject to be broken at

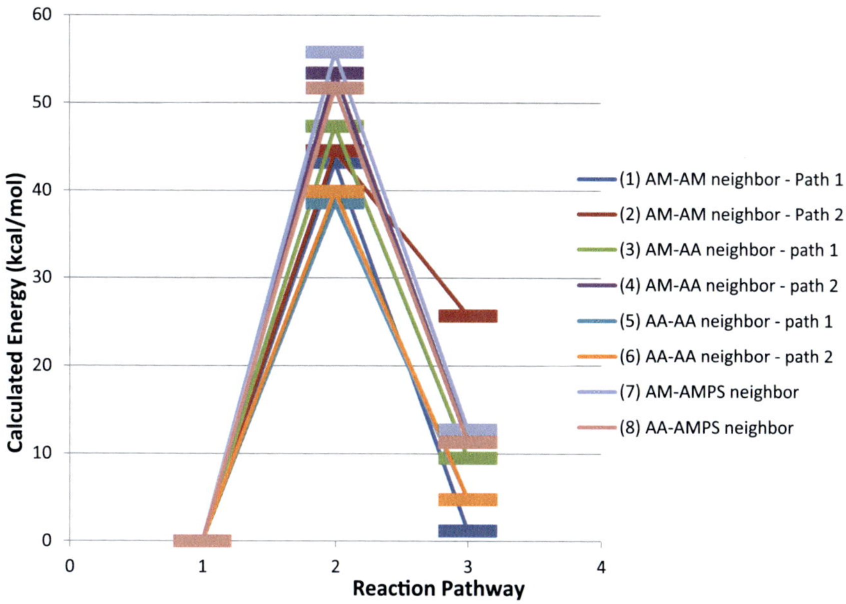

FIGURE 5.6

Reaction pathway acrylamide–acrylamide, acrylamide–acidic acid, acidic acid–acidic acid, acrylamide–2-acrylamido-2-methylpropane sulfonic acid, and acidic acid–2-acrylamido-2-methylpropane sulfonic acid neighbors.

elevated temperatures. Because the molecular systems of our interest only involve covalent bonds of simple nonmetal elements (H, C, O, N, and S), wherein for these types of chemical bonds, bond interaction parameters have been well-developed and calibrated for the semiempirical methods. Also considering the molecular systems of the polymeric gel are often too complex to be accurately calculated using high-level density functional theory (DFT) approaches, we selected AM1 (Austin method) as our primary computational method to calculate the formation and decomposition energies of different polymeric gels.

Two special cross-linked monomers, namely, *N*,*N*-alkylenebisacrylamide (for instance, *N*,*N*,-methylenebisacrylamide) and polyethylene glycol diacrylate (for instance, triethyleneglycol diacrylate) are used in synthesis. When forming the copolymer of AM monomer, these cross-linking monomers bridge various lengths of PAM polymers [30], as illustrated in Fig. 5.7.

Without catalyzing, the cross-linked PAM gel undergoes thermal decomposition with the insertion of the water molecule into the dangling bonds on the bridge. Two possible dissociation pathways were studied, and the transition barriers are compatible in the range of 37 ~ 39 kcal/mol (Fig. 5.8 and 5.9). Note that at the semiempirical AM level, substantial large energy errors exist, in particular those for the transition configurations. A high level of computational studies will

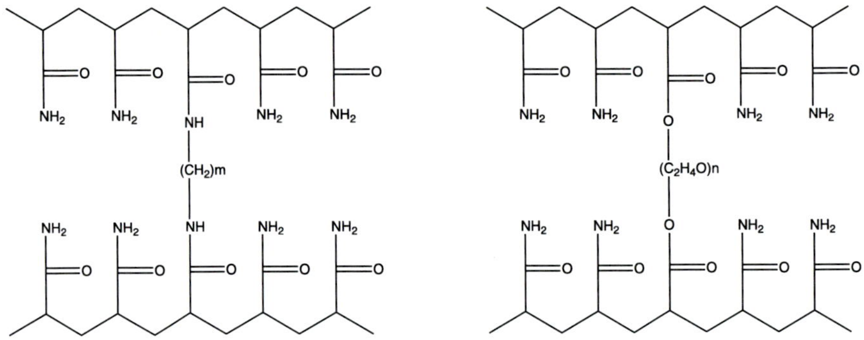

FIGURE 5.7

Molecular structures of cross-linked polyacrylamide polymeric gel by *N,N*,-methylenebisacrylamide (A) and triethyleneglycol diacrylate (B).

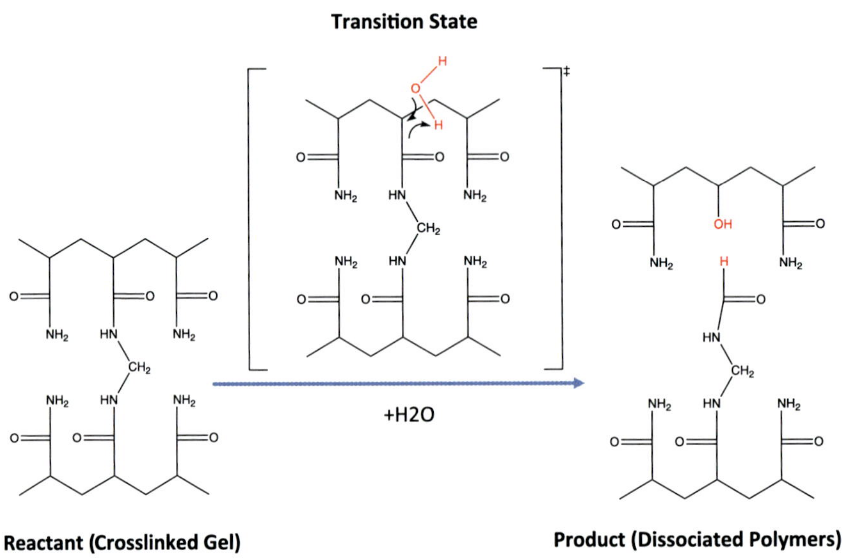

FIGURE 5.8

Possible dissociation path #1 of the cross-linked gel.

be needed if different cross-linking agents are investigated. Further simplifications of the modeling systems might also be needed.

5.2.5 Rheological property predictions

Rheological properties of the polymer solution play key roles in the EOR operations. Commonly, the poor oil recovery rate is occasioned by differences between

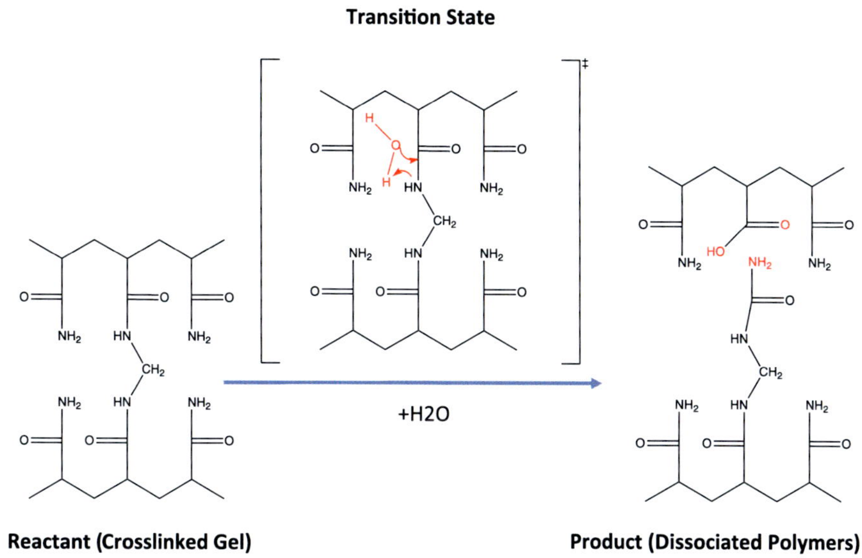

FIGURE 5.9

Possible dissociation path #2 of the cross-linked gel.

the mobility of the injected displacing medium and the in situ reservoir oil, and equally importantly, by permeability variation within the reservoir. Rheological behaviors are related to the response of the polymers in solution with respect to the environmental changes, for example, flow rate and shear stress. A robust molecular prediction model holds the potential to significantly accelerate the development of polymer-based rheological modifiers.

The recent development of our multiscale predictive framework allows the successful prediction of rheological properties from the chemical structure for polymers of experimentally relevant molecular weights and provides an in silico machine learning engine for screening novel compositions and structures with the optimized non-Newtonian response, required for both shear-thinning and shear-thickening applications. The recent development of our multiscale predictive framework allows the successful prediction of rheological properties from the chemical structure for polymers of experimentally relevant molecular weights and provides an in silico machine learning engine for screening novel compositions and structures with the optimized non-Newtonian response, required for both shear-thinning and shear-thickening applications.

Our framework provides:

- Procedures and tools for systematic coarsening from atomistic models and reverse mapping of coarse-grain models to atomistic,
- Unique ab initio methods to characterize the atomistic origin of colloidal and interfacial interactions and phenomena,

- Systematic structure and composition builders based on practical descriptors that drive rheological changes in polymer melts and diluted polymer mixtures,
- A rheological properties engine capable of predicting viscosity in the zero-shear limit and under realistic dynamic conditions (for shear rates commensurate with experiments) for large heterogeneous systems,
- Coarse-grain FFs with improved nonbond descriptions based on accurate quantum mechanics,
- An in silico screening machine learning engine that feeds from the systematic model builders to cover the descriptors search space, computes the rheological properties from converged trajectories spanning submilliseconds, and ranks them for each structure/composition using an automated viscosity-versus-shear rate fitness function that can be tuned for shear-thickening, shear-thinning, and other rheological responses.

While steps 1–3 are related to the system building block for the construction of the close-to-real polymeric molecular ensemble subject to investigation by integrating the practical molecular descriptors associated with the rheological properties of the diluted polymer mixture, step 4 is designed to directly monitor the system response under typical environmental changes, for example, the shear stress acting on the system. Specifically, the stress autocorrelation function is used to estimate the zero-shear viscosity of the system via the Green–Kubo equation. The procedure involves relating the ensemble average of autocorrelation of the pressure cross-term, p_{xz}, to η as follows,

$$\eta = \frac{V}{k_{B}T}\int_{0}^{\infty}\langle p_{xz}(0)p_{xz}(t)\rangle dt \tag{5.4}$$

where p_{xz} is computable from stress virials (Pressure tensor), k_B is the Boltzmann constant, T is the average temperature, and the autocorrelation function is averaged over MD trajectories. The asymptotic behavior of the pressure autocorrelation is $t^{3/2}$, which makes an accurate estimation of the viscosity difficult, or in practical terms, convergence limited.

The general methodology implemented consists of solving for η in the equation,

$$J = -\eta\frac{\partial V_x}{\partial z} = -\eta\nabla V_{xz} \tag{5.5}$$

using the following algorithm:

- Build coarse-grain, bead-spring models of polymer components using structural and chemical cues from the equivalent molecular systems.
- Equilibrate the system at a temperature of 300K.
- Apply a shear deformation to the simulation box and integrate the SLLOD [31] equations of motion at a controlled temperature and fixed volume, at a shear rate relative to the perpendicular direction. This requires that the bead velocities should be remapped to the sheared box at every timestep.

- Monitor the velocity profile in the fluid mixture (should be consistent with the applied shear strain rate, for a diluted solvated polymer mixture), which determines ∇V_{xz} in the equation above. ∇V_{xz} is the derivative in the z-direction of the V_x component of fluid motion (i.e., $\nabla V_{xz} = \frac{\partial V_x}{\partial z}$).
- Monitor the off-diagonal component of the pressure or stress tensor, p_{xz} to determine the J (momentum flux in units of momentum per area per time) term in the equation above.

5.3 Computational simulation of enhanced oil recovery surfactants

Various types of surfactants have been studied and employed in the chemical EOR processes [32]. New candidates are proposed to target different reservoir conditions and various EOR processes [33,34]. With thousands of different surfactant products available in the marketplace, the searching and selection of chemical candidates suitable for a particular oilfield have become more complicated. Especially, the performance of chemicals is greatly affected by the complex reservoir conditions, such as the local water chemistry and the crude oil composition under high-temperature and high-pressure formation. Traditionally, the formulation of EOR chemicals has been primarily based on the laboratory trial-and-error approach. There is the hope for the modern computational modeling to make a contribution to the smarter selection of the EOR chemicals. Molecular modeling is the investigation of molecular structure, properties, and their interactions by the collective use of fundamental theory and computational methods and graphical visualization techniques. The advantages of using molecular modeling to study the EOR chemicals are its capacity to predict the properties and behavior at the atomic level. If properly used, molecular modeling can provide important insights to guide laboratory studies and validation.

Surfactant is a class of special chemical compounds that are capable of altering the interfacial properties between two phases (two liquids, a liquid and a solid, a liquid and a gas, etc.). Structurally, surfactants are composed of at least an oleophilic nonpolar tail and a dissimilar hydrophilic polar head. Based on the type of the hydrophilic head, surfactants can be classified into several categories, including nonionic, anionic, cationic, and zwitterionic surfactants. In the petroleum industry, surfactants are used for specific purposes such as increasing the miscibility of oil in water, inducing wettability alternation in the porous formation, mobilizing trapped oils, or forming bobbles or microemulsion. All of these applications are governed by a single-core underlying mechanism, which is to take advantage of the special properties of surfactants to participate in two phases, rather than staying in a single phase. And, the main reason for this biphasic partition arises from the molecular composition and the structural variation of the surfactant. Consequently, there exist certain correlative relationships between the microscopic structural properties and the macroscopic physical and chemical behaviors of surfactants. One of the key objectives of the molecular design with the assistance of computational modeling is to identify such correlative relationships, built from the existing and known data, which can be extended to unknown candidates. Another important contribution of the computational simulation is to provide an in-depth understanding, at the molecular level, the fundamental basis of the effect of the microscopic variation of surfactants on the macroscopic behaviors, ideally, under different application environments.

5.3.1 Quantitative structure–activity/property relationship

5.3.1.1 Computational simulation enhanced quantitative structure–activity relationship/quantitative structure–property relationship modeling of surfactants

The fundament of the quantitative structure–activity relationship/quantitative structure–property relationship (QSAR/QSPR) method is based on the observation that there are certain types of correlative relationships between the appearance properties and the molecular structure of organic molecules; and importantly, such correlative relationships can be expressed in a more systematic and quantitative way. The appearance properties can be macroscopic (physical, chemicals, or biological) measurements; or microscopic molecular reactivity when participating in biological or chemical reactions. The task of the QSAR/QSPR study is to search for the definitive mathematical formula of correlative relationships, from the measured property/activity and a set of descriptive parameters (descriptors) derived from the molecular structure of a group of organic molecules. In principle, the derived relationship can be further applied to predict properties/activities of similar organic molecules that are not included in the original group, if such a quantitative relationship is properly correlated.

Attempts to establish the QSAR/QSPR models were made much early even before the modern computational simulation methods are widely applied. The linear free energy relationships (LFERs) established by Hammett [35,36] and Taft [37,38] provided important insights into many organic reaction mechanisms. Wiener index [39,40] is considered the first topological molecular descriptor developed for the purposes of QSAR/QSPR. The Wiener index is defined as the sum of the lengths of the shortest paths between all pairs of vertices in the chemical graph representing the non-hydrogen atoms in the molecules [41]. Unlike the constitutional descriptors that are mainly derived from the chemical composition in an organic compound, the topological descriptors are obtained from the graphic connectivity of the organic compounds (without all hydrogen bonds). For example, the *n*-butane molecule has four vertical carbon atoms. There are three pairs of vertices at a distance of one from each other, two pairs of vertices at a distance of two, and one pair of vertices at a distance of three. Therefore, the Wiener index of *n*-butane is summed as $3\cdot1+2\cdot2+1\cdot3=10$; or more generally, $(N^3-N)/6$, where N is the number of vertices for the *n*-alkanes.

Computational molecular calculations add a new dimension to generate a large amount of molecular descriptors in a more systematic and speedy way. On top of the constitutional and topological descriptors which are derived from empirical approaches, more descriptors can be generated directly from computations. Many new molecular descriptors including those related to the geometric, electronic, and thermodynamic characteristics of the molecule can be computed from the first principle. At least three main contributions from modern computational simulation have rejuvenated the QSAR/QSPR method: (1) the ability to obtain the optimized molecular geometry of any molecular systems because this is the key step in the QSAR/QSPR study. The quantum chemistry calculations either from the Hartree–Fock theory or DFT start from searching for the geometry of the system corresponding to the total ground state energy of the electronic motion; that is, to achieve the lowest system energy by continuously optimizing the structure of molecules; (2) the ability to compute additional molecular descriptors at the atomic level and to systematically manage the large database system. Through decades of development, the number of molecular descriptors has increased dramatically [42,43]. Computer-aided simulations with the development of new software and database tools allow more effective data management; (3) the ability to compute the molecular descriptors of compounds, including those properties unmeasured or not even be synthesized. The inspiration of the molecular design is to predict the properties of chemicals by using these molecular descriptors through the optimum quantitative relationships developed from the QSAR/QSPR model (Fig. 5.10).

Building a reliable QSAR/QSPR model is to search for the mathematic function (F) linking the measured property (P_i) of chemicals with a set of selected molecular descriptors ($m_{i,j}$):

$$P_i = F(m_{i1},\ m_{i2}, \ldots m_{iN}) \tag{5.6}$$

Multilinear regression (MLR) is by far the most widely used approach to search for the QSAR/QSPR models, in which the function F is just a linear addition of a set of selected descriptors (weighted); and the weighted factors are optimized, in term of the square correlation coefficients (R^2), Fisher criterion value (F), and standard deviation (S^2). Other correlation schemes, such as principal component regression and partial least-squares methods, have also been used. More comprehensive nonlinear regression methods have been applied. Among these, the artificial neural network (ANN) and the support vector machine approaches have shown to provide improved models compared to the standard MLRs, with the increasing complicity and number of databases.

For decades, numerous QSAR/QSPR studies have been conducted to establish correlation models of many properties of surfactants, including boiling point, melting point, viscosity, refractive index, density, dielectric constant, polarizability, vapor pressures, critical temperatures, critical pressure, heat of vaporization, heat of formation, entropy, solubility, octanol–water partition coefficient (logP), hydrophilic–lipophilic balance, critical micelle concentration (CMC), surface

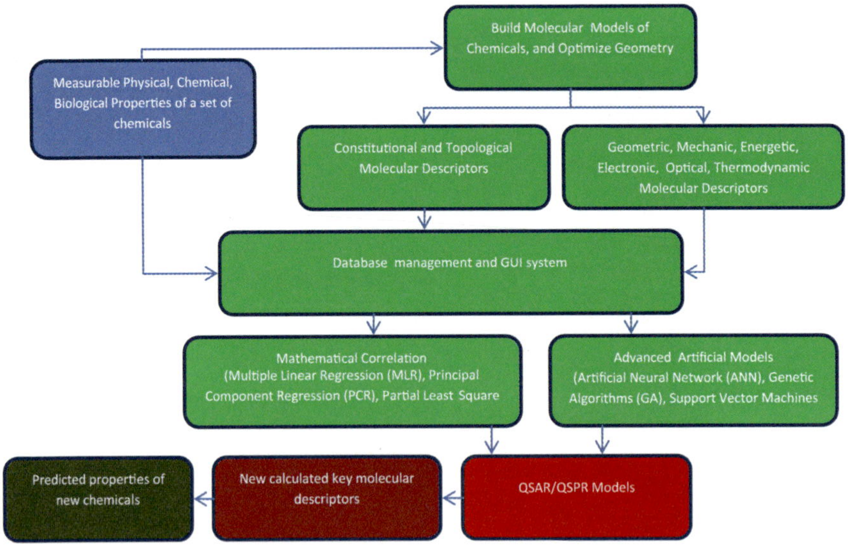

FIGURE 5.10

Schematic flow diagram of using quantitative structure–activity relationship/quantitative structure–property relationship models for the molecular designs.

tension, cloud point, acid dissociation constant, chain transfer constant, glass transition temperature, and more. A few good reviews on QSAR/QSPR modeling of surfactants have been published [44,45]. In this section, two parameters more relevant to the chemical EOR applications of surfactants, CMC and IFT, are discussed.

5.3.1.2 Critical micelle concentration prediction

CMC is an important property of surfactants, with the increase of concentration in solution, the hydrophobic portion of surfactants start to arrange themselves by forming certain micelle structures. It has been found that the many important properties of surfactant solutions, such as surface tension, IFT, conductivity, emulsification, and foaming, are significantly different crossing the CMC point. In 1976 Rosen [46] initially proposed a two-term equation to correlate two molecular descriptors with CMC of selected nonionic and anionic surfactants. A nonlinear term was added by Becher in 1984 [47] and Ravey in 1988 [48]. For nonionic polyoxethylene alcohols C_ME_N

$$\log \text{CMC} = 1.77 - 0.52\,\text{N} + 0.032\,\text{M} + 0.002\,\text{N}\cdot\text{M} \tag{5.7}$$

QSAR/QSPR models based on the constitutional and topological descriptors using MLR approaches have been developed by Huibers [49,50]. For nonionic surfactants, the best model obtained is:

$$\log \mathrm{CMC} = -(1.80 \pm 0.16) - (0.567 \pm 0.009)\cdot \text{c-KH}_0 + (1.054 \pm 0.048)\cdot \text{c-AIC}_2 + (7.5 \pm 1.0)\mathrm{RNNO} \quad (5.8)$$

c-KH_0 stands for the Kier and Hall molecular connectivity index of zeroth order for a hydrophobic fragment, the second-order average structural information index (c-AIC_2) reflects the basic chemical characteristics of hydrophobic moiety; and RNNO is related to the numbers of nitrogen and oxygen atoms.

For anionic surfactants, the model developed is:

$$\log \mathrm{CMC} = (1.89 \pm 0.11) - (0.314 \pm 0.010)\cdot \text{t-sum-KH}_0 - (0.034 \pm 0.003)$$

$$\mathrm{TDIP} - (1.45 \pm 0.18)\cdot \text{h-sum-RNC} \quad (5.9)$$

t-sum-KH_0 represents the Kier and Hill index of zeroth order for the whole hydrophobic domain; TDIP is the molecular total dipole moment; and h-sum-RNC is the sum of the relative numbers of carbon atoms over all hydrophilic fragments.

These models have been tested with various surfactant classes, including nonionic [51–57]; cationic [58–60], anionic [52,61–66], and Gemini [67–69] and sugar-based surfactants [70], or combination [71]. For MLR modeling, three parameters, namely, coefficient of determination (R^2), Fisher ratio for the regression (F), and the standard error of the regression (S^2) are used as criteria. Good correlations have been reported in previous studies.

5.3.1.3 Interfacial tension prediction

Unlike other properties, the applications of QSAR/QSPR methods to investigate the effects of surfactant on the IFT alternation are more challenging. In general, the value of IFT of the oil–water interface is dependent on many factors. Internal parameters such as the type of oil, salinity of water, and solutes in the water; as well as external parameters such as temperature, pressure, and concentration all play a role in determining the actual IFT of a particular interface. The early QSAR/QSPR models of IFT were built based on the correlation with the surface tension values at CMC, Υ_0, that is, the lowest surface tension (or free energy) that a surfactant can mostly reach. From the chosen data set of Υ_0 of 30 diverse structures of nonionic surfactant, Wang [72,73] proposed a two-descriptor model as:

$$\Upsilon_0 = 12.66 - 0.01529\cdot \Delta H_f + 0.6310\cdot \mathrm{KH0} + 0.0006576\cdot \Delta H_f \times \mathrm{KH0} \quad (5.10)$$

where ΔH_f is the heat of formation and KH0 is a topological descriptor, which represents the size of the hydrophobic segment and contains group contribution from all nonhydrogen atoms in the fragment:

$$\mathrm{KH0} = \sum_{i=1}^{N} \left(\delta_i^v\right)^{-1/2}, \text{where } \delta_i^v = \frac{Z_i^v - H_i}{Z_i - Z_i^v - 1}, \quad (5.11)$$

Here, Z_i is the total number of electrons in the ith atom, Z_i^v is the number of valence electrons, and H_i is the number of hydrogen directly attached to the ith atom.

A new model for the same dataset was also proposed to return an improved correlation as:

$$\Upsilon_0 = 11.98 - 0.01053 \cdot \Delta H_f + 0.5848 \cdot \mathrm{KH0} + 0.09734 \cdot D + 0.4780 \cdot \mathrm{NO} - 0.0007763 \cdot E_T + 0.1345 \cdot \mathrm{NO} \cdot \mathrm{KH0} \quad (5.12)$$

where E_T is the total energy, D is the dipole moment, and NO is the number of oxygen atoms of the hydrophilic segment.

Wang [73] also reported a QSAR/QSPR model of Υ_0 of the anionic surfactant at different temperatures as:

$$\Upsilon_0 = 27.71 - 0.005239 \cdot \Delta H_f + 1.080 \cdot \mathrm{KH0} - 0.2130^{*} D + 17.16 \cdot C - 0.1520 \cdot T \quad (5.13)$$

Here, C is the concentration in a unit of mol/L and T is the temperature in °C.

Fini et al. [74] studied the IFT of a set of ionic surfactants on the interface of the n-decane and water and proposed a QSAR/QAPR model as:

$$\log \mathrm{IFT} = 1.989 - 0.167 \cdot \mathrm{Jhetv} - 0.059 \cdot \mathrm{RDF110u} + 0.061 \cdot \mathrm{RDF140m} \quad (5.14)$$

Here, Jhetv is the van der Waals' weighted distance matrix, RDF110u is the radial distribution function −11.0/unweighted, and the RDF140m is the radial distribution function −14.0/weighted by atom masses.

Instead of IFTs of the oil–water interface, the surface tension of the surfactant solution representing the liquid–air interfaces is more readily determined by experiments. A large amount of the QSAR/QSPR works has been published for different types of surfactants [72,75–81]. The ANN-based machine learning technology has been applied to train the model of the surface tension values from a large database containing 1602 pure compounds from a large number of data points (18,298) at different temperatures [82]. The Sequential Search mathematical algorithm was used to handle such a large data set. In this study, a three-layer ANN has been optimized using the Levenberg–Marquardt optimization strategy. The model has a wide range of applicability by the prediction capability of the model is restricted to the compounds, which are similar to those one applied to develop the model. Application of the model for the totally different compounds than the investigated ones is not recommended although it may be used for a rough estimation of the surface tensions of these kinds of compounds.

5.3.2 Packing parameter of surfactants

Surfactants consist of at least one hydrophobic and one hydrophilic counterpart of molecular species in one molecule. When added to the aqueous solution, the hydrophobic parts of the surfactants turn to gather together, forming the so-called micelles. Depending on the molecular structure and the concentration of surfactant, different types of micelle can be formed, such as spherical, worm-like, vesicle, and planar layer. The Packing Parameter is one of the commonly used numbers to relate the molecular structure to its partition in the aqueous solution.

By definition, the packing parameter of a surfactant that consists of at least one hydrophobic and one hydrophilic group can be determined from:

$$P = \frac{1}{a_e} \cdot \frac{V_0}{l_0} \tag{5.15}$$

where V_0 is the volume of the surfactant hydrophobic (tail) portion, l_0 is the length of the tail, and a_e is the equilibrium area per surfactant molecule at the aggregate interface.

For example, for the spherical close-packed micelle with a radius R_s consisting of g surfactants, we have the area of the core $A = g \cdot a_e = 4\pi R_s$, and the volume of the micelle, $V = g \cdot V_s = 4\pi\ R_s^3/3$, $R_s = 3V_s/A_e$. In the close-pack model, $R_s < \mathrm{l}_0$; then we have the packing parameter for the spherical micelle, $P < 1/3$ (for the close-packed model).

The "hydrophobic effect" of surfactants can be described by the fundamental theory related to the self-assembly of the surfactant tails (hydrophobic portions) to avoid contact with water. When the surfactants are added to the water solution, this often leads to the formation of different shapes of micelle. The oil-in-water emulsion or microemulsion can be formed when the surfactants are added to the oil–water mixture. The statistical thermodynamic models for detailed descriptions of the micelle shape and distribution in aqueous solution, determined from the calculation of the packing parameters have been extensively studied previously [83–86]. With these models, the chemical potential changes from the infinitely diluted state to the self-assemble state are determined by different terms, such as the energy gained by arranging the surfactant tails to avoid the contact with water, and the energy penalty by packing the surfactant heads (head–head interaction), as well as deformation of tails when packing, the interfacial term because of the nonideal close-packed model, etc. Based on these models, a number of molecular descriptors have been identified and can be calculated with calibration. In addition, computational approaches to improve the ability to obtain more accurate molecular descriptors have also been proposed.

Betaine is a class of special surfactants with more complicated molecular structures, and charge distributions (Fig. 5.11). Betaine surfactant typically has both cationic and anionic centers attached to the same molecule, rendering much stronger head–head interactions when the packing parameter is calculated.

Fig. 5.12 illustrates various energy terms calculated for a cocamidopropyl betaine in a spherical micelle. Compared to that of the regular ionic surfactant, we can see that the dipole–dipole interaction in the Betaine surfactant is significantly different from that in the ionic (and nonionic) surfactants (Fig. 5.13). It is worth noting that the current calculation using the classical Debye–Huckel solution might overestimate the interaction energy term, and more detailed investigations and calibration will be needed to obtain a more accurate dipole interaction energy term. Nevertheless, these comparisons clearly indicated the effect of the dipole interaction will have a more meaningful impact on the surfactant distributions in the aqueous solution.

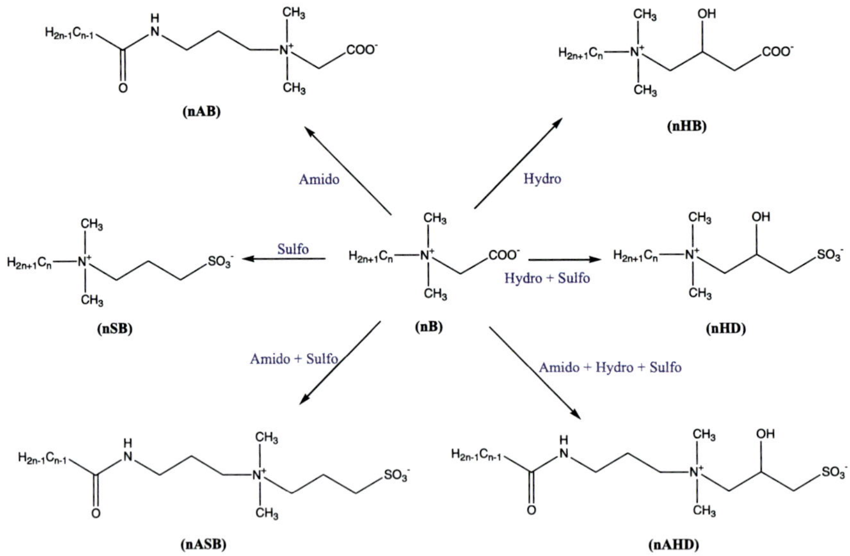

FIGURE 5.11

Some variations in the betaine surfactant molecules.

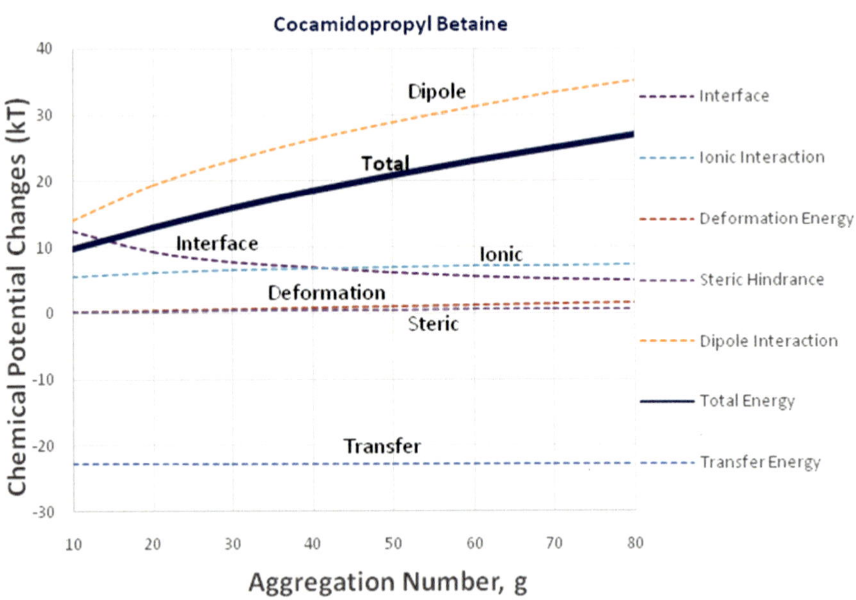

FIGURE 5.12

Calculated energy terms of the cocamidopropyl betaine packed in the spherical micelle.

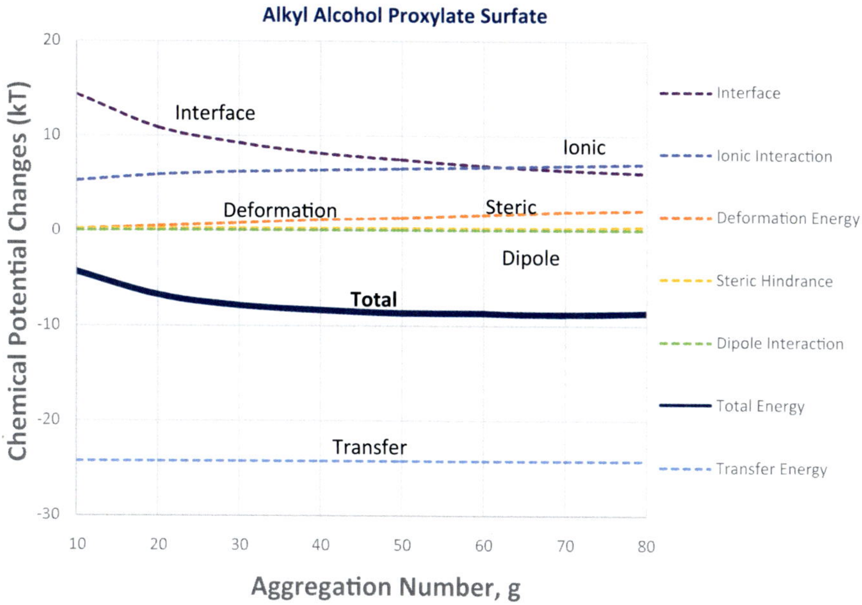

FIGURE 5.13

Calculated energy terms of the ionic surfactant packed in the spherical micelle.

5.3.3 Molecular dynamics and Monte Carlo simulations

The increasing computational power and the massively parallelized codes have enhanced our ability to deal with larger and more complex molecular systems, enabling the multiscale modeling based on the molecular dynamics (MD) and MC calculations. Molecular MD/MC modeling studies of surfactant systems are very rich, including, but with no limits to those on the phase and micelle modeling [87–90]; structural information [91–95]; adsorption on surfaces [96,97]; self-assembly [98–101]; interaction [102]; and Mass transport [103,104].

The dependence of the superficial energy and entropy of a model heptanes/water system as a function of a surfactant concentration was studied [105]. MD Calculations with polarizable potentials were conducted to investigate the *n*-alkane/water interface, including sodium–halide ion distributions [106]. Constant surface tension and constant volume (NVT) MD simulations with fixed IFT for a series of bolaamphiphilic poly-(fluorooxetane)s with perfluoroalkyl chain lengths ranging from $-CF_3$ to $-(CF_2)_4F$ were conducted to understand their behavior on the water/air interface [107]. The effects of the surfactants on the gas–liquid interface have been an important issue related to the CO_2-EOR operations. A number of MD calculations have been applied to provide the molecular view of the CO_2–water interface [108,109].

The fundamental basis for the molecular dynamic modeling of IFT calculations starts from the mechanic law through pressure tensor calculations. The definition of IFT accordingly to the Kirkwood and Buff formula [110]:

$$\gamma = \frac{1}{2}\int_0^{Lz} dz[P_N(z) - P_T(z)] \tag{5.15}$$

where P_N and P_T are the normal and tangential components of the pressure with respect to the planar interface, respectively. P_N and P_T are the same in the bulk phase because the structure is isotropic in any direction, but are different near the interface as it becomes very anisotropic. P_N and P_T were calculated from each slab of the simulation box during the simulation as a time average as:

$$P_N(z) = \rho(z)k_B T - \frac{1}{V_{\text{slab}}}\left\langle \sum_{i,j} \frac{z_{ij}^2}{r_{ij}} \frac{du(r_{ij})}{dr_{ij}} \right\rangle \tag{5.16}$$

$$P_T(z) = \rho(z)k_B T - \frac{1}{V_{\text{slab}}}\left\langle \sum_{i,j} \frac{x_{ij}^2 + y_{ij}^2}{2r_{ij}} \frac{du(r_{ij})}{dr_{ij}} \right\rangle \tag{5.17}$$

where $\rho(z)$ denotes the density of the slab at Z and V_{slab} denotes the slab volume. k_B and T are the Boltzmann constant and the absolute temperature. Angle bracket means an ensemble average of all atoms located in the slab at z. r_{ij}, x_{ij}, y_{ij}, and z_{ij} are distance between atoms and its components, respectively, and $u(r_{ij})$ is the potential energy of the atomic pair i and j.

The origin of IFT between two contact phases is often attributed to the difference in their cohesive energy densities (CEDs). CED, which is defined as the energy per unit volume, is a measure of the extent of atomic interactions in a system. The inhomogeneity in CED leads to an excess of free energy and a tensile net force at the interface, which are related to the IFT. The IFT of a decane($C_{10}H_{22}$)−water interface mediated by a family of alky benzene sulfonates surfactants has been computed theoretically using MD simulations [111]. The surfactants, denoted by *m*-C16, indicate a benzene sulfonate group attached to the *m*th carbon in the hexadecane backbone. Four different isomers ($m = 2$, 4, 6, and 8) were studied to understand the effect of the molecular structure of surfactants on IFT. The general procedure of the MD calculation is described as:

1. Atomic interaction models—in this simulation, decane and water molecules are described separately, with the united atom model for decane and the F3C model for water, whereas the surfactant molecules are described using the DREIDING FF.
2. The system configuration—all systems consist of 120 decane, 800 water, and 32 surfactant molecules.
3. System construction—surfactant, decane, and water systems were first optimized separately in the box with the same *x* and *y* dimensions; then combined into one simulation box.

4. NVT and NPT MD simulation—both NVT (200 ps at 300K) and NPT (400 ps, 300K, and 1 atm) MD calculations were conducted sequentially to obtain final equilibrium configurations.

The calculated IFT value showed a trend with respect to the different surfactant isomers, in which the 4-C16 is corresponded to the lowest IFT value, in general, consistent with the experimental observation [112].

A number of important insights can be obtained from these MD calculations.

1. The conformation of surfactants at the interface is characterized by calculating the moments of inertia (I_1, I_2, and I_3) of each surfactant and the tilt angle between the largest principal moment and the interface normal vector (Fig. 5.2). It has been found the 4-C16 has the largest $I_3/I_{1,2}$ ratio and the smallest tilt angle, indicating that the 4-C16 surfactant is aligned more vertically with the smaller molecular cross-sectional area at the interface, thus more compact packing than the others.
2. Miscibility and effective alkyl tail length are studied by comparing the end-to-end length of the decane (as oil phase) and the difference between the end-to-end length of long and short alkyl chains, $r_{effective} = r_{long} - r_{short}$. The 9.53 Å of the effective alkyl tail length in 4-C16 is the closest to that of the decane (9.97 Å). This rationalizes that the size similarity of the hydrophobic group and the oil-phase solvent is responsible for the lowest IFT of 4-C16.
3. The density profiles of the liquid–vapor interface can be analyzed based on the 10–90 criteria, which define the interfacial thickness to be the distance between two positions where the density varies from 10% to 90% of the density of the bulk phase. The oil/surfactant/water system has a more complex phase, so the 90–90 criteria are proposed instead. The distance between two positions where the densities of decane and water are 90% of their own bulk densities (0.997 for water and 0.725 g/cm^3 for decane at 300K and 1 atm). As shown in Fig. 5.3, for the pure decane/water system, the interfacial layer is between 90%W and 90%D; whereas for the decane/surfactant/water, the water phase is within 90%W and 10%W, and the decane phase is within 90%D and 10%D; a middle phase wherein the surfactants are mainly located is within 10%D and 10%W.
4. Interface formation energy (IFE) is defined as IFE = $\{E_{total} - (n \cdot E_{surfactant,single} + E_{decane-water})\}/n$ is a measure of average intermolecular interaction per surfactant molecule arising from the insertion of that surfactant into the decane–water interface. Here, E_{total}, $E_{surfactant,single}$, and $E_{decane-water}$ are the energies of the whole system, the single surfactant molecule that is calculated from a separate 100 ps MD simulation in vacuum at the same temperature, and a bare decane–water system, respectively.
5. The total energy of the system can be analyzed from the main field theory, which is defined as:

$$E(\text{total}) = E(\text{interface}) + E(\text{elastic}) + E(\text{mixing}) \tag{5.18}$$

where E (interface), E (elastic deformation), and E (mixing) are individual energy term of contributions, which can be described as:

$$E(\text{interface}) = A \cdot \gamma \tag{5.19}$$

where A is the interface area and γ is the IFT of interest

$$E(\text{elastic}) = \frac{3}{2} k_B T n_{\text{chain}} \left(\frac{t_{\text{surf}}^2}{N_{\text{long}} {}^* a^2} + \frac{N_{\text{long}} {}^* a^2}{t_{\text{surf}}^2} - 2 \right) \tag{5.20}$$

And

$$E(\text{mixing}) = \frac{A^*(t_{\text{total}} - t_{\text{water}})}{N_{\text{decane}} {}^* a^3} {}^*(1-\eta)^* \ln(1-\eta) \tag{5.21}$$

Here, n_{chain} is the number of chains, N_{long} is the number of beads, N_{decane} is the number of bead in decane, a is the van der Waals diameter, t_{surf} is the fraction of alkyl tail of surfactant in the volume of interphase, and the volume of the interphase is $A \times (t_{\text{total}} - t_{\text{water}})$. All parameters and energy terms can be extracted from the MD trajectory analyses.

5.3.4 Continuous progresses in molecular dynamics/Monte Carlo simulations

In essence, MD simulations are based on the many-body equation of motions and dynamic evolution of a collection of molecules. The molecules are allowed to interact with each other following certain predefined relations, for example, potential functions or FF for atomic interactions. Because the molecular systems typically consist of a large number of different molecules, MD simulations will have the potential to provide theoretical predictions of the molecular properties at macroscopic or at least mesoscopic levels. However, the extremely large and complex systems also employ significant challenges for the simulation, in particular, to achieve the optimum configurations of the molecular system that best represent the actual meso/macroscale formations. The conventional MD/MC simulations circumvent this problem by sacrificing lengthy numerical calculations. The advances in high-speed computer powers and the network-type parallel computational codes have made MD/MC simulation more competent for many applications. In addition, significant efforts have been conducted to improve the methodologies of the MD/MC simulation.

5.3.4.1 Statistical mechanics

The fundamental basis of MD/MC simulation is based on the statistical mechanics which provides the rigorous mathematical expressions that relate macroscopic properties to the distribution and motion of atoms and molecules of the N-body system.

- Microcanonical ensemble (NVE): An isolated system with a fixed number of atoms (N), a fixed volume (V), and a fixed total energy (E). The system is characterized as the exchange of the potential and kinetic energy, with total energy being conserved.
- Canonical ensemble (NVT): An isolated system with a fixed number of atoms (N), a fixed volume (V), at a fixed temperature (T). The system is characterized as the exchange of energy of endothermic and exothermic processes with a thermostat.
- Isobaric-isothermal ensemble (NPT): An isolated system with fixed pressure (P), a fixed number of atoms (N), and a fixed temperature (T). This corresponds most closely to laboratory conditions with a flask open to ambient temperature and pressure.
- Grand canonical ensemble (μVT): A system with fixed chemical potential (μ), a fixed volume (V), and a fixed temperature (T). The system is considered an open system because it can exchange the atoms and energy with a reservoir, so that different states of the system could have different total system energy and a total number of atoms in the system.
- Grand canonical MC/MD methods.

5.3.4.2 Potentials in molecular dynamics calculation

The ability to describe molecular interactions through a definitive potential function is one key element of the MD simulation to be distinguished from conversion MC modeling. In chemical and biologic simulations, such potential functions are often represented by the FF, which defines the interaction forces between a pair of atoms in the system. Basically, the FFs are characterized by different functional terms, such as covalent bonds, nonbonded, long-range electrostatic and van der Waals, and so on. Additional terms are added to describe special interactions, such as polarizable FFs and reactive FFs. Numerous FF parameters have been developed and integrated into most commercial and academic computational software. A number of such FF include, but with not limited to:

1. Classical FFs—such as Assisted Model Building and Energy Refinement [113], Chemistry at HARvard Molecular Mechanics [114], Universal Force Field [115], and DREIDING, a generic FF for molecular simulation [116], and so on. Basically, there are two major types of potential energy forms in the molecular system, bonded and nonbonded terms. Parameterized FF terms are used to describe these energy forms. The bonded terms for atomic interaction typically include covalent bond, distortion energies of angle blending, and dihedral constraints; and the nonbonded terms often include those long-range electrostatic and van der Waals forces and so on. Therefore, the most common FF terms will have:

$$E = E_{\text{bonded}} + E_{\text{nonbonded}} = \left(E_{\text{bond}} + E_{\text{angle}} + E_{\text{dihedral}}\right) + \left(E_{\text{electrostatic}} + E_{\text{vdWaals}}\right) \tag{5.22}$$

2. Polarizable FFs—developed to account for charge distribution variation under a dielectric environment. Since the first introduction of the CFF/ind polarizable FF in 1976 [117], many types of polarizable FFs have been proposed, including the expansion of the classical FFs to include these electrostatic terms. Basically, the polarizable FF terms are included to better describe the nonbonded electrostatic interactions beyond the fixed charge model. Typically, three additional terms including "induced point dipole," "Drude oscillator" and "Fluctuating charge" are used, whereas the first two terms are originated from the dipole induction, and the last term is included to represent the redistribution of charges within a molecule, so,

$$E_{\text{electrostatic}} = E_{\text{Coulomb}} + E_{\text{Induced}} + E_{\text{Drude}} + E_{\text{FQ}} \tag{5.23}$$

3. Reactive Force Fields (ReaxFF)—special classes of FFs to account for the transition of the molecular systems. In addition, to use the atomic parameters from their equilibrium configuration, ReaxFF uses quantum chemistry DFT to calibrate those parameters during transitions. ReaxFF is a versatile and expendable FF method. Once the parameters are calibrated, they can provide fast and reliable dynamic properties. The original ReaxFF was designed to describe interactions during the atomic transition, that is, to establish a general relationship between bond distance and bond order on one hand and between bond order and bond energy on the other hand that leads to proper dissociation of the bond to separated atoms, so the bonded energy terms are expanded to include:

$$E = E_{\text{bond}} + E_{\text{over}} + E_{\text{under}} + E_{\text{val}} + E_{\text{pen}} + E_{\text{tors}} + E_{\text{conj}} + E_{\text{vdWaals}} + E_{\text{Coulomb}} \tag{5.24}$$

where E_{over} and E_{under} are the energy penalties due to the under or over coordination of the total bond order (i.e., <4 for the carbon atom and >1 for a hydrogen atom); E_{val} is the three-body energy contribution from valence angle terms goes to zero as the bond orders in the valence angel goes to zero and the energy penalties of over/under-coordination; E_{pen} is the additional penalty energy caused by two double bonds sharing an atom in a valence angle; E_{tors} is the four-body energy for torsion rotation barriers; and E_{conj} is the contribution of conjugation effects to the molecular energy [118]. Some details of the continuous development and applications of ReaxFF have been summarized [119].

4. Coarse-grained (CG) FFs—another type of FFs to reduce the computational cost by reducing the degree of freedom in the model for large systems, especially the biological processes, such as DNA and protein folding. A few examples of the CG FFs include MARTNI [120], SIRAH [121], and VAMM [122]. The key of the CG FF is to reduce the complexity of the simulated system by grouping atoms together into virtual particles, such that to smooth the energy landscape and to reduce the number of interactions in the system. In general, the CG simulation is depended on the details of the models related to the scientific question to be addressed, so it is critical to preserve the

features that drive the phenomenon of interest and not to oversimplify the description of the system [123].

Advanced FFs based on machine learning [124–126], neural network [127,128], and big data [129] approaches have been also applied to generate field fields.

5.4 Conclusion

The designs and selections of suitable chemicals, for example, polymers and surfactants, for the chemical EOR applications, have been mainly relied on the trial-and-error experimental approaches. Applications of the advanced computational simulation methods are expected to provide a new dimension of understanding to assist and accelerate the chemical optimization processes. In this section, we attempt to highlight some applications of molecular modeling to deal with special aspects of issues typically related to the applications in the chemical EOR operations. Computational simulations of complex systems are often time consumed and demand higher computer power. The proper selection of suitable computational methods is important to provide meaningful modeling results to any special systems. The computational methods involve high-level DFT calculations on reaction mechanisms; semiempirical energy calculations, classical thermodynamic models, FF interactions and MD, multidimensional modeling the advanced machine learning algorithm. The level of computational accuracy needs to constantly be calibrated. To provide more meaningful theoretical prediction, the molecular modeling should be integrated with laboratory experiments and application data.

Nomenclature

AA acidic acid
AM acrylamide
AMPS 2-acrylamido-2-methylpropane sulfonic acid
ANN artificial neural network
CED cohesive energy density
CMC critical micelle concentration
DFT density functional theory
EOR enhanced oil recovery
FDA fluid-diverting agent
FVA fluid viscosification agent
HLB hydrophilic–lipophilic balance
HT/HS high temperature/high salinity
HPAM hydrolyzed polyacrylamide
IFT interfacial tension
LFER linear free energy relationship

MC Monte Carlo
MD molecular dynamics
MLR multilinear regression
NPT isothermal–isobaric ensemble
NVE microcanonical ensemble
NVP *N*-Vinyl pyrrolidone
NVT canonical ensemble
OOIP original oil-in-place
PAM polyacrylamide
PCR principal component regression
ps picosecond
QSAR quantitative structure–activity relationship
QSPR quantitative structure–property relationship
SVM support vector machines
VSE volumetric sweep efficiency
μVT grand canonical

References

[1] L.C. Uren, E.H. Fahmy, Factors influencing the recovery of petroleum from unconsolidated sands by waterflooding, Transactions of the AIME 77 (1927) 318–335.
[2] R.D. Shupe, Chemical stability of polyacrylamide polymers, Journal of Petroleum Technology 33 (1981) 1513–1529.
[3] A. Zaitoun, B. Potie, Limit conditions for the use of hydrolyzed polyacrylamides in brines containing divalent ions, in: SPE-11785-MS, SPE Oilfield and Geothermal Chemistry Symposium, June 1–3, 1983, Denver, Colorado, USA.
[4] H.A. Nasr-El-Din, B.F. Hawkins, K.A. Green, Viscosity behavior of alkaline, surfactant, polyacrylamide solutions used for enhanced oil recovery, in: SPE-21028-MS, SPE International Symposium on Oilfield Chemistry, February 20–22, 1991, Anaheim, California, USA.
[5] A.A. Bonapasta, F. Buda, P. Colombet, Interaction between Ca ions and poly(acrylic acid) chains in macro-defect-free cements: a theoretical study, Chemistry of Materials 13 (2001) 64–70.
[6] R.S. Seright, A.R. Campbell, P.S. Mozley, et al., Stability of partially hydrolyzed polyacrylamides at elevated temperatures in the absence of divalent cations, SPE Journal 15 (2010) 341–348.
[7] A. Moradi-Araghi, P.H. Doe, Hydrolysis and precipitation of polyacrylamides in harsh brines at elevated temperature, SPE Reserve Engineering 2 (1987) 189–197.
[8] D. Kahne, W.C. Still, Hydrolysis of a peptide bond in neutral water, Journal of the American Chemical Society 110 (1988) 7529–7534.
[9] R.S. Brown, A.J. Bennet, H. Ślebocka-Tilk, Recent perspectives concerning the mechanism of H_3O^+- and OH^--promoted amide hydrolysis, Accounts of Chemical Research 25 (1992) 481–488.
[10] R.S. Brown, Studies in amide hydrolysis: the acid, base, and water reactions, in: A. Greenberg, C.M. Breneman, J.F. Liebman (Eds.), The Amide Linkage: Selected Structural Aspects in Chemistry, Biochemistry and Materials Science, John Wiley Sons, 2000, pp. 85–114.

[11] J.P. Guthrie, Hydration of carboxamides. Evaluation of the free energy change for addition of water to acetamide and formamide derivatives, Journal of the American Chemical Society 96 (1974) 3608–3615.

[12] H. Ślebocka-Tilk, A.J. Bennet, H.J. Hogg, et al., Predominant ^{18}O exchange accompanying base hydrolysis of a tertiary toluamide: N-ethyl-N-(trifluoroethy1)toluamide. Assessment of the factors that influence partitioning of anionic tetrahedral intermediates, Journal of the American Chemical Society 113 (1991) 1288–1294.

[13] R.L. Schowen, G.W. Zuorick, Amide hydrolysis. Superimposed general base catalysis in the cleavage anilides, The Journal of American Chemical Society 88 (1966) 1223–1225.

[14] J.F. O'brien, J. Pranata, Ab initio study of the base-promoted hydrolysis of N-methylformamide, Journal of Physical Chemistry 99 (1995) 12759–12763.

[15] D. Bakowies, P.A. Kollman, Theoretical study of base-catalyzed amide hydrolysis: gas- and aqueous-phase hydrolysis for formamide, Journal of the American Chemical Society 121 (1999) 5712–5726.

[16] H. Ślebocka-Tilk, A.A. Neverov, R.S. Brown, Proton inventory study of the base-catalyzed hydrolysis of formamide. Consideration of the nucleophilic and general base mechanisms, Journal of the American Chemical Society 125 (2003) 1851–1858.

[17] Y. Xiong, C.-G. Zhan, Theoretical studies of the transition-state structures and free energy barriers for base-catalyzed hydrolysis of amides, The Journal of Physical Chemistry A 110 (2006) 12644–12652.

[18] T. Cohen, J. Lipowitz, Acid-catalyzed amide hydrolysis assisted by a neighboring amide group, Journal of the American Chemical Society 86 (1964) 5611–5616.

[19] J.P. Krug, P.L.A. Popelier, R.F.W. Bader, Theoretical study of neutral and of acid and base promoted hydrolysis of formamide, Journal of Physical Chemistry 96 (1992) 7604–7616.

[20] D. Zahn, Theoretical study of the mechanisms of acid-catalyzed amide hydrolysis in aqueous solution, The Journal of Physical Chemistry B 107 (2003) 12303–12306.

[21] T.K. Manojkumar, S.B. Suh, K.S. Oh, et al., Theoretical studies on the mechanism of acid-promoted hydrolysis of N-formylaziridine in comparison with formamide, The Journal of Organic Chemistry 70 (2005) 2651–2659.

[22] P.D. Bolton, Hydrolysis of amides. II. Substituent effects in dilute acid and alkali, Australian Journal of Chemistry 19 (1966) 1013–1021.

[23] P.D. Bolton, G.I. Jackson, Hydrolysis of amides. V. Alkaline hydrolysis of alkyl-substituted amides, Australian Journal of Chemistry 24 (1971) 969–974.

[24] Q. Ma, P.J. Shuler, C.W. Aften, et al., Theoretical studies of hydrolysis and stability of polyacrylamide polymers, Polymer Degradation and Stability 121 (2015) 69–77.

[25] L.I. Tolstikh, N.I. Akimov, I.A. Golubeva, et al., Degradation and stabilization of polyacrylamide in polymer flooding conditions, International Journal of Polymeric Materials 17 (1992) 177–193.

[26] W.M. Leung, D.E. Axelson, Thermal degradation of polyacrylamide and poly(acrylamide-co-acrylate), Journal of Polymer Science: Part A: Polymer Chemistry 25 (1987) 1825–1846.

[27] J.D. Van Dyke, K.L. Kasperski, Thermogravimetric study of polyacrylamide with evolved gas analysis, Journal of Polymer Science: Part A: Polymer Chemistry 31 (1993) 1807–1823.

[28] J.J. Maurer, G.D. Harvey, Thermal degradation characteristics of poly(acrylamide-co-acrylic acid) and poly(acrylamide-co-sodium acrylate) copolymers, Thermochimica Acta 121 (1987) 295–306.
[29] C. Jiang, X. Xia, S. Kang, et al., Neighboring group effect on the thermal degradation of polyacrylamide and its derivatives, Journal of Polymer Engineering 39 (2019) 239–247.
[30] C. Jiang, B. Yu, Q. Ma, et al., Crosslinked polymers as "smart" viscosifiers used in hostile environments, Journal of Petroleum Science and Engineering 173 (2018) 1332–1339.
[31] D.J. Evans, G.P. Morriss, Nonlinear-response theory for steady planar Couette flow, Physical Review A 30 (1984) 1528–1530.
[32] C. Negin, S. Ali, Q. Xie, Most common surfactants employed in chemical enhanced oil recovery, Petroleum 3 (2017) 197–211.
[33] A.M. Johannessen, K. Spildo, Enhanced oil recovery (EOR) by combining surfactant with low salinity injection, Energy Fuels 27 (2013) 5738–5749.
[34] N.N. Zulkifli, S.M. Mahmood, S. Akbari, et al., Evaluation of new surfactants for enhanced oil recovery applications in high-temperature reservoirs, Journal of Petroleum Exploration and Production Technology 10 (2020) 283–296.
[35] L.P. Hammett, Some relations between reactions rates and equilibrium constants, Chemical Reviews 17 (1935) 125–136.
[36] L.P. Hammett, Linear free energy relationship in rate and equilibrium phenomena, Transactions of the Faraday Society 34 (1938) 156–165.
[37] J.R.W. Taft, Linear free energy relationship from rates of esterfication and hydrolysis of aliphatic and ortho-substituted benzoate esters, Journal of the American Chemical Society 74 (1952) 2729–2732.
[38] J.R.W. Taft, The general nature of the propotionality of polar effects of substituent groups in organic chemistry, Journal of the American Chemical Society 75 (1953) 4231–4238.
[39] H. Wiener, Structural determination of paraffin boiling points, Journal of the American Chemical Society 69 (1947) 17–20.
[40] H. Wiener, Correlation of heats of isomerization and differences in heats of vaporization of isomers among the paraffin hydrocarbons, Journal of the American Chemical Society 69 (1947) 2636–2638.
[41] D.H. Rouvray, The rich legacy of half a century of the Wiener index, in: D.H. Rouvray, R.B. King (Eds.), Topology in Chemistry: Discrete Mathematics of Molecules, Horwood Publishing, 2002, pp. 16–37.
[42] M. Karelson, Molecular Descriptors in QSAR/QSPR, John Wiley & Sons, New York, 2000.
[43] R. Todeschini, V. Consonni, Handbook of Molecular Descriptors, Wiley-VCH, Weinheim, 2000.
[44] A.R. Katritzky, M. Kuanar, S. Slavov, et al., Quantitative correlation of physical and chemical properties with chemical structure: utility for prediction, Chemical Reviews 110 (2010) 5714–5789.
[45] J. Hu, X. Zhang, Z. Wang, A review on progress in QSPR studies for surfactants, International Journal of Molecular Sciences 11 (2010) 1020–1047.
[46] M.J. Rosen, The relationship of structure to properties in surfactants. IV. Effectiveness in surface or interfacial tension reduction, Journal of Colloid and Interface Science 56 (1976) 320–327.

[47] P. Becher, Hydrophilic-lipophile balance: history and recent developments, Langmuir Lecture—1983, Journal of Dispersion Science and Technology 5 (1984) 81–96.
[48] J.C. Ravey, A. Gherbi, M.J. Stebe, Comparative study of fluorinated and hydrogenated nonionic surfactants. I. Surface activity properties and critical concentrations, Progress in Colloid & Polymer Science 76 (1988) 234–241.
[49] P.D.T. Huibers, Prediction of critical micelle concentration using a quantitative structure-property relationship approach. 1. Nonionic surfactants, Langmuir 12 (1996) 1462–1470.
[50] P.D.T. Huibers, V.S. Lobanov, A.R. Katritzky, et al., Prediction of critical micelle concentration using a quantitative structure–property relationship approach 2. Anionic surfactants, Journal of Colloid and Interface Science 187 (1997) 113–120.
[51] Z. Wang, G. Li, X. Zhang, et al., A quantitative structure-property relationship study for the prediction of critical micelle concentration of nonionic surfactants, Colloids and Surfaces A: Physiochemical and Engineering Aspects 197 (2002) 37–45.
[52] S. Yuan, Z. Cai, G. Xu, et al., Quantitative structure-property relationships of surfactants: prediction of the critical micelle concentration of nonionic surfactants, Colloid and Polymer Science 280 (2002) 630–636.
[53] R.A. Saunders, J.A. Platts, Correlation and prediction of critical micelle concentration using polar surface area and LFER methods, Journal of Physical Organic Chemistry 17 (2004) 431–438.
[54] E.A.M. Gad, QSPR for nonionic surfactants, Journal of Dispersion Science and Technology 28 (2007) 231–237.
[55] A.R. Katritzky, L.M. Pacureanu, S.H. Slavov, et al., QSPR study of critical micelle concentrations of nonionic surfactants, Industrial & Engineering Chemistry Research 47 (2008) 9687–9695.
[56] A. Mozrzymas, B. Rozycka-Roszak, Prediction of critical micelle concentration of nonionic surfactants by a quantitative structure-property relationship, Combinatorial Chemistry & High Throughput Screening 13 (2010) 39–44.
[57] K. Roy, H. Kabir, QSPR with extended topochemical atom (ETA) indices: modeling of critical micelle concentration of non-ionic surfactants, Chemical Engineering Science 73 (2012) 86–98.
[58] M. Jalali-Heravi, E. Konouz, Multiple linear regression modeling of the critical micelle concentration of alkyltrimethylammonium and alkylpyridinium salts, Journal of Surfactants and Detergents 6 (2003) 25–30.
[59] A.R. Katrizky, L.M. Pacureanu, S.H. Slavov, et al., QSPR study of the first and second critical micelle concentrations of cationic surfactants, Computers and Chemical Engineering 33 (2009) 321–332.
[60] A. Mozrzymas, B. Rozycka-Roszak, Prediction of critical micelle concentration of cationic surfactants using connectivity indices, Journal of Mathematical Chemistry 49 (2011) 276–289.
[61] M. Jalali-Heravi, E. Konouz, Prediction of critical micelle concentration of some anionic surfactants using multiple regression techniques: a quantitative structure–activity relationship study, Journal of Surfactants and Detergents 3 (2000) 47–52.
[62] D.W. Roberts, Application of Octanol/Water partition coefficients in surfactant science: a quantitative structure-property relationship for micellization of anionic surfactants, Langmuir 18 (2002) 345–352.

[63] X. Li, G. Zhang, J. Dong, et al., Estimation of critical micelle concentration of anionic surfactants with QSPR approach, Journal of Molecular Structure (Theochem) 710 (2004) 119–126.

[64] A.R. Katrizaky, L. Pacureanu, D. Dobchev, et al., QSPR study of critical micelle concentration of anionic surfactants using computational molecular descriptors, Journal of chemical information and modeling 47 (2007) 782–793.

[65] D.T. Stanton, On the importance of topological descriptors in understanding structure-property relationships, Journal of Computer-Aided Molecular Design 22 (2008) 441–460.

[66] S. Rahal, N. Hadidi, M. Hamadache, In silico prediction of critical micelle concentration (CMC) of classic and extended anionic surfactants from their molecular structural descriptors, Arabian Journal for Science and Engineering, 45, 2020, pp. 7445–7454.

[67] G. Absalan, B. Hemmateenejad, M. Soleimani, et al., Quantitative structure–micellization relationship study of Gemini surfactants using genetic-PLS and genetic-MLR, QSAR & Combinatorial Science 23 (2004) 416–425.

[68] Z. Kardanpour, B. Hemmateenejad, T. Khayamian, Wavelet neural network-based QSPR for prediction of critical micelle concentration of Gemini surfactants, Analytica Chimica Acta 531 (2005) 285–291.

[69] C. Guo, P. Zhou, J. Shao, et al., Integrating statistical and experimental protocols to model and design novel Gemini surfactants with promising critical micelle concentration and low environmental risk, Chemosphere 84 (2011) 1608–1616.

[70] T. Gaudin, P. Rotureau, I. Pezron, et al., New QSPR models to predict the critical micelle concentration of sugar-based surfactants, Industrial and Engineering Chemistry Research 55 (2016) 11716–11726.

[71] N. Anoune, M. Nouiri, Y. Berrah, et al., Critical micelle concentrations of different classes of surfactants: a quantitative structure property relationship study, Journal of Surfactants and Detergents 5 (2002) 45–53.

[72] Z. Wang, D. Huang, G. Li, et al., Effectiveness of surface tension reduction by anionic surfactants - Quantitative structure–property relationships, Journal of Dispersion Science and Technology 24 (2003) 653–658.

[73] Z. Wang, J. Feng, H. Wang, et al., Effectiveness of surface tension reduction by nonionic surfactants with quantitative structure-property relationship approach, Journal of Dispersion Science and Technology 26 (2005) 441–447.

[74] M.F. Fini, S. Riahi, A. Bahramian, Experimental and QSPR studies on the effect of ionic surfactants on n-decane - water interfacial tension, Journal of Surfactants and Detergents 15 (2012) 477–484.

[75] G.W. Kauffman, P.C. Jurs, Prediction of surface tension, viscosity, and thermal conductivity for common organic solvents using quantitative structure-property relationships, Journal of Chemical Information and Computer Science 41 (2001) 408–418.

[76] E.J. Delgado, G.A. Diaz, A molecular structure based model for predicting surface tension of organic compounds, SAR and QSAR in Environmental Research 17 (2006) 483–496.

[77] J. Wang, H. Du, H. Liu, et al., Prediction of surface tension for common compounds based on novel methods using heuristic method and support vector machine, Talanta 73 (2007) 147–156.

[78] R.L. Gardas, J.A.P. Coutinho, Applying a QSPR correlation to the prediction of surface tensions of ionic liquids, Fluid Phase Equilibria 265 (2008) 57–65.

[79] A. Khajeh, H. Modarress, Quantitative structure–property relationship for surface tension of some common alcohols, Journal of Chemometrics 25 (2011) 333–339.

[80] A.K. Khajeh, H. Modarress, QSPR prediction of surface tension of refrigerants from their molecular structures, International Journal of Refrigeration 35 (2012) 150–159.

[81] F.S. Mjalli, K. Shabbaz, M.A. Hashim, et al., Surface tension of ionic liquids analogues using the QSPR correlation, International Journal of Chemical Engineering and Applications 4 (2013) 96–100.

[82] F. Gharagheizi, A. Eslamimanesh, B. Tirandaze, et al., Handling a very large data set for determination of surface tension of chemical compound using quantitative structure-property relationship strategy, Chemical Engineering Science 66 (2011) 4991–5023.

[83] E. Ruckenstein, R. Nagarajan, Thermodynamics of amphiphilar aggregation into micells and vesicles, in: K.L. Mittal (Ed.), Micellization, Solubilization, and Microemulsions, Vol. 1, Plenum Press, New York and London, 1977.

[84] R. Nagarajan, E. Ruckenstein, Critical micelle concentration: a transition point for micellar size distribution: a statistical thermodynamics approach, Journal of Colloid and Interface Science 62 (1977) 221–231.

[85] R. Nagarajan, E. Ruckenstein, Theory of surfactant self-assembly: a predictive molecular thermodynamic approach, Langmuir 18 (1991) 31–38.

[86] R. Nagarajan, E. Ruckenstein, Molecular packing parameter and surfactant self-assembly: the neglected role of the surfactant tail, Langmuir 7 (2002) 2934–2969.

[87] H. Chen, A.Z. Panagiotopoulos, Molecular modeling of surfactant micellization using solvent-accessible surface area, Langmuir 35 (2019) 2443–2450.

[88] H. Liu, Y. Liu, Y. Shang, et al., Molecular simulation and experimental studies on the interfacial properties of a mixed surfactant SDS/C_4mimBr, Molecular Simulation 45 (2019) 223–229.

[89] M. Mannoor, S. Kang, Y.K. Suh, Molecular dynamics simulation of aggregates in the Dodecane/spen80 system and their behavior in an electric field, Advanced in Condensed Matter Physics 2015 (2015) 739458.

[90] D. Viduna, A. Milchev, K. Binder, Monte Carlo simulation of micelle formation in block copolymer solutions, Macromolecular Theory and Simulation 7 (1998) 649–658.

[91] R.G. Larson, Monte Carlo lattice simulation of amphiphilic systems in two and three dimensions, The Journal of Chemical Physics 89 (1988) 1642–1650.

[92] D. Lu, Z. Liu, Z. Liu, Z., et al., Molecular simulation of surfactant-assisted protein refolding, The Journal of Chemical Physics 122 (2005) 134902.

[93] S. Vierros, M. Sammalkorpi, Hybrid atomistic and coarse-grained model for surfactants in apolar solvents, ACS Omega 4 (2019) 15581–15592.

[94] E.J.A. Suys, D.B. Warren, A.C. Pham, et al., A nonionic polyethylene oxide (PEO) surfactant model: experimental and molecular dynamics studies of kolliphor EL, Journal of Pharmaceutical Sciences 108 (2019) 193–204.

[95] P. Shi, H. Zhang, L. Lin, et al., Molecular dynamics simulation of four typical surfactants at oil/water interface, Journal of Dispersion Science and Technology 39 (2017) 1258–1265.

[96] G. Lu, X. Zhang, C. Shao, et al., Molecular dynamics simulation of adsorption of an oil-water-surfactant mixture on calcite surface, Pet. Sci. 6 (2009) 76–81.

[97] I. Moncayo-Riascos, C.A. Franco, F.B. Cortes, Dynamic molecular modeling and experimental approach of fluorocarbon surfactant-functionalized SiO_2 nanoparticles for gas wettability alternation on sandstones, Journal of Chemical and Engineering Data 64 (2019) 1860–1872.

[98] P.K. Maiti, Y. Lansac, M.A. Glaser, et al., Self-assembly in surfactant oligomers: a coarse-grained description through molecular dynamics simulations, Langmuir 18 (2002) 1908–1918.
[99] M. Luo, L.L. Dai, Molecular dynamics simulations of surfactants and nanoparticles self-assembly at liquid-liquid interfaces, Journal of Physics: Condensed Matter 19 (2007) 375109.
[100] W. Shinoda, R. De Vane, M.L. Klein, Coarse-grained molecular modeling of non-ionic surfactant self-assembly, Soft Matter 4 (2008) 2454–2462.
[101] G.M.C. Silva, P. Morgado, M. Goldmann, et al., Spontaneous self-assembly and structure of perfluoroalkylalkane surfactant hemimicelles by molecular dynamics simulations, Proceedings of the National Academy of Sciences of the United States of America 116 (2019) 14868–14873.
[102] S. Choe, R. Chang, J. Jeon, et al., Molecular dynamics simulation study of a pulmonary surfactant film interacting with a carbonaceous nanoparticles, Biophysical Journal 95 (2008) 4102–4114.
[103] A. Gupta, A. Chauhan, D.I. Kopelevich, Molecular modeling of surfactant covered oil-water interfaces: dynamics, microstructure, and barrier for mass transport, The Journal of Chemical Physics 128 (2008) 234709.
[104] X. Tang, S. Xiao, O. Lei, et al., Molecular dynamics simulation of surfactant flooding driven oil-detachment in nano-silica channels, The, Journal of Physical Chemistry 123 (2019) 277–288.
[105] G. Urbina-Villalba, R.M. Landrove, J.A. Guaregua, Molecular dynamics simulation of the interfacial behavior of a heptane/water system in the presence of nonylphenol triethoxylated surfactants. 1. Surface energy, surface entropy, and interaction energies as a function of temperature and surfactant concentration, Langmuir 13 (1997) 1644–1652.
[106] C.D. Wick, T.M. Chang, J.A. Slocum, et al., Computational investigation of the n-alkane/water interface with many-body potentials: the effect of chain length and ion distributions, Journal of Physical Chemistry 116 (2012) 783–790.
[107] B.C. Stephenson, K.J. Beers, Determination of the interfacial characteristics of a series of bolaamphiphilic Poly(fluorooxetane) surfactants through molecular dynamics simulation, The Journal of Physical Chemistry B 110 (2006) 19363–19405.
[108] S.R. da Rocha, K.P. Johnston, P.J. Rossky, Surfactant-modified CO_2-water interface: a molecular view, The Journal of Physical Chemistry 106 (2002) 13250–13261.
[109] M.T. Stone, S.R. da Rocha, P.J. Rossky, et al., Molecular differences between hydrocarbon and fluorocarbon surfactants at the CO_2/water interface, The Journal of Physical Chemistry B 107 (2003) 10185–10192.
[110] J.H. Irving, J.G. Kirkwood, The statistical mechanical theory of transport processes. IV. The equations of hydrodynamics, The Journal of Chemical Physics 18 (1950) 817–829.
[111] S.S. Jang, S.T. Lin, P.K. Maiti, et al., Molecular dynamics study of a surfactant-mediated decane-water interface: effect of molecular architecture of alkyl benzene sulfonate, The Journal of Physical Chemistry B 108 (2004) 12130–12140.
[112] P.H. Doe, W.H. Wade, R.S. Schechter, Alkyl benzene sulfonate for producing low interfacial tensions between hydrocarbons and water, Journal of Colloid and Interface Science 59 (1977) 525–531.
[113] W.D. Cormell, P. Cieplak, C.I. Bayly, et al., A second generation force field for the simulation of proteins, nucleic acids, and organic molecules, Journal of the American Chemical Society 117 (1995) 5179–5197.

[114] B.R. Brooks, R. Bruccoleri, B.D. Olafson, et al., CHARMM: a program for macromolecular energy, minimization, and dynamics calculations, Journal of Computational Chemistry 4 (1983) 187–217.

[115] A.K. Rappe, C.J. Casewit, K.S. Colwell, et al., UFF, a full periodic table force field for molecular mechanics and molecular dynamics simulation, Journal of the American Chemical Society 114 (1992) 10023–10035.

[116] S.L. Mayo, B.D. Olafson, W.A. Goddard III, DREIDING: a generic force field for molecular simulations, The Journal of Physical Chemistry 94 (1990) 8897–8909.

[117] A. Warshel, M. Levitt, Theoretical studies of enzymic reactions: dielectric, electrostatic and steric stabilization of the carbonium ion in the reaction of lysozyme, Journal of Molecular Biology 103 (1976) 227–249.

[118] A.C. van Duin, S. Dasgupta, F. Lorant, F., et al., ReaxFF: a reactive force field for hydrocarbons, The Journal of Physical Chemistry, A 105 (2001) 9396–9409.

[119] T.P. Senftle, S. Hong, M.M. Islam, et al., The ReaxFF reactive force-field: development, applications and future directions, Nature: Computational Materials 2 (2016) 15011.

[120] S.J. Marrink, H.J. Risselada, D.P. Yefimov, et al., The MARTINI force field: coarse grained model for biomolecular simulation, The Journal of Physical Chemistry B 111 (2007) 7812–7824.

[121] L. Darre, M.R. Machado, P.D. Dans, et al., Another coarse grain model for aqueous solvation: WAT FOUR? Journal of Chemical Theory and Computation 6 (2010) 3793–3807.

[122] A. Korkut, W.A. Hendrickson, A force field for virtual atom molecular mechanics of proteins, Proceedings of the National Academy of Sciences of the United States of America 106 (2009) 15667–15672.

[123] J. Barnoud, L. Monticelli, Coarse-grained force fields for molecular simulations, Methods in Molecular Biology 1215 (2015) 125–149.

[124] F. Guo, Y.S. Wan, S.Q. Feng, et al., Intelligent-ReaxFF: evaluating the reactive force field parameters with machine learning, Computational Materials Science 172 (2019) 109393.

[125] Z.E. Hughes, J.C. Thacker, A.L. Wilson, et al., Description of potential energy surfaces of molecules using FFLUX machine learning models, Journal of Chemical Theory and Computation 15 (2019) 116–126.

[126] R. Jinnouchi, J. Lahnsteiner, F. Karsai, et al., Phase transitions of hybrid perovskites simulated by machine-learning force fields trained on the fly with Bayesian inference, Physical Review Letters 122 (2019) 225701.

[127] Q. Hu, M. Weng, X. Chen, X, et al., Neural network force fields for metal growth based on energy decompositions, The Journal of Physical Chemistry Letter 11 (2020) 1364–1369.

[128] J.S. Smith, O. Isayev, A.E. Roitberg, ANI-1: an extensible neural network potential with DFT accuracy at forcefield computational cost, Chemical Science 8 (2017) 3192–3203.

[129] R. Ramakrishnan, P.O. Dral, M. Rupp, et al., Big data meets quantum chemistry approximations: the machine learning approach, Journal of Chemical Theory and Computational 11 (2015) 2087–2096.

CHAPTER

Gas flooding: Gas Enhanced Oil Recovery (G-EOR) to CO_2 sequestration

6

Birol Dindoruk[1] and Russell Johns[2]

[1]*Department of Petroleum Engineering and Chemical and Biomolecular Engineering, University of Houston, Houston, TX, United States*

[2]*Department of Energy and Mineral Engineering, Pennsylvania State University, University Park, PA, United States*

6.1 Introduction

Gas flooding involves many aspects of engineering, from surface facilities to production engineering to reservoir engineering. All elements of such a broad scope are not possible to capture in detail in this chapter and therefore we focus largely on reservoir engineering aspects of the problem. A very detailed examination of all aspects of classical gas flooding from the surface to the reservoir can be found in the SPE monograph by Jarrell et al. [1] as well as the SPE monograph by Stalkup [2]. There are also newer theories and their applications, especially in the context of miscibility development and assessment as summarized by Dindoruk et al. [3]. Some of the renewed momentum in gas flooding also comes from the potential for subsurface CO_2 sequestration and other gases (H_2, CH_4, etc.) storage to lower the overall footprint of greenhouse gases. More than ever, there is a wider envelope of applications for petroleum engineers that includes coupled CO_2 sequestration and geothermal energy production [4].

6.2 What is gas flooding and what does it achieve?

Increasing oil recovery using improved oil recovery (IOR) and/or enhanced oil recovery (EOR) has two main components. The first is to increase the contacted area and/or volumetric sweep, while the second is to reduce the residual oil in the contacted area. We can view this as the aggregate of the macroscopic and microscopic sweep. This is especially important in terms of interpreting the laboratory experiments. Often, these two terms IOR and EOR are used interchangeably, even though there are conceptual differences like how the remaining oil will be recovered or whether the process will lead to altering the residual oil saturation. For most EOR methods, the main objective is to reduce/mobilize the residual oil

Recovery Improvement. DOI: https://doi.org/10.1016/B978-0-12-823363-4.00005-4

saturation by changing the contacted in situ fluid properties by interacting with the in situ fluids. For example, in many thermal methods, among other aspects, in situ oil viscosity is changed by the thermal energy delivered. Gas flooding can be implemented in both ways, as an IOR and EOR method. If the gas is injected to provide energy to the reservoir to recover the oil while not appreciably changing the residual oil, which normally happens at lower pressure and/or where the phase behavior between the injected fluid and the oils is minimal, then this flood is more of an IOR process than an EOR process. If gas is injected at higher pressures or if the gas and oil have high levels of mutual solubilities in a favorable way to reduce the residual oil, which is normally quantified in the laboratory with miscibility pressure, then the flood is an EOR process. Clearly, there could be a succession of various flooding stages between these two application limits of IOR and EOR while the main objective remains to recover the maximum amount of oil in practical ways.

Fluids injected into reservoirs using EOR are physically and chemically different from in situ oil to be displaced. In general, a mixture of the injectant and in situ oil forms more than one phase. When more than one phase exists, capillary forces affect the flow in the porous media. Capillarity is related to the interfacial tension (IFT), for example, for a capillary tube, capillary pressure, P_c, is defined by $P_c = 2\sigma \cdot \cos\theta/r$, where σ is the IFT, θ is the contact angle, and r is the pore or tube radius [5]. High capillarity causes low local microscopic displacement efficiency. The obvious approach to a total or high oil recovery on a pore-to-pore basis would be to reduce the IFT to zero [2,5,6]. Fig. 6.1 conceptually explains how the residual oil saturation can be reduced, through the capillary number, where the capillary number N_{Ca} is defined by: $N_{Ca} = u \cdot \mu/\sigma$, where u is the interstitial velocity, μ is the viscosity of the injectant, and σ is the IFT.

According to Fig. 6.1, lowering IFT while keeping the other variables, velocity and viscosity, constant increases the capillary number and at the limit (IFT $\rightarrow$ 0) capillary number becomes infinite, resulting in very low or zero residual oil saturation for the contacted oil (bypassed oil can remain). A similar plot using the data from Berea sandstone and a Rocky Mountain Rock samples is also shown by Stalkup [2]. When the injected fluid and the reservoir oil are miscible, the IFT vanishes, so that a miscible displacement has high displacement efficiency. To achieve high displacement efficiency is the main goal of EOR. As IFT is one of the fundamental fluid properties, many laboratory studies measure either IFT as a direct measurement or as the outcome of low IFT for the displacement processes under consideration.

In a simplistic yet conceptual way, one may seek conditions that can lead to high capillary number conditions for various processes. In terms of the constituents of the capillary number, Fig. 6.1 shows that the capillary number has to be increased exponentially to lower the residual oil to zero or near zero. In terms of physical limitations of the constituents of the capillary number, IFT is the only physical parameter that can be modified this significantly. If an EOR process can lower the IFT to zero or near zero, nearly all trapped oil contacted can be displaced.

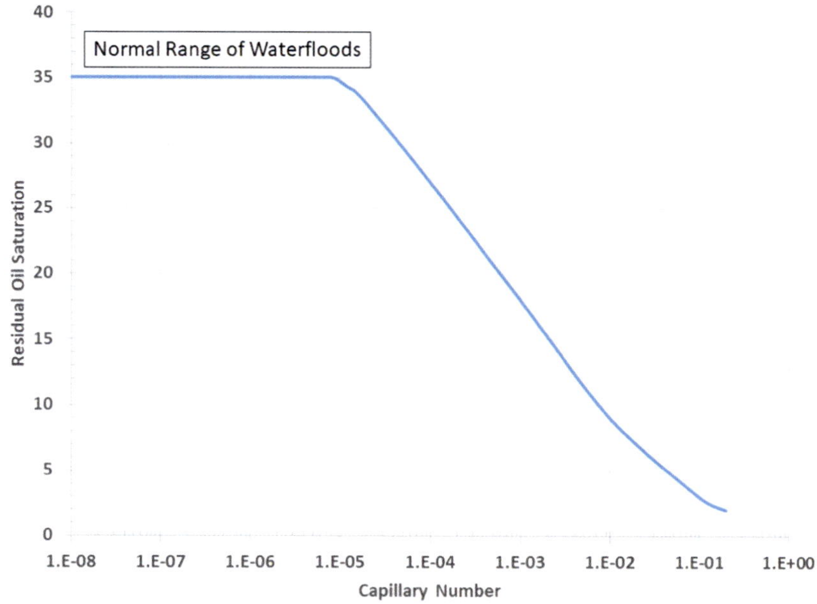

FIGURE 6.1

Dependence of residual oil saturation on capillary number (N_{Ca}).

Gas flooding is the injection of hydrocarbon or nonhydrocarbon components into oil reservoirs, and sometimes even into gas/gas-condensate reservoirs. Injected components are usually in a vapor form (gas phase) at separator and/or atmospheric temperature and pressure and may include mixtures of hydrocarbons from methane to butane, and nonhydrocarbon components, such as carbon dioxide, nitrogen, and even toxic but naturally occurring gases, such as hydrogen sulfide. Such gaseous streams can also include low levels of heavier hydrocarbon compounds especially if separator gases and/or recycled gas streams are considered. One of the key measures of success in injecting fluids into reservoirs is to displace the maximum amount of contacted oil (high displacement efficiency). Therefore we look for the combinations of conditions that can lead to such displacement efficiencies. In doing that laboratory studies will often be the starting point while considering various analog systems, if available.

In general terms, gas flooding is the injection of hydrocarbon or nonhydrocarbon components into oil reservoirs that are typically waterflooded to residual oil while it can also be applied, but with more real or perceived financial risk, in combination with the primary production. Injected components are usually gases at atmospheric temperature and pressure and may include mixtures of hydrocarbons from methane to propane, and also carbon dioxide, nitrogen, and even hydrogen sulfide. Miscible agents are also known as solvents. In some cases, certain gaseous solvents, such as dimethyl ether compounds, can also be considered.

These components are typically referred to as novel solvents [7–9]. Each gaseous injectant, whether found in nature or not, often has a qualitative similar impact on phase behavior in various ways. However, their interactions with in situ fluids may have significant quantitative differences for a given pressure, temperature, and reservoir fluid. Although these components are usually gases at atmospheric temperature and pressure, they typically are supercritical fluids at reservoir temperature and pressure in that some of their properties may be more liquid like. Carbon dioxide, for example, has a density similar to oil, but a viscosity more like gas at pressures greater than 1000 psia. Gas injection today often means CO_2 injection to recover residual oil and potentially to store or sequester CO_2 streams from various processes or atmospheric CO_2.

6.3 Symptomatic improvement approaches

The key to gas flooding is to contact as much of the reservoir with the gas as possible and to recover nearly all oil once contacted. Thus gases are designed to be miscible with the oil so that oil previously trapped by capillary forces is released into the injected or contacted "gas" phase. The gas phase then carries the oil components to the production well. Ideally, flow is piston like in that whatever gas volume is injected displaces an approximately equal volume of in situ reservoir fluid. Unfortunately, in real fields, miscibility is not always possible and reservoir heterogeneities can cause gas to cycle through one or more layers, which results in poor sweep efficiency (which is relevant for any type of flooding exposed to reservoir heterogeneities). A proper gas flood design will consider both microscopic and macroscopic displacement efficiency, as they both impact the profitability of the project (see Chapters 1–3 for discussion on key factors that affect the displacement and sweep efficiency). Each reservoir is unique so that an engineer must have a good understanding of process basics.

In many EOR methods or methodologies, it is very common to improve the performance of a particular method using symptomatic approaches. For example, for gas injection processes, high mobility of the injected fluid (gas) with respect to the in situ fluids, especially in the presence of reservoir heterogeneities, can lead to lower levels of sweep in the reservoir, especially in combination with the gravity effects that may push the gas toward the top of the reservoir. Mobility contrast between the injectant and the in situ fluids is usually quantified with the term called, mobility ratio. Mobility ratio is expressed as mobility of displacing fluid divided by mobility of displaced fluid. Mobility is the ratio of relative permeability to viscosity of one phase.

$$\lambda_j = \frac{k_j}{\mu_{ij}} \tag{6.1}$$

where

λ_j is the mobility of phase j (1/cP),

k_j is the relative permeability of phase j (dimensionless), and
μ_{ij} is the viscosity of phase j (cP)

$$M = \frac{\lambda_D}{\lambda_d} \tag{6.2}$$

where

M is the mobility ratio, fraction (dimensionless),
λ_D is the mobility of displacing fluid (1/cP), and
λ_d is the mobility of displaced fluid (1/cP).

If we are injecting gas to displace oil, the mobility ratio equation will be:

$$M = \frac{\lambda_g}{\lambda_o} = \left(\frac{k_g}{\mu_g}\right) / \left(\frac{k_o}{\mu_o}\right) = \frac{k_g \mu_o}{k_o \mu_g} \tag{6.3}$$

Various options are tried to reduce M including foams (with and without nanoparticles), direct thickening and polymers and gels. Here, we will discuss two distinct approaches briefly. When gas is injected into a reservoir, due to gravitational differences with respect to the in situ oil it has a tendency to rise and segregate to the top of the reservoir, thereby bypassing some of the remaining oil. While the density difference between the gases and the in situ oils is the main factor for this, injected gas can also finger through the oil because of its high mobility and lower viscosity. One way to control the gas mobility and improve its sweep efficiency (and thus improve the incremental oil recovery) is by injecting the gas in the foam form, or emulsion form. The foam, because of its high viscosity, diverts the gas away from the fast paths (thief zones) already created. Foam will block the flow channels in a relative sense (with respect to gas it creates much higher resistance) while forcing gas to the unswept areas and therefore increasing the reservoir sweep.

- *Viscosification of the injected gas*

  One of the ways of controlling the mobility of the gas is to increase its viscosity by adding various chemicals and/or nanoparticles. This particular topic has been an interest to researchers for a long time and a recent review paper by Enick et al. [10] covers various details of mobility and conformance control aspects of CO_2 EOR via thickeners, foams, and gels. Although the subject paper [10] mainly focuses on CO_2 as the injected gas, the topic is also applicable to other injections. One of the most discussed and studied approach has been foam injection into the reservoir to promote the displacement process, especially by remediation of the adverse mobility ratios while some of the applications mainly focus on fracturing and as well as conformance control aspects of the injected fluids. Overall, there is no known economic field scale, displacement type of additives as gas thickeners for gas injection processes. Gels and foams have been practically applied for various treatment, fracturing, etc., operations because they can be injected in the aqueous phase.

- *Gas injection design and water-alternating gas (WAG)*

  One of the classical symptomatic improvement approaches is to alternate gas and water injection. Fig. 6.2 illustrates a typical gas flooding process where CO_2 or another solvent is injected into several slugs alternating with water. This process, which is known as WAG, can significantly improve sweep efficiency since water is less mobile (greater viscosity) than CO_2 and hence improves the average mobility ratio of the flood. More favorable mobility ratios mean that less gas cycles through the high permeability layers than it otherwise would. One advantage of WAG outside of mobility control is that less expensive gas is injected in favor of cheaper water or chase fluid. WAG has proven to be very effective as a mobility control measure and is nearly always used.

Although WAG increases recoveries, due to gravitational differences gas can still tongue upwards in the formation during the gas injection cycle, while water can move downward in the water cycle (Fig. 6.3). This segregation of fluids will occur when there is sufficient vertical permeability (large aspect ratio) and

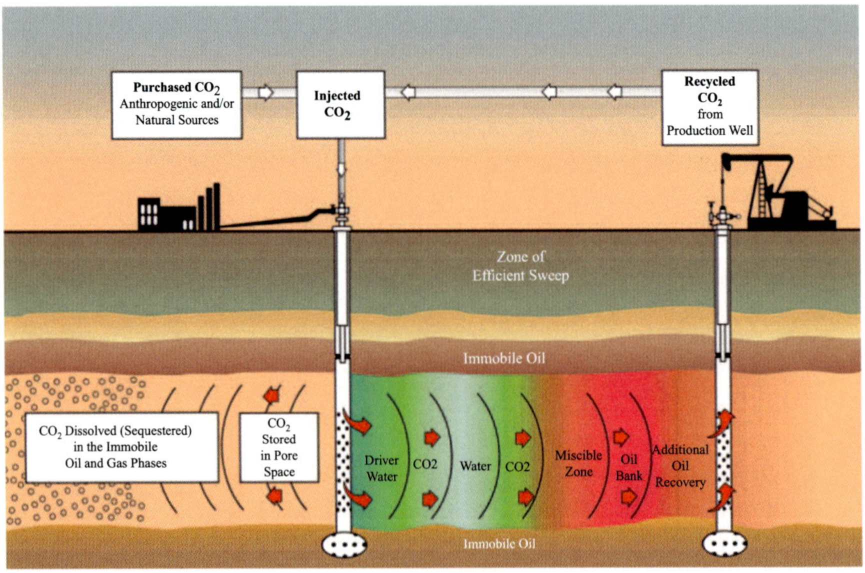

FIGURE 6.2

Illustration of gas flooding process showing a typical water-alternating gas process where carbon dioxide is injected in several slugs followed by water (right) and the illustration of CO_2 injection for CO_2 sequestration purposes into an oil reservoir (left) [11].

Reproduced with permission from L.W. Lake, M. Lotfollahi, S.L.Bryant, CO_2 Enhanced Oil Recovery Experience and its Messages for CO_2 Storage, in: P. Newell, A.G. Ilgen (Eds.), Science of Carbon Storage in Deep Saline Formations: Process Coupling across Time and Spatial Scales, Elsevier, 2018, pp. 15–31.

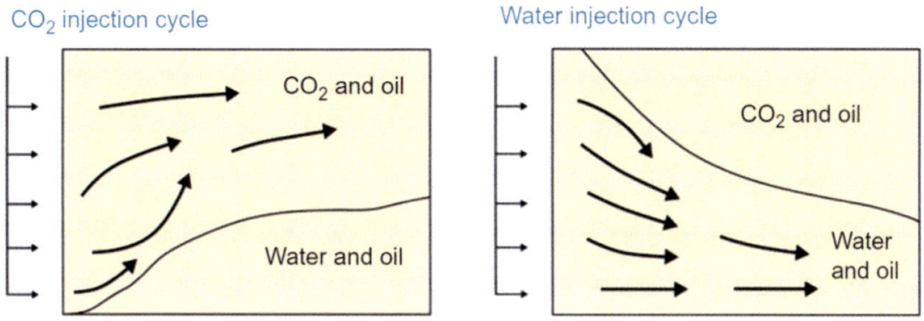

FIGURE 6.3

During water-alternating gas, gas can move upwards owing to its low density, while injected water can move downwards [12].

Reproduced with permission from R.T. Johns, B. Dindoruk, Gas flooding, in: J. Sheng (Ed.), Enhanced Oil Recovery Field Case Studies, Elsevier, 2013, pp. 1–22.

sufficient gravity forces (large gravity number). It is important to adjust the volume of water and gas injected so that a maximum in the displacement efficiency is achieved. Too much water or too much gas can result in poor sweep efficiency. Reservoirs that have smaller permeability at the top generally perform better than those with larger permeability at the top since low-density gas wants to gravitate to the top of the reservoir overcoming low permeability there.

Gas flooding can be implemented in a variety of ways either as continuous gas injection, continuous gas injection chased with water (or possibly another gas), conventional WAG, or tapered WAG (TWAG) as illustrated in Fig. 6.4. For WAG, the total volume of gas injected should be determined along with the frequency (number of gas–water cycles) and the water–gas ratio (volume of water divided by the volume of gas injected in each cycle). In West Texas, gas injection is sometimes tapered so that more gas is delivered up front but is quickly changed to water injection once gas breaks through a production well. Tapering ensures that the gas slug size is sufficient to sweep between wells. Simultaneous water–gas injection (SWAG) has also been attempted, although this requires significant monitoring owing to injection of multiple phases. How gas is delivered must be examined for each reservoir under consideration since each reservoir and oil is unique. Detailed parametric analysis of WAG was studied by Babayev [13].

Other mobility control methods include foams, which are injected primarily into the highest permeability layers. Unlike WAG, as mentioned earlier in the context of viscosification of injected gas, foam is still in the research stage for large-scale application. The idea is to create stable CO_2 foam by injection of a small amount of surfactant in either the gas or more typically the water phase. If a stable foam can be generated within the "thief zones" or high permeability layers it can restrict flow through them causing the gas to be diverted to other less-permeable layers. Other potential conformance methods include gels as

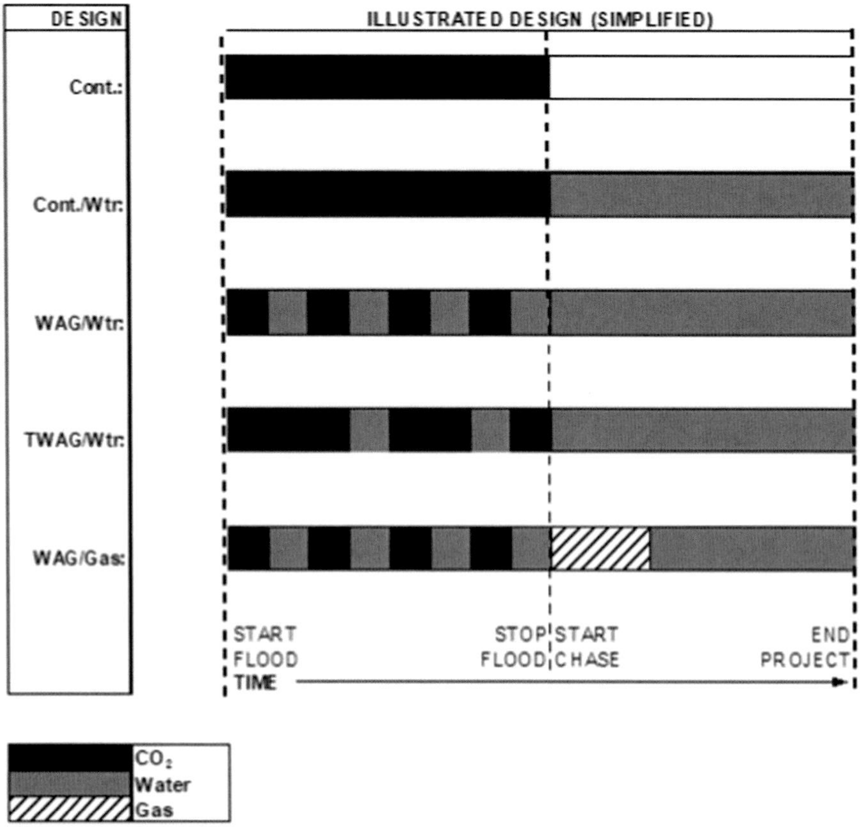

FIGURE 6.4

Water-alternating gas floods can take on many forms. "Cont." = continuous gas injection; "Cont./Wtr" = continuous gas injected chased with water; "WAG/Wtr" = conventional WAG flood chased with water; "TWAG/Wtr" = tapered alternating gas and water chased with water; and "WAG/Gas" = conventional WAG chased with gas.

outlined in detail by Green and Willhite [14]. All conformance measures (and gas injection in general) can suffer from injectivity problems in that water or gas may not be injected in sufficient quantities and in some cases existence of additional phases can cause impediment to flow for the others. Injectivity tests or pilot tests are often performed in the field to reduce the risk associated with these methods.

WAG injection for various schemes was studied by Babayev [13]. In the past expensive solvents have been tried during miscible WAG injection. At the present time cheaper and relatively more abundant solvents, such as N_2, CO_2, and methane, are more attractive to achieve more cost-effective recoveries. All these gases have their own MMP (minimum miscibility pressure) for each oil composition and reservoir temperature. For light or lean reservoir gases, reservoir pressures tend to be less than

MMP, especially when the reservoir is under depletion (pressure decline), A side benefit of gas and water injection is to repressurize the reservoir to reach higher pressure and hopefully to MMP (pressurization). The pressure-up phase is also important to increase the reservoir energy (and also drive the oil to producers) to produce as much of oil volume as possible [15]. Comparison of widely used pure gases for miscible injection, CO_2, is generally the best choice for several reasons. First of all, the availability of CO_2 makes this gas cheaper than propane, butane, etc., which are normally refined (separated from streams) products in large quantities. Also, the molecular weight of CO_2 is 44 [as the mass density is $P \cdot \mathrm{MW}/(Z \cdot R \cdot T)$] according to the real gas equation (where P is pressure, MW is molecular weight, T is temperature, R is the gas constant, and Z is gas compressibility factor), which sufficiently large (i.e., part of the reason behind the relatively higher CO_2 densities), for miscible processes, since it has lower minimum miscibility pressure as compared to methane, or any other light hydrocarbon gases, for the same oil composition. Moreover, the viscosity and density of CO_2 under reservoir conditions are more favorable than methane. But along with CO_2 other gases or mixtures of gases (blends) can also be efficiently used, and a detailed study is needed to find the best or optimum gas composition. Having said that CO_2 is the best relative to the N_2 or hydrocarbon gases, there are some other gases that can be applied locally that can be even better agents than the CO_2. For example, H_2S, while being very toxic, has a better miscibility character with respect to CO_2. In the following, we will briefly explain various types of WAG processes.

A variant of WAG injection is the immiscible WAG injection. This type of WAG process has been applied with the aim of improving frontal stability or for more efficiently contacting unswept zones. The main difference between the immiscible WAG and miscible WAG injection is that there is no significant mass exchange between the in situ fluid (oil) and the injectant (gas).

Hybrid WAG injection is based on a large slug of gas injection, followed by a number of smaller slugs of water and gas. Generally, the first gas slug is expensive hydrocarbon gas or carbon dioxide. The main purpose of this slug is to contact the oil phase. The following gas slugs consist of cheaper and lighter gases. This type of WAG injection has been used in the Dollarhide Devonian field [16], San Andres formation, West Texas [15].

Simultaneous water and gas injection is the process when water and gas are injected simultaneously, at the same time. There are two SWAG injection techniques. In one technique, water and gas are mixed on the surface and then are injected into the formation via one injection well. Another technique assumes no mixing of the surface; the two phases are injected separately into the selected formation's layers. This type of injection has been used on Siri Field, Kuparuk River field, Joffre Viking field, etc. [17].

Gas flooding can also be implemented in a single well. In a standard gas flood, gas is transported through the formation from one well to another. In a cyclic single-well treatment, gas is injected into a single well and then shut in. That same well is then produced after the gas is allowed to soak. This process may be repeated several times. This single-well method is generally not as good in recovering oil as multiwell gas flooding but is increasingly being used in an immiscible process to reduce viscosity and swell heavier oils [1], and more recently for EOR in tight shale reservoirs.

6.4 Phase behavior and experiments

Phase behavior controls whether an oil and gas mixture will be miscible at reservoir temperature and pressure. Thus an important step in any gas flood design is to properly characterize the fluid over a wide range of reservoir pressures and temperatures. This is also a key to optimizing production as the fluid properties change from the reservoir to the surface facilities.

6.4.1 Standard pressure volume temperature data

Pressure volume temperature (PVT) experiments are nearly always performed to determine the relationships between pressure, volume, and temperature for the fluid of interest. There are four key experiments that are done on most fluids, independent of whether gas is to be injected or not. These are constant-mass expansion tests, constant-volume depletion tests, differential liberation tests, and separator tests. These tests duplicate in some limited way the oil recovery process that may be occurring in a reservoir. A complete description of these tests can be found in McCain [18] and Pedersen and Christensen [19]. These classical black-oil experiments are still very important and the primary, secondary, and as well as uncontacted area flows are controlled by those parameters.

A constant-mass expansion test gives important information on the saturation pressures of the reservoir, which are the pressures at which a second phase will form. For black-oil reservoirs, we are most interested in the bubble-point pressure, where the bubble-point pressure is the pressure at which a bubble of gas first forms [20]. For gas-condensate reservoirs, dew-point pressures are more important because as pressure is reduced liquid forms (drops out), which can constrict the flow of gas to the production well. For purposes of this chapter, we consider only black or volatile oils where the reservoir temperature is smaller than the mixture critical temperature. They are the main targets of gas injection processes while there are various applications of enhanced gas recovery (EGR) processes and gas reinjection processes for condensate recovery optimization (see [21] for a field case implementation). Some of the aspects of the gas–gas displacement from an experimental point of view are studied, for example, by Turta et al. [22] and Sim et al. [23] for dry gas systems. Some of the nonequilibrium aspects of gas injection in condensate reservoirs are studied in detail by Huang et al. [24]. Furthermore, one of the condensate mature case analysis for the Lian 4 fault block was studied by Feng [25]. A comprehensive review on CO_2 EGR and sequestration in depleted gas reservoirs is completed recently by Hamza et al. [26].

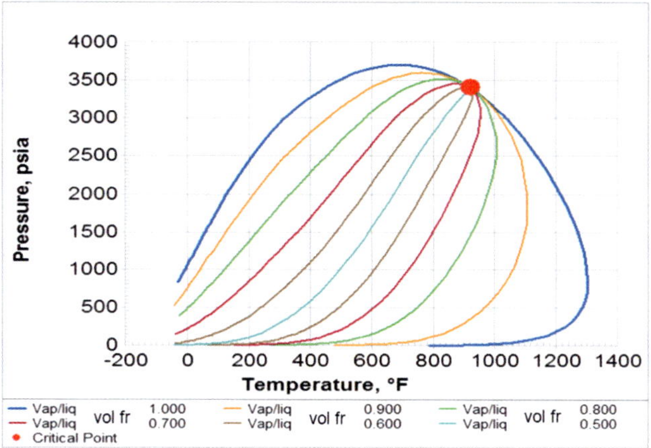

FIGURE 6.5

Pressure–temperature diagram for a typical black oil.

Fig. 6.5 illustrates a PT diagram for typical black oil at a reservoir temperature of T_r. As shown the bubble-point pressure P_b is reached as the pressure is reduced from the reservoir pressure P_r. Vapor is formed below the bubble-point pressure as given by the quality lines (constant % of vapor). The location and the magnitude of bubble-point pressure plays important role in gas injection processes, especially in the context of miscibility development.

6.4.2 Swelling test

When the gas injection is planned, a swelling test is often done where the selected injection gas is mixed with the oil. A swelling test is used to determine how much gas dissolves in the oil, how much the oil will swell as a result, and the resulting bubble-point pressures as injection gas is added. Equilibrium phase compositions can also be measured for several overall compositions (or per mixing ratio of oil and added gas) during the test.

A swelling test begins with reservoir oil at its bubble-point pressure. At the reservoir temperature, a fixed amount of gas is mixed with the oil. The pressure is then increased maintaining the same reservoir temperature until all gas is solubilized into the oil. When the last gas bubble is dissolved, the oil–gas mixture is at its new bubble-point pressure (as a result of the added gas). The pressure and volume of the mixture are then measured. The new volume is called the "swollen" volume. Although swelling is of secondary importance in miscible gas flooding, some oils can nearly double their volume when contacted by CO_2. Swelling could therefore cause some oil previously trapped by capillary forces to flow within the reservoir. This process of adding gas and measuring the new bubble-point pressure is repeated several times at different percentages of gas injected.

If a lot of gas is added, a dew point could be reached as pressure is increased. Obtaining a dew-point pressure is ideal, as the mixture critical point would lie between the last bubble-point pressure measured and this dew-point pressure. This brackets the critical point for fluid characterization using a cubic equation of state (EOS). Gas and oil viscosities are often also measured in swelling tests as these are very important in determining fluid mobilities and mobility ratios.

Fig. 6.6 illustrates pressure–temperature and pressure–composition diagrams for gas A and crude oil. The measured bubble-point pressure for each composition is shown with a solid dot and is given as P_{b1}, P_{b2}, and so forth. The pressure–composition diagram is easily derived from the saturation pressures of the PT diagram at a fixed temperature. The composition axis of the pressure-composition diagram is the mole fraction of the added injection gas A in the oil–gas mixture. Many times, the overall structure of the envelope is captured with 5–6 sparse experimental data points with at least two points being at high mixing ratios leading to a point to the right of the critical point (dew-point mixtures). In terms of direct use cases, this experiment can yield first-contact miscibility pressure for systems with closed envelopes. However, the use of the data in combination with

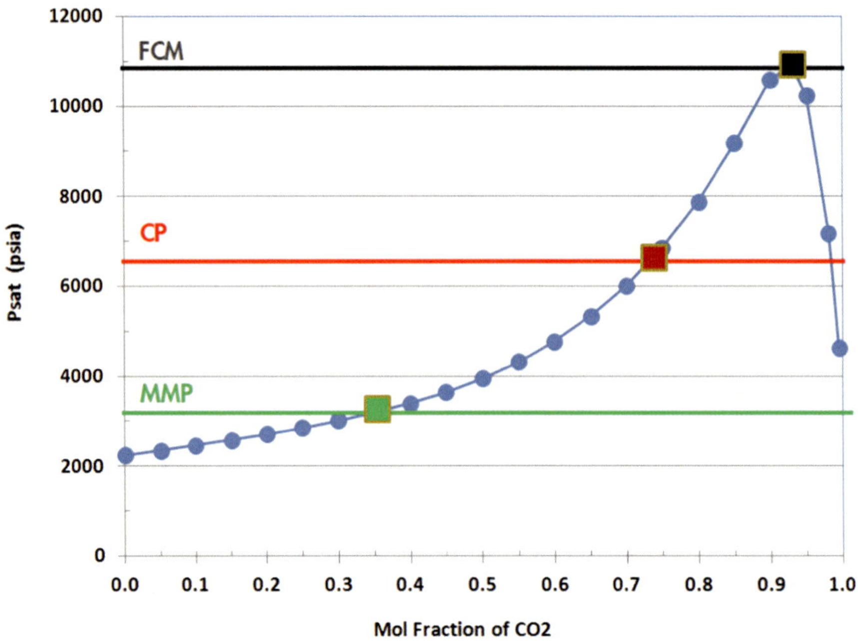

FIGURE 6.6

Phase envelope generated for a single-contact experiment for a CO_2 reservoir oil system. First-contact miscibility pressure is the apex of the system indicated with the black square. Any pressure at or above first-contact miscibility pressure will lead to a single-phase system for any mixture formed between the reservoir oil and the injectant (CO_2).

the basic PVT data is extremely useful to constrain the EOS for the subject gas and its interactions with the in situ hydrocarbons, especially when there are no slim-tube data is available.

6.4.3 Slim-tube test

Slim-tube experiments are very useful to define the minimum pressure for miscibility using crude oil obtained from the field. The slim tube consists of a very slim (small ID) coiled tube filled with core or glass bead material. The tube can be quite long to allow miscibility to develop some distance from the injection point. The tube is first saturated with a known volume of oil. The temperature is then usually kept fixed at the reservoir temperature. Gas is injected through the tube to displace the oil and the amounts of gas and oil are recorded with time. The recovery is defined as the ratio of the volume of oil produced to the initial oil volume. This is known as the pore volumes recovered. Once a very large recovery is obtained, the procedure is repeated for higher pressure. Typically, the pore volumes recovered are plotted against the pressure at a fixed time, say at 1.2 pore volumes injected. The lowest pressure at which the recovery is about 90 to 95% is often used to define the MMP for any oil–gas system (see Fig. 6.7) or when a "break" in recovery versus pressure plot is observed.

Sometimes MMP is referred to as "dynamic miscibility pressure" as it is obtained from a system in the presence of flow. The presence of flow will have an impact on the composition of the mixtures (or grading/transition) between the

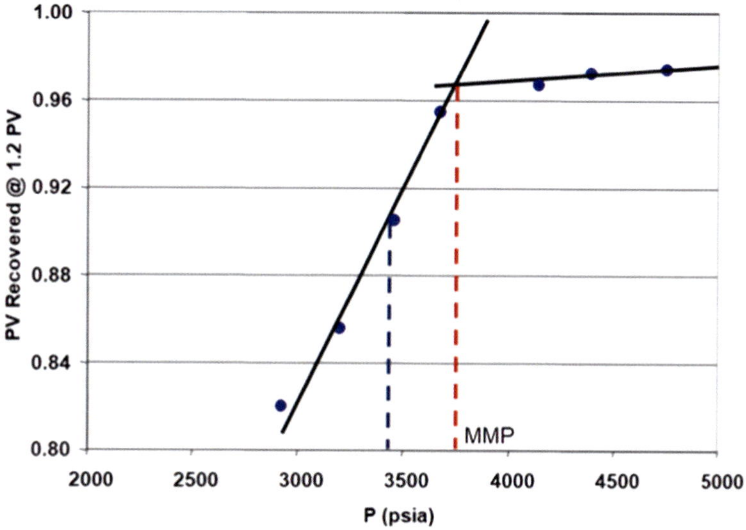

FIGURE 6.7

Example slim-tube recoveries for an oil displaced with pure CO_2.

injected fluid and the initial fluid. Interpretative definition of MMP is arbitrary depending on what criteria are adopted, however, so there is some uncertainty in the MMP. For further discussion of the MMP criteria see the review article by Dindoruk et al. [3]. Nevertheless, if the reservoir pressure exceeds this "MMP" good recovery of contacted oil will likely be obtained. Dispersion in the field, however, can degrade the oil recovery by reducing the effective strength of the gas [27]. Slim-tube experiments are also very time-consuming and expensive so that few of them can be made. Many times, they are not performed and as a result, many correlations and numerical models were developed. Again, readers can look at the details presented in Dindoruk et al. [3] for potential mechanistic (mass transfer mechanisms) issues of some of the correlative techniques and as well as why CO_2 is treated somewhat different than many of the hydrocarbon gases as injectants.

6.4.4 Multicontact test

The interaction of flow and phase behavior is critical to determine the MMP for real fluids. The multicontact test attempts to mimic this interaction within the reservoir. Gas is again mixed with oil, but this time the pressure and temperature are fixed so that as gas and oil are mixed two equilibrium phases may result. The two phases can have different mobilities so that one would flow ahead of the other within the reservoir. The liquid phase that forms is nearly always of lower mobility so that it is contacted by fresh gas in what is termed a backward contact. Alternatively, the gas phase that forms will then contact fresh oil in a forward contact. This process is repeated over again in repeated steps. Forward and backward contacts are well described in both Stalkup [2] and Lake et al. [28]. In the multicontact test, however, only forward or backward contacts are measured, but not both simultaneously. The multicontact test is very useful for vaporizing displacements since miscibility for vaporizing drives is developed using forward contacts as is discussed in the following section. A few equilibrium phase compositions can also be measured in these tests for potential use in fluid characterization.

6.4.5 Fluid characterization using an equation of state

PVT and gas injection experiments do not provide the full range of data that are required by compositional simulation. For example, swelling tests are conducted only along the mixing line between the injected gas and oil fluid compositions. Thus mathematical models called EOS are used to fill in these data gaps. These models are tuned to the available experimental data in a very complicated and often subjective process. The difficulty arises because oils consist of hundreds of components and isomers, many of which cannot be precisely identified experimentally. In addition, compositional simulation with EOS is computationally intensive so that fewer components are required, generally less than fifteen. Many components and isomers are then lumped into pseudocomponents. Each pseudocomponent must be assigned values for EOS properties, such as critical temperature and pressure. Because the components are not known precisely, these properties are tuned to match the measured data to achieve as good of a fit as possible without adjusting too many parameters or changing their values outside of a reasonable range. Over fitting can reduce the predictability of the EOS model for data outside the range of the experiments. Fluid characterization and EOS are outlined in detail by Pedersen and Christensen [19], Danesh [29], and Firoozabadi [30]. Egwuenu et al. [31] further showed that tuning to the MMP can significantly improve the fluid characterization. However, many of these models are still somewhat limited in their methodology to the use of basic PVT data. Regardless of the approaches, any adopted model should at least honor the measured data consistently using a logical and physically constrained recipe. Generally, these models will predict differently when they are not constrained to the same data or analog system behavior.

6.5 Minimum miscibility pressure and displacement mechanisms

The primary means of recovery in gas floods is by means of extraction of oil components. There are three types of displacements that can occur depending on the level of extraction. The first type is first-contact miscibility (FCM) where the oil and gas are miscible when mixed in all proportions. Minimum pressure to achieve FCM can be obtained from single-contact experiments and it corresponds to the apex of the phase envelope as shown in Fig. 6.6. This type is very difficult to achieve with many solvents, but if FCM occurs the displacement can extract all contacted oil, leaving no residual oil. Miscibility can also be developed within the reservoir by a multicontact miscibility (MCM) or dynamic process where gas and oil mix in contacts that are forward, backward, or a combination of both. Miscibility in MCM floods is developed when the phase compositions that form in each contact move toward a critical point (see advanced reading box).

The last type of gas flood is an immiscible one, which has more limited, but varying degrees of extraction. The term "immiscible gas flood" is really a misnomer because gas will always extract some oil components. Recovery (or displacement efficiency) improves as a gas flood becomes more miscible (more dominant phase behavior between oil and gas).

6.5.1 Simplified ternary representation of displacement mechanisms

Fig. 6.8 shows a simplified phase behavior at constant reservoir temperature and pressure where only three pseudocomponents are present (light, intermediate, and heavy weight components). In plots of this kind, all possible compositions that form as gas and oil are mixed must lie on a line segment drawn between them. This line is called a mixing or dilution line.

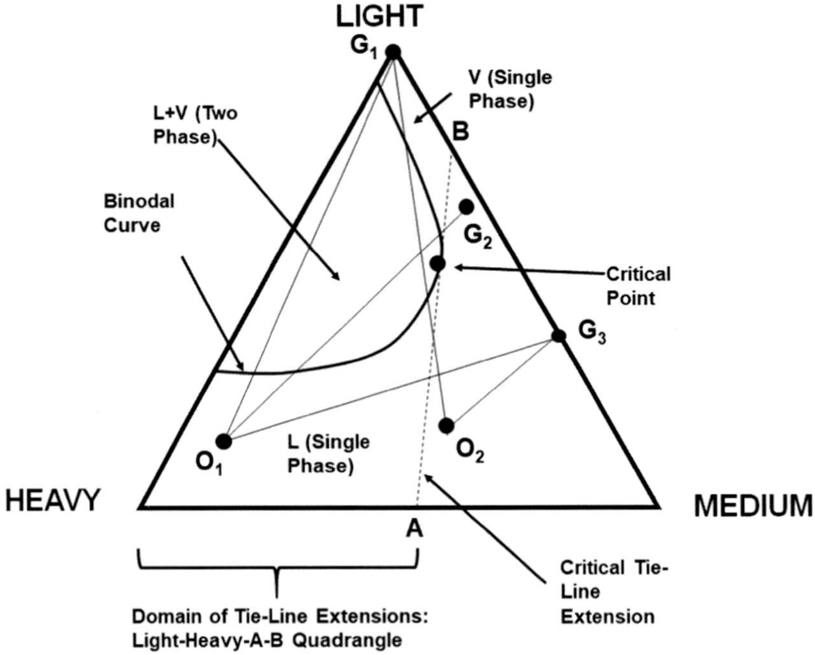

FIGURE 6.8

Ternary representation of gas flooding.

Injection of lean gas G_1 with the heavier oil O_1 (relative to O_1 as O_2 has more heavier components in it) in Fig. 6.8 is immiscible because the mixing line between them traverses a large section of the two-phase region. There is some extraction of oil components in this case, but recoveries in a slim-tube experiment would be low at 1.2 PVI. Gas G_2, however, is multicontact miscible because it lies outside the tie-line extensions as shown in Fig. 6.8 (it is to the right of the critical tie-line/limiting tie-line extension). Contacts of this gas with equilibrium oil would eventually reach the critical point. In addition, such a displacement causes the intermediate component to condense into the equilibrium oil, causing the trapped oil to swell, which helps recovery. This process is known as a condensing drive, where miscibility is developed at the trailing edge of the displacements by backward contacts.

Gas G_3 is first-contact miscible with oil O_1 as its mixing line does not intersect the two-phase region at all. This case is not achievable for most reservoirs and injection gases.

Displacement of a lighter oil, O_1, is multicontact miscible even for the lean gas G_1 because that oil lies outside the critical tie-line extension as is shown. In this type of displacement, known as a vaporizing drive, the intermediate component in the oil is vaporized into the equilibrium gas phase. Vaporizing drives develop miscibility at the leading edge of the displacement because two-phase compositions that result from these forward contacts approach the critical point there. Both gases G_2 and G_3 are FCM with the oil O_2 (as the dilution lines between G_2 and G_3 and O_2 do not intersect the two-phase region).

6.5.2 Displacement mechanisms for field gas floods

The simplified model of Fig. 6.8 does not adequately represent field displacements where more than three components are present. Most real CO_2 and enriched-gas displacements have features of both condensing and vaporizing (CV) drives, in that a complex combination of both forward and backward contacts intersects a critical locus first. In CV displacements, miscibility is developed somewhere between the condensing and vaporizing portions as shown in Fig. 6.9. The relevant analytical theory was developed by Johns et al. in the early 1990s [32].

Swelling of oil occurs near the front of the displacement followed by a series of vaporizing fronts, where heavier and heavier components in the oil are vaporized in a chromatographic-like separation. Individual contributions of each mass transfer mechanism (vaporizing and/or condensing) as the relative weight of condensing and vaporizing segments of the displacement, can be quantified using the methodology developed by Johns et al. [33]. Injection of very lean gases, such as nitrogen or methane, however, are primarily vaporizing drives. In addition, the sensitivity of numerical solutions to each mass transfer mechanism or their combinations is shown in the same study.

6.5.3 Determination of minimum miscibility pressure

The MMP is one of the most important design considerations for a gas flood. Orr et al. [34] showed that the displacement mechanism significantly affects the MMP. For a simple four-component CO_2 displacement, the MMP assuming a vaporizing displacement was 4500 psia, while the MMP for the combined CV displacement was only 1750 psia. Thus incorrectly determining the displacement mechanism could lead to large errors in the MMP (and thus any correlations developed using such outcomes). This is also true for the minimum miscibility enrichment (MME), where the MME is the solvent enrichment required for the reservoir to be at the MMP. For a simple four-component example, the MME calculated assuming a condensing gas drive was 53% solvent enrichment (propane), while the MME for a combined condensing/vaporizing displacement required only 39% propane [32]. Thus the requirement for miscibility in terms of the MMP or MME is less when the correct displacement mechanism of a combined condensing and vaporizing process is accounted for.

There are several proven methods that can capture the correct physics to determine the MMP:

- Slim-tube and multicontact experiments;
- Mixing cell methods;
- Empirical correlations (with the right imbedded physics and data);
- Data-driven models [35];

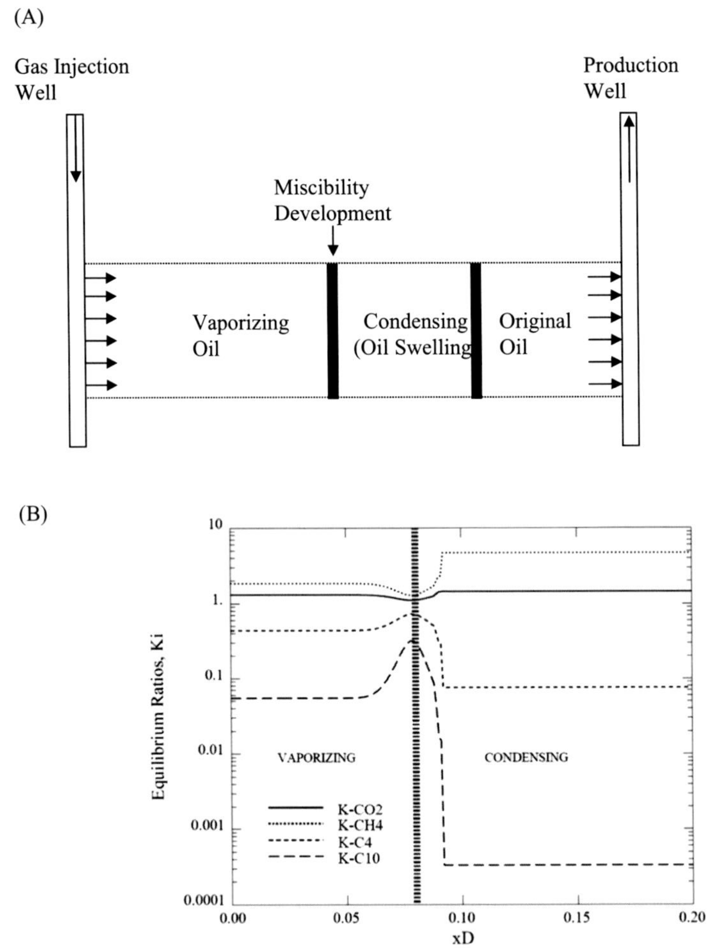

FIGURE 6.9

(A) Location of developed miscibility within the formation for MCM gas floods (a sketch of a condensing/vaporizing drive). (B) Miscibility development can also be seen in terms of converging K-values ($K_i \rightarrow 1$, for all components) for an analog four component systems.

- Compositional simulation of slim-tube displacements; and
- Analytical methods using EOS and the method of characteristics (MOC).

Some of the key points of the first two sets of experiments were explained earlier briefly. A very detailed analysis of MMP (experimental and numerical) is covered recently in Dindoruk et al. [3]. Therefore we will only cover some of the key elements here.

As discussed earlier, experiments are very useful, but are costly and time-consuming to perform. Mixing cell methods for MMP determination can yield good

MMP estimates if the proper displacement mechanism is accounted for and the fluid characterization is accurate [36]. Empirical correlations for determining the MMP can be good depending on whether the reservoir fluid is like those used to develop the correlation [1]. Analytical methods using MOC rely on an accurate EOS that is properly tuned to available PVT data [34]. The advantage of these methods is that they are quick and cheap, but the calculation methodology must be done carefully [37]. Compositional simulation to estimate the MMP is also cheap; yet its result is dependent on correcting for the size of the grid blocks (numerical dispersion error).

Some fluids are more sensitive to dispersion, whether physical or numerical when the mixing line between gas and oil intersects tie lines at large angles. For gas–oil systems that are sensitive to physical dispersion, the gas flood should be operated well above the MMP to achieve good displacement efficiency [27]. See Johns et al. [33] for the sensitivity to the relative weight of each displacement mechanism (what percentage is condensing and what percentage is vaporizing drive). The advanced reading box discusses what is mixing and dispersion.

The MMP increases with increasing temperature and with less gas enrichment. Figures 6.10 and 6.11 give examples of the MMP calculated using MOC for a lean gas (methane) and solvent (CO_2). Figures 6.10 and 6.11 are computed using the approach presented by Yuan et al. [37].

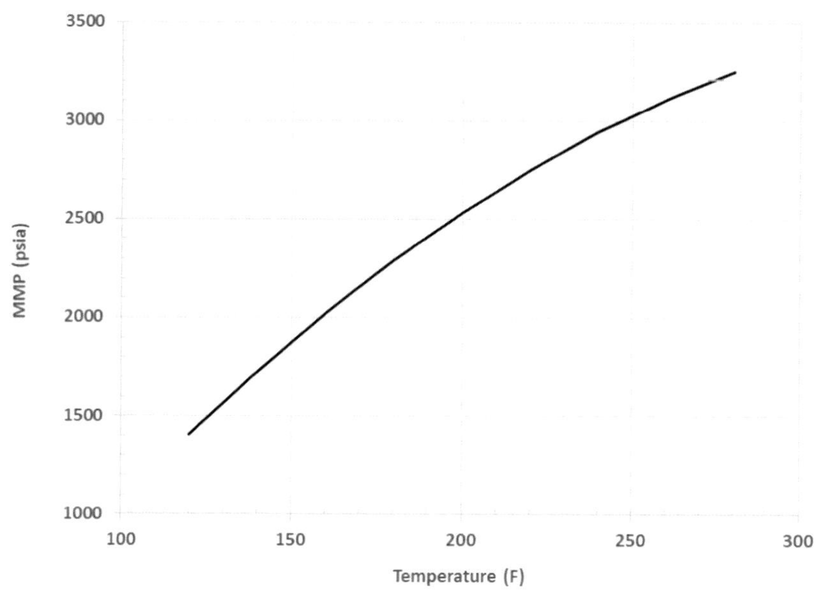

FIGURE 6.10

Effect of temperature on the minimum miscibility pressure for pure CO_2 injection. Subject oil has a molecular weight of C7 + = 148.92 with mole % C2–C6 = 14.11.

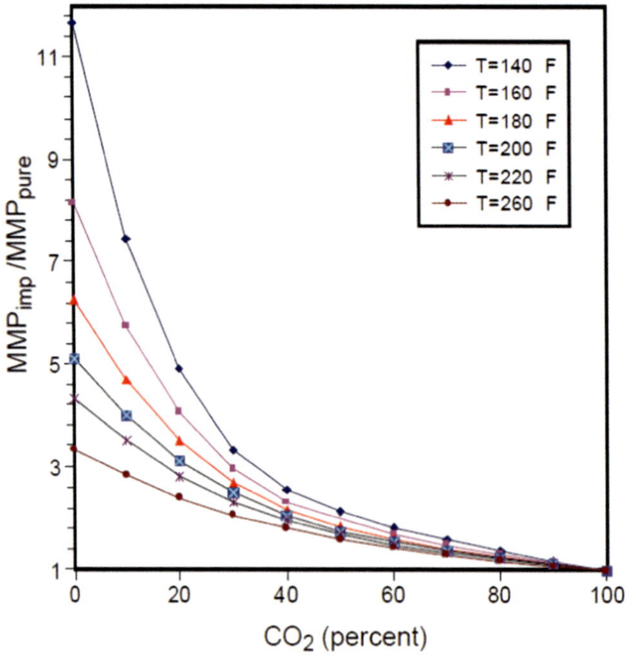

FIGURE 6.11

Effect of temperature and enrichment of lean gas (methane) with solvent (CO_2) on minimum miscibility pressure. The minimum miscibility pressure ratio is the ratio of the actual minimum miscibility pressure at that enrichment and temperature (MMP_{imp}) to the minimum miscibility pressure for injection of pure CO_2 (MMP_{pure}). Subject oil has a molecular weight of C7 + = 148.92 with mole % C2–C6 = 14.11.

As is shown, hot reservoirs will have larger MMPs, while adding solvent to lean gases can significantly decrease the MMP. The MMP with lean gas injection is nearly 11 times higher for cold reservoirs than the MMP with pure CO_2 injection. Although hot reservoirs have much higher MMPs, they show less variation in the MMP with solvent gas enrichment.

6.5.4 Development of miscibility, mixing processes, and the role of dispersion

Most of the fundamental theory of miscibility is developed by Johns [38] and Dindoruk [39] and the aggregate of all the development in that area by Orr and his coworkers is summarized in the book by Orr [40]. Here we will mainly utilize the elements of those studies and studies by the others cited in the literature without going into the mathematical details as summarized by Orr [40]. Miscibility is often developed by a multicontact process where phases that form upon contact with injected gas and oil move toward a critical point. Fig. 6.12 illustrates these contacts starting with fresh oil and injection gas (i.e., analog of the mixing zone shown in Fig. 6.9). A proportion of the gas is mixed with the oil until two equilibrium phases form between them. This is the first contact with the reservoir. These phases are denoted by *X* and *Y*, where *X* is the equilibrium compositions for the liquid (equilibrium oil) phase and *Y* are the equilibrium compositions for the vapor (equilibrium gas) phase. Clearly, if the gas and oil are FCM two phases can never form, so this illustration is only for MCM flow. We assume here that the gas mobility is greater than the liquid mobility so that equilibrium gas always moves ahead of the equilibrium oil as shown below. Flow is from left to right in the diagram, which simulates one-dimensional flow from the injector well to the production well. The brackets indicate which compositions are mixed at each contact. The colors indicate that these *X* and *Y* compositions lie on the same tie line so that if they are mixed, the same *X* and *Y* result. Hence, we do not mix them.

After the first contact, a second contact occurs, where fresh gas contacts equilibrium oil from the 1st contact, and equilibrium oil contacts fresh oil. This process is repeated as shown for additional contacts as gas moves further into the reservoir. After *N* contacts, anyone of the equilibrium gas and oil pairs that form may approach or intersect a critical point. If one of them intersects a critical point, miscibility is developed at that *X* and *Y* pair, because at the critical point there is no IFT and only one phase exists. For a condensing drive, the *X* and *Y* pair closest to the gas approaches miscibility first. For a vaporizing drive, the *X* and *Y* pair closest to the oil approaches miscibility first. For a combined

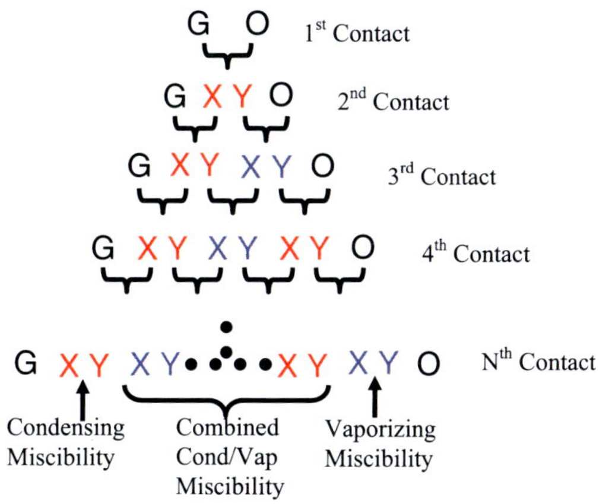

FIGURE 6.12

Illustration of mixing cell contacts between a gas G and oil O. Miscibility is developed from backward contacts for condensing drives, while miscibility is developed from forward contacts for vaporizing drives. The condensing and vaporizing process develops miscibility between these two limits.

condensing/vaporizing drive, miscibility is achieved anywhere between these two pairs. The degree to which a displacement is condensing or vaporizing in a CV drive depends largely on the relative position of the critical *X* and *Y* pair with respect to fresh gas and oil (and the relative proportionality of each segment can be quantified as shown in Johns et al. [33]). The relative location of where miscibility is developed per mass transfer mechanism is shown in Fig. 6.13.

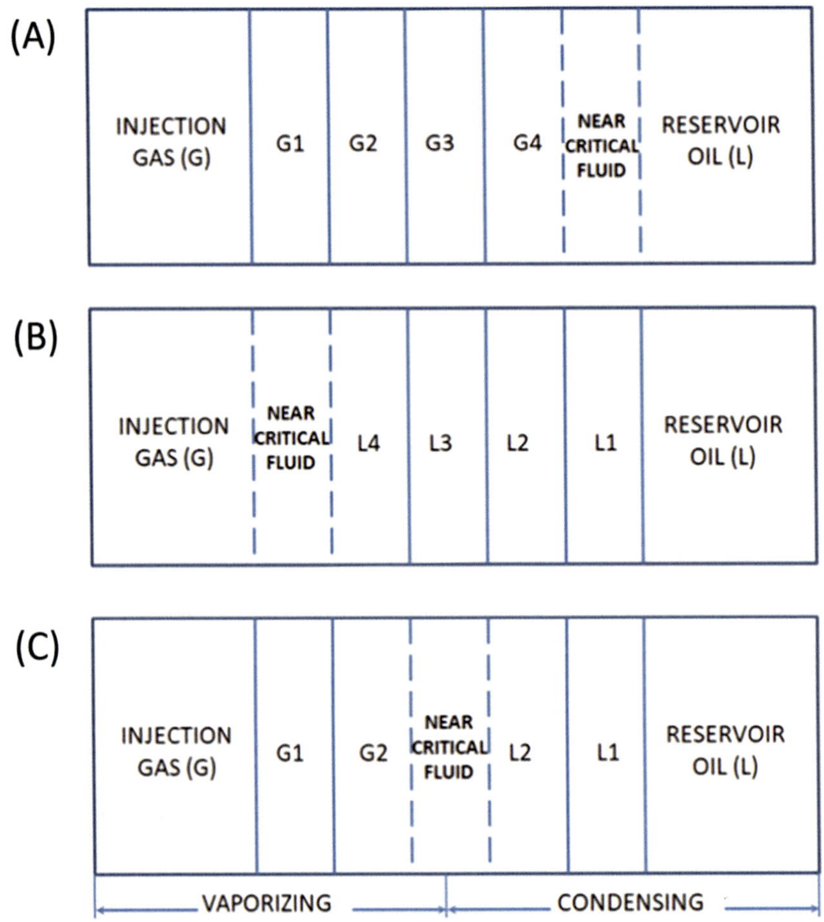

FIGURE 6.13

Relative location of the critical point achieved by displacement of reservoir oil by injection gas, where L = liquid phase, G = gas/vapor phase: (A) Vaporizing Gas Drive (VGD), (B) Condensing Gas Drive (CGD), and (C) Condensing and Vaporizing Gas Drive (CVD). Reproduced with permission from Ref. [12].

Reproduced with permission from R.T. Johns, B. Dindoruk, Gas flooding, in: J. Sheng (Ed.), Enhanced Oil Recovery Field Case Studies, Elsevier, 2013, pp. 1–22.

MCM and minimum miscibility concept and relevant experiments are all developed assuming theoretically no dispersion or minimum dispersion (large Peclet number) when experiments are designed. However, in field-scale implementation, it is hard to implement such flows, especially in complex systems and/or complex flow geometries. Even in numerical solutions, we inherently introduce numerical dispersion aligned with the grid structures and the direction of the flow.

6.5.5 Role of dispersion

Fig. 6.14 shows displacement paths for an MCM displacement using two different conceptual models. The solid thicker line indicates the dispersion-free path for MCM displacement of 48% C4 and 52% C10 mixture by pure CO_2 at 1600 psia and 160°F while the thinner line with symbols indicates the impact of dispersion on the displacement path. In this case, physical dispersion was represented using

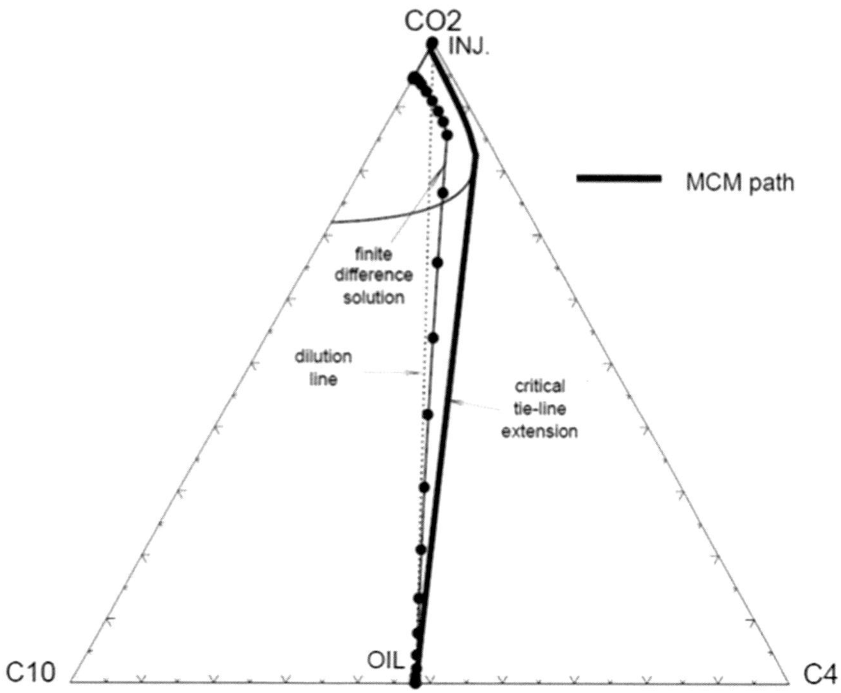

FIGURE 6.14

Effect of dispersion on multicontact miscibility 48% C4 and 52% C10 mixture by continuous injection of pure CO_2 at 1600 psia and 160°F. A finite difference solution at 0.2 PV injection is obtained using 100 grid blocks (corresponding to numerical Peclet number of 250 as defined by Lantz [41]).

coarse grid blocks, which yields numerical Peclet number of 250 (as defined by Lantz [41]). Generally, dispersion will pull the miscible compositional path toward the trajectory of the dilution line in the two-phase region. The impact of physical and numerical dispersion was studied by many authors [28,38–40,42,43] in the past and key points related to the impact of dispersion in the context of gas injection are summarized in Orr [40]. Some of the details in terms of how to keep miscibility sustained in the presence of dispersion for larger-scale domains are studied by Solano et al. [27] and Johns et al. [44]. In that study going beyond the MMP or MME to offset the loss of miscibility due to dispersive forces was recommended.

6.5.6 What is the mixing of oil and gas?

The term dispersion is used loosely in the petroleum industry, but its impact on the gas flooding process can be critical. Some important questions are what is the "true" mixing of oil and gas, and what level does it occur at the field scale compared to the laboratory scale? What grid-block sizes should be used in a compositional simulation? What value of miscible residual oil saturation (S_{orm}) should be used in a simulation? The answers to these questions depend on what causes true mixing.

True mixing of oil and gas is caused only by the diffusion of molecules within a single phase. Diffusion occurs to equilibrate chemical potentials that exist owing to a concentration gradient in the phase. Diffusion itself is relatively slow so that under some circumstances cannot yield significant mixing of oil and gas. For example, in the laboratory, there is not sufficient time for diffusion because fluids are produced quickly. Thus at the laboratory scale, there is little mixing and displacements are controlled by advection.

Consider diffusion when mixing different colored paints as shown in Fig. 6.15. Red paint is initially placed in a circular area as shown. The area is relatively small, however, so that diffusion, as illustrated by the white region, is relatively small. Strong reds and blues remain indicating that mixing is not significant. The paint is then stirred mechanically so that the highs and lows (red and blue) are displaced significantly. This is not mixing but is rather related to velocities variations that occur during the mixing process.

Becomes of the mechanical stirring, the area contacted between the reds and blues has increased by several magnitudes from the original unstirred paint can. Thus with time, diffusion can enhance mixing causing true mixing. Diffusion is enhanced here for two reasons, first that the area of contact has increased, but also that the distance a molecule must now travel is substantially less (i.e., the concentration gradients are increased as evidenced by the smaller distance between reds and blues). Now the paint can effectively become homogenized resulting in a more uniform color.

True mixing of oil and gas occurs in the same way in the field. If oil and gas are miscible, then there is a concentration gradient within a hydrocarbon phase. Diffusion can equilibrate those concentrations resulting in a reduced gas

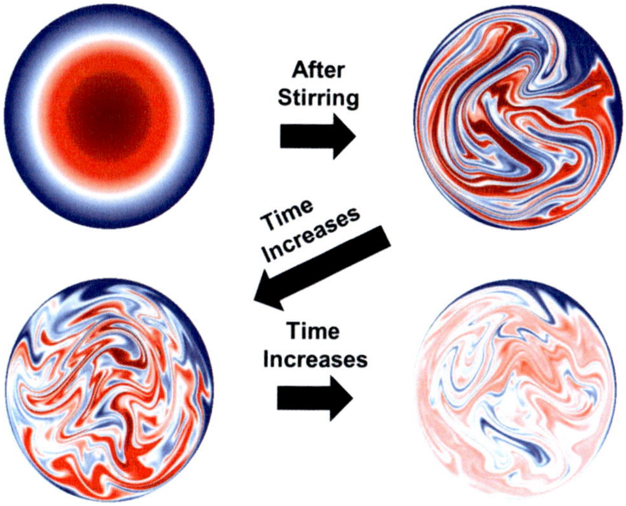

FIGURE 6.15

Mixing of red and blue paint in a paint can. Red paint is initially placed in the paint can in a circular area as shown in the upper left plot. The white region shows the transition zone as diffusion occurs. Then, the paint is stirred so that "highs" and "lows" are displaced greatly increasing the contact area. Now with time diffusion can enhance true mixing, causing the paint color to be homogenized.

Courtesy of Jacques Vanneste, University of Edinburgh.

"strength" (or peak concentration). Stirring of the gas and oil occurs in an oil reservoir owing to variations of velocity by heterogeneity. Gas will channel through the larger permeability pathways, increasing the contact area between the gas and oil. Diffusion then enhances mixing owing to the increased contact area. The scale of the displacement is critical to the level of mixing found in the field. If the well spacing is increased, then for a given heterogeneity, the level of true mixing will be increased (Fig. 6.16). At the laboratory scale, the distance traveled is small so that true mixing is small, owing to a small contact area and time involved. At the field scale, however, mixing can be accelerated by several magnitudes so that mixing can be quite important.

The gas slug size injected depends on many factors, including mixing. If the level of true mixing is large, then the gas slug size should be increased to maintain miscibility as the peak concentration (say of CO_2) will be reduced significantly with time. The impact of diffusion is generally overlooked in the industry, yet it can be very significant for larger well spacing, greater reservoir heterogeneity (especially small permeability correlation lengths vertically and large correlation lengths horizontally), and larger mobility ratio [45].

Upscaling a model to obtain larger grid-block sizes should consider mixing [46]. Very large grid blocks homogenize the reservoir resulting in less "stirring"

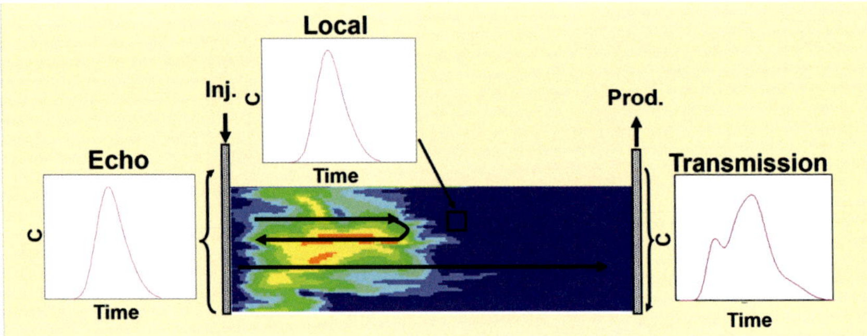

FIGURE 6.16

Illustration of three types of mixing. Transmission mixing is not representative of true mixing, while echo and local are similar.

or true mixing, while at the same time increasing the level of numerical dispersion (or mixing caused by large grid blocks that are at one concentration). This implies there is an optimum grid-block size to achieve the same level of mixing. Ideally, however, very small grid blocks will be used to properly represent reservoir heterogeneity [47]. The level of true mixing in the field is best measured by using single-well tracer tests (SWTT), where a tracer is injected outwards from a well, and then produced back. This accounts for convective spreading through layers of different permeability. True mixing is irreversible, while convective spreading is reversible. That is, tracer that moved outward quickly in a high permeability layer will also come back quickly. Diffusion is irreversible though, and so the concentration profile at the well will be an indication of the level of mixing at the scale of the SWTT measurement (typically on the order of tens of meters). This level of true mixing is sometimes referred to as echo dispersion (Fig. 6.16).

The level of true mixing can also be determined by modeling a tracer in compositional simulation. Here, fine grid blocks are used and an estimate of the dispersion is obtained in each grid block. The values of dispersion obtained can be averaged at the same length scale (distance traveled). That level of dispersion, which is called local dispersion, is similar to the level of true mixing at that length scale.

The final type of dispersion is called transmission dispersion. This is "false" or "apparent" mixing induced primarily by convective spreading through layers of differing permeability. At the production well, the concentrations are then averaged resulting in an unrealistically large value of dispersion. This type of dispersion should not be used to model the true level of mixing in a reservoir and, therefore, in determining the gas slug size required.

A simple example of FCM displacement illustrates the impact of mixing on two-dimensional oil recovery. Fig. 6.17 shows four reservoirs, each of which has the same mean permeability, but the permeabilities are arranged differently within

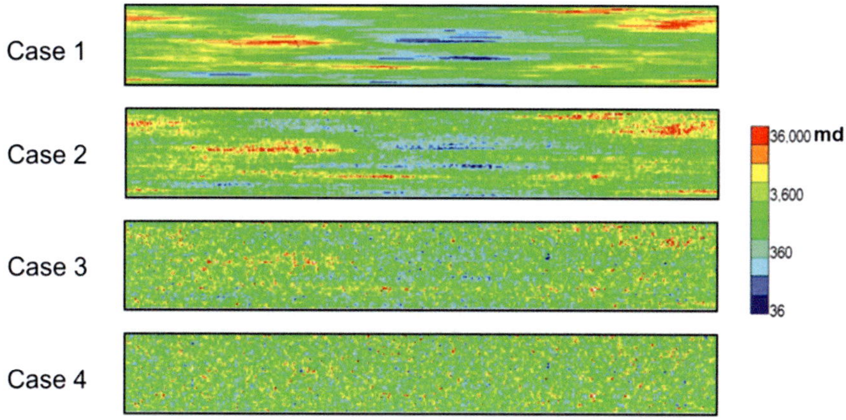

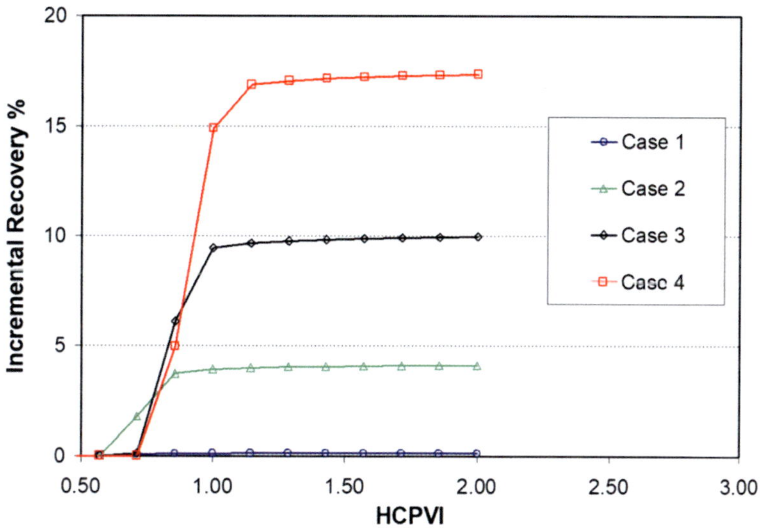

FIGURE 6.17

Effect of mixing on oil recovery from a 0.33 PVI gas slug for four reservoirs with the same mean permeability, but different permeability arrangements. Case 1 has the greatest horizontal correlation length, while case 4 has more random permeability (small correlation length). Case 1 has greater "true mixing," while case 4 has little mixing. The final recovery is much higher for the case with less mixing.

the reservoir. Case 1 has the greatest horizontal correlation length, followed by case 2, then case 3, and then case 4 where permeability is nearly random within the reservoir. An FCM gas slug is injected for 0.33 PVI in all cases, where the injection well is on the left and the production well is on the right of Fig. 6.18.

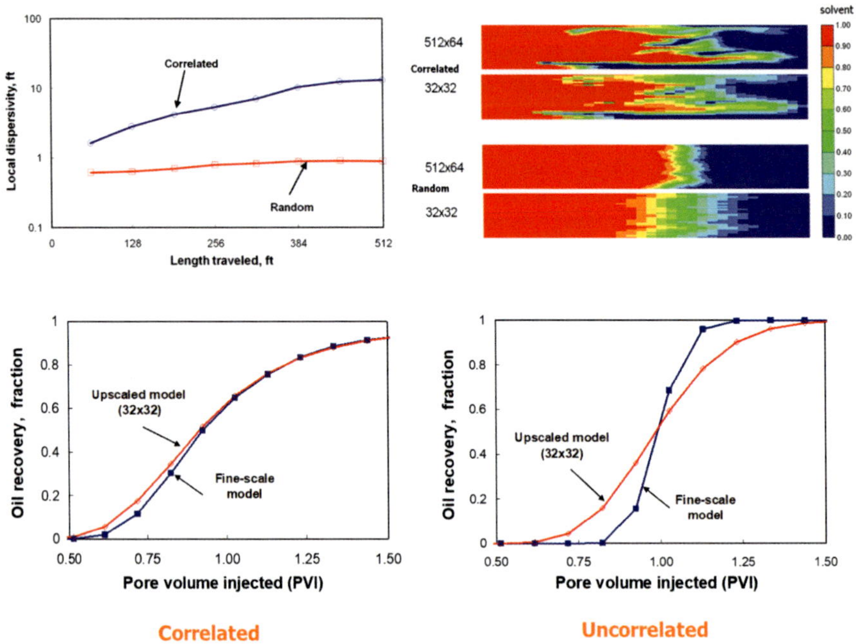

FIGURE 6.18

Upscaling of case 1 (correlated) and case 4 (uncorrelated) reservoirs. Case 1 can be upscaled accurately because the mixing level is large, while case 4 cannot be upscaled and if done would result in a faster breakthrough.

Case 1 induces more "stirring" owing to its greater correlation length and contact area with time, while case 4 shows no increase in contact area with time (flow is piston like). True mixing is therefore much greater in case 1 so that recovery is near zero even at a late time. That is, the injected slug completely mixes with oil before oil can be displaced. Case 4 shows significant incremental oil recovery as mixing between oil and gas is low and flow is piston like. Case 1 requires more gas to be injected to achieve good recoveries (Fig. 6.18).

To match the recoveries in case 4, very small grid blocks would have to be used. Case 4 however can be upscaled to larger grid blocks owing to its greater level of mixing. This is typically the opposite of what most believe. Fig. 6.18 shows the correlated (case 1) and uncorrelated (case 4) saturation profiles. Increasing the grid-block size in case 4 would always cause more mixing than is present in the reservoir and cannot capture the breakthrough time. Case 4 does a good job in matching the recoveries for the upscaled case 1 because its level of mixing is much larger. Most real reservoirs are highly correlated so that mixing levels are large and a reasonable amount of upscaling can be done. While upscaling can be done, case 4 requires a much larger gas slug size than case 1.

6.6 Hydrocarbon gas injection

When dry gas/separator gas is injected into an oil reservoir, phase equilibrium is reached between contacted oil and the gas. Generally, intermediate components (i.e., C3–C4) will be vaporized into the gas phase. As more contacts continue, progressively heavier components can transfer into the gas phase and the magnitude of this transfer depends on the composition of the oil, gas, reservoir temperature, and pressure of the system. In short, the process is dictated by the phase behavior and the dynamic mixing of the components in different phases. Storage of the natural gas in oil reservoirs can also lead to additional oil recovery via the vaporization process. Generally, hydrocarbon gas injection can be considered when there is a nearby source or excess hydrocarbon gas available. Overall there are three categories of hydrocarbon gas injection schemes:

1. Locally hydrocarbon gas(es) is available due to pipeline access or excess field gas is available. This is probably one of the most common situations. The most classical case is the consideration of locally available separator gases. Generally, methane is the most dominant hydrocarbon species in separator gases. Depending on the character and the geochemistry of the oil and its depositional environment and history, the associated gas may have more or less wetness. Generally, the abundance of C2 + components (wetness) can be beneficial for flood. However, often associated gases will require high pressures to make the process miscible (i.e., high MMP values). One obvious extension is to use components heavier than methane. This is generally known as an enrichment process.
2. There is gas to be purchased either to mix with the locally available gas or to be injected in limited quantities. Generally, additional hydrocarbon gas is purchased either to make up the difference for voidage replacement or to improve the quality of the gas injected to make it more favorable for miscibility. Often the addition of richer (heavier) hydrocarbon gases is considered to obtain the maximum benefit for miscibility under compression or reservoir integrity considerations. In general, we like to inject at a lower pressure with the maximum miscibility benefit, however, lean gases normally require high pressures to achieve miscibility. Adding some heavier/richer gases (i.e., C2 +) into the associated gases may lead to better performance at lower pressures. The second reason that we may like to stay at a lower pressure is related to the integrity of the wells and the caprock. Even though we may feel that we can inject at a high pressure economically, the reservoir itself and/or the wells can dictate what pressure ranges should be used. Therefore taking the flexibility in the pressure domain away will only leave flexibility in terms of manipulating the injection compositions.
3. Hydrocarbon gas injection is due to local stranded gas that contains condensates. Stranded gas is another issue faced, as some discoveries are far away from the markets and the infrastructure for sales and transportation (i.e.,

pipelines). In many of those cases, the main economic driver is to strip the liquids from the gas stream and reinject the gas stream (leaner gas) back to the reservoir as venting or flaring the gas is no longer permitted. In such cases, the gas that is being injected could be what is produced and may contain some CO_2. Often, this particular topic is captured under the topic of EGR.

6.7 N_2/flue gas

Nitrogen (N_2) is abundant in the air and is therefore available everywhere. Thus the cost of using nitrogen is primarily associated with the cost of separation from the air and processing. However, nitrogen has adverse properties in terms of its interaction with the oil. If we want N_2 to interact with the oils the same way as CH_4, it will take much higher pressures, especially for oils with APIs less than 25 at or below moderate reservoir temperatures ($<180°F$).

A variation of the process using CO_2 is possible from flue gas, which can contain on the order of 87–88% nitrogen while the rest is CO_2 and other inert gases in the air. The technique was initially proposed to take advantage of CO_2 in the flue or exhaust gases, especially when other CO_2 sources are not available. However, one of the issues, for an injection gas that contains such a high concentration of N_2, is again the adverse impact of N_2 on the miscibility behavior of the system. Elevated amounts of N_2 can lead to much higher miscibility pressures. This will cause higher levels of compression if the flood is to be implemented as a near miscible flood. The behavior of N_2 with CH_4 is studied in detail by Dindoruk et al. [48] and while not having a systematic analysis available, several aspects of N_2 with CO_2 are studied by Gajbhiye [49] in terms of IFT measurements, by Vahidi and Zargar [50] for the calculation of MMP, and Wang et al. [51] for asphaltene precipitation experiments. As expected, because N_2 is the leanest gas in terms of its partitioning into the liquid hydrocarbon phases, adding CH_4 or CO_2 into the injected N_2 stream has a significant effect at some threshold concentration of CH_4 or CO_2. Such threshold values where miscibility is improved vary depending on the nature of the oil and the system temperature [48]. Overall, there is a need for a good predictive correlation for N_2-CH_4, and N_2-CO_2 dilution ratios on the MMP of the system.

Overall, flue gas or nitrogen may seem attractive, however, the assumed benefit in terms of oil recovery may not be there for any oil–gas system. Thus systems of interest should be screened accordingly. For example, having a submissive system leading to the early breakthrough of injected fluid may lead to additional costs of gas handling and separation even though the acquisition cost of N_2 or flue gas may seem attractive. Additional complexities may also stem from the onset of asphaltenes within the operational envelope or some of the impurities, such as SO_x and NO_x, in the flue gases.

6.8 CO_2 sequestration

In terms of the theory of the gas injection process, CO_2 exhibits hybrid condensing/vaporizing drive behavior due to its similarities to the rich gas injection systems. As early as the 1940s high-pressure CO_2 injection was suggested for the recovery of the residual oil [52]. One of the most commonly considered gas is CO_2 due to naturally occurring CO_2 reservoirs and as a result developed pipeline infrastructure in such areas, such as West Texas. Nowadays, CO_2 injection is considered for a wider spectrum of reservoirs and other subsurface assets as explained below as it can be acquired from anthropologic sources and as well as from the air via direct air capture (DAC). The concern about the rise in carbon dioxide (CO_2) concentrations in the atmosphere and its impact on climate change have been increasing over past decades. While one of the primary sources of CO_2 emissions is power or manufacturing plants (large point sources) that burn fossil fuels to generate electricity, the cumulative emissions from vehicles and humans/animals (small but distributed point sources) can be significant. This necessitates the development of carbon capture technologies not only from flue gases but also directly from the atmosphere [53].

CO_2 emissions have increased significantly since the industrial revolution and parallel the growth in power requirements to sustain the demand and as well as current lifestyles. It is evident from Fig. 6.19 that the energy generated is almost linear with the CO_2 generation.

During the last two decades and especially during the current time frame, there is significant momentum in terms of CO_2 abatement technologies. Just due to the size of the problem, and its urgency many options are being considered. However, the urgency coupled with cost and as well as the size of the problem highlights the importance of one of the key storage solution options: subsurface storage. The petroleum industry is one of the few industries that deal with large-scale systems with high-pressure high-temperature conditions involving complex fluid behavior, rocks and their interactions, and flow dynamics in the subsurface along with the subsurface uncertainties. This expertise along with the subsurface storage volumes makes the aggregate experience gained in the industry quite important. Moreover, the petroleum industry as a whole has the critical domain expertise to reach and manage deep reservoirs both onshore and offshore. These capabilities and know-how put the petroleum industry at the forefront to capture, utilize, and store carbon dioxide and other greenhouse gases in the subsurface. More details on carbon capture, utilization, and storage options for the petroleum industry can be found in the study by Magzymov et al. [54].

Although CO_2 injection has been practiced in enhancing oil recovery at the field scale since the 1970s, geological CO_2 sequestration known as geological carbon storage as a proposed technology to store CO_2 has not been implemented at a large scale. CO_2 from large point sources or even with distributed dilute sources as in the case of DAC can be permanently stored in permeable subsurface structures. These geological formations are required to have certain characteristics: (1) they should

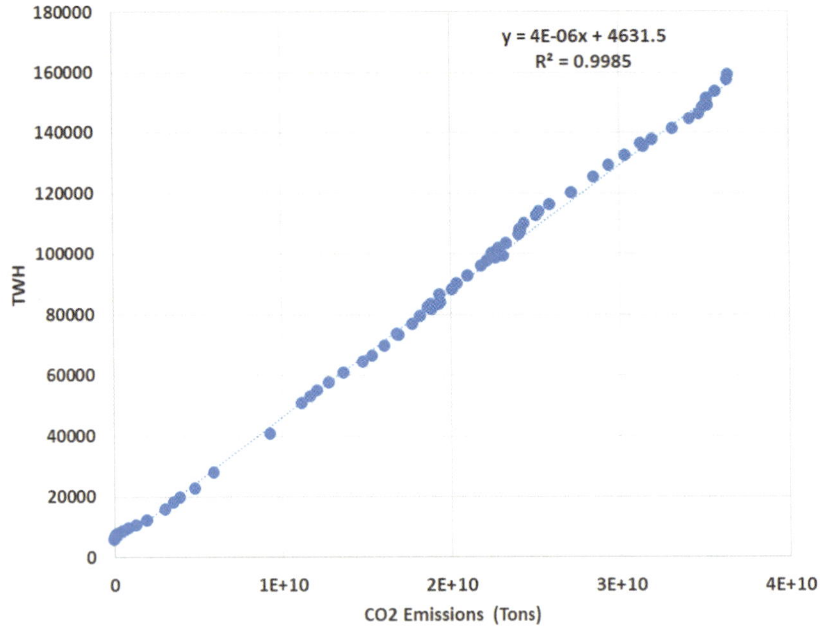

FIGURE 6.19

Global energy consumption (terawatt hours) versus global CO_2 emissions till the end of 2020.

have a caprock or multiple impermeable formations (the equivalent of an impermeable caprock) to prevent upward migration and potential leakage of CO_2 (containment element) and (2) the subject formation should have sufficient volumetric storage (amount stored per unit bulk volume), defined by porosity, reactivity, and phase behavior/solubility of CO_2 in the in situ fluids to hold large amounts of CO_2. Main geological storage units are:

- Depleted oil and gas reservoirs,
- Deep saline formations,
- (Unmineable) coal seams,
- Basalt formations, and
- Deep aquifers with geothermal potential.

We explain these categories below and then compare the efficiency of the two most common storage options.

1. *Depleted oil and gas reservoirs*

   One of the most important categories for CO_2/gas storage falls in this category. There are two key reasons for this: (1) The storage and the characteristics of the reservoirs are already known since they have been produced to their final stage. As result, at least intrinsically, the more they are produced, the more we know about their reservoirs in terms of their capacity and flow characteristics along with the connectivity, both being extremely important for project executions, (2) the sealing capacity of their caprocks are known to be very good, and (3) large volumes of fluid can be stored per bulk volume owing to water and oil already being extracted.

2. *Deep saline formations*

   Deep saline formations are widespread so in theory, significant volumes of CO_2 can be stored in them. Unfortunately, they are not as efficient as depleted oil and gas reservoirs because relatively incompressible water fills the pore space so that less CO_2 can be stored per bulk volume. This approach relies on CO_2 displacing water from the pores without fracturing the caprock near the well and having CO_2 being trapped in a variety of ways. Trapping can occur through hysteresis by water coming in behind the CO_2, by CO_2 dissolving into the water phase, and by precipitation of CO_2 into a solid phase. Dissolved CO_2 in the water phase is typically only on the level of 2%–4% by mole fractions, however. Trapping should last for thousands of years, although the time scale for this trapping has not been explicitly defined. Furthermore, unlike depleted oil and gas reservoirs, the sealing capability of the caprock is not well known. Furthermore, deep saline formations often outcrop far from the injection source, so that CO_2 may migrate and reach the atmosphere. Regional groundwater flow is well known to exist in most aquifers, so that groundwater flows to various outcrops, rivers, and other water surface features.

3. *Coal seams, basalt formations, and deep aquifers with geothermal potential*

   This group of potential CO_2 storage formations suffers from even less bulk volume storage efficiency than the first two. If storage is reported on a pore volume basis, these techniques look promising to store CO_2, but that is not the case when the bulk volume is considered.

   There is some confusion in the literature on the potential for significant volumes of CO_2 to be stored in the various types of reservoirs. Part of this confusion is that the mass of CO_2 that can be stored is often reported based on the pore volume, and not bulk volume. On a pore volume basis, a reservoir with little porosity could look very promising for storage when reported on a pore volume basis, but in fact, it would not be due to the small reservoir pore volume for CO_2 storage. A simple calculation can illuminate the differences in the potential for CO_2 storage in the different categories.

   Consider that CO_2 is stored solely in its pure CO_2 state as a separate phase and also solubilized with water to its maximum possible value. Other storage mechanisms are possible, such as CO_2 stored as a precipitated phase, but for purposes of this calculation, we will ignore that amount as it could occur in all formation types. Table 6.1 shows that EOR reservoirs after depletion can store twice as much CO_2 as deep brine reservoirs on a bulk volume basis. Because there is less risk with permeable depleted oil reservoirs (known geological quantities), depleted reservoirs should be the first priority for

Table 6.1 Relative amounts of CO_2 stored in three types of formations (densities: 1.0 for oil and water, 0.6 gm/cc for CO_2, and 0.04 mole fraction CO_2 in water).

		Phase saturation			Mass CO_2 Storage	
Reservoir	**Porosity**	**CO_2**	**Water**	**Oil**	**Ratio (PV)**	**Ratio (BV)**
Brine	0.35	0.2	0.8	0	0.6	4.0
EOR	0.35	0.6	0.3	0.1	1.2	8.1
Tight rock	0.05	0.5	0.2	0.3	1	1

EOR, Enhanced oil recovery.

storage. Tight rocks, while giving high CO_2 storage on a pore volume basis, show poor storage efficiency on a bulk volume basis. For example, CO_2 in the calculation in Table 6.1 shows four times less mass storage of CO_2 than in deep brine reservoirs and eight times less storage efficiency in EOR reservoirs on a bulk volume basis.

There are ways to increase the amount of CO_2 to be stored per bulk volume from that calculated here. For example, horizontal wells could be drilled in a staggered line drive pattern, while liquids below (water and/or oil) can be produced. CO_2 injected continuously is lighter than water and will become buoyant. CO_2 would then build up in a gravity stable or enhanced process [55]. Estimated recoveries when injected into an oil reservoir are on the order of 80%, while storage is also maximized. The process relies however on the ability to produce the liquids at the bottom of the reservoir, so that the reservoir must be isolated (no water influx from outside so that all reservoir boundaries must be no flow). Further, the process requires nonzero vertical permeability. A significant advantage of the process of injecting in vertical wells is that gravity can overcome large heterogeneities resulting in a good sweep from top to bottom.

6.9 Field cases

We consider four field cases that provide important lessons on the application of gas flooding for EOR. The recovery in the field depends on both volumetric sweep and displacement efficiency. Summaries of gas floods performed can be found in Christensen et al. [15] and Manrique et al. [56].

6.9.1 Slaughter estate unit CO_2 flood

This miscible flood pilot is in the West Texas San Andres dolomite and is an example of a CO_2 gas flood with very good oil recovery. The permeability is low averaging around 4 mD at a depth of about 5000 ft. Good recovery was obtained in a water flood in the early 1970s prior to gas flooding (see Fig. 6.20).

The gas injected contained 72 mol.% CO_2 and 28 mol.% H_2S. The MMP of approximately 1000 psia with this gas and moderately viscous oil (32 degrees API) is substantially less than the average reservoir pressure of 2000 psia. Thus this flood is MCM. The acid gas was eventually replaced with a chase gas consisting of mostly nitrogen and then water. Water was injected alternately with the acid gas with a WAG ratio of about 1.0. A 25% hydrocarbon pore volume (HCPV) slug of acid gas was injected. The chase gas was also alternated with water, and eventually the gas–water ratio was reduced to 0.7 to improve vertical sweep. The cycles shown in Fig. 6.20 for the chase gas correspond to the cycles in WAG injection. During WAG, water injectivity losses of about 50% were experienced. Furthermore, there is evidence of gas channeling as the GOR began to climb at the same time or just prior to the oil production (see Fig. 6.20).

Incremental tertiary recovery was 19.6% original oil in place (OOIP), which is largely the result of good WAG management and the use of H_2S in the gas [57]. H_2S, although very dangerous, is a very good miscible agent. When added to the primary and secondary recovery (water flood) of about 50%, the total recovery in this pilot is expected to be around 70% OOIP, which is well above the average for most fields. Because of the great success of the pilot, the unit was gas flooded field wide. The field-wide flood has also been successful although a higher gas–water ratio was used, which may lead to more channeling and poorer sweep than in the pilot.

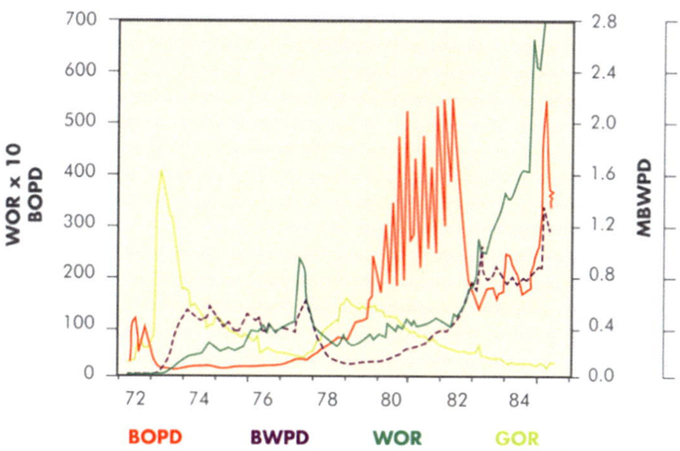

FIGURE 6.20

Oil and water rates and water and gas–oil ratios for Slaughter Estate Unit Pilot.

6.9.2 Weeks Island gravity stable CO_2 flood

A pilot test of the "S" sand at the Weeks Island, Louisiana field was performed in the early 1980s. This sand, which is up against a salt dome, is highly dipping (see Fig. 6.21) and is very permeable both vertically and horizontally. Following a water flood, gas was injected up dip so that gravity would stabilize the front in a relatively horizontal interface. The idea is that as gas is injected the interface will move down vertically recovering oil and displacing it to the down dip production wells. This process can be highly efficient (good volumetric sweep) if there is good vertical permeability, the gas interface is stable, and moves vertically downward. The gas injected was a mixture of CO_2 and about 5% plant gas. The plant gas was used to "lighten" the CO_2 so that the gas–oil interface is more stable.

At the reservoir temperature of 225°F, the flood is immiscible, not miscible. Nevertheless, a pressure core taken in zones where the gas traversed were nearly "white," indicating oil saturations of approximately 2% or less. The excellent recovery of over 90% demonstrates that even immiscible floods when properly designed can achieve sufficient extraction of oil components. Immiscible gas floods in general can achieve better displacement efficiency than water floods owing to decreased oil viscosity, oil swelling, IFT lowering, partial extraction, and the potential for good gravity drainage. One disadvantage of immiscible gas flooding over waterflooding is the potential for poor sweep due to its adverse mobility ratio.

A subsequent commercial test of the gravity stable process was not as successful largely because of significant water influx down dip of the production wells. Thus injection of CO_2 largely pressurized the gas cap but did not cause the gas–oil interface to move vertically downward. A gravity stable process like this would be very effective as long as the water influx is relatively small.

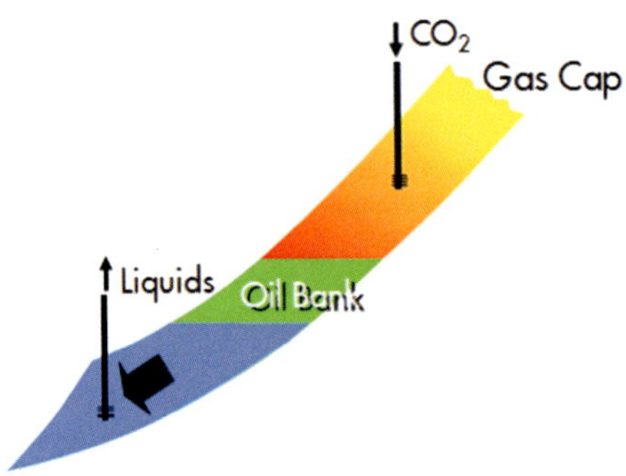

FIGURE 6.21

Illustration of oil bank movement during gravity stable CO_2 flood at Weeks Island.

6.9.3 Jay little Escambia creek nitrogen flood

The Jay field near the Alabama–Florida border is one of a few nitrogen floods ever conducted. The reservoir is in the Smackover carbonate at a depth of 15,000 ft. Nitrogen is a good miscible gas in this reservoir because of its very light sour crude (50-degree API), and high reservoir pressure around 7850 psia. The formation permeability averages 20 mD and the oil is a very light, sour crude, with an API gravity of 51 degrees. A significant advantage of nitrogen is that it is readily available via separation from the air. Nitrogen was injected in a WAG ratio near 4.0, which is greater than typical.

The overall recovery at Jay is expected to be near 60% OOIP. Incremental recovery from miscible nitrogen injection is forecast to be around 10% OOIP (see Fig. 6.22). The high primary and secondary recovery of around 50% OOIP is likely the result of low vertical permeability coupled with good horizontal permeability in the dolomite facies [58]. Low vertical permeability caused by shale lenses or in this case cemented zones associated with thin stylolites, is an ideal candidate for both water and gas flooding as fluids are less able to segregate vertically.

6.9.4 North Cross Devonian unit (Crossett field)

The Crossett field, located in west Texas, was discovered in 1944. The productive Devonian Main Pay consists primarily of siliceous limestone and tripolitic chert

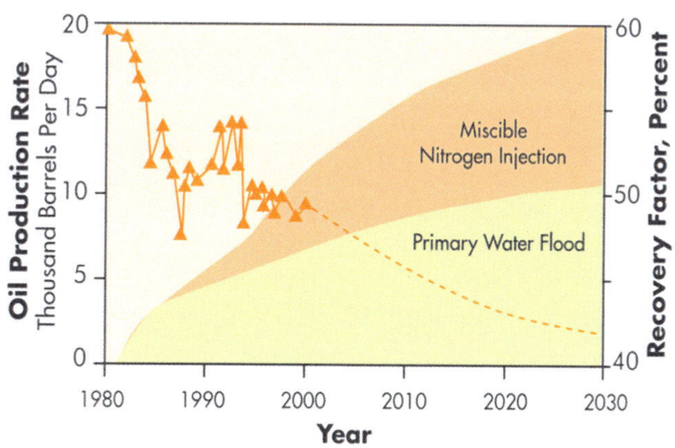

FIGURE 6.22

Oil recovery from Jay's miscible nitrogen flood. Miscible nitrogen injection is expected to give an incremental recovery of 10% of original oil in place over waterflooding alone.

Reproduced with permission from J.J. Lawrence, N.K. Maer, D. Stern, et al., Jay nitrogen tertiary recovery study: managing a mature field, in: SPE-78527-MS, Abu Dhabi International Petroleum Exhibition and Conference, October 13–16, 2002, Abu Dhabi, UAE.

with effective porosities ranging from 9% to 35%, and corresponding permeabilities range from less than 1 mD up to about 30 mD, with an average permeability of 5 mD. The average reservoir pressure and temperature are estimated to be approximately 2500 psi—well above minimum miscibility pressure (MMP)—and 105°F, respectively. The crude oil has an API gravity of 44. In these conditions, relatively low reservoir temperature plays a positive role in the miscibility of CO_2 while it can also lead to complex phase behavior [59]. The field was developed using regular 40 acre spacing and produced under primary recovery until 1964. North Cross is the oldest operating field-scale CO_2 injection project in the world. Large-scale continuous CO_2 injection began in April 1972 and was expanded over time to provide CO_2 support throughout the field. The field has been under continuous CO_2 flood for over four decades. Currently, over 200% HCPV of CO_2 has been injected, and over 60% of the latest estimate of its target OOIP has been recovered. The field continues to perform well with a gross utilization (MCF CO_2 injected/bbl oil recovered) of approximately 30 MCF/BO (20 MCF/BOE). North Cross is an excellent example of a successful mature continuous CO_2 flood [60–62].

6.10 Overview of field experience

Gas flooding technology is well developed and has demonstrated good recoveries in the field [56]. Recoveries from both immiscible and miscible gas flooding vary from around 5%–20% OOIP, with an average of around 10% incremental OOIP for miscible gas floods [15]. Immiscible gas flooding recoveries are less on average, around 6% OOIP. Although recovery by gas flooding is very economic at these levels, 55% OOIP still remains on average postmiscible gas flooding assuming 65% OOIP prior to gas flooding. The significant amount of oil that remains is largely the result of gas channeling through the formation owing to large gas mobility. Channeling will also result in the early breakthrough of the gas (solvent), typically around the same time that oil is produced. This is in contrast to surfactant/polymer flooding, which almost always exhibits an oil bank prior to surfactant breakthrough. Poor volumetric sweep is not as much of a problem for surfactant–polymer floods or water floods, which have more favorable mobility ratios. Nevertheless, miscible flooding is generally very economic and less complex than chemical flooding, especially for deeper reservoirs that are more technically challenging for surfactant/polymer floods (see Chapter 4). Owing to the high cost of chemicals, surfactant floods have not yet been commercially successful.

Besides poor sweep and early breakthrough of gas, several problems have been observed in implementing gas floods [1]. Corrosion of injection and production facilities can occur when CO_2 is injected. Asphaltenes, hydrate, and scaling formation have also been a problem, but these problems have often been overcome. Reduced water injectivity following a gas cycle during WAG has also been reported to be a significant problem in some fields. This is likely the result of relative permeability changes as fluids are injected (hysteresis). Gas injectivity is generally not as much of a problem, and in some cases, gas injectivity has increased with time owing to the dissolution of carbonate rocks with CO_2.

Most gas floods use WAG. Of those, the WAG ratio is typically around 1.0 [15]. Total volumes of gas injected (slug sizes) are mostly in the range of 0.1–0.3 PVI. Injection of more gas, however, can yield larger recoveries, although at a cost of diminishing returns (less oil recovered per volume of gas injected). The effectiveness of gas floods is often measured in terms of gas utilization factors, which are the volume of gas injected per volume of oil recovered. Utilization factors less than 10 MCF/STB are very efficient.

One of the most important aspects of operating gas floods is reservoir management. Reservoir management of a field is a life-cycle process that requires good data acquisition and monitoring. Extensive core data, geologic descriptions, and good reservoir simulation models are necessary to understand how to best implement a gas flood process.

Nomenclature

BOPD	barrels of oil per day
CV	combined condensing and vaporizing drive
DAC	direct air capture
EOR	enhanced oil recovery
EOS	equation of state
EGR	enhanced gas recovery
FCM	first-contact miscibility
ft	feet
GOR	gas-to-oil ratio
HCPV	hydrocarbon pore volume
IFT	interfacial tension
IOR	improved oil recovery
MBWPD	thousand barrels of water per day
MCM	multicontact miscibility
mD	millidarcies
MME	minimum miscibility enrichment
MMP	minimum miscibility pressure
MOC	method of characteristics
OOIP	original oil in place
PVI	pore volumes injected
PVT	pressure volume temperature
SWAG	simultaneous water–gas injection
SWTT	single-well tracer tests
TWAG	tapered water-alternating gas
WAG	water-alternating gas
WOR	water-to-oil ratio

References

[1] P.M. Jarrell, C.E. Fox, M.H. Stein, et al., Practical aspects of CO_2 flooding, in: SPE Monograph Series, Society of Petroleum Engineers, Richardson, 2002.
[2] F.I. Stalkup Jr., Miscible Displacement, SPE Monograph Series, Society of Petroleum Engineers, Richardson, 1983.
[3] B. Dindoruk, R.T. Johns, F.M. Orr Jr., Measurement and calculation of minimum miscibility pressure: a state of the art review, SPE Reservoir Evaluation & Engineering 24 (2021) 367–389.
[4] J.B. Randolph, O.M. Saar, J. Bielicki, Geothermal energy production at geologic CO_2 sequestration sites: impact of thermal drawdown on reservoir pressure, Energy Procedia 37 (2013) 6625–6635.
[5] C.A. Hutchinson Jr., P.H. Braun, Phase relations of miscible displacement in oil recovery, AIChE Journal 7 (1961) 64–72.
[6] C.R. Smith, Mechanics of Secondary Oil Recovery, fourth ed., Krieger Publishing, Malabar, 1985.
[7] R. Ratnakar, B. Dindoruk, I. Wilson, Experimental investigation of DME-water-crude oil phase behavior and PVT modeling for the application of DME-enhanced waterflooding, Fuel 182 (2016) 188–197.
[8] R. Ratnakar, B. Dindoruk, I. Wilson, Development of empirical correlation for DME-partitioning between brine and crudes for enhanced waterflooding applications, Journal of Petroleum Science and Engineering 157 (2017) 264–272.
[9] R. Ratnakar, B. Dindoruk, I. Wilson, Phase behavior experiments and modeling of DME-brine-crude oil mixture based on Huron-Vidal mixing rule for EOR applications, Fluid Phase Equilibria 434 (2017) 49–62.
[10] R.M. Enick, D. Olsen, J. Ammer, et al., Mobility and conformance control for CO_2 EOR via thickeners, foams, and gels - A literature review of 40 years of research and pilot tests, SPE-154122-MS, SPE Improved Oil Recovery Symposium, April 14–18, 2018, Tulsa, Oklahoma, USA.
[11] L.W. Lake, M. Lotfollahi, S.L. Bryant, CO_2 enhanced oil recovery experience and its messages for CO_2 storage, in: P. Newell, A.G. Ilgen (Eds.), Science of Carbon Storage in Deep Saline Formations: Process Coupling across Time and Spatial Scales, Elsevier, 2018, pp. 15–31.
[12] R.T. Johns, B. Dindoruk, Gas flooding, in: J. Sheng (Ed.), Enhanced Oil Recovery Field Case Studies, Elsevier, 2013, pp. 1–22.
[13] J. Babayev, Investigation of Water Alternating Gas (WAG) Injection Recovery in Layered Reservoirs (MS thesis), University of Houston, Houston, 2010.
[14] D.W. Green, G.P. Willhite, Enhanced Oil Recovery, second ed., Society of Petroleum Engineers, Richardson, 2018.
[15] J.R. Christensen, E.H. Stenby, A. Skauge, Review of WAG field experience, SPE Reservoir Evaluation & Engineering 4 (2001) 97–106.
[16] E.C. Lin, E.S. Poole, Numerical evaluation of single-slug, WAG, and hybrid CO_2 injection processes, Dollarhide Devonian Unit, Andrews County, Texas, SPE Reservoir Engineering 6 (1991) 415–420.
[17] M.T. Barnawi, A Simulation study to Verify Stone's Simultaneous Water and Gas Injection Performance in a 5-Spot Pattern (MS thesis), Texas A&M university, College Station, 2008.

[18] W.E. McCain, The Properties of Petroleum Fluid, third ed., PennWell Books, Tulsa, 2017.

[19] K.S. Pedersen, P.L. Christensen, Phase Behavior of Petroleum Reservoir Fluids, CRC Press, Boco Raton, 2007.

[20] L.P. Dake, The Practice of Reservoir Engineering, Elsevier, Amsterdam, 1994.

[21] W.J. Justice, Review of cycling operations in the La Gloria field, Journal of Petroleum Technology 2 (1950) 281–286.

[22] A.T. Turta, S.S.K. Sim, A.K. Singhal, et al., Basic investigations on enhanced gas recovery by gas-gas displacement, in: Paper 2007-124, Proceedings of the Petroleum Society's 8th Canadian International Petroleum Conference (58th Annual Technical Meeting), June 12–14, 2007, Calgary, Alberta, Canada.

[23] S.S.K. Sim, A.T. Turta, A.K. Singhal, et al., Enhanced gas recovery: factors affecting gas-gas displacement efficiency, Journal of Canadian Petroleum Technology 48 (2009) 49–55.

[24] K. Huang, L. Zhang, W. Zhu, et al., Nonequilibrium override phenomenon between gases in condensate gas development, Journal of Petroleum Science and Engineering 205 (2021) 108894.

[25] W.Y. Feng, Recovery enhancement at the later stage of supercritical condensate gas reservoir development via CO_2 injection: a case study on Lian 4 fault block in the Fushan sag, Beibuwan Basin, Natural Gas Industry B 3 (2016) 460–466.

[26] A. Hamza, I.A. Hussain, M.J. Al-Marri, et al., CO_2 enhanced gas recovery and sequestration in depleted gas reservoirs: a review, Journal of Petroleum Science and Engineering 196 (2021) 107685.

[27] R. Solano, R.T. Johns, L.W. Lake, Impact of reservoir mixing on recovery in enriched-gas drives above the minimum miscibility enrichment, SPE Reservoir Evaluation & Engineering 4 (2001) 358–365.

[28] L.W. Lake, R.T. Johns, W. Rossen, et al., Fundamentals of Enhanced Oil Recovery, Society of Petroleum Engineers, Richardson, 2014.

[29] A. Danesh, PVT and Phase Behavior of Petroleum Reservoir Fluids, Developments in Petroleum Science No. 47, Elsevier, 1998.

[30] A. Firoozabadi, Thermodynamics of Hydrocarbon Reservoirs, McGraw-Hill, 1999.

[31] A.M. Egwuenu, R.T. Johns, Y. Li, Improved fluid characterization for miscible gas floods, SPE Reservoir Evaluation & Engineering 11 (2008) 655–665.

[32] R.T. Johns, B. Dindoruk, F.M. Orr Jr., Analytical theory of combined condensing/vaporizing gas drives, SPE Advanced Technology Series 1 (1993) 7–16.

[33] R.T. Johns, H. Yuan, B. Dindoruk, Quantification of displacement mechanisms in multicomponent gas floods, SPE Journal 9 (2004) 314–321.

[34] F.M. Orr Jr., R.T. Johns, B. Dindoruk, Development of miscibility in four component CO_2 floods, SPE Reservoir Engineering 8 (1993) 135–142.

[35] U. Sinha, B. Dindoruk, M. Soliman, Prediction of CO_2 minimum miscibility pressure (MMP) using machine learning techniques, SPE Journal 26 (2021) 1666–1678.

[36] K. Ahmadi, R.T. Johns, Multiple mixing-cell model for MMP determination, SPE Journal 16 (2011) 733–742.

[37] H. Yuan, R.T. Johns, A.M. Egwuenu, B. Dindoruk, Improved MMP correlation for CO_2 floods using analytical gas flooding theory, SPE Reservoir Evaluation & Engineering 8 (2005) 418–425.

[38] R.T., Johns, Analytical Theory of Multicomponent Gas Drives with Two-Phase Mass Transfer (Ph.D. dissertation), Stanford University, Stanford, 1992.

[39] B. Dindoruk, Analytical Theory of Multiphase, Multicomponent Displacement in Porous Media (Ph.D. dissertation), Stanford University, Stanford, 1992.

[40] F.M. Orr Jr., Theory of Gas Injection Processes, Tie-Line Publications, Stanford University, 2007.
[41] R.B. Lantz, Quantitative evaluation of numerical dispersion (truncation error), SPE Journal 11 (1971) 315–320.
[42] A. Arya, T.A. Hewett, R.G. Larson, et al., Dispersion and reservoir heterogeneity, SPE Reservoir Engineering 3 (1988) 139–148.
[43] B.W. Walsh, F.M. Orr Jr., Prediction of miscible flood performance: the effect of dispersion on composition paths in ternary systems, In Situ 14 (1990) 19–47.
[44] R.T. Johns, P. Sah, S. Subramanian, Effect of gas enrichment above the MME on oil recovery in enriched gas floods, in: SPE-56826-MS, SPE Annual Technical Conference and Exhibition, October 3–6, 1999, Houston, Texas, USA.
[45] M. Connolly, R.T. Johns, Scale-dependent mixing for adverse mobility ratio flows in heterogeneous porous media, Transport in Porous Media 113 (2016) 29–50.
[46] O.O. Adepoju, L.W. Lake, R.T. Johns, Anisotropic dispersion and upscaling for miscible displacement, SPE Journal 20 (2014) 421–432.
[47] Gholamreza Garmeh, Russell T. Johns, Upscaling of miscible floods in heterogeneous reservoirs considering reservoir mixing, SPE Reservoir Evaluation & Engineering 13 (5) (2010) 747–763. Available from: https://doi.org/10.2118/124000-PA.
[48] B. Dindoruk, F.M. Orr Jr., R.T. Johns, Theory of multicontact miscible displacement with nitrogen, SPE Journal 2 (1997) 268–279.
[49] R. Gajbhiye, Effect of CO_2/N_2 mixture composition on interfacial tension of crude oil, ACS Omega 5 (2020) 27944–27952.
[50] A. Vahidi, G. Zargar, Sensitivity analysis of important parameters affecting minimum miscibility pressure (MMP) of nitrogen injection into conventional oil reservoirs, in: SPE-111411-MS, SPE/EAGE Reservoir Characterization and Simulation Conference, October 28–31, 2007, Abu Dhabi, UAE.
[51] P. Wang, F. Zhao, J. Hou, et al., Comparative analysis of CO_2, N_2, and gas mixture injection on asphaltene deposition pressure in reservoir conditions, Energies 11 (2018) 2483.
[52] S.J. Pirson, Tertiary recovery of oil, paper presented at the Central Appalachian Section, in: AIME Meeting, June 26, 1941.
[53] R. Ratnakar, S. Santhosh, R. Agrawal, et al., Modeling and experimental study on CO_2 adsorption in fixed-bed columns: applications to carbon capture and utilization, Journal of Natural Gas Science and Engineering 94 (2021) 104111.
[54] D. Magzymov, B. Dindoruk, R.T. Johns, Carbon capture, utilization, and storage in the context of petroleum industry: a state-of-the-art review, in: SPE-209368-MS, SPE Improved Oil Recovery Conference, April 25–29, 2022, Virtual.
[55] L. Li, S. Khorsandi, R.T. Johns, et al., CO_2 enhanced oil recovery and storage using a gravity-enhanced process, International Journal of Greenhouse Gas Control 42 (2015) 502–515.
[56] E.J. Manrique, V.E. Muci, M.E. Gurfinkel, EOR field experiences in carbonate reservoirs in the United States, SPE Reservoir Evaluation & Engineering 10 (2007) 667–686.
[57] M.H. Stein, D.D. Frey, R.D. Walker, Slaughter Estate Unit CO_2 flood: comparison between pilot and field-scale performance, Journal of Petroleum Technology 44 (1992) 1026–1032.
[58] J.J. Lawrence, N.K. Maer, D. Stern, et al., Jay nitrogen tertiary recovery study: managing a mature field, in: SPE-78527-MS, Abu Dhabi International Petroleum Exhibition and Conference, October 13–16, 2002, Abu Dhabi, UAE.

[59] X. Wang, A. Strycker, Evaluation of CO_2 injection with three hydrocarbon phases, in: SPE-64723-MS, SPE International Oil and Gas Conference and Exhibition, November 7–10, 2000, Beijing, China.
[60] S.A. Aryana, C. Barclay, S. Liu, North Cross Devonian Unit—a mature continuous CO_2 flood beyond 200% HCPV injection, in: SPE-170653-MS, SPE Annual Technical Conference and Exhibition, October 27–29, 2014, Amsterdam, The Netherlands.
[61] L.E. Henderson, The use of numerical simulation to design a carbon dioxide miscible displacement project, Journal of Petroleum Technology 26 (1974) 1327–1334.
[62] G.J. Mizenko, North Cross (Devonian) Unit CO_2 flood: status report, in: SPE-24210-MS, SPE/DOE Enhanced Oil Recovery Symposium, April 22–24, 1992, Tulsa, Oklahoma, USA.

CHAPTER

Enhanced oil recovery in unconventional reservoirs

7

B. Todd Hoffman
Montana Technological University, Butte, MT, United States

7.1 Introduction

Enhanced oil recovery (EOR) in unconventional reservoirs is in its early stages, but the potential is huge given there are hundreds of billions of barrels of oil in each of the major unconventional oil basins in North America [1]. To put that in context, the largest conventional oil field, Ghawar in Saudi Arabia, is around 200 billion barrels of oil [2], and the total oil production from the United States from the first Drake wells in 1859 through the end of 2020 is around 235 billion barrels [3]. While the volumes of oil in place are large, the recovery factors for primary production are small, mostly in the single digits [4,5]; thus there is a strong need for some type of injection strategy to improve recovery from these types of reservoirs.

Over the last 10–15 years, there has been significant progress in implementing EOR in unconventional reservoirs. There have been over 30 field trials by more than 15 different operators in at least a half dozen basins. While the success of the field pilots has varied widely, significant additional oil has been produced, and more importantly, significant additional knowledge has been gained.

This chapter begins by describing unconventional reservoirs and how they differ from conventional reservoirs and how these differences impact EOR. Then, the different physical processes and recovery mechanisms for each of the major methods are discussed including how each one induces additional oil production in unconventional reservoirs. We will explore the field trials that have occurred to date and discuss where each method has been successful. There are significant lessons learned from these projects that will impact projects going forward. This chapter ends with the operational aspects and considerations needed to implement a project in the field.

7.2 Unconventional reservoirs

The term unconventional reservoir is somewhat vague, and these types of reservoirs have been called many different names, such as shale oil reservoirs, source

Recovery Improvement. DOI: https://doi.org/10.1016/B978-0-12-823363-4.00004-2

rock reservoirs, light-tight reservoirs, hard-to-produce reservoirs, and others. The unconventional reservoir is the most commonly used name to refer to them; thus that term is also used in this chapter. To ensure clarity, however, unconventional reservoirs are fully defined in this section including the features that all of them have in common.

First, unconventional reservoirs have extremely low permeability in the micro- and nano-Darcy range (0.01–0.00001 mD). This is due in part because the producing intervals are also source rocks, which are formed by small-sized particles. Due to the low permeability, unconventional reservoirs require development with long horizontal wells and multistage hydraulic fracturing. This development strategy combined with the low-permeability source rocks defines unconventional reservoirs.

Because they are the source rocks for conventional production, potential unconventional reservoirs are located across the globe (Fig. 7.1). The exploitation of unconventional reservoirs started in the United States and Canada around 2000 and has become fully developed in a number of basins. In the last 5–10 years, the process has expanded to other parts of the world including Argentina, Russia, China, and other locations.

7.2.1 Formation and geology

When sediment is eroded and washed out to sea, only the smallest particles stay in suspension and make it out to deep marine environments (Fig. 7.2). If this

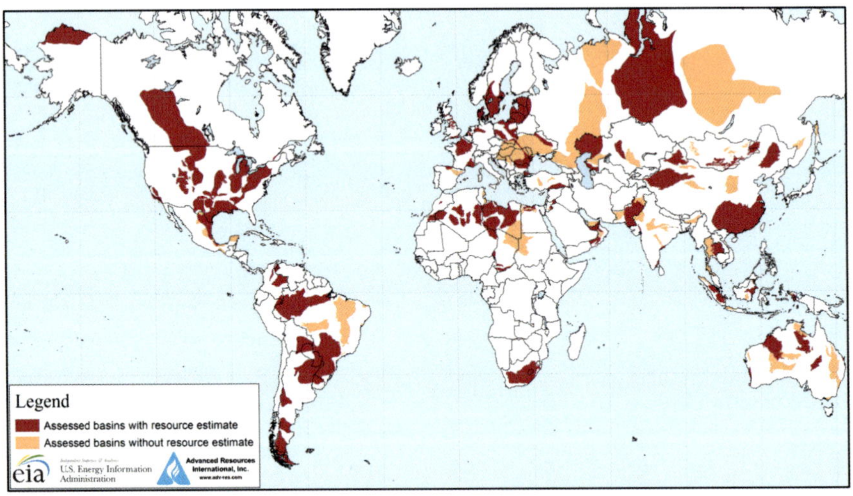

FIGURE 7.1

Location of potential unconventional reservoirs across the globe [6].

Reproduced from V.A. Kuuskraa, S.H. Stevens, K.D. Moodhe, World Shale Gas and Shale Oil Resource Assessment, US Energy Information Administration, Washington, DC, 2013.

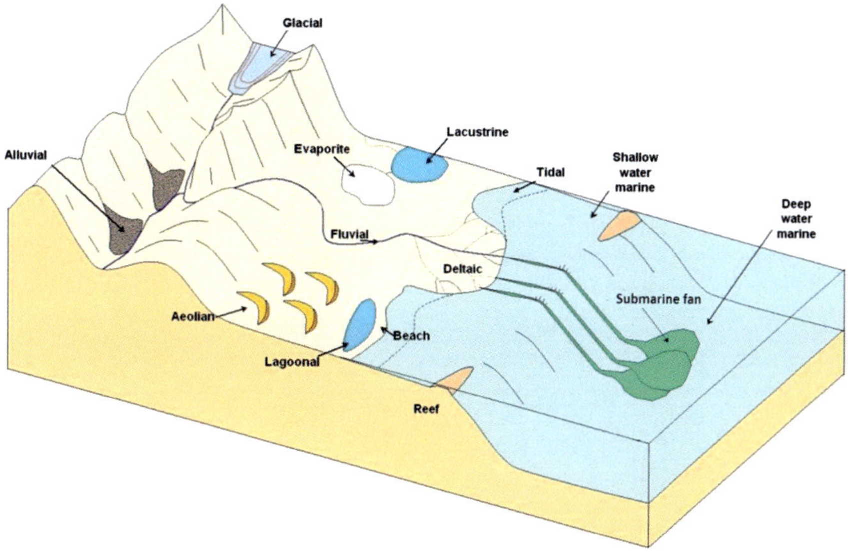

FIGURE 7.2

Terrestrial and oceanic depositional environments.

From https://en.wikipedia.org/wiki/Depositional_environment.

occurs near the equator or in a warm climate, then organic material would have flourished and been deposited on the sea floor with the small inorganic particles. The organic material is slowly transformed into kerogen, and over time, it gets buried deeper where the pressure and temperature increase. Some of the kerogen is converted into producible oil and natural gas. These buried, organic-rich sediments become hydrocarbon source rocks.

The process of converting kerogen to hydrocarbons causes a volume expansion resulting in an increase in the source rock pore pressure. Unconventional reservoirs therefore are often overpressured. The higher pressure may also create microfractures around the kerogen and an increase in formation permeability. The pressure rise may also cause some hydrocarbons to be expelled, and migrate to more permeable formations. If there is a trapping mechanism to stop the migration, the hydrocarbons form a conventional reservoir (Fig. 7.3). Regardless of how much hydrocarbon has left the source rock, significantly more remains; consequently, the amount of oil and gas in unconventional reservoirs far exceeds even the largest conventional reservoirs.

In addition to microfractures, many unconventional reservoirs have larger natural fractures caused by uplift and tectonics that may be feet to tens of feet in length. These may play a role in hydrocarbon production depending on the direction, size, and open/sealed nature of the fractures. Unconventional reservoirs also tend to have low matrix porosity due to the small pore spaces and mixtures of organic and inorganic particles.

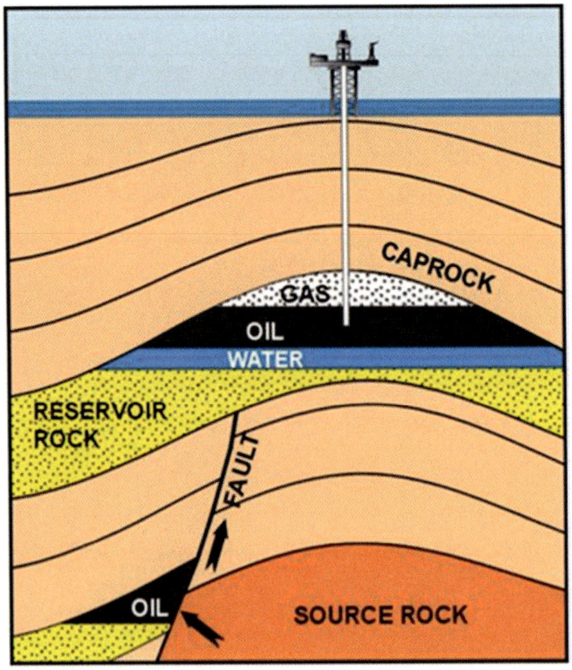

FIGURE 7.3

Simple representation of a hydrocarbon system with source rocks, migration, formation seals, and permeable reservoirs.

From https://en.wikipedia.org/wiki/petroleum.

7.2.2 Development

Due to the unique characteristics of unconventional reservoirs, techniques to exploit these types of reservoirs had to be established, namely, drilling long horizontal wells and completing multistage hydraulic fractures.

While other parts of the petroleum industry are drilling more complex and difficult horizontal wells, the sheer number of horizontal wells drilled in unconventional reservoirs makes their development unique. There have been tens of thousands of horizontal wells drilled in unconventional basins over the last 15 years with the main improvement being the time it takes to drill a well. By 2020, it is typical to drill a 20,000 ft well (10,000 ft deep and 10,000 ft lateral length) in 10 days, which is at least 3–4 times faster than in 2010. This was achieved through numerous incremental changes in both the tools developed and the techniques used.

While the improvements in horizontal drilling are impressive, the advancements in hydraulic fracturing are even more remarkable. Due to the very low permeability, hydraulic fracturing was needed for these types of formations, and some of the early wells in the Bakken used a single-stage fracture, where oilmen simply pumped sand and fluid down a well and hoped for the best. With this method, it is likely that only one or two fractures are created at the weakest points along the horizontal, but even with these simple fracture treatments, production rates improved significantly.

The idea of fracturing multiple stages in a single well started in the tight gas plays, such as the Denver-Julesburg, Uinta, and Piceance Basins, during the 1990s [7]. These were vertical wells drilled through multiple sand lenses. Each sand body needed its own fracture, so each lens (starting at the bottom) was isolated with a bridge plug. The well would be perforated and the sand lens would be fractured; then, another plug was set above it and the next zone was fractured. This was repeated until all the zones were fractured. At first, this was difficult to do in horizontal wells because once the well goes lateral, gravity no longer helps move tools to the bottom of the well. Early horizontal wells only had two or four stages, but improvements in technology allowed more and more stages to be fractured. Now it is common to do 40–50 stages in a single 10,000 ft lateral, and the record is over 100 stages [8].

Another improvement in hydraulic fracturing is the amount of fracture surface area that is created in each stage. Early attempts created bi-wing fractures that extended out perpendicular to the wellbore, but with improvements in design and execution, modern completions tend to create more fractures near the wellbore in a stimulated reservoir volume. More fracture surface area is better for primary production because there are more places for oil in the matrix to flow to fractures. This is also much better for EOR in unconventional reservoirs as there is more area for the EOR fluids to interact with the oil-in-place and recover more of it.

7.3 Unconventional enhanced oil recovery

The technology employed to develop these reservoirs is extraordinary, but primary recovery factors are not. Even with the most advanced drilling and fracturing techniques, single-digit recovery factors are typical. Due to the pore structure and low permeability, these reservoirs are hard to produce; thus an injection strategy or some types of EOR are needed to increase the recovery.

7.3.1 Screening criteria

There are numerous methods to increase recovery in oil reservoirs, but not all methods work for all types of reservoirs. Taber et al. [9] proposed an initial screening process to identify options appropriate for the characteristics of the reservoir under consideration. These are called screening criteria, and a similar process is completed here for unconventional reservoirs.

Table 7.1 Enhanced/improved oil recovery techniques.

Miscible gas	Chemical	Thermal	Other
CO_2	Polymer	Steam drive	Waterflood
Nitrogen	Surfactant	Cyclic steam	Immiscible gas
Lean gas HC	Alkaline	High-pressure air	Microbial
Rich gas hydrocarbon	Alkaline/polymer surfactant	Steam-assisted gravity drainage	Low salinity or nanofluid

Table 7.1 lists options for EOR in conventional reservoirs—many are discussed in earlier chapters of this book. Some can be quickly dismissed for unconventional reservoirs because they are simply the wrong process or physically cannot work. Unconventional reservoirs are too deep for steam injection methods; the steam would not make it to the reservoir in the vapor phase. Moreover, unconventional reservoirs have light, high-API gravity oils, so viscosity reduction, which is the main mechanism for steam injection, is not that important for unconventional reservoirs. Polymers are also problematic because of the small pore spaces in unconventional reservoirs; injecting polymers would likely damage the matrix formation and hinder production.

Other methods, such as microbial, high-pressure air, and immiscible gas injection, may be possible but are likely to be ineffective for various reasons. Microbial methods have not found commercial success in conventional reservoirs, so that makes them less likely to work for unconventional reservoirs. While there have been a few high-pressure air injection cases in light oils, this thermal method is mostly used for viscosity reduction in heavy oils. Immiscible gas methods mostly provide pressure support where gas cap injection can be used. Unconventional reservoirs do not have those structural components, so immiscible gas is inappropriate.

Water, surfactant, and miscible gas injection methods have emerged from the screening process with the highest chance of success; therefore they are the main focus of this chapter. For the remainder of this section, we discuss how these methods work, and physically why they produce additional oil. These three methods also are the ones primarily tested in the field, and those pilots are covered in an upcoming section.

7.3.2 Continuous versus cyclic injection

Besides selecting the fluids that are going to be injected, the strategy for injection also must be determined. In this subsection, two injection strategies are considered: continuous injection and cyclic injection. In the first method, fluid is continuously injected down dedicated "injection" wells and continuously produced from other dedicated "production" wells. The goal of this strategy is to (1) provide pressure support by creating a pressure gradient and (2) sweep oil and lower the residual oil saturation between the injection and production wells. This method requires patterns of wells to most efficiently recover the oil, and it is the most common injection strategy for conventional reservoirs whether EOR or waterflooding.

The other main strategy is a cyclic injection, which is often referred to as a huff-n-puff injection. In this method, the same well is used for both injection and production, and the switch from one operational mode to the other occurs rather frequently—weeks to months. The goal of this method is to increase recovery in the near-well region. The idea originated in early heavy-oil steam injection projects. Highly viscous oils make it difficult to initiate a well-to-well steam flood, but the project can be developed more efficiently if the cyclic injection is initially performed on all the wells.

With the cyclic injection, steam is injected down a well for some time (~30–60 days), then the well is shut in for a soak phase (~5–15 days) to let the heat transfer from the steam into the heavy oil, which causes its viscosity to decrease. Finally, the well is put into production (~60–180 days). After a few repeated cycles, the reservoir has higher injectivity and producibility, so a continuous steam injection project can be implemented. Most heavy-oil steam floods use cyclic injection at the beginning to get things started. In some fields, however, cyclic steam injection is used for the entire life of the field [10].

In conventional reservoirs, most miscible gas injection developments use continuous injection strategies, but there are some cyclic gas injection projects [11] where CO_2 is used as the injectant. The main advantage of cyclic injection is that the increase in production occurs sooner. Conversely, the benefit of continuous injection is that overall recovery is higher, especially between wells where much lower residual saturation can be achieved.

For unconventional reservoirs, there are advantages and disadvantages of both continuous and cyclic injection strategies. Most unconventional horizontal wells are drilled in the direction of minimum horizontal stress (Sh_{min}) so that the hydraulic fractures extend away from the wells as transverse fractures. This is good for creating lots of fracture surface area and improving primary production, but in many cases, these fractures extend to the offset wells and cause well-to-well interference. During continuous injection, this communication causes problems with rapid breakthrough and poor sweep.

If wells are drilled in the direction of maximum horizontal stress (SH_{max}), then the fractures tend to propagate along the wellbore—both above and below—and create longitudinal fractures. Injecting into these kinds of wells in a continuous fashion allows more flat-front displacement and improved sweep efficiencies. There are some projects in Canada [12] and Alaska [13] where this has been successfully implemented with waterflooding and water-alternating gas (WAG) injection, respectively. While these fields have been developed with long horizontal wells and multistage hydraulic fracturing, the permeability of these reservoirs is around 1 mD, so they are not considered unconventional by the definition in this chapter. Due to the very low matrix permeability in unconventional reservoirs, injecting into wells with longitudinal fractures may not be viable; however, since it has not been attempted, the results are unknown.

There are a few field pilots using wells with transverse hydraulic fractures where the continuous injection was tested, but these projects had major breakthroughs and conformance control issues and are discussed during the Field Trials section of this chapter. Because of these issues, most injection projects in unconventional reservoirs have used cyclic (huff-n-puff) injection strategies. Thus most of the remainder of this chapter focuses on cyclic injection methods.

7.3.3 Cyclic gas injection—conceptual model

The most common EOR method in unconventional reservoirs is cyclic gas injection. This subsection covers how the process works in the reservoir and discusses the physical recovery mechanisms that are causing additional oil to be produced.

Prior to injection, the well is produced on primary recovery to lower the local pressure and to capture the high initial rates from the wells. Most unconventional reservoirs are overpressured, so this early production aids injection by lowering the pressure. In some cases, the initial production is required to get the injection pressure below the fracture gradient. The longer the primary production, however, the further the reservoir pressure declines, so a greater volume of injected gas is needed to build pressure back up. This requires a balancing act—some initial production is necessary, but not too much. The length of time varies for different situations, but typically there are months to years of primary production.

Once injection starts, gas is sent down the wellbore and fills the highly permeable fractures; however, the volume of fractures is small, so the pressure builds rapidly. The interface between the fracture and matrix is where the injected gas and reservoir oil interact. Fig. 7.4 shows a simplified rendering of this system.

The matrix permeability is so low that gas has difficulty entering the matrix pore spaces. There are two driving forces to get gas into the matrix: diffusion and advection. Advection can be thought of as Darcy flow or pressure-driven flow where gas is forced from the higher-pressure fractures into the lower-pressure matrix. Diffusion causes gas to enter the fluids in the matrix space due to molecular

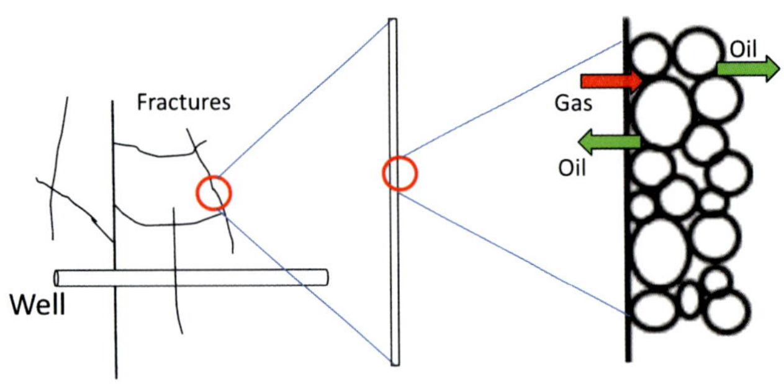

FIGURE 7.4

Conceptual model of gas injection into an unconventional reservoir system.

interactions and compositional differences. In conventional reservoirs, diffusion is small compared to advection, and it can be safely ignored. In low-permeability unconventional reservoirs, however, advection can be very slow, and in some instances, diffusion makes a meaningful contribution.

Analytical equations can be used to estimate the contribution from each process (advection vs diffusion). Darcy's Law for one-dimensional (1D) linear flow (Eq. 7.1) calculates the quantity of gas flowing into the matrix by advection, while Fick's Law (Eq. 7.2) is used to estimate the volume of gas entering due to diffusion.

$$q = 0.345\frac{(P_{\text{inj}} - P_{\text{r}})}{B}\left(\frac{k\phi c_t}{\mu t}\right)^{\frac{1}{2}} \tag{7.1}$$

$$C(x,t) = \text{erfc}\left(\frac{x}{\sqrt{4Dt}}\right) \tag{7.2}$$

where q is the flow rate of gas entering the matrix (MSCF/day), P_{inj} and P_{r} are the injection and reservoir pressures respectively (psi), B is the formation volume factor (res. ft^3/SCF), k is the matrix permeability (mD), ϕ is the matrix porosity (%), c_t is the total system compressibility (1/psi), μ is the viscosity (cp), and t is the time in hours. The constant, 0.345, accounts for unit conversion. $C(x,t)$ is the concentration of gas at a given location and time, erfc() is the complementary error function, x is the distance from the fracture, and D is the diffusion coefficient.

Darcy's Law and Fick's Law are simple models that do not fully represent the true physics of the system; however, they are reasonable approximations for comparison and they can provide an estimate of the relative contribution of both processes. The impact of advection is highly dependent on the matrix permeability, which varies significantly across unconventional reservoirs; likewise, the influence of diffusion is strongly related to the diffusion coefficient, which depends on the injection gas, reservoir fluids, and pore systems [14,15].

Using the methodology developed by Hoffman and Rutledge [16], Fig. 7.5 shows how much advection is contributing (with diffusion providing the remaining support) as a function of matrix permeability and diffusion coefficient. For high permeability systems, advection provides almost all the contribution, but for systems with lower matrix permeability and higher diffusion coefficients, over half of the gas entering the matrix may be due to diffusion.

For both advection and diffusion, the amount of gas entering the matrix is proportional to the fracture surface area, which is the main reason that the primary completion (hydraulic fracturing) is so important for successful EOR in unconventional reservoirs. There are several ways to estimate fracture surface area: volumetric, analytical, and experimental. For completions with 40–50 stages in a 10,000 ft horizontal lateral, the surface area is millions to tens of millions of square feet [16,17], which provides a lot of area for gas/oil interactions to occur. With this amount of fracturing, millions of standard cubic feet (SCF) per day of gas can be taken up by the matrix.

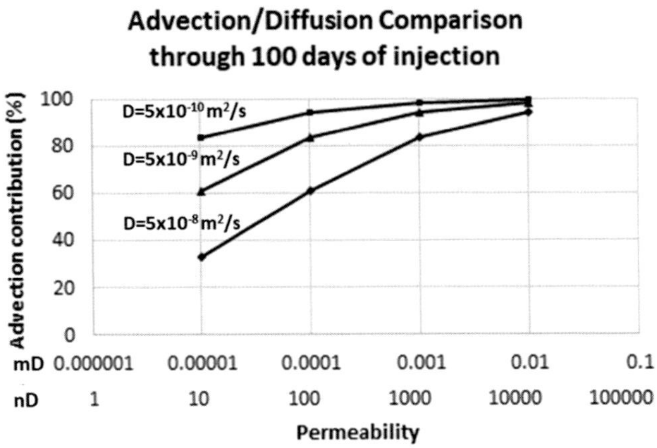

FIGURE 7.5

The relative amount of injection gas entering unconventional reservoir matrix due to advection and diffusion.

Gas entering the matrix is only the first step in the recovery process. If the injection pressure is high enough, gas will dissolve into the oil (the liquid hydrocarbon phase) and cause it to swell. Under the same conditions, the injected gas may vaporize some of the reservoir liquids into the gas phase; therefore it is important to understand the potential contributions of different recovery mechanisms.

7.3.4 Cyclic gas injection—recovery mechanisms

Recovery mechanisms are important for recognizing what is happening in the reservoir and for improving the overall reservoir performance. If the process is better understood, procedures can be developed to maximize recovery efficiency, so more oil can be produced per unit of gas injected. Without understanding the mechanisms, improving the process becomes an expensive trial-and-error endeavor. There have been numerous recovery mechanisms proposed for gas injection into liquid-rich unconventional reservoirs. Table 7.2 lists some of the proposed mechanisms, which should be considered a thorough but not exhaustive list.

To quantitatively evaluate how much each mechanism is contributing, Hoffman and Reichhardt [18] proposed a method to isolate each mechanism and to determine how much each one is contributing. They constructed a numerical model where each mechanism could be turned on and off, such that the contribution from each one could be resolved. They were able to evaluate the first five mechanisms listed in Table 7.2 with their method.

The work was carried out for reservoir fluids that ranged from black oils to retrograde gas condensates; the initial gas–oil ratios (GORs) varied from 500 SCF/STB to 20,000 SCF/STB. Two main conclusions were drawn from this

Table 7.2 Proposed recovery mechanisms for gas injection into unconventional oil reservoirs.

Recovery mechanisms	
Vaporization	Viscosity reduction
Oil swelling (expansion)	Pressure support
Secondary solution gas drive	Wettability change
Injection-induced fracturing	Interfacial tension reduction

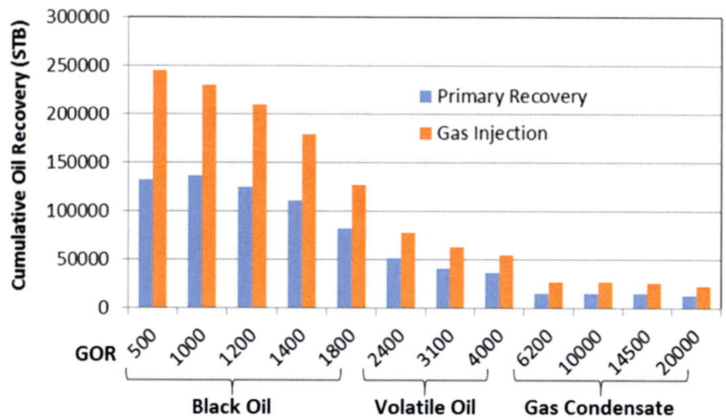

FIGURE 7.6

Cumulative oil recovery from primary production and from the gas injection as a function of initial GOR [18].

Reproduced from B.T. Hoffman, D. Reichhardt, Recovery mechanisms for cyclic (huff-n-puff) gas injection in unconventional reservoirs: a quantitative evaluation using numerical simulation, Energies 13 (2020) 4944.

work. Their first conclusion is that recovery is a strong function of initial GOR, with the highest total recoveries for black oils (Fig. 7.6). This makes sense because black oils have more liquids per bulk reservoir volume than gas condensates, so there is more oil available to recover.

Fig. 7.6 also demonstrates that additional recovery is observed for all reservoir fluid types, and the incremental production for each fluid is about $40\% - 60\%$ more than primary recovery. The potential for gas condensate reservoirs is still high, but if a company has acreage in both black oil and gas condensate regions, it may be more economical to perform EOR in the black oil area first.

The second main conclusion, summarized in Fig. 7.7, is that there is a large variance in how important each mechanism is for different reservoir fluids. The y-axis on Fig. 7.7 (Efficiency) represents the fraction of EOR recovery compared to primary recovery.

For reservoirs with lower initial GOR (black oils), the main contribution is gas dissolving into the oil phase, which causes swelling and secondary solution gas

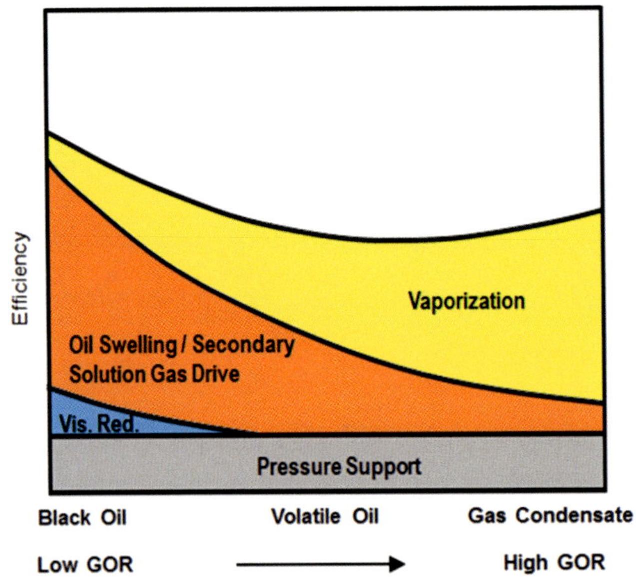

FIGURE 7.7

Contribution of different recovery mechanisms for cyclic gas injection in unconventional reservoirs [18].

Reproduced from B.T. Hoffman, D. Reichhardt, Recovery mechanisms for cyclic (huff-n-puff) gas injection in unconventional reservoirs: a quantitative evaluation using numerical simulation, Energies 13 (2020) 4944.

drive. The increased volume, due to swelling forces some of the oil from the matrix into the fractures. In addition, when the pressure is reduced during production, the injected gas comes out of the solution and provides pressure to push oil toward the well, similar to the primary solution gas drive. The authors were unable to isolate the relative amounts these two mechanisms provided, so they are presented as one category.

For higher initial GOR areas (gas condensates), vaporization of liquid hydrocarbons into the gas phase is the most significant mechanism, which makes sense because liquid dropping out near the wellbore is a significant problem for gas condensate reservoirs. Vaporization solves this problem by transporting the liquids in the gas phase. Pressure support provides a small contribution for all reservoir fluid types, and viscosity reduction plays, at most, a minor role in black oil reservoirs.

Other mechanisms, such as wettability changes or additional fracturing, may also contribute, but they could not be evaluated with this methodology. Sensitivity analysis was completed on operational parameters (e.g., injection rates and cycle times) and reservoir properties (e.g., matrix permeability). These showed that higher injection rates improve production by shortening injection times and lengthening production cycles. Also, as patterns become more mature, they tend to be more affected by vaporization than less developed patterns. Later in the project life, there is more free gas present because the reservoir liquid composition has changed and because more injected gas has remained in the reservoir, which allows more vaporization.

7.3.5 Cyclic gas injection—injection gases

The miscible gas injection has been successfully implemented on a large number of conventional oil reservoirs; mostly carbon dioxide (CO_2) floods in places, such as West Texas and Wyoming, as well as natural gas injection projects on the North Slope of Alaska. In these cases, it has been almost exclusively continuous injection strategies between injectors and producers.

Miscibility is the degree that two fluids mix and is a function of pressure, temperature, and the properties of the two fluids. The minimum miscibility pressure (MMP) is the lowest pressure that these two fluids mix. It is an important parameter in conventional gas injection because once this pressure is attained, a single-phase mixture of reservoir and injection fluids can travel through the reservoir and result in very low residual oil saturation (Fig. 7.8). Increasing pressure above the MMP does not result in higher recovery.

In cyclic gas injection, the importance of miscibility is different. Instead of creating a miscible bank, injected gas mixes with oil and causes it to swell. Alternatively, for gas condensate systems, as the pressure reaches the dew point, it vaporizes liquid into the gas phase. Also, some research [20] shows that as pressure is increased above MMP, the amount of recovery also increases. The relationship between MMP and cyclic gas injection in unconventional reservoirs is not fully understood at this point in time.

There are a number of different gases that can be injected for EOR, yet so far in this chapter, only a generic injection gas has been discussed. The type of injection gas has a large impact on the effectiveness of recovery. Table 7.3 lists six gases that have been proposed for EOR in unconventional reservoirs. The advantages and disadvantages of each gas are considered below.

Lean hydrocarbon gases are 90%–95% methane, and typically, they have gone through a gas plant where the heavier components have been stripped off and sold separately. These gases can be used for EOR where they occur, but depending on where the gas plant is located, it may require long flowlines to get the gas to the injection wells. It will cost less per unit than other hydrocarbon gases, as the more valuable parts have already been extracted and sold. Because it is mostly methane, lean gas requires higher pressures and thus more volume to achieve MMP. For gas condensate reservoirs, where vaporization is the primary recovery mechanism, lean gas likely performs better than other potential injection gases.

Rich hydrocarbon gas is mostly field gas that is produced with oil. It may be lightly treated at the well location, but it contains a significant fraction of ethane and higher-weight hydrocarbon components. It varies by location and over time,

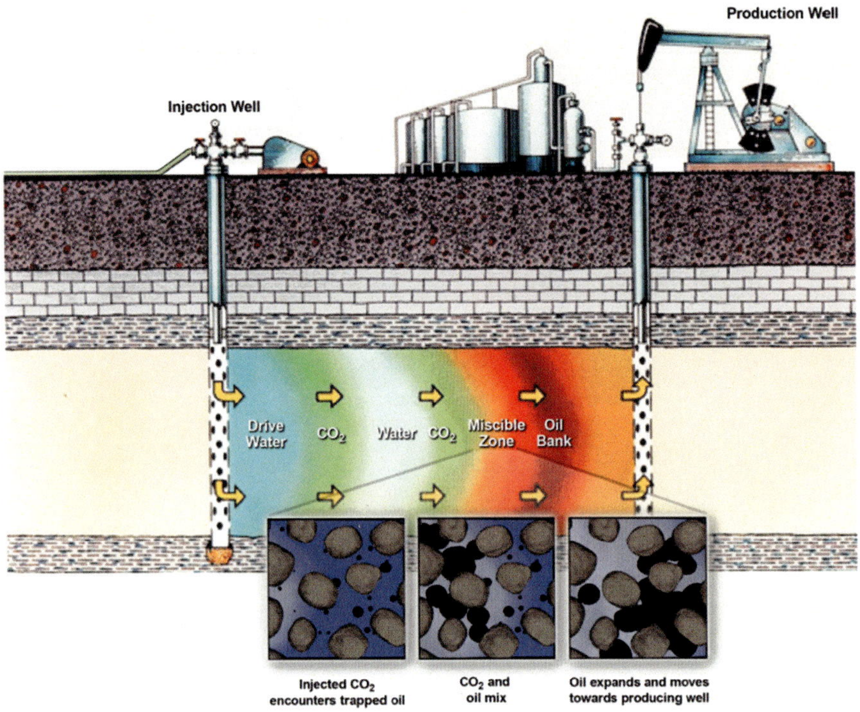

FIGURE 7.8

Cross-section illustrating how a miscible bank is formed in a CO_2 flood [19].

Reproduced from National Energy Technology Laboratory, Carbon Dioxide Enhanced Oil Recovery: Untapped Domestic Energy Supply and Long-Term Carbon Storage Solution, US Department of Energy, Washington, DC, 2010.

Table 7.3 Potential gases for injection into unconventional enhanced oil recovery (EOR) projects.

EOR injection gases	
Lean hydrocarbon	Rich hydrocarbon
Carbon dioxide	Ethane
Liquified petroleum (propane/butane)	Nitrogen

but most unconventional oil basins have field gas with 60%–80% methane and 20%–40% heavier hydrocarbon components. One advantage is that rich gas can dissolve in the oil at a lower pressure. In many unconventional reservoir locations, there is more produced gas than can be treated, so a significant portion is being flared [21]. If produced gas can be captured and injected, it is not wasted and can be produced again later. The injection equipment required is similar to lean gas injection, but the compressors need to be able to manage liquids dropping out between each stage, which most modern equipment can do.

Carbon dioxide is the most common gas used in conventional reservoir gas floods, but its use in unconventional gas injection has been limited. It is superior to lean gas for miscibility, but depending on the composition of rich gas, CO_2 may be only slightly better when mixing with unconventional reservoir oils. The two main issues with CO_2 are availability and corrosion. Most unconventional basins do not have a CO_2 source nearby. The Permian basin is the exception since CO_2 flooding occurs at neighboring conventional fields, and a few CO_2 pilots in unconventional reservoirs have been performed there. The other concern is when CO_2 and water are in contact, carbonic acid is formed, which causes operational issues. Downhole and surface equipment must be replaced with materials that can handle corrosive environments; this can be a considerable added expense. CO_2 does dissolve easily into reservoir oils, and when available, CO_2 is less expensive per unit than hydrocarbon gas, so there are advantages of CO_2. So far, however, the disadvantages of CO_2 outweigh the advantages for unconventional EOR.

Ethane is only available from gas plants, but recently in some of the major unconventional basins, the supply of ethane has surpassed the demand, leaving a potential low-cost supply. Ethane is more miscible than CO_2 or rich hydrocarbon gas [22], so it may be a very effective injectant, but long-term availability may be an issue if ethane demand increases. Research with ethane has been limited to laboratories and modeling with no field trials of ethane injection to date.

Liquified petroleum gases (LPGs), such as propane and butane, have the potential to recover additional hydrocarbons from unconventional reservoirs. LPGs are more expensive per unit, but less volume may be needed to see the same benefits as other hydrocarbon gases. LPGs mix with in situ hydrocarbons more completely and easily because LPGs are liquid at injection pressures and temperatures. This may also reduce operational expenses because LPGs can be pumped downhole instead of requiring a large compressor. There has been at least one pilot using propane for injection [23], and it performed well.

Nitrogen has been proposed for injection, but it has not been studied as much as the other gases. It requires a very high pressure for nitrogen to go into the solution, so its effectiveness for swelling oil is limited. There are some immiscible nitrogen floods in conventional continuous injection projects [24], but in cyclic gas injection, the advantages of immiscible flooding are not present. Nitrogen has been used in combination with CO_2 or methane in conventional retrograde gas condensate reservoirs with success [25], so it may have potential in unconventional condensate reservoirs, but this has not been studied.

7.3.6 Water injection

In addition to gas injection methods, other EOR methods for unconventional reservoirs have been reported and tested in the field. Two of them will be discussed in this chapter: water injection and surfactant injection.

Waterflooding is by far the most common injection strategy for conventional reservoirs, with most reservoirs receiving water injection at some point in their life. It is no surprise therefore that water injection was one of the first methods considered for unconventional reservoirs. Due to the low permeability of unconventional reservoirs, however, there was concern early on that injectivity would be too low to achieve a meaningful rate. This is potentially significant because water is much more viscous than gas. Nevertheless, field pilots from the early 2010s showed that injection rates were sufficiently high, which was attributed to hydraulic and natural fractures that provided permeability.

Similar to gas injection in unconventional reservoirs, two different injection strategies were considered for water injection: continuous and cyclic. Most wells are drilled in the direction of Sh_{min} and have transverse fractures. For continuous injection, the injected water broke through to the production wells very quickly and did not sweep the reservoir oil; therefore most continuous waterfloods pilots in unconventional reservoirs did not perform well.

The cyclic injection strategy has been the most common method implemented in unconventional reservoirs. In this case, water is injected down the wellbore and easily fills the hydraulic fracture networks. Water enters the matrix by spontaneous imbibition and advection (forced imbibition). For water-wet rocks, the solid surface has more attraction to water than oil, so the formation pulls water from the fracture and displaces oil in the matrix. During injection, the pressure in the fractures is higher than the pressure in the matrix, so additional water is forced into the matrix based on Darcy's Law. In cyclic injection, there is counter-current flow; as water is entering the matrix from the fracture, oil is exiting the matrix to the fracture. Hatiboglu and Babadagli [26] present experimental results that show this phenomenon. When the well is put back into production, the oil will be produced from the fractures.

There has been extensive research on the behavior of water with oil-saturated unconventional rocks [27,28]. Kurtoglu [29] showed an Amott cell experiment with a sidewall core from the Bakken formation and microlayers can be observed where oil is expelling out along the striations (Fig. 7.9).

In addition to the experimental data, anecdotal evidence demonstrated the potential for cyclic water injection. In some formations, oil rates increased in offset wells when they were shut in for longer periods after the initial completion, and there was less fracture-water flow back in these wells. This indicates that completion fluids were being imbibed into the formation and oil was being transported out of the matrix [30]. A well in the Niobrara formation actually produced over 13,000 STB of oil during the drilling process as drilling mud was imbibing into the formation.

This effect only occurs in strongly water-wet rocks, but many unconventional reservoirs have oil-wet tendencies since they often contain a large amount of organic material and clay particles. Surfactants and other injectants can cause a formation to become more water wet, so there has been significant work using these processes to increase recovery from unconventional reservoirs.

FIGURE 7.9

Amott cell experiment on Bakken sidewall core showing oil expelling along microlayer beds [29].

Reproduced from B. Kurtoglu, Integrated Reservoir Characterization and Modeling in Support of Enhanced Oil Recovery for Bakken Effects of CO_2 Flooding Horizontal Wells in a Tight Oil Reservoir: Elm Coulee Field Case Study (Ph.D. dissertation), Colorado School of Mines, 2013, with permission.

7.3.7 Altering wettability

Surfactants injected with water can effectively increase oil recovery from unconventional reservoirs. The surfactants alter the interfacial tension, thereby modifying the wetting and spreading properties of the reservoir. The rock–fluid and fluid–fluid interactions change, causing the system to become more water wet.

For unconventional reservoirs, cyclic injection is likely to be employed for wettability alteration. Surfactant-laden water quickly fills the extensive fracture network during the injection. Some of the injected fluid enters the matrix due to Darcy flow. At the molecular scale, surfactants interact with the solid pore walls, where they attach to the walls and/or release molecules from the walls [31,32]. Fig. 7.10 demonstrates this process.

These interactions change the wettability and allow more injected fluid to imbibe into the reservoir. At the same time, some of the surfactant molecules are working at the oil–water interface to reduce the interfacial tension. This allows the oil and water to flow more easily within the pore spaces. When the well is put back into production, more oil is able to go from the matrix to the wellbore.

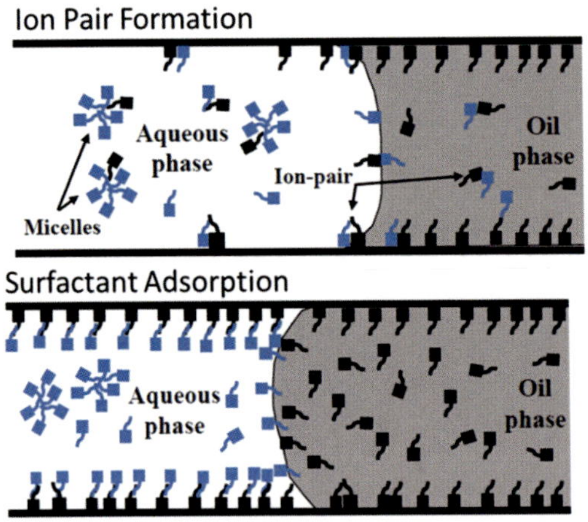

FIGURE 7.10

Conceptual model of surfactant molecules causing a change in wettability and interfacial tension [32].

Reproduced from J.O. Alvarez, Wettability Alteration Using Surfactants to Improve Oil Recovery From Unconventional Liquid Reservoirs (Ph.D. dissertation), Texas A&M University, 2017, with permission.

Laboratory studies show that contact angles—and thus the wettability—are altered when surfactants contact unconventional rocks [33,34]. Furthermore, these and other studies [35] have shown that oil recovery is increased due to imbibition with surfactants. Experiments with cores from a number of different unconventional basins and formations have produced 3–4 times as much oil with surfactants than injecting water. Fig. 7.11 shows one of these experiments for a Bakken core.

Field trials of surfactant injection are limited, but there are a few published cases that indicate additional oil recovery in unconventional reservoirs [36]. There are potential issues with surfactants, as identified by Alfrage et al. [37]. These issues include: (1) adsorption of surfactant onto solid rock surfaces, (2) imbibition may be uneconomically slow at field scales, (3) high salinity formation waters may reduce effectiveness, and (4) cost of surfactant may not be offset by additional oil. The first three of these concerns can be overcome by injecting a larger quantity of surfactants; however, due to the cost of surfactants, injecting more may not be economically feasible.

Additional methods have been suggested to address this situation; for example, one proposal is to inject additional surfactants during the hydraulic fracturing process. Some surfactants are used during the hydraulic fracturing process, but to increase oil recovery, a greater volume and different types of surfactant are needed. These would enter the formation with the proppant and fracturing fluids [38]. After fracturing, stages and wells may be shut in for days or weeks while the well is waiting to be brought online. This time would allow for larger soak times and may get the surfactant to go further into the reservoir.

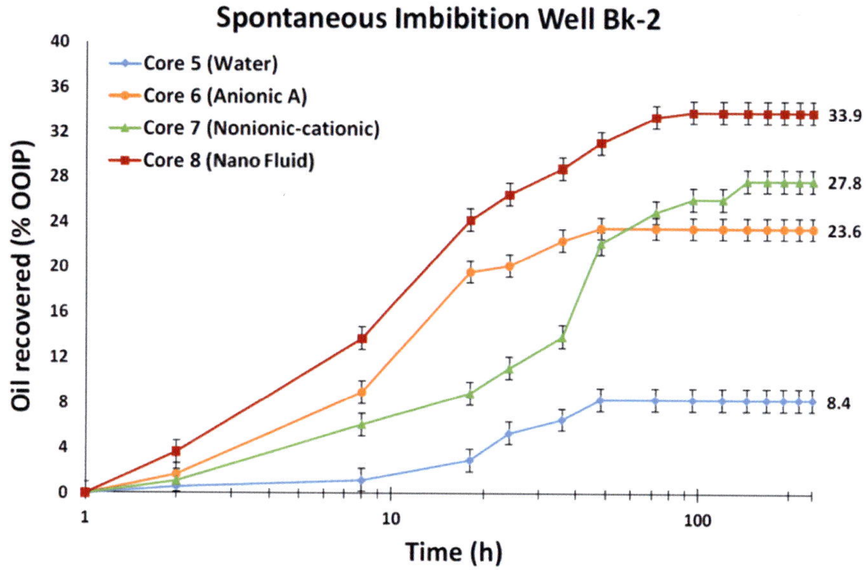

FIGURE 7.11

Bakken core flood results showing increased recovery when surfactants are used instead of pure water [32].

Reproduced from J.O. Alvarez, Wettability Alteration Using Surfactants to Improve Oil Recovery From Unconventional Liquid Reservoirs (Ph.D. dissertation), Texas A&M University, 2017, with permission.

A promising EOR technique in conventional reservoirs is low-salinity waterflooding (LSW). The method is discussed in earlier chapters of this book. Due to a number of mechanisms, LSW alters the wettability of rocks towards water wet. Additionally, LSW could be combined with surfactants where the amount of surfactant could be significantly reduced. There have been some studies to apply LSW in unconventional reservoirs [39,40] that have demonstrated promise.

The last technique discussed is using nanofluids for EOR. Nanofluids have microscopic particles that provide an alternate and possibly more effective method to change the contact angle and wettability in oil reservoirs. Fig. 7.12 is a conceptual drawing showing how nanofluids could impact the contact angles. Laboratory trials [42] have shown its effectiveness, but published field tests are limited.

7.3.8 Summary

In the current section, a number of potential EOR methods for unconventional reservoirs are discussed, including gas injection, water injection, and surfactant injection. Each method is described in detail including conceptually how they work and the potential recovery mechanisms. Different injection strategies, such

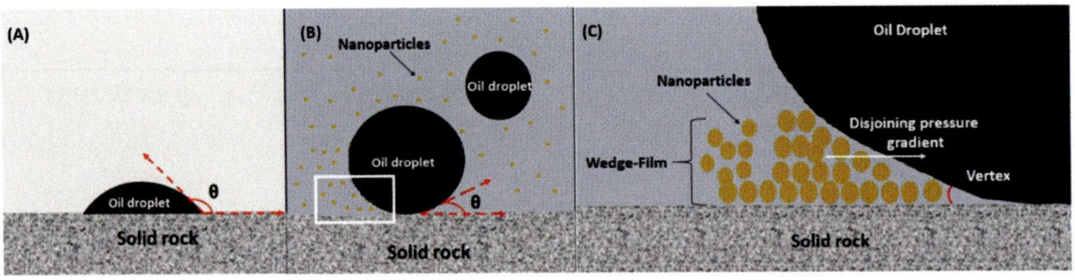

FIGURE 7.12

Enhanced oil recovery mechanism of a nanofluid (A) without nanoparticles, (B) in the presence of nanoparticles, and (C) the wedge film of nanoparticles leads to a disjoining pressure gradient [41].

Reproduced from F. Sagala, T. Montoya, A. Hethnawi, et al., Nanopyroxene-based nanofluids for enhanced oil recovery in sandstone cores at reservoir temperature, Energy & Fuels 33 (2019) 877–890, with permission.

as continuous and cyclic, were also compared. In the next section, field trials (or pilot tests) are examined for the various methods. Learnings and conclusions are drawn from each one, and best practices are provided.

7.4 Field trials

There have been over 30 EOR pilot tests within at least 6 unconventional basins in the United States and Canada with substantial variability in the types of fluids injected and injection strategies. From these pilot tests, significant knowledge has been gained regarding unconventional EOR processes. Pilot tests from the Bakken, Eagle Ford, and Permian Basin are highlighted in this chapter.

7.4.1 Early Bakken trials

The Bakken formation was the earliest unconventional oil development; thus it was the first area to see the need for EOR. There were at least seven pilot injection projects in the Bakken between 2008 and 2014. Two of the pilots were performed in the Elm Coulee field on the Montana side of the Bakken [43], and five of the pilots were in North Dakota [44]. Fig. 7.13 shows the general locations where the pilots took place in Richland County, Montana, and Mountrail County, North Dakota (green ovals).

A summary of the locations, the dates, the type of injection fluids, and the injection strategy is given in Table 7.4.

Of the seven field trials, three used water as the injecting fluid and four used gas. Of the gas injection pilots, three used carbon dioxide (CO_2) and one used enriched natural gas. The earliest tests were designed as injectivity tests with a cyclic injection strategy. The objectives of these tests were to see if any fluid could be injected into such low-permeability formations and if any additional oil

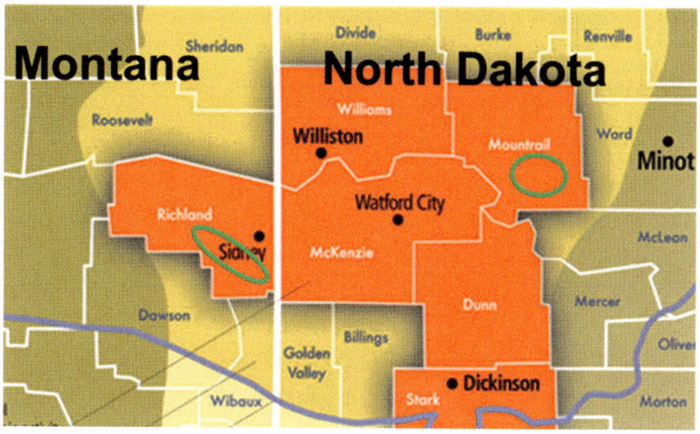

FIGURE 7.13

General locations (*green circles*) of seven injection pilot tests in the Bakken formation.

Table 7.4 Summary of early pilot tests in the Bakken.

Name	State	Year	Fluid	Type
Pilot Test #1	ND	2008	CO_2	Huff-n-puff
Pilot Test #2	MT	2009	CO_2	Huff-n-puff
Pilot Test #3	ND	2012	Water	Huff-n-puff
Pilot Test #4	ND	2012–13	Water	Flood
Pilot Test #5	ND	2014	CO_2	Vertical inj.
Pilot Test #6	MT	2014	Water	Flood
Pilot Test #7	ND	2014	Nat. gas	Flood

would be produced once the injection well was converted back to production. These included two CO_2 pilots and one water injection pilot.

Three of the remaining trials examined continuous injection strategies where increased production was expected at wells offset from the injector. Two of these pilots were water injection projects, and the third was an enriched natural gas injection project that followed one of the water injection pilots. The other test was a CO_2 injection pilot where the injection occurred in a vertical well that was near a horizontal well.

7.4.1.1 Injectivity cyclic pilot tests

The first attempts at fluid injection in the Bakken were designed to evaluate the feasibility of injecting into the submillidarcy rock. The projects used CO_2 mostly likely because (1) most conventional gas floods at the time used CO_2 instead of hydrocarbon gas and (2) it is less viscous and easier to inject than water. Two operators tested two different areas of the Bakken about 6 months apart with similar results. Both were able to easily inject the gas. In one project, 1000 MSCF/day was injected for approximately 30 days and the other project injected 1500–2000 MSCF/day for 45 days at 2000–3000 psi, which is well below the fracture pressure.

Both projects were successful from an injectivity standpoint. They were able to inject at the planned rates of 1–2 million SCF/day, and the injection pressures did not exceed the limits. In the first case, an increase in CO_2 concentration was observed at an offset well about 5000 ft away in less than 2 weeks. The increased CO_2 rate at the offset well was calculated to be about 4 MSCF/day. There was no increased oil production at that or any other offset well.

The second objective of these early projects was to determine if CO_2 would stimulate and ultimately increase production from the wells. Unfortunately, this second objective was not realized for either project. Fig. 7.14 shows the production response for both wells. In Pilot Test #1, there was no observed increase in oil production. For Pilot Test #2, the oil rate increased once the well went back into production; however, the increase did not fully offset the production lost during injection and shut-in times

A third cyclic pilot test was performed in the Bakken using water as the injection fluid to determine if counter-current imbibition would cause additional recovery. In this case, there were two cycles of injection. Once again, injectivity did not appear to be an issue; around 1200 barrels of water were injected per day. Each injection cycle lasted just over a month, and it seemed that there were no problems injecting that amount of fluid into the well. After the first injection cycle, the well was shut in for a soak period of around 2 weeks. The well was put back on production for about 3–4 months before the second injection period started. After the second injection cycle, the well was shut in again for about 2 weeks and then allowed to produce.

Fig. 7.15 shows the fluid production responses from the cyclic water pilot test. The two injection cycles are indicated in the figure, and there does not appear to be any additional oil produced after either cycle.

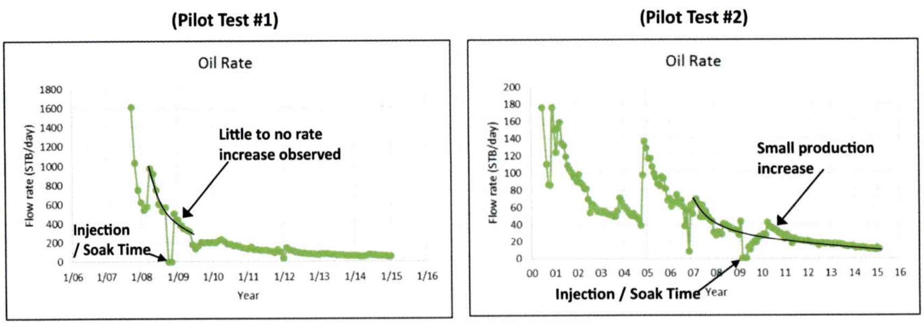

FIGURE 7.14

Production from two Bakken wells performing huff-n-puff CO_2 injection.

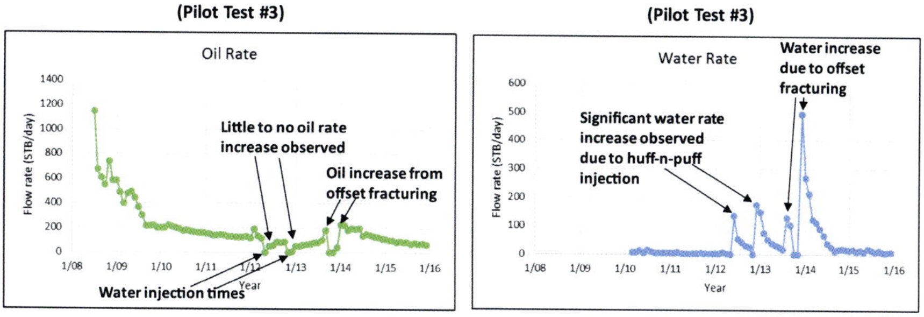

FIGURE 7.15

Oil and water production from water huff-n-puff operation in Bakken.

Interestingly, there is a point in early 2014 where an increase in oil production is observed after a shut-in period, so one might speculate that this was a third injection cycle. This occurred, however, when newly drilled offset wells were being hydraulically fractured. The offset wells were around 1000–3000 ft away from the original well. The increase in oil and especially water production in early 2014 was likely due to well interference or "frac-hits" where hydraulic fracturing at a nearby well causes a response (sometimes positive and sometimes negative) in an offset well [45]. For this well, it actually happened twice, which is more noticeable in the water rate plot (Fig. 7.15). Offset well communication during fracturing is not a focus of this chapter, but it is important to understand whether well responses are caused by injection or something else, and frac-hits affected many of the early Bakken pilots.

Pilot Test #5 was a planned CO_2 cyclic operation with a vertical well that had about 60 ft of middle Bakken pay thickness. The plan was to inject at 300–500 MSCF/day for 20 to 30 days, shut the well in for 20 days and then turn the well back on production; however, CO_2 was observed at an offset well about 900 ft away in less than 24 hours, and operations were shut down shortly thereafter. The vertical well had only about one-hundredth of the pay zone compared to the horizontal well pilots, but the gas rates were only about two-tenths less than the horizontal well projects. Therefore the per foot injectivity is about 20 times higher for this case than in the previously discussed pilots, which is presumably the reason gas broke through so quickly in this pilot.

7.4.1.2 Continuous injection pilot tests

The results of the cyclic injection projects revealed that both gas and water could easily be injected into the low-permeability Bakken formation with hydraulically fractured wells; however, there was little to no additional oil production from these tests. The first pilot where the goal was to increase recovery at an offset well occurred in North Dakota, and water was used as the injectant (Pilot Test #4).

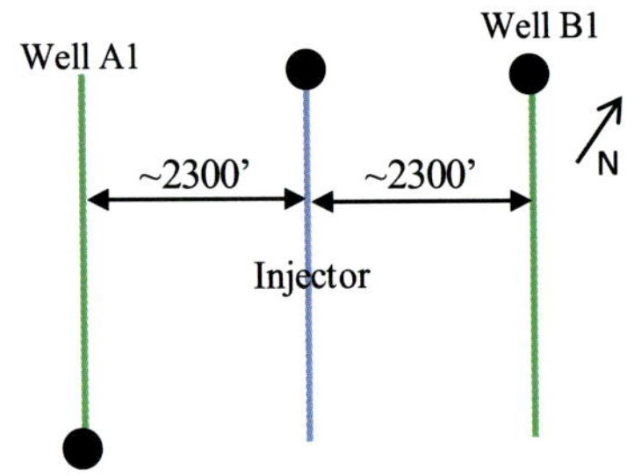

FIGURE 7.16

General layout of Pilot Test #4.

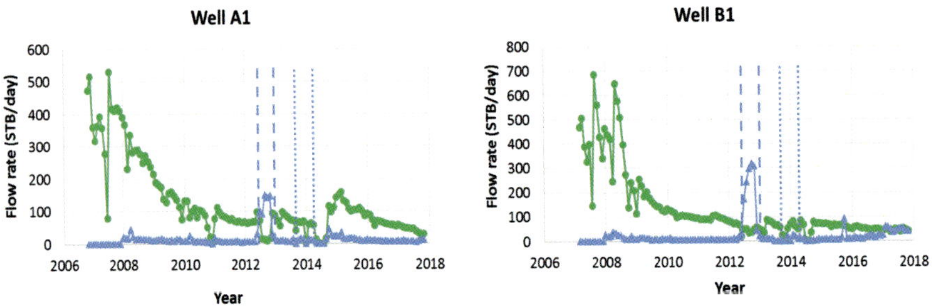

FIGURE 7.17

Oil and water production from wells A1 and B1 in Pilot #4.

Fig. 7.16 shows the general layout of the pilot. There were offsets to the east and west of the injector that was approximately 2300 ft away (A1 and B1).

Figure 7.17 shows oil and water production for wells A1 and B1. Around 1350 STB/day of water was injected for 8 months in 2012, which caused the bottom hole pressure to increase by 1500 psi—to around 6000 psi. The offset wells both had a significant increase in water production during this time, and a breakthrough occurred in less than a month for both wells. Unfortunately, no material improvement in oil productivity was realized. If anything, the oil rates may have decreased, but this could be due to the high volumes of water choking off some of the oil.

After the first round of injection, the well stopped injecting for about 6 months. This was due to the early water breakthrough and resulting high water cuts observed at the two offset wells. During this time, the water cuts in the offset wells returned to preinjection levels. The injection was started again for another 8 months, but this time the rate was much lower, around 380 STB/day, and the bottom hole pressure did not increase. In addition, the water production at the offset wells did not increase as before. This showed that injection can occur and maintain pressure while not quickly breaking through to the offset producers;

nonetheless, there was still no increase in oil production from offset wells. Overall, there was about 440,000 STB of water injected but only about 65,000 STB of additional water produced, with the rest of the water remaining in the reservoir.

After the water injection ended, the injection well was produced intermittently (about 1/3 of the time) the for next 4 months but very little oil was produced. The well was then converted into a gas injector; this is listed as Pilot Test #7 in Table 7.4. For this pilot, enriched natural gas was used as the injectant. The gas was sourced from surrounding production wells and was approximately 55% methane, 10% nitrogen, and 35% C_{2+} fractions. Gas was injected for 55 days in the middle of 2014 at an average injection rate of 1600 MSCF/day. The target surface injection pressure was 3500 psi.

There was increased production in the offset wells immediately after the gas injection, but the interpretation is complicated by the fact that wells further to the west were being hydraulically fractured during the same time, so while clear increases were observed in well A1, it is less clear if this was caused by gas injection. After 1 week of injection, an increase in gas was observed at well B1 by about 160 MSCF/day; hence, around 10% of the injected gas was being produced at B1. The oil rates may have flattened for a short time, but there was no sustained oil increase, and it quickly returned to its previous decline (Fig. 7.18).

While the gas flood was discontinued after only 55 days, this pilot indicates that rich natural gas can be used as an injection fluid, and since it is more available than CO_2 in unconventional oil plays, it may be a more appropriate injection gas. Furthermore, using natural gas will prevent flaring, which continues to be an issue in some parts of the Bakken and Permian.

The final early Bakken pilot is a waterflood in Montana (Pilot Test #6). There is one injection well and at least four offset wells were monitored for increases in pressure and flow rate. For the first 3 months, the well was injecting around 1700 STB/day of water; then, the rates were reduced to just under 1000 STB/day for the last 5 months of the project. Just like the North Dakota waterflood pilot, the

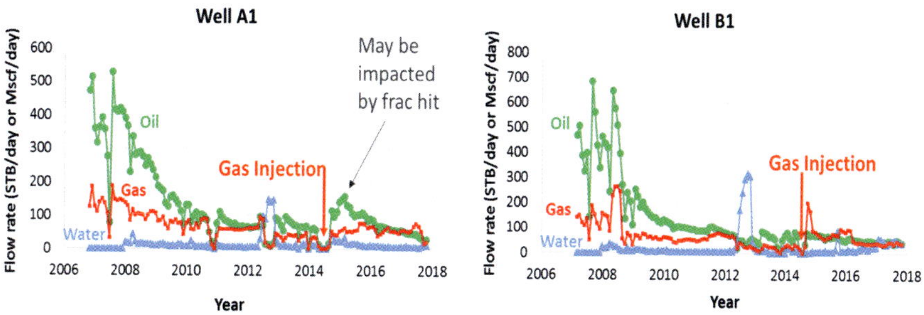

FIGURE 7.18

Oil and gas production from well Offset-North of Pilot #4/7.

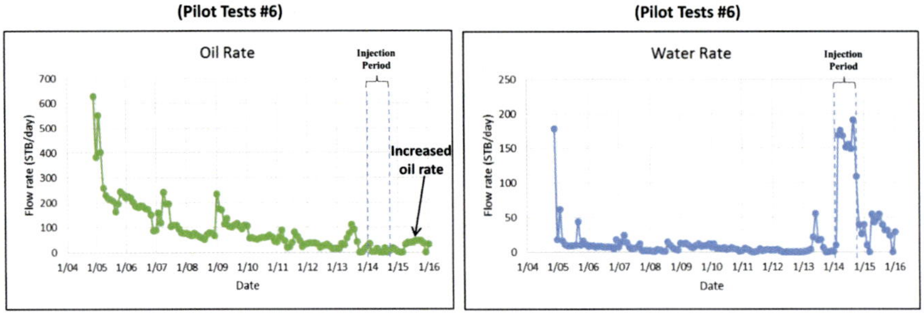

FIGURE 7.19

Oil and water production rates from offset well in Montana waterflood pilot.

results of the Montana pilot are obscured because nearby wells were being hydraulically fractured at the same time water was being injected. Almost all of the surrounding wells had increases in oil and water rates after the water injection started; however, most of these instances could be attributed to frac-hits.

The closest offset well (about 880 ft away) was clearly impacted by the injection well indicated by the increase in water production, but the oil rate did not increase during this time (Fig. 7.19). The breakthrough happened about 1 week after the injection began. Injection rates decreased after the breakthrough possible because of the water production. In addition, the average tubing pressure on the injection well started to increase, which might have contributed to the decrease in injection rate.

During the last month of injection, the operator ran an injection profile log and determined that half of the injection was going in the two frac stages closest to the heel of the well while the rest of the water was spread out over the other nine stages. They made a couple of attempts over the next few months to bypass these two stages during injection, but they were unsuccessful and the injection well remained shut-in.

7.4.1.3 Summary

There were several pilots in the Bakken formation between 2008 and 2014. While they were carried out by multiple operators, in many different locations, and used various injection fluids, there are some similarities among all of them. Thus some general conclusions can be drawn from the results of the projects.

Injectivity does not appear to be an issue for either gas or water. Initially, there was concern that it would be difficult to inject into these very low-permeability reservoirs but that does not seem to be a problem.

All the pilots were performed in stimulated wells, so it is unclear whether injectivity is assured because there is enough increase in conductivity due to the hydraulic fractures or if the multithousand feet of reservoir contact from a horizontal well would be enough for easy injection even if the well was unstimulated.

Conformance control is a major issue. Breakthrough times in days and weeks seemed to be the normal case in these early pilots. This not only caused fluid handling issues in the production wells, but in some instances, it also caused the oil rates to be inhibited.

There could be several reasons for the conformance control issues. The hydraulic fractures in both the producers and injectors may have caused the wells to link up resulting in early breakthrough times and poor sweep efficiencies. It could also be that natural fractures in the Bakken caused the poor performance. Another possibility is that some combination of both hydraulic and natural fractures led to the problems.

Increased oil recovery was insignificant from these early trials. Most pilots did not show any increase in oil production due to the injection, and for the ones that did, the increase was minimal or possibly caused by frac-hits. The volumes of gas injected in the cyclic process are small compared to other locations, such as the Eagle Ford, which may have influenced the recovery as well.

While these early pilots were valuable and much was learned from them, they were not necessarily designed to optimize oil rates, so it is not surprising that higher recovery factors were not realized. Because there is still so much oil remaining in these reservoirs, it seemed prudent to try a second round of pilots using what was learned from these ones, and that was done in the Eagle Ford.

7.4.2 Eagle Ford pilots

The unconventional basin with the most EOR development is the Eagle Ford. In the Bakken, many different injection fluids and injection strategies were tried, but in the Eagle Ford, only cyclic injection with hydrocarbon gases has been implemented. EOG Resources, Inc. (EOG) executed the first pilot in 2013 with a single well, and based on its success, they expanded with three multiwell pilots in 2015. The results of these field tests were reported in an investor relations presentation in May 2016 [46]. They indicated that wells could produce 30% to 70% more oil than primary production if the cyclic gas injection was deployed (Fig. 7.20).

The potential recovery EOG envisioned spurred them and other operators to test further areas in the Eagle Ford. By 2020, there have been over 20 projects by at least 6 companies from Dimmit County in the Southwest to Madison County in the Northeast (Fig. 7.21) with over half of the projects being executed by EOG.

The green background in Fig. 7.21 represents wells completed in the most liquid-producing regions of the Eagle Ford, and the red represents wells in the gas portion. The circles highlight the general area where the pilots took place, and the different colors represent the different operators.

All the pilots used hydrocarbon gas, but the composition of the gas varied across the locations with some using almost pure methane and others with a significant fraction of heavier hydrocarbons. In addition, all of the field trials used a cyclic injection scheme. Table 7.5 lists the pilots and when each pilot started as well as the operator and lease name.

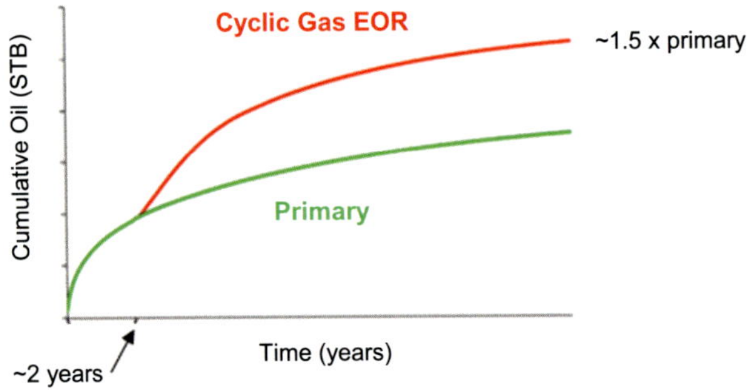

FIGURE 7.20

Expected recovery from cyclic gas injection into Eagle Ford.

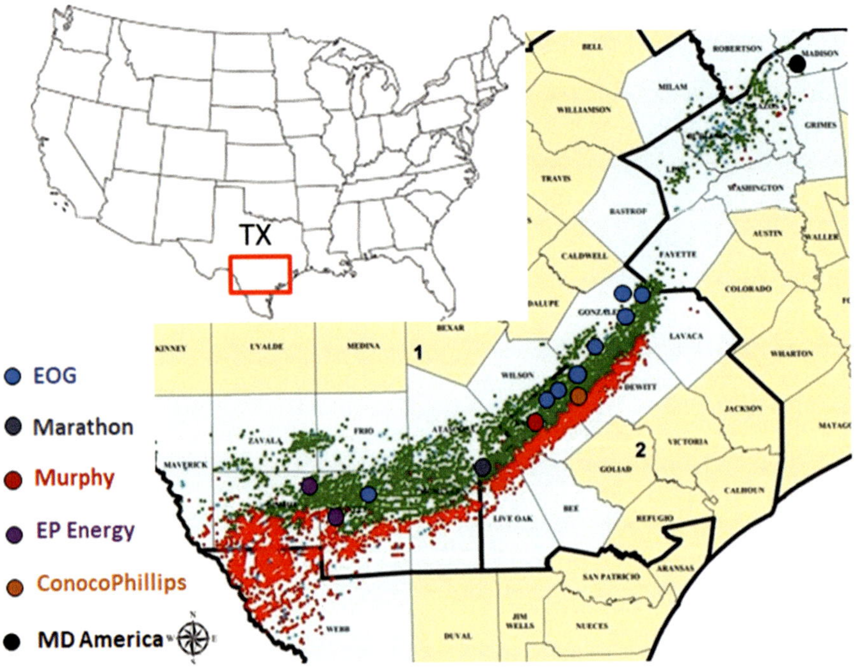

FIGURE 7.21

Eagle Ford wells and approximate locations of cyclic gas injection pilots.

7.4.2.1 *Steen-Scruggs*

The first pilot carried out by EOG was a single isolated well where the nearest offset producers were over 2 miles away. This provided some assurance that the injected gas would not migrate to other wells. This pilot started in early 2013 and

Table 7.5 General information for Eagle Ford enhanced oil recovery field trials.

Lease	Operator	County	Year
Steen-Scruggs	EOG	Gonzales	2013
Whyburn	EOG	Gonzales	2015
Henkhaus	EOG	Gonzales	2015
Martindale	EOG	La Salle	2015
Middle McCowen	Marathon	Atascosa	2015
Baker Deforest	EOG	Gonzales	2016
Mitchell	EOG	Gonzales	2016
Drees	Murphy	Karnes	2017
Ritchie Farms	EP Energy	Dimmit	2017
Maltsberger	EP Energy	La Salle	2018
Klein	ConocoPhillips	DeWitt	2018
Montana	MD America	Madison	2019
12 + others	EOG	Karnes Gonzales	2016–2020

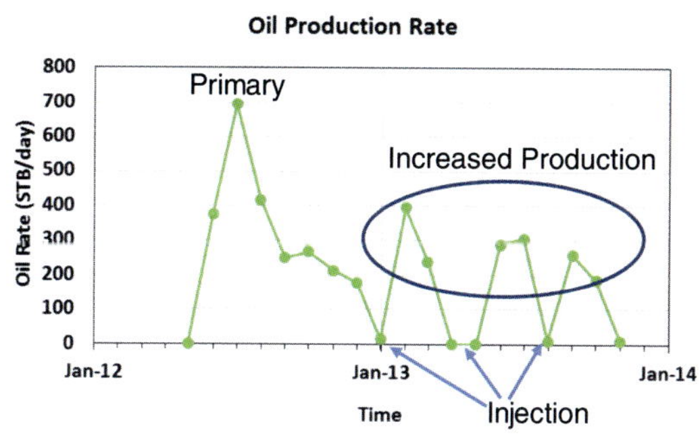

FIGURE 7.22

Oil production rate for single-well Eagle Ford Pilot Steen-Scruggs.

had three distinct cycles during that year. The oil rates for the well (Fig. 7.22) indicated a positive response to the process. Each cycle increased the production rate to approximately half of the well's initial rate. Once the oil rate started to drop, another injection cycle was started. This is encouraging because not only the first cycle, but subsequent cycles were able to increase production.

EOG has not publicly released data related to this pilot; however, there is production data reported by the state agency in Texas (Railroad Commission—RRC). The Texas RRC data are more limited than other states [47], which prevents some evaluation of the projects. For example, the oil production rates are reported per lease and not per well, so if there is more than one well in a lease, the individual well production is unknown. For the Steen-Scruggs area, this

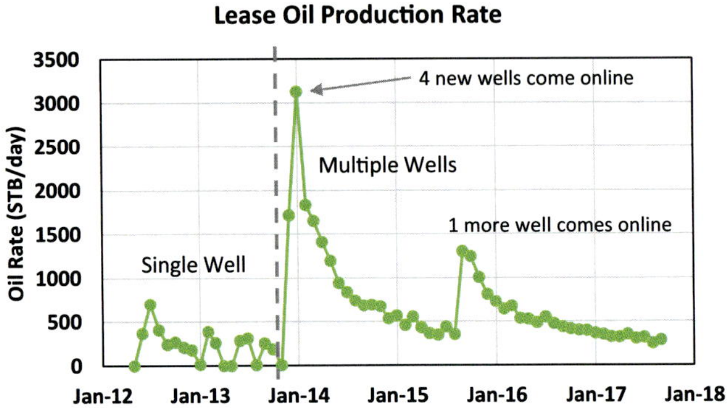

FIGURE 7.23

Steen-Scruggs lease production including up to six wells.

causes a problem in late 2013 when four more wells came online in the lease and hid the cyclic injection results (Fig. 7.23). Also, gas injection rates are not reported to the RRC for most of the pilots.

Even with these data limitations, some important observations can be made. For this first Eagle Ford pilot, the injection plus soak time was approximately 4–6 weeks, which is time the well was not producing. The injectant was a lean natural gas [46] that has been processed through a nearby gas plant to remove the heavier natural gas components. Thus the gas for injection is likely 90% – 95% methane, and the other 5% – 10% is made up of C_{2+} fraction.

7.4.2.2 Whyburn and Henkhaus

The next two pilots are discussed together due to the similarities in the development and results. These are both multiwell pilots near the Steen-Scruggs where injection only occurs in some of the wells in the lease. Throughout the project life, the wells that are injected changed—approximately every month. Some wells were dedicated producers, especially at the lease boundaries, to ensure gas did not leave the lease. Many of the wells continued to produce while other wells were injecting. For Whyburn, there are 8 wells in the lease, and for Henkhaus, there are 14 wells in the lease with some injecting and some producing.

While information is limited due to the publicly available data, it may not matter which wells are injected, as all wells in the lease are likely to be improved by the process. Fig. 7.24 shows the oil production rates for the two leases. It is easy to see when injection started for both pilots, as the production rates increase meaningfully from the baseline trend. The purple line is a harmonic decline curve fit through the preinjection oil rates. Much like Steen-Scruggs, there is incremental oil production resulting from the pilots.

From the time injection started in early 2015 through 2017, the average lease oil production rate was 370 STB/day for Whyburn and 1065 STB/day for

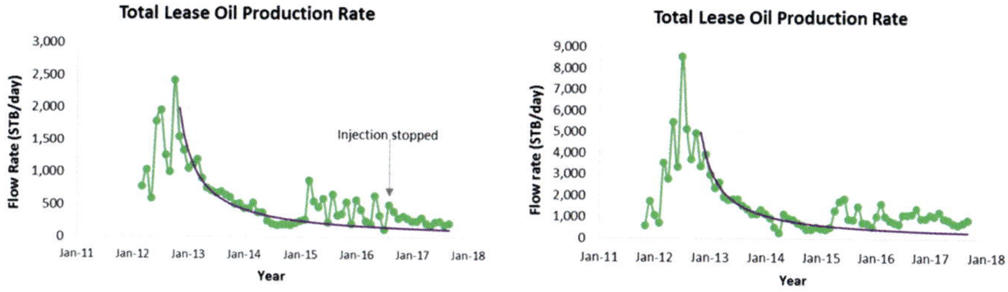

FIGURE 7.24

Oil production rates from the Whyburn and Henkhaus leases in Eagle Ford.

Henkhaus. If the injection project had not been implemented, the average oil rates would have been 170 and 420 STB/day for the leases, respectively. They produced more than double the rate that they would have based on the preinjection decline curves.

The injection stopped around the middle of 2016 for Whyburn. This is apparent not only from the production data in Fig. 7.24, but satellite images confirmed that the compressors had been moved off of location during this time. Although injection stopped in 2016, the production rate does not go back to the preinjection rates but starts to follow a new decline. This indicates that the injection process does not simply accelerate oil production, but it causes incremental oil to be produced. In the 2 ½ years reported, Whyburn cumulatively produced 200,000 STB and Henkhaus produced 550,000 STB of additional oil compared to the primary decline estimates.

7.4.2.3 Martindale

This pilot is somewhat different from the three Eagle Ford pilots previously discussed. It is located in La Salle county over 100 miles to the southwest of the other three pilots. Since Martindale is not close to the gas plant, it is assumed that this pilot used rich hydrocarbon gas (field gas) for injection. Also, all four wells in the lease are being injected at the same time, so it is easier to interpret what is happening. Fig. 7.25 displays the production rates and the cumulative production for this lease. Note that the rate shows the average well rate; since all four wells are cyclic injectors, the lease rates were divided by the number of wells. The cumulative production is for the entire lease.

The timing of each cycle is very apparent from the production data. For example, the last three injection cycles are shorter (around 1 month) than the first three (around 2 months). The reduced injection/soak times made a positive difference in the pattern recovery, which is particularly noticeable on the cumulative plot where an increase in the slope of the green line occurs. The production following the first injection cycle peaks at approximately 65% of the initial rate, which is higher than the other pilots. Because the Whyburn and Henkhaus pilots have multiple wells producing and injecting at the same time, the individual well peak

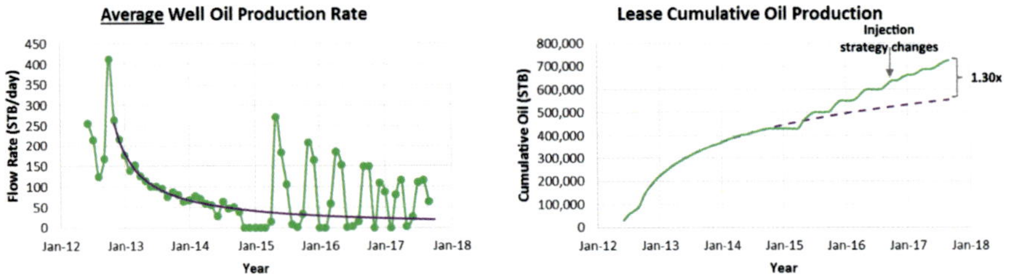

FIGURE 7.25

Average well rate and cumulative lease production for Martindale cyclic gas injection pilot.

rates cannot be discerned. The Martindale pilot clearly performs better than Steen-Scruggs, and there could be a number of reasons including: different rates or volumes of injected gas, various reservoir rock and/or fluid properties, or using rich gas instead of lean gas.

Over the 3-year lifespan of this project, the four wells produced 300,000 STB. Without the injection process, it is estimated the lease would have only produced 130,000 STB during this timeframe. The gas injection process therefore has more than doubled the incremental production during that time. When considering all 6 years of the lease's production history, the injection case has already produced 1.30 times more than the predicted preinjection cumulative production.

Because the data quality is high for this pilot, future production for these wells can be predicted. A decline curve was fit through the peak production rates with the cyclic process reproduced by having the well produce at the peak rate for two months and then shutting in the wells for one month (to replicate the injection stages). The predicted rates in Fig. 7.26A are only shown for the first 4 years, but the models predict for 10 years. The cumulative production in Fig. 7.26B displays 10 years of production. For the four-well lease, the process is expected to recover over 1.2 million STB and incrementally 480,000 STB or 65% more than the pre-injection model calculates.

7.4.2.4 Mitchell and Baker Deforest

Due to the success of Henkhaus, gas injection projects were started in two offset leases: Mitchell and Baker Deforest. There are 14 wells in each of these leases. Gas injection rates are reported for these leases as shown in Fig. 7.27. The Mitchell lease injected around 24 million SCF/day, and the Baker Deforest lease had a peak injection rate of 40 million SCF/day, but it is unclear exactly how many of the 14 wells these rates are being spread across.

The shaded area between the oil production rates and the predicted primary rate represents the incremental production from each lease. The increases are 600,000 STB for the Mitchell and 900,000 STB for the Baker Deforest. The injection was shut in for about 6 months for both leases in 2018. Oil production dropped off over that time, but rates increased when the injection was restarted.

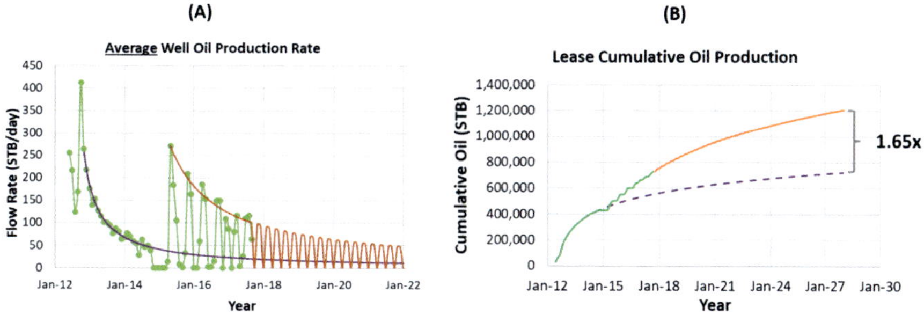

FIGURE 7.26

Predicted (A) rates and (B) cumulative production for Martindale cyclic gas injection project.

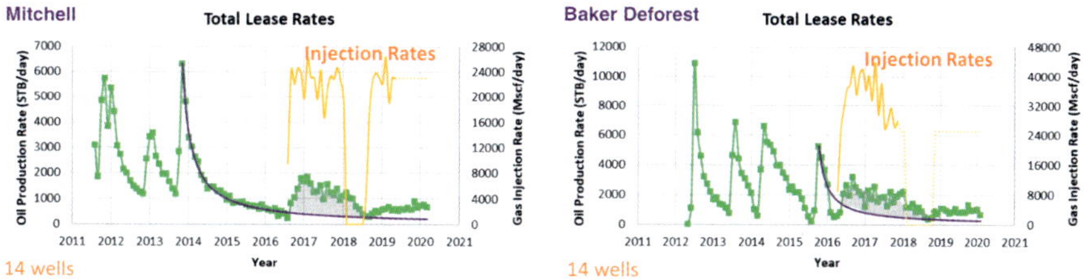

FIGURE 7.27

Oil production and gas injection rates from the Mitchell and Baker Deforest leases.

7.4.2.5 Drees A-179

The injection pilot operated by Murphy Oil in Karnes County was the first to be implemented in the Eagle Ford after the EOG field tests were reported. It was conducted in late 2016/early 2017, and it appears that rich hydrocarbon gas (field gas) was used. There were four wells in the project (Fig. 7.28), which are isolated from other Eagle Ford wells. Gas was injected into one of the boundary wells and its immediate offset was shut in while the other two wells were produced. After 30 days of injecting, the pattern was inverted and the other boundary well was used for injection; its offset well was shut in and the other two wells were produced [48]. This process was repeated for a number of cycles over 6–8 months. The combined oil production rates from all four wells in the lease are shown in Fig. 7.28.

The increase in oil production does not appear as dramatic as the previous pilots, and this is likely due to a couple of reasons: (1) due to the size of the compressor, the gas injection rates were lower, probably around 1–2 million SCF/day and (2) the length of time for injection was only a few months. Rogers et al. [48] used a reservoir model—history matched to the pilot data—to estimate what the

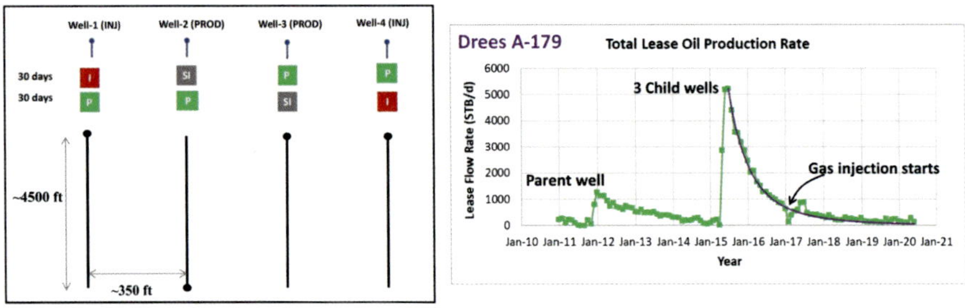

FIGURE 7.28

General layout and production response from field gas injection project in Karnes Co.

expected recovery would have been if the injection had lasted longer and with a higher injection rate. If the existing injection rate would have continued for 4 years, the cumulative production was estimated to be 25% higher than primary production, and if the injection rate had been increased by 6 times, the cumulative production would have been 70% higher than the primary production, which is in line with the expectations of the earlier pilots.

7.4.2.6 Klein D

ConocoPhillips performed at least two pilots in 2018–2019 on their liquid-rich acreage in DeWitt County. There was no technical publication on this pilot, but there were reports in investor relations presentations [49] that shed light on the projects. There were 11 wells in the two pilots and Fig. 7.29 presents the publicly available production data from one of the pilot areas.

There were four wells in this pattern, and at least two cycles of injection occurred in 2018. The oil response is evident even if it is not large. Like most of the other pilots in Texas, the amount of gas injected in this project is not known, so a larger injection rate may have caused more oil production.

7.4.2.7 Other Eagle Ford pilots

There are tens of wells in some of the leases in Texas, so if the injection pilot is one or even a few wells, the production from the other wells masks the response from the pilot wells. This is evident from one of the earlier pilots performed in the Eagle Ford. In 2015, Marathon Oil Company planned a pilot in the Middle McCowen lease in Atascosa County, where RCC permits indicated they planned to inject 2.5 million SCF/day into a well for 2 months. The pilot was performed, but the results are unknown. The actual injection amount is not reported to the state agency, and there are over 60 wells in the lease, so the oil production response could not be determined from public data.

There are a few other pilot locations in Texas where the number of wells in the lease obscured the results, and while some of the pilot data can be found

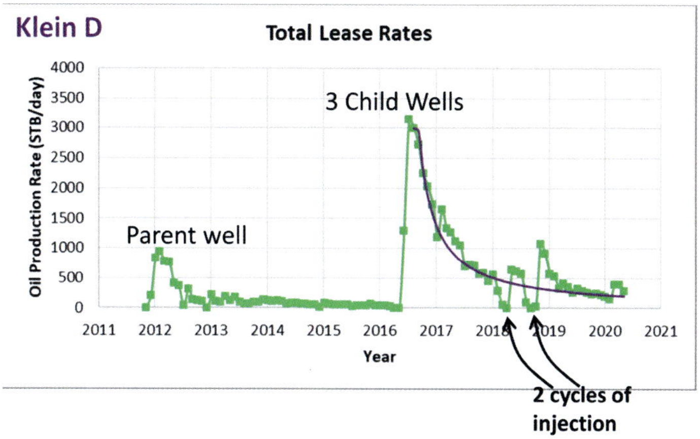

FIGURE 7.29

Oil response from gas field gas injection project in DeWitt Co.

elsewhere (investor relations or technical papers), this information is mostly excluded from the public domain.

EP Energy has performed gas injection pilots in two different locations in their Eagle Ford acreage: one is in the black oil region and the other is in the gas condensate region. In the leases surrounding the pilot areas, there were at least 50 wells, so the production data are not known. There were a few conference presentations tangentially related to these projects [50–52], so there is some basic information about these field trials. For the black oil pilot, the gas injection rates ranged from 5 to 15 million SCF/day. The injected gas quickly traveled to the nearby wells, so those wells were shut in, but the gas continued to the next offset wells, which were also shut in. Over 10 wells had to be closed during the first cycle of injection. When the wells were put back into production, additional oil was produced, but the amount was not publicly reported.

The second pilot is one of the few performed in a gas condensate region. There were 11 wells in the project and 4 were cyclic injectors/producers, while the other 7 were dedicated producers. Two injection wells were on the east side of the project and two were on the west. Injection times were initially around 30 days, but over the project life, injection times were reduced to around 10 days. Significant additional liquid hydrocarbon production occurred; over a year twice as much was produced than if wells had stayed on primary [53].

In addition to EOG's six leases discussed earlier in this subsection, EOG also injected gas in other Eagle Ford leases in Gonzales and Karnes County, which total around 200 wells. This is more developed than the combined efforts of all of the other companies in all the unconventional basins. In the most successful lease, they have doubled the expected primary production from the wells, and most leases are expected to recover 30% – 50% more oil than primary. The clear technical success is overwhelming and based on available data, the projects were economically successful as well [54].

7.4.2.8 Summary

Cyclic natural gas injection pilots have clearly shown that this process will recover additional oil in the Eagle Ford. This is confirmed with multiple pilots in multiple areas with multiple operators. While recoveries vary significantly, it appears that 30%–70% of incremental oil recovery—first reported by EOG in 2016—is consistent with the most recent pilot results. At least five other companies have injected into the Eagle Ford, and many more are at various levels of planning.

One advantage of cyclic injection is that the oil response happens quickly compared to continuous injection methods. Access to CO_2 in the Eagle Ford is limited, so natural gas has worked well, and based on early data, it appears that field gas richer in heavier end hydrocarbons performs better, which is consistent with the miscibility of rich gas being better than lean gas.

Pilots with smaller gas injection volumes have increased oil production; however, the pilots with more injection have much higher oil recoveries There have been issues with conformance control (keeping gas contained) and building pressure, particularly where the wells are closely spaced and where there are strong natural and hydraulic fracture connections between wells. Techniques to deal with these issues need to be developed.

7.4.3 Permian injection projects

The Permian Basin in west Texas has been a leader in conventional CO_2 injection, but the development of the Permian's unconventional formations occurred later than the Bakken or the Eagle Ford, so EOR also has been slower to develop there. However, there have been a number of projects to date, and Fig. 7.30 shows the locations of some of these pilots.

At a minimum, five operators have implemented projects in the Permian, but because they have happened more recently, less information is available about them than the Eagle Ford or Bakken. Furthermore, because they are mostly in Texas, much of the production and injection data are not submitted to the state agency or are limited to the lease level.

Reportedly, both Occidental and Chevron performed cyclic CO_2 injection pilots in 2017–2018, but nothing has been published. Additional pilots with natural gas injection were started in 2019 by Occidental and EP Energy, but the results are obscured by the number of wells in the lease. Two recent pilots with available data are discussed here.

7.4.3.1 Hydraulic fracture test site

A consortium of companies along with the US Department of Energy (DOE) was formed to study hydraulic fracture development in an area of the Permian [55]. The wells were instrumented to measure fracture growth, and drill back studies

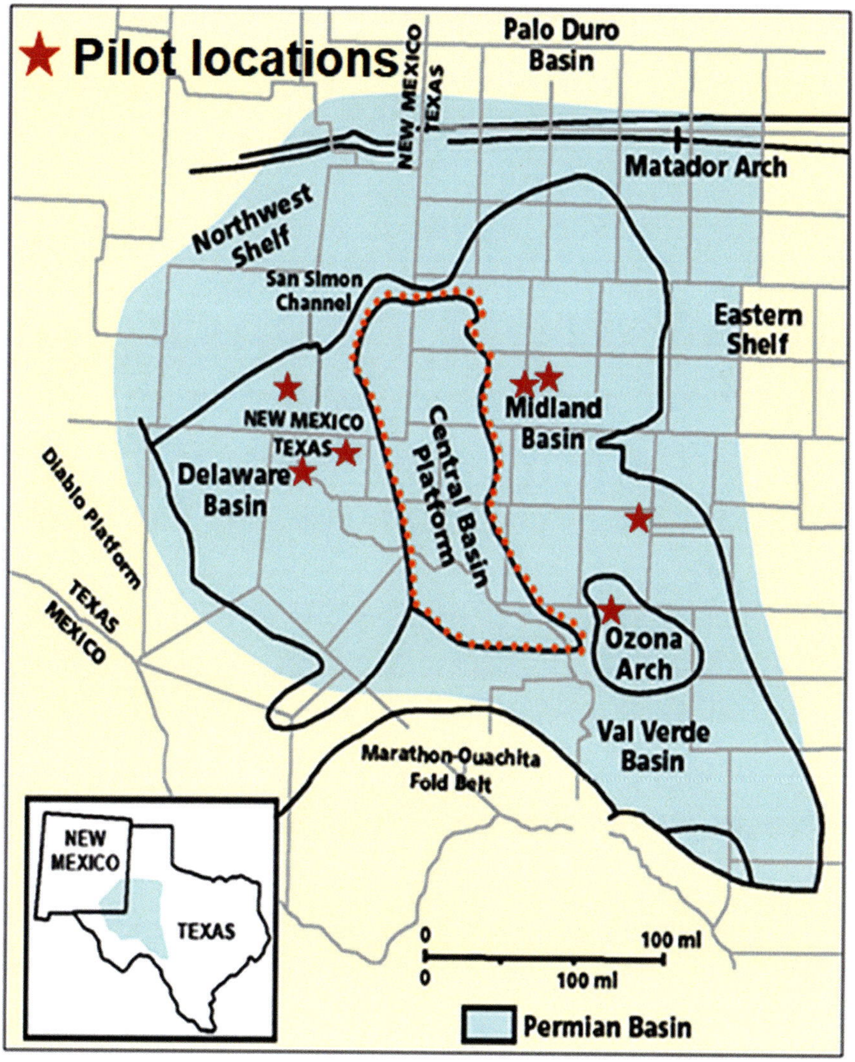

FIGURE 7.30

Location of Permian enhanced oil recovery pilot projects in unconventional formations.

and coring operations were performed through the original fractures to visualize them. An EOR project with hydrocarbon gas injection was planned for Phase 2 of the study, but the original location was not suitable for gas injection, so the Phase 2 area was moved to the northwest by a couple of miles (Fig. 7.31).

Five wells were closely monitored—the injection well and two wells on each side of it. Gas was injected into the central well at 2.5 million SCF/day for 4

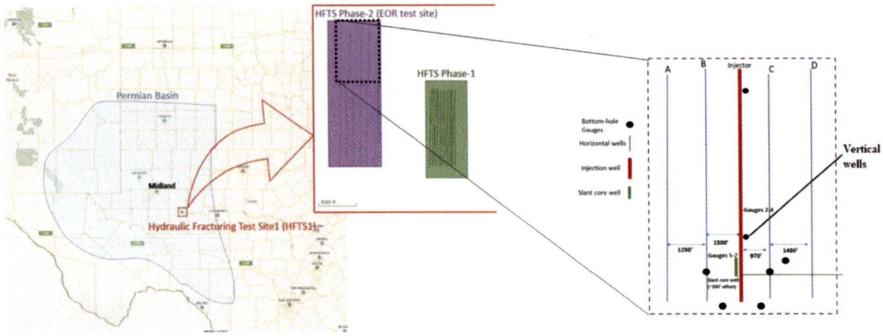

FIGURE 7.31

Location of enhanced oil recovery test site for US Department of Energy-funded project [56].

Reproduced from S. Salahshoor, J. Ciezobka, Hydraulic fracturing test site HFTS Phase-2 EOR pilot: huff-and-puff pilot in the Permian-Midland Basin, in: SPE-200468-MS, SPE Improved Oil Recovery (Virtual), 31 August–4 September 2020, with permission.

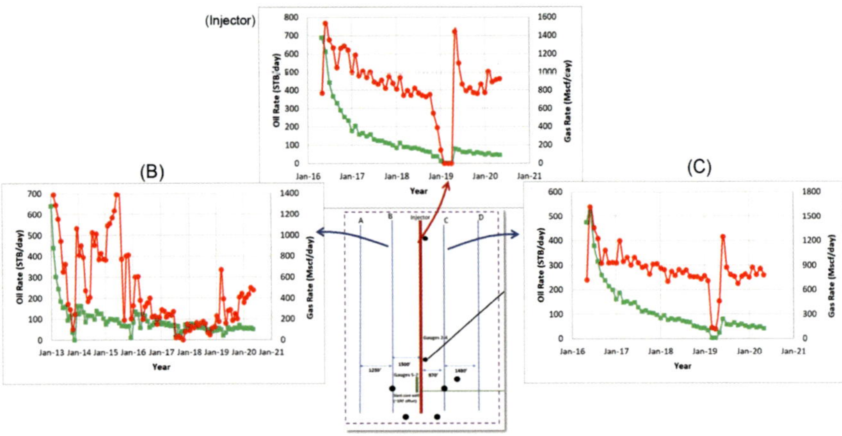

FIGURE 7.32

Oil and gas rates for injection well and two closest offset wells.

months in early 2019. Gas and oil rates for the three central wells are shown in Fig. 7.32. Because this was a DOE-funded study, the production rates are reported per well instead of per lease. There was no noticeable response in the outer wells (A and D) so their production rates are not included. The injection well was put into production in April of 2019 and an increase in oil and gas rates was observed in it and the two adjacent wells (B and C). Well C is about 970 ft to the east of the injection well and saw the injected gas rather quickly, so it was shut in during the injection. It was turned on in April and has a well-defined increase in oil and gas rates. Well B is about 1500 ft to the west of the injection well. Injected gas was observed here also, but it was a much lower quantity than Well C, so this well continued to produce throughout the injection process. Oil rates appear to have flattened or increased slightly while gas rates increased more notably.

Gas was mostly contained with gas breakthrough only occurring for the closest wells. This is encouraging for implementing EOR in other parts of the Permian. The increase in oil rates was not large, but neither were the gas injection rates. Results from the Eagle Ford suggest that higher injection rates could make a difference for the Permian.

7.4.3.2 Nail unit

An additional Permian gas injection pilot was carried out in Martin County in 2020. There were three wells in this area that are isolated from offset development. These wells are about 600 ft apart and have lateral lengths around 7500 ft. State permits indicate that gas injection rates can be up to 5 million SCF/day and injection pressures can be as high as 5000 psi. The oil production is reported as a total for these three well, so Fig. 7.33 shows the average well production rates—dividing the lease production by 3.

After the first couple of injection cycles, additional oil may be produced, but the long-term results are unclear since the project has been going on for less than a year and has only reported two full cycles.

7.4.4 Recent Bakken trials

EOR in unconventional reservoirs started in the Bakken in the late 2000s, and while there was not a lot of additional oil produced, there was a significant amount learned about the process. More recent pilot projects have furthered the knowledge. In this subsection, three recent Bakken pilots are discussed.

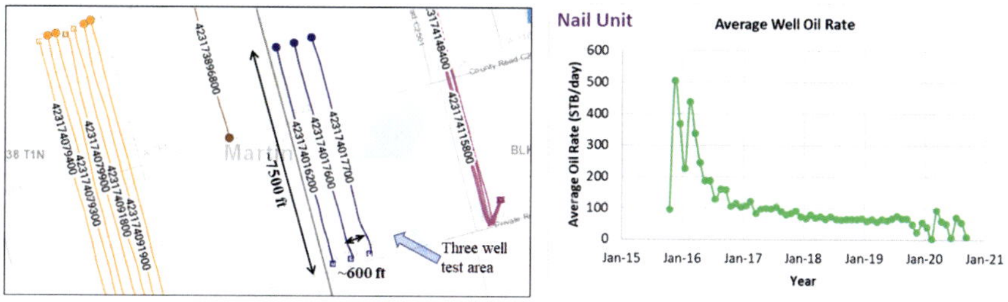

FIGURE 7.33

Location and oil production from gas injection pilot in Martin Co. operated by ExxonMobil.

7.4.4.1 Stomping horse

After the success of the cyclic natural gas injection pilots in the Eagle Ford, there was an attempt to replicate that in the Bakken. Liberty Resources implemented a project on their Stomping Horse lease in Williams County ND [56]. There were 12 wells in the project located in the Middle Bakken and the Upper Three Forks in an alternating (wine rack) pattern with half in each zone. Two boundary wells on each side of the pattern ensured that no injected gas would escape. Five of the eight wells in the middle were used for injection at different times.

To prevent allocation issues, only gas from the lease was used for injection, but that volume of gas was insufficient to increase pressure. The injection rate ranged from 0.8 to 1.6 million SCF/day. During all injection cycles, injected gas was observed in the offset wells of both formations. This represents a significant volume to fill up, so the pressure was not able to increase much. The MMP was 2450 psi, but the highest bottom hole pressure achieved was 1800 psi in one well. Other wells did not even achieve that pressure, so injected gas did not mix with the reservoir oil. The result is that little to no additional oil was produced. The operators recommended that new pilots should (1) have a larger supply of gas, (2) design for higher injection rates and volumes, (3) consider geologic complexity and well spacing in site selection, (4) start earlier when reservoir pressures are near MMP, and (5) consider WAG injection [56].

7.4.4.2 Red Sky—Nelson farms

A propane injection pilot with a vertical injector and a horizontal producer was conducted in North Dakota in 2017–2018 [23]. The objective was to demonstrate the technical feasibility of using propane to enhance oil recovery from unconventional reservoirs. An advantage of propane is that it is first-contact miscible at low pressures—around 650 psi for the Bakken oil and temperature. Furthermore, propane is a liquid under field conditions and can be pumped into the injection well without the need for a compressor. These factors make propane a high-quality EOR agent for unconventional reservoirs; however, it must be handled carefully in the field for safety reasons.

The horizontal producing well was drilled in 2005; its lateral length is around 4400 ft, and before the test, it was producing around 30 STB/day on the primary. The vertical well was drilled in late 2016 about 1000 ft from the horizontal well and was completed with a small hydraulic fracture. Around 19 million SCF of propane was injected into the vertical well over an 8-month period. The injection rate varied from 100 to 150 MSCF/day. The first 5 months of injection are shown in Fig. 7.34A.

The horizontal well was shut in during injection, plus another 9 months due to operational issues. The well came back on production at around 70 STB/day; the blue line in Fig. 7.34B shows the measured production for the next 8 months while the black lines show the expected primary production if the well continued on primary production after being shut-in for 17 months. This amounts to about

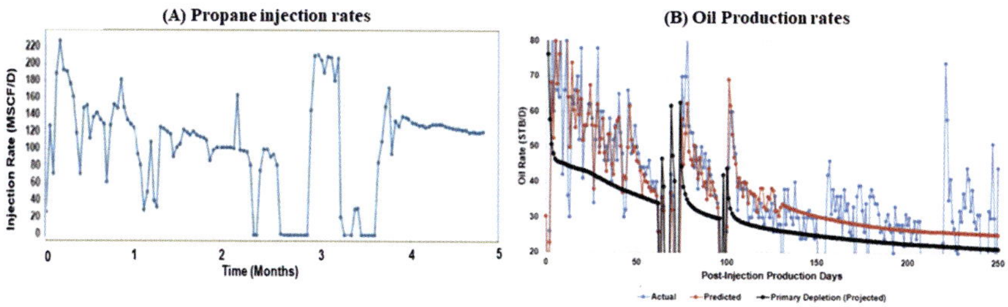

FIGURE 7.34

(A) Propane injection rates and (B) measured oil production rates in blue and predicted primary oil production rates in black [23].

Reproduced from N.R. Nagarajan, D. Stoll, M.L. Litvak, et al., Successful field test of enhancing Bakken oil recovery by propane injection: part I. Field test planning, operations, surveillance, and results, URTeC-2020-2768, in: SPE/AAPG/SEG Unconventional Resources Technology Conference (Virtual), 20–22 July 2020, with permission.

17% more oil than would have been recovered on primary [23]. At best, the vertical well was impacting only about one-eighths of the 4400-ft lateral well, so the increase in production was impressive. If a larger volume of propane was injected into a horizontal well, a much larger increase in oil production might have occurred.

7.4.4.3 Surfactant flood

Due to the oil-wet nature of most unconventional reservoirs, water injection strategies are not as effective at recovering oil. Field trials using a chemical treatment were carried out in the Niobrara and Bakken formations [36]. Fig. 7.35 shows the response from a single-well trial in the Bakken.

The 10,000 ft lateral well had 30 stages of hydraulic fractures, and it was in production for about two and a half years before surfactants were injected. The chemical concentration was around 1,500 ppm in a low-salinity brine. The injection continued for 2 weeks—below the fracture pressure—and was followed by 4 months of a shut-in. When the well was put back on production, the oil rate increased by about 100 STB/day and sustained the increase for at least one and a half years. By predicting the increased production for 5 years, Kazempour et al. showed that about 25% more oil will be produced in this well [36].

7.4.5 Summary

There has been significant EOR activity in the three basins primarily discussed in this section (Bakken, Eagle Ford, and Permian), and EOR projects have been carried out in at least three other unconventional resource basins. The Niobrara in Colorado had a surfactant injection project and possibly a small gas injection project. In 2019–2020 there was a hydrocarbon gas injection project in the gas

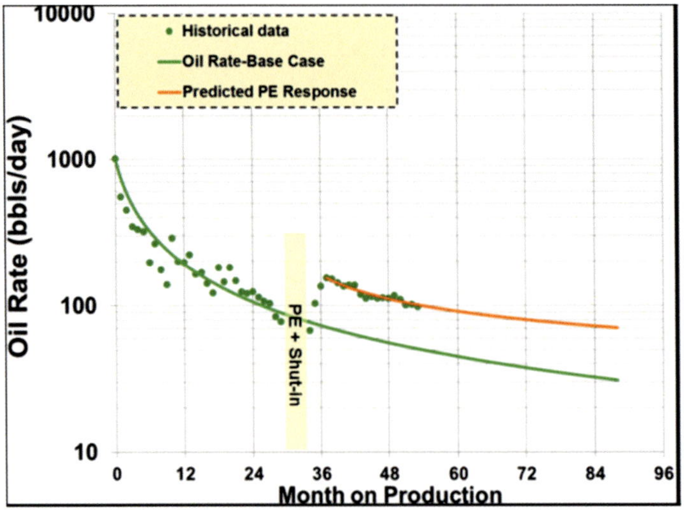

FIGURE 7.35

Oil rate increase due to surfactant injection in a Bakken well [36].

Reproduced from M. Kazempour, M. Kiani, D. Nguyen, et al., Boosting oil recovery in unconventional resources utilizing wettability altering agents: successful translation from laboratory to field, in: SPE-190172-MS, SPE Improved Oil Recovery Conference, 14–18 April 2018, Tulsa, Oklahoma, USA, with permission.

condensate region of the Montney formation in British Columbia. Also, there was a single-well natural gas injection project in the SCOOP area of Oklahoma. Similar to the basins with more EOR activity, these pilots also had varying degrees of success. Most produced additional oil so there was a technical success, and the economic outcomes are being evaluated.

In field trials that have been completed, there has been significant knowledge gained and lessons learned. The early pilots showed that injectivity was not a problem, but conformance control and early breakthrough were potentially challenging. This led subsequent pilots to focus on cyclic injection strategies.

The next set of projects showed that building pressure is important for the success of cyclic gas injection. If the reservoir does not achieve sufficient pressure (above the bubble or dew point pressure), then the injected gas is not able to dissolve into the oil phase or vaporize the hydrocarbon liquids. In this case, minimal additional hydrocarbons are produced. There are several reasons why pressure may not increase above the saturation pressure, including insufficient gas volumes, gas leakage (poor conformance), and/or highly depleted reservoir pressures.

Higher injection rates can be beneficial for cyclic gas injection because less time is required to inject a given volume of gas and pressure builds more rapidly. This allows the wells to be put back into production sooner and for longer periods, which helps the project economics. Higher injection rates, however, require larger compressors, which are more expensive, so there are trade-offs for higher injection rates.

Other injection fluids (water, surfactants, propane) have been used to increase recovery in unconventional oil reservoirs, and these fluids may be more appropriate for some situations and basins. More research is needed to understand when and where each injectant performs best.

In this section, various injection field trials were discussed. The next section discusses best practices for implementing a successful unconventional EOR project.

7.5 Project development

Previous parts of this chapter discussed how EOR in unconventional reservoirs works, and some of the field trials were evaluated to report lessons learned from the projects. The operational aspects that should be considered are covered in this subsection.

While EOG has successfully expanded past the pilot stage to large-scale development, most companies are in the early planning and development stages of EOR for unconventional reservoirs. Therefore a short description of the important characteristics of pilot design is presented. This is followed by details that apply to both pilot design and large-scale development.

7.5.1 Pilot design

Pilots are performed to reduce the technical and economic risks of implementing a new project or idea and to learn some of the potential pitfalls and correct them before implementing the process on a larger scale. The most important part of designing a pilot is choosing the location. While there are many aspects that should be considered, there is no perfect location. There are pros and cons for all areas, which need to be balanced against the requirements of the specific pilot. As a minimum, potential considerations include:

- Geology
- Natural fractures
- Completion/hydraulic fractures
- Well spacing
- Well quality
- Available data (logs, cores, fluids, etc.)
- Primary production
- Pressure depletion
- Injection fluids source
- Working interest.

It might be tempting to put the pilot in a poor performing area since these wells are not doing well and EOR might "improve" them. EOR, however, cannot overcome geology. If the pilot is placed in the poorest acreage, it may hide the true potential of the method. Alternatively, if the pilot is placed in the best acreage, it may mask problems with the process that need to be addressed before developing other areas. Therefore it is best to find a pilot area with average properties for the region.

Natural fractures or hydraulic fractures may cause conformance problems, but if fractures are prevalent throughout the field or if the main goal of the project is to study how to mitigate early breakthrough, then the pilot needs to be placed in an area with representative fractures.

It is better to use a location with contiguous operated acreage that extends outside the pilot. This may prevent problems with injection fluids from leaving the pattern where they are difficult to monitor. It is preferred that a pilot be located where the operator has a 100% working interest, but that is often impractical or does not fit with other criteria for selecting a location. In those cases, it is important to select a location where partners are committed and willing to work together on the project.

If wells have hydraulic communication with each other, then all wells in communication must be considered part of the pilot, even if fluid is not being injected into all of them. In some cases, the closest wells to the injectors may be shut for some time to allow the pressure to build. All offset wells in the pilot must be monitored for the presence of the injectant. Vertical containment also needs to be considered; if injected fluid is lost to the overburden, the economics of the project are reduced significantly.

A further consideration when choosing a pilot location is the physical integrity of the wells. Wells with damaged casing or cement have a disastrous effect on injection pilots. Workovers to clean out sand or remove scale often are required to make sure that all wells are in proper condition before starting a pilot.

Before injection starts, contingency plans should be formulated for when things do not go as intended. There are numerous scenarios where things go wrong, and as many of these problems as possible need to be considered. For example, if the injected fluid starts being produced at the offset wells, should those wells be shut in, or should the injection be reduced or moved to different wells? Thinking through possible scenarios is the best way to prevent prematurely halting the project due to operational issues.

During these current early days of EOR in unconventional reservoirs, pilot projects are vital to expanding the technology so that more efficient and successful basin-wide development occurs.

7.5.2 Laboratory measurements

For conventional EOR, standard core and fluid measurements are well established. For EOR in unconventional reservoirs, laboratory measurements are still being refined. For cyclic gas injection, standard pressure, volume, and temperature (PVT) tests, such as differential liberation and constant volume depletion, are necessary. In addition, swelling tests should be completed, where injection gas is added to the reservoir oil and pressurized until all of the gas is in solution. This is repeated at increasing concentrations, and the volume of the combined solution is measured. Fluid expansion is one of the more important recovery mechanisms for EOR in unconventional reservoirs, and the swelling test provides information about how much expansion occurs.

The miscibility of injected gas and reservoir oil is a strong function of pressure, and for conventional EOR, the MMP is one of the defining characteristics of a gas injection project. The slim tube test is the standard experiment to determine the MMP, but other lab methods and correlations are used as well. For cyclic gas injection in unconventional reservoirs, the MMP may not be as significant as in conventional reservoirs. Research has shown that recovery continues to increase as the pressure increases above MMP [20]. That being said, miscibility is still important to determine how easily the two substances mix, and slim tube tests are often used for gas injection into unconventional oils.

There have been unconventional core experiments completed for cyclic gas injection, but they have been completed mostly at universities and in research labs. The mechanical setup varies, but the general procedure for several core huff-n-puff studies is the same [20,57,58]. The core is saturated with oil and aged, and afterward, gas is added to a pressurized container with the core inside. The gas containing the liquid hydrocarbons is pulled out of the vessel and fresh gas is replaced for a number of cycles. The oil recovered is measured as a function of injection cycles. While similar in procedure, these experiments have not been incorporated into a standard technique.

More recently, cores from unconventional rocks had slits cut through the slabbed face to represent hydraulic fractures. The slits were approximately 1 mm wide by 10 mm high and extend through the entire core plug. These slits are filled with frac sand. This core setup was used for a range of cyclic gas injection studies [15]. Gas was injected around the outside of the core in a cyclic manner, and fluid was produced through the end of the core with the slits.

Laboratory core studies are needed to evaluate cyclic gas injection in unconventional reservoirs, and there are a number of proposed core experiments—with varying strengths and limitations. Most methods are still at the research level, but as improved techniques are developed, standardization may become possible.

7.5.3 Reservoir simulation

Reservoir simulation studies are important to perform before implementing projects in the field. With simulation, different development strategies are tested and optimized, such as injection rates, cycle times, and well strategies. The models are used to evaluate potential problems that might occur and plan contingencies for the various situations. It is also nice to have a working model during the project so that it can be used to assess incidents that arise during the pilot. This subsection discusses the various aspects of reservoir simulation that are unique to EOR in unconventional reservoirs; basic knowledge of reservoir simulation is assumed.

An entire conventional reservoir can be modeled, but due to the vast size of unconventional systems, only a portion of the reservoir needs to be modeled. The sizes of the model vary from a multiwell pattern model to a single-well model or even a smaller segment model, which can be as small as a quarter of a fracture. Fig. 7.36 shows some typical model sizes.

The smaller segment models require that the small volumes can represent the whole system, and in most situations that assumption is valid. Flow rates from these models are scaled up to the wellbore scale for validation. The number of gridblocks also plays a pivotal role in the overall model size. The gridblocks need to be small enough to capture the physics of the EOR process, which is more complex than primary production. If there are too many gridblocks, however, the models run exceedingly slow or do not converge. The smaller segment models can allow for finer gridblock sizes without exceeding the total gridblock limit.

Modeling the petrophysical properties, such as porosity and permeability, is vital to have a reliable model to study EOR in unconventional reservoirs. Porosity tends to be small in unconventional reservoirs—mostly in the single digits (5% − 8%)—but log-measured values, if evaluated properly, are as accurate as conventional reservoirs and can be used in unconventional flow models. Permeability, on the other hand, needs some special treatment compared to conventional reservoirs. There are three systems that need to be considered: (1) matrix and small-scale natural fractures, (2) large-scale natural fractures, and (3) stimulated hydraulic fractures.

Core measured permeability values tend to be very small, in the micro-Darcy and nano-Darcy range. These values may include minor natural fractures that are

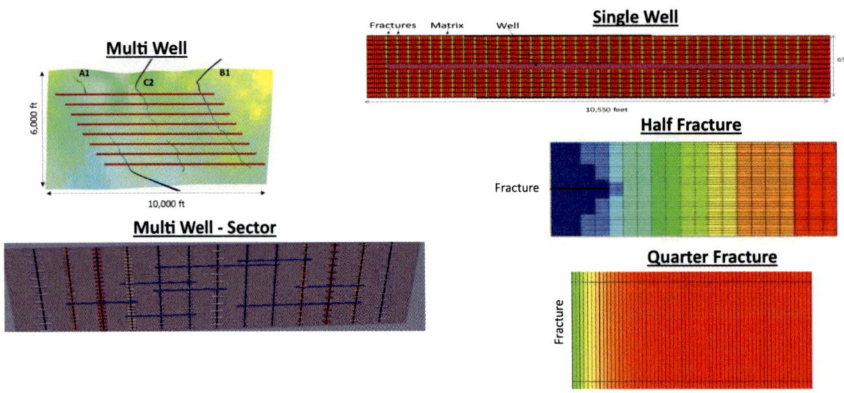

FIGURE 7.36

Flow simulation model sizes for unconventional enhanced oil recovery projects.

smaller than the core, such as microfractures caused by hydrocarbon generation. Natural fractures that are larger than the core but smaller than the gridblocks (i.e., small-scale natural fractures), should be incorporated into the gridblock permeability and cause it to be higher. Fractures that are larger than the gridblocks may also contribute to the overall system in a meaningful way. If they are important, then a dual-porosity/permeability model needs to be used to account for their contribution, where the fracture porosity and permeability are included in a colocated second grid.

For simulation models of unconventional reservoirs, the stimulated hydraulic fractures must always be included and many methods have been proposed to incorporate them. In early modeling of unconventional systems, fine-scale gridding of the fractures (either through logarithmic gridding or local grid refinement) was used [59], but that practice has decreased over time. Another option is to place the fractures into the models as a separate object that interacts with the matrix and well [60,61]. A third option is to include these fractures directly into the gridded model. With this method, the porosity of the gridblock is reduced to account for the small volume the hydraulic fracture is occupying. Eq. (7.3) can be used to reduce the gridblock porosity:

$$\phi_{gb} = \frac{w_f \phi_f}{w_{gb}} \tag{7.3}$$

where ϕ_{gb} is the porosity for the gridblock, w_f is the width of the actual fracture, ϕ_f is the porosity for the fracture, and w_{gb} is the width of the gridblock. The gridblock permeability is equal to the fracture conductivity divided by the fracture height. This method can be used to model the fractures in a single-porosity model or a dual-porosity/permeability model. If this is done for a single porosity system, the overall model size has to account for the loss of volume where the fractures are included. The dual-porosity/permeability system already accounts for this in the secondary grid [18].

For cyclic gas injection, the simulation fluid model should be a fully compositional model where an equation of state is used to represent the PVT behavior of the reservoir and injected fluids. The in situ fluid make-up changes each injection cycle, and the black oil fluid model is not able to represent this kind of fluid mixture.

Due to the ultra-low permeability in unconventional rocks, it is nearly impossible to measure relative permeability in the lab; therefore standard Corey-type curves are typically used and can be modified during the history matching process. Capillary pressures are also difficult to measure, and due to the small pore spaces in unconventional reservoirs, capillary pressure may be higher than in conventional reservoirs. The initial pressure and saturation are defined the same as in conventional models, but many unconventional reservoirs are overpressured. Wells are modeled as horizontal wells, and the well constraints, either rate or bottom hole pressure, are defined similarly to conventional reservoir models.

Once the basic model is constructed but before it is used for prediction scenarios, the model must be validated against historical production rates and pressures.

This process is called history matching, and it is important to achieve the goals of the study. History matching assures that the model provides reasonable predictions that are used to make decisions about how to best develop the field. Pressures and primary production rates are always matched, and when modeling EOR processes, the breakthrough times and volumes of injected gas also must be reproduced.

Before EOR projects begin, there is no injection behavior to match; however, understanding subsurface connections between wells is important for the accurate prediction of injection performance. Injection results from other parts of the basin can be used to substantiate the model. Furthermore, sensitivity cases should be run with both strong and weak subsurface connections to establish the range of possible outcomes. Once injection starts, the results will demonstrate which situation is most suitable. For all modeling studies, sensitivity analysis of different parameters and operating conditions should be employed to account for uncertainty and determine a range of possible outcomes.

Modeling is important for the successful implementation of EOR into unconventional reservoirs and should be employed for most projects. The time and effort spent on putting together a high-quality model is returned when the EOR project is efficiently implemented.

7.5.4 Injection gas

One of the first considerations for a cyclic gas injection project is to determine where the gas will be sourced. Most locations currently use hydrocarbon gases, but typically there is not enough gas produced from the wells in the pattern to supply the entire project. The additional gas could come from produced gas at nearby wells operated by the same company or be purchased from a nearby gas plant or gas pipeline. Once the project gets going, much of the needed gas can be recycled from wells in the project, but some make-up gas is required for the entire project life. There have been some projects that were delayed and even canceled because the companies could not get access to enough gas to run the project.

Most projects today use hydrocarbon gas, but if CO_2 is going to be used as the injection gas, it needs to be secured for the project. CO_2 is available in fewer locations than hydrocarbon gas, so sourcing it may be more challenging. In addition, corrosion needs to be mitigated if CO_2 gas is used. When CO_2 mixes with water, carbonic acid is formed. Corrosion protection is needed for the downhole equipment, the tubing, surface flowlines, and surface separation equipment until the CO_2 goes through a dehydrator. Common corrosion protection methods include (1) alloy metals, such as stainless steel, (2) carbon steel internally coated with a corrosion-resistant material, such as plastic, or (3) chemical corrosion inhibitors. A combination of all three is often required.

A decision may be made about what to do with the produced CO_2. The CO_2 can be separated from hydrocarbon gas using an anime tower. In this case, the produced hydrocarbon gas can be sold and the CO_2 can be reinjected. Another option is to inject the produced hydrocarbon gas with the CO_2 stream, which saves the cost of the amine facility—but then the hydrocarbon gas cannot be sold. An economic evaluation determines the best option for each project.

Other important considerations are the volume of gas injected each cycle and the injection time. The volume injected influences how high the pressure is, and the pressure must be high enough for gas to go into the solution. Analytical models and flow simulation tools can be used to estimate the volume of gas needed to achieve the desired pressure for each cycle. The injection time depends on the required volume and the rate that can be achieved by the compressor. The higher the rate, the less time is needed for injection, but higher rates require more up-front costs for the compressor.

Unlike cyclic steam injection, there is little to no need for a soak phase during gas injection into unconventional reservoirs; once the injection is completed, the wells are returned to production as soon as possible. There is no evidence from field data to suggest that longer soak times result in any more production, and the longer the wells are not producing, the more it damages the economics.

7.5.5 Equipment

For the cyclic gas injection, the largest capital cost is the compressor, and there is a wide range of compressors in terms of size, performance, and costs. For the field trials discussed in previous sections, the compressor size had a considerable impact on the results of the pilots. The largest compressors had an outlet pressure of 6000–9000 psi and could output 5–15 million SCF/day. The smaller compressors were injecting 1–2 million SCF/day at much lower pressures. The pilots displaying the biggest increase in oil rates and recovery used the larger compressors.

With the larger compressors, the pressure builds to higher values in less time, and more gas is able to dissolve into the oil, which in turn allows more oil to be produced. The ratio of injection time to production time is much less for the larger compressors, so the wells produce for longer intervals. The downside to a large compressor is that they cost significantly more than the smaller one, and from 2017 to 2019 there were long lead times for delivery (up to 24 months). Recently, the supply of compressors has increased and some rental options are available, but securing a compressor is still one of the most important features for planning a gas injection project. The pilots with smaller compressors were able to show, as a proof of concept, that incremental oil is produced, but the economic viability of large-scale development demands the larger compressors.

These compressors typically call for around 1000 psi of inlet suction pressure, so if the gas is sourced from field gas, a booster compressor is necessary to get to the inlet pressure. If the gas is sourced from a gas plant or pipeline, the need for a booster compressor depends on the line pressure.

Regardless of the compressor size, the efficiency of the operations should be maximized to ensure a profitable project. This necessitates that the compressors do not sit idle. After injecting into one or more wells, the gas would be injected into the next set of wells, while the original wells are producing. Fig. 7.37 shows two possible well injection patterns.

If there are wells that are running both north and south from a single pad location, then the injection can be switched between wells in a repeating manner. Alternatively, if there are a large number of wells going in one direction from a single location, the injection can be switched between them in a cyclic manner. If all of the wells are not located on the same pad, additional high-pressure piping is required to move the gas from one pad to another. If the number of days producing is much larger than the number of days injecting, then there is an opportunity for a third injection location. Finally, if the offset wells are operated by a different

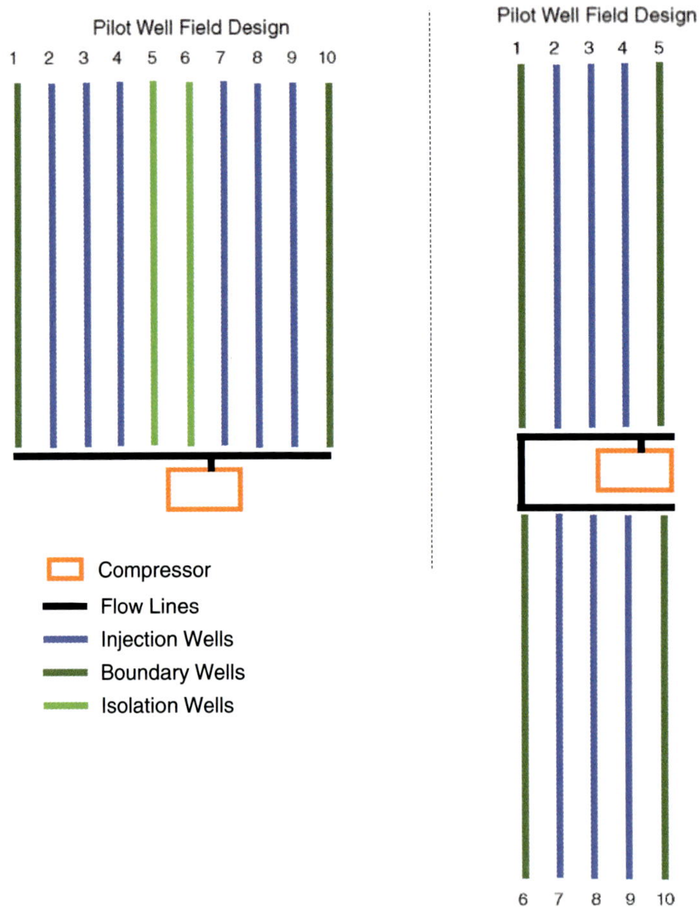

FIGURE 7.37

Possible cyclic gas injection well patterns to ensure compressor utility. Inject into wells 2, 3, and 4, then while those wells are producing, inject into wells 7, 8, and 9.

company, then a partnership may be necessary to maximize the usage of the compressor.

Besides the compressor, additional equipment is needed for a successful cyclic gas injection project. For example, a wellhead rated for 5000 psi working pressure may be sufficient for primary production, but during gas injection, surface pressures may exceed that and a larger wellhead is required. Before injection, the larger wellhead can be installed, but if the gas injection is planned from the beginning, it might be less expensive to connect the higher-rated wellhead from the start. A similar situation could arise for gas-tight tubing, which is needed for injection. Other surface and subsurface equipment should be checked to ensure they are appropriate for gas injection.

At the surface, the separation equipment must handle large volumes of produced gas during the huff-n-puff injection. Typically, wells in unconventional reservoirs come on at high rates and drop off quickly, so the surface equipment often has spare capacity when gas injection starts, but sufficient capability should be verified.

7.5.6 Primary completions and conformance control

Methods to drill and hydraulic fracture wells are very influential in the success of the injection strategy. The wells need to stay in the zone as much as possible, and the hydraulic fractures need to create as much surface area as possible. Matrix permeability is so low that the injectant cannot push oil through the matrix like a conventional flood, so the injected fluid has to access the remaining oil at the interface of the matrix and fractures. Therefore, the more fracture complexity and fracture surface area, the better chance of success for EOR.

Large, bi-wing fractures are not needed in unconventional reservoirs. The limiting factor is not the permeability of the fractures, but the amount of flow that can go from the matrix to the fractures. More fractures that are less permeable will perform better than fewer more permeable fractures. The focus of hydraulic fracturing throughout the 2010s was to create more fracture surface area. Better drilling and completing techniques not only lead to better EOR, but these same ideas maximize primary production as well.

A second problem with long bi-wing fractures is that they tend to connect wells and cause an early breakthrough. Even during cyclic injection, the injected fluids easily move to the offset production wells and prevent pressure build-up and gas going into the solution. Hydraulic fractures are often designed to prevent these kinds of well-to-well connections, yet they exist—possibly due to natural fractures. Regardless of the reason, some method to reduce the connection between wells may need to be implemented.

Different methods are proposed to reduce well-to-well connections in unconventional reservoirs. Many techniques are similar to ones in conventional EOR projects, which often have the same problem. First, the location of the problem along the wellbore is determined using some form of production log. Then, the offending locations can be isolated and shut off. Physical solutions using downhole movable sleeves are highly effective [62], but they have to be installed with the initial completion of the well. Other physical solutions, such as using cement to squeeze the perforations, can also be effective if the well interval causing the problem is isolated. High viscosity fluid diverters are also a common solution. A more recent proposed technique is to precipitate calcite in the worst offending high perm fractures [63].

All of the previous methods can add significant cost, but in some situations, the injectant can be used to counteract the conformance problem without adding costs. A portion of the injection gas can be redirected to the well or wells where gas is migrating, thus turning them into injection wells [64]. The offending wells can also be choked back or simply shut-in while the injection is occurring.

Maintaining the injected fluid close to where it is injected significantly improves the economics of EOR. A multipronged strategy for injection conformance control should be planned and ready to deploy before a breakthrough occurs. Many solutions with various expenses and effectiveness are possible.

7.5.7 Monitoring and optimization

Surveillance or monitoring is simply acquiring data or information about a project on a regular basis. The information is evaluated—on a less frequent but also regular timeframe—to optimize the project's overall operational efficiency. Because EOR projects tend to be more complex and costlier than other operations, more rigorous monitoring and extensive evaluations are done. EOR in unconventional reservoirs requires the same high level of rigor.

Every project is unique so there are no universal data collection reports, and the format and information collected should be tailored to the specific project. Typically, both physical and economic data are gathered. Physical data should be measured for the injection and production wells as well as any nearby wells that might be in communication with the pilot area. The physical data may include:

- Production/injection rates,
- Breakthrough times and volumes,
- Gas and fluid compositions, and
- Production/injection profiles along the wellbore.

Other information is collected as needed.

Economic data shows when and where inefficiencies are occurring in the process. If costs are increasing or revenues are decreasing, there needs to be a reasonable explanation. Typical economic information to collect include:

- Compression/pumping costs,
- Treatment facilities expenses,
- Injection gas/fluid costs, and
- Well maintenance/workover costs.

Additional economic information may also be collected.

During the pilot, more data are collected than at any other time in a project. This is when the most information is learned about the EOR process in the specific field. For example, potential problems will be identified, and methods to correct them will be implemented. This will reduce risks and costs associated with the large-scale field implementation. Pilots may require more expensive data collection techniques (e.g., fiber optic lines or passive seismic) that would not be used during regular field development, but it is worth the added expense in the pilot to more fully understand the process.

A consistent data collection process during the life of the field is critical, so the process can be continuously improved, and injection can be stopped once the pattern is mature. In conventional EOR, there are standard assessment tools to determine how well the process is working. For example, in CO_2 floods, gas utilization (Eq. 7.4) is a key metric to determine if a flood is efficient, and in steam floods, a steam–oil ratio is often used to determine how a pattern is behaving.

$$\text{Gas utilization} = \frac{\text{Gas injected (MSCF)}}{\text{Incremental oil produced (STB)}} \tag{7.4}$$

The monitoring metrics for EOR in unconventional reservoirs have not been standardized, but there are a couple of options. Similar to conventional CO_2 injection, gas utilization may be used. Low numbers represent more efficient processes, and for conventional reservoirs, a value less than 10 MSCF/STB is considered desirable, although this may be different for unconventional reservoirs. The advantage of gas utilization is that there is a historical connection with conventional gas injection projects to provide context.

Alternatively, Carlsen et al. [51] proposed using incremental oil production divided by the amount of gas injected as the standard monitoring tool—in a term they called EOR efficiency (Eq. 7.5).

$$\text{EOR efficieny} = \frac{\text{Incremental oil produced (STB)}}{\text{Gas injected (MMSCF)}} \tag{7.5}$$

The EOR efficiency is the inverse of the gas utilization with the exception that the injected gas in the denominator has units of millions of SCF. The advantage of EOR efficiency is that as the values increase, the project is more efficient. Typical values for some of the Eagle Ford projects are around 20–50 STB/MMSCF.

For both gas utilization and EOR efficiency, it is important to use incremental oil in the calculations. The oil that would have been produced on primary is excluded from the calculations.

7.5.8 Economics

The final topic to address is the economics of EOR in unconventional reservoirs. During the planning stage, economic calculations and estimations are needed to help the decision-making process. While this is similar to economics for any petroleum project, there are some unique features of EOR and unconventional reservoirs that are addressed. An example from an Eagle Ford pilot is used to explain a few key aspects; basic knowledge of economic analysis is assumed.

As with any economic project, it can be broken down into revenues and expenses. For EOR (or any petroleum-producing project), the revenues come from selling oil and gas. For EOR and waterfloods, the oil volumes need to be the incremental oil. To estimate incremental oil, a baseline of production is determined. Fig. 7.38A shows typical oil production rates from a well in an unconventional formation. The measured oil rates, in green, start high and drop off quickly.

The purple line is a decline curve that has been fit through the primary production and represents the baseline production. The orange line starting in 2021 is the predicted oil production for cyclic gas injection. The rates above the baseline are the incremental production. For EOR, it is sometimes easier to see the incremental oil on a cumulative production plot (Fig. 7.38B).

The estimated oil price multiplied by the estimated oil production for the month (or any time interval being used—day, year, etc.) gives the revenue for the month. Because oil prices are so volatile, it is good practice to use multiple price scenarios, which provides some range of uncertainty for the revenues.

For most projects, the natural gas produced is also included in the revenue stream. This is done by multiplying the expected gas rate by the different gas price scenarios and adding it to the oil revenues. For cyclic gas injection, however, it may not be included if all of the produced gas is being reinjected.

The expenses or costs are further broken down into two parts: capital and operating. Capital costs are the fixed, up-front money that is spent to get the project started. For EOR, the largest capital costs are the compressors or pumps used to inject the gas or fluid downhole. Additional capital costs include any well workovers, flowlines, or other new equipment that is needed.

The largest operating cost is usually the gas or chemicals that are injected. A second and usually smaller cost is the fuel and maintenance to run the compressors or pumps. Similar to revenues, only incremental costs should be included in the economic analysis. For example, if an employee is checking on the wells

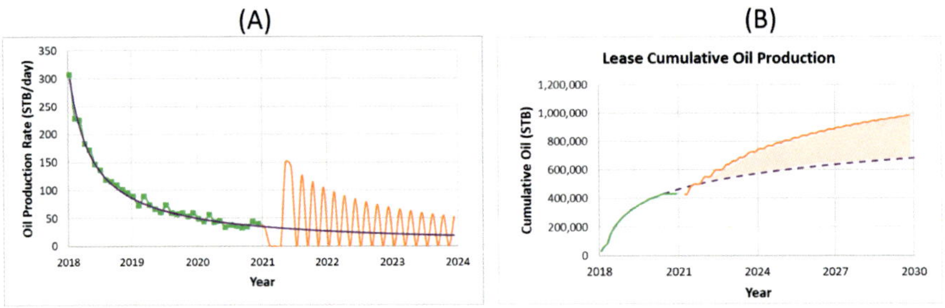

FIGURE 7.38

(A) Measured and predicted oil rates for one of four wells in an unconventional reservoir lease. (B) The cumulative oil produced for all four wells in the lease.

Table 7.6 Simple economic evaluation of an Eagle Ford cyclic gas injection project with four wells in US dollars.

Oil prices	CapEx=$2 million		CapEx=$4 million		CapEx=$6 million	
$60/STB	$7.78M	$6.16M	$5.81M	$4.81M	$3.84M	$1.74M
$45/STB	$4.76M	$3.13M	$2.78M	$1.15M	$0.81M	$-1.28M
$30/STB	$1.73M	$0.10M	$-0.25M	$-1.87M	$-1.21M	$-4.31M
Break even	$21/STB	$29/STB	$31/STB	$39/STB	$41/STB	$51/STB

NPV (rows $60/STB to $30/STB)

OpEx (first column of each CapEx pair): 5% fuel gas, 10% make up gas

OpEx (second column of each CapEx pair): 10% fuel gas, 20% make up gas during primary production, this expense is the same for EOR, so it does not need to be included in the operating costs. Also, like revenues, there are uncertainties when forecasting the capital and operating expenses, which should be incorporated into the analysis.

Once the costs and revenues have been established, the economic calculations are done to find the discounted cash flow, net present value, rate of return, breakeven price, payback time, and any other economic metrics a company uses.

Using the production forecasts in Fig. 7.38, the results of a simplified economic study for a four-well Eagle Ford lease are shown in Table 7.6. This economic model assumes that capital costs are $1 million per well. The operating costs were based on injection rates of 3 million SCF/day for a 3-month fill-up time. Then, it is assumed that 80% of the gas is recycled, so that an additional 20% is needed for make-up gas each month. The fuel and associated operating costs are assumed to equal 10% of the injected gas volume. The 20% make-up gas and 10% fuel gas assumptions come from rules of thumb for conventional gas injection projects. This may be larger than expected for unconventional EOR, so a second case using 10% make-up gas and 5% fuel gas is also evaluated. Table 7.6 includes three different oil price scenarios and three different capital costs along with the two operating costs. Only net present value and breakeven oil price are shown but other metrics could be calculated. At the lowest oil prices and highest costs, the project does not make money, but for most scenarios, the economics are positive.

EOR in unconventional reservoirs is still in its infancy, and operational efficiencies will improve over time, as injection rates, cycle times, and other aspects are optimized. As EOR is fully integrated into the lifecycle of wells and field development, further improvements will be observed; thus project economics will continue to improve.

7.6 Summary and conclusions

One of the most important messages from this chapter is the huge potential that EOR has for unconventional reservoirs. There are hundreds of billions of barrels of oil in each unconventional producing basin, and the primary recovery factors are in the single digits. Implementing EOR into these types of reservoirs has the potential to produce billions of additional barrels of oil. The development of unconventional reservoirs is in the early stages, so ongoing improvements in drilling and hydraulic fracturing should positively impact recovery from EOR.

There are many types of EOR that have been proposed for unconventional reservoirs including water, surfactants, and miscible gas, and both continuous and cyclic delivery methods have been considered. Field and lab tests have occurred for all of these methods, and most have some viability, but as of 2020, cyclic miscible gas injection with field hydrocarbon gases appears to have the most potential. For cyclic injection, the additional recovery is driven by the interaction of the injectant and the reservoir oil at the fracture–matrix interface. For cyclic gas injection, the mechanisms for recovery include oil swelling, secondary solution gas drive, vaporization, and pressure support.

Over the last 15 years there have been tens of different field trials with various injection fluids, locations, operators, and the results from these pilots have varied greatly as well. Much was learned from the field tests:

- Injectivity is not a problem; as much gas or fluid could be injected as possible.
- Conformance control or early breakthrough is an issue that needs to be considered for each project.
- Maximizing the deliverability of injection fluid minimizes injection time and improves efficiency.
- If not enough volume is injected to build pressure, less incremental oil is produced.

The Eagle Ford has had the most EOR development—more than all of the other basins combined—and cyclic gas injection appears to be the method of choice there. As most testing occurs, however, other methods may also turn out to be viable.

For successful pilot design and field development, extensive planning must be completed. Compressors are the biggest expense, but all equipment needs to be properly sourced and installed. The well layout, contingency planning, and surveillance are all important aspects of having a successful EOR project.

Since it is early in the life of EOR for unconventional reservoirs, standardization has not yet occurred in laboratory experiments and reservoir simulations. For example, core experiments need to be fully developed for cyclic gas injection. For flow simulation, the most important aspect is incorporating hydraulic fractures into the model as accurately as possible. However, even as new techniques are developed, EOR projects still require laboratory work and flow models, so tasks continue to be completed as skillfully as possible.

The economics of EOR in unconventional reservoirs looks positive. The field pilots clearly show that additional oil can be produced, and the revenues appear to be greater than the costs for most oil price scenarios. As the process becomes more efficient over time, the economics should continue to improve.

The future of EOR in unconventional reservoirs is promising but lots of work still needs to be done. More field trials with new experiments and simulation models need to be completed, but the size of the prize is so enormous that the industry will keep working on it. This could be the cusp of something really significant.

Nomenclature

1D One dimensional
b/d Barrels per day
C_2+ Ethane plus fraction

CO_2	Carbon dioxide
DOE	US Department of Energy
EOG	EOG Resources, Inc.
EOR	Enhanced oil recovery
ft	Feet
GOR	Gas–oil ratio
LPG	Liquified petroleum gas
LSW	Low-salinity waterflooding
mD	Milli-Darcy
MMP	Minimum miscibility pressure
MSCF	Thousands of standard cubic feet
PVT	Pressure, volume, and temperature
RRC	Texas Railroad Commission
SCF	Standard cubic feet
SHmax	Maximum horizontal stress
SHmin	Minimum horizontal stress
STB	Stock tank barrel
WAG	Water-alternating gas

References

[1] A.M. Jaffe, K.B. Medlock, R. Soligo, The Status of World Oil Reserves: Conventional and Unconventional Resources in the Future Supply Mix, James Baker Institute for Public Policy, Rice University, Houston, 2011.

[2] R. Sorkhabi, Ghawar, Saudi Arabia: the king of giant fields, Geo ExPro: Geoscience and Technology Explained 7 (2010) 24–29.

[3] EIA, U.S. Field Production of Crude Oil, U.S. Energy Information Administration, Washington, DC, 2020.

[4] A.J. Clark, Determination of recovery factor in the Bakken Formation, Mountrail County, ND, in: SPE Annual Technical Conference and Exhibition, 4–7 October 2009, New Orleans, Louisiana, USA.

[5] S.A. Gherabati, U. Hammes, F. Male, et al., Assessment of hydrocarbon in place and recovery factors in the Eagle Ford shale play, SPE Reservoir Evaluation & Engineering 21 (2018) 291–306.

[6] V.A. Kuuskraa, S.H. Stevens, K.D. Moodhe, World Shale Gas and Shale Oil Resource Assessment, U.S. Energy Information Administration, Washington D.C., 2013.

[7] D.P. Craig, C.E. Odegard, W.C. Pearson, et al., Case history: observations from diagnostic injection tests in multiple pay sands of the Mamm Creek Field, Piceance Basin, Colorado, in: SPE-60321-MS, SPE Rocky Mountain Regional/Low-Permeability Reservoirs Symposium and Exhibition, 12–15 March 2000, Denver, Colorado, USA.

[8] K. Vidma, P. Abivin, D. Fox, et al., Fracture geometry control technology prevents well interference in the Bakken, in: SPE-194333-MS, SPE Hydraulic Fracturing Technology Conference and Exhibition, 5–7 February 2019, The Woodlands, Texas, USA.

[9] J.J. Taber, F.D. Martin, R.S. Seright, EOR screening criteria revisited—part 1: introduction to screening criteria and enhanced recovery field projects, SPE Reservoir Engineering 12 (1997) 189–198.

[10] C.H. Bowman, S. Gilbert, A successful cyclic steam injection project in the Santa Barbara field, Eastern Venezuela, Journal of Petroleum Technology 21 (1969) 1531–1539.

[11] K. Bybee, Screening criteria for carbon dioxide huff 'n' puff operations, Journal of Petroleum Technology 59 (2007) 55–59.

[12] C. Carpenter, Water-injection operations and gas-injection sensitivities in the Bakken formation, Journal of Petroleum Technology 69 (2017) 80–82.

[13] A. Bond, R. Jasser, V. Johnson, et al., Completion optimization provides step changes in horizontal well performance in the Oooguruk field, Alaska, SPE Drilling & Completion 32 (2017) 1–12.

[14] M. Cronin, H. Emami-Meybodi, R.T. Johns, Diffusion-dominated proxy model for solvent injection in ultratight oil reservoirs, SPE Journal 24 (2019) 660–680.

[15] B.F. Thomas, M. Piwowar, M. Noroozi, et al., Insight gained from experimental gas-cycling EOR in the unconventional Montney and Duvernay formations, in: SPE-200359-MS, SPE Improved Oil Recovery Conference (Virtual), 1–4 September 2020.

[16] B.T. Hoffman, J.M. Rutledge, Mechanisms for huff-n-puff cyclic gas injection into unconventional reservoirs, in: SPE-195223-MS, SPE Oklahoma City Oil and Gas Symposium, 9–10 April 2019, Oklahoma City, Oklahoma, USA.

[17] K.T. Raterman, H.E. Farrell, O.S. Mora, et al., Sampling a stimulated rock volume: an Eagle Ford example, SPE Reservoir Evaluation & Engineering 21 (2018) 927–941.

[18] B.T. Hoffman, D. Reichhardt, Recovery mechanisms for cyclic (huff-n-puff) gas injection in unconventional reservoirs: a quantitative evaluation using numerical simulation, Energies 13 (2020) 4944.

[19] National Energy Technology Laboratory, Carbon Dioxide Enhanced Oil Recovery: Untapped Domestic Energy Supply and Long-Term Carbon Storage Solution, US Department of Energy, Washington, DC, 2010.

[20] F.D. Tovar, M.A. Barrufet, D.A. Schechter, Gas injection for EOR in organic rich shale. Part I: operational philosophy, in: SPE Improved Oil Recovery Conference, 14–18 April 2018, Tulsa, Oklahoma, USA.

[21] A.J. Calderón, N.J. Pekney, Optimization of enhanced oil recovery operations in unconventional reservoirs, Applied Energy 258 (2020) 1–21.

[22] P.L. McGuire, R. Okuno, T.L. Gould, et al., Ethane-based enhanced oil recovery: an innovative and profitable enhanced-oil-recovery opportunity for a low-price environment, SPE Reservoir Evaluation & Engineering 20 (2017) 42–58.

[23] N.R. Nagarajan, D. Stoll, M.L. Litvak, et al., Successful field test of enhancing Bakken oil recovery by propane injection: part I. Field test planning, operations, surveillance, and results, URTeC-2020-2768, in: SPE/AAPG/SEG Unconventional Resources Technology Conference (Virtual), 20–22 July 2020.

[24] D. Denney, Enhanced oil recovery with high-pressure nitrogen injection, Journal of Petroleum Technology 53 (2001) 55–56.

[25] P.J. Sanger, J. Hagoort, Recovery of gas-condensate by nitrogen injection compared with methane injection, SPE Journal 3 (1998) 26–33.

[26] C.U. Hatiboglu, T. Babadagli, Experimental and visual analysis of co- and counter-current spontaneous imbibition for different viscosity ratios, interfacial tensions, and wettabilities, Journal of Petroleum Science and Engineering 70 (2010) 214–228.

[27] S. Takahashi, A.R. Kovscek, Spontaneous countercurrent imbibition and forced displacement characteristics of low-permeability, siliceous shale rocks, Journal of Petroleum Science and Engineering 71 (2010) 47–55.

[28] S. Morsy, J.J. Sheng, Effect of water salinity on shale reservoir productivity, Advances in Petroleum Exploration and Development 8 (2014) 9–14.

[29] B. Kurtoglu, Integrated Reservoir Characterization and Modeling in Support of Enhanced Oil Recovery for Bakken Effects of CO_2 Flooding Horizontal Wells in a Tight Oil Reservoir: Elm Coulee Field Case Study (Ph.D. dissertation), Colorado School of Mines, 2013.

[30] T. Jacobs, Improving shale production through flowback analysis, Journal of Petroleum Technology 67 (2015) 37–42.

[31] A. Telmadarreie, J.J. Trivedi, Post-surfactant CO_2 foam/polymer-enhanced foam flooding for heavy oil recovery: pore-scale visualization in fractured micromodel, Transport in Porous Media 113 (2016) 717–733.

[32] J.O. Alvarez, Wettability Alteration Using Surfactants to Improve Oil Recovery From Unconventional Liquid Reservoirs (Ph.D. dissertation), Texas A&M University, 2017.

[33] D. Wang, R. Butler, J. Zhang, et al., Wettability survey in Bakken Shale with surfactant-formulation imbibition, SPE Reservoir Evaluation & Engineering 15 (2012) 695–705.

[34] K. Prateek, K.K. Mohanty, EOR in tight oil reservoirs through wettability alteration, in: SPE-166281-MS, SPE Annual Technical Conference and Exhibition, 29 September–2 October 2013, New Orleans, Louisiana, USA.

[35] J.O. Alvarez, D.S. Schechter, Wettability alteration and spontaneous imbibition in unconventional liquid reservoirs by surfactant additives, SPE Reservoir Evaluation & Engineering 20 (2017) 107–117.

[36] M. Kazempour, M. Kiani, D. Nguyen, et al., Boosting oil recovery in unconventional resources utilizing wettability altering agents: successful translation from laboratory to field, in: SPE-190172-MS, SPE Improved Oil Recovery Conference, 14–18 April 2018, Tulsa, Oklahoma, USA.

[37] D. Alfarge, M. Wei, B. Bai, Fundamentals of Enhanced Oil Recovery Methods for Unconventional Oil Reservoirs, Volume 67 of Developments in Petroleum Science, Elsevier Science, Amsterdam, 2020.

[38] K. He, L. Xu, Y. Gao, et al., Evaluation of surfactant performance in fracturing fluids for enhanced well productivity in unconventional reservoirs using Rock-on-a-Chip approach, Journal of Petroleum Science and Engineering 135 (2015) 531–541.

[39] M.K. Valluri, J.O. Alvarez, D.S. Schechter, Study of the rock/fluid interactions of sodium and calcium brines with ultra-tight rock surfaces and their impact on improving oil recovery by spontaneous imbibition, in: SPE Low Perm Symposium, 5–6 May 2016, Denver, Colorado, USA.

[40] A. Katende, F. Sagala, A critical review of low salinity water flooding: mechanism, laboratory and field application, Journal of Molecular Liquids 278 (2019) 627–649. Available from: https://doi.org/10.1016/j.molliq.2019.01.037.

[41] F. Sagala, T. Montoya, A. Hethnawi, et al., Nanopyroxene-based nanofluids for enhanced oil recovery in sandstone cores at reservoir temperature, Energy & Fuels 33 (2019) 877–890.

[42] B. Peng, L. Zhang, L. Luo, et al., A review of nanomaterials for nanofluid enhanced oil recovery, Royal Society of Chemistry Advances 51 (2017) 32246–32254.
[43] Montana Board of Oil and Gas (MBOG), Live Data Access, 2020.
[44] North Dakota Industrial Commission (NDIC), Oil and Gas Division, 2020. Online Production Data.
[45] B. Kurtoglu, A. Salman, How to utilize hydraulic fracture interference to improve unconventional development, in: SPE-177953-MS, International Petroleum Exhibition and Conference, 9–12 November 2015, Abu Dhabi, UAE.
[46] W.R. Thomas, L.W. Helms, T.K. Driggers, et al., EOG Resources (EOG) Earnings Call, 2016.
[47] Texas Railroad Commission (RRC), Oil and Gas Division, 2020. Online Production Data.
[48] J.B. Rogers, B. Basbug, T. Firincioglu, et al., A simulation case study of gas injection pilot in Eagle Ford, in: SPE-200430-MS, SPE Improved Oil Recovery Conference (Virtual), 18–20 August 2020.
[49] A. Hirshberg, G. Leveille, H. Harding, Eagle Ford Investor Field Tour (2018). 3–5 April ConocoPhillips, Houston.
[50] M. Torres, W. Yu, R. Ganjdanesh, et al., Modeling interwell fracture interference and huff-n-puff pressure containment in Eagle Ford using EDFM, in: SPE-195240-MS, SPE Oklahoma City Oil and Gas Symposium. 9–10 April 2019, Oklahoma City, Oklahoma, USA.
[51] M. Carlsen, C. Whitson, M.M. Dahouk, et al., Compositional tracking of a huff-n-puff project in the Eagle Ford, in: URTeC-2019-539-MS, SPE/AAPG/SEG Unconventional Resources Technology Conference, 22–24 July 2019, Denver, Colorado, USA.
[52] E. Kerr, K.K. Venepalli, K. Patel, et al., Use of reservoir simulation to forecast field EOR response—an Eagle Ford gas injection huff-n-puff application, in: SPE-199722-MS, SPE Hydraulic Fracturing Technology Conference and Exhibition, The Woodlands, Texas, USA, 4–6 February, 2020.
[53] M. Carlsen, S. Mydland, M.M. Dahouk, et al., Compositional tracking of a huff-n-puff project in the Eagle Ford: a second take, in: URTeC-2020–2869-MS, SPE/AAPG/SEG Unconventional Resources Technology Conference (Virtual), 20–22 July 2020.
[54] G. Grinestaff, C. Barden, J. Miller, et al., Evaluation of Eagle Ford cyclic gas injection EOR: field results and economics, in: SPE-200427-MS, SPE Improved Oil Recovery Conference (Virtual), 18–20 August 2020.
[55] S. Salahshoor, J. Ciezobka, Hydraulic fracturing test site HFTS Phase-2 EOR pilot: huff-and-puff pilot in the Permian-Midland Basin, in: SPE-200468-MS, SPE Improved Oil Recovery (Virtual), 31 August–4 September 2020.
[56] G. Pospisil, P. Weddle, S. Strickland, et al., Report on the first rich gas EOR cyclic multiwell huff n puff pilot in the Bakken tight oil play, in: SPE-201471-MS, SPE Annual Technical Conference & Exhibition (Virtual), 26–29 October 2020.
[57] S.B. Hawthorne, C.D. Gorecki, J.A. Sorensen, et al., Hydrocarbon mobilization mechanisms from upper, middle, and lower Bakken reservoir rocks exposed to CO_2, in: SPE-167200-MS, Unconventional Resources Conference Canada, 5–7 November 2013, Calgary, Alberta, Canada.

[58] L. Li, J.J. Sheng, Optimization of huff-n-puff gas injection to enhance oil recovery in shale reservoirs, in: SPE180219-MS, SPE Low Perm Symposium, 5–6 May 2016, Denver, Colorado, USA.

[59] W. Pu, B.T. Hoffman, EOS modeling and reservoir simulation study of Bakken gas injection improved oil recovery in the Elm Coulee field, Montana, in: URTeC-2014–1922538, SPE/AAPG/SEG Unconventional Resources Technology Conference, 25–27 August 2014, Denver, Colorado, USA.

[60] L. Li, S.H. Lee, Efficient field-scale simulation of black oil in naturally fractured reservoir through discrete fracture networks and homogenized media, SPE Reservoir Evaluation & Engineering 11 (2008) 750–758.

[61] Y. Xu, J.S. Cavalcante Filho, W. Yu, et al., Discrete-fracture modeling of complex hydraulic-fracture geometries in reservoir simulators, SPE Reservoir Evaluation & Engineering 20 (2017) 403–422.

[62] C. Bell, B. Cheeseman, B. Gowisnock, et al., New horizontal producer in existing waterflood: mitigating water breakthrough, in: SPE-201540-MS, SPE Annual Technical Conference and Exhibition (Virtual), 27–29 October 2020.

[63] C.M. Kirkland, R. Hiebert, R. Hyatt, R., et al., Direct injection of biomineralizing agents to restore injectivity and wellbore integrity, SPE Production & Operations 36 (2020) 216–223.

[64] A. Baldwin, L. Lasecki, D. Mohrbacher, et al., A methodological workflow for assessment and design of a huff-n-puff hydrocarbon gas injection pilot test as an EOR technique for Eagle Ford shale oil reservoirs, in: SPE-200415-MS, SPE Improved Oil Recovery Conference (Virtual), 31 August–4 September 2020.

CHAPTER

Microbial enhanced oil recovery

8

Hakan Alkan[1], Soujatya Mukherjee[2] and Felix Kögler[3]
[1]*TU Bergakademie Freiberg, Freiberg, Germany*
[2]*Wintershall Dea AG, Kassel, Germany*
[3]*University of Strasbourg, Strasbourg, France*

8.1 Introduction

With its positive and negative impacts, oil is one of the most significant elements of modern human civilization. It is still the major source of energy for the world; its demand rising steadily, being a critical resource for the petrochemical industry. However, a big part of the oil remains in the reservoir after primary recovery. This remaining oil can comprise between 50% and 95% of the original oil in place. To increase oil recovery, the reservoir can be waterflooded in a secondary recovery process. Enhanced oil recovery (EOR) processes are mostly applied after waterflooding to produce even more of the remaining oil. Various grouping efforts are made for EOR methods, such as thermal, gas, and chemical [1,2]. However, one unique method cannot be assigned to any of these classifications: the microbial enhanced oil recovery (MEOR).

"Microbe" or microorganism is a general and loose term that comprises a variety of organisms too small to be seen with the naked eye. They are mostly unicellular, that is, consist of only one cell as the smallest possible unit of a living organism but can also be multicellular. Examples include bacteria, fungi, algae, and many others. In his book "The Invisible Kingdom," Ben Barak mentioned that "until about 20 years ago looking at microbes as social organisms wasn't too popular. Lately, aided by technological advances, observations of organisms have caused scientists to view microbial interactions more diversely. It now appears as if microbes can display a host of quite complex behavior patterns involving communication, cooperation, and deception" [3]. This is often referred to as quorum sensing and one of the best examples of this microbial communication is perhaps biofilm formation: Microbes produce a regulating substance, *N*-acyl homoserine lactone (AHL), in this case. AHL induces gene expression, that is, retrieving the genetic information stored as DNA in the cells and producing the respective biomolecules, when it reaches a threshold concentration. In effect, cells "communicate" via the concentration of AHL and will collectively form a biofilm upon a certain abundance. MEOR can be interpreted as a result of many aspects, such as

Recovery Improvement. DOI: https://doi.org/10.1016/B978-0-12-823363-4.00009-1

this, but is foremost driven by microbial reproduction and metabolic activity. Arguably, bacteria are the most important microorganisms in MEOR. There are more than 30,000 formally named species of bacteria that are in pure culture and for which the physiology has been investigated [4].

8.1.1 History and milestones

The idea of using microorganisms to increase oil production from reservoirs was first proposed by Beckmann [5]. Over the following decades until the 1990s, several milestones and development further established MEOR as an EOR method, as listed in Table 8.1. From the table, it can be concluded that Claude E. Zobell

Table 8.1 Essential milestones of microbial enhanced oil recovery.

Milestone	Source
1926—Idea that microorganisms could be used to release oil from porous media	[5]
1936—The existence of a rich microflora inhabiting ocean bottom deposits; uncontaminated sediment sampling using a modified Ekman coring device	[6]
1946—The first MEOR patent (Patent US2413278): "Bacteriological process for treatment of fluid-bearing earth formations"	[7,8]
1950—Biopolymer "Xanthan gum" discovered at the Northern Regional Research Laboratories of the US in 1950	[9]
1953—Zobell's second patent application describing an anaerobic process for removing sulfur from petroleum hydrocarbons (Patent US2641564)	[10]
1954—The first field test was carried out in the Lisbon field, Union County, AR, USA	[11]
1961—First report on the use of MEOR in former USSR	[12]
1962—Bioplugging as an MEOR mechanism experimentally proved	[13]
1983—Patent search revealed over one hundred patents based on the use of microorganisms in oil production by 1983	[14]
1984—MEOR is mentioned for the first time with explicit classification "in situ" and "ex situ"; reference to the use of biotechnological techniques, involving rDNA	[15]
1985—Experimental investigations on the potential use of microorganisms in oil recovery	[16]
1984—Growth of sulfate-reducing bacteria in reservoirs and their contribution to hydrogen sulfide souring	[17]
1985—The lipopeptide-producing Bacillus strain JF-2 has been used in coreflood experiments and in field trials (ex situ surfactant production)	[18]
1986—First reporting from field applications; first published listing of screening criteria	[19,20]
1988—First MEOR numerical reservoir simulator	[21]
1990—Design of a field pilot with a well-described experimental and numerical study	[22]

MEOR, *microbial enhanced oil recovery.*

dominates the first decades of MEOR development. His studies and contribution to reservoir microbiology and MEOR are summarized in [6]. Many of his publications from the 1940s and 1950s discussed the key role of microorganisms in the hydrocarbon biogeochemistry. At a time when the industry considered the importance of the sustainable/reliable oil supply, Zobell's thoughts were turning to improved oil recovery using bacterial products, such as acids and gases, to aid the mobilization of oil in the reservoir. He also recognized the inevitable problems of oil spill pollution and the potential for use of bacteria as a mitigation measure. Zobell worked in most of these areas and published a total of 66 papers between 1938 and 1978 on various aspects of petroleum microbiology. He described almost all the EOR mechanisms, which are still valid today in his patent from 1946 "Bacteriological Process for Treatment of Fluid-bearing Earth Formations" [7]. Among other things, he suggested to increase the recovery of oil or gas by increasing the porosity of formations (solubilization of calcareous material), releasing fluids from rock surfaces (acid and gas production), facilitating flow in formations due to surfactant generation, increasing gas pressure (carbon dioxide production) and reducing the viscosity of oils (biotransformation of oils) [6]. He is also one of the pioneers on investigating the effect of microbial souring. Simultaneously, studies on MEOR were started in the same period as Zobell's works in the former Russia (USSR). Kuznetsov's work validated the technology of activating microbial community of a reservoir, which was later further developed by Ivanov and his research group [8].

In the 1960s and 1970s, significant research activity also took place in some European countries, such as former Czechoslovakia, Hungary, Romania, and Poland [8]. The field trials developed in these countries were based on the injection of mixed anaerobic or facultative anaerobic bacteria selected because of their ability to generate high quantities of gases, acids, solvents, polymers, surfactants, and cell biomass.

The first MEOR field test was carried out in the Lisbon field of Union County, USA, in 1954 [11]. The classification being in situ and ex situ was first made by Saxman [15]. One of the first attempts at ex situ surfactant production by using Bacillus strain JF-2 is reported by Javaheri et al. [18]. The first reports on the screening criteria for MEOR were made by Bryant and Douglas [19]. The same reference also provides preliminary information on the design of the MEOR field pilots as well as operational economics.

The growth of sulfate-reducing bacteria (SRB) in reservoirs and their contribution to hydrogen sulfide-induced souring, corrosion, and plugging of formations was first studied by Iverson and Olson [17]. Knapp et al. [22] reported a field MEOR design covering the result of experimental, numerical as well as operational efforts. The report describes coreflood experiments performed with defined substrates to unlock responsible EOR mechanisms based on simple modeling approaches. The numerical model developed by the authors describes microbial transport in a multidimensional porous medium. This description is completed with a kinetic model of bacterial growth, which leads to either bacterial plugging and mobility control or alteration of oil properties, such as reduction of oil viscosity and interfacial tension (IFT) [22]. Notably, this is the first attempt of its kind.

The quantity and quality of the studies on MEOR are increased significantly after the 1990s, due to technological developments in microbiology, with publications reporting both in situ and ex situ MEOR. With the oil price increase after 2005, the number of publications soared in parallel and was fourfold higher compared to the 1990s. Well-prepared peer-reviewed publications as well as field reports dominate the last decades [23–28]. Interestingly, most of the comprehensive reviews about MEOR field trials report positive results in terms of an increase in oil recovery.

It can be concluded that MEOR is one of the oldest EOR methods with continuing interest worldwide. Note that from widely applied EOR methods, the first industrially significant thermal EOR project was conducted in 1931 in Texas, USA [29]. Chemical methods (surfactant and polymer) are applied since the 1960s and the first CO_2 injection test as an oil recovery method was performed in 1964 [30]. With the advent of MEOR applications, microbiology became an additional pillar to be considered in reservoir/EOR engineering. It was perceived as essential that reservoir microbiology needs to provide valuable inputs to reservoir engineering for MEOR as will be discussed in the following sections.

The inherent value of MEOR and its theoretical advantages compared to other EOR methods lie in the fact that MEOR is always characterized as an easy to apply and a cost-effective operational oil recovery method. Therefore it has a significant impact on the oil industry from an EOR perspective. However, MEOR is often performed in the form of single-well tests or in marginal "stripper" wells with a low productivity, where the expectation for incremental recovery is low to begin with [31,32]. Accordingly, the number of MEOR applications remained low, not exceeding even 1% of the total EOR applications and oil recovered by EOR. Yet, within a larger perspective and especially considering the start with Zobell''s works, it can be argued that MEOR is even more than just a method to increase incremental oil production: MEOR is the introduction of modern microbiology to reservoir engineering and should be considered from exploration to field management.

8.1.2 Definitions and classifications

One of the earliest attempts to classify the MEOR was made by Saxman and Crull by grouping six processes occurring in MEOR [15]. Bryant et al. defined the MEOR as the "injection of selected microorganisms into the reservoir and subsequent stimulation and transportation of their in situ growth products in order that their presence will aid in further reduction of residual oil left in the reservoir after secondary recovery is exhausted" [33]. Maudgalya et al. developed individual classifications according to reservoir types and application strategies, type of recovery mechanisms, microbes used either injected or in situ, and nutrients applied [34].

Recently, MEOR has been defined as "a biological based technology consisting in manipulating function and/or structure of microbial communities in oil reservoirs" [35]. A classification according to the oil production challenges in the field is provided as follows:

- Wellbore clean up removes mud and other debris blocking the channels where oil flows through
- Well stimulation improves the flow of oil from the drainage area into the wellbore
- Waterfloods are enhanced via stimulating microbial activity by injecting selected nutrients and sometimes indigenous microbes

For microbial well stimulation, a fluid/nutrient medium defined by using similar techniques applied in MEOR is injected into a producing well and a shut-in period of a few days is allowed as incubation (soak) time and produced from the same well. Such huff-n-puff (HnP) type single well treatments (SWT) are intended to stimulate the microbial activity of indigenous microbes in the near-wellbore region and to change the petrophysical properties via generated metabolites and/or biomass. The microbial metabolites, predominantly organic acids in this case, reduce near-wellbore skin and enhance oil flow to the well. The reduced interfacial tension and increased water wettability of rock in the wellbore region can also help to assure more oil flow by reducing the fractional flow of water. This treatment method showed significant improvement in oil production in some applications [36,37]. These processes are certainly due to one or multiple MEOR mechanisms occurring near the wellbore and are mostly considered an MEOR application. In addition, these treatments also provide simultaneous removal of asphaltene and/or paraffin and scale from a well, by consuming the heavier carbon fractions or components in the wellbore, potentially enhancing crude oil or gas recovery [38,39]. Although the MEOR background expertise is partly applicable in such applications, these could be interpreted as a part of well stimulation processes rather than an EOR method.

It can be concluded from the above description that there is no consensus on the classification of MEOR types. In a relatively recent publication, Patel et al. divided MEOR once again into two main categories, that is, ex situ and in situ [32]. This classification is based on whether the principal aspect of MEOR, the metabolic activity of microorganisms, occurs inside or outside the reservoir and will also be used in this chapter (Fig. 8.1).

8.1.2.1 Ex situ microbial enhanced oil recovery

In ex situ MEOR, as a general application, cultivable bacterial strains or consortia, potentially isolated from an oil reservoir and/or genetically engineered, produce a specific microbial metabolite, such as a biopolymer or a biosurfactant. This microbial product is prepared in an industrial facility and is then transported and injected into the reservoir. Ex situ MEOR can be quite costly because the bacterial community targeting a specific metabolite as well as the metabolite itself

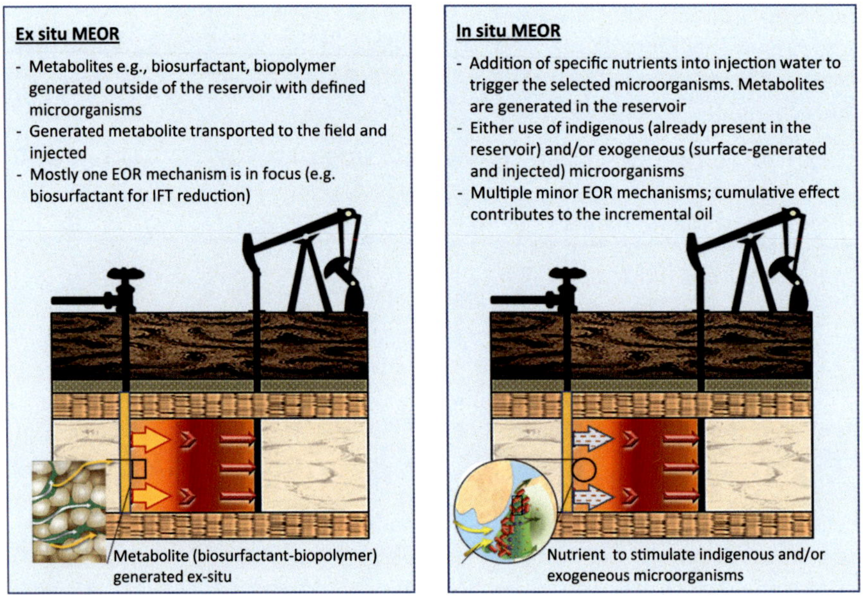

FIGURE 8.1

Schematic description and definition of ex situ and in situ microbial enhanced oil recovery.

must be produced in high amounts outside of the field, transported to the field, has to be stabilized, has to be compatible with reservoirs fluids, etc. Similar to the production of any chemical at an industrial scale, upscaling, purification, and formulation efforts further increase the costs of this approach. On the other hand, ex situ MEOR is more appealing to operators because the injection of a defined product with an anticipated biochemical reaction in the reservoir appears to be a manageable and controllable process [32]. However, the full potential of MEOR mechanisms is unlikely to be achieved with this method because not all mechanisms, namely related to biomass, biofilm, and gas production, are incorporated in the process.

8.1.2.2 In situ microbial enhanced oil recovery

In in situ MEOR, metabolites are generated in the reservoir as a reaction of the injected nutrients with targeted microorganisms. This kind of application can be further subdivided depending on the origin of the targeted microorganisms: indigenous (already present in the reservoir) or exogenous (isolated from other sources). The state-of-the-art is the first option: defining the beneficial microorganisms in situ (in the reservoir) and feeding them with selected nutrients, resulting in the growth of the microorganisms and in the generation of the metabolites, both acting on EOR. It is frequently affirmed that, in comparison to the injection of ex situ microorganisms, the in situ application has more chances to be successful as the microorganisms are already adapted to the reservoir conditions and can grow if a well-selected nutrient is injected for targeted EOR mechanisms. In contrast, it is highly unlikely that the injected ex situ microorganisms can out-compete the indigenous microbes, which have acclimated to the harsh conditions over geological time scales. Rather than one specific effect, in situ MEOR targets several effects, possibly each to a minor extent, for enhancing total oil production.

8.1.3 Workflow: from screening to field application

The workflow of an MEOR operation is similar to workflows applied in other EOR projects; however, the contribution of microbiology in all steps makes an important difference. Obviously, with a pregenerated injectant, the design of ex situ MEOR applications is comparable to chemical EOR methods. One addition, however, is that microbiological monitoring plays a major role. In situ MEOR projects, on the other hand, require an integral cycle with the interaction of an interdisciplinary team of microbiologists, reservoir engineers and operational staff. Therefore the resulting workflow for in situ MEOR projects is unique and can be described as follows.

The design of an MEOR operation starts with the selection of a suitable field. For this decision, screening criteria are applied first to conclude whether a field in question can be a suitable candidate for an MEOR application. Well-integrated sampling and laboratory activities of the microbiology and reservoir teams are needed to define the microbiological impact and resulting EOR effects in the field. The objective here is to tailor a nutrient to trigger the metabolic activity/growth of the selected microorganisms and the definition and quantification of the EOR mechanisms as a result of this triggered microbial activity. This needs an effective working cycle between microbiology and reservoir engineering to optimize the nutrient for maximum EOR effect, therefore maximum incremental oil recovery. The data provided during the laboratory phase is used in the numerical simulation of MEOR at a pilot scale (if any) and/or further in the field scale. This necessary exercise, which is valid for all EOR methods, helps to design the operational parameters to estimate the incremental oil and thus the overall technical and economic feasibility of the operation. In the operational phase, the microbiology and reservoir engineering teams should work in an integrated manner with the operational team on designing and on setting the surface equipment as well as the surveillance program, including sampling and evaluation. For example, operational staff might need to be trained in taking samples for microbiological analysis without any contamination, for example, intrusion of oxygen, and with sterile equipment. A general work breakdown structure (WBS) that can be applied in MEOR projects is shown in Fig. 8.2.

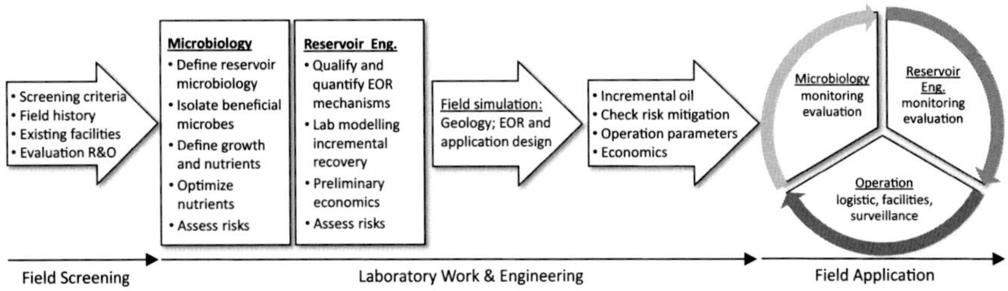

FIGURE 8.2

The general work breakdown structure for microbial enhanced oil recovery projects. *R&O*, risks and opportunities.

8.2 Primary screening criteria

To perform preliminary screening of reservoirs, a defined list of screening criteria for MEOR is required, similar to any other EOR method. A well-known example of a defined list of tabulated data is published by Taber et al. [40]. Here, all criteria for the application of common EOR methods, based on applications in selected oil fields, are summarized. Therefore a relatively quick decision for the way forward can be easily made if the relevant characteristics of a reservoir are known. In Taber's screening criteria all EOR methods are listed against the relevant reservoir parameters. Interestingly, screening criteria for MEOR are not listed in Taber's publication.

8.2.1 Reservoir screening

The first available screening for MEOR was compiled by Bryant and Douglas and contained five criteria [19]. The application ranges given for four of these, namely permeability, salinity, temperature, and oil gravity were updated numerous times by several authors, based on the results of new applications and/or new technologies [41–45]. The fifth states that the reservoir should not have been flushed with a biocide for at least 2 weeks before the MEOR starts. Considering that biocide can inhibit the reservoir microbiology completely, this is correct.

One of the most recent screening tables is provided by Patel et al. under the title of "Standardized Certain Well Selection Criteria for MEOR Jobs" [32]. It largely refers to the US Department of Energy (DOE) reservoir database and the Institute of Reservoir Studies (IRS, India) which have standardized certain well-selection criteria for MEOR jobs. Safdel et al. further extended this table by reviewing some other studies and applications [46]. Based on these most recent screening tables, the following should be considered:

According to both DOE and IRS, a suitable formation type for MEOR is sandstone [32]. However, there are studies on the successful applications of MEOR in carbonates as well [47–49]. As early as 1988, Lazar even stated that the presence of carbonates is a prerequisite for successful MEOR [50]. Recent studies support this claim and concluded that calciferous reservoirs could be particularly suitable for certain in situ MEOR approaches [51].

According to IRS, the reservoir temperature should be lower than 90°C; while the DOE limit is 71°C [32]. There are applications and validated presence of microbial communities beneficial for MEOR at temperatures slightly higher than 80°C. However, this range should be considered as the maximum limit for MEOR, considering that reservoirs with a temperature of > 80°C showed less/no microbial activity as indicated by a lack of biotransformation [52]. Depth can be applied as a criterion in cases where an estimation of the reservoir temperature is necessary.

The reservoir pressure is given in IRS list as < 300 bar as the only example among screening tables. Generally, unless the pressure does not change abruptly [53,54], pressure effects impairing microbial growth, such as conformation changes of biomolecules, are not expected below 500 bar [55].

Permeability is a limiting factor for the applicability of MEOR. Various studies confirm the importance of permeability on the movement of the bacteria in porous media [56–58]. Although microorganisms used in MEOR have various geometries, it is reasonable to assume a general determinant scale of 1–5 μm for bacteria. This dimension is critical for low-permeability formations with narrow pore throats. Davis and Updegraff stated that the pore/throat sizes of the rock should be at least twice the cell sizes to allow the passage of microbes through the rock without substantial plugging [59]. Therefore the reservoir permeability is and should be one of the most important selection criteria for MEOR. Current recommendations range from a lower limit between 50 mD (IRS) to 100 mD (DOE), but even successful field trials from formations with a permeability of ≥5 mD were reported recently [60]. However, reservoir permeabilities are mostly given as averaged values of petrophysical measurements and/or well tests analyses, also for fractured reservoirs with a very heterogeneous permeability. This means that even low-permeability reservoirs will have certain sections where the permeability is high enough to allow microbial growth and transport. Here, bioplugging is one of the most promising mechanisms. Therefore the permeability criteria should be considered for fractured heterogeneous reservoirs carefully; in such cases, fracture geometry is of high importance for MEOR.

The oil density and viscosity of the reservoir are interdependent on each other. Low density and viscosity oils are preferred in in situ MEOR as the efficiency of EOR mechanisms is higher, except for cases where the biotransformation plays the major oil recovery role. As the activation of the other EOR mechanisms is difficult for viscous heavy oils, this biotransformation mechanism of MEOR, that is, microbial conversion of heavy oil components to light components to induce a potential decrease in oil viscosity to increase its mobility, should be the principal focus in heavy oil reservoirs.

MEOR is typically applied in mature reservoirs in which the water cuts (WC) of the producers are high, and this does play a significant role in the selection of the reservoir; yet, this is a criterion for all EOR methods. Generally, oil saturation for EOR should be higher than 25% except for gas and steam flooding (>40%) [40].

The pH range in the reservoir is given as 6–9 in the IRS screening criteria. In none of the other screening criteria table, pH is cited as a parameter. The activity of many bacteria generally decreases at a very low and high pH, and for fermentative microbes commonly used in MEOR, microbial growth can stop at a pH below 5.

The maximum formation brine salinity (NaCl) is given as 10% in both recommendations (IRS and DOE) and corresponds to most of the previous screening criteria for MEOR. However, experience from recent MEOR applications shows that there are cases in which the MEOR can be successful at higher salinities up to 200 g/L [61,62].

In conclusion from the above discussion, the most relevant parameters for the preliminary screening of MEOR are temperature (or depth), permeability, salinity, density (or viscosity), and pH. Other parameters are less relevant in the primary EOR screening stage and do not need to be considered in further detail. In addition, two parameters, which have not been mentioned yet, are of particular importance for in situ applications where fermenters are used with carbon sources as nutrients.

The first one is the carbonate content of sandstone formations as carbonate acts as a pH buffer and prevents a decrease in pH and microbial activity in the reservoir upon the generation of organic acids.

The second one is the presence of sulfate in the reservoir because SRB may use MEOR metabolites to produce hydrogen sulfide (H_2S). Such biogenic reservoir souring can be a showstopper for any MEOR operation and is discussed later under the section on risk management. On the other hand, as mentioned previously, the oil density is not crucial in in situ MEOR cases where the biotransformation is the main recovery mechanism.

In conclusion, Fig. 8.3 depicts updated screening criteria based on various recent publications for the application of in situ MEOR [26,32,45,46,61,62]. Many of these parameters should be considered as smooth ranges rather than hard

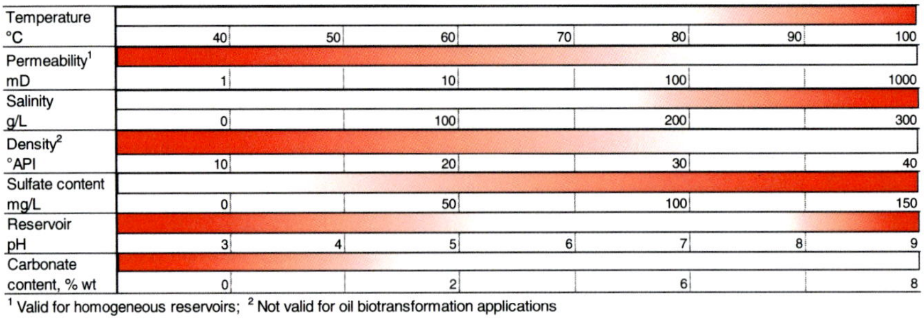

FIGURE 8.3

In situ microbial enhanced oil recovery screening parameters and applicable ranges (light to dark = feasible to not feasible; in situ microbial enhanced oil recovery based on biotransformation is not included).

cut-offs as MEOR research and proof of concepts for its application are being developed continuously.

Obviously, the main requirement related to the reservoir in the case of in situ MEOR is the presence of beneficial bacteria that can create EOR effects after having been fed with selected nutrients. After selecting the potential reservoir for an MEOR application, confirming the presence of such bacteria or bacterial communities is the next step for a decision toward a field operation. This step is time-consuming and does not fit the scope of a quick screening exercise.

8.2.2 Field application criteria

Before passing the second screening, which comprises sampling and microbiological analysis, some other requirements regarding the feasibility of MEOR operations should be answered via a quick but careful look at the history and existing facilities in the field:

- Is the geology and the reservoir well characterized? The geology and petrophysical properties are of high importance in any EOR operation and the availability of related information is helpful in the decision of a field operation. The heterogeneity of a reservoir can affect the performance of the EOR methods considerably. If this is a well-defined heterogeneity (determined by using, e.g., the coefficient of permeability variation described by Dykstra and Parsons [63]), this might be in support of emphasizing specific effects of MEOR, such as bioplugging (see field application example in Vizcacheras Field discussed in Section 8.5.1).
- Does a validated reservoir model exist? This information is crucial as the design and performance prediction can be better achieved with the numerical simulations with an existing model. Obviously, the more reliable the model, the more consistent the operational design and performance predictions will be by reducing the technical and economic uncertainties.
- Can the existing facilities in the field be used or easily upgraded for a field operation? As will be discussed in the next sections, the main attractiveness of MEOR is that it is cost effective and easy to apply. MEOR is mostly applied in waterflooded fields and the quality and quantity of the water preparation, distribution, and pumping facilities are important factor facilitating MEOR applications.
- Does the preliminary-high level economics based on reservoir screening criteria and field availabilities support the idea of an MEOR application?

8.3 Field-specific tailoring: physical modeling

The application of screening criteria to the reservoirs or wells in consideration may give an idea as to whether MEOR is applicable or not for any individual case. Combining the field requirements and high-level (screening) economics decides the way forward in the field operation. From a project management point of view, considering the generalized breakdown structure given in Fig. 8.2, the importance of a well-integrated team covering the microbiology, reservoir engineering, and operational aspects in all work packages is extremely critical/essential. If no dedicated MEOR team is already present, a common and clear terminology between science/technology and engineering must be established, which might take considerable time. The data transfer between microbiology and physical/numerical modeling with upscaling of all information to field scale necessitates the appropriate-common metrics and communication. In particular, the deliverables expected from the microbiology team must be clear, because data commonly obtained by microbiological methods might be of little use as input data for the modeling team (e.g., parameters that cannot be directly or easily implemented in models and/or simulators, such as numbers of microbial cells).

8.3.1 Microbiology and enhanced oil recovery evaluation

8.3.1.1 Microbiology workflow

Whether ex situ or in situ application, reservoir microbiology is the essential starting point in MEOR. The physicochemical characteristics of the reservoirs determine the diversity and activity of the microorganisms. The temperature is the main limiting factor, but salinity, pH, density, and permeability and resulting steric hindrances can limit the microbial activity.

Based on available data and considering that any molecular oxygen is rapidly removed by microbial and/or chemical processes, it is suggested that only strict anaerobic bacteria can be considered as truly indigenous to oil reservoirs [64]. The widespread anaerobic biosphere in oil reservoirs was confirmed in many studies [65,66]. Hence, preferred metabolic pathways include fermentation, anaerobic hydrocarbon degradation, acetogenesis, methanogenesis, and other anaerobic respiration processes such as nitrate or iron reduction and reduction of oxidized sulfur species, including sulfate. Especially when the in situ temperature or salinity of the studied oilfield is high, a careful examination of the physiological properties of sampled bacteria, that is, their capability of growing under reservoir conditions, can be a good indication of their possible indigenous nature [66]. In contrast, the presence of aerobic bacteria should be considered a potential indicator of contamination from outside of the reservoir. Of course, indigenous facultative bacteria might also have capabilities for an aerobic metabolism, but under in situ conditions with depleted oxygen levels, these are of little or no relevance for MEOR. Since fermentative microbial metabolization of nutrients (i.e., carbon sources or substrates) is arguably the most-approved and most-often applied MEOR approach, an in-depth assessment of this approach from an application-oriented reservoir engineering point of view is more focused.

From a pragmatic reservoir engineering point of view, the ideal way forward for an MEOR application would be to characterize the reservoir microbiology and identify species with known EOR capabilities in a catalog. Such catalogs are published from time to time but need to be reviewed often as new MEOR microorganisms are reported; a recent collection is provided by [46]. All of them are bacteria species and most of them are used for ex situ MEOR to produce biosurfactants and biopolymers. Among the biosurfactant-producing bacteria, for instance, *Rhodococcus* sp., *Bacillus subtilis*, *Bacillus licheniformis*, *Acinetobacter* sp., and *Pseudomonas* sp. can be mentioned [67–71]. Some others mainly used for in situ MEOR is collected under the group name fermentative bacteria and include *Enterobacter* and *Clostridium* species [72,73].

The microbiological workflow that can be followed in MEOR applications is given in Fig. 8.4 [62,74–77]. MEOR investigations start with retrieving water and oil samples from the selected reservoirs. The next steps are to define the microbiological community of the reservoir in terms of taxonomy and their relative abundance and to isolate the microorganisms that can be beneficial for oil recovery. The cultivation of the isolated microorganisms and/or communities with preselected nutrients and testing them in terms of EOR effects then needs the input of reservoir engineering. Obviously, the optimization of the nutrient for the beneficial communities for maximizing the EOR effects needs an efficient iteration between the disciplines, namely microbiology and reservoir engineering.

a. *Sampling:*

The representativity of the samples is crucial as the sampling is mostly done at the wellheads of the fields. Oil and water samples need to be free of

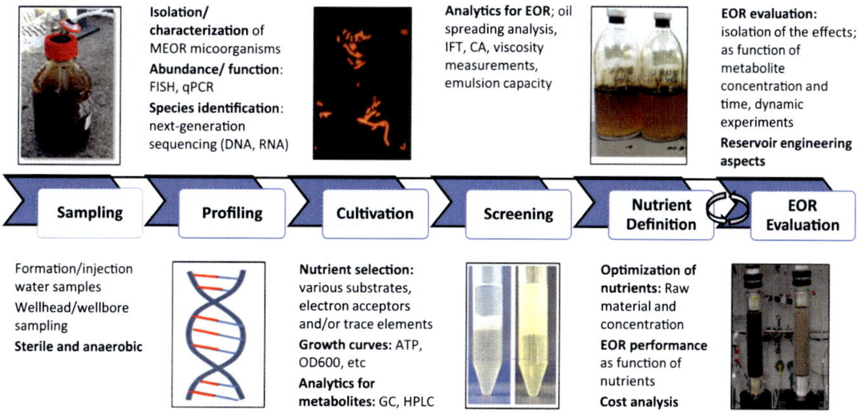

FIGURE 8.4

Microbiology workflow than can be applied for microbial enhanced oil recovery applications.

contamination to represent original reservoir microbiology. One way to ensure this is to retrieve samples directly from the wellbore, which is a safe way for reservoirs not already contaminated by the injected water. Besides, many studies report a distinct difference between the microbial diversity of planktonic (i.e., dispersed in the water phase) communities and their sessile counterparts (attached to surfaces, e.g., as biofilms) [78], making direct subsurface sampling the method of choice. However, this procedure is costly. Krüger et al. address this issue by comparing wellhead samples with samples taken directly from the wellbore using a commercial single-phase reservoir sampler [54]. Interestingly, they conclude that the wellhead water samples are representative for MEOR development. Still, special care should be taken when sampling at the wellhead: the samples should be collected under a constant flush of nitrogen and filled into sterile glass bottles until completely full and all used equipment should be sterile.

b. *Profiling:*

"Profiling" requires a microbiology laboratory and refers to characterizing the microbial reservoir community on a taxonomic level, for example, on the family level, and to assign relative abundances to each taxon. In addition, classical microbiological parameters, such as cell numbers, physiology, and morphology of microorganisms, are determined by microscopy and other analytics. These methods include a direct count of specific groups of bacteria, for example, with microscopy using fluorescent probes binding to specific nucleic acid sequences to detect specific species or functions (fluorescence in situ hybridization, FISH). Additional methods, most importantly sequencing of the 16 S ribosomal RNA gene (16 S rRNA) for taxonomic classifications and qPCR (quantitative polymerase chain reaction) for quantification of microbes are based on direct extraction of cell genetic material (DNA/RNA). Since the microbial biomass present in oil reservoirs is typically low, it is often most straight forward to filter several liters of water samples with sterile cell-retaining filters (at least 0.2 μm, preferably below) and subject the filters to subsequent analysis, including direct DNA isolation from those filters. The isolated DNA can be used as a template for a range of downstream biomolecular analytical methods. Very important among those are sequencing techniques in which a specific part of the microbial genome (i.e., the total amount of genes in a cell) is amplified and then analyzed or "read" for its nucleic acid sequence. For a taxonomic classification, that is, a "profiling" of the microbes present in the original samples, the most common gene analyzed is the 16 S rRNA gene. The corresponding sequencing technology has experienced a tremendous boost in the last 15 years: So-called second-generation sequencing (including 454 pyrosequencing and Illumina sequencing) has become a very feasible and affordable analysis method offered by many specialized service laboratories and even third-generation sequencing methods (which offer even longer reads of the genetic information and hence a higher taxonomic resolution on family or even species level) are routinely available (see, e.g., [79–81] for comprehensive reviews). The isolated microorganisms that can be beneficially used for in situ MEOR are defined as a result of this analysis. The measurements and analysis used for profiling are listed in Table 8.4 under the topic monitoring in conjunction with results from incubation screening experiments (Section 8.5.3).

c. *Cultivation of microorganisms:*

Microbial communities obtained and isolated from oil and water samples are incubated with various nutrient media. In general, these nutrient media include vitamins, trace elements, yeast extract, and carbon sources, such as peptone or meat extracts. For the initial cultivations, various electron donor and acceptor combinations and redox conditions can be tested, including fermentative, sulfate-reducing, nitrate-reducing, and methanogenic conditions. Oil from the field can also be added to some cultivations for the enrichment of oil-degrading microorganisms. Salt composition, salinity, pH, and incubation temperature are to be kept as close to field conditions as possible. The growth of the enrichments can be monitored by simple optical density (OD) and adenosine triphosphate (ATP) measurements. For the sake of practicability, initial incubations can be performed under atmospheric pressure conditions. The types of microorganisms present in the enrichments should be determined by 16S rRNA analysis using the sequencing technologies mentioned above and the results can then be compared with the initial or pre-MEOR profiling results.

d. *Characterization of enrichments; screening:*

The microbial enrichment cultures obtained during the cultivation step are analyzed for liquid and gaseous metabolites that are produced by the microorganisms. In this step various analytics and equipment of conventional EOR methods, such as determination of gas composition and amount (gas chromatography, absolute pressure), acids and solvents (pH, HPLC), and surface-active substances (HPLC, oil spreading analysis, IFT measurements, contact angle, viscosity measurements), can be used for the characterization of the enrichments in terms of EOR effects. Consortium or consortia, that is, a community of several different species or "types" of bacteria, that show good performance in growth and in the production of metabolites, are selected for further tests. Several growth experiments should be performed with the defined isolated consortium using various compositions and concentrations of nutrients. The growth characteristics including the incubation time, growth phases, such as stationery, and decline phases are important as they dictate the generation of metabolites and biomass (living and death cells) that are governing the EOR effects. These characteristics are also important for the numerical modeling of MEOR processes in a laboratory or field scale. An example of a growth curve is given in Fig. 8.5 where growth is assessed by measuring the optical density (OD600), that is, the turbidity of the culture, using a spectrophotometer. Thus both active and dead cells are considered (biomass). Fermentative

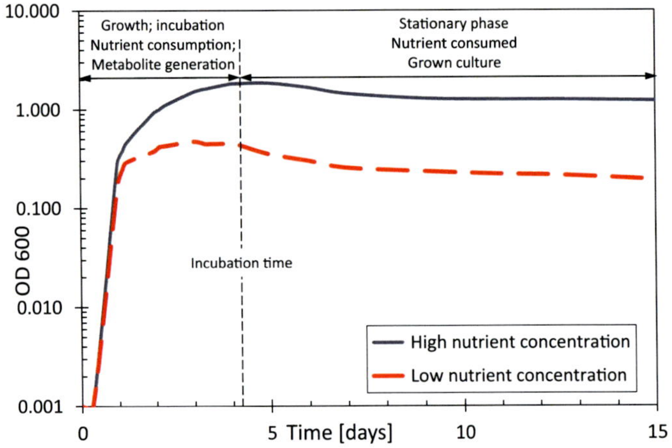

FIGURE 8.5

Example growth curves for microorganisms selected for microbial enhanced oil recovery and isolated from an oil field, fed by nutrients with two different concentrations.

growth typically starts with a lag phase in which the organisms adjust to the incubation conditions and cell number increase is relatively low. Then, the microorganisms grow exponentially. Growth ceases after all the nutrient is consumed or when the growth conditions become unfavorable, for example, because the produced metabolites (e.g., ethanol) reach a toxicity threshold and act as inhibitors. At this point, cell numbers in the culture are highest and a stationary phase begins in which the number of active cells stays constant. After the stationary phase, the cell death (i.e., cell lysis) exceeds cell division (i.e., forming of new cells), leading to a decay phase with slowly decreasing cell numbers (not shown in the figure). The incubation time is perhaps the most crucial information for planning the operation in the field as discussed in more detail in Section 8.4. Another observation from the figure is the obvious effect of the nutrient concentration on the growth. As can be expected, the growth is a function of the nutrient concentration and also the ambient temperature. Therefore it is important to perform all growth tests under reservoir temperature.

e. *Nutrient definition and optimization:*

The production of metabolites correlates with the composition and amount of the nutrients added. Theoretically, the highest optical density is reached with the highest amount of selected nutrients added. However, in an MEOR application, this maximum amount of nutrients is commonly not applied due to economical (costs), physicochemical (adsorption to formation rock, dilution in the reservoir), and petrophysical (excess of biomass in limited pore space) limitations. The incubation of microorganisms (typically bacteria) with nutrients in the formation water and the generation of metabolites can be described with the following general expression:

$$\text{Bacteria}\left(<1.10^4\text{cells/mL}\right) + \text{Nutrient(liquid)} \rightarrow \text{Metabolites(liquid)} + \text{Metabolites(gas)} + \text{Bacteria}\left(>1.10^6\text{cells/mL}\right) \quad (8.1)$$

For the planned in situ MEOR application, the promising consortia and nutrient(s) should be further investigated with respect to reservoir engineering. The aim of this investigation is to optimize the nutrient on one hand and to evaluate the EOR effects and their relationship with tested nutrients on the other hand. Therefore the simple analogy of the in situ MEOR injectant under laboratory conditions is the nutrient mixed into the formation water at a given concentration. The nutrient is consumed by the in situ and/or ex situ microorganisms and the displacing fluid in MEOR consists of metabolites and a higher number of microorganisms in the formation water.

8.3.1.2 Enhanced oil recovery evaluation

EOR evaluation aims at the identification of MEOR mechanisms and their quantification for given field cases and is performed based on standardized EOR experimentation and numerical methods. In addition to the analytics for an initial assessment of MEOR cited in the previous section, dynamic experiments can be applied to further investigate individual EOR mechanisms. These can be used for screening MEOR in general, but also to investigate various feeding strategies and the impact of reservoir characteristics on it.

Dynamic experiments are laboratory settings designed to assess potential microbe–fluid–rock interactions as well as MEOR mechanisms and their quantification, as they provide a relatively realistic fluid to rock surface ratio. Another aspect of dynamic experiments is that bacteria are exposed to rheological (e.g., shear) forces comparable to porous media. Also, if properly designed, both advective- and diffusion-dominated transport properties and realistic adsorption characteristics can be accounted for.

a. *Micromodels:*

   Nowadays, micromodels used for EOR studies are capable of state-of-the-art visualization of the flow and EOR mechanisms at the pore scale. Micromodel applications can be used to complement well-established technologies, such as sandpack and coreflooding experiments, both very important in an EOR screening process before field implementation. Various micromodel studies were published using microchips with various geometries and capabilities [82–87]. An important limitation of the use of micromodels is the difficulty to upscale the fluid flow behavior from idealized microchips to real reservoir rocks, as pore structure is one main factor that governs the fluid flow behavior in reservoir rocks. However, new approaches to construct the micromodels based on the pore structure in real porous media are emerging [88].

b. *Sandpacks:*

Sandpack or column experiments have been performed numerously, for example, for investigating the transport of microbes in aquifers or for soil bioremediation [89,90]. Another application of sandpack experiments is the evaluation of EOR methods. Especially in the cases where a consolidated core from the reservoir cannot be obtained, sandpack experiments with unconsolidated porous media (formation material) are convenient to investigate the impact of the reservoir mineralogical composition on MEOR. However, design parameters in sandpack experiments need special attention and expertise to avoid/reduce wall effects, preferential flow, and fines migration leading to pore blocking. Additionally, recreating the representative petrophysical properties, such as porosity, pore size, and permeability, is challenging. Moreover, several effects complicate the interpretation of the results, such as parallel in situ gas and biofilm formation, potential micro heterogeneities, and the geometry of the selected columns (which impacts capillary end effects). Especially the in situ gas generation and its effect on the flow and production behavior can be relevant for MEOR experiments, as the gases are one of the main metabolites generated in microbial fermentation. To mitigate this risk, sandpacks that are often designed for experiments at atmospheric pressure, should be upgraded to function at higher pressures to avoid the formation of free gas. Because of their relatively simplistic nature, sandpacks are widely used for oil recovery evaluations and in studies for souring assessment [91–94].

c. *Corefloods:*

Among laboratory efforts to simulate the flow and EOR effects in porous media, corefloods are the most preferred. This is because the experiments can be conducted in real porous media, ideally with cores retrieved from the reservoir itself. Particularly the testing under reservoir pressure makes the use of corefloods paramount for EOR studies. In MEOR, this aspect is mainly relevant for establishing reservoir-representative gas solubility. For MEOR, the need for coreflood experiments prior to any decision on a field application is a prerequisite because of the representative reservoir mineralogy and porous media characteristics that only corefloods provide. There are numerous examples of the use of corefloods for the performance prediction of MEOR [87,95–99].

Compared to other EOR setups, MEOR dynamic experiments need a special expertise in their design. To avoid microbiological contaminations, the setups should be sterile and in most of the cases, anaerobic before any experiment starts. This obviously needs significant knowhow and experience. The other requirement is that auxiliary equipment that allows proper MEOR sampling and analysis might differ compared to other EOR experiments. Table 8.2 shows a compilation of the analytics and dynamic experiments that can be used in MEOR studies.

Table 8.2 List of enhanced oil recovery experiments that are applicable for static and dynamic screening of microbial enhanced oil recovery.

Experimental technique/equipment	Rationale/advantage/disadvantage
Monitoring microbial growth in batch experiments OD600, ATP concentration, atmospheric pressure, and reservoir temperature	To define the effect of nutrient characteristics on growth, experiments at high pressure also possible; growth under various nutrient conditions [100]
Gas generation measurements GC and absolute pressure	Especially important in in situ MEOR in which the fermentation is the main metabolic pathway, mostly CO_2 is produced [73,101,102]
Viscosity measurements Rotational type viscometer mostly under static conditions for in situ MEOR can be performed at T_{res}, P_{res}.	To define the rheology and viscosity of the cell broth containing the grown culture and metabolites; for the ex situ MEOR standard procedures of polymer flooding are applied [99,103,104]
Emulsion activity Emulsification capacity of a two-phase system	Can be applied as a quick assessment of surface activity and emulsification property [76]
Oil spreading; surface tension (ST)/ interfacial tension (IFT) ST with ring–pendant drop tensiometer; in the case of lower IFTs (e.g., <1 mN/m.) spinning drop is used; oil spreading needs no specialized equipment	Spreading experiment can be used as a quick test for surface activity. Surface tension measurement as an indication of water–oil IFT; IFT should be preferred, with pendant drop-spinning drop under reservoir conditions possible [87,105–107]
Wettability measurements Contact angle measurements on the formation rock aged in oil	To define the wettability alteration through MEOR. Possible under P_{res}, T_{res}, dynamic measurements help to determine the change of wettability during MEOR [108,109]
Spontaneous imbibition (SI) Standard SI equipment via Amott-Harvey wettability index (AI)	SI as an indication and quantification of wettability change; the impact of wettability alteration on oil recovery [87,99,110]
Micromodels In glass/silicon micromodels designed to mimic the petrophysics; T_{res} is possible; P_{res} is challenging	Observation of EOR effects at pore scale; visualization of the bacteria and its movement, and metabolites, such as biofilm; results difficult to upscale to reservoir [85–88]
Sandpacks Columns filled with model sands as well as crushed formation material; modeling of reservoir porous media is challenging, potential over-under estimations due to experimental artifacts	Measuring dynamic growth and metabolite generation, dynamic material balance nutrient/metabolites, observation of incremental oil recovery; EOR effects at the core/Darcy scale; can also be used for the dynamic evaluation of souring due to MEOR [92,94,111]

(*Continued*)

Table 8.2 List of enhanced oil recovery experiments that are applicable for static and dynamic screening of microbial enhanced oil recovery. *Continued*

Experimental technique/equipment	Rationale/advantage/disadvantage
Corefloods Standard coreflooding setups with the emphasis on the sterility and anaerobic conditions, T_{res}, P_{res}, and use of reservoir cores are possible	Modeling the flow and MEOR dynamics in a real porous system; observation of EOR effects at core/Darcy scale; if available reservoir cores can be used for the best approximation of reservoir conditions [87,97,98]

ATP, *adenosine triphosphate;* EOR, *enhanced oil recovery;* MEOR, *microbial enhanced oil recovery.*

8.3.1.3 Isolated community versus reservoir community

In the literature, the results of in situ MEOR laboratory studies are mostly provided as the experiments performed with isolated microorganisms and/or consortia that are known as beneficial for their EOR effects [112–114]. For these experiments, isolated microorganisms are often mixed with the sterilized formation water (synthetic or original) to investigate the EOR effects. However, the microbiology of the formation water is much more complex, and the feeding chemistry and strategy influence the overall microbiological composition. For instance, SRB can be present in the entire in situ microbial community but may not necessarily constitute a part of a bacterial consortium isolated for MEOR assessment. Hence, potential adverse effects, such as souring, would not be considered during the experimental assessment of MEOR. Such shortcomings may create essential issues in terms of overall risk assessment. Therefore the performance of the isolated consortia should always be tested against the overall microbial community; in other words, by using original (unsterile) formation water instead of sterile formation water where only the isolated communities are added. Alkan et al. highlighted the difference between the performances of isolated and reservoir microbial communities in terms of EOR effects in an MEOR project where the differences in water phase viscosity, wettability, and bioplugging effects are significant [99]. Accordingly, relying on experiments performed only by using isolated microbial communities (as mostly done) for deciding on the fate of an application can yield over-optimistic conclusions on the MEOR potential and therefore on the design of a field operation.

8.3.2 Recovery mechanisms

8.3.2.1 Basics

The objective of an EOR application is to increase oil production by overcoming the physical forces and/or changing the petrophysical properties responsible for keeping the oil in the reservoir. In ex situ MEOR, one or two EOR mechanisms, such as IFT reduction and wettability alteration, in case of biosurfactants or improved mobility ratio and reservoir conformance in case of biopolymer are targeted and these are qualified and quantified similarly to the corresponding chemical EOR agents. However, the precise mechanism of in situ MEOR to recover oil in the reservoir cannot be modeled in detail, as is the case for other EOR methods. Since the microbiology is reservoir specific, the in situ MEOR mechanisms are field specific as well. There exist some studies relating MEOR to new theories: for example, Sunde et al. argued that, in the application of MEOR, the residual oil is connected as continuous oil strands extending from injector to producer and blocked only by water in pores close to the producer [115]. However, the consensus in most studies in the literature confirms the conventional concept, that is, that MEOR consists of various minor EOR mechanisms and that the summation of all individual mechanisms leads to a significant total contribution to incremental oil. These EOR mechanisms are briefly elaborated in the following.

EOR is applied in the secondary mode just after primary production or in the tertiary mode after waterflooding. The major aspect of EOR is to increase oil recovery from the reservoir by improving the displacement efficiency (E). E can be increased by increasing the volumetric or the macroscopic (E_v) and/or microscopic sweep (E_m) efficiencies:

$$E = E_v \times E_m \tag{8.2}$$

Both E_v and E_m can be positively impacted as follows.

a. *Mobility ratio improvement:*

The volumetric displacement efficiency (E_v) is high if the mobility of the displacing phase (in most of the cases water phase) is similar to or lower than the mobility of the displaced phase (oil); therefore a mobility ratio (M) equal to or less than unity defines an ideal displacement. From the equation below, it can be interpreted that the relative permeabilities of the displacing (k_{rw}) and displaced phases (k_{ro}) are the parameters relating the volumetric displacement efficiency to microscopic displacement:

$$M = \frac{\text{displacing phase mobility}}{\text{displaced phase mobility}} = \frac{M_w}{M_o} = \frac{k_{rw}\mu_o}{k_{ro}\mu_w} \tag{8.3}$$

As can be concluded from Eq. (8.3), mobility improvement is possible by changing the phase viscosities and/or the relative permeabilities. Assuming that the relative permeabilities remain constant (e.g., no relevant improvement in microscopic displacement), better mobility control can be achieved simply by adjusting the water (μ_w) and oil phase (μ_o) viscosities.

b. *Increase in capillary number:*

For a higher microscopic displacement efficiency (E_m), the capillary forces (i.e., interfacial tension between oil and water phase, $\sigma_{w\text{-}o}$) dominating in the reservoir should be decreased. Whether the decrease is sufficient for mobilizing additional oil is mostly expressed by the dimensionless capillary number, N_{Ca}:

$$N_{\text{Ca}} = \frac{\text{viscous forces}}{\text{capillary forces}} = \frac{\mu_w \cdot v}{\sigma_{w-o}} \tag{8.4}$$

To recover the oil trapped in the pores of the reservoir, the interfacial tension between water and oil phases (the denominator of Eq. 8.4) should be reduced, and/or the viscous forces (the numerator of Eq. 8.4, water viscosity μ_w and velocity of the displacing phase v) should be increased. As the extent of the viscous forces in the reservoir is limited by the injection rate and the viscosity of the water phase, the reduction of the capillary forces is one of the main parameters for increasing the capillary number, and thus oil recovery. For a substantial decrease in oil saturation, the capillary number should increase above the critical capillary number in the capillary desaturation curve (CDC). The critical capillary number in the CDC depends mainly on the porous medium (e.g., the reservoir rock) and the nonwetting phase and is in the range of 10^{-6} to 10^{-5} for the nonwetting phase in typical sandstones [26,116,117]. As the interfacial tension is linked to wettability as described in the Laplace-Young equation, the capillary number is also dependent on wettability. Many authors also include the contact angle (CA, θ) as a measure of the wettability in the capillary number definition [26,117,118]. For example, in the case of a greater contact angle, oil largely spreads on the rock surface and the capillary forces to overcome are much higher, therefore most of the oil remains in the reservoir.

c. *Petrophysical modification:*

Modification of petrophysics via changes in fracture and/or matrix properties, such as plugging of the pore space and/or dissolution of the rock, can provide a significant contribution to oil recovery by improving both macroscopic and microscopic sweep efficiencies. Conformance control with selective plugging is an important mechanism in reservoirs showing heterogeneous permeability distribution. Conformance treatments are usually applied to block high permeability channels and existing water highways with higher flux. As a result, the flow is diverted toward the lower permeability areas containing mobile unswept oil.

In laboratory investigations the permeability reduction factor (PRF) is mostly used to estimate the reduction in permeability:

$$\mathrm{PRF} = \frac{k_f}{k} \tag{8.5}$$

where k is the original permeability and k_f is the modified (final) permeability calculated via the measured pressure difference across the length of the porous medium.

On the other hand, if the rock contains carbonates or other acid-soluble minerals, acidic components of reservoir/injected fluids can dissolve the matrix. This causes a pore volume increase, which in turn increases the rock permeability and facilitates the flow, and frees oil attached to the rock.

8.3.2.2 Mobility improvement

a. *Increase of water viscosity:*

Polysaccharides produced by various microorganisms can be added as biopolymers in the injection water for ex situ MEOR applications. Xanthan gum (XG) and Scleroglucan produced from *Xanthomonas* sp. and *Sclerotium* sp., respectively, are well-known examples of such ex situ biopolymers. *Xanthomonas* species are used to ferment carbohydrates and produce a thermally stable heteropolysaccharide called XG. Its physical properties of viscosity, shear resistance, temperature, and salt tolerance in conjunction with its mobility controlling properties of the displacing phase make it almost an ideal polymer for use in EOR. Studies investigated the effect of concentration, solution ionic strength and composition, interactions with reservoir rocks, sediments, and polymer degradation on the rheology of the shear-thinning solution [119–121]. Fig. 8.6 shows the viscosity of XG at 600 mg/L concentration in water with two different Na^+ concentrations as a function of shear rate. Another biopolymer, Scleroglucan, is produced by a species of the fungus *Sclerotium* and was examined for possible use in EOR [122,123]. Curdian (*Alcaligeness* sp.), Dextran (*Leuconostoc* sp.) Pullukan (*Aureobasidium* sp.), and Levan (*Bacillus sp.*) are also reported for industrial production [46,124]. Schizophyllan (from the fungus *Schizophyllum commune*) has been reported to exhibit significant viscosifying effects under high salinity–temperature reservoir conditions [125,126]. The engineering and use of the biopolymers can be classified under chemical EOR expertise with the addition of special attention to the microbiological interactions with the biopolymers in the reservoirs, for instance, the degradation of the biopolymers itself as a nutrient for in situ microorganisms, souring, etc.

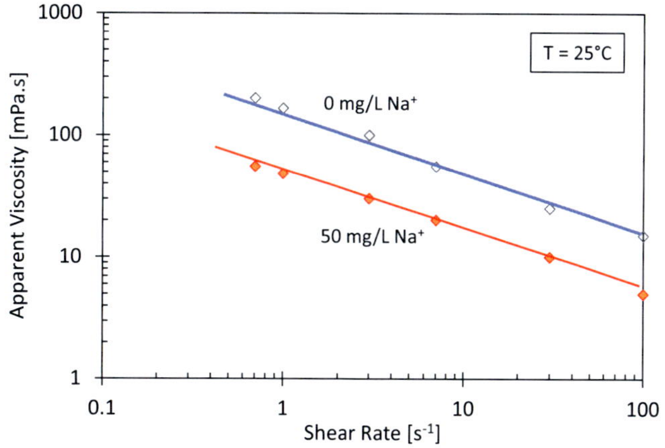

FIGURE 8.6

Measured rheology and viscosity of Xanthan gum with a concentration of 600 mg/L.

Reproduced with modifications from L. Zhong, M. Oostrom, M.J. Truex, et al., Rheological behavior of Xanthan gum solution related to shear thinning fluid delivery for subsurface remediation, Journal of Hazardous Materials 244–245 (2013) 160–170.

Similar to ex situ case, for in situ MEOR, the mechanisms of increasing the water phase viscosity are due to the generation of exopolymeric substances (EPS) which are natural polymers of high molecular weight secreted by microorganisms into their environment as a result of their metabolism. EPS is mostly composed of polysaccharides (exopolysaccharides) and proteins but includes other macromolecules, such as DNA or lipids. The generation of EPS in the injection/formation brine increases its viscosity. It should be noted that although the increased water viscosity can be one of the relevant mechanisms of MEOR, almost all literature data is on the ex situ production of biopolymers with isolated microbes. A compilation of the in situ MEOR water phase viscosifying effect is given by Alkan et al. [99]. In available studies, measured rheologies are shear thinning and viscosities measured at shear rate of 7/s are in the range of 2–3 mPa·s, a range that is significantly lower than the reported viscosities of ex situ biopolymers; few studies provide higher values up to 10 mPa·s [99,103,104]. It is also reported that in the absence of the cells, for example, after filtration through porous media, the measured viscosities of the in situ cultures are lower. Thus microbial cells themselves contribute to the viscosity of the aqueous phase and cell-containing metabolite dispersions often exhibit a higher viscosity than their cell-free analogs. Currently, no reported apparent viscosity values measured in porous media (corefloods) exist, which are specific for in situ MEOR [99].

b. *Decrease in oil viscosity:*

The decrease in oil viscosity through MEOR is related to at least two phenomena: Biodegradation of oil and gas generation. Biodegradation with respect to MEOR can be defined as a series of natural processes resulting from the activities of microorganisms by which organic materials are transformed into other chemical structures. In most cases, complex organic molecules (polymers or hydrocarbons) are broken down into smaller compounds. This has two different impacts: Firstly, the degradation of the ex situ MEOR products, such as biopolymers, results in the at least partial loss of its EOR capacities (i.e., viscosity decrease) as discussed in the previous section. This degradation can obviously be a challenge in in situ MEOR but has a rather minor physical and economic impact as the EPS amount generated is rather low. Secondly, biodegradation implies a positive effect in in situ MEOR representing one of the oil recovery mechanisms. To differentiate it from the previous definition, the term biotransformation is used herein. In this case, certain bacteria metabolize hydrocarbon molecules with a high molecular weight, converting them into smaller chain molecules that generally have lower viscosities [127]. Various studies investigated the biotransformation of the oils due to grown microbes under anaerobic as well as aerobic conditions [33,128–130]. Under in situ conditions, anaerobic hydrocarbon biotransformation should be the more relevant process by far. Although the exact mechanisms are not yet well understood, it is currently believed that anaerobic biotransformation is a syntrophic process [131] and the bacteria involved include sulfate-, nitrate-, and iron-reducing bacteria [132]. Hydrocarbons that can potentially be metabolized anaerobically include alkanes, aromatics, and cyclic hydrocarbons. In this context, it should be noted that under anoxic conditions, "high-quality hydrocarbons" may be used preferentially by microbes [133,134], which can lead to adverse effects of biotransformation including an increase in acidity and/or sulfur content of the oil [135].

On the other hand, Sabut et al. reported on the laboratory as well as field applications relying on the bacterial transformation of oil [136]. The rheological measurements before and after the treatments exhibit favorable improvement in oil viscosity; the oil viscosity decreased 20%–35% relative to its initial value. This indication of a change in the oil composition was also confirmed by carbon analyses of the oil before and after the treatment. Hou et al. [137] presented the results of oil hydrocarbon degradation by the selected bacteria *Brevibacillus brevis* and *Bacillus cereus* based on metabolite analysis and field results. After bacteria injection, 44 from 60 SWTs in China showed a significant increase in oil production, mainly due to the viscosity reduction induced by oil biotransformation [137]. Recent studies discussed various laboratory and field cases with biotransformation as the main MEOR mechanism determined by reduction in oil viscosity and related oil recovery [129,138–141].

This mechanism is also used for the microbial well stimulation. Besides the reduced flow of highly viscous crude oil within reservoirs, precipitated paraffinic and asphaltic fractions from the crude oil can block the main flow pathways. Microbial transformation of precipitated crude oil fractions can therefore be expected to help re-establish oil flow. A summary of such efforts is provided by Bachmann et al. [124]. Lazar et al. used strains of *Pseudomonas* in a laboratory study on solid and semisolid paraffins and demonstrated that significant degradation of paraffins occurred [142]. Microbial culture products were first used in 1986 in the Austin Chalk formation in Texas, USA, to control paraffin deposition. This technology was reported to successfully improve oil production from several individual wells and commercial paraffin control is reported to slow the rate of decline in oil production and extend the operational life of marginal oil fields [143–148]; the hydrocarbon degraders were injected into the well along with nutrients, and the wells are shut for several weeks up to months for microbial growth and metabolism to proceed. The microbial treatments are repeated on a regular schedule (every several weeks or monthly). Youssef et al. compiled procedures and field results of such treatments to control paraffin deposition [148]. Using a mixture of indigenous microorganisms *Arthrobacter* sp., *Pseudomonas* sp., and *Bacillus* sp., Jinfeng et al. showed a reduction of paraffin from 29.8% to 25.5% in 9 months in a field trial [149].

The second mechanism via which MEOR reduces the oil viscosity is the dissolution of the gas generated as the result of microbial fermentation. Alkan et al. provide a compilation of the CO_2 amounts generated by fermentative bacteria fed with glucose or sucrose, based on the available related literature [99]. Behlülgil and Mehmetoglu report a viscosity decrease from 80 mPa·s to 50 mPa·s due to dissolved CO_2 produced in situ by *Clostridium acetobutylicum*, an anaerobic microorganism that utilized 12% molasses and 5 g/L yeast extract in an aqueous solution as a substrate [72]. However, it should be noted that this nutrient concentration is high and possibly not economic for use in a full-scale MEOR application. Similar gas (acid- and solvent-based) mechanisms were attributed to *C. acetobutylicum* in reviews by Van Hamme et al. [150] and Youssef et al. [148].

Kobayashi et al. showed that yeast extract could enhance the gas production by the indigenous microorganisms of a depleted petroleum reservoir [112]. The results suggest that reservoir microorganisms and proteinaceous substrates could be considered for a new MEOR procedure, in which proteinaceous substrates (e.g., agro-industrial waste waters) would be injected into the reservoir to stimulate gas production by the indigenous microorganisms of the target reservoir. They reported up to 250 mM CH_4 production per liter of the nutrient after 224 days of incubation, which is significantly higher than the values reported by Alkan et al. [99]. Apart from the above-cited notable exceptions, the reported values are derived from batch experiments and are in the range of 10–45 mM. This amount of CO_2 can only lead to a small decrease in oil viscosity. Correspondingly, the limited number of reported measurements are in the range of only 0.5–1.5 mPa·s as a decrease in oil viscosity.

The above-mentioned viscosity changes in both water and oil phases in in situ MEOR are relatively low in comparison to polymer or CO_2 flooding EOR cases; yet this might be still sufficient for an improvement in mobility ratio. For example, in an application using fermentative bacteria an increase in water phase viscosity due to EPS production of 1 to 3 mPa·s and a decrease in oil viscosity of 1–2 mPa·s seems to be possible based on literature data [99]. For a given reservoir with initial water and oil viscosities of 1 and 22 mPa·s, respectively, this may result in a mobility ratio improvement from approximately 22 down to 7 under reservoir conditions.

8.3.2.3 Increase in capillary number

a. *Decrease in IFT:*

Numerous studies present and discuss the use of biosurfactants in MEOR as ex situ or in situ applications. A list of the biosurfactants produced industrially is provided by Safdel et al. [46]; the microorganisms and their products, such as *B. subtilis* (Emulsan), *Pseudomonas* sp. (Alasan), *Arthrobacter* sp. (Viscosin), and *Acinetobacter* sp. (Rhamnolipid), are listed. Their biodegradability, low toxicity, emulsification activity, and their stability in different environments are reported to be their advantages in comparison to chemical surfactants. Safdel et al. further classify five major types of biosurfactants in detail, including glycolipids (e.g., Rhamnolipids), lipopeptides and lipoproteins (e.g., Surfactin), fatty acids, neutral lipids, and phospholipids, polymeric surfactants (e.g., Emulsan) and particulate biosurfactants (e.g., cells themselves or cell debris) [46]. For instance, Surfactin produced by *B. subtilis* during fermentation is known as one of the most effective biosurfactants and has been reported to be successfully used in MEOR [151,152]. Various studies investigate the effects of the physicochemical reservoir conditions, such as pH, salinity, and temperature on the solubility, stability, and activity of such biosurfactants. For example, Olsen and Janshekar [153] report optimized surfactant systems formulated with biosurfactants that have been found to lower oil/water IFT to as low as 5.10^{-3} mN/m in an earlier attempt.

Al-Sulaimani et al. [154], Samir et al. [155], and Amani et al. [106] reported the performance of biosurfactants produced by various species isolated from oil-contaminated soils or oil fields by measuring IFTs and oil recoveries in coreflood tests. Ghojavand et al. [156] evaluated the potential of biosurfactant produced by *Bacillus* strains for oil recovery in dolomite cores. A *Bacillus mojavensis* strain that was isolated from an Iranian oil field was used to produce a lipopeptidic biosurfactant and the IFT between this aqueous phase and *n*-hexadecane is reported to be 0.1 mN/m. The biosurfactant broths prepared ex situ were injected in dolomite cores to quantify the oil recovery after the waterflooding and up to 7% more recovery is reported. Rabiei et al. [157] report on biosurfactant generated by the isolates *Enterobacter cloacae* and *Enterobacter hormaechei*. Coreflooding tests are reported to investigate the effects of IFT reduction and wettability alteration on oil recovery during the MEOR process; dynamic interfacial tension at the interface of the cell-free broth and crude oil reached its final value in less than 15 min, decreasing from 29 to 3.2 mN/m. Comparing the results of oil recovery from in situ and ex situ tests showed slightly higher incremental oil recovery during the ex situ process than in situ process due to the partially purified biosurfactant that was used in the ex situ test. Sarafzadeh et al. [110] used *Enterobacter cloacae*, a facultative biosurfactant-producing bacterium isolated from heavy crude oil contaminated soil of Iran. They also compared the results for in situ and ex situ coreflood experiments showing that the oil recovery efficiencies are similar for both in situ and ex situ biosurfactant cases performed using dolomite-based cores. In both cases, IFT measurements showed a reduction to 3.0 mN/m from the original value of 32 mN/m. In addition, the wettability test showed a significant reduction of the Amott-Harvey index (AI) from 0.89 to 0.11 for the ex situ case. They affirm that during the two-stage flooding in both the in situ and ex situ cases, IFT reduction shows its most pronounced effect in the first stage while in the second stage wettability alteration is the dominating mechanism having a more significant effect on the total microbial oil recovery in ex situ case. Based on the results, the biosurfactant from *E. cloacae* is a good candidate for the EOR because of its high capability to reduce IFT and alter rock wettability.

Song et al. [87] reported crude lipopeptides produced by a strain isolated from an indigenous community in an oil reservoir. It was found that crude lipopeptides can effectively reduce the IFT to $10^{-1} \sim 10^{-2}$ mN/m under high salinity conditions without forming stable emulsions, and the wettability of natural sandstone could be enhanced (AI, from 0.36 to 0.48). The results of coreflooding experiments indicated that an additional 5.2% of OOIP can be recovered with a 9.5% reduction in injection pressure.

A compilation of the biosurfactant effect in in situ MEOR was given by Alkan et al. [99]. The reductions in IFT are low to moderate except in the cases reported by Kowalewski et al. [98] and Song et al. [87] which can be categorized rather as ex situ biosurfactant generation efforts. Based on literature data it is reasonable to assume a best-case IFT reduction in in situ MEOR case from 30 to 1 mN/m. With $\mu_w = 1.1$ mPa s. and $v = 1.10^{-7} - 1.10^{-6}$ m/s (which are representative of waterflooding in average reservoir conditions), the resulting capillary number during waterflooding is between 10^{-7} and 10^{-6}. In an in situ MEOR application considering a potential increase in water viscosity in addition to a decrease in IFT as mentioned above, the capillary number might increase slightly beyond 1.10^{-6}, which remains one order of magnitude below the critical capillary number in CDC given for sandstones in general. Therefore a significant decrease in the residual oil saturation cannot be expected due to a decrease in the IFT effect (unless significant) of in situ MEOR.

b. *Wettability alteration:*

The wettability alteration through MEOR is one of the most highlighted oil recovery mechanisms of MEOR. In ex situ MEOR, the wettability alteration is mostly discussed as one of the recovery mechanisms of biosurfactants with a similar mode of action as chemical surfactants [110,157].

In in situ MEOR wettability alteration is also considered to be one of the strongest mechanisms, especially in the cases where the original wettability is "intermediate to oil wet." Afrapoli et al. [158] assumed that the hydrophilic bacterial attachment on the oil-covered surface is driven by the consumption of the oil by bacteria using it as a carbon source. According to Karimi et al. [108], aged glass surfaces regained their initial water wetness because the bacteria could remove the adsorbed polar and asphaltene compounds. Khajepour et al. [86] examined the effects of the MEOR process on wettability and residual oil saturation using microscopic visualization of two-phase flow in transparent glass micromodels. A biosurfactant-producing bacterial strain (*E. cloacae*) was utilized to investigate the impacts of both biofilm and biosurfactant on the wettability of the micromodel pore walls by measuring the relative permeabilities before and after microbial treatment. Results indicated that the wettability of the pores was altered toward more water wetness. Hajibagheri et al. [159] investigated the synergic effects of anionic and cationic chemical surfactants and bacterial solutions on the wettability alteration of carbonate rock. Moreover, the experiments reported are based on the use of isolated microorganisms prescreened for their biosurfactant properties (*Rhodococcus sp 094* and *E. cloacae*) rather than in situ communities found in the reservoirs. Alkan et al. [160] reported an experimental study performed using reservoir fermentative communities without any isolation or enrichments. Spontaneous imbibition (SI) experiments as well as contact angle (CA) measurements conducted using Bentheimer cores and original reservoir core slices aged with dead oil showed the change of the wettability from intermediate to water wet after the incubation with sucrose and yeast extract. Supporting measurements conducted with an oscillating pendant drop (OPD) apparatus confirmed quick biofilm formation at the oil–water interface. The viscoelasticity observed in OPD tests may affect the SI behavior in two ways: the first one is the wettability change, believed to be the main effect. Secondly, the biofilm enveloping the oil droplet affects the mechanical equilibrium at the oil–water–rock interface beneficially, whereby the oil droplet can be contracted toward the center of the droplet, thus decreasing the CA. Although assumed to be a minor effect, this elasticity decreases the snap-off effects in porous media by minimizing the capillary blockage and enabling oil mobility through narrow pores [161]. Alkan et al. [160] provided the typical SI curves in the case of an in situ MEOR in comparison to control (sterile) experiments as well as contact angles measured with an HPHT (high pressure, high temperature) pendant drop apparatus on rock slices initially aged in dead oil. The SI curve of MEOR deviates clearly from the control experiments after 2–3 days, which corresponds to the exponential growth phase of the bacteria. The CA measured on quartz, sandstone, and reservoir slices aged in dead oil decreased from 120 to 130 degrees to 70 to 60 degrees indicating the wettability alteration toward more water wet due to MEOR (Fig. 8.7).

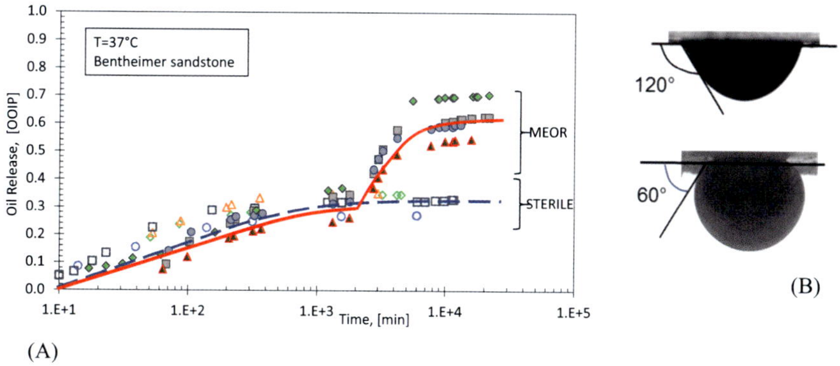

FIGURE 8.7

(A) Spontaneous imbibition curves for microbial enhanced oil recovery (red) compared with control/sterile (blue) experiments; (B) contact angle on quartz slice aged in dead oil; top: control/sterile; bottom (B): microbial enhanced oil recovery case.

8.3.2.4 Petrophysical modification

a. *Bioplugging:*

As up to 20% of the global oil reservoirs are fractured [162], bioplugging bears significant potential for incremental oil production. In ex situ MEOR, no bacterial growth is triggered purposely in the reservoir; therefore biofilm and/or biomass generation for effective selective plugging is low. However, swelling and gel formation might result in minor plugging. It is reported that the XG with a Cr^{3+} system can be used as an effective chemical due to its swelling characteristics. The effect of various pregelled biopolymers on the permeability of sandpacks was also studied and a decrease of 74.8%, 99.8%, 96.3%, 92.3%, and 96.5% of permeability with 1 g/L of XG, polyhydroxy butyrate, guar gum, polyglutamic acid, and chitosan, respectively, was reported after 11 days of injection [163]. It is reported that this is due to the development of a large flow resistance caused by in situ gelation occurring over a short length of the sandpack column. Conformance control using biopolymers also occurs from the high viscosities resulting in low shear rates. As the shear rate decreases, the apparent viscosity of the biopolymer solution increases making an additional but relatively lower plugging effect.

In contrast to the conventional conformance control mechanisms, bioplugging due to biofilm and/or biomass formation is mentioned as one of the most important in situ MEOR mechanisms throughout the literature [33,163,164]. Note that a biofilm is defined as microorganisms and extracellular polymers associated with a substratum or solid surface [89].

A review of the literature reporting bioplugging was provided by Alkan et al. [99]. Dynamic experiments performed by using cores and sandpacks with injecting microbial cultures as well as nutrients to stimulate the defined microorganisms in the formation water previously injected in the porous media results in decreasing permeability with PRFs varying between 0.03 and 0.75. The magnitude of the effect is depending on the amount injected as well as on the distribution of the oil in the system. In an early attempt the permeability reduction caused by the formation of biofilm in sandpacks was expressed with the following equations as a function of bacterial organic carbon (BOC), measured as nonsoluble organic carbon:

$$\mathrm{PRF} = e^{\left(a[\mathrm{BOC}]+b[\mathrm{BOC}]^2\right)} \quad [\mathrm{BOC}] \leq 0.4\mathrm{mg/cm^3} \tag{8.6}$$

$$\mathrm{PRF} = c \quad [\mathrm{BOC}] > 0.4\mathrm{mg/cm^3} \tag{8.7}$$

where [BOC] is the mass of BOC per unit volume of the porous medium and a, b and c are constants depending on the type of porous media. In the study, the following values are given for $a = -37$, $b = 45$, and $c = 5.10^{-4}$ [89].

Torbati et al. [165] proved experimentally that microorganisms preferentially plug the larger pores rather than the smaller pores. In an experiment performed using Berea sandstone cores a permeability decrease of three times of magnitude is reported after 234 days of continuous nutrient injection [89]. Rittmann [166] investigated the effect of continuous and discontinuous biofilm formation on permeability, concluding that the pore throat size reduction is most important when the biofilms are continuous and relatively thick. Gullapalli et al. [167] reported a microbial profile modification process consisting of sequential injection of spores and nutrients, which was designed for a carbonate reservoir (Eunice Monument South Unit, Lea County, New Mexico, USA). They used *B. licheniformis* isolates that generated biofilm and biomass in situ when fed with sucrose as the main nutrient. Coreflood experiments performed in Berea cores showed permeability reductions up to 80% of its initial value (PRF = 0.2) resulting in incremental oil recovery in the single well field application.

Gandler et al. found that the permeability reductions varied between 0.2 and 0.7 [164] when a *Pseudomonas aeruginosa* culture was inoculated into columns of glass beads and Berea sandstone cores and a growth substrate was injected continuously. Al-Hattali et al. [168] investigated the bioplugging performance of an isolated *B. licheniformis* strain with a sucrose-containing medium in Indiana limestone core plugs where a fracture was simulated by slicing the cores vertically into two sections. The results were promising with an enhancement of oil recovery observed after bacterial injection. A total of 27% − 30% of the remaining oil was produced after 11 hours of incubation. Gaurav [169] reported permeability decreases down to 0.39 of the original permeability of 434 mD in the coreflood experiments performed with sandstones after a static incubation of 10 days.

Bioplugging experiments performed in Bentheimer cores and sandpacks were reported by Alkan et al. [99]. In one phase experiments performed with cores of 1400 mD, a quick decrease in permeability resulting in almost complete plugging was recorded after 15 PV continuous injection of nutrient. The magnitude of the permeability decrease in sandpacks was lower being in the range of 0.4–0.5 of original permeability [99]. Moreover, the experiments performed using oil showed significantly lower plugging in corefloods. No permeability decrease was reported in sandpacks. Notably, this also demonstrated the importance of using corefloods as dynamic EOR screening tools. The optical examination of the cores used in bioplugging experiments showed slime as an indication of biofilm accumulated especially on the inlet side, whereas the outlet side was free of any slime. A similar phenomenon was also indicated by Davey et al. [170]. Fluorescent microscopy observations (DNA stained with a fluorescent nucleic acid stain, SYTO™ 9) from sandpack samples (one phase experiments) supported this information from the cores, that is, a high cell density and strong adsorption of the cells, especially in the inlet side of the sandpacks are observed.

Micromodel studies provide a comprehensive insight into the mechanisms of bioplugging in porous media. Stewart and Fogler [171] investigated the bioplugging mechanisms generated by *Leuconostoc mesenteroides* under nutrient-rich conditions in idealized micromodels of porous media. They affirmed that, as the nutrient flows through the micromodels, the initial bioplugging occurs at the nutrient–inoculation interface due to growth in the larger pore throats, while the smaller pores remain free of cells. If the nutrient injection continues, the pressure drop in this bioplugged region increases to a point where a channel opens, and biomass advances deeper in the micromodel. The biomass-plug propagation and channel breakthrough are described well using the Bingham yield stress model also representing the stability of biofilm against shear stress.

Gaol et al. [88] used glass–silicon–glass micromodels to simulate the idealized and real porous structures and observed the formation and the effects of the biofilms in in situ MEOR applications. In one-phase experiments, allowing ten days for the incubation of a fermentative community after the injection of the nutrient solution, the biofilm and cells were observed by using fluorescence microscopy after the injection of a fluorescence marker solution of SYTO™ 9 to stain the DNA of microorganisms in the micromodel. Fig. 8.8 shows biofilm formation in the vicinity of the grain surface. However, the biofilm was not distributed uniformly in the micromodel. The biofilm thickness, based on visual inspection, varies from 5 to 140 μm. The morphology of biofilm at several positions was observed to follow the grain shape, particularly toward the main diagonal of the artificial structure micromodels. As a result, the permeability reduction was observed in both micromodels mimicking idealized and real porous media, where the permeability of real structure was reduced from 1.2 to 0.67 D compared to 13.5 to 10.9 D in artificial structure. These results are equal to the PRF values of 0.56 and 0.81 for real and artificial structures, respectively. These PRF values were then compared to a mathematical model based on the Kozeny–Carman (K–C) approach as illustrated in Fig. 8.9. As can be expected the K–C approach can better estimate the idealized case rather than the real structure case. In the two-phase experiments performed using microchips with an analogy to real heterogeneous porous systems with two low (porosity and permeability of 0.18 and

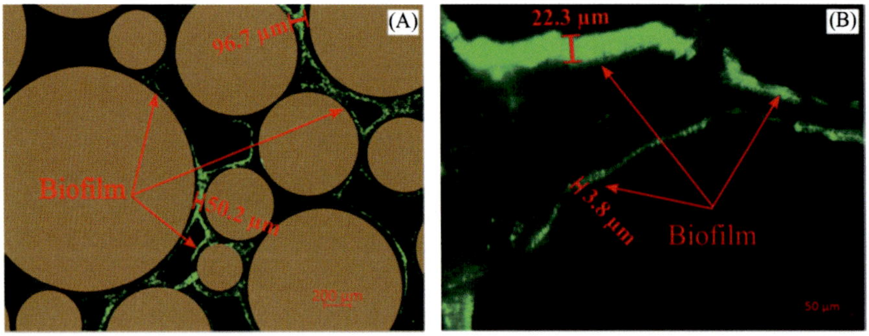

FIGURE 8.8

Images of biofilms after injection of SYTO™ 9 in the (A) artificial structure and (B) real-structure micromodels (note the different scales in the images).

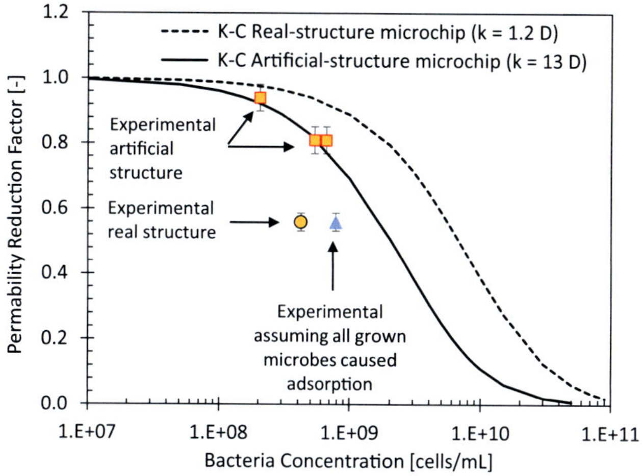

FIGURE 8.9

Permeability reduction factor in the experiments with micromodels based on differential pressure data compared to the Kozeny–Carman model.

Reproduced with modifications from C.L. Gaol, L. Ganzer, S. Mukherjee, et al., Investigation of clogging in porous media induced by microorganisms using a microfluidic application, Environmental Science: Water Research & Technology 7 (2021) 441–454.

800 mD, respectively) and one high permeability (porosity and permeability of 0.22 and 4 D, respectively), potential effects of biofilm formation on the oil recovery are reported. During the flooding process, all the MEOR experiments showed an additional oil recovery between 2.2% and 7.1% of OOIP, with the extra oil being mainly produced from low permeability zones of micromodels. This result combined with visual detection of the changes in oil saturations indicates that the plugging due to microbe attachment and straining, occurred in the high permeability zone; thus the water flow diverges to the low permeability zones and mobilizes the remaining oil as seen in Fig. 8.10.

These findings support established assumptions that this MEOR mechanism is most effective in heterogenous reservoirs. In contrast, very homogenous reservoirs might even be adversely affected by bioplugging. For example, this undesired bioplugging effect was reported to cause injectivity issues [172,173]. This bioplugging process may reduce the pore throat diameter, thus reducing the permeability and restricting the fluid flow within the reservoir. The growth of microbes that leads to the bioplugging process can be prevented by treating the reservoir with biocides (antimicrobial agents) or common well stimulation techniques, such as acidizing.

b. *Carbonate dissolution:*

Organic acids are very common metabolites of in situ MEOR and Zobell's pioneering work already made early remarks on the decomposition of inorganic

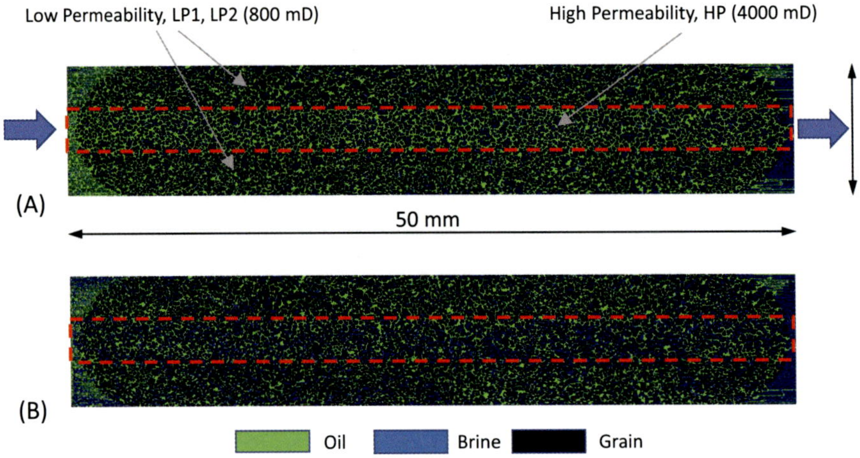

FIGURE 8.10

Heterogeneous microchip used for the visualization, saturation distribution before (a) and after microbial enhanced oil recovery (b).

Reproduced with modifications from C.L. Gaol, L. Ganzer, S. Mukherjee, et al., Investigation of clogging in porous media induced by microorganisms using a microfluidic application, Environmental Science: Water Research & Technology 7 (2021) 441–454.

carbonates [7]. Yet, the interaction between organic acids and reservoir minerals remained insufficiently investigated for a long time. The presence of organic acids implies two important aspects in terms of reservoir engineering of MEOR. If the reservoir rock contains carbonates, organic acids may dissolve them, causing a porosity and a corresponding permeability increase. Another very important implication of the presence of carbonates is their buffer capacity to keep the formation pH constant if organic acids are generated as metabolites during MEOR. Roldan-Carrillo et al. [113] indicated that in sandpack experiments the amount of oil released was higher in porous media systems composed mainly of carbonated granular media (41% wt. of total) compared to quartz systems. However, no discussion is provided on the reasons for this difference between the carbonate and quartz materials used for the porous media.

As one of the rare studies directly relating rock type to MEOR and pH buffering, Rinanti [174] reported the result of an experimental work conducted using isolated indigenous bacteria from a reservoir with low permeability (<40 mD) limestone rock samples. The porosity of the cores increased from 23% to 56% as an average of four experiments after MEOR treatment. An interesting result of their study is that despite the expected buffering effect of carbonate they measured a significant reduction in pH from neutral (7.0) to acidic range (5.2 ± 0.5). Alkan et al. [99] conducted the MEOR studies on a reservoir containing a considerable amount of siderite ($FeCO_3$, 10% wt. at average). In corresponding sandpack experiments performed with original formation rock, carbonate dissolution was confirmed by a significant decrease in surface iron content after the MEOR experiments by X-ray photoelectron spectroscopy (XPS) [51]. The same study also reports that even minor amounts of noninert minerals (1% pyrite and calcite) facilitated microbial fermentative growth significantly and the authors suggest to introduce the acid neutralization capacity of the reservoir rock as an additional screening parameter for in situ MEOR relying on fermentative microbial growth. It is also indicated that the permeability measurements cannot give the magnitude of carbonate dissolution, since the permeability decrease due to bioplugging is concurrent with the permeability increase due to carbonate dissolution. But estimations calculated with K–C by using the decrease in the specific surface area measured before and after the experiments show clear evidence of porosity and permeability increase due to carbonate dissolution [51].

Coreflood experiments are reported to investigate the cumulative effects of in situ MEOR [97,101,175,176]. Coreflood experiments with strain *B. licheniformis* BNP29 and a sucrose-based nutrient were performed with lime-free and lime-containing, oil-bearing sandstone cores. Oil recovery increased from 9.3% to 22.1% of the waterflood residual oil saturation [101]. It is reported that acid production along with selective plugging, are important mechanisms leading to increased oil recovery, presumably through resulting changes in rock porosity and alteration of wettability. Sugihardjo et al. [97] performed coreflood studies with isolated in situ bacteria (*Bacillus* sp.), formation water, and a defined nutrient. They observed higher recoveries after successive nutrient injection into the high permeability (793 − 1052 mD) sandstone cores (IRF = 27.5%, after three injection steps) and lower recoveries (max 12.6%) in native limestone cores with a low permeability (16.5 − 18.4 mD), mainly due to biosurfactant production. SEM and X-ray analyses revealed no damage to the core sample. However, permeability tests showed a significant decrease in permeability after injection of microbes of up to 50.1% of the initial value. Feng et al. [176] reported the results of corefloods performed in synthetic and reservoir cores with a recovery of 13.9% after fermentative growth of the NG80 community in the core. Displacement tests performed using bacterial single metabolite and NG80 metabolites show that the main factors affecting recovery are type and concentrations of biosurfactant generated. The highest incremental oil effect is supposed to be obtained from the injection of bacteria at first and then displaced with metabolites.

8.4 Mathematical modeling of microbial enhanced oil recovery

Modeling of the MEOR processes involves a complex combination of physical, chemical, and biological processes. However, in addition to laboratory experiments, numerical modeling represents the second pillar of the field-specific tailoring of MEOR. By reducing the uncertainties, it paves the way for a well-founded decision process on whether MEOR can be applied or not. Numerous studies have been performed to model the MEOR processes in porous media using pore-scale, core/laboratory scale, conceptual/sector scale, and full-field reservoir scale simulation.

In the literature, there are several attempts for mathematical modeling of the MEOR mechanisms in porous media, including the prediction of the growth of bacteria. Several numerical simulators, such as Eclipse, CMG-Stars, UTCHEM, TOUGH2, BUGSIM, and COMSOL, have been used to model and evaluate MEOR performance. Numerical modeling of ex situ MEOR is generally performed by using state-of-the-art chemical analogs of the MEOR byproducts (biopolymers and biosurfactants) for laboratory-scale and field-scale simulations [175,177].

A recent overview of the state-of-the-art of numerical modeling of in situ MEOR was provided by Patel et al. [32]. They list the numerical efforts from simplistic laboratory to complex numerical models including the modeling of microbial growth, the transport of the metabolites and microorganisms in the reservoir, and related physicochemical processes and interactions of MEOR with reservoir components. A revised table of the numerical efforts for modeling in situ MEOR is provided in Table 8.3.

Despite offering new solutions and progress for the modeling of in situ MEOR processes, all these studies also indicate the complexities of numerical efforts. However, as is the case in all EOR processes, numerical modeling is of paramount importance for validating the MEOR mechanisms and improving the prediction of the outcome of the process, thereby increasing the chances for successful application of MEOR. One other important aspect is the identification of potential risk parameters having the greatest influence on the uncertainty of the MEOR predictions and mitigating the uncertainty accordingly by determining specific input ranges. From a pragmatic point of view of reservoir engineering, the following modeling approaches can be considered for designing in situ MEOR operations and estimating their performance numerically.

8.4.1 Estimating the incubation radius

Crucial information in in situ MEOR is the portion of the reservoir in which the reaction (metabolism) between injected nutrient and targeted microorganisms takes place. Similar to the tools routinely used for microbial-derived products in biochemical engineering, the portion of the reservoir containing the injected volume can be referred to as a bioreactor. After injection, the nutrient solution moves through the reservoir radially assuming that incubation follows the growth curves similar to that defined in the laboratory and if no other hydrodynamic effect exists. Fig. 8.11 shows the growth in the reservoir schematically. Injected nutrient (N) meets reservoir microorganisms as soon as it reaches the reservoir and the incubation starts. The injection water contains the targeted microorganisms likely as well and, in this case, theoretically, the incubation starts already at the surface where the nutrient is mixed with the injection water. In the reservoir, the incubation continues as the nutrient advances from the wellbore deeper into the

Table 8.3 Numerical efforts to model in situ microbial enhanced oil recovery.

Models	Source
Preliminary attempts	
1983—Fundamental conservation laws along with growth and retention kinetics of biomass to predict porosity reduction in porous media	[178]
1985—A review of modeling kinetics of nutrient-limited transport and microbial growth	[179]
1988—Modeling of growth and transport of microorganisms in porous formations	[21]
1991—Model validated with laboratory data describing adsorption, growth, and decay of microorganisms, consumption of nutrients, and other physical processes	[95]
Analytical models	
1992—Advection-dispersion equation modified to account for the deposition and entrainment to breakthrough curves from sandpacks for organisms with two different sizes	[57]
1996—Model for the changes in porosity, specific surface area, and permeability caused by biomass accumulation in porous media	[180]
2001—A biofilm growth model suitable for microscale simulations of biofilm activity incorporating a growth-hydrodynamics model in porous media	[181]
2002—Estimation of the bioreactor radius, reaction time versus residence time, and nutrient consumption in the reservoir	[182]
1 D laboratory models	
1994—A modeling method for mixed type rock consisted of matrix blocks and fractures tested and validated on experimental data	[183]
2006—MEOR is simulated using MATLAB and tested	[184]
2010—Modeling partitioning of surfactant (metabolite) between oil and water phases as a novel approach in the context of MEOR	[185]
2016—Modeling injection of surfactant-producing bacteria into the reservoir, different types of bacteria attachment and growth scenarios are studied	[186]
3D Reservoir models	
1990—Microbial transport completed with a kinetic model of bacterial growth leading to either plugging, mobility control, or change of IFT	[187]
1993—Scaling criteria and similarity rules derived from a new approach for modeling microbial plugging in combination with bacterial growth and nutrient transport models	[188]
2008—Multicomponent transport model used to simulate the effects of physicochemical parameters on OOIP in a hypothetical reservoir	[189]
2017—Modeling MEOR under the effect of reservoir temperature, pH, and microbial sorption kinetics	[173]
2019—Performing both core and field-scale simulations of MEOR compared with conventional waterflooding	[190]
2020—Modeling the influence of effective porosity, microbial kinetics, and operational parameters within a sandstone formation	[191]
Reactive transport models	

(Continued)

Table 8.3 Numerical efforts to model in situ microbial enhanced oil recovery. *Continued*

Models	Source
2016—For the biophysiochemical effects on multispecies involving *Pseudomonas putida* for MEOR application	[173]
2018—UTCHEM-PHREEQC was used to simulate reactions between the reservoir brine and nutrient for MEOR	[192]
2018—Modeling the influence of interaction between injection and formation brine salinities on in-situ MEOR by coupling of multiple-ion exchange transport model	[193]
Streamline simulation	
2010—The reaction and transport processes are solved simultaneously along each streamline. Gravity effects are implemented using an operator splitting technique	[194]
Adaptation of commercial simulators	
2014—CMG STARS chemical option to model in situ MEOR including the bacteria lifetime, transport, metabolite generation, and EOR effects	[195]
2014—ECLIPSE software to model MEOR using anaerobic gas-producing bacteria	[196]
2016—CMG STARS for estimating the growth of SRB and H_2S formation during an MEOR application	[175]
Coupling with uncertainty analysis	
2018—Sensitivity analysis in modeling MEOR including the uncertainties on the use of Petrotoga species for MEOR and biotransformation	[197]
2020—Sensitivity analysis based on the statistical distributions in MEOR experiments	[99]
Nonstandard models	
1998—A network model to determine porosity and permeability changes in a porous medium as a result of changes in the amount of biomass	[198]
2002—An improvement of the 3D-3P simulator UTCHEM including a microbial population model with various options for feeding and transport	[199]
2008—Computational pore network modeling of the influence of biofilm on bioplugging in porous media	[200]
2011—Modeling population patterns of chemotactic bacteria in homogeneous porous media	[201]
2011—Pore network simulator for modeling biofilm evolution in porous media including a reaction term to account for nutrient consumption	[84]
2012—One-dimensional isothermal model, comprising displacement of oil by water containing bacteria and nutrients for a quantitative prediction of recovery factor	[202]
2017—A model to include the interfacial area between oil and water to be used for the optimization of the related laboratory works	[203]
2018—Utilizing genomic information to constrain the biological parameters needed for modeling souring, providing a pathway for using microbial data from oil reservoir studies	[204]

(*Continued*)

Table 8.3 Numerical efforts to model in situ microbial enhanced oil recovery. *Continued*

Models	Source
2019—Model by incorporating a temperature-dependent biokinetic for simulating the cold-water MEOR calibrated and confirmed using the experimental results	[205]

IFT, *interfacial tension;* MEOR, *microbial enhanced oil recovery.*

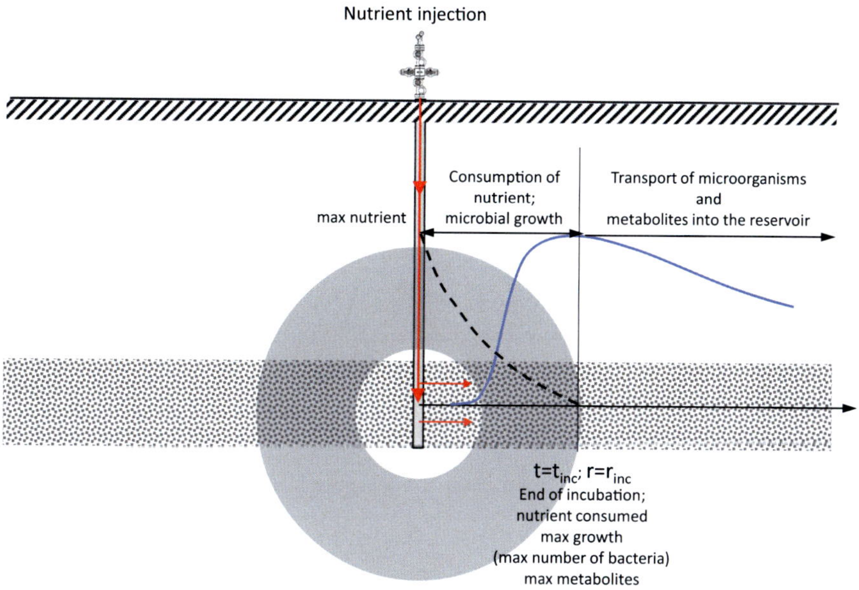

FIGURE 8.11

Schematic representation of the nutrient injection, incubation with microorganism distribution around the wellbore, nutrient consumption (black dashed line), and concentration of biomass and metabolites (solid line).

reservoir, stimulating the growth of the targeted microorganisms. At the incubation radius r_{inc}, the nutrient is completely metabolized at time t_{inc}; the nutrient is theoretically consumed and at this point, growth of the targeted microorganisms stops, resulting in the maximum amount of metabolites in the formation water; along with an increased number of cells, the metabolites continue to move deeper into the reservoir as dictated by the hydraulic gradient.

Byrant and Lockhart [182] estimated the relationship between residence time t_{res}, which is the time the injected water spends within the flooded part of the reservoir, and the characteristic reaction time or incubation time t_{inc}, the time required to complete the metabolism. They used the half-life, λ in their original equation, which is the inverse of the incubation time (t_{inc}):

$$\lambda = \frac{1}{t_{inc}} \tag{8.8}$$

According to their definition, reaction time is the incubation time of the bacteria with the injected N and is in turn dependent on the growth characteristics of the bacteria, a laboratory input from the microbiological analysis. The incubation radius (r_{inc}) in which the complete consumption of the nutrients takes place within time t_{inc} can be estimated with the following equation:

$$r_{inc} = \sqrt{\frac{q_{inj}}{\lambda \cdot \pi \cdot h \cdot \varnothing (1 - S_{or})}} \tag{8.9}$$

In this equation, q_{inj} is the injection rate, h is formation thickness, ϕ is porosity and S_{or} is residual oil saturation. Obviously, for a given half-life λ, the r_{inc} is a function of the injection rate, the geometry of the reservoir, and the porosity considering that the oil saturation around an injector well should be constant and close to zero. The λ which is the inverse of t_{inc} is depending on the type of the microorganisms, the nutrient, and the thermodynamic data, especially temperature, as defined in the laboratory. For example, colder injection water cools down the zone around the injection well, therefore the t_{inc} is increased due to the existing temperature profile of the reservoir, hence extending the r_{inc}.

The nutrient concentration, n, can be calculated assuming the following simple stoichiometric relationship:

$$n = \quad \varepsilon_N c \tag{8.10}$$

In this equation, c is the metabolite concentration and the stoichiometric coefficient ε_N defines the conversion efficiency of nutrient N into metabolite C.

To estimate the reaction time and the change in the concentrations of both nutrient (n) and metabolites (c), an isothermal plug flow through the reactor is assumed, with the consumption of N being first order and irreversible, and with N being injected at an initial concentration $n(0) = n_0$. The rate equation is:

$$\frac{dc}{dt} = -\varepsilon_N \frac{dn}{dt} = \lambda \varepsilon_N n \tag{8.11}$$

When integrated subject to the initial condition $n(0) = n_0$:

$$n = n_0 e^{-\lambda t} \quad \rightarrow \quad \frac{dn}{dt} = -\lambda n_0 e^{-\lambda t} \tag{8.12}$$

The nutrient concentration n can be expressed in terms of the radius by inserting the corresponding term for t:

$$n = n_0 e^{-\lambda \frac{\pi . h . \varnothing (1 - S_{or}) . r^2}{q}} \tag{8.13}$$

This term gives the nutrient concentration in the reservoir as a function of various operational and reservoir parameters, such as injection rate (q), porosity (ϕ), distance from the injector (r), and incubation time. With an assumption that ε_N is equal to unity, which means the total conversion of N to C, the rate of change of metabolite concentration c is inversely proportional to n.

Although it can be argued that this above given approach is an oversimplification, it provides the basics of designing an in situ MEOR operation in terms of metabolism kinetics and its extent in the reservoir.

For a better understanding of this approach, an example calculation can be made for a generic reservoir with $h = 20$ m; $\phi = 0.23$; and $S_{or} = 0.25$. Fig. 8.12 depicts the nutrient and metabolite concentration profile as a function of the radius for two different incubation times $t_{inc} = 2.5$ days, (equivalent to $\lambda = 0.4$/day) and $t_{inc} = 10$ days, (equivalent to $\lambda = 0.1$/day). The nutrient solution is injected at a constant rate ($q = 100$ m^3) with a stoichiometric coefficient (ε_N) of unity and dispersion is neglected, which is an acceptable assumption in comparison to the convective influence at the near wellbore.

For the case where the incubation time is short ($t_{inc} = 2.5$ days; $\lambda = 0.4$/day), the full MEOR potential is reached at an 11-m distance from the wellbore. In the case where the incubation time is longer ($t_{inc} = 10$ days; $\lambda = 0.1$/day), a larger radius of about 22 m for the given example is calculated. This simplified approach is repeated with a higher injection rate, namely $q_{inj} = 300$ m^3/day. The change in the metabolite concentration theoretically calculated (100%) is given in comparison to the case with $q_{inj} = 100$ m^3/day in Fig. 8.13. As can be concluded from the figure, in the case $q_{inj} = 300$ m^3/day the metabolite front extends deeper into the reservoir than in the case of $q_{inj} = 100$ m^3/day. In the case of $q = 300$ m^3/day, the incubation is complete, that is, delivering the maximum amount of metabolite, at approximately 40 m. This means also that until this point the metabolite is extensively

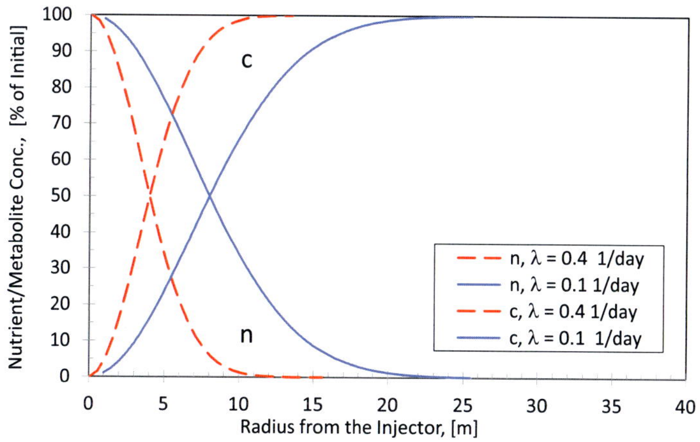

FIGURE 8.12

Profiles of the nutrient and metabolite fronts for various half-life λ (inverse of incubation times).

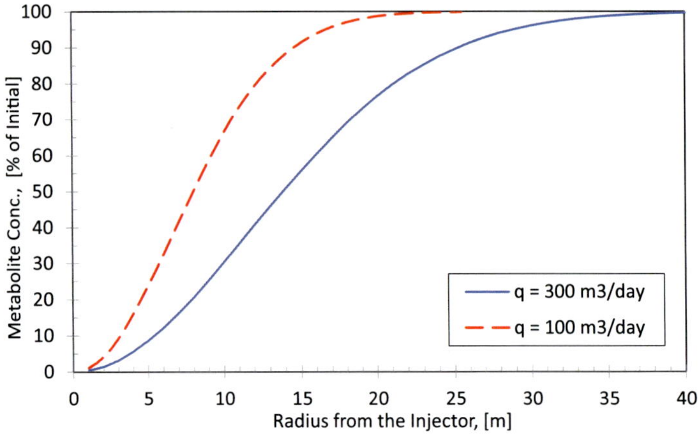

FIGURE 8.13

Metabolite concentration as a function of distance from the wellbore due to in situ microbial enhanced oil recovery for two different injection rates ($\lambda = 0.1$/day).

dispersed within the reservoir. To conclude, the fundamental design criterion given by Bryant and Lockhart [182] is that to be effective, the residence time of the injected fluid should be greater than the incubation time so that the reaction takes place in the right place in the reservoir:

$$t_{inc} \ll t_{res} \tag{8.14}$$

As the t_{inc} can be changed through the nutrient concentration for given microorganisms, this condition is satisfied for large values of n_0, for smaller values of r_{inc}, and q_{inj}.

Economics may favor higher injection rates, however; for an effective MEOR, injected nutrient solution should have time to incubate with the bacteria selected for MEOR without being diluted too much in the reservoir. Higher flow rates imply lower residence times and hence require higher rate constants (λ), which means lower incubation times (t_{inc}). However, from another perspective, a shorter residence time can be beneficial to transport the nutrient deeper into the reservoir whereby the incubation starts far from the injection well. This fundamental design criterion can be used to estimate the distribution of nutrients as well as incubation radius with produced metabolites at any given time in the reservoir.

The radius of the microbial slug increases with injected volume; the activated part of this volume, that is, bioreactor, depends primarily on the incubation time as exemplarily shown in Fig. 8.12. However, some common practices may change this situation. For instance, if the injection stops, the complete injected volume will totally be activated in place as in situ (or ex situ) microorganisms find sufficient time for metabolizing the nutrients; the complete growth might also occur in the wellbore in this case. In the second practice, if the nutrient solution slug is followed by water injection, this would also give the nutrient slug sufficient time to be activated at a given radius; the MEOR/metabolite front is displaced by the water injection into the reservoir.

Obviously, the above-presented model is related to the estimation of the incubation radius, in other words, location of the metabolite formation in the reservoir. Numerical modeling using the discretisation of the reservoir as well as time is the most commonly used one for simulating the EOR processes.

8.4.2 Transport of microorganisms in porous media

Despite the general assumption that the microorganisms are in motion in the aqueous phase as suspended cells, this is not always true. Frequently, microorganisms of various species adhere to a solid surface, forming biofilms. Therefore it is assumed that the microorganisms live in the form of biofilms, an assumption that is valid for porous media as well. However, it is also experimentally validated that in the reservoir, after the in situ application of MEOR, the number of microorganisms increases several orders of magnitude and the suspended portion starts to move in the reservoir due to general hydrodynamic laws. This movement is preferable for MEOR, as the microorganisms and their metabolic activity are needed for the MEOR mechanisms throughout the reservoir. The transport of microorganisms in porous media has been investigated both physically and numerically in recent decades. Corapcioglu and Haridas [172] modeled the diffusivity of the bacteria by application of the Stokes–Einstein equation, which effectively treats the microbes as particles undergoing Brownian motion. But in MEOR, this effect is likely to be insignificant compared to fluid advection. The remaining parameters needed to characterize the transport of the bacteria are the longitudinal and transverse dispersivities in the porous medium obeying the processes illustrated in Fig. 8.14. A mass conservation equation for microorganisms can be established considering the following effects:

- Microbial growth and decay: Microorganisms are included as the source and sink terms in the mass-conservation equation, which is based on the Monod equation [206–208].
- Chemotaxis: It allows bacteria to move in a directed manner along chemical concentration gradients toward nutrient molecules; however, in porous media under injection–production cycle, this can be neglected in comparison to advection and dispersion [201].
- Interphase transfer: If not present within the biofilm on the rock surface, the bacteria prefer mostly to dwell in the water phase and the water–oil interface. The transfer of bacteria from water to the oil phase can be neglected. The preference of bacteria toward biofilm present on the rock surface or water–oil interface is also depending on bacteria type, specifically on hydrophilic or hydrophobic cell membrane characteristics [110]. Making this distinction is important because adhering to the oil/water interface or to the rock surface

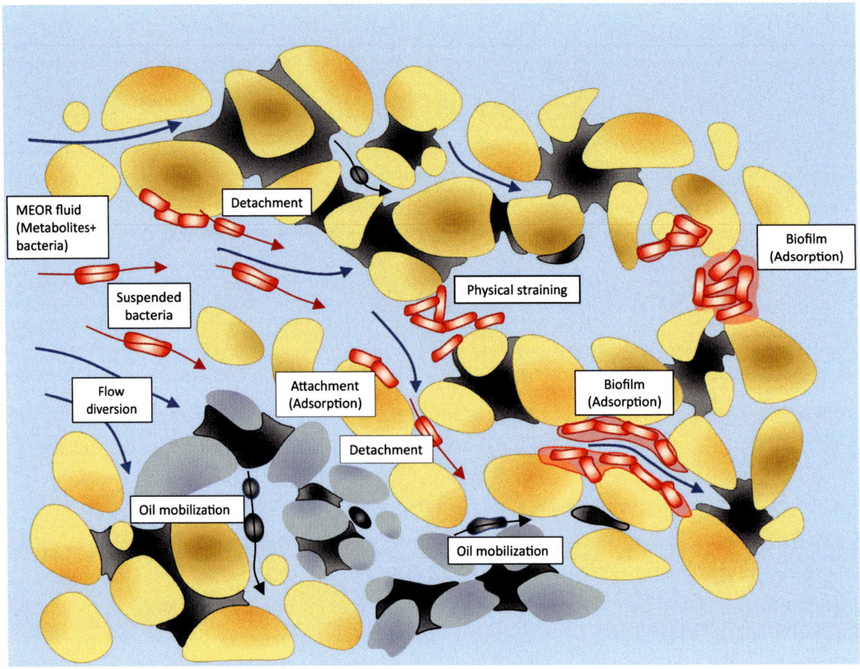

FIGURE 8.14

Schematic of the transport of microorganisms in porous media (not to scale).

improves the oil recovery via different mechanisms, for example, reduction of IFT versus wettability alteration.

- Adsorption: Several studies investigated the physical adsorption of bacterial cells to surfaces. Often, such adsorption is the first step of biofilm formation and greatly influenced by the zeta potentials of the two surfaces (e.g., cells and rock). In the second step, EPS segregated by bacteria attaches the cells to each other and to the pore walls. For modeling purposes, the resulting biofilm could be considered a kind of chemisorption, although the actual molecular adhesion mechanisms are various and complex. The attachment or filtration process can be assumed to be irreversible or reversible. This can be modeled with known equilibrium isotherm models, such as the Langmuir, Freundlich, or the linear isotherms such as Henry, but need to be calibrated with experimental data.
- Biofilm formation and its effect on bioplugging: Biofilm accumulation in porous structures result from the complex interplay of nutrient mass transfer, metabolic conversion rates, and biofilm detachment. Cell detachment from the biofilm depends on the surrounding shear field. The detached fragments enter the fluid stream and move along the streamlines. During the past decade several different approaches for multidimensional modeling of biofilm growth

as well as for biofilm detachment have been developed. These approaches are roughly grouped into discrete biomass-based models [209], discrete individual-based models [210,211], and continuum biofilm models [212]. However, all these models are of laboratory scale and difficult to upscale to the whole reservoir. The knowledge of macroscopic laws, such as the permeability–porosity relationship is not sufficient to account for the coupling between hydrodynamics, biomass growth, and biofilm cells in porous media as bacterial biofilms are formed of populations able to adapt their metabolism to the evolving microenvironment, and in particular to hydrodynamic conditions. It appears that the predictive character of such models needs to be largely improved, if closure laws were not postulated a priori, but rather deduced either from upscaling techniques, or from ad hoc experiments performed at the global scale [213]. Biofilms can also be adsorbed onto the rock surface as shown in Fig. 8.14.

- Physical straining, filtration: This is especially important for larger microorganisms or rods and clumps of bacterial cells, which have dimensions in the range of prevailing pore throats. Straining takes place when a particle in suspension flowing through a pore is larger than the pore opening, so that suspended microorganisms accumulate on the rock grains. Experimental studies reveal a linear relationship between the straining rate of spherical bacteria and the ratio of bacterial diameter to average grain size. Gruesbeck and Collins [214] suggested that the ratio between bacteria diameter to the pore size must be in the range of $0.17 - 0.25$ for straining-induced plugging to occur, which corresponds to the claim made by Davis and Updegraff [59] that the pore throat sizes of the rock should be at least twice the cell sizes to allow passage of microbes through the rock without substantial plugging. Additionally, an absolute pore-throat plugging could occur when bacteria formed a filter cake on the porous media's top face when the ratio between bacteria diameter to the pore size is bigger than 0.38. Generally, the major axis of rod-shaped bacteria tends to adopt a preferred orientation in direction of the flow [208].
- Gravity settling: A comparison of the gravitational forces and the surface forces on almost neutrally buoyant Brownian particles clearly shows that the surface forces are several orders of magnitude larger [215]. Gravity should therefore play a relatively minor role in determining the extent of bacterial retention.

8.4.3 Upscaling to the field

In MEOR literature it is hard to find studies on the upscaling procedures of laboratory works to field scale. A reason for that is the complexity of MEOR because of the additional components, that is, the microorganisms. The other reason is the routine of EOR projects in which an upscaling is done by using validated numerical reservoirs models for both laboratory and field cases. However, due to the lack of such representative models, this routine is difficult for MEOR.

Islam and Gianetto [188] presented a set of scaling groups based on analyzing governing partial differential equations at laboratory and field scales. These included three-phase transport equations as well as microbial transport, nutrient propagation, and growth equations.

The upscaling of bacterial growth rate and time can be done based on the principles discussed in previous sections. However, the effects of the metabolite and bacterial dispersion in formation water observed in laboratory experiments are difficult to scale up. One way is to conduct experiments with various nutrients and metabolite concentrations for estimating the dispersion of metabolites during their flow and their EOR effects.

One discrepancy is the aspect of slug sizes and concentrations: these should be the same in laboratory experiments and in the field, but often, laboratory experiments require higher slug sizes/concentrations to simulate the same effect as in the field. Obviously, this should be done in conjunction with geometric similarity so that bacteria and nutrient distributions can be scaled properly. Another important practice is, that reservoir fluids and porous media (instead of artificial brines, model rock, etc.) should be used in the experiments, which is a difficult condition to be fulfilled. In general, only dead oil (stock tank barrel, STB) is available under atmospheric conditions, which means only a part of the reservoir fluid can be represented in the experiments. Sampling oil under wellbore conditions or creating reservoir fluid via recombination is a costly and difficult procedure. On top of that, conducting successful representative experiments under reservoir pressure requires time-consuming preparations and set-ups.

Scaling groups relating to geometry dictate that the model and field prototype have geometric similarities. However, flooding experiments are often performed in linear cores. These models are mostly insufficient in terms of geometry and petrophysical representativity, especially for studying bioplugging where isotropy is essential.

Moreover, it is crucial to obtain and use representative rock samples in dynamic experiments. Cores and/or plugs retrieved from the reservoir can be used, but this does not always guarantee an acceptable representation of the reservoir porous media in terms of geometry, petrophysics and rock mineralogy as most of the reservoirs exhibit micro- and macroheterogeneities. A combination of experimental approaches (micromodels, sandstones, and corefloods) should be used and the results should then be interpreted for more reliable physical modeling of MEOR under reservoir conditions.

The concentration and the slug size can be optimized in the field scale by considering the metabolite yield of the nutrient injected and the simplified approach proposed by Bryant and Lockhart [182]. In relation to the generated metabolites and their effective concentrations, nutrient slug size can then be estimated for creating EOR effects. A simplified example can be given by assuming a reservoir with one injector corner and one producer at a 250-m distance with natural water influx in the direction from injector to producer as schematized in Fig. 8.15. The

r [m]=	250
h [m]=	10
ϕ [-]=	0.23
S_o [-]=	0.5
Swept/total [-]=	0.25
q_{inj} [m^3/day]=	150
N (sucrose+yeast) [g/L]=	5 g/L
t_{inj} [months]=	8

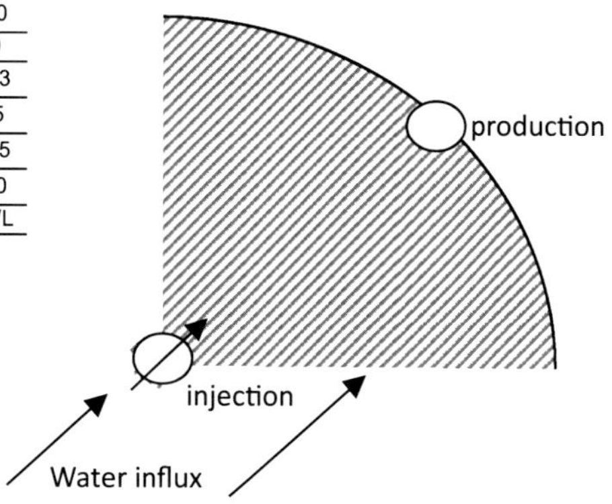

FIGURE 8.15

Simplified example for the estimation of the nutrient slug size and metabolite concentration in the reservoir.

thickness and the porosity are 10 m and 0.23, respectively, with an oil saturation of 0.5. The ratio of the swept zone to total is 0.25. After 8 months of nutrient injection with a rate of 150 m^3/day, a total of approximately 36,000 m^3 nutrient with a concentration of 5 g/L sucrose and yeast extract is injected to yield a metabolite concentration of 2.55 g/L liquid metabolite in formation water and 20 mM CO_2. The EOR effect of this amount can then be estimated, based on experimental data generated in the laboratory.

EOR field applications need performance prediction before and during the operation. This routine part of all field applications is necessary to reduce the uncertainties of the operation and therefore to better forecast the outcome of the application. The EOR modules of commercial reservoir simulators are supporting thermal, gas, and chemical EOR operations. However, because of the above-mentioned difficulties of modeling MEOR in the field scale, no commercial simulator provides a dedicated option for modeling/supporting in situ MEOR. Section 8.4.4 represents an attempt to model in situ MEOR using the mathematical analogies provided by the commercial reservoir simulator CMG as an example.

8.4.4 Discretized modeling using commercial reservoir simulators

A MEOR modeling approach using CMG STARS™ (or any other reservoir simulator offering similar modeling capabilities) can be applied for reservoir simulation [195]. The most relevant features for MEOR modeling, which can be modeled using CMG STARS™ are: (1) The modeling of dispersed components (stabilized dispersion, such as droplets, bubbles, and lamellae), which can be treated as components in the carrying phase that provides the flexibility of modeling and quantification of bacteria, nutrients, biopolymers, biosurfactants and CO_2 for specific cases. This concept can be coupled with the component property input package capabilities, such as adsorption, blockage, nonlinear viscosity, dispersion, and nonequilibrium mass transfer, to model complex phenomena via input data choices. (2) The reaction kinetics module allows the calibration of the bacteria concentration as it changes with time and nutrient availability, as an analogy to the Monod Equation. It also enables the generation of metabolites as a product of the biochemical reactions [216].

As discussed in the previous sections, for in situ MEOR, a nutrient (substrate) is added to a fluid containing targeted microorganisms (in most of the cases bacteria). Then, during the bacterial growth, the nutrient is consumed and is converted into additional cells and metabolites. This process can be formulated as a chemical reaction as follows:

$$\sum n\text{Reac}, i \cdot M\text{Reac}, i = \sum n\text{Prod}, i \cdot M\text{Prod}, i \tag{8.15}$$

In which n and M are the stoichiometric coefficients of the ith reactant "Reac" and the product "Prod." Simulation of the chemical processes requires well-defined reaction kinetics. There is no standard input option defined, so the user can define the reaction components arbitrarily making the usage of flexible modeling options.

The stoichiometric coefficients are responsible for the material balance between bacterial number (concentration) and the nutrient supply. As many reactions as necessary can be applied for the investigated MEOR process if experimental evidence allows. For example, if bacteria use multiple energy sources and grow at corresponding different rates, multiple reactions must be defined. Several equations are used to express substrate uptake ability for transport and growth of microorganisms [179,217]. In microbiology, the Monod equation is generally applied for modeling the S-shaped growth and growth kinetics [218,219]. The simplest form of the Monod equation is given for the bacterial growth rate, g_g as follows:

$$g_g = g_{\max} \frac{S}{K_S + S} \tag{8.16}$$

In which S is substrate (nutrient, n) concentration, $g_{\max}$ is the maximum growth rate of the microorganism, and K_S is the half-rate constant representing at which the rate is $g_{max}/2$. It should be mentioned that the Monod equation is half-empirical; however, it can successfully reproduce the literature data with an average error of less than 5% [220,221].

Bacteria decay is modeled with a second linear equation using the decay constant, K_d:

$$g_d = X \cdot K_d \tag{8.17}$$

The resulting change in the bacterial concentration (X) can be expressed as:

$$\frac{dX}{dt} = g_g - g_d \tag{8.18}$$

The Monod approach is not implemented in STARS. However, the Arrhenius Equation can be used as a replacement giving the necessary freedom for modeling the growth and reaction rate (how many reactions/bacteria per time unit):

$$r = F_{\text{Freq}} \cdot e^{\left(\frac{-E_{act}}{R.T}\right)} \prod_{i=1}^{n_c} c_i \tag{8.19}$$

The reaction rate, r, which replaces the Monod kinetics is calculated depending on the frequency factor, F_{Freq}, temperature, T, and concentration of reacting components (n), that is, bacteria and nutrients (c_i), with the Arrhenius Equation. The exponential term (Arrhenius Factor) is responsible for the temperature dependency. By setting the activation energy equal to zero, an isothermal reaction is obtained. This assumption is valid in most of the cases; an MEOR application can be assumed as isothermal because the nutrient solution injected into the reservoir is heated up in a short time as it advances into porous media. If the assumption would not be valid due to significantly lower injection temperatures, the thermal option can be included in the simulation also considering the effect of temperature gradient on the incubation time. The reactant and product concentration factors are calculated depending on their concentration in a reference phase and reactants and products are considered as tracers in water and oil phases. Their EOR effects are given as a function of their concentration in the existing phases.

The EOR effects can be modeled by applying related standard procedures. The amounts of the metabolites (products) that can be classified as polymer or surfactant are reproduced by using the material balance like Eq. (8.15) and the related effects (polymer viscosity and IFT) are defined as functions of the metabolites generated. Polymer is modeled using a mixing rule (linear or nonlinear) and can be expanded for shear-thinning/thickening rheology. Surfactant reduces the oil−water IFT and/or changes the wettability, which is modeled using an interpolated relative permeability curve based on two different sets of relative permeability curves (as inputs) within the simulator. The adsorption and partitioning effects in the water and oil phases of these products can be considered as usual in a numerical simulation if data is available. Simulation of selective and nonselective plugging due to bacterial activity can be modeled using the adsorption option allowing the mechanical blockage in terms of adsorption. The adsorption is generally assessed by the Langmuir isotherm correlation. Permeability alteration often accompanies adsorption (especially if adsorption exits as mechanical entrapment, which is mostly the case for MEOR). The simulator accounts for this via region-dependent residual resistance factors (RRF) which allow correlation of local permeability with local adsorption levels. Microbial gas generation is considered another product (metabolite) of a "reaction" between bacteria and nutrient(s) according to Eq. (8.15). Dissolved gas reduces oil viscosity and density as calculated by mixing properties of both components. The partitioning of gas between two phases at different pressures and temperatures is modeled via K-values, for example, gas−liquid K-value for CO_2 partitioned between oil and gas. Gas partitioning between two liquid phases, that is, oil and water, can be defined via liquid–liquid K-values. The properties of the oil–CO_2 mixtures should be provided in the PVT inputs. The generated data should preferentially be based on the results of a swelling test in which the swelling of oil (density) and the viscosity are measured. For modeling purposes, the relative permeabilities corresponding to wettability changes can be integrated into the model and interpolated as a function of the intermediate alterations. Among all MEOR mechanisms, the modeling of porosity and permeability increase due to carbonate dissolution can exhibit a challenge as this may cause instabilities in calculating the changes in fluid saturations and relative permeabilities. The issue can be solved by using numerical models developed for investigating the phenomena and upscaling the results for full-field simulations.

The calibration and validation of the model can be conducted with dynamic experiments, namely sandpacks and corefloods; the geometry of these setups can be discretized and the output data from these experiments can be used to calibrate and regress the EOR properties by inverse modeling technique (history matching). The established and calibrated model can be further used for MEOR field modeling if a reservoir model of the field exists. This deterministic phase of MEOR modeling can be shortened by eliminating the modeling of the growth, which is the first step of modeling. This might be a sound suggestion considering the range of uncertainties in the laboratory phase as well as the time-consuming numerical process of matching the growth. Eliminating the growth means neglecting incubation time and microbial activity area during the growth phase with respect to metabolite transport in the rest of the reservoir. In other words, growth and resulting metabolite generation occur in a short time (a few days) and within a short radius from the injection well. In the case of neglecting the growth, the model establishes a material balance of metabolites injected from the surface while the following modeling phases remain the same. A test to compare both modeling approaches shows that neglecting growth during modeling is a pragmatic way resulting in an insignificant difference compared to the full cycle.

All steps of numerical modeling need input from the laboratory, including EOR static and dynamic experiments. However, all these experiments are subject to uncertainties due to various factors. One of these uncertainties is the bacterial growth and growth rate changing w.r.t. experimental environments as well as initial bacterial concentration in the formation sample. For instance, the bacterial concentration in a given sample can vary significantly depending on the location of the sampling point (e.g., injection vs production water), sampling season, or history of the sampling point (e.g., previous biocide and/or cleaning treatments, such as pigging). In the reservoir, these uncertainty ranges grow, considering thermodynamic as well as physicochemical factors affecting also related EOR effects. A good way of including these uncertainties is to perform an uncertainty analysis yielding a probabilistic distribution of oil, gas, and water production profiles for a given field. As also suggested by Ansah et al. [192], a Monte Carlo approach can be used for this purpose. In this case, the EOR mechanisms introduced into the model can be assigned as a base, best and worst cases with corresponding statistical distributions based on the experience from the laboratory works.

Fig. 8.16 depicts the steps of numerical simulation of in situ MEOR by using a commercial reservoir tool. The outcome is a model, incorporating the growth reaction and the EOR effects as a function of the metabolite concentration validated by dynamic experiments. The next step is the use of this model in field applications if a reservoir model of the studied field exists. Obviously, the reservoir model has to be validated before this step, to reduce the geological, petrophysical, and operational uncertainties.

A result of a simplified in situ MEOR simulation using a sector reservoir model is given in Fig. 8.17 as an example. The sensitivity analysis is performed using Monte Carlo simulation by assigning statistical distribution functions for the water and oil viscosity changes as well as wettability alteration and permeability reduction due to biofilm formation. In the sector model of a heterogeneous reservoir with intermediate to oil rock wettability using one injector and two producers, a selected nutrient was injected for 6 months to model the in situ MEOR effect. All steps stated in Fig. 8.16 were fulfilled based on experimental studies previously conducted, including the validation process by using dynamic experiments. Fig. 8.17 shows the production profiles in one of the producers (Producer-1) with numerous proxy runs as suggested in the Monte Carlo method as a function of the time for the worst, base, and best MEOR cases and the waterflooding case as a comparison. The cumulative oil production is also given in the same figure as a statistical distribution. By further extending this approach, potential contributions of individual MEOR mechanisms to the overall oil recovery could be investigated by a sensitivity

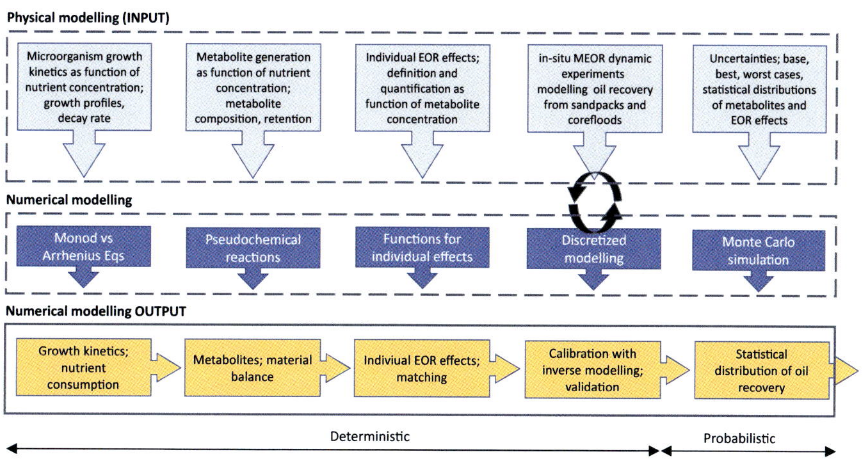

FIGURE 8.16

Simplified algorithm of microbial enhanced oil recovery numerical modeling by using CMG STARS reservoir simulator.

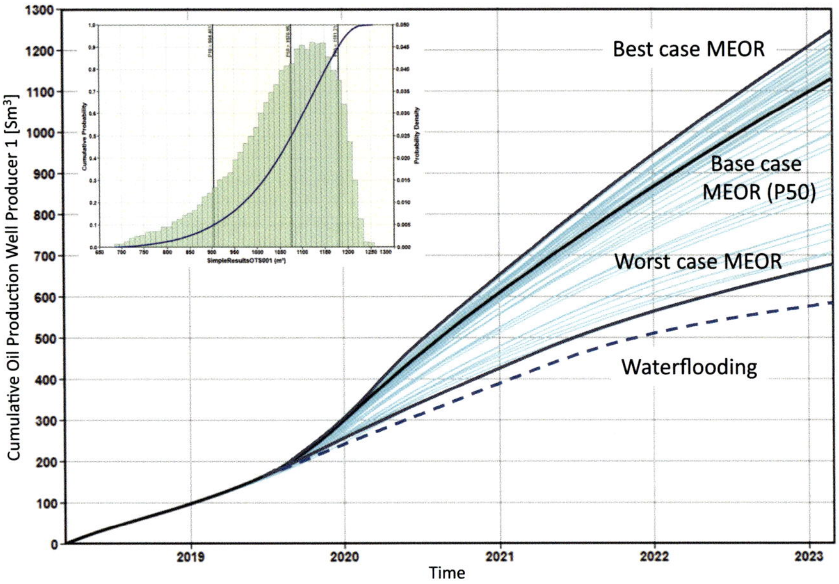

FIGURE 8.17

Simulation of oil production with probabilistic approach (Monte Carlo) for the well Producer-1 of the sector reservoir model applied for microbial enhanced oil recovery case versus waterflooding (microbial enhanced oil recovery effects are given as statistical distributions based on laboratory data; BEST case stands for all MEOR mechanisms work perfectly; WORST case stands for all microbial enhanced oil recovery mechanisms work at their minimum).

analysis; the main isolated EOR effects were determined as bioplugging and wettability alteration with individual incremental recoveries up to $2-3\%$ of OOIP at the end of 20 years of application. The other MEOR mechanisms (biosurfactant, changes in fluid viscosities) played a minor role in the total recovery. The effect of carbonate dissolution is also determined to be high, if the formation rock contains carbonate minerals.

8.5 Field applications and operational aspects

8.5.1 Field applications

In the last two decades the MEOR operations are reported with more detail and some of them have documented results based on well-documented monitoring work.

Safdel et al. presented a comprehensive collection of MEOR field applications including 47 cases in 21 countries including applied MEOR technologies, reservoir formation names, utilized microorganisms, and nutrients and obtained effects relevant to each case [46]. It was reported that only 10% of the worldwide field trials resulted in no incremental oil recovery performance. Most of the applications were performed using MEOR flooding (33%) followed by cyclic HnP (26%) technologies with an increase in oil production within the range of 8% to 350%.

Besides the review of Safdel et al., country-specific MEOR field applications were also reported. Among them, Ibragimov et al. presented the information on the MEOR applications in 15 fields in Azerbaijan emphasizing the availability of raw materials [24]. In India, a field application supported by experimental works was given as an example for the case study Bentayan field [25]. Najafi-Marghmaleki et al. investigated the MEOR for potential applications in reservoirs in Iran [27]. Successive papers highlighted the development of MEOR in China and applications in various fields as in situ and ex situ MEOR [26,28]. Gao summarized 31 MEOR field projects carried out in China since 2000 and also provided the reservoir characteristics as well as some insights on the applications [28]. The following conclusions can be drawn from the MEOR experience in China:

- From 31 cases, 25 operations report increase in the production of 75% at average, with minimum and maximum 33% and 100%, respectively.
- Most of the reservoirs have a relatively low temperature being in the range of 23°C – 65°C. Three reservoirs were reported to be warmer with 80°C.
- Except one case with 160 g/L, the formation salinities were low, in the range of 4–46 g/L.
- The permeabilities are generally moderate to high ranging from 100 to 1800 mD. Two cases are reported to be successful even in tight reservoirs with a permeability range of 19 – 25 mD.
- The oils are light to medium heavy with a dead oil viscosity range of 21–188 mPa·s. Three cases with high oil gravity (low degrees API) with viscosity range of 1885–2135 mPa·s are reported.
- In most of the cases, in situ MEOR with selected exogeneous microbes and successive nutrient injection were applied. In few cases only nutrients were injected.
- In most of the cases, the applications were performed as HnP applications. In 13 applications, flooding was done between injector and producer well pairs.
- Reduction of oil viscosity, emulsification, and biotransformation are the most reported MEOR mechanisms, especially in HnP applications. In some cases, it is reported that the MEOR application eased the wax-paraffin scale problem in the wells. Other potential mechanisms are reported to be IFT reduction and CO_2 generation. In few applications, the bioplugging of high permeability channels is reported as potential reason for reduction of WC, thus increasing oil production.
- No operational risks are reported.

One of the fields reported is the Baolige Field and the full field application is well documented [222]. The formation has an average porosity and permeability of 0.18 and 144 mD, respectively. The reservoir temperature is 50°C. The strains used in this study were *B. subtilis*, *Arthrobacter* sp., *Geobacillus subterraneus*, *P. aeruginosa*, *B. licheniformis*, and *Rhodococcus* sp. and they were isolated from the production fluid of the field. Laboratory based investigations showed that all six strains were able to decrease the oil viscosity by biotransformation. Fermentation liquids of these six strains significantly reduced the IFT between oil and water by 70%. The two selected mixtures of strains were tested in laboratory, coreflooding confirming their MEOR performance, that is, with IRFs of 9.1% and 13.2%, respectively, whereas IRFs obtained using either single strain ranged from 7.0% to 8.7%. Based on these preliminary studies, the composition of the nutrient used in the field test consisted of (wt.%) glucose 0.4%, peptone 0.1%, yeast extract 0.08%, ammonium chloride 0.2%, and potassium dihydrogen phosphate 0.1%. Using the strain mixture selected under the laboratory conditions, four cycles of microbial injection were carried out by the mid of 2015, (indicated by the arrows in Fig. 8.18). The field application achieved a significant oil recovery with 210,000 tons of crude oil produced over 43 months from 169 production wells, reporting oil viscosity reductions from 188 to 80 cP. This is a well-documented

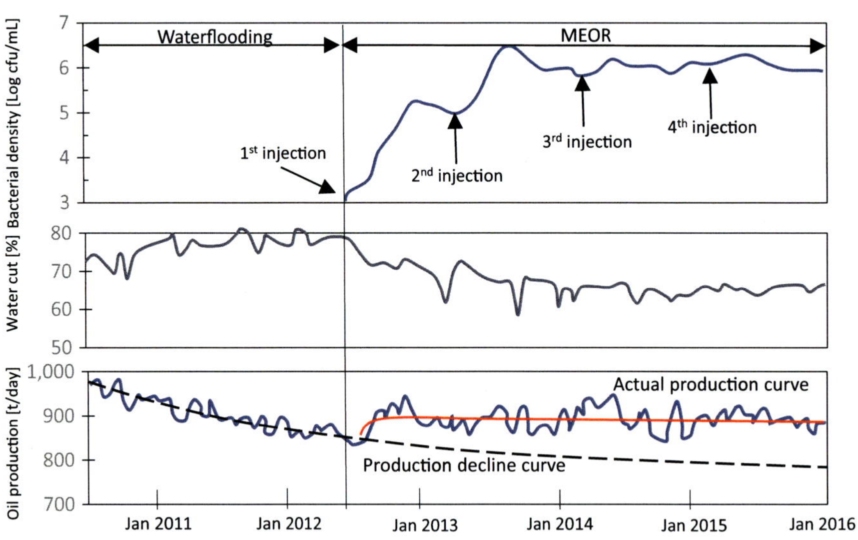

FIGURE 8.18

Daily based oil production, water cut, and bacterial density during microbial enhanced oil recovery in Baolige Oilfield, showing average values of 169 oil production wells.

Reproduced with permission from C. Ke, G. Lu, Y. Li, et al., A pilot study on large-scale microbial enhanced oil recovery (MEOR), Baolige oilfield, International Biodeterioration & Biodegradation 127 (2018) 247–253.

example of in situ MEOR acting with multiple mechanisms and operated in flooding mode.

In Vizcacheras Field of Argentina an MEOR program was started in September 1999 with a systematic inoculation from one injector using hydrocarbon degrading anaerobic-facultative microorganisms [223]. The reservoir has a temperature of 92°C, an average permeability of 1 Darcy, an effective porosity of 0.25 and residual oil saturation of 0.25. The project was developed for 19 months of uninterrupted treatment by injecting batches of nutrients and microorganisms periodically into the reservoir to colonize the contacted pore volume. It is reported that fractional flow alteration in produced fluids has shown a clear improvement: oil production increased, and water retracted in neighboring producers of the treated injector. Three dominant combined mechanisms explain this behavior: (a) Further reduction of residual oil saturation by in situ biosurfactants lowering IFT, (b) improved oil mobility from short-chain solvents originated by biotransformation of heavy paraffinic ends, and (c) conformance efficiency improvement created by the colonizing biomass in active zones that diverts fluids to unflooded zones. MEOR increased the oil production rate by 20% in average over baseline. This is a typical in situ MEOR application relying on multiple EOR mechanisms acting for recovery enhancement. Fig. 8.19 shows the production increase in one of the producers as well as the compositional alterations observed in this well due to MEOR application. This change in oil composition caused a decrease of approximately 58% in oil viscosity as one of the oil recovery mechanisms.

In an application in Brazil, the objective was the bioplugging of the high permeability zones of the reservoir [224]. In the laboratory phase the beneficial microorganisms were screened from the sampled formation water. The reservoir temperature is 45°C and formation water salinity is reported to be low. A nutrient consisting of sucrose and ammonium phosphate was selected to feed the selected microbes. After 6 PV of nutrient injection in sandstone cores representing the formation rock, a sixfold permeability decrease was observed after bacterial incubation of ten days followed by 9 PV of water injection. The biomass and biopolymer were detected in the pores via SEM studies. The operational phase was started with one injector. The selected microorganisms were injected as a slug to ensure that these are present in the high permeability zone. After this stage the nutrient was dosed continuously to the injection water. During the nutrient injection it was observed that a threefold increase in injection pressure as well as a reduction of the tracer concentration at the production well as a strong indication of the plugging of the high permeability zone occurred. After the successful pilot, the field application was extended to ten other injection wells. The injection was targeted to the high permeability zone with layer thickness of 1.5–7.5 m. After the decrease in the injectivity the nutrient injection was continued to all the zones in the reservoir. The results showed a sustainable pressure increase in eight injectors due to the plugging of the high permeability zone. In five of these wells the injection rate decreased significantly. Unfortunately, there are no further

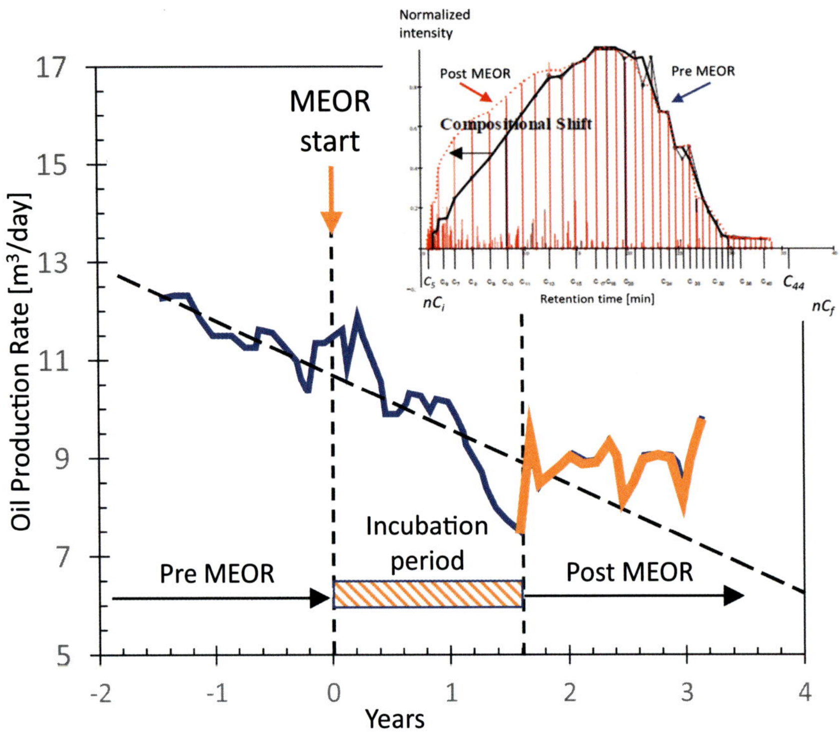

FIGURE 8.19

Oil production profile and shift of oil composition (high-performance liquid chromatography, upper right corner) due to in situ microbial enhanced oil recovery application in Vizcacheras Field.

Reproduced with modifications from L.A. Strappa, J.P. De Lucia, M.A. Maure, et al., A novel and successful MEOR pilot project in a strong water-drive reservoir Vizcacheras Field, Argentina, in: SPE-89456-MS, SPE/DOE Symposium on Improved Oil Recovery, 17–21 April, 2004, Tulsa, Oklahoma, USA.

reports on the influence of the operation on oil recovery, however this is an exemplary application aiming at one specific MEOR mechanism, namely bioplugging. A similar field application with effective and specific bioplugging by injecting exogenous microorganisms with nutrients is also reported by Jackson et al. [225]. They observed an increase of 20% in oil production rate.

The number of published field studies on ex situ MEOR applications is even more limited.

In an early attempt to use biopolymer for conformance control and mobility improvement, an area within the North Alma Penn Unit (USA) had been selected to evaluate the effects of injecting a biopolymer at various concentrations into the Sims Sands reservoirs beginning during the early stages of waterflooding [226]. The polymer "slug" was started after 0.31 PV of the reservoir had been filled with brine. In total approximately 0.39 PV of the biopolymer solution was injected at a polymer concentration of 304 ppm and this caused no apparent damage to the reservoirs. Five of nine producing wells responded to water injection, and three of these showed a reduction in water production.

A biopolymer pilot project had been performed in the Eddesse-Nord sandstone reservoir (Germany) with a depth of 350 m, a temperature of 22°C and a thickness of 5 m [227]. A XG solution with a concentration of 800 ppm was injected from 1985 until 1988 into 2 wells. The xanthan was provided as a 2% fermentation broth and no severe problems were encountered during mixing of the polymer and injection of the polymer solution. There was no plugging tendency in the observed wells, which is attributed to a thorough laboratory quality testing prior to the field application. A response in the production wells occurred earlier than predicted by reservoir simulation and the incremental oil production was somewhat higher than the predicted 6% of the OOIP. The xanthan produced in one well after a residence time in the reservoir of about three years, showed no degradation, neither chemical nor bacterial.

A successful biosurfactant application is reported for Daqing Field in China with a biosurfactant RH (rhamnolipid fermentation liquor) which contains 20%–25% rhamnolipid [228]. They studied the synergism of RH, and mixture of biosurfactant and alkylbenzene sulfonate (ORS, an anionic chemical surfactant) systems at different ratios and evaluated the effect of RH as sacrificial agent on reducing the loss of surfactant. When the ratio of RH/ORS is 2:1 and the total concentration of the two surfactants is over 0.06 wt.% the IFT value can reach the order of 1.10^{-3} mN/m. A recent paper studied the feasibility of using biosurfactant surfactin–fermentation liquor (SF) as a complementary agent of alkaline–surfactant–polymer (ASP) flooding [229]. The results show that the activity of surfactin is tolerant toward high salinity (up to 12 wt.%), high temperature (50°C–100°C) and high Ca^{2+}/Mg^{2+} concentrations. The results show interfacial characteristics of surfactin and significant synergic effects between surfactin and ORS. In an ASP-flooding pilot test, the application of surfactin could reduce the amount of ORS by 50%, resulting in the cutting the total cost of the ASP system by 30%.

8.5.2 Surface engineering

A consensus on MEOR is that this is an easy to apply and cost-effective method, especially benefitting mature waterflooded marginal fields greatly. However, this does not hold true for ex situ MEOR, because the application of biopolymers and biosurfactants at industrial scale requires complex surface facilities similar to their chemical analogs. In situ MEOR, on the other hand, can be described as easy to apply and cost-effective method because the facilities to inject the nutrients in an already waterflooded field are easily compatible with existing injection facilities; there is no need for expensive filtration, pumping and mixing units and the operational costs remain low.

Knapp et al. reported a preliminary design of the surface facilities used in an MEOR field pilot [22]. Also, they provide the mixing and injection protocol/procedure. Tanks used in fracking operations were reused to mix the nutrient slug and hold it prior to injection. Oil trucks transported molasses to the injection site and provided a parallel heating/temperature control for a lower working viscosity. A triplex pump was used to mix the nutrient-brine injection slugs and to inject this mixture and subsequent brine slugs into the formation. A filter system reduced the particulate matter in the injection fluid. Filter and recirculation pumps were used to move the liquids between tanks and recirculate (roll) the fluids in the tanks to ensure mixing. Similar surface set-ups used in MEOR field applications are described in various studies; tanks with various volumes are used for storing the nutrients, which are mixed with the injection water and mixture nutrient and injection water is injected into the reservoir a procedure comparable to waterflooding [230–232].

In a recent paper a skid to mix the nutrient and injection water in an offshore platform has been described [233]. It is mentioned that, if necessary, each injection line can be split to numerous injection points using orifice chokes. Alternatively, the two points can be combined to increase total nutrient injection into a single point. The nutrient is stored in intermediate bulk containers or other available tanks.

What can be seen from the literature survey is that the applied surface facilities are straight forward to install and operate in comparison to ex situ MEOR as well as other EOR methods in which cost extensive separation and pumping units (e.g., CO_2 injection) or large generation units (e.g., thermal methods) or fine filtration and mixing units (chemical EOR) are required [234]. In general, a nutrient or a nutrient-microorganism solution is mixed with injection water at a predefined mixing ratio and injected to the reservoir by using the existing injection pump.

Inspired by the basic concepts and designs from existing publications, Alkan et al. [235] recently described the surface facilities for state-of-the-art in situ MEOR pilot applications (single and multiwell tests) (Fig. 8.20).

In in situ MEOR, the nutrient solution to be injected into the reservoir can be prepared in various ways. The raw material (glucose, sucrose, molasses, etc.) can be procured and a concentrate can be prepared directly on site to be mixed with injection water. The raw material can be delivered to the field in liquid or solid (powder) form and prepared by the operator using a proper mixer. Nevertheless, this way needs a carefully designed surface set-up and dedicated personnel to maintain the set-up and the mixing operation. Service companies can also be commissioned to prepare premixed liquid nutrient concentrates, based on a recipe provided and developed by the client. However, such externally prepared nutrient concentrates should also be tested by the client under similar laboratory conditions as the original laboratory-scale formulation to ensure consistency and effectiveness of the upscaled formulation. For instance, changing the manufacturer of specific raw materials, such as sucrose or yeast extract, during the upscaling can influence the microbial growth. The concentrate is generally transported with tank trucks to the injector well location as shown with #1 in Fig. 8.20. The head space

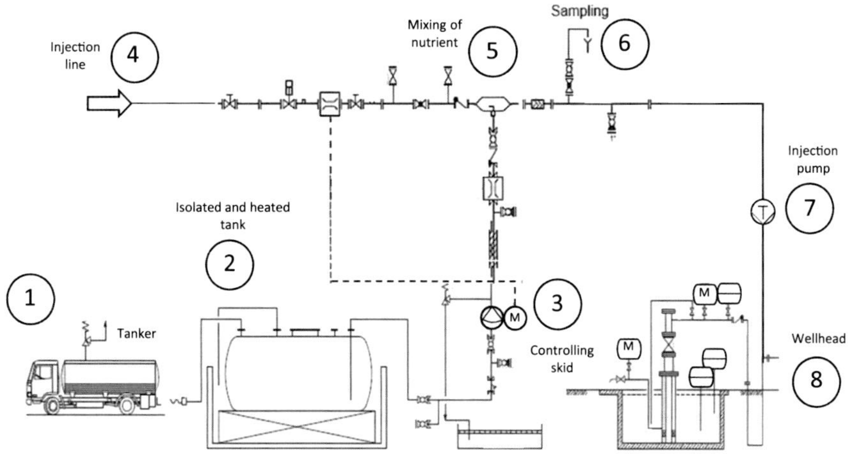

FIGURE 8.20

Schematic representation of the microbial enhanced oil recovery facilities designed for field operations.

Reproduced with modifications from H. Alkan, S. Mukherjee, W. Jelinek, et al., Front-End Engineering Practice of In Situ MEOR Applications, Petroleum Science and Engineering. Article in Press. (2022).

of the tank trucks is filled with N_2 to minimize O_2 exposure to the concentrate. Tank trucks with isolation and heating equipment should be preferred in colder regions/times to keep the concentrate at a constant temperature assuring continuous flow to the tanks in the injector well location, which are also isolated and heated (#2 in Fig. 8.20). An engineered and manufactured skid for field operations is needed for regulating and monitoring the mixing of the nutrient concentrate with injection water (#3 in Fig. 8.20); a flow meter measures the water flow coming from the injection line (#4 in Fig. 8.20) continuously and based on this flow rate the concentrate is pumped by a dosage pump (membrane pump) installed in the skid automatically into the injection line. This equipment and the necessary electronics can be set in the skid, making the complete set-up transportable and applicable in all similar injection set-ups. To assure a homogeneous distribution and delivery of the concentrate in the flowline, an orifice can be placed directly in the injection pipeline in counter current direction (#5 in Fig. 8.20). During the field application, injection samples can be taken via a sampling port (#6 in Fig. 8.20) at the wellhead to confirm the nutrient́s uniformity over time and to check that the injected concentrations remain constant and in a range to achieve the laboratory-confirmed EOR effects. The selection of filtering system is depending on the quality of the injection water in terms of solid particle content; mostly a standard 10-μm filter is used just before the injection pump (#7 and #8 in Fig. 8.20). At the production side, the wellhead is connected to a set-up with a sampling port to monitor the improvement in the oil production as well as indicators for operational parameters, microbiological performance and risks including souring.

8.5.3 Monitoring

Proper and timely monitoring before, during and after any EOR operation is necessary to ensure the success of the operation by mitigating risks, which are the main determinants for the way forward. The monitoring before the operation aims at establishing a baseline, which facilitates the identification of changes due to the EOR operation and the evaluation of its outcome by comparing monitoring data before, during and after the EOR operation. The importance of monitoring should be the main emphasis in MEOR applications compared to other EOR methods. However, to date only few studies were published on the topic. Bryant et al. pointed out that monitoring is an important criterion for a proof of concept of MEOR field tests [41]. Knapp et al. reported monitoring works in a field pilot applied in Oklahoma aiming especially at confirming the bioplugging effect after the MEOR treatment [236]. After the MEOR project was initiated, its status was determined by monitoring MEOR bacteria population, metabolite concentration in produced water as well as oil production. Portwood proposed various monitoring techniques for evaluating the microbiological activity and increase in oil production [237]. Maure et al. summarizes the surveillance plan applied in an MEOR pilot test in Piedras Coloradas field in Argentina [238]. Ohno et al. listed the monitoring activities performed in an MEOR application in addition to standard surveillance in the field [231]. Sabut et al. presented the results of an MEOR field performance based on a pre- and postmonitoring program. The program emphasizes the oil production quantification and in situ mechanisms [136]. Jackson et al. also presented monitoring tools and methods they used as indicators of the effectiveness of the MEOR well treatment [239], and they emphasize the importance and ability of gathering meaningful field data before and after any application to correctly evaluate the results and plan the next steps. Cui et al. conducted monthly sampling of the production wells for observing the response of an in situ MEOR [240]; chemical analysis, DNA extraction, and pyrosequencing of partial 16S rRNA genes as well as evaluation of community abundance by quantitative PCR were described alongside the water-cut measurements for determining the oil recovery from the producers.

Wang et al. presented the results of monitoring works to evaluate the MEOR application in Fuyu oilfield in China [241]. Ke et al. reported on the performance of a large scale MEOR application, which is evaluated with a monitoring program consisting mainly of observing the bacterial density [222].

In all field applications, a well-planned and scheduled monitoring program should be applied before, during, and after the operations comprising of the microbiological, chemical and petrophysical analyses.

Microbiological monitoring is mainly based on the experiences and all state-of-the-art methods and technologies applied in laboratory works. In addition to microbiological monitoring, chemical and petrophysical analyses are crucial for determining the success of the operation in terms of incremental oil produced and of the mitigation of risks. The chemical composition of the injected and produced brine as well as of the oil can give important evidences on the performance of MEOR. Routine surveillance techniques of the reservoir engineering, such as well tests, logs, and core analysis, may provide data supporting the microbiological and chemical data. The methods that can be applied in monitoring programs as well as in laboratory studies preceding the field operations are listed in Table 8.4.

To confirm an in situ MEOR process developed in the laboratory, the baseline monitoring is particularly important and should start ideally 3–6 months earlier than the operation. Only with a well-documented baseline, changes due to MEOR operations can be detected. During the monitoring period, sampling can start with once every two weeks and the sampling intervals can be increased to two times per week a month before the operation starts. The frequency of monitoring and the list of the analysis required can be updated during and after the operation based on the obtained results. Often, samples can be taken with a high frequency with only a fraction of them being analyzed immediately. The remaining samples can be stabilized, stored, and analyzed as required. If, for instance, any analytical response (such as the detection of a tracer in a water sample) is observed in any of the initial samples, the retained samples can be used to further define the exact timing and magnitude of such a response.

In addition to the listed analyses in Table 8.4, some additional measurements and related evaluations can also be used for required observations. For instance, stable isotope-labeled water (D_2O) can be used as natural, nonradioactive tracer especially in SWTs. It only requires determination of the background concentration of D_2O in the injection water. Thus the recovered fraction of the injected brine during the back production can be quantified, allowing detection of fluid losses due to dilution und subsurface water flow.

To obtain the maximum level of information from field pilots performed in various modes (SWT or MWT), full microbiological and chemical analysis should be performed. In the case of SWT, back production data whereas in case of MWT production data needs to be collected and monitored and the results need to be compared with the baseline and laboratory data. The potential variance of the injection water in terms of reservoir microbiology should also be investigated prior to any field operation. In an SWT-HnP reported, the cell numbers of different types of microbes generally increased (fermentative, methanogenic bacteria/archaea) as expected and was a proof of the MEOR concept developed under laboratory conditions [235]. After accounting for the dispersion in the reservoir, metabolites (gases, organic acids, alcohols) were detected in concentrations and ratios similar to laboratory experiments. The results of the formation water microbiology as well as the compositions of the metabolites are furnished as an example of monitoring activities in this SWT in Fig. 8.21. It can be seen that, after feeding, the number of target bacteria *Halanaerobium* sp. significantly increased compared to other bacteria. Moreover, the total abundance increased both in laboratory and in the reservoir at a comparable level, which is a proof of concept that the MEOR in the reservoir can work similarly to the laboratory. The comparison of the metabolite concentrations depicted in the same figure shows that the

Table 8.4 Analytical methods applied and most relevant analytes measured in microbial enhanced oil recovery applications.

Microbiology	Explanation
Isolation of DNA	Genomic DNA (gDNA) isolation from filters of ≤0.2-μm-filtered water samples; gDNA can be stored and is used for various downstream biomolecular analysis, including quantifying (qPCR) and identifying (sequencing) microbes
16S rRNA sequencing	The 16S rRNA gene can be used to identify different microorganisms in one sample; for this, genomic DNA (i.e., the complete DNA of all microorganisms in a given sample) is used as template and the corresponding region is amplified in a polymerase chain reaction (PCR); obtained amplicons can be sequenced by various means (DGGE, second-generation sequencing, such as Illumina and/or third-generation sequencing) to identify the present microbes and their relative abundance
Fluorescence microscopy (e.g., FISH/CARD-FISH; DAPI or SYTO staining)	Fluorescence microscopy can be used to quantify microorganisms; in FISH (fluorescent in situ hybridization), so-called "probes" (fluorescent molecules binding to specific DNA region) can be modified to target specific functional microbial groups; FISH can be applied to quantify active (via RNA-binding probes) or all (via DNA-binding probes) microbes present; other DNA-binding fluorescent stains to quantify all microbial cells (both living and dead) include DAPI and SYTO; additionally, stains/dyes targeting EPS-specific proteins allow biofilm detection
qPCR	Quantitative polymerase chain reaction, a PCR method for quantifying organisms based on DNA sequences; different groups of microorganisms can be targeted and quantified (e.g., sulfate reducing bacteria)
OD_{600}	The optical density at 600 nm using a photometer, a simple way of determining bacterial growth by turbidity of a sample
ATP	Adenosine triphosphate: ATP content is measured using the luciferin/luciferase system for determining bacterial growth; ATP is essential for all metabolic cellular processes and only detected in active cells due to its rapid environmental degradation

(Continued)

Table 8.4 Analytical methods applied and most relevant analytes measured in microbial enhanced oil recovery applications. *Continued*

Microbiology	Explanation
MPNs	Most probable number (MPN) is a dilution method to quantify microbes, based on the statistical random dispersion of microorganisms per volume in a sample; also used for determining bacterial growth as it targets only living cells cultivable under laboratory conditions; with the design of the employed nutrient medium, specific bacterial groups (e.g., fermenters, SRB) can be targeted
Chemical	
HPLC	High-performance liquid chromatography for determining the chemical composition, to separate, identify, and quantify each component in a mixture
μGC	Micro gas chromatography, to determine and quantify the gases in a gas mixture
DOC/DIC TOC/TIC	Dissolved organic/inorganic carbon (DOC/DIC); Total organic/inorganic carbon (TOC/TIC)
Fe(II) + Fe(III)	Detection of Fe(II) + Fe(III); concentrations in the formation water. (implications of dissolution of iron-bearing minerals)
H_2S	Using gaseous samples and μGC as standard analytical method or methods for measuring hydrogen sulfide (including test strips) for determining the souring
SO_4^{2-}	Measured as part of water analysis or spectrophotometrically; important parameter for souring-related aspects
pH	To specify the acidity or basicity of an aqueous solution
Molybdenum/nitrate	Molybdenum/nitrate concentrations if they are used as souring inhibitor
Salt content	Water chemistry analyses, might indicate dissolution processes
Redox potential	Redox potential (also known as oxidation/reduction potential), indicates which microbial biochemical reactions can be expected
Formaldehyde	One of the most common biocides used in top side facilities; concentration affects microbial growth
Ionic composition	Chemical composition of formation water
Oil composition	The components and their mol/mass composition are determined with GC/HPLC; essential of biotransformation of oil is to be assessed
Petrophysical	
Density	Density of the oil/water phase

(*Continued*)

Table 8.4 Analytical methods applied and most relevant analytes measured in microbial enhanced oil recovery applications. *Continued*

Microbiology	Explanation
Viscosity	Dynamic viscosity of the oil/water phase mostly measured with rotational double coaxial cylinder viscometer
Particle content	Particle amount and size can be determined by sieve analysis, light scattering, etc.
XRD analysis	For determining the mineral composition of the reservoir rocks
XPS	X-ray photoelectron spectroscopy, superficial (<50 to 100 nm) characterization of surfaces, including oxidation and binding states of atoms
Water cut	Water amount of the produced fluid at the wellhead, given as volume %
Oil production	Oil production, as a reciprocal of water cut. Measured at the wellhead
T, P	Temperature, pressure with standard methods of petroleum engineering,
Zeta-potential	Measure of the surface charge of a sample (solid, bacterium)
Contact angle	Measure for the wettability of a sample

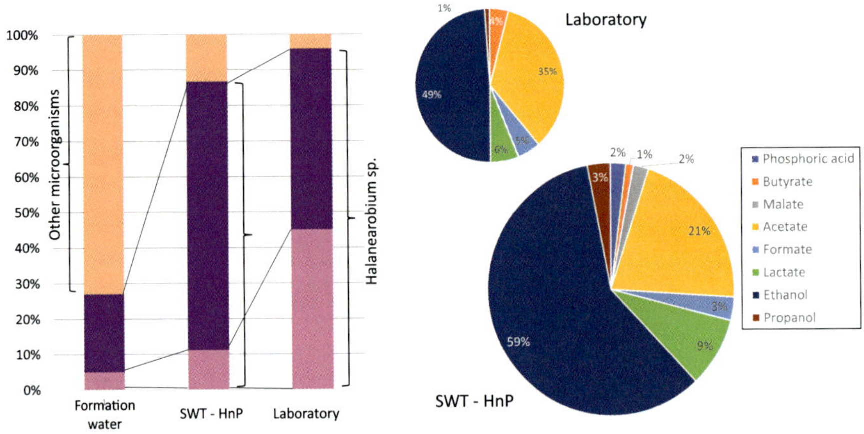

FIGURE 8.21

Results of a microbiological monitoring. Left: relative abundance of microbes in the field before (formation water) and after feeding with optimized nutrients (single well test huff-n-puff), compared with microbial enhanced oil recovery in batch experiments (laboratory); right: metabolites measured in the production fluid compared with laboratory observations (mass %).

Reproduced with modifications from H. Alkan, S. Mukherjee, W. Jelinek, et al., Front-End Engineering Practice of In Situ MEOR Applications, Petroleum Science and Engineering. Article in Press. (2022).

metabolism occurring in the reservoir is comparable to the laboratory results. Gas analysis showed that amounts of H_2S were minor and comparable to the baseline (<2 ppm). This indicated that no measurable reservoir souring occurred in the reservoir during the application of MEOR in this HnP.

8.5.4 Risk assessment and mitigation

Detailed risk assessments from beginning and efforts for mitigating the risks should be considered in the workflow of MEOR projects. For assessing the potential risks, a risk register should be created by listing the probability of the risks and their impact on schedule, scope, and cost before and after the mitigation. The risk parameters can also be classified as surface, subsurface and time risks, which is important during the economical calculation for MEOR. The assessment should be updated in each phase of the project by actualizing the risks based on the mitigation results and by adding the risks associated to upcoming phases. The impacts on the schedule, scope, and costs with quantitative impacts (such as time delay and additional costs) should be provided in a risk register with clearly defined changes on time, scope, and cost inputs on the way forward to field operations. This process of risk identification and mitigation is vital in the decision quality of MEOR projects similarly to all comparable initiatives.

An example of the risk register that is applied in an MEOR project at its operational phase is provided in Table 8.5. As can be concluded from the table, some apparently relevant items were categorized with low to very low probability; due to previously performed mitigation measures considered during the R&D and engineering phases [235]. Good examples of such cases were the potential presence of pathogenic bacteria and the uncontrolled bacterial activity in the reservoir. It was experimentally confirmed that no pathogenic bacteria existed in microcosmos of the reservoir and uncontrolled bacterial activity upon leaching of nutrient broth or cell-containing fluids was not possible. While evaluating the risks prior to the operational phase, the impact of the proof of MEOR concept developed in previous phases was weighted higher than the impacts on time and cost for the MEOR application.

Major risks classified in Table 8.5 are described in the following:

- Adsorption of the nutrient and metabolites (#1): A stronger retention of the injected nutrient and in situ generated metabolites, irrespective of the mechanism, is a high risk on the scope of the operation as it decreases the active concentration of the MEOR products. Based on related studies [99], this is expected to be a minor concern in the design of the operation however included in the R&O list for better understanding and mitigation measures with low probability as its impact can be important, an unsuccessful operation in its maximum.
- Plugging of near wellbore (#2): From the literature it can be deduced that the bioplugging especially around the wellbore can be an issue [99]. Also based

Table 8.5 Major risks determined prior to operational phase of the project.

#	Risk description	Probability	Time	Impact on scope	Cost
1	High adsorption of the injected nutrients on the reservoir rock	L	M	M	H
2	Plugging of the near wellbore by extensive growth of bacteria	L	L	H	L
3	Sterile mixing, sterile well head conditions	VL	M	H	L
4	Anaerobic mixing	VL	H	H	M
5	Dosage of biocide into injection water	H	L	H	L
6	Regulatory approval	L	M	L	M
7	Contamination of the groundwater reservoirs	VL	L	H	M
8	Microbially induced corrosion (MIC)	M	L	M	H
9	Changed thermodynamics in the reservoir	M	L	H	L
10	Insufficient growth of bacteria in the reservoir	M	M	H	M
11	Insufficient resolution by detecting the increase in production due to MEOR	H	M	H	L
12	Location and well not suitable for the pilot	H	H	M	H
13	Reservoir souring due to the potential activation of SRB	H	M	H	M

H, *high;* L, *low;* M, *medium;* MEOR, *microbial enhanced oil recovery;* SRB, *sulfate reducing bacteria;* VL: *very low.*
Source*: Reproduced with modifications from H. Alkan, S. Mukherjee, W. Jelinek, et al., Front-End Engineering Practice of In Situ MEOR Applications, Petroleum Science and Engineering. Article in Press. (2022).*

on the literature the mitigation is relatively straight forward, however its field specific nature is subject to be considered in all field operations.

- Sterile and anaerobic mixing conditions (#3, #4): Preservation of sterility and anaerobicity is not only one of the most important issues in laboratory experiments, but also for operations to ensure that samples are not microbially contaminated or exposed to O_2. However, "microbial contamination" during a field trial rather affects the shelf life and potential deterioration of surface fluids, for example, the nutrient solution, and not the reservoir itself (it is substantially impossible that small numbers of nonnative bacteria would be very active under reservoir conditions, let alone causing detrimental effects on the reservoir). Thus the risk of contamination of the injection fluid by other microbes (other than in situ) is ranked generally as very low. For instance, in the case of using nutrient concentrate consisting of sucrose or molasses, using a concentration of higher than 60 wt.% makes the solution self-sterile. As the nutrient concentrate is mixed with injection water (basically the formation water in a closed circulation system) no contamination is expected. The anaerobic character of the nutrient concentrate can be assured by filling head space of the tanks and tanker trucks with N_2 during the logistics.

- Dosage of biocide into injection water (#5): A biocide can be applied periodically in order to mitigate the fouling and souring in the processing loop. In terms of MEOR, this can result in delayed microbial growth in the reservoir due to a lower initial number of cells in the injection water. This risk can be mitigated by reducing the concentration of the biocide to a minimum-acceptable level but also by stopping its injection sufficiently before any MEOR operation starts.
- Regulatory approval (#6): As is the case for all operational projects a regulatory approval is needed for any MEOR field operation, both for pilot and full field operations. This approval is required for fulfilling the requirements on geological, operational and safety issues, which are documented in an operational plan for submission to the respective regulatory authorities. Typically, it is a standard procedure. However, for a nonstandard operation, such as MEOR case, a more comprehensive preparation might be needed. Detailed and good quality documentation informing on the process to be applied, on the field characteristics and on the operational preparations is among the mitigation measures. In addition, supplementary information might be required to be sent to the environmental, water and wastewater regulatory bodies, for a complete project approval.
- Contamination of the groundwater reservoirs (#7): This can be related to the risks associated to the regulatory approval and should be studied under laboratory conditions with a worst-case operational scenario, for example, leakage in casing or from the annulus of the injector, leading to a spill of nutrient solution into groundwater horizons, feeding the microbes in these formations. However, it is an unlikely risk that a mixture of nutrients in groundwater would lead to the growth of pathogenic bacteria.
- Microbially induced corrosion (MIC) (#8): As a well-known fact in oil industry, microbial corrosion, also called microbiologically influenced corrosion (MIC) can harm the material of top facilities or the wellbore significantly [242]. This is due to the presence and activity of microorganisms in biofilms on the surface of the corroding material. Common MIC indicator microorganisms are the sum of SRA and SRB (sulfate reducing archaea and bacteria). A standard procedure of MIC assessment is based on empirical relations between microbial data from injection and production systems and associated problems.
- Changed thermodynamics in the reservoir (#9): The potential negative effect of changed thermodynamics in the wellbore and near wellbore during the nutrient injection on the efficiency of MEOR is one of the major risks. It is known that the pressure changes do not make any significant impact on the microbial activity unless these changes happen abruptly [54].

  The effect of temperature change on the other hand could be significant knowing that the temperature is one of the major factors affecting the growth rate of the bacteria. As the injected water is normally cooler than reservoir temperatures it can cool down the reservoir starting from the near injector wellbore. This temperature drop should be accounted in designing any MEOR operations. As an example, Fig. 8.22 shows the effect of temperature on growth behavior (ATP) of an isolated in situ bacterium. As can be concluded from the figure the maximum bacterial activity occurs between the second and third days in the range of reservoir temperature (37°C–43°C); it delays as the temperature decreases. At 20°C the maximum activity is significantly lower than the higher temperatures and took place around day 14. This means the radius in which incubation takes place may increase as the cooled region increases and this fact should be considered in risk management as well as operational designs.

- Insufficient growth of bacteria in the reservoir (#10): It is a well-known fact that the laboratory experiments, such as sandpacks and corefloods, cannot completely represent the reservoir conditions. This fact is much more important in the MEOR case in which the interactions microbiology–geochemistry–petrophysics in the reservoir are excessively diverse. Therefore this risk should be mitigated by performing a sensitivity analysis and thereafter by testing and validating the growth of the bacteria within the reservoir under the reservoir conditions. This should be a success criterion of the pilot applications prior to full field operations.

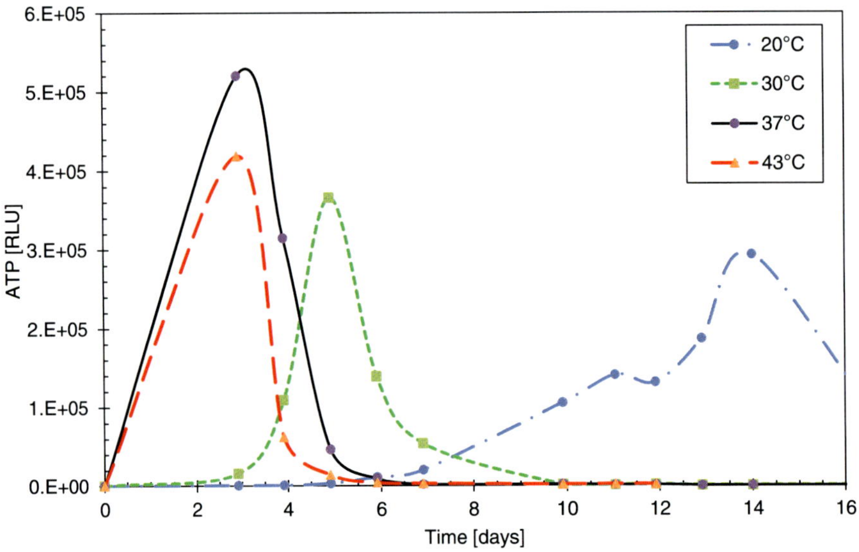

FIGURE 8.22

Bacterial activity during growth at different incubation temperatures in a field studied for a microbial enhanced oil recovery application. *ATP*, adenosine triphosphate.

- Insufficient resolution in detecting the increase in production due to MEOR (#11): As the fields are mostly mature with high WC at producers and as the oil recoveries expected are relatively low in comparison to other EOR methods the risk of missing the potential increase in oil production due to MEOR application is high. Combining this risk with natural fluctuations in water/oil productions may put a considerable impact on the scope of any field application. Mitigating this risk requires more reliable and accurate WC measurements and the increase of the monitoring frequency of WC to decrease the influence of measurement biases on the value precision. A water sampler designed for improvement in the precision and accuracy is discussed in [235]. Using this electronically controlled mobile sampler, a precision of $\pm 0.2\%$ can be achieved in WC measurements.
- Location and well not suitable for the pilot (#12): As MEOR is mostly applied in mature fields the risks related to the well/location selected are one of the major issues of the operational phase. Table 8.6 provides a list of potential issues in this risk frame.

  An unsuitable location for an EOR operation is one of the main concerns in the operational planning of any EOR project. Especially in the preliminary phases of an MEOR operation in which the process is in the test phase, a location with well-defined geological boundaries and

Table 8.6 Risk issues related to well/location not suitable for microbial enhanced oil recovery operation.

Risk issue	Aim	Mitigation measures
Field suitable for proof of concept (PoC)	To screen and select the best location for reducing the uncertainties in terms of PoC	Apply screening criteria for the best location by analyzing available data and cooperating with field management
Failing well integrity	To conduct a safe operation in short and long term to fulfill the requirements for the license to operate	Check/analyze all available/existing data on the well; test the integrity with pressure and USIT, PMIT; repair the well when needed
Uncertainties in reservoir geology/petrophysics	To know the reservoir better for a safe and aim-oriented operation; to estimate the connectivity of the injectors and producers	Check/analyze all available/existing data on the reservoir; perform and analyze the well tests; perform and analyze the tracer tests
Reservoir model not representative	To model the MEOR process; evaluate and predict the mechanisms and souring mitigation, the economics of the operation and the development scenarios	Improve/update existing reservoir model with new data; calibrate the MEOR model with laboratory and field data

MEOR, *microbial enhanced oil recovery;* PMIT, *Platform Multifinger Imaging Tool;* USIT, *Ultra Sonic Imager Tool.*

production history is a suitable primary target. The following three main criteria below can be applied to select a location for any multiwell pilot EOR operation in a mature field and are not exclusive to MEOR:

- Low shale content (the effect of clay content on the microbial activity is not clearly defined in the literature)
- Relatively good defined geological structure (a requirement for all similar projects)
- High enough remaining oil after waterflood to be mobilized with favorable economics.

In addition to these criteria, the distance between injectors and producers is one of the critical issues in MEOR, but not necessarily a showstopper. The distance should be short enough to receive the reservoir response as soon as possible, which is a relevant criterion for planning the next field applications. However, as MEOR requires an incubation time in the given reservoir volume, the distance between injector and producer should be sufficiently long to allow complete metabolization of the nutrients in a reasonable radius before the producers. A location with a sufficiently large area to install and operate the surface facilities is also a facilitating factor in deciding on the field operations. To evaluate, design and predict the MEOR operation and reduce the uncertainties, all available data and information on the operational locations as well as on the full field needs to be collected and analyzed. The monitoring/surveillance data for establishing a baseline for the upcoming operations are also to be included into the reservoir analyses.

In addition, uncertainties involving well integrities and factors, such as injectivity, pressure communication, and flow boundaries, w.r.t. the MEOR location needs to be mitigated. Mature, waterflooded fields are often operated with old wells. Therefore a proper well integrity check before any EOR operation is mandatory. Usually, pressure tests are performed in the casing and tubing to check for leakages. Imaging tools, such as USIT (Ultra Sonic Imager Tool) or PMIT (Platform Multifinger Imaging Tool), are primarily used for checking additional integrity aspects of mature wells. Casing aspects, such as casing corrosion damage, mechanical damage, cement evaluations, and casing thickness analysis, for determination of collapse and burst pressures prove to be extremely helpful for checking integrity of mature wells. Pressure communication between wells in the test area can be determined by carrying out interwell pulse tests, whereas analyses of multirate and fall-off tests can be used to determine important parameters, such as injectivity index, transmissibility, skin factor, and reservoir flow boundaries. Such measures and desired information help to reduce the uncertainties of the petrophysical characteristics of the selected location. Tracer tests are especially appropriate if there are doubts on the hydrodynamic connectivity of the injectors and producers. The results from the tracer tests can be interpreted to reduce the modeling uncertainties by providing further information on in situ heterogeneities (channels, fractures, preferred paths, etc.) and geological features.

One of the critical issues in the application of ex situ or in situ MEOR is the consideration and mitigation of the risk of microbial souring (#13 in Table 8.5) in the reservoir due to MEOR application.

8.5.5 Reservoir souring due to the activation of sulfate-reducing bacteria

The reservoir souring due to the activation of sulfate-reducing bacteria (SRB) remains always one of the major technical risks in MEOR applications [243,244]. If sulfate is present, injection of water or mixtures of water with various organic materials leads to an increasing potential of reservoir (bio)souring. In the case of injection of nutrients in MEOR, the metabolites generated during the in situ activity of microorganisms, such as hydrogen or organic acids, may serve as potential SRB substrates. SRB are very common in many environments and can be active as long as the temperature is in the biological relevant range (maximum ~80°C). Even if a reservoir has no documented history of neither elevated sulfate levels nor souring, souring can occur once sulfate is introduced. Acetate (CH_3COO^-), for instance, is a common substrate for SRB and a potential metabolite produced during MEOR and the reaction of H_2S generation in the presence of SO_4^{2-} can be expressed as follows:

$$CH_3COO^- + SO_4^{2-} + 3H^+ \rightarrow 2CO_2 + H_2S + 2H_2O \tag{8.20}$$

The H_2S produced due to this microbial interplay should be considered in designing all MEOR studies.

Prior to MEOR operations, an assessment of the souring risk should be incorporated in laboratory analysis of the injection and formation water together with microbial growth to specify the following important aspects:

- The chemical and microbiological composition of the injection and formation water: The sulfate content of the injection and formation water is the critical parameter in terms of microbially induced souring. Presence of SRB (e.g., *Desulfovibrionales*) should be checked in field samples.
- The presence of organic acids: Organic acids are products of the fermentation in MEOR processes. Acetate, lactate and formate are among the metabolites generated and the amount generated is generally high enough for complete reduction of the available sulfate by SRB. Other relevant SRB substrates may also include hydrogen and hydrocarbons.
- The thermodynamics in the reservoir: Since SRB are indigenous they can be active in a range valid for the reservoir studied. The reservoir pressure has no known detrimental effect on the SRB [245].

- pH: pH is critical for the activity of SRB and it is known that the optimum pH range for most SRB is around $6-9$. But even if the pH is reduced to a range of 5–6 by organic acids during MEOR, SRB are most likely still active.
- Permeability: It is known that SRB can differ in shape and size with an average length of 1 μm, similar to most other bacteria. The permeability or better formulated the average pore size of the reservoir studied ($k_{avr} > 100$ mD, pore diameter >10 μm) should be high enough for the free movement and transport of the SRB throughout the reservoir.

If the souring is defined as a major risk based on this above assessment, efforts should be initiated to quantify and mitigate it. The magnitude of microbial induced souring can be theoretically calculated with some basic but valid assumptions. As the organic acids during MEOR-related fermentation processes are present in excess, the resulting H_2S in the reservoir only depends on the amount of sulfate. Another conservative assumption in this calculation is that the SRB are fully active to be able to completely metabolize all available sulfate. Hence, the theoretically calculated amount is based on the stoichiometry with a given SO_4^{2-} concentration. However, with practical considerations during the field implementation, the H_2S produced is expected to be much lower due to several natural mitigation mechanisms of reservoir souring:

- Partitioning of H_2S in oil and water phases
- Retention of sulfate and H_2S reacting with reservoir rock/water
- Limitation of H_2S with injected mitigation chemicals

If any H_2S is produced in the reservoir, this can exist in dissolved form in the formation water. On the other hand, the reservoir pressure is generally higher than the bubble point and much higher than the partial pressure of generated H_2S. Hence, it can be assumed that no H_2S will be present in the gas phase under reservoir conditions. Except at high pH, H_2S is generally considerably more soluble in organic compounds than in aqueous solutions. Its solubility in oil is increasing with increasing concentrations of aromatic compounds and decreasing with increased concentrations of paraffinic or polar compounds in oil. Therefore as implied and quantified in many studies, which considered partitioning coefficients (K_{ow}), most of the H_2S generated is found as dissolved form in the oil phase. The definition of K_{ow} is not straight forward, but the literature data is around 30 at P and T specific to most oil reservoirs; however, this value is also depending on the oil composition and water salinity [246,247]. When the formation water is highly saline and the oil is viscous in nature, it can be assumed that the K_{ow} would not be lower than 30, but in fact even higher. This implies that potentially generated H_2S is dissolved in oil if it is not in gas phase.

Second, the H_2S retention in the reservoir is thought to be significant due to precipitation reactions with dissolved iron Fe(II) and iron-rich minerals, such as siderite ($FeCO_3$). If, for instance, siderite is present in reservoir rock this would react with H_2S to form pyrrhotite (FeS) or other amorphous sulfidic minerals:

$$FeCO_3 + H_2S \rightarrow H_2O + CO_2 + FeS\downarrow \tag{8.21}$$

A simplistic approach gives an idea on the magnitude of the retention of H_2S due to above given precipitation reaction with following realistic assumptions: an initial sulfate content of 0.5 mM in formation with a reservoir volume of 1 m^3; with a porosity of 0.25, the pore volume is 0.25 m^3. The H_2S generated by full consumption of aqueous SO_4^{2-} contained in brine filling this pore volume fully (with $S_w = 1.0$) is approximately 0.5 mM (or 0.125 mol per m^3 of reservoir volume. In accordance to above given chemical reaction this reacts with the same molar amount of siderite to precipitate as FeS. For a conservative arbitrary case, the siderite content of one m^3 reservoir rock might be 1% or approximately 20 kg (172.6 mol) given a rock density of 2,650 kg/m^3. If only a small portion (1%, 1.72 mol) of the siderite is available to react with H_2S according to Eq. (8.21), there is still more than enough reactive siderite in excess to neutralize the resulting 0.5 mM H_2S. This simple approach coupled with strongly conservative assumptions shows the relevance of retention mechanisms to mitigate the H_2S. In addition to the above given retention mechanisms, the divalent cation content of the formation water causes the sulfate to precipitate as Ca^{2+}, Ba^{2+}, or Sr^{2+} sulfates (e.g., $CaSO_4$, $BaSO_4$, and $SrSO_4$). Parallel and in combination with the experimental works numerical modeling can be used as a tool to estimate the magnitude of souring due to MEOR applications. With proper modeling tools and experimental data (if any) the partitioning of H_2S in water and oil phases as well as its retention in the reservoir with various precipitation mechanisms can be modeled [246–249]. This step can reduce the uncertainties on souring greatly and provides strong background information for additional mitigation measures.

Parallel to the works on determining the natural mitigation mechanisms and their magnitude, chemical mitigation methods must be conducted to mitigate the remaining souring risk, if any. In related literature, specific attention is given to the concept of nitrate injection (mostly as calcium-nitrate), which is claimed to be a cheap and environmentally friendly alternative to biocide injections. There are already some examples available for the use of nitrate in oil fields reporting mixed positive effects on corrosion and souring [250,251]. It is known however that the formation and subsequent accumulation of nitrite can be challenging due to its reactivity with many metals.

It is reported that when nitrate reducers are present and active, H_2S is effectively inhibited [252]. The duration of inhibition is dependent on the added amount of nitrate, but as soon as nitrate is depleted, and electron donors are still present in the medium, sulfate reduction continues. This is especially problematic given that some bacteria can reduce both nitrate and sulfate: Hence, bacteria leading to souring can be stimulated by nitrate injection. Consequently, a permanent addition of nitrate into the injection water would be needed for H_2S inhibition. It is also reported that during nitrate addition corrosion rates increased due to microbial influenced corrosion caused by nitrate-reducing activity with formation of nitrite or elemental sulfur [253,254]. Another side effect during nitrate reduction is the reported formation of microbially induced mineral precipitations. Two different types of minerals were observed: an amorphous iron mineral phase, which was assumed to be ferrihydrite or a similar Fe(III) hydroxide [253,254]. Furthermore, in the literature it is described that some nitrate-reducing bacteria (NRB) can couple denitrification to Fe(II) oxidation to gain energy, which would also lead to mineral precipitations [255]. Concurrent to iron reduction, calcite minerals are formed, which is known as microbially induced calcite precipitation [256]. In addition to the inhibition effect, nitrate relies on the presence and activity of NRB. It is possible that the applied injection water does not always contain active NRB, which may be another limiting factor for using nitrate [251].

Kögler et al. mentioned that molybdate inhibited the activity of sulfate-reducing microorganisms (SRMs) very efficiently, while still enabling MEOR due to its specific inhibition mechanism [257]. As shown in Fig. 8.23, in batch experiments, 0.5 mM molybdate inhibited SRM growth and preflushing of molybdate delayed microbial sulfate reduction under dynamic conditions more efficiently than using 100 mM nitrate. However, significant molybdate adsorption onto reservoir rock occurred and maximum Langmuir saturation was estimated to be ≥ 34 μmol/g of rock. Residual oil allowed a further propagation of molybdate in

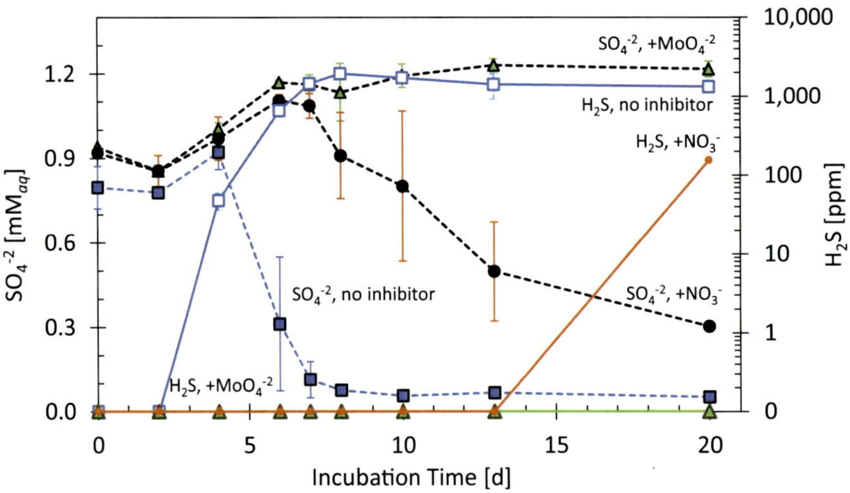

FIGURE 8.23

Effect of nitrate ($+NO_3^-$, 100 mM) and molybdate ($+MoO_4^{2-}$, 0.5 mM) on the H_2S generation for the field studied. The experiments are performed as batch. H_2S concentrations are measured at the head space or in the liquid phase, respectively.

Reproduced with permission from F. Kögler, F.S.F. Hartmann, D. Schulze-Makuch, et al., Inhibition of microbial souring with molybdate and its application under reservoir conditions, International Biodeterioration & Biodegradation 157 (2021) 105158.

sandpack experiments, but a pH < 6 limited the efficiency of molybdate due to its rapid adsorption. Under favorable souring conditions, they also observed the localized formation of macroscopic iron sulfide precipitations in sandpacks. These resulted in a fourfold permeability decrease after the injection of SRM substrates for 40 days and a calculated mean sulfate reduction rate of 52 μM per hour. However, delayed sulfate reduction in molybdate-preflushed sandpacks suggests an inhibitory effect even if molybdate is partially adsorbed. Sulfate reduction was not detected when molybdate was continuously injected with MEOR nutrients into sandpacks demonstrating its inhibitory efficiency in case it is applied in early phases of field operations with a potential risk of souring.

8.6 Economics

For an MEOR application, similar to economic considerations for other EOR applications, one should consider risk factors and rate of return on investment as, among others, the most important factors required for a robust economic evaluation. While the risk entity is included in the calculation of net present value (NPV), the rate of return on investment can be calculated by setting the NPV factor to zero, and solving for the appropriate discount rate, which results in the project's internal rate of return (IRR). Thus while NPV provides the difference between the present value of cash inflow and outflow over a period, the IRR is used to estimate the profitability of investments made for the project.

For MEOR, the risk entity can be further broken down primarily into surface, subsurface and time risks as outlined in Section 8.5.4. Time risks include risk factors, such as regulatory approval, time for application, and payout time, which can be solved relatively easily. Surface risks, such as sterile mixing conditions and operational integrity of surface facilities, can be included in the economics considering some realistic probability ranges and costs and their incorporation in the economic calculations is not too difficult. However, the highest risk probability is for the subsurface, since the input to quantify the probability of a risk associated with a given EOR mechanism or process could be extremely sensitive to the economics. The greater the difference in the ranges of input, the greater will be the range of possible outcomes. Therefore great care should be taken to determine the input ranges and identify potential risk parameters having the greatest influence on the uncertainty of the outcome. As mentioned in Section 8.4.4, numerical modeling using both deterministic and probabilistic approaches are used to determine oil production for MEOR application, which serves as an input to the project economics and hence to the NPV calculations.

MEOR is generally known as a cost-effective EOR application and costs reported from field applications confirm the theory. However, due to the existing competition between the E&P operators and associated confidentiality, only simplified economical estimates have been cited in various studies. For example, Bryant et al. reported the results of two field applications in which molasse is injected and gives cost estimates up to 5 USD_{2020}/STB [230]. In a recent paper,

Cui et al. concluded a cost of approximately 11 USD_{2020}/STB based on the results of MEOR tests in Chinese fields [240].

In view of the above discussions and challenges, a simplified overview of economical estimates for MEOR is provided in this section, starting with some additional factors influencing the NPV calculations and their value in MEOR as follows:

- Oil prices: Crude oil prices are determined mainly by global supply and demand as well as speculative changes in world economics and politics. Economic growth is one of the biggest factors affecting petroleum product—and therefore crude oil—demand. Growing economies increase demand for energy in general and especially for transporting goods and materials from producers to consumers. The speculative influence is also important, depending on the global politics and international conflicts. MEOR is known to be less influenced by oil prices than any chemical EOR processes as the nutrient used for MEOR are mostly natural products, such as sugar, and therefore independent of changes in the oil prices. Moreover, higher oil prices increase the revenue of any EOR operation, thus increasing the economic feasibility. The same is valid for MEOR as well.
- Capital expenditures (CAPEX): As per definition, CAPEX are major purchases a company makes that are designed to be used over the long term. As discussed, the investment costs of in situ MEOR is minor in most cases. The major needs are tanks and dosage pumps to inject the nutrients into the injection water. If the injection-mixing facility is rented, then this could be categorized as operational expenses as well. In a mature oil field under waterflooding the injection pumps can be used for MEOR to inject water–nutrient mixture into the reservoir. The investment costs of ex situ MEOR can be significantly high mostly comparable to the CAPEX costs of other chemical methods.
- Operating expenses (OPEX): As per definition, OPEX are the day-to-day expenses a company incurs to keep its business operational including operational costs of injection and production. OPEX may have following components in MEOR:
  - *Injectant:* In in situ MEOR, the nutrient solution is the chemical mixed with injection water and injected into the reservoir. Considering the sugar (glucose, sucrose, molasse) as the main and most costly chemical of in situ MEOR, the price of the chemical is approx. 2 USD_{2020}/m^3 of the injected nutrient solution with a concentration of 5 g/L. Assuming that Xanthan has the same oil recovery yield with a concentration of 1000 ppm in injection water, its cost is about 3 USD_{2020}/m^3 in polymer solution, which is approximately the same. If Scleroglucan is used, then the prices increase to about 10 USD_{2020}/m^3. The chemicals to be used potentially for the mitigation of the reservoir/process related risks, such as mitigation of souring in the cases of ex situ and in situ and the degradation in case of ex situ MEOR obvious part of the injectant cost. The costs of such additional treatments should also be considered. Biocide application in the case of Xanthan (or any other biopolymers/biosurfactants) to reduce the

biodegradation or use of nitrate or molybdate to mitigate the souring in the reservoir in the presence of ex situ or in situ MEOR metabolites are mostly a part of the injectant composition. For instance, the use of nitrate or molybdate in the nutrient composition with an effective concentration to mitigate the potential souring increases the cost of injectant by 1.2 USD_{2020} for molybdate and 5 USD_{2020} for nitrate per m^3 solution [235].

  - *The injection and production costs* of in situ MEOR are not significantly different from the waterflooding costs. Especially in a mature field, maintenance of the existing wells is needed, which includes planned activities, such as workovers. Testing the injectors and producers before any EOR activity is crucial for the long-term safe operability and therefore the additional workover/field testing should be considered as a significant part of OPEX. Similar to conventional waterflooding no major expenses are needed on the production side, a fact differing from the chemical EOR methods.
  - *Engineering*: The microbiology and related analyses are part of the process design works and engineering of the operation. This is not the biggest part of the operational expenses; however, its magnitude is strongly depending on the level of the available technology and expertise in the company. If a well-established, experienced microbiology laboratory team is a part of the engineering workflow and the reservoir is well studied including a validated model, a period of 3–9 months of engineering activities should be included in operational costs. Since MEOR is highly interdisciplinary, additional costs might arise because external experts must be recruited, for example, for ensuring the microbiological knowhow.
  - *Testing & monitoring:* These are the activities to reduce the uncertainties, such as well tests, tracer injections and monitoring activities before, during, and after the operation including sampling and microbiological surveillance.
- Process effectiveness: This can be defined as the incremental oil production during the operational time period. In NPV calculation, this is a main part of revenues on the top of the sales of the oil produced. The incremental recovery expressed as the fraction of the original oil place depends on the process applied. In surfactant and polymer applications this can increase up to 10%–15% of OOIP, the same ratio can be theoretically possible in ex situ MEOR with well compatible biopolymers and/or biosurfactants; however, considering missing expertise and current challenges, it seems to be difficult to reach this amount. In in situ MEOR incremental oil expectations are less. As discussed in related chapters the IRF of in situ MEOR is in the range of 4%–10%.
- Taxation, discount rate, etc.: Other NPV related considerations affect the economics of the process considerably. Taxation is a country-specific item in economic considerations. These taxes can make up between 25% and 35% of the total costs, depending on the load of CAPEX and OPEX costs. In many cases, the taxation regime has a huge impact on the overall project returns. The real discount rate depends on the nominal rates and inflation and is also one of the parameters of the economic predictions.

Table 8.7 gives an overview of the in situ MEOR costs provided by several references. Most of these costs are based on simplified considerations of operational (nutrient) and investment costs (injection facility) and do not reflect the prices of full field applications. To keep it simple, the revenue is compared against the simplified costs in USD_{2020}/STB in most cases. This list also confirms the low-cost character of MEOR within a range of 2 − 11 USD_{2020}/STB. However as concluded from the oversimplifications of the calculations these values are worth to be further discussed. The most recent data given in the table is slightly higher than many other examples [235]. It is also noted that to have a fair comparison among the different MEOR case studies, simplistic economical estimates consisting of the nutrient costs as the only OPEX and the injection facility as the only CAPEX are made in this example as well.

Most of the in situ MEOR applications described in these studies are reporting the use of molasses as the main nutrient injected to feed the microorganisms. In the study by Robertson et al. [262], a potato base medium was even used initially to feed the bacteria because of the availability as a cheap carbon source (potato starch). Such cheap but low-quality nutrients decrease the cost of the operation but also the overall recovery efficiency; however no sufficient information is provided on the application of these substances, for example, whether any chemical as well as petrophysical interactions occurred in the reservoir that may exhibit short and long-term restrictions on oil production.

Table 8.7 Cost of microbial enhanced oil recovery application from various references adjusted with inflation (USD_{2020}/STB, CPI).

Type of application	USD_{2020}/STB	Source
In situ, single-well application	2	[258]
Injection of molasses	3.5 − 5.2	[230]
Various projects, SWT, MEOR flooding	4	[237]
Various flooding projects average	5.2	[259]
Hydrocarbon-degrading anaerobic microorganism	5 − 11	[238]
HnP treatments	12	[260]
Nutrient injection, MEOR flooding	2.3	[104]
MEOR flooding, field pilot	4	[223]
Various applications	3 − 8	[34]
Piedras Coloradas field, HnP	6.6	[261]
Papagayos reservoir, bacteria solution flooding	4.0	[261]
Various Chinese fields	8	[23]
Various Chinese fields	4	[23]
Not mentioned	11	[240]
In situ MEOR, fermentative bacteria	10 − 15	[235]

HnP, *huff-n-puff;* MEOR, *microbial enhanced oil recovery;* STB, *stock tank barrel;* SWT, *single well test.*
Source: *Modified from H. Alkan, S. Mukherjee, W. Jelinek, et al., Front-End Engineering Practice of In Situ MEOR Applications, Petroleum Science and Engineering. Article in Press. (2022).*

As mentioned above, these costs do not consider costs for the mitigation of operational and reservoir related risks. As MEOR is mostly planned for mature fields, the repair and related extended maintenance of the wells can be an issue. The chemicals to be used for the mitigation of the souring adds potential costs on operational costs. Such additional costs are mentioned in none of these applications.

Fig. 8.24A shows a comparison of the unit cost of the MEOR applications with other EOR methods as values from the available literature adjusted to the averaged inflation and available max–min limiting bars for each. As can be concluded, MEOR cost is slightly higher than waterflooding costs, but still significantly lower than other EOR costs. The costs for chemical flooding applications includes an average for both polymer and surfactant applications; however, few available references on the cost of surfactant flooding indicate considerably higher costs than polymer injection being in the range 50–70 USD_{2020}/STB. Fig. 8.24B shows the split of a field pilot (MWT) cost considering the data from literature as an example [235]. As can be seen the injectant costs including logistics is about 66% of the total costs and

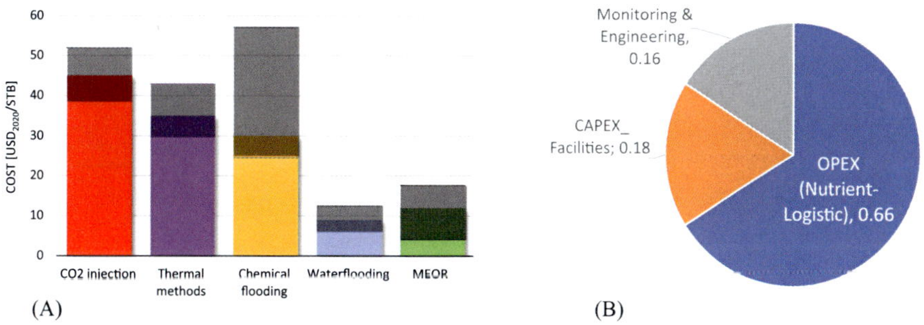

FIGURE 8.24

(A) Comparison of the microbial enhanced oil recovery costs with other enhanced oil recovery costs, Gray columns as data variance (B) Splitting of the costs in a microbial enhanced oil recovery multiple well test; a rough estimation without considering well repair and research and development costs.

(A) Averaged with data from H. She, D. Kong, Y. Li, et al., Recent advance of microbial enhanced oil recovery (MEOR) in China, Geofluids 2019 (2019) 1871392; P. Simandoux, D. Champlon, E. Valentin, Managing the cost of enhanced oil recovery, Oil & Gas Science and Technology 45 (1990) 131–139; M. Perera, R.P. Gamage, T.D. Rathnaweera, et al., A review of CO2-enhanced oil recovery with a simulated sensitivity analysis, Energies 9 (2016) 481; S. Roussanaly, A. Grimstad, The economic value of CO2 for EOR applications, Energy Procedia 63 (2014) 7836–7843; C. Nikolova, T. Gutierrez, Use of microorganisms in the recovery of oil from recalcitrant oil reservoirs: current state of knowledge, technological advances and future perspectives, Frontiers in Microbiology 10 (2020) 2996; E. Dean, J. French, M. Pitts, et al., Practical EOR agents—there is more to EOR than CO2, in: SPE-190424-MS, SPE EOR Conference at Oil and Gas West Asia, 26–28 March, 2018, Muscat, Oman; R. Rommerskirchen, B. Jakobs-Sauter, R. Ng, et al., A holistic approach to identify the optimum cost-performance surfactants for chemical EOR projects, in: SPE-196353-MS, SPE/IATMI Asia Pacific Oil & Gas Conference and Exhibition, 29–31 October, 2019, Bali, Indonesia; M.T. Al-Murayri, H.E. Al-Mayyan, K. Moudi, et al., Chemical EOR economic evaluation in a low oil price environment: Sabriyah Lower Burgan reservoir case study, in: SPE-190337-MS, SPE EOR Conference at Oil and Gas West Asia, 26–28 March, 2018, Muscat, Oman.

only 18% of the total cost are spent for facilities. A monitoring and engineering cost of 16% is shown separately, to highlight the cost required for workover and integrity checks for various existing surface facilities. In principle, this is a part of the OPEX as defined above, thus bringing the total OPEX cost to 82% for this specific example. Knowing that the main constituents of the nutrient are sucrose and yeast for the given example, their contribution to OPEX costs is strongly dependent on the market prices of these substances, especially on the price of the sugar. Considering the proportion of these substances, sucrose is the main driver of OPEX costs in this field test example. As discussed above, molasses were used in many applications worldwide as cost effective alternative. However, using molasses and cheaper yeast may decrease the performance of the processes considerably as the metabolite yield decreases [235].

Fig. 8.25 shows the historical course of the sugar prices adjusted with inflation rate. In the same figure the nutrient cost derived from [235] with corresponding sucrose prices and with the assumption of 5% IRF is given for arbitrarily selected dates again adjusted to inflation rate. As can be concluded as the sugar prices increases the operational costs are also increasing, sometimes even to a point beyond their economic feasibility. The economics of the project is also strongly depending on the oil prices. In the same figure the oil prices are also shown to give an idea on the feasibility of MEOR. Considering all this information it can be argued that MEOR has not been always the cheapest technology and even sometimes a technology that could not be applied economically. However, it should be noted that the above cost comparison, was calculated based on sucrose, that is, a very expensive option, as main nutrient.

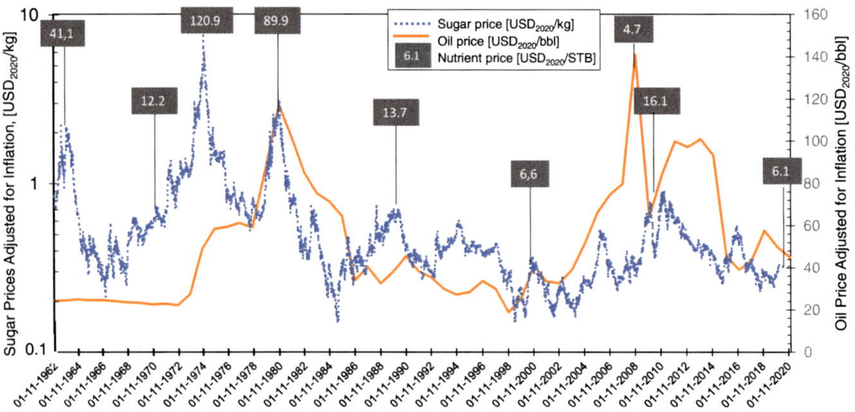

FIGURE 8.25

Historical course of sugar (blue) and oil prices (orange). Nutrient prices per incremental oil (as USD_{2020}/stock tank barrel; calculated on the basis sugar prices and amounts needed for the assumed 5% incremental recovery factor) are shown for some arbitrarily selected dates to compare with current day (2020) costs (6.1 USD_{2020}/stock tank barrel).

Operations in mature fields under waterflooding have obvious disadvantages as well. Based on MEOR operational experience the first disadvantage is the mostly not updated geology of the field, which is especially important when the field is highly heterogeneous and faulted. As the production–injection plans are managed based on a preliminary numerical assessment, new and updated data is often missing, and this makes the planning and execution of the operation difficult. The other issue is the old facilities, such as wells and pumps. The integrity of the wells assigned for injection and production must be checked. In a known mature field case, for instance, an integrity confirmation performed by USIT failed reflecting the need for workover of all related wells before the start of operations. Such unexpected operations, the costs of workovers as well as additional tests for reducing the uncertainties of the application can make a significant part of the operational costs. This fact should obviously be considered in the economic considerations; however, such issues are valid in all EOR projects, in general.

The other issue on the costs are the R&D works, which is not included in the above provided overall cost calculations. Considerable time and budget are needed for the development of the expertise and methodology for in situ MEOR. This expertise and methodology can be applied for the front-end engineering of follow-up MEOR operations. Still, an MEOR start-up in a mature field requires 3–9 months of time in a microbiology-reservoir engineering laboratory with full capability including static and dynamic compatibility tests.

8.7 Concluding remarks

Although oil recovery performance of MEOR is discussed in several case studies, a general conclusion on the effectiveness of the "EOR mechanisms" of MEOR cannot be drawn easily. This is primarily due to the fact that the available data from field applications vary extensively depending on geology, microbiology and operational aspects and is therefore difficult to quantity. Table 8.8 attempts to summarize and, more importantly to quantify the oil recovery performance (IRF as % of OOIP) of standalone MEOR mechanisms based on related expertise, laboratory data, experimentally calibrated numerical works, and the available field results. The MEOR mechanisms are evaluated for ex situ and in situ MEOR separately, as applicable.

Ex situ MEOR with industrially produced biopolymers has a significant potential if applied as mobility control and/or conformance control agents. It is suggested that biopolymers exhibit a higher tolerance of salinity compared to their chemical counterparts, and additional chemicals can further improve their salt stability. However, temperatures higher than 80°C are challenging for biopolymers, such as Xanthan. Both Scleroglucan and Schizopyllan are known to be more resistant at high temperature (<120°C) and high salinity (<130,000 ppm)

Table 8.8 Evaluation of microbial enhanced oil recovery mechanisms ex situ and in situ.

MEOR mechanism	Comments/conclusion
Mobility improvement increase of water viscosity	Ex situ biopolymers (e.g., Xanthan gum and Scleroglucan) are produced using specific medium and/or genetically engineered bacteria in bioreactors and have comparable effect to that of synthetic polymers. IRF is comparable to synthetic polymers in laboratory studies, however limited field data/application as PoC. Microbial degradation is a challenge to be mitigated.
	In situ generation of exopolymeric substances (EPS) increasing the water viscosity (shear thinning rheology with viscosity in the range of 3 – 10 mPa·s at typical shear rates of 7 to 10/s valid in reservoirs away from the wellbore); low and/or negligible effects are reported for in situ MEOR; Expected standalone IRF; 0.5% < IRF < 1.5%
Mobility improvement decrease of oil viscosity due to gas generation	In situ generation and dissolution of gas (CO_2) in oil; generally due to fermentation; with economic amounts of nutrients gas generation results only in a limited decrease in oil viscosity, in the range of 1 – 3 mPa·s. Expected standalone IRF; 0.5% < IRF < 1%.
Mobility improvement decrease of oil viscosity due to biotransformation	In situ biotransformation of the oil by oil transforming microorganisms, given mostly as standalone effect. In laboratory and field studies significant changes in oil viscosity down to 40%. Reports of promising field applications with increase in oil production in HnP mode especially in China.
Increase in capillary number IFT reduction	Ex situ: industrial biosurfactants, such as Emulsan, Alasan and Viscosin, generating ultra-low IFTs (10^{-3} – 10^{-5} mN/m) that can improve the microscopic displacement efficiency at a range similar to chemical analogs; IRF up to 20% in experiments; reports of significant increase in oil recovery with moderate decreases in IFT potentially due to wettability alteration. Microbial degradation is the challenge to mitigate.
	In situ: low-to-moderate changes in IFTs with an expected standalone IRF < 1%
Increase in capillary number wettability alteration	Ex situ: biosurfactants are reported to alter the wettability and changing the IFT as well in a similar way as chemical surfactants
	In situ: strong wettability alterations are confirmed based on spontaneous imbibition and contact angle measurements. The explanations are different however the effect on oil recovery is confirmed with corefloods. Expected standalone IRF range 1%–2%.

(Continued)

Table 8.8 Evaluation of microbial enhanced oil recovery mechanisms ex situ and in situ. *Continued*

MEOR mechanism	Comments/conclusion
Conformance control via bioplugging	Ex situ: conformance control by various industrial biopolymers produced using specific medium or genetically engineered bacteria in bioreactors results in swelling and flow diversion phenomenon. Significant controllable permeability reductions observed in dynamic experiments. However, well documented field experiences are missing. In situ: various bacterial communities are reported to be successfully used in heterogeneous reservoirs with bioplugging. High permeability reductions are reported from the sandpack and coreflood experiments. In homogeneous reservoirs this can cause undesired plugging. Expected standalone IRF; 1% < IRF < 2%
Pore volume increase; porosity—permeability increase	In situ: if carbonates are present in the reservoir rock, these can be dissolved by organic acids generated as metabolites. Expected standalone IRF; 1% < IRF < 2%

IFT, *interfacial tension;* IRF, *incremental recovery factor;* MEOR, *microbial enhanced oil recovery;* PoC, *Proof of concept.*

conditions [122,125]. A crucial aspect of field applications is the biodegradation of the biopolymers. The biopolymers often serve more readily as nutrient to in situ microorganisms, possibly due to their microbial origin. On the other hand, it is reported that biopolymers are more resistant to mechanical degradation compared to synthetic polymers, due to their more rigid backbone structure of the biopolymer. The main challenge is to find a tradeoff between temperature and biodegradation, that is, balance the higher temperature tolerance of biopolymers versus the reduced biodegradation due to reduced microbial activity at high reservoir temperatures. Apart from these side effects, the performance of biopolymers can be estimated similarly to state-of-the-art polymer flooding with synthetic polymers. The same conclusion is valid for biosurfactants as well. Laboratory studies report significant IFT reduction and wettability alteration effects of biosurfactants, although the biodegradation appears to be a challenge for field applications. However, there are not well documented field cases in the literature. Industrial biosurfactants exhibiting low to ultra-low IFTs (10^{-3}–10^{-5} mN/m) can obviously increase capillary number for reducing the residual oil saturation similarly to chemical surfactants. It is mostly speculated that, alongside IFT reduction, wettability alteration is the main contributor on capillary number. In the case of successful application using biosurfactants, an IRF comparable to surfactant applications (IRFs up to 20%) can be expected.

The catalogs of microorganisms creating EOR effects are listed and actualized frequently in related publications [44,266]. Fermentative bacteria represent the main target microorganisms of in situ MEOR. The fermentative metabolism creates multiple MEOR mechanisms in the reservoir. The definition of the in situ fermenters as well as the selection of the nutrients needs a microbiology workflow, which should be interconnected with reservoir engineering to quantify the EOR mechanisms in the reservoir. The field application, in turn, needs the contribution of the operational expertise.

As can be concluded from Table 8.8, in situ biopolymer as well as gas generation yield insignificant and/or low oil recoveries; measured increase in water phase viscosities and the gas (CO_2) generations are low; however, these small changes can make a noteworthy improvement in mobility ratio.

It seems that the in situ biotransformation is the only standalone mechanisms causing a relevant increase in oil mobility by decreasing its viscosity. However, its effect is dependent on the crude oil compositions as well as sufficient dispersion and activity of oil-utilizing bacteria throughout the reservoir for an economic viable feasibility. Since anaerobic hydrocarbon biotransformation is a potentially slow and unpredictable process, the economic benefit should be tested in SWTs to a limited extent. However, field reports indicate successful application of biotransformation for reducing oil viscosity as well as for reducing the near wellbore paraffin precipitation. Especially the field experiences reported from China give promising hints on the successful applications, often by injecting exogeneous bacteria known for their oil biotransformation properties.

Another MEOR effect, which is likely minor, based on available laboratory studies and especially considering the retention in the reservoir, is the in situ biosurfactant production. During in situ fermentation, biosurfactant generation is reported to be insignificant. Therefore measured decreases in IFTs are not sufficient for mobilizing the residual oil. However, the wettability alteration is concluded to be an important mechanism of in situ MEOR, especially the change from intermediate oil to strongly water wet. But since the presence of both cells and metabolites are required, the sustainability of the mechanism throughout the reservoir is questionable. Nevertheless, the oil recovery by IFT reduction and wettability alteration is evaluated to be in the range $2\% - 3\%$.

The bioplugging effect via biofilm and biomass generation can be considerable if the reservoir is heterogeneous, that is, exhibiting high permeability "thief" zones. The in situ bioplugging can be effective to divert the flow into the lower permeability oil containing matrix. Obviously, the magnitude of this effect depends on many factors, but numerical evaluation shows that effective bioplugging can create IRFs up to 2%.

The carbonate dissolution is another potential mechanism of in situ MEOR in the reservoirs containing carbonates in the rock mineralogy. However, quantifying the effect of pore volume increase is challenging. Yet, permeability increases as well as related mobilization of rock-attached oil may result in an IRF in the range of $1 - 2\%$.

With the exception of biotransformation, the incremental oil from the in situ MEOR is due to the cumulative effects of multiple mechanisms as mentioned in many studies. The individual incremental recoveries are estimated to be small depending on various factors. Based on field experiences as documented by Alkan et al. [62,99], the main MEOR mechanisms are bioplugging, wettability alteration and carbonate dissolution. If all the mechanisms work at the same time and effectively, the in situ MEOR can increase the oil recovery up to 10% of OOIP.

In conclusion, MEOR is even more field-specific than other EOR methods. Firstly, this is because the very reservoir-specific microbiology, which depends on temperature, salinity and various other factors. Secondly, the success of the above-mentioned mechanisms strongly depends on the reservoir characteristics. Bioplugging is effective in heterogeneous reservoirs with preferential flow channels. As the wettability alteration can change the wettability from intermediate-oil wet to water wet, a reservoir with strongly water wet rock will not produce additional oil from this mechanism. To create a pore volume increase due to dissolution of carbonates, the obvious prerequisite is the presence of carbonates in the reservoir rock. Other mechanisms namely increase in water phase viscosity by generation of EPS, gas generation and its dissolution in oil to reduce its viscosity and IFT reduction to increase the capillary number and emulsification of the oil are minor and can only be accounted for additionally, if the other major mechanisms are active. Obviously, the activation of the in situ or injected microorganisms with well-selected and quantified nutrient is the paramount condition for in situ MEOR. The growth of the beneficial bacteria, the generation of the metabolites and their distribution in the reservoir with effective concentration is needed for the technical success of the process.

In addition to the EOR concerns, the success of an in situ MEOR application is also depending on various other field-related aspects, which need to be considered, including:

- Reservoir characterization: In order to design the subsurface processes with a material balance between the injected nutrients and EOR effects, a careful reservoir engineering is a prerequisite. One of the conditions for this task is the sufficient characterization of the reservoir in consideration. The petrophysical properties (porosity, permeability distributions, mineralogy, wettability, etc.) and the heterogeneity issue (DP coefficient for the anisotropy) as well as a validated geological and reservoir model are critical for the reservoir engineering design. Also, knowledge of the field history with previously conducted operations should be collected and evaluated before the operation starts. Numerical modeling is a tool that should be used to reduce the technical risks. Commercial simulators can be used with appropriate calibration; however, the validation and predictive capabilities of the models are based on significant inputs from laboratory work conducted on microbiology and EOR capacities.
- Field facilities: One of the advantages of in situ MEOR is the easy to apply characteristic. However, for a realization of the field operation, the use of existing facilities is necessary. Obviously, a mature field with well-maintained surface facilities and wells facilitates the MEOR operation and reduce its cost. If wells and pumps for waterflooding are operational without any need for major workovers and there is no substantial oxygen ingress in the operational facilities (pipelines), the only additional equipment for the in situ MEOR operation is a pump for dosing the nutrient stored in tanks to the injection water with a given concentration.

- Risk definition and mitigation: A risk assessment is a required part of an in situ MEOR project; the risks should be defined at the beginning of the project, monitored continuously, and mitigated properly. In in situ MEOR projects souring is always the main risk to be considered. The presence of sulfate in the formation water and/or its injection into the reservoir is a factor to be controlled. The souring should be assessed properly, and mitigation measures should be taken if it arises to be a critical issue.
- Economics: In situ MEOR is relatively cheap in comparison to other EOR methods. This general assumption is mostly based on the theoretical background and field applications. However, the economics should be revisited depending on the nutrient and oil prices; if the nutrient prices are high and the oil price is low economics can yield quite challenging OPEX. The situation in the field in consideration also plays a crucial role on the economics; mature fields with worn out wells and operational equipment, pumps requiring severe workovers might increase the project costs and therefore decrease the economic viability of the application.

To summarize, in situ MEOR has the actual potential to live up to the often-associated assumptions of being an easy-to-apply low-cost EOR method, whereas ex situ MEOR is similar to chemical EOR, that is, potentially more expensive. Following closing recommendations can be deducted for potential in situ MEOR considerations:

As simple as it is in its application, in situ MEOR is perhaps one of the most complicated and demanding EOR methods in terms of initial planning and screening considerations. In order to prevent economically nonfeasible or even reservoir-damaging operations, the utmost importance of a reservoir-specific field screening before a potential field application, a representative evaluation of EOR effects during the research phase and ongoing risk management during every phase of an in situ MEOR project cannot be stressed enough.

Laboratory and field experiences raise the question whether in situ MEOR and related mechanism can be sustained in the reservoir for a sufficient time. To be effective, most of the in situ MEOR mechanisms (e.g., bioplugging, wettability alteration, and biopolymer effects) require both the presence of the bacterial metabolites as well as the bacterial cells as an integrated system in the porous medium. This implies that the success of this "system" is strongly depending on its activity and advance in the reservoir. It seems that this is not straightforward considering the "straining" of bacteria in porous media as well as the dilution of the metabolites.

The laboratory as well numerical models can capture only a part of the "real" physics of EOR methods. Considering the complexity of the process, this general aspect of EOR engineering is much more relevant in MEOR. Therefore a pragmatic approach based on various inputs and experiences should be preferred in designing the field applications.

Considering the above cited challenges and the cost-effective and easy-to-apply aspects of in situ MEOR, a field pilot is only necessary to validate the risk mitigation measures if any. Pilots are run to down-scale the field-wide application to eventually assess the technical and economic viability of the project. In in situ MEOR, the economic viability is more obvious in comparison to other EOR applications. But for any meaningful number, incremental recovery from the field application is needed. Unfortunately, it is not only difficult but impossible to capture this number from any downscaled MEOR pilot. Therefore also considering the complexity and cost involved, the gain from an MEOR field pilot is not always guaranteed.

Most of the successful field applications of in situ MEOR worldwide are single-well operations in HnP mode. Considering the above discussed challenges of in situ MEOR this can be comprehensible. On the other hand, multiwell applications might often be the preferred way, because the potential for certain MEOR mechanisms is increased, for example when bioplugging is expected in a heterogeneous reservoir.

For several reasons, the oil industry is sometimes reluctant to deploy EOR projects. This hesitation particularly pertains to MEOR: Despite numerous successful R&D projects and evidence for low CAPEX and OPEX, the number of field applications is limited due to hesitations in face of potential risks and missing ad hoc success perspectives. On the other hand, an experienced approach coupled with a pragmatic and quick decision process is the key to success, as demonstrated in the recent worldwide MEOR operations.

Nomenclature

16s rRNA	16S ribosomal RNA, for gene coding
AHL	*N*-Acryl homoserine lactone
AI	Amott-Harvey wettability index, dimensionless
ASP	Alkaline–surfactant–polymer
ATP	Adenosine triphosphate (for microbial growth, activity)
bbl	Barrels
BOC	Bacterial organic carbon (mg/cm^3)
CA	Contact angle (degrees)
CAPEX	Capital expenditure
CDC	Capillary desaturation curve
DGGE	Denaturing gradient gel electrophoresis
DOC/DIC	Dissolved organic/inorganic carbon;
DOE	US Department of Energy
EC	Emulsification capacity (fraction)
EOR	Enhanced oil recovery
EPS	Exopolymeric substances, exopolysaccharides
FISH	Fluorescence in situ hybridization
GC	Gas chromatography
GOR	Gas–oil ratio
HnP	Huff-n-puff
HPLC	High-performance liquid chromatography
HPHT	High pressure, high temperature

IFT	Interfacial tension (mN/m)
IRF	Incremental recovery factor (% of OOIP)
IRR	Internal rate of return
IRS	Institute of reservoir studies (India)
K−C	Kozeny−Carman
M	Mobility, mobility ratio
MEOR	Microbial enhanced oil recovery
MIC	Microbiologically influenced corrosion
MPN	Most probable number (to quantify microbes statistically)
MWT	Multiple well test, multiple well treatment
NPV	Net present value
NRB	Nitrate reducing bacteria
OD600	Optical density at a wavelength of 600 nm (for microbial growth)
OOIP	Original oil in place
OPEX	Operational expenditures
OPD	Oscillating pendant drop (apparatus)
PCR	Polymerase chain reaction
PMIT	Platform multifinger imaging tool (log)
PoC	Proof of concept
PRF	Permeability reduction factor (fraction)
PV	Pore volume (fraction)
R&O	Risks and opportunities
R&D	Research and development
rDNA	Ribosomale deoxyribonucleic acid
RH	Rhamnolipid fermentation liquor
RLU	Relative luminescence units
RNA	Ribonucleic acid
RRF	Residual resistance factors
qPCR	Quantitative polymerase chain reaction
S	Saturation (fraction)
SEM	Scanning electron microscope
SF	Surfactin−fermentation liquor
SI	Spontaneous imbibition
SRB	Sulfate-reducing bacteria
SRM	Sulfate-reducing microorganisms
ST	Surface tension (mN/m)
STB	Stock tank barrel
SWT	Single well test, treatment
SYTO™	(DNA) stained with a fluorescent nucleic acid stain
TOC/TIC	Total organic/inorganic carbon
USD_{2000}	Cost adjusted to US Dollars in year 2020
USIT	Ultrasonic imager tool (log)
USSR	Union of Soviet Socialist Republics
WBS	Work breakdown structure
WC	Water cut
wt	Weight
XG	Xanthan gum
XPS	X-ray photoelectron spectroscopy

References

[1] E.M.A. Mokheimer, M. Hamdy, Z. Abubakar, et al., A comprehensive review of thermal enhanced oil recovery: techniques evaluation, Journal of Energy Resources Technology 141 (2019) 030801.

[2] S.H. Shafiai, A. Gohari, Conventional and electrical EOR review: the development trend of ultrasonic application in EOR, Journal of Petroleum Exploration and Production Technology 10 (2020) 2923–2945.

[3] I. Ben-Barak, The Invisible Kingdom: From the Tips of Our Fingers to the Tops of Our Trash, Inside the Curious World of Microbes, Basic Books, New York, 2009.

[4] D. Dykhuizen, Species numbers in bacteria, Proceedings of the California Academy of Sciences 56 (2005) 62–71.

[5] J. Beckman, Action of bacteria on mineral oil, Industrial and Engineering Chemistry, News Edition 4 (1926) 23–26.

[6] C. Bass, Zobell's contribution to petroleum microbiology, in: C.R. Bell, M. Brylinsky, P. Johnson-Green (Eds.), Proceedings of 8th International Symposium on Microbial Ecology, Atlantic Canada Society for Microbial Ecology, Halifax, 1999, pp. 51–56.

[7] C. Zobell, Bacteriological process for treatment of fluid-bearing earth formation, Patent US24133278, 1946.

[8] I. Lazar, I.G. Petrisor, T.F. Yen, Microbial enhanced oil recovery (MEOR), Petroleum Science and Technology 25 (2007) 1353–1366.

[9] A. Margaritis, J.E. Zajic, Mixing, mass transfer, and scale-up of polysaccharide fermentations, Biotechnology and Bioengineering 20 (1978) 939–1001.

[10] C.E. Zobell, Process of removing sulfur from petroleum hydrocarbons and apparatus, Patent US2641564, 1953.

[11] H. Yarbrough, V. Coty, Microbially enhanced oil recovery from the upper cretaceous nacatoch formation, Union County, Arkansas. Technology Transfer Branch, Bartlesville in: E.C. Donaldson, J.B. Clark (Eds.), Proceedings of 1982 International Conference on Microbial Enhancement of Oil Recovery, 1983, pp. 149–153.

[12] S.I. Kuznetsov, Geologic Activity of Microorganisms, first ed., Springer, New York, 1962.

[13] P.B. Crawford, Water technology: continual changes observed in bacterial stratification rectification, Producers Monthly 26 (1962) 10–11.

[14] T. Jack, B. Thompson, Patents employing microorganisms in oil production, in: J.E. Zajic, D.G. Cooper, T.R. Jack, et al. (Eds.), Microbial Enhanced Oil Recovery, PennWell Publishing Co., Tulsa, 1983, pp. 14–25.

[15] D. Saxman, A. Crull, Biotechnology and enhanced petroleum production, in: SPE-13146-MS, SPE Annual Technical Conference and Exhibition, 16–19 September, 1984, Houston, Texas, USA.

[16] G.E. Jenneman, R.M. Knapp, M.J. McInerney, et al., Experimental studies of in-situ microbial enhanced oil recovery, SPE Journal 24 (1984) 33–37.

[17] W. Iverson, G. Olson, Problems related to sulfate reducing bacteria in the petroleum industry, in: R.M. Atlas (Ed.), Petroleum Microbiology, Macmillan, New York, 1984, pp. 619–641.

[18] M. Javaheri, G.E. Jenneman, M.J. McInerney, et al., Anaerobic Production of a Biosurfactant by Bacillus licheniformis JF-2, Applied and Environmental Microbiology 50 (1985) 698–700.

[19] R. Bryant, J. Douglas, State-of-the-Art Review: Microbial Field Technology, NIPER-138, National Institute for Petroleum and Energy Research, Bartlesville, 1986.

[20] D. Hitzman, Petroleum microbiology and the history of its role in enhanced oil recovery, in: E.C. Donaldson, J.B. Clark (Eds.), Proceedings of International Conference on Microbial Enhancement of Oil Recovery, The National Technical Information Service, Springfield, 1982, pp. 162–218.

[21] R. Knapp, F. Civan, M. McInerney, Modeling growth and transport of microorganisms in porous formations, in: R. Vichnevetsky, P. Borne, J. Vignes (Eds.), Proceeding of 12th World Congress on Scientific Computation, IMACS, Paris, 1988, pp. 676–679.

[22] R. Knapp, M. McInerney, D. Menzie, et al., Microbial Field Pilot Study (Final Report), Oklahoma University, Norman, 1993.

[23] C.H. Gao, A. Zekri, Applications of microbial-enhanced oil recovery technology in the past decade, Energy Sources, Part A: Recovery, Utilization, and Environmental Effects 33 (2011) 972–989.

[24] K.M. Ibragimov, F. Abdullayeva, N. Guseynova, Experience of microbial enhanced oil recovery methods at Azerbaijan fields, in: SPE-177377-MS, SPE Annual Caspian Technical Conference & Exhibition, 4–6 November, 2015, Baku, Azerbaijan.

[25] D. Astuti, T. Ariadji, P. Aditiawati, et al., A comprehensive preparation study for microbial nutrient injection of microbial enhanced oil recovery: reservoir screening and laboratory analysis—case study Bentayan Field, in: SPE-186249-MS, SPE/IATMI Asia Pacific Oil & Gas Conference and Exhibition, 17–19 October, 2017, Jakarta, Indonesia.

[26] H. Guo, Y. Li, Z. Yiran, et al., Progress of microbial enhanced oil recovery in China, SPE-174697-MS, in: SPE Asia Pacific Enhanced Oil Recovery Conference, Society of Petroleum Engineers, 11–13 August, 2015, Kuala Lumpur, Malaysia.

[27] A. Najafi-Marghmaleki, S. Kord, A. Hashemi, et al., Experimental investigation of efficiency of MEOR process in a carbonate oil reservoir using Alcaligenes faecalis: impact of interfacial tension reduction and wettability alteration mechanisms, Fuel 232 (2018) 27–35.

[28] C. Gao, Experiences of microbial enhanced oil recovery in Chinese oil fields, Journal of Petroleum Science and Engineering 166 (2018) 55–62.

[29] S. Stoval, Recovery of oil from depleted sands by means of dry steam, Oil Weekly 73 (1934) 17–24.

[30] A.V. Kane, Performance review of a large-scale CO_2-WAG enhanced recovery project, SACROC Unit Kelly-Snyder field, Journal of Petroleum Technology 31 (1979) 217–231.

[31] J. Portwood, F. Hiebert, Mixed culture microbial enhanced waterflood: tertiary MEOR case study, in: SPE-24820-MS, SPE Annual Technical Conference and Exhibition, 4–7 October, 1992, Washington, DC, USA.

[32] J. Patel, S. Borgohain, M. Kumar, et al., Recent developments in microbial enhanced oil recovery, Renewable and Sustainable Energy Reviews 52 (2015) 1539–1558.

[33] R.S. Bryant, E.C. Donaldson, T.F. Yen, et al., Microbial enhanced oil recovery, in: E.C. Donaldson, G.V. Chilingarian, T.F. Yen (Eds.), Developments in Petroleum Science, Elsevier, Amsterdam, 1989, pp. 423–450.

[34] S. Maudgalya, R.M. Knapp, M. McInerney, Microbially enhanced oil recovery technologies: a review of the past, present and future, in: SPE-106978-MS, Production and Operations Symposium, 31 March 31 3 April, 2007, Oklahoma City, Oklahoma, USA.

[35] U. Saikia, R. Bharanidharan, E. Vendhan, et al., A brief review on the science, mechanism and environmental constraints of microbial enhanced oil recovery (MEOR), International Journal of ChemTech Research 5 (2013) 1205–1212.

[36] S.J. Nelson, P. Launt, Stripper well production increased with MEOR treatment, Oil & Gas Journal 89 (1991) 114–118.

[37] G. Trebbau, G. Nunez, R. Caira, et al., Microbial stimulation of Lake Maracaibo oil wells, in: SPE-56503-MS, SPE Annual Technical Conference and Exhibition, 3–6 October, 1999, Houston, Texas, USA.

[38] L.P. Streeb, F.G. Brown, MEOR-Altamon/Bluebell field project, in: SPE-24334-MS, SPE Rocky Mountain Regional Meeting, 18–21 May, 1992, Casper, Wyoming, USA.

[39] K. Raney, New microbial approach advances enhanced oil recovery, well stimulation and paraffin dispersal, The American Oil & Gas Reporter (May 2018).

[40] J.J. Taber, F.D. Martin, R.S. Seright, EOR screening criteria revisited—part 1: introduction to screening criteria and enhanced recovery field projects, SPE Reservoir Engineering 12 (1997) 189–198.

[41] R.S. Bryant, T.E. Burchfield, Review of microbial technology for improving oil recovery, SPE Reservoir Engineering 4 (1989) 151–154.

[42] R.S. Bryant, T.E. Burchfield, D.M. Dennis, et al., Microbial enhanced waterflooding: a pilot study, in: E.C. Donaldson (Ed.), Developments in Petroleum Science, Elsevier, Amsterdam, 1991, pp. 399–419.

[43] G.C. Segovia Nolasco, V.A. Huerta Quinones, G. Guitierrez, Improving MEOR performance by a selection methodology in mature oilfields, in: SPE-123072-MS, Latin American and Caribbean Petroleum Engineering Conference, 31 May – 30 June, 2009, Cartagena de Indias, Colombia.

[44] D. Thrasher, D.A. Puckett, A. Davies, et al., MEOR from lab to field, in: SPE-129701-MS, SPE Improved Oil Recovery Symposium, 24–28 April, 2010, Tulsa, Oklahoma, USA.

[45] R.L. Zahner, S.J. Tapper, B. Marcotte, et al., Lessons learned from applications of a new organic-oil-recovery method that activates resident microbes, SPE Reservoir Evaluation & Engineering 15 (2012) 688–694.

[46] M. Safdel, M.A. Anbaz, A. Daryasafar, et al., Microbial enhanced oil recovery, a critical review on worldwide implemented field trials in different countries, Renewable and Sustainable Energy Reviews 74 (2017) 159–172.

[47] R. Almehaideb, A.Y. Zekri, Laboratory investigation of parameters affecting optimization of microbial flooding in carbonate reservoirs, Petroleum Science and Technology 20 (2002) 377–392.

[48] G. Castorena-Cortés, I. Zapata-Peñasco, T. Roldán-Carrillo, et al., Evaluation of indigenous anaerobic microorganisms from Mexican carbonate reservoirs with potential MEOR application, Journal of Petroleum Science and Engineering 81 (2012) 86–93.

[49] M. Khisametdinov, D. Sokolova, A. Fedorov, et al., Development of microbial enhanced oil recovery technology for carbonate formations: stage 1, in: SPE-171263-MS, SPE Russian Oil and Gas Exploration & Production Technical Conference and Exhibition, 14–16 October, 2014, Moscow, Russia.

[50] I. Lazar, S. Dobrota, M. Stefanescu, Some considerations concerning nutrient support injected into reservoirs subjected to microbiological treatment, in: T.E. Burchfield, R.S. Bryant (Eds.), Proceedings of the Symposium on Applications of Microorganisms to Petroleum Technology, National Technical Information Service, Springfield, 1988. XIV 1–XIV 6.

[51] F. Kögler, N. Dopffel, E. Mahler, et al., Influence of surface mineralogy on the activity of Halanaerobium sp. during microbial enhanced oil recovery (MEOR), Fuel 290 (2021) 119973.
[52] I.M. Head, D.M. Jones, S.R. Larter, Biological activity in the deep subsurface and the origin of heavy oil, Nature 426 (2003) 344–352.
[53] A.Y. Halim, U.D. Fauzi, S. Siregar, et al., Microbial enhanced oil recovery: an investigation of bacteria ability to live and alter crude oil physical characteristics in high pressure condition, in: SPE-123506-MS, Asia Pacific Oil and Gas Conference & Exhibition, 4–6 August, 2009, Jakarta, Indonesia.
[54] M. Krüger, N. Dopffel, J. Sitte, et al., Sampling for MEOR: comparison of surface and subsurface sampling and its impact on field applications, Journal of Petroleum Science and Engineering 146 (2016) 1192–1201.
[55] P.M. Oger, M. Jebbar, The many ways of coping with pressure, Research in Microbiology 161 (2010) 799–809.
[56] J.T. Raleigh, D.L. Flock, A study of formation plugging with bacteria, Journal of Petroleum Technology 17 (1965) 201–206.
[57] G.M. Hornberger, A.L. Mills, J.S. Herman, Bacterial transport in porous media: evaluation of a model using laboratory observations, Water Resources Research 28 (1992) 915–923.
[58] A. Halim, A. Shapiro, A.E. Lantz, et al., Experimental study of bacterial penetration into chalk rock: mechanisms and effect on permeability, Transport in Porous Media 101 (2014) 1–15.
[59] J.B. Davis, D.M. Updegraff, Microbiology in the petroleum industry, Bacteriological Reviews 18 (1954) 215–238.
[60] H. She, D. Kong, Y. Li, et al., Recent advance of microbial enhanced oil recovery (MEOR) in China, Geofluids 2019 (2019) 1871392.
[61] A.R. Awan, R. Teigland, J. Kleppe, A survey of North Sea enhanced-oil-recovery projects initiated during the years 1975 to 2005, SPE Reservoir Evaluation & Engineering 11 (2008) 497–512.
[62] H. Alkan, N. Klueglein, E. Mahler, et al., An integrated German MEOR project update: risk management and Huff'n Puff design, in: SPE-179580-MS, SPE Improved Oil Recovery Conference, 11–13 April, 2016, Tulsa, Oklahoma, USA.
[63] J.L. Jensen, I.D. Currie, A new method for estimating the Dykstra-Parsons coefficient to characterize reservoir heterogeneity, SPE Reservoir Engineering 5 (1990) 369–374.
[64] M. Magot, B. Ollivier, B. Patel, Microbiology of petroleum reservoirs, Antonie van Leeuwenhoek 77 (2000) 103–116.
[65] G.S. Grassia, K.M. McLean, P. Glénat, et al., A systematic survey for thermophilic fermentative bacteria and archaea in high temperature petroleum reservoirs, FEMS Microbiology Ecology 21 (1996) 47–58.
[66] M. Magot, Similar bacteria in remote oil fields, Nature 379 (1996). 681-681.
[67] Q. Cui, S. Sun, Y. Luo, et al., Comparison of in-situ and ex-situ microbial enhanced oil recovery by strain Pseudomonas aeruginosa WJ-1 in laboratory sand-pack columns, Petroleum Science and Technology 35 (2017) 2044–2050.
[68] S. Dastgheib, M. Amoozegar, E. Elahi, et al., Bioemulsifier production by a Halothermophilic Bacillus strain with potential applications in microbially enhanced oil recovery, Biotechnology Letters 30 (2008) 263–270.
[69] S. Geetha, I. Banat, S. Joshi, Biosurfactants: production and potential applications in microbial enhanced oil recovery (MEOR), Biocatalysis and Agricultural, Biotechnology 14 (2018) 23–32.

[70] N. Youssef, D.R. Simpson, M.J. McInerney, et al., In-situ lipopeptide biosurfactant production by Bacillus strains correlates with improved oil recovery in two oil wells approaching their economic limit of production, International Biodeterioration & Biodegradation 81 (2013) 127–132.
[71] F. Zhao, C. Guo, Q. Cui, et al., Exopolysaccharide production by an indigenous isolate Pseudomonas stutzeri XP1 and its application potential in enhanced oil recovery, Carbohydrate Polymers 199 (2018) 375–381.
[72] K. Behlülgil, M.T. Mehmetoğlu, Bacteria for improvement of oil recovery: a laboratory study, Energy Sources 24 (2002) 413–421.
[73] Y. Xu, M. Lu, Microbially enhanced oil recovery at simulated reservoir conditions by use of engineered bacteria, Journal of Petroleum Science and Engineering 78 (2011) 233–238.
[74] P.F. Almeida, R.S. Moreira, R. Almeida, et al., Selection and application of microorganisms to improve oil recovery, Engineering in Life Sciences 4 (2004) 319–325.
[75] E.J. Gudiña, L.R. Rodrigues, J.A. Teixeira, et al., Biosurfactant producing microorganisms and its application to enhance oil recovery at lab scale, in: SPE-154598-MS, SPE EOR Conference at Oil and Gas West Asia, 16–18 April, 2012, Muscat, Oman.
[76] M. Shavandi, G. Mohebali, A. Haddadi, et al., Emulsification potential of a newly isolated biosurfactant-producing bacterium, Rhodococcus sp. strain TA6, Colloids and Surfaces B: Biointerfaces 82 (2011) 477–482.
[77] M.J. Veshareh, S. Ayatollahi, Microorganisms' effect on the wettability of carbonate oil-wet surfaces: implications for MEOR, smart water injection and reservoir souring mitigation strategies, Journal of Petroleum Exploration and Production Technology 10 (2019) 1539–1550.
[78] H.J. Smith, A.J. Zelaya, K.B. De, León, et al., Impact of hydrologic boundaries on microbial planktonic and biofilm communities in shallow terrestrial subsurface environments, FEMS Microbiology Ecology 94 (2018) 1–16.
[79] S. Goodwin, J.D. McPherson, W.R. McCombie, Coming of age: ten years of next-generation sequencing technologies, Nature Reviews Genetics 17 (2016) 333–351.
[80] J. Jo, J. Oh, C. Park, Microbial community analysis using high-throughput sequencing technology: a beginner's guide for microbiologists, Journal of Microbiology 58 (2020) 176–192.
[81] J.S. Johnson, D.J. Spakowicz, B. Hong, et al., Evaluation of 16S rRNA gene sequencing for species and strain-level microbiome analysis, Nature Communications 10 (2019) 5029.
[82] R.S. Bryant, J. Douglas, Evaluation of Microbial Systems in Porous Media for EOR, SPE Reservoir Engineering 3 (1988) 489–495.
[83] M. Shabani Afrapoli, S. Alipour, O. Torsaeter, Fundamental study of pore scale mechanisms in microbial improved oil recovery processes, Transport in Porous Media 90 (2011) 949–964.
[84] C.C. Ezeuko, A. Sen, A. Grigoryan, et al., Pore-network modeling of biofilm evolution in porous media, Biotechnology and Bioengineering 108 (2011) 2413–2423.
[85] L. Wang, Y. Tang, P. Guo, et al., Microbial functioning on crude oil in a gas-permeable single microfluidic channel, Journal of Petroleum Science and Engineering 104 (2013) 38–48.
[86] H. Khajepour, M. Mahmoodi, D. Biria, et al., Investigation of wettability alteration through relative permeability measurement during MEOR process: a micromodel study, Journal of Petroleum Science and Engineering 120 (2014) 10–17.

[87] Z. Song, H. Han, W. Zhu, Morphological variation and recovery mechanism of residual crude oil by biosurfactant from indigenous bacteria: macro- and pore-scale experimental investigations, Journal of Microbiology and Biotechnology 25 (2015) 918–929.
[88] C.L. Gaol, L. Ganzer, S. Mukherjee, et al., Investigation of clogging in porous media induced by microorganisms using a microfluidic application, Environmental Science: Water Research & Technology 7 (2021) 441–454.
[89] S.W. Taylor, P.R. Jaffé, Biofilm growth and the related changes in the physical properties of a porous medium: 3. Dispersivity and model verification, Water Resources Res. 26 (1990) 2171–2180.
[90] D.G. Brown, J.R. Stencel, P.R. Jaffé, Effects of porous media preparation on bacteria transport through laboratory columns, Water Research 36 (2002) 105–114.
[91] V. Somasundaram, L. Philip, S.M. Bhallamudi, Laboratory scale column studies on transport and biotransformation of Cr(VI) through porous media in presence of CRB, SRB and IRB, Chemical Engineering Journal 171 (2011) 572–581.
[92] R.T. Armstrong, D. Wildenschild, Microbial enhanced oil recovery in fractional-wet systems: a pore-scale investigation, Transport in Porous Media 92 (2012) 819–835.
[93] S. Rellegadla, G. Prajapat, A. Agrawal, Polymers for enhanced oil recovery: fundamentals and selection criteria, Applied Microbiology and Biotechnology 101 (2017) 4387–4402.
[94] F. Kögler, E. Mahler, N. Dopffel, et al., The microbial enhanced oil recovery (MEOR) potential of Halanaerobiales under dynamic conditions in different porous media, Journal of Petroleum Science and Engineering 196 (2021) 107578.
[95] M. Chang, F. Chung, R. Bryant, et al., Modeling and laboratory investigation of microbial transport phenomena in porous media, in: SPE-22845-MS, SPE Annual Technical Conference and Exhibition, 6–9 October, 1991, Dallas, Texas, USA.
[96] B. Rouse, F. Hiebert, L.W. Lake, Laboratory testing of a microbial enhanced oil recovery process under anaerobic conditions, in: SPE-24819-MS, SPE Annual Technical Conference and Exhibition, Society of Petroleum Engineers, 4–7 October, 1992, Washington, DC, USA.
[97] Sugihardjo, E.H. Legowo, Partomo, et al., Microbial coreflooding experiments using indigenous microbes, in: SPE-57306-MS, SPE Asia Pacific Improved Oil Recovery Conference, Society of Petroleum Engineers, 25–26 October, 1999, Kuala Lumpur, Malaysia.
[98] E. Kowalewski, I. Rueslåtten, E. Gilje, et al., Interpretation of microbial oil recovery from laboratory experiments, in: Proceedings of 13th European Symposium on Improved Oil Recovery, 2005 25–27 April, Budapest, Hungary.
[99] H. Alkan, S. Mukherjee, F. Kögler, Reservoir engineering of in-situ MEOR: impact of microbial community, Journal of Petroleum Science and Engineering 195 (2020) 107928.
[100] H. Alkan, E. Biegel, M. Kruger, et al., An integrated MEOR project; Workflow to develop a pilot in a German field, in: SPE-169151-MS, SPE Improved Oil Recovery Symposium, 12–16 April, 2014, Tulsa, Oklahoma, USA.
[101] M.M. Yakimov, M.M. Amro, M. Bock, et al., The potential of Bacillus licheniformis strains for in situ enhanced oil recovery, Journal of Petroleum Science and Engineering 18 (1997) 147–160.
[102] P. Arora, D.R. Ranade, P.K. Dhakephalkar, Development of a microbial process for the recovery of petroleum oil from depleted reservoirs at 91–96°C, Bioresource Technology 165 (2014) 274–278.

[103] S.M. Desouky, M.M. Abdel-Daim, et al., Modelling and laboratory investigation of microbial enhanced oil recovery, Journal of Petroleum Science and Engineering 15 (1996) 309–320.
[104] D.O. Hitzman, A.K. Stepp, D.M. Dennis, et al., Innovative MIOR Process Utilizing Indigenous Reservoir Constituents, National Petroleum Technology Office, Tulsa, 2003.
[105] E. Kowalewski, I. Rueslåtten, K.H. Steen, et al., Microbial improved oil recovery—bacterial induced wettability and interfacial tension effects on oil production, Journal of Petroleum Science and Engineering 52 (2006) 275–286.
[106] H. Amani, M.H. Sarrafzadeh, M. Haghighi, et al., Comparative study of biosurfactant producing bacteria in MEOR applications, Journal of Petroleum Science and Engineering 75 (2010) 209–214.
[107] X. Wei, K. Liu, D. Li, Laboratory investigation of the effect of microbial metabolite on crude oil-water interfacial tension under reservoir conditions, in: SPE-165216-MS, SPE Enhanced Oil Recovery Conference, 2–4 July, 2013, Kuala Lumpur, Malaysia.
[108] M. Karimi, M. Mahmoodi, A. Niazi, et al., Investigating wettability alteration during MEOR process, a micro/macro scale analysis, Colloids and Surfaces B: Biointerfaces 95 (2012) 129–136.
[109] Y. Al-Wahaibi, S. Joshi, S. Al-Bahry, et al., Biosurfactant production by Bacillus subtilis B30 and its application in enhancing oil recovery, Colloids and Surfaces B: Biointerfaces 114 (2014) 324–333.
[110] P. Sarafzadeh, A.Z. Hezave, M. Ravanbakhsh, et al., Enterobacter cloacae as biosurfactant producing bacterium: differentiating its effects on interfacial tension and wettability alteration Mechanisms for oil recovery during MEOR process, Colloids and Surfaces B: Biointerfaces 105 (2013) 223–229.
[111] S. Rellegadla, S. Jain, A. Agrawal, Oil reservoir simulating bioreactors: tools for understanding petroleum microbiology, Applied Microbiology and Biotechnology 104 (2020) 1035–1053.
[112] H. Kobayashi, H. Kawaguchi, K. Endo, et al., Analysis of methane production by microorganisms indigenous to a depleted oil reservoir for application in microbial enhanced oil recovery, Journal of Bioscience and Bioengineering 113 (2012) 84–87.
[113] T. Roldán-Carrillo, G. Castorena-Cortés, J. Reyes-Avila, et al., Effect of porous media types on oil recovery by indigenous microorganisms from a Mexican oil field, Journal of Chemical Technology & Biotechnology 88 (2013) 1023–1029.
[114] P. Hosseininoosheri, H. Lashgari, K. Sepehrnoori, Characterization of in-situ biosurfactant in microbial enhanced oil recovery, in: SPE-180877-MS, SPE Trinidad and Tobago Section Energy Resources Conference, Society of Petroleum Engineers, 13–15 June, 2016, Port of Spain, Trinidad and Tobago.
[115] E. Sunde, B. Lillebø, T. Torsvik, Towards a new theory for improved oil recovery from sandstone reservoirs, in: SPE-154138-MS, SPE Improved Oil Recovery Symposium, 14–18 April, 2012, Tulsa, Oklahoma, USA.
[116] L.W. Lake, R. Johns, B. Rossen, et al., Fundamentals of Enhanced Oil Recovery, Society of Petroleum Engineers, Richardson, 2014.
[117] K.J. Humphry, B. Suijkerbuijk, H.A. van der Linde, et al., Impact of wettability on residual oil saturation and capillary desaturation curves, Petrophysics 55 (2014) 313–318.
[118] I. Chatzis, N.R. Morrow, Correlation of capillary number relationships for sandstone, Society of Petroleum Engineers Journal 24 (1984) 555–562.

[119] L. Zhong, M. Oostrom, M.J. Truex, et al., Rheological behavior of Xanthan gum solution related to shear thinning fluid delivery for subsurface remediation, Journal of Hazardous Materials 244–245 (2013) 160–170.
[120] H.Y. Jang, K. Zhang, B.H. Chon, et al., Enhanced oil recovery performance and viscosity characteristics of polysaccharide xanthan gum solution, Journal of Industrial and Engineering Chemistry 21 (2015) 741–745.
[121] S. Ghoumrassi-Barr, D. Aliouche, A rheological study of xanthan polymer for enhanced oil recovery, Journal of Macromolecular Science, Part B 55 (2016) 793–809.
[122] A.L. Compere, W.L. Griffith, Scleroglucan biopolymer production, properties, and economics, in: M. Moo-young, C. Vezina, K. Singh (Eds.), Fermentation Products, Pergamon, Oxford, 1981, pp. 441–446.
[123] N.A. Castillo, A.L. Valdez, J.I. Fariña, Microbial production of scleroglucan and downstream processing, Frontiers in Microbiology 6 (2015) 4606123.
[124] R.T. Bachmann, A.C. Johnson, R.G.J. Edyvean, Biotechnology in the petroleum industry: an overview, International Biodeterioration & Biodegradation 86 (2014) 225–237.
[125] B. Leonhardt, B. Ernst, S. Reimann, et al., Field testing the polysaccharide schizophyllan: results of the first year, in: SPE-169032-MS, SPE Improved Oil Recovery Symposium, 12–16 April, 2014, Tulsa, Oklahoma, USA.
[126] D. Prasad, B. Ernst, G. Incera, et al., Field testing the polysaccharide schizophyllan—single well test design and current results, in: Proceedings of 19th European Symposium on Improved Oil Recovery, 2017, April 24–27, Stavanger, Norway.
[127] E.J. Gudiña, J.F.B. Pereira, R. Costa, et al., Biosurfactant-producing and oil-degrading Bacillus subtilis strains enhance oil recovery in laboratory sand-pack columns, Journal of Hazardous Materials 261 (2013) 106–113.
[128] D. Deng, C. Li, Q. Ju, et al., Systematic extensive laboratory studies of microbial eor mechanisms and microbial EOR application results in Changqing oilfield, in: SPE-54380-MS, SPE Asia Pacific Oil and Gas Conference and Exhibition, 20–22 April, 1999, Jakarta, Indonesia.
[129] C. Uzukwu, D. Dionisi, Biodegradation of hydrocarbons as a mechanism of microbial enhanced oil recovery, in: SPE-180137-MS, 78th EAGE Conference and Exhibition, 30 May – 2 June, 2016, Vienna, Austria.
[130] J. You, G. Wu, F. Ren, et al., Microbial community dynamics in Baolige oilfield during MEOR treatment, revealed by Illumina MiSeq sequencing, Applied Microbiology and Biotechnology 100 (2016) 1469–1478.
[131] N.D. Gray, A. Sherry, R.J. Grant, et al., The quantitative significance of Syntrophaceae and syntrophic partnerships in methanogenic degradation of crude oil alkanes, Environmental Microbiology 13 (2011) 2957–2975.
[132] S.M. Mbadinga, L.-Y. Wang, L. Zhou, et al., Microbial communities involved in anaerobic degradation of alkanes, International Biodeterioration & Biodegradation 65 (2011) 1–13.
[133] I.M. Head, D.M. Jones, W.F.M. Röling, Marine microorganisms make a meal of oil, Nature Reviews. Microbiology 4 (2006) 173–182.
[134] D.M. Jones, I.M. Head, N.D. Gray, et al., Crude-oil biodegradation via methanogenesis in subsurface petroleum reservoirs, Nature 451 (2008) 176–180.
[135] S. Larter, H. Huang, J. Adams, et al., The controls on the composition of biodegraded oils in the deep subsurface: part II—geological controls on subsurface biodegradation fluxes and constraints on reservoir-fluid property prediction, AAPG Bulletin 90 (2006) 921–938.

[136] B. Sabut, M. Salim, A. Hamid, et al., Further evaluation of microbial treatment technology for improved oil production in Bokor field, Sarawak, in: SPE-84867-MS, SPE International Improved Oil Recovery Conference in Asia Pacific, 20–21 October, 2003, Kuala Lumpur, Malaysia.

[137] Z. Hou, P. Han, J. Le, et al., The application of hydrocarbon-degrading bacteria in Daqing's low permeability, high paraffin content oilfields, in: SPE-114344-MS, SPE Symposium on Improved Oil Recovery, 20–23 April, 2008, Tulsa, Oklahoma, USA.

[138] C. Yao, G. Lei, J. Ma, et al., Laboratory experiment, modeling and field application of indigenous microbial flooding, Journal of Petroleum Science and Engineering, 90– 91 (2012) 39–47.

[139] Y. Sugai, K. Komatsu, K. Sasaki, et al., Microbial-induced oil viscosity reduction by selective degradation of long-chain alkanes, in: SPE-171850-MS, Abu Dhabi International Petroleum Exhibition & Conference, Society of Petroleum Engineers, 10–13 November, 2014, Abu Dhabi, UAE.

[140] R. Chandankere, J. Yao, M. Cai, et al., Properties and characterization of biosurfactant in crude oil biodegradation by bacterium Bacillus methylotrophicus USTBa, Fuel 122 (2014) 140–148.

[141] B. Shibulal, S.N. Al-Bahry, Y.M. Al-Wahaibi, et al., The potential of indigenous Paenibacillus ehimensis BS1 for recovering heavy crude oil by biotransformation to light fractions, PLoS One 12 (2017) e0171432.

[142] I. Lazar, A. Voicu, C. Nicolescu, et al., The use of naturally occurring selectively isolated bacteria for inhibiting paraffin deposition, Journal of Petroleum Science and Engineering 22 (1999) 161–169.

[143] S.A. Bailey, T.M. Kenney, D.R. Schneider, Microbial enhanced oil recovery: diverse successful applications of biotechnology in the oil field, in: SPE-72129-MS, SPE Asia Pacific Improved Oil Recovery Conference, 8 9 October, 2001, Kuala Lumpur, Malaysia.

[144] J.W. Pelger, Microbial enhanced oil recovery treatments and wellbore stimulation using microorganisms to control paraffin, emulsion, corrosion, and scale formation, in: E.C. Donaldson (Ed.), Developments in Petroleum Science, Elsevier, Oxford, 1991, pp. 451–466.

[145] L. Nelson, D.R. Schneider, Six years of paraffin control and enhanced oil recovery with the microbial product, Para-Bac™, in: E.T. Premuzic, A. Woodhead (Eds.), Developments in Petroleum Science, Elsevier, Oxford, 1993, pp. 355–362.

[146] L.K. Oberding, L.M. Gieg, Methanogenic paraffin biodegradation: alkylsuccinate synthase gene quantification and dicarboxylic acid production, Applied and Environmental Microbiology 84 (2018) e01773–17.

[147] B. Zahner, A. Sheehy, B. Govreau, MEOR success in Southern California, in: SPE-129742-MS, SPE Improved Oil Recovery Symposium, 24–28 April, 2010, Tulsa, Oklahoma, USA.

[148] N. Youssef, M.S. Elshahed, M.J. McInerney, Microbial processes in oil fields: culprits, problems, and opportunities, Adv. Appl. Microbiol. 66 (2009) 141–251.

[149] J. Liu, L. Ma, B. Mu, et al., The field pilot of microbial enhanced oil recovery in a high temperature petroleum reservoir, Journal of Petroleum Science and Engineering 48 (2005) 265–271.

[150] J.D. van Hamme, A. Singh, O.P. Ward, Recent advances in petroleum microbiology, Microbiology and Molecular Biology Reviews 67 (2003) 503–549.

[151] K.D. Schaller, S.L. Fox, D.F. Bruhn, et al., Characterization of surfactin from bacillus subtilis for application as an agent for enhanced oil recovery, in: M. Finkelstein,

J.D. McMillan, B.H. Davison, et al. (Eds.), Proceedings of the Twenty-Fifth Symposium on Biotechnology for Fuels and Chemicals, 2003. May 4–7, Totowa, New Jersey, USA.

[152] M. Jahanbani Veshareh, E. Ganji Azad, T. Deihimi, et al., Characterization, production optimization and wetting effect on carbonate surfaces, Journal of Petroleum Exploration and Production 9 (2019) 233–245.

[153] D. Olsen, H. Janshekar, Biosurfactant production and laboratory application tests for heavy crude oil, in: Third International UNITAR Heavy Crude & Tar Sands Conference, 22–31 July, 1985, Long Beach, California, USA.

[154] H. Al-Sulaimani, Y. Al-Wahaibi, S. Al-Bahry, et al., Experimental investigation of biosurfactants produced by bacillus species and their potential for MEOR in Omani oil field, in: SPE-129228-MS, SPE EOR Conference at Oil & Gas West Asia, 11–13 April, 2010, Muscat, Oman.

[155] M. Samir, M. Abu El Ela, S. El Marsafy, et al., New incubated Pseudomonas Aeruginosa bacteria for increasing oil recovery under reservoir conditions, in: SPE-126104-MS, North Africa Technical Conference and Exhibition, 14–17 February, 2010, Cairo, Egypt.

[156] H. Ghojavand, F. Vahabzadeh, A.K. Shahraki, Enhanced oil recovery from low permeability dolomite cores using biosurfactant produced by a Bacillus mojavensis (PTCC 1696) isolated from Masjed-I Soleyman field, Journal of Petroleum Science and Engineering 81 (2012) 24–30.

[157] A. Rabiei, M. Sharifinik, A. Niazi, et al., Coreflooding tests to investigate the effects of IFT reduction and wettability alteration on oil recovery during MEOR process in an Iranian oil reservoir, Applied Microbiology and Biotechnology 97 (2013) 5979–5991.

[158] M.S. Afrapoli, S. Alipour, O. Torsaeter, Analysis of microscopic displacement mechanisms of a MIOR process in porous media with different wettability, Transport in Porous Media 93 (2012) 705–719.

[159] F. Hajibagheri, M. Lashkarbolooki, S. Ayatollahi, et al., The synergic effects of anionic and cationic chemical surfactants, and bacterial solution on wettability alteration of carbonate rock: an experimental investigation, Colloids and Surfaces A: Physicochemical and Engineering Aspects 513 (2017) 422–429.

[160] H. Alkan, M. Szabries, N. Dopffel, et al., Investigation of spontaneous imbibition induced by wettability alteration as a recovery mechanism in microbial enhanced oil recovery, Journal of Petroleum Science and Engineering 182 (2019) 106163.

[161] M.M. Bidhendi, G. Garcia-Olvera, B. Morin, et al., Interfacial viscoelasticity of crude oil/brine: an alternative enhanced-oil-recovery mechanism in smart waterflooding, SPE Journal 23 (2018) 803–818.

[162] A. Saidi, Simulation of naturally fractured reservoirs, in: SPE-12270-MS, SPE Reservoir Simulation Symposium, 15–18 November, 1983, San Francisco, California, USA.

[163] R. Khachatoorian, I.G. Petrisor, C.C. Kwan, et al., Biopolymer plugging effect: laboratory-pressurized pumping flow studies, Journal of Petroleum Science and Engineering 38 (2003) 13–21.

[164] G. Gandler, A. Gbosi, S.L. Bryant, et al., Mechanistic understanding of microbial plugging for improved sweep efficiency, in: SPE-100048-MS, SPE/DOE Symposium on Improved Oil Recovery, 22–26 April, 2006, Tulsa, Oklahoma, USA.

[165] H.M. Torbati, R.A. Raiders, E.C. Donaldson, et al., Effect of microbial growth on pore entrance size distribution in sandstone cores, J. Ind. Microbiol. 1 (1986) 227–234.

[166] B.E. Rittmann, The significance of biofilms in porous media, Water Resources Research 29 (1993) 2195–2202.

[167] I.L. Gullapalli, J.H. Bae, K. Hejl, et al., Laboratory design and field implementation of microbial profile modification process, SPE Reservoir Evaluation & Engineering 3 (2000) 42–49.

[168] R. Al-Hattali, H. Al-Sulaimani, Y. Al-Wahaibi, et al., Improving sweep efficiency in fractured carbonate reservoirs by microbial biomass, in: SPE-154679-MS, SPE EOR Conference at Oil and Gas West Asia, 16–18 April, 2012, Muscat, Oman.

[169] A. Gaurav, Microbial profile modification MPM: a noble application of biotechnology in improving oil recovery, in: SPE-191277-STU, SPE Annual Technical Conference and Exhibition, Society of Petroleum Engineers, 9–11 October, 2017, San Antonio, Texas, USA.

[170] M.E. Davey, D. Gevertz, W.A. Wood, et al., Microbial selective plugging of sandstone through stimulation of indigenous bacteria in a hypersaline oil reservoir, Geomicrobiology Journal 15 (1998) 335–352.

[171] T.L. Stewart, H.S. Fogler, Pore-scale investigation of biomass plug development and propagation in porous media, Biotechnology and Bioengineering 77 (2002) 577–588.

[172] M. Yavuz Corapcioglu, A. Haridas, Transport and fate of microorganisms in porous media: a theoretical investigation, Journal of Hydrology 72 (1984) 149–169.

[173] P. Sivasankar, G.S. Kumar, Influence of pH on dynamics of microbial enhanced oil recovery processes using biosurfactant producing Pseudomonas putida: mathematical modelling and numerical simulation, Bioresource Technology 224 (2017) 498–508.

[174] A. Rinanti, The potential of indigenous bacteria to increase porosity-permeability of reservoir rock: a preliminary study, International Journal 12 (2017) 71–75.

[175] M. Saito, Y. Sugai, K. Sasaki, et al., Experimental and numerical studies on EOR using a biosurfactant, in: SPE-183496-MS, Abu Dhabi International Petroleum Exhibition & Conference, 7–10 November, 2016, Abu Dhabi, UAE.

[176] Q. Feng, J. Zhou, Z. Chen, et al., Study on EOR Mechanisms by microbial flooding, in: SPE-79176-MS, Annual International Conference and Exhibition, 5–7 August, 2002, Abuja, Nigeria.

[177] E.W. Al-Shalabi, Numerical modeling of biopolymer flooding in high-temperature high-salinity carbonate cores, in: OTC-28447-MS, Offshore Technology Conference Asia, 2018,20–23 March Kuala Lumpur, Malaysia.

[178] R. Knapp, M.J. McInerney, D.E. Menzie, et al., The Use of Microorganisms in Enhanced Oil Recovery, US Department of Energy, Washington, DC, 1983.

[179] D.K. Button, Kinetics of nutrient-limited transport and microbial growth, Microbiological Reviews 49 (1985) 270–297.

[180] T.P. Clement, B.S. Hooker, R.S. Skeen, Macroscopic models for predicting changes in saturated porous media properties caused by microbial growth, Groundwater 34 (1996) 934–942.

[181] S.K. Tiwari, K.L. Bowers, Modeling biofilm growth for porous media applications, Mathematical and Computer Modelling 33 (2001) 299–319.

[182] S.L. Bryant, T.P. Lockhart, Reservoir engineering analysis of microbial enhanced oil recovery, SPE Reservoir Evaluation & Engineering 5 (2002) 365–374.

[183] A.A. Sitnikov, N.A. Eremin, R.R. Ibattulin, A mathematical model of microbial enhanced oil recovery (MEOR) method for mixed type rock, in: SPE-28903-MS, European Petroleum Conference, 25–27 October, 1994, London, UK.
[184] M.R. Ghadimi, M. Ardjmand, Simulation of microbial enhanced oil recovery, in: 101734-MS, International Oil Conference and Exhibition in Mexico, 31 August – 2 September, 2006, Cancun, Mexico.
[185] S.M. Nielsen, A.A. Shapiro, M.L. Michelsen, et al., 1D simulations for microbial enhanced oil recovery with metabolite partitioning, Transport in Porous Media 85 (2010) 785–802.
[186] S.M. Nielsen, I. Nesterov, A.A. Shapiro, Microbial enhanced oil recovery—a modeling study of the potential of spore-forming bacteria, Computers & Geosciences 20 (2016) 567–580.
[187] M.R. Islam, Mathematical modeling of microbial enhanced oil recovery, in: SPE-20480-MS, SPE Annual Technical Conference and Exhibition, 23–26 September, 1990, New Orleans, Louisiana, USA.
[188] M.R. Islam, A. Gianetto, Mathematical modelling and scaling up of microbial enhanced oil recovery, Journal of Canadian Petroleum Technology 32 (1993) 30–36.
[189] M. Behesht, R. Roostaazad, F. Farhadpour, et al., Model development for MEOR process in conventional non-fractured reservoirs and investigation of physico-chemical parameter effects, Chemical Engineering & Technology 31 (2008) 953–963.
[190] E.O. Ansah, Mechanistic modeling of MEOR as a sustainable recovery technology: coreflooding validation, sensitivity and field application, in: SPE-199770-STU, SPE Annual Technical Conference and Exhibition, 30 September – 2 October, 2019, Calgary, Canada.
[191] S. Chakraborty, S.K. Govindarajan, S.N. Gummadi, Numerical modeling on the influence of effective porosity, microbial kinetics, and operational parameters on enhanced oil recovery by microbial flooding within a sandstone formation, SPE Journal 25 (2020) 2932–2961.
[192] E.O. Ansah, Y. Sugai, K. Sasaki, Modeling microbial-induced oil viscosity reduction: effect of temperature, salinity and nutrient concentration, Petroleum Science and Technology 36 (2018) 1113–1119.
[193] P. Sivasankar, G. Suresh Kumar, Modelling the influence of interaction between injection and formation brine salinities on in-situ microbial enhanced oil recovery processes by coupling of multiple-ion exchange transport model with multiphase fluid flow and multi-species reactive transport models, Journal of Petroleum Science and Engineering 163 (2018) 435–452.
[194] S.M. Nielsen, K. Jessen, A.A. Shapiro, et al., Microbial enhanced oil recovery: 3D simulation with gravity effects, in: SPE-131048-MS, SPE EUROPEC/EAGE Annual Conference & Exhibition, Society of Petroleum Engineers, 14–17 June, 2010, Barcelona, Spain.
[195] H. Büeltemeier, H. Alkan, M. Amro, A new modeling approach to MEOR calibrated by bacterial growth and metabolite curves, in: SPE-169668-MS, SPE EOR Conference at Oil and Gas West Asia, 31 March – 2 April, 2014, Muscat, Oman.
[196] P. Spirov, Y. Ivanova, S. Rudyk, Modelling of microbial enhanced oil recovery application using anaerobic gas-producing bacteria, Petroleum Science 11 (2014) 272–278.
[197] E.O. Ansah, Y. Sugai, R. Nguele, et al., Integrated microbial enhanced oil recovery (MEOR) simulation: main influencing parameters and uncertainty assessment, Journal of Petroleum Science and Engineering 171 (2018) 784–793.

[198] B.J. Suchomel, B.M. Chen, M.B. Allen, Network model of flow, transport and biofilm effects in porous media, Transport in Porous Media 30 (1998) 1–23.

[199] M. Delshad, K. Asakawa, G.A. Pope, et al., Simulations of chemical and microbial enhanced oil recovery methods, in: SPE-75237-MS, SPE/DOE Improved Oil Recovery Symposium, 13–17 April, 2002, Tulsa, Oklahoma, USA.

[200] M. Thullner, P. Baveye, Computational pore network modeling of the influence of biofilm permeability on bioclogging in porous media, Biotechnology and Bioengineering 99 (2008) 1337–1351.

[201] F. Centler, I. Fetzer, M. Thullner, Modeling population patterns of chemotactic bacteria in homogeneous porous media, Journal of Theoretical Biology 287 (2011) 82–91.

[202] E.C. Lacerda, V.I. Priimenko, A.P. Pires, Microbial EOR: a quantitative prediction of recovery factor, in: SPE-153866-MS, SPE Improved Oil Recovery Symposium, Society of Petroleum Engineers, 14–18 April, 2012, Tulsa, Oklahoma, USA.

[203] D. Landa-Marbán, F.A. Radu, J.M. Nordbotten, Modeling and simulation of microbial enhanced oil recovery including interfacial area, Transport in Porous Media 120 (2017) 395–413.

[204] Y. Cheng, C.G. Hubbard, L. Zheng, et al., Next generation modeling of microbial souring—parameterization through genomic information, International Biodeterioration & Biodegradation 126 (2018) 189–203.

[205] E. Hong, M.S. Jeong, T.H. Kim, et al., Development of coupled biokinetic and thermal model to optimize cold-water microbial enhanced oil recovery (MEOR) in homogenous reservoir, Sustainability 11 (2019) 1–19.

[206] R.W. Harvey, S.P. Garabedian, Use of colloid filtration theory in modeling movement of bacteria through a contaminated sandy aquifer, Environmental Science & Technology 25 (1991) 178–185.

[207] J. Ryan, M. Elimelech, Colloid mobilization and transport in groundwater, Colloids and Surfaces A: Physicochemical and Engineering Aspects 107 (1996) 1–56.

[208] N. Tufenkji, Modeling microbial transport in porous media: traditional approaches and recent developments, Advances in Water Resources 30 (2007) 1455–1469.

[209] G.E. Kapellos, T.S. Alexiou, A.C. Payatakes, Hierarchical simulator of biofilm growth and dynamics in granular porous materials, Advances in Water Resources 30 (2007) 1648–1667.

[210] J.-U. Kreft, G. Booth, J.W. Wimpenny, BacSim, a simulator for individual-based modelling of bacterial colony growth, Microbiology 144 (1998) 3275–3287.

[211] J.-U. Kreft, C. Picioreanu, J.W.T. Wimpenny, et al., Individual-based modelling of biofilms, Microbiology 147 (2001) 2897–2912.

[212] E. Alpkvista, I. Klapper, A multidimensional multispecies continuum model for heterogeneous biofilm development, Bulletin of Mathematical Biology 69 (2007) 765–789.

[213] M. Karrabi, P. Séchet, C. Morra, et al., Investigation of hydrodynamic/biomass growth coupling in a pilot scale granular bioreactor at low pore Reynolds number, Chemical Engineering Science 66 (2011) 1765–1782.

[214] C. Gruesbeck, R.E. Collins, Entrainment and deposition of fine particles in porous media, SPE J. 22 (1982) 847–856.

[215] M.M. Sharma, Y.C. Yortsos, Transport of particulate suspensions in porous media: model formulation, AIChE J. 33 (1987) 1636–1643.

[216] STARS, Thermal & Advanced Processes Reservoir Simulator, Computer Modelling Group Ltd., Calgary, 2014.

[217] J.C. Merchuk, J.A. Asenjo, The Monod equation and mass transfer, Biotechnology and Bioengineering 45 (1995) 91–94.

[218] K. van't Riet, R.G.J.M. van der Lans, Mixing in bioreactor vessels, in: M. Moo-Young (Ed.), Comprehensive Biotechnology, second ed., Academic Press, Burlington, 2011, pp. 63–80.

[219] W.H. Kampen, Nutritional requirements in fermentation processes, in: H.C. Vogel, C.L. Todaro (Eds.), Fermentation and Biochemical Engineering Handbook, second ed., William Andrew Publishing, Westwood, 1996, pp. 122–160.

[220] J.A. Robinson, J.M. Tiedje, Nonlinear estimation of Monod growth kinetic parameters from a single substrate depletion curve, Applied and Environmental Microbiology 45 (1983) 1453–1458.

[221] M. Kanna, Y. Matsumara, Applicability of Monod equation to growth curves of various microorganisms, Journal of the Japan Petroleum Institute 55 (2012) 236–240.

[222] C. Ke, G. Lu, Y. Li, et al., A pilot study on large-scale microbial enhanced oil recovery (MEOR) in Baolige oilfield, International Biodeterioration & Biodegradation 127 (2018) 247–253.

[223] L.A. Strappa, J.P. De Lucia, M.A. Maure, et al., A novel and successful MEOR pilot project in a strong water-drive reservoir Vizcacheras Field, Argentina, in: SPE-89456-MS, SPE/DOE Symposium on Improved Oil Recovery, 17–21 April, 2004, Tulsa, Oklahoma, USA.

[224] R. Reksidler, A.G. Torres Volpon, L.C. Barbosa, et al., A microbial enhanced oil recovery field pilot in a Brazilian onshore oilfield, in: SPE-129930-MS, SPE Improved Oil Recovery Symposium, 24–28 April, 2010, Tulsa, Oklahoma, USA.

[225] S.C. Jackson, A.W. Alsop, R. Fallon, et al., Field implementation of DuPont's microbial enhanced oil recovery technology, in: SPE-159128-MS, SPE Annual Technical Conference and Exhibition, 8–10 October, 2012, San Antonio, Texas, USA.

[226] B. Perry, Performance history on use of biopolymer in springer sand waterflood in Southern Oklahoma, in: SPE-4085-MS, Fall Meeting of the Society of Petroleum Engineers of AIME, 8–11 October, 1972, San Antonio, Texas, USA.

[227] W. Littmann, W. Kleinitz, B.E. Christensen, et al., Late results of a polymer pilot test: performance, simulation adsorption, and xanthan stability in the reservoir, in: SPE-24120-MS, SPE/DOE Enhanced Oil Recovery Symposium, 22–24 April, 1992, Tulsa, Oklahoma, USA.

[228] D. Li, S. Lu, Y. Liu, et al., The effect of biosurfactant on the interfacial tension and adsorption loss of surfactant in ASP flooding, Colloids and Surfaces A: Physicochemical and Engineering Aspects 244 (2004) 53–60.

[229] D. Wang, Y. Zhang, Y. Liu, et al., The application of surfactin biosurfactant as surfactant coupler in ASP flooding in Daqing oil field, in: SPE-119666-MS, SPE Middle East Oil and Gas Show and Conference, 15–18 March, 2009, Manama, Bahrain.

[230] R.S. Bryant, A.K. Steep, K.M. Bertus, et al., Microbial Enhanced Waterflooding Mink Unit and Phoenix Field Pilots (Final Report), National Institute for Petroleum and Energy Research, Bartlesville, 1993.

[231] K. Ohno, S. Maezumi, H.K. Sarma, et al., Implementation and performance of a microbial enhanced oil recovery field pilot in Fuyu Oilfield, China, in: SPE-54328-MS, SPE Asia Pacific Oil and Gas Conference and Exhibition, 20–22 April, 1999, Jakarta, Indonesia.

[232] C. Li, Y. Li, X. Li, et al., The application of microbial enhanced oil recovery technology in Shengli oilfield, Petroleum Science and Technology 33 (2015) 556–560.

[233] M. Samuel, K. Nimitz, B. Gaubert, C. Swider, Offshore chemical injection skid for use in biological enhanced oil recovery BEOR, in: OTC-27173-MS, Offshore Technology Conference, 2016, Houston, Texas, USA.

[234] A.Y. Zekri, K.K. Jerbi, Economic evaluation of enhanced oil recovery, Oil and Gas Science and Technology 57 (2002) 259–267.

[235] H. Alkan, S. Mukherjee, W. Jelinek, et al., Front-End Engineering Practice of In Situ MEOR Applications, Petroleum Science and Engineering. Article in Press. (2022).

[236] R.M. Knapp, M.J. McInerney, J.D. Coates, et al., Design and implementation of a microbially enhanced oil recovery field pilot, Payne County, Oklahoma, in: SPE-24818-MS, SPE Annual Technical Conference and Exhibition, 4–7 October, 1992, Washington, DC, USA.

[237] J.T. Portwood, A commercial microbial enhanced oil recovery technology: evaluation of 322 projects, in: SPE-29518-MS, SPE Production Operations Symposium, 2–4 April, 1995, Oklahoma City, Oklahoma, USA.

[238] M.A. Maure, F.L. Dietrich, V.A. Diaz, et al., Microbial enhanced oil recovery pilot test in Piedras Coloradas Field, Argentina, in: SPE-53715-MS, Latin American and Caribbean Petroleum Engineering Conference, 21–23 April, 1999, Caracas, Venezuela.

[239] S.C. Jackson, J. Fisher, A. Alsop, et al., Considerations for field implementation of microbial enhanced oil recovery, SPE-146483-MS, in: SPE Annual Technical Conference and Exhibition, 30 October – 2 November 2, 2011, Denver, Colorado, USA.

[240] Q. Cui, L. Yu, H. Dong, et al., Pilot test of indigenous microorganism flooding in a heavy oil reservoir, in: SPE-169145-MS, SPE Improved Oil Recovery Symposium, 12–16 April, 2014, Tulsa, Oklahoma, USA.

[241] J. Wang, Z. Wang, J. Tian, et al., An evaluation of microbial flooding for enhanced oil recovery, Fuyu oilfield, China, in: SPE-189001-MS, SPE Annual Caspian Technical Conference and Exhibition, 1–3 November, 2017, Baku, Azerbaijan.

[242] T.L. Skovhus, D. Enning, J.S. Lee, Microbiologically Influenced Corrosion in the Upstream Oil and Gas Industry, CRC Press, Boca Raton, 2017.

[243] A. Visser, I. Beeksma, F. van der Zee, et al., Anaerobic degradation of volatile fatty acids at different sulphate concentrations, Applied Microbiology and Biotechnology 40 (1993) 549–556.

[244] L.M. Gieg, T.R. Jack, J.M. Foght, Biological souring and mitigation in oil reservoirs, Applied Microbiology and Biotechnology 92 (2011) 263–282.

[245] B. Herbert, P. Gilber, H. Stockdsle, et al., Factors controlling the activity of sulphate-reducing bacteria in reservoirs during water injection, in: Proceedings of European Offshore Conference, 10 September, 1985, Aberdeen, UK.

[246] D.J. Ligthelm, R.B. de Boer, J.F. Brint, et al., Reservoir souring: an analytical model for H2S generation and transportation in an oil reservoir owing to bacterial activity, in: SPE-23141-MS, SPE Offshore Europe, 3–6 September, 1991, Aberdeen, UK.

[247] I. Skjælaaen, A. Ebigbo, M. Espedal, et al., A model for transport of hydrogen sulfide in oil-and water-saturated porous media, Computing and Visualization in Science 13 (2010) 265–273.

[248] D. Coombe, C. Hubert, G. Voordou, Mechanistic modelling of H2S souring treatments by application of nitrate or nitrite, in: PETSOC-2004-292, Canadian International Petroleum Conference, 8–10 June, 2004, Calgary, Canada.

[249] H. Alkan, G. Namazova, S. Hatscher, et al., Modelling approaches to assess biogenic souring during waterflooding and EOR operations, in: Proceedings of First EAGE/IFPEN Conference on Sulfur Risk Management in Exploration and Production, 18–20 September, 2018, Rueil-Malmaison, France.

[250] J.J. Arensdorf, K. Miner, R. Ertmoed, et al., Mitigation of reservoir souring by nitrate in a produced-water reinjection system in Alberta, in: SPE-121731-MS, SPE International Symposium on Oilfield Chemistry, 20–22 April, 2009, The Woodland, Texas, USA.

[251] G. Voordouw, A. Agrawal, H.S. Park, et al., Souring treatment with nitrate in fields from which oil is produced by produced water reinjection, in: SPE-141354-MS, SPE International Symposium on Oilfield Chemistry, 11–13 April, 2011, The Woodland, Texas, USA.

[252] N. Klueglein, F. Kögler, I.J. Adaktylou, et al., Understanding selective plugging and biofilm formation of a halophilic bacterial community for MEOR application, in: SPE-179620-MS, SPE Improved Oil Recovery Conference, 11–13 April, 2016, Tulsa, Oklahoma, USA.

[253] E.A. Vik, A.O. Janbu, F.K. Garshol, et al., Nitrate based souring mitigation of produced water—side effects and challenges from the Draugen produced water re-injection pilot, in: SPE-106178-MS, International Symposium on Oilfield Chemistry, 28 February – 2 March, 2007, Houston, Texas, USA.

[254] D. Martin, K. Dodds, I.B. Butler, et al., Carbonate precipitation under pressure for bioengineering in the anaerobic subsurface via denitrification, Environmental Science & Technology 47 (2013) 8692–8699.

[255] K.L. Straub, M. Benz, B. Schink, et al., Anaerobic, nitrate-dependent microbial oxidation of ferrous iron, Applied and Environmental Microbiology 62 (1996) 1458–1460.

[256] B.M. Mortensen, M.J. Haber, J.T. DeJong, et al., Effects of environmental factors on microbial induced calcium carbonate precipitation, Journal of Applied Microbiology 111 (2011) 338–349.

[257] F. Kögler, F.S.F. Hartmann, D. Schulze-Makuch, et al., Inhibition of microbial souring with molybdate and its application under reservoir conditions, International Biodeterioration & Biodegradation 157 (2021) 105158.

[258] A.J. Sheehy, Field studies of microbial EOR, in: SPE-20254-MS, SPE/DOE Enhanced Oil Recovery Symposium, 22–25 April, 1990, Tulsa, Oklahoma, USA.

[259] F.L. Dietrich, F.G. Brown, Z.H. Zhou, et al., Microbial EOR technology advancement: case studies of successful projects, in: SPE-36746-MS, SPE Annual Technical Conference and Exhibition, 6–9 October, 1996, Denver, Colorado, USA.

[260] A. Yusuf, S. Kadarwati, Nurkamelia, Sumaryana, Field test of the indigenous microbes for oil recovery, Ledok Field, Central Java, in: SPE-57309-MS, SPE Asia Pacific Improved Oil Recovery Conference, 25–26 October, 1999, Kuala Lumpur, Malaysia.

[261] C.H. Gao, A. Zekri, K. El-Tarabily, Field cases illustrate MEOR effectiveness, Oil and Gas Journal 107 (2009) 41–46.

[262] E.P. Robertson, G.A. Bala, S.L. Fox, et al., An Overview of Field Specific Designs of Microbial EOR, US DOE Scientific and Technical Information, Washington, DC, 1995.

[263] P. Simandoux, D. Champlon, E. Valentin, Managing the cost of enhanced oil recovery, Oil & Gas Science and Technology 45 (1990) 131–139.

[264] M. Perera, R.P. Gamage, T.D. Rathnaweera, et al., A review of CO_2-enhanced oil recovery with a simulated sensitivity analysis, Energies 9 (2016) 481.
[265] S. Roussanaly, A. Grimstad, The economic value of CO_2 for EOR applications, Energy Procedia 63 (2014) 7836–7843.
[266] C. Nikolova, T. Gutierrez, Use of microorganisms in the recovery of oil from recalcitrant oil reservoirs: current state of knowledge, technological advances and future perspectives, Frontiers in Microbiology 10 (2020) 2996.
[267] E. Dean, J. French, M. Pitts, et al., Practical EOR agents—there is more to EOR than CO_2, in: SPE-190424-MS, SPE EOR Conference at Oil and Gas West Asia, 26–28 March, 2018, Muscat, Oman.
[268] R. Rommerskirchen, B. Jakobs-Sauter, R. Ng, et al., A holistic approach to identify the optimum cost-performance surfactants for chemical EOR projects, in: SPE-196353-MS, SPE/IATMI Asia Pacific Oil & Gas Conference and Exhibition, 29–31 October, 2019, Bali, Indonesia.
[269] M.T. Al-Murayri, H.E. Al-Mayyan, K. Moudi, et al., Chemical EOR economic evaluation in a low oil price environment: Sabriyah Lower Burgan reservoir case study, in: SPE-190337-MS, SPE EOR Conference at Oil and Gas West Asia, 26–28 March, 2018, Muscat, Oman.

CHAPTER

Heavy oil and extra heavy oil (bitumen) recovery

9

Ian D. Gates, Jingyi Wang, Tara D. Gates, Yi Su, Ranjani Kannaiyan and Ran Luo
Department of Chemical and Petroleum Engineering, University of Calgary, Calgary, AB, Canada

9.1 Introduction

Heavy oil and extra heavy oil (referred to as bitumen) production has increased globally [1]. This is due to a variety of reasons, including the increasing markets for heavy oil components for fuels, such as diesel and jet fuel, as well as new technological developments that have led to more economic and less environmentally impactful recovery processes for obtaining these oils from reservoirs [2,3]. Furthermore, in many oil jurisdictions, the location and volumes of the resource are well established, for example, in the western provinces of Alberta and Saskatchewan, nearly all heavy oil and extra heavy oil resources have been mapped extensively [4]. Clastic-based extra heavy oil resources are commonly referred to as oil sands reservoirs.

Heavy oil generally refers to oil that has density between 920 and 1,000 kg/m^3 (22.3–10 degrees API) and viscosity between 100 and 50,000 cP [5,6]. Extra heavy oil, referred to as bitumen, typically has density greater than 1,000 kg/m^3 (lower than 10 degrees API) with viscosities typically greater than 100,000 cP [5,6]. Heavy oil and extra heavy oil prices are often set by not only its density and viscosity but also its sulfur content, acid number, and olefin content [3]. Viscosities of an example liquid including heavy oil and extra heavy oil at room temperature are displayed in Fig. 9.1.

The key challenges faced by bitumen recovery processes operating in heavy oil and oil sands reservoirs are typically embodied by two issues: (1) the viscosity of the bitumen is high, typically in the hundreds to thousands of cP (for heavy oil) and tens of thousands to millions of cP (for bitumen) and (2) the reservoir is heterogeneous with the variability of geological properties associated with the numerous kinds of facies within the reservoir as well as reservoir features, such as top and bottom thief zones (top gas and water or bottom water zones) and intraformational water zones. Recovery processes, to be successful, must be capable of dealing with these two issues—the process must mobilize the heavy oil or bitumen such that it can be produced under reasonable drive mechanisms and move the mobilized bitumen to a production well through the heterogeneity of

Recovery Improvement. DOI: https://doi.org/10.1016/B978-0-12-823363-4.00008-X

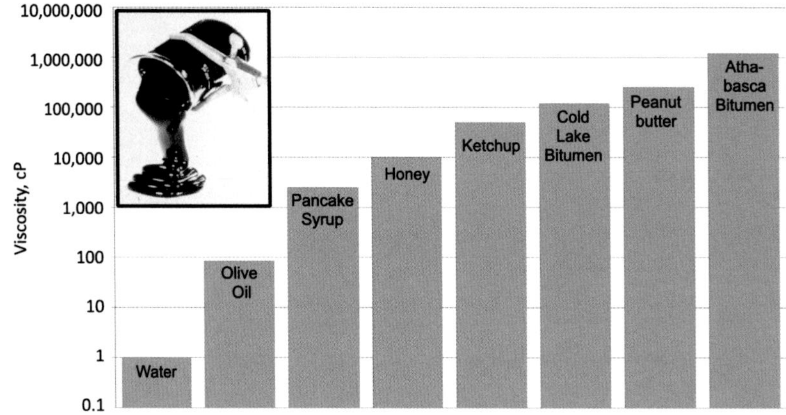

FIGURE 9.1

Comparison of liquid viscosities and those of heavy oil and extra heavy oil (bitumen).

the formation. The recovery process, ideally, would be able to mobilize and produce bitumen reliably no matter the heterogeneity of the formation—this is a tall order and in practice, is likely hard to achieve with a single recovery process.

As described by Gates and Wang [7], technology for in situ recovery of heavy oil and extra heavy oil has been developing for over 50 years. From their analysis, they determined that the features of technologies that persisted were those that were relatively simple and cost effective. Many of the processes they list that were complex did not persist and evolve into commercial technologies [7]. Prior to about 2010, the significant feature of the methods that were developed appeared to focus on technology that could raise the recovery factor, accelerate production rates, and minimize the cost intensity of the recovery process. Beyond about 2010, with increased awareness and policy pressures for lowering carbon dioxide emissions (sometimes referred to as carbon emissions or intensity), much of the focus on technology development has been on less emissive recovery processes that are also cost effective. In some jurisdictions, motivation to develop lower cost and less emissive recovery processes is further reinforced by carbon taxes imposed on emissions associated and generated from recovery processes [8]. Gates and Wang [7] found that the major technological developments that enabled the production of heavy and extra heavy oil were directional and horizontal well drilling, completion design, improved pump technology, operating strategy, and better characterization of the reservoir itself. In recent years there has been major developments on the design of flow control devices for heavy and extra heavy oil production [9,10]. This includes devices for sonic and supersonic steam injection, multiphase oil and water flow control, and stabilization of steam-trap control for thermal (steam) recovery techniques [11].

For heavy and extra heavy oils, there are generally two broad recovery process approaches that are being used. For lower viscosity heavy oil with sufficiently high solution gas, the oil is produced by using cold production technologies where the solution gas in the heavy oil is sufficient to mobilize the oil within the reservoir and move it to the production well. Examples are found in Canada, Venezuela, and Albania where the heavy oil can be produced in a primary recovery method manner to the surface [12]. In cold production processes, the key is to maintain the pressure in the reservoir such that a large amount of oil is recovered from the reservoir. In most of these operations, the recovery factor of the recovery process is typically between 5% and 15% [12,13]. This implies that majority of the oil remains in the reservoir after the process is deemed uneconomic and thus there is a search for new methods to produce this oil after cold production. Several postcold production processes have been explored including solvent-based processes, carbon dioxide injection-based processes, and air injection-based processes [14–17].

For extra heavy oils, the most commonly used methods use thermal stimulation and this can be done by several different methods [6,18]. This includes steam-based recovery process and air injection-based recovery methods. Steam-based methods include recovery processes, such as cyclic steam stimulation (CSS) and steam-assisted gravity drainage (SAGD) [6,18]. In these methods, steam is injected into the reservoir, which then lowers the viscosity of the oil, which enables it to move from the reservoir to the production well where it is then transported to the surface. The steam required in these processes is typically generated from the combustion of natural gas [19,20]. This implies that these methods have large carbon dioxide emissions to the atmosphere [21,22]. In an attempt to lower the carbon emissions, one variant of steam-based recovery processes that has been explored in oil sands reservoirs have been thermal-solvent hybrid recovery processes [6]. This includes expanding solvent SAGD (ES-SAGD) and rich solvent processes, such as VAPEX and NSolv [6,23–25]. Another variant of steam-based recovery processes that has been explored recently to lower the amount of steam required, which in turn reduces the amount of carbon emissions, are those where noncondensable gas (NCG) is coinjected with the steam into the reservoir [26,27]. Examples of these processes include the steam and gas push (SAGP) and eMSAGP processes [26]. In these processes, in some operations, the steam requirement has been reduced by over 80% [28]. This implies that the carbon emissions have also been reduced significantly.

Emerging recovery processes for extra heavy oil are rich solvent-based methods and electromagnetic based methods [29–33]. Rich solvent methods have the advantage that they avoid steam use and thus avoid carbon dioxide emissions to the atmosphere. Electromagnetic based methods have the advantage that they can also be less submissive to the atmosphere, especially if the power required by the method can be generated from renewable power sources. Other recovery methods that continue to be developed are air injection methods, such as the toe-to-heel air injection (THAI), where oxygen injected into the reservoir reacts with the in situ oil generating heat, which enables greater oil mobilization enabling production to the surface [34–36].

The literature is rich with information and science and technical contributions on the production of heavy and extra heavy oil reservoirs. These include:

- Properties of heavy and extra heavy oil (bitumen) [18,37–47],
- Reservoir characterization [44,48–63],
- Reservoir modeling studies [13–16,18,57–61,64–70],
- Micromodel experiments [71–74],
- Physical model experiments [75–79],
- Pilot data analysis [80–85], and
- Commercial operation data analysis [86–93].

The studies also include aspects of geology and geological heterogeneity including studies on the structure of oil sands deposits, recovery process energy efficiency, environmental emissions, economics, and the search for new processes and variants of current existing commercial processes that improve energy efficiency (raising economic efficiency) or lower environmental impacts including greenhouse gas (GHG) emissions and water consumption.

In this chapter, we review the technical basis for the recovery of heavy and extra heavy oil with a focus towards what the key technical barriers and potential technological solutions are required to enable the processes.

9.2 Technical basis for heavy and extra heavy oil recovery processes

The flow rate of oil within a porous reservoir rock is given by:

$$\text{Oil Flow Rate} = (\text{Oil Phase Mobility})\text{x}(\text{Driving Pressure Gradient}) = \left(\frac{k_{ro}k_{abs}}{\mu}\right)\left(\frac{\Delta P}{L}\right) \tag{9.1}$$

where $k_{ro}k_{abs}$ is the effective permeability, μ is the viscosity of the oil phase, ΔP is the pressure drop, and L is the distance over which the pressure drop applies. The technical basis for the production of heavy or extra heavy oil (bitumen) from the reservoir is set by the mobility of the oil phase and the ability of the driving pressure gradient to move the mobilized oil to the production well. This leads to the two requirements for production of heavy oil and bitumen systems:

1. Mobilize the oil (by effective permeability or viscosity or both) and
2. Move the mobilized oil to a production well (by any drive mechanism possible).

If any one of these two requirements are not satisfied, oil production to surface cannot happen and the recovery process will fail. Thus the design of recovery processes for these systems centers around these two requirements. In a gravity drainage process, the key success requirement of the process is the mobilization of the oil; the driving pressure gradient term in Eq. (9.1) becomes $\Delta\rho g$ (where: $\Delta\rho$ is the density difference between phases, and g is the gravity of Earth) [6].

The potential for enhancing the oil phase effective permeability is small since this would mean raising the oil phase relative permeability or absolute permeability of the reservoir or both. Unless the reservoir is fractured as is done in hydraulic fracturing operations, this is difficult. However, in cyclic steam stimulation, as practiced in the Cold Lake oil sands deposit in Alberta, Canada, steam fracturing is done, which can enhance the permeability of the reservoir by several times [94,95]. For enhancing the relative permeability, this would mean reducing the interfacial tension between the oil and water phases, which can be done via surfactants or by raising the temperature of the system. However, the relative permeability cannot be enhanced significantly even with both surfactant and elevated temperatures.

9.2.1 Cold production of heavy oil

Fig. 9.2 displays a schematic representation of the onset of foamy oil flow within the reservoir. For cold producible heavy oil with sufficient solution gas content, viscosity reduction is achieved when the reservoir pressure drops and solution gas exsolves due to the opening and pumping at the production well. More specifically, when the production at the well is dropped, gas dissolved in the oil comes out of solution to form microbubbles [12,13]. As the pressure declines further, the microbubbles grow in size and expand, and they do so with flow consequent in the direction towards the lowest pressure point in the reservoir, which happens to be the production well [12]. The foamy oil that is generated during the gas exsolution process yields are overall oil phase with lower viscosity than that of the original heavy oil [12]. This is what provides the enhanced mobility of the heavy oil within the reservoir. One key feature of the system is that because the oil phase is viscous, the gas bubbles do not tend to merge and form a single connected gas phase (which if connected could then move as a single phase if it reached the critical gas saturation depleting the reservoir of the solution gas "energy"). Since the bubbles remain separate, this enhances the ability of the system to mobilize the heavy oil and maintain the gas content in the reservoir for extended time. The pressure difference that occurs from the reservoir to the production well and the expansion of the solution gas bubbles down the path of the pressure gradient are the driving mechanisms that move the mobilized heavy oil to the production well where it is then pumped to the surface. This means that the recovery process is a displacement recovery process. In cold heavy oil production with sand, not only is the foamy oil and formation water produced to the surface but also sand leading to the creation of wormholes within the reservoir [13]. Wormholes are parts of the reservoir where the sand has been produced to the surface, in other words, these are parts of the reservoir where the sand has been depleted creating void space within the reservoir [13,16]. In this manner, the two

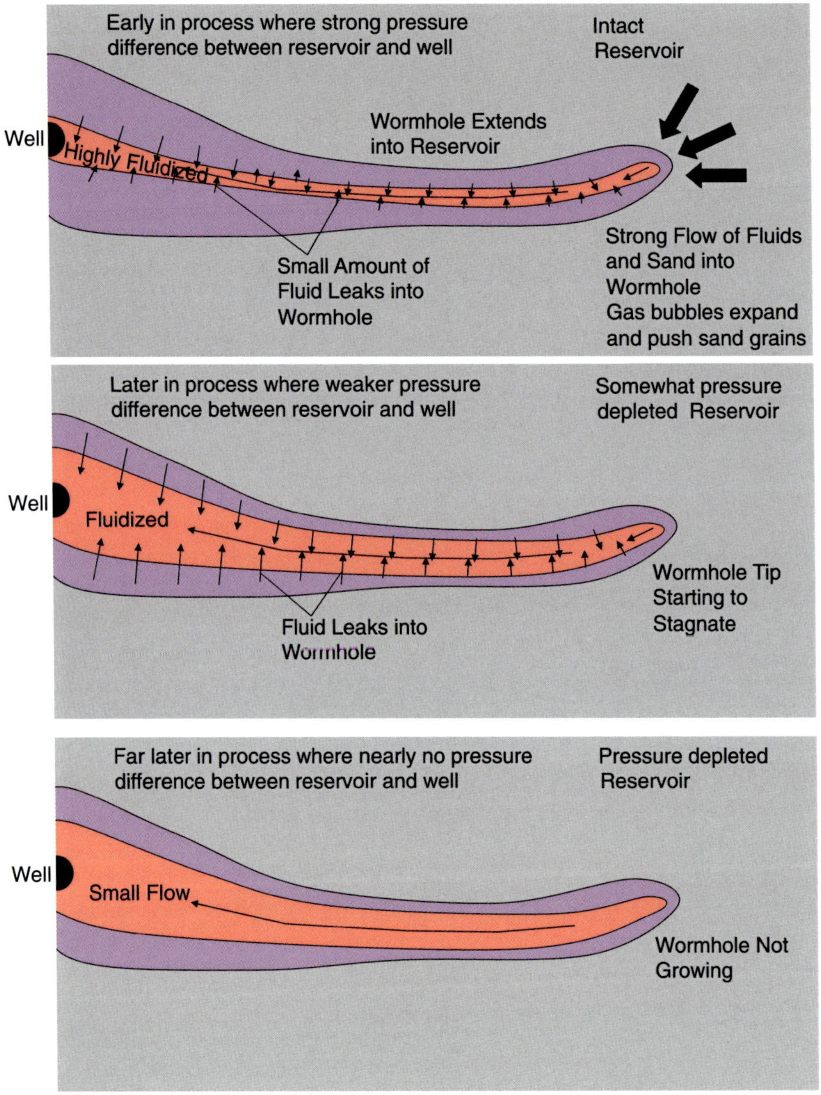

FIGURE 9.2

Cold production of heavy oil under solution gas drive and sand production.

requirements for the heavy oil recovery process as described above are met in a cold heavy oil recovery process. The key challenge faced by cold production processes for heavy oil is that the recovery factors tend to be low typically achieving between 5% and 15%. This means that a significant amount of the oil remains in the reservoir after the process becomes uneconomic and thus there is a search for postcold production processes that both yield more oil but with low emissions intensity.

Fig. 9.3 displays the evolution of solution gas in a viscous oil core. The sand grains are displayed as white–gray objects whereas the gas bubbles are depicted in various shades of blue. The images show that some gas bubbles remain essentially stationary but grow within the pore space whereas other bubbles migrate through the pore space. The disconnected nature of the solution gas bubbles is clear.

9.2.2 Recovery of extra heavy oil (bitumen) via stimulation

For extra heavy oil (bitumen) systems, there are several knobs that can be used to lower the viscosity of the oil phase—this includes temperature, solvent, and asphaltene precipitation (or in situ oil upgrading). The effect of temperature on the viscosity is well known: as shown in Fig. 9.4, the higher the temperature, the lower the viscosity of the oil. At original reservoir temperature of 10°C, oil sands bitumen has a viscosity of order of hundreds of thousands to millions of cP but at 200°C, it drops to below 10 cP reflecting an enhancement of the oil mobility by up to five orders of magnitude when the oil's temperature is raised to 200°C. When bitumen is mixed with solvent, for example, diluent (natural gas condensate) or butane, the viscosity of the solvent–bitumen mixture is lowered and if sufficient solvent is added, the viscosity can be orders of magnitude lower than that of the original bitumen [6,18]; Fig. 9.5 provides an example. With some solvents, such as propane, butane, and hexane, if the concentration is high enough, asphaltenes will precipitate yielding a heavy very viscous or solid-like phase and a lower viscosity de-asphalted oil phase [96–98].

9.2.2.1 Temperature-based processes

Focusing on temperature as a knob for the mobilization of heavy oil or bitumen, this implies a need for raising the temperature within the reservoir. The underlying technological requirement to do this either means that a hot agent must be injected into the reservoir to heat it or that an agent must be injected into the reservoir, which has the action of creating heat within the reservoir. The two

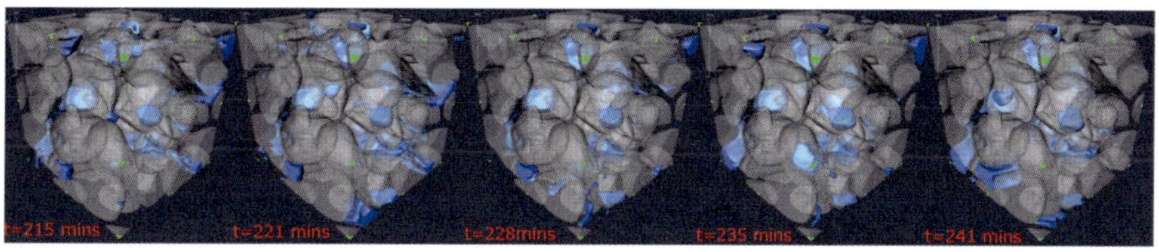

FIGURE 9.3

Dynamics of solution gas bubbles resulting from pressure depletion in an oil core as observed using a CT scanner.

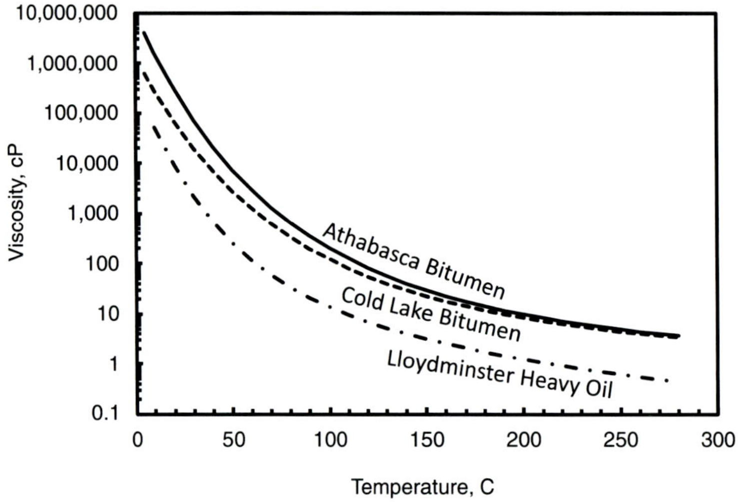

FIGURE 9.4

Effect of temperature on the viscosity of heavy oil and bitumen.

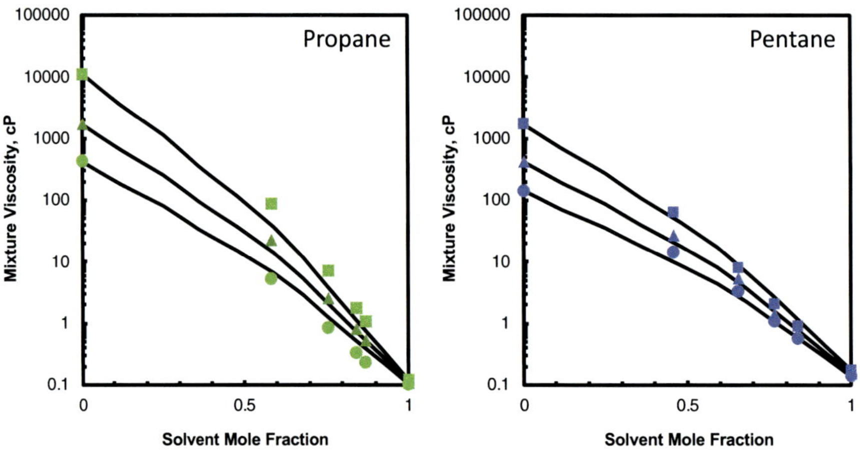

FIGURE 9.5

Effect of solvent on the viscosity of Athabasca bitumen (dots are experimental data and lines are nonlinear fits to Arrhenius model).

technological routes that are commonly used to do this are either steam injection or air injection. In steam injection, the steam carries its latent and sensible heat into the reservoir and after condensing within the reservoir releases its latent heat to the surrounding reservoir, heating both the reservoir rock and formation water and bitumen within the pore space. Once heated, the viscosity of the hot oil is reduced, enabling its mobilization within the reservoir after which the mobilized oil is moved to the production well and then produced to the surface. Often the condensed steam, referred to as condensate, is also produced to the surface with the mobilized oil. In this manner, the condensate can be recycled and used for steam injection once again. In an air injection process, oxygen after being injected reacts with the oil (combustion) within the reservoir creating heat. This heat raises the temperature of and mobilizes the bitumen, which enables its movement to the production well.

Both steam injection and air injection processes imply the use of technology within the reservoir. This includes directional drilling and placement of the well, completion design, which includes the nature of the perforations, and the pumps used to move the fluids to surface. The well and completion design is critical to ensure that there is uniform conformance of the injectant within the reservoir surrounding the wells. The pump technology must be able to withstand high temperature, multiphase flow, and potentially fine sand production, which can erode parts in the pump. Another aspect of the in situ technology is also the operating strategy for the injection itself: for example, there are benefits for starting steam injection at high pressure and then lowering it as the process evolves [99]. However, conformance control underground is always hard to achieve and due to the heterogeneities of the reservoir, the depletion zone in the reservoir is nonuniform [58]. This implies that utilization of the injectant for the production of the heavy oil or bitumen will be suboptimal and this connects to the requirement for the injectant to contact the oil within the reservoir so that the oil is mobilized and the driving pressure gradient moves the mobilized oil to the production well.

In steam-assisted gravity drainage (SAGD), depicted in the cross section in Fig. 9.6, two horizontal wells, both parallel and one atop the other, are placed in the reservoir. The lower production horizontal well is placed between 1 and 2 m above the base of the reservoir whereas the upper injection well is positioned between 5 and 7 m above the lower well [6,18]. Initially, steam circulation through the two wells is done to heat the near well region around the two wells. When sufficiently hot, steam is injected into the upper well and fluids are produced from the lower well and a steam chamber is created as the mobilized bitumen is removed from the reservoir. This creates the steam depletion chamber in the reservoir around the upper well. Steam is then injected into the steam chamber convecting its heat to the edge of the chamber where on release of its latent heat, the bitumen is heated consequently lowering its viscosity. Due to the difference of the densities of the steam and the mobilized bitumen, the bitumen then drains under gravity to the base of the steam chamber where the production well is located. To protect from having the injected steam be directly produced by the lower well, the operation is controlled such that a liquid layer remains above the lower production well. This is referred to as steam-trap control and it is an effective way to ensure that injected steam is distributed out into the depletion chamber rather than being produced directly by the lower well [100]. The SAGD

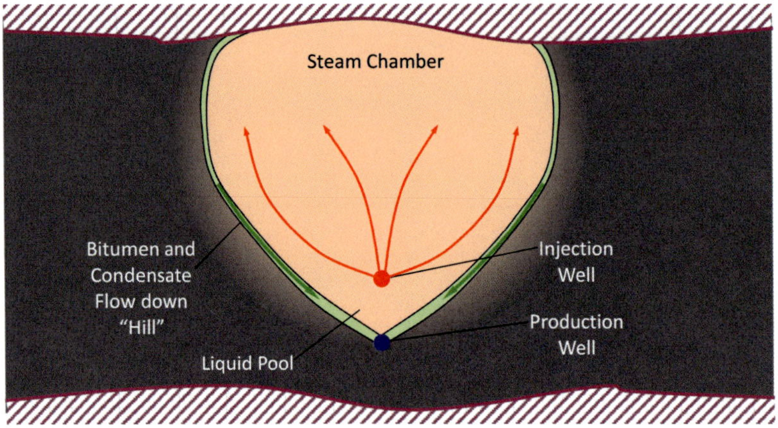

FIGURE 9.6

Cross-sectional view of steam-assisted gravity drainage; horizontal wells project into the page.

process achieves the two requirements as listed above: first, the steam injected in the system raises the temperature and enhances the mobility of the bitumen, and second, the density difference between the steam and the mobilize bitumen enables gravity drainage to deliver that mobilized bitumen to the production well where it is then pumped to the surface. As the bitumen drains from the pore space, the steam replaces it in that pore space. In typical practice, SAGD achieves recovery factors between 35% and 70%.

In cyclic steam stimulation (CSS), as depicted in Fig. 9.7, steam is injected into the reservoir at high pressure to create a heated zone within the reservoir [101–103]. As practiced in Cold Lake, Alberta, steam injection is done above the fracture pressure of the reservoir. After the required steam volume has been injected into the reservoir, the system soaks (no injection or production) enabling heat transfer from the hot zone of steam into the bitumen beyond that hot zone. Thereafter, the well is then put on production and a mixture of mobilized bitumen, steam condensate, formation water, and gases are produced to the surface. When the oil production rate drops below a threshold value, steam injection is started again and the cycle repeats. In common practice, between 10 and 15 cycles are done over a period of up to 15 years for a CSS operation. In CSS, the two requirements as stated above are also met. First, the injected steam heats the bitumen lowering its viscosity and enhancing its mobility. Second, the pressurization that occurs due to the injected volume plus the thermal expansion of the fluids plus exsolved solution gas due to the elevated temperature, all lead to a set of drive mechanisms enabling the movement of the bitumen from the reservoir to the production well where it is then pumped to surface. In the early stages of the process, another important recovery mechanism is formation recompaction: as the

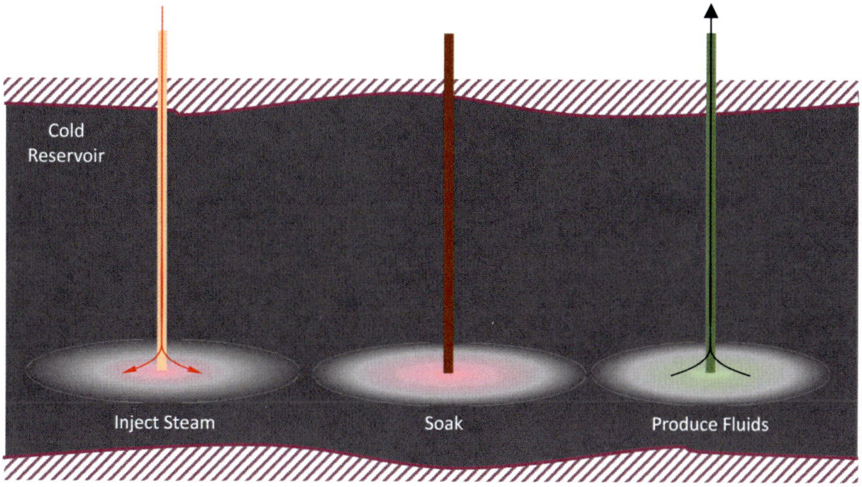

FIGURE 9.7

Illustration of cyclic steam stimulation.

steam is injected into the formation, it lifts the overburden and thus on production as the overburden resets, it drives fluids from the reservoir. In typical practice, the recovery factor from cyclic steam stimulation is between 20% and 35% depending on the reservoir thickness and adjacency to bottom water [6,18].

Many researchers have examined options to improve the thermal efficiency of steam-based recovery processes and in particular, how to operate the reservoir process to reduce the SOR [19,58,59,99,104–106]. The SOR, the steam-to-oil ratio, also reflects the greenhouse gas impact per unit volume of produced bitumen as well as the economic efficiency of the process at a given time (this can be thought of as an instantaneous value typically representing the value over a day). Over an extended period of time (months to years or from the start of the process to a given point of time), the SOR is expressed as the cumulative steam-to-oil ratio (cSOR), defined as the ration of the cumulative steam (expressed as cold water equivalent) injected and cumulative oil produced over that period of time. With respect to environmental impact, steam generation is done by combusting fuel (normally natural gas) which leads to carbon dioxide emissions to the atmosphere. The literature shows that the emissions intensity (carbon dioxide emitted per unit volume of bitumen) depends on the cSOR [19]. One example of the relationship is displayed in Fig. 9.8.

With respect to the SOR and economic efficiency, the steam represents a cost (the cost of fuel for steam generation as well as the handling, treatment, make-up, and disposal of water) and the bitumen indicates the revenues. Thus the SOR is a measure of the cost-to-revenue ratio. The larger is the SOR, the larger is the emissions intensity and the cost-to-revenue ratio. Thus there is a strong motivation in

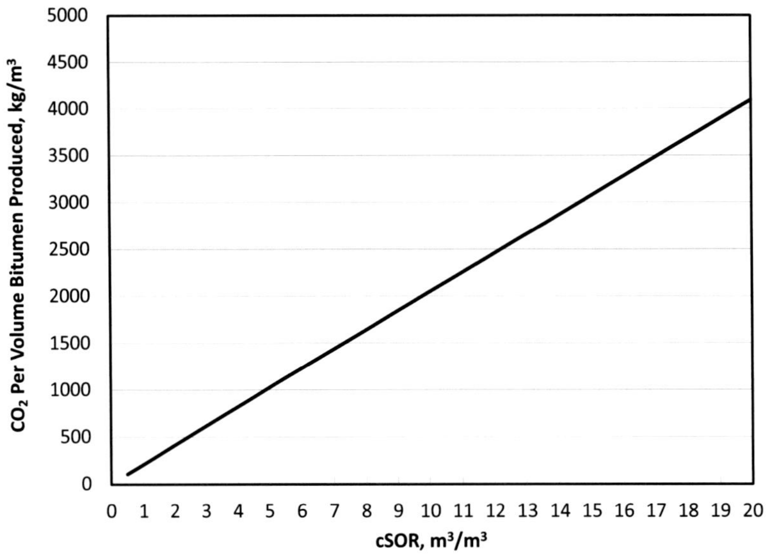

FIGURE 9.8

Emitted CO_2-to-oil ratio versus cumulative steam-to-oil ratio.

Reproduced with permission from I.D. Gates, S. Larter, Energy efficiency and emissions intensity of SAGD, Fuel 115 (2014) 706–713.

SAGD operations to lower the SOR. Several researchers have demonstrated that the control of the injection pressure can improve the SOR of the recovery process and this has been used by industry to improve the SOR [87,99]. However, most research has focused on variants of SAGD where solvents are added to the steam or changes to the well completion design [75,76,83,105,107–109].

There are several variants of air injection processes that have been used for heavy oil and bitumen reservoirs including cyclic injection, continuous injection, dry and wet combustion processes, and toe-to-heel air injection (THAI) [34–36,110]. In THAI, shown schematically in Fig. 9.9, air is injected through a vertical well into the reservoir and fluids are produced from the system through a horizontal production well that is positioned near the base of the reservoir. At sufficiently high pressure to initiate combustion of the heavy oil within the reservoir. As combustion occurs through the many reactions that occur within the system including low temperature oxidation, high temperature oxidation, thermal cracking (pyrolysis), aquathermolysis (hydrous pyrolysis), heat is generated within the reservoir, which then leads to the mobilization of oil. A horizontal well whose toe is placed near the vertical well gathers the mobilized oil and produces it to the surface. THAI accomplishes the previously stated requirements of a recovery process for viscous oils: first, air injection leads to heat generation, which leads to the mobilization of the oil. Second the density difference between the mobilized

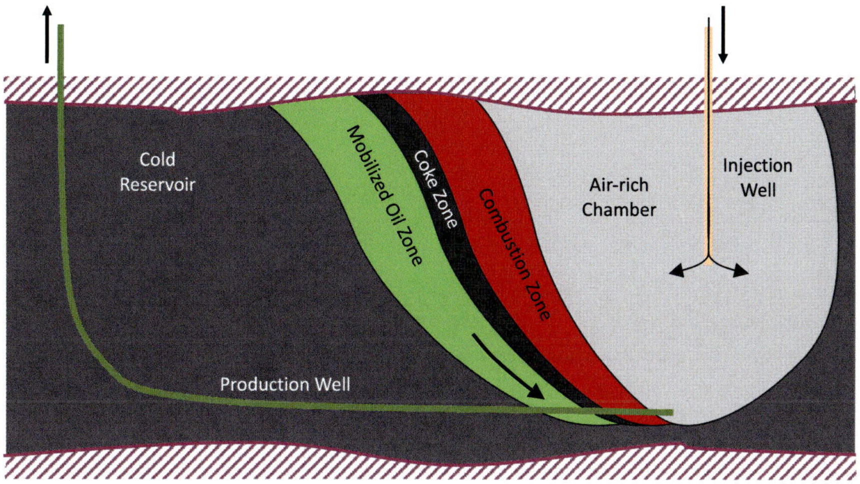

FIGURE 9.9

Illustration of the toe-to-heel air injection.

oil and the air zone leads to gravity-induced drainage of the mobilized oil to the production well that sits along the base of the reservoir. One key feature of air injection systems is that they generate steam not only from the reactions themselves but also from formation water or adjacent water zones (such as bottom water) which are vaporized generating steam. This in situ generated steam also transmits heat within the reservoir heating and mobilizing oil within the system. In this manner, air injection processes can be viewed as in situ steam generation recovery processes.

Both steam injection and air injection processes imply the use of technology on the surface. For steam, this means the use of steam generators, such as once-through steam generators (OTSGs) or heat recovery steam generators (HRSGs), which in turn implies the requirement for fuel to generate the steam [111–113]. For steam from OTSGs and HRSGs, there is a requirement for water treatment, handling, and processing so that it can be recycled for steam reinjection with minimal deposition of solid scale within the tubes of the OTSGs and HRSGs [112]. A new type of steam generator that is emerging for these processes are direct contact steam generators (DCSGs). In these systems, the water is directly injected into the combustion flame and a mixture of steam and carbon dioxide (and nitrogen if air is used in the process) is injected into the formation [114,115]. The key benefit offered by DCSGs is that the energy lost in the flue gases as is the case in OTSGs and HRSGs does not occur in DCSGs. This results in DCSGs having far higher efficiency than that of OTSGs and HRSGs. Another advantage of DCSGs is that the water used does not have to be treated and cleaned as is the case in OTSGs and HRSGs where the water must be softened and cleaned of all organics prior to being vaporized to avoid the deposition of solids within the steam generator [114]. In DCSGs, all the organic matter in the water is combusted as fuel and any solids formed in the process can be separated from the steam-carbon dioxide mixture prior to its injection into the reservoir [114,115]. Another key advantage of DCSGs is that some fraction of the carbon dioxide injected into the formation is sequestered in the reservoir—this, together with its overall higher thermal efficiency, lowers the GHG intensity of the recovery process [115]. One concern that arises with DCSGs is corrosion issues arising from the presence of carbon dioxide with steam in surface lines, underground well liners, and tubing strings. Currently, DCSGs are still at research stage and are being explored as an alternative to OTSGs and HRSGs.

For air injection, this implies the use of a compressor to pressurize and motivate gas flow into the reservoir. Although gas compression use is routine, it has its own inefficiencies and GHG emissions intensity. One issue that is confronted by air injection systems is oxygen flow within surface lines, well liners, and tubing strings. This is especially the case if enriched oxygen is used in the system [116,117]. This can lead to oxidation of metal leading to corrosion.

9.2.2.2 Solvent-based processes

Focusing on solvent dilution as a knob for the mobilization of heavy oil or bitumen implies a need to inject solvent into the reservoir. There are several technical requirements that must be specified with respect to the solvent—this includes the choice of solvent, the temperature and pressure (conditions) of the solvent injected, and potential changes of the solvent as the process evolves. One key benefit from solvent injection is that if no steam is injected, then the GHG emissions associated with steam generation is avoided [25,31]. In some processes, such as ES-SAGD and solvent-assisted SAGD (SA-SAGD), the solvent is coinjected with steam into the reservoir. In these so-called thermal-solvent processes, the amount of steam injected is reduced below that which would have been injected if SAGD was practiced [29,108].

With respect to solvent injection, if the solvent is injected into the reservoir in a vapor phase with elevated temperature, this implies the use of solvent pressurization and heating/vaporization on surface. This means that the technology for vaporizing solvent is required. For steam-solvent coinjection, the solvent, if its volume fraction in the steam is not too large, can be simply injected directly into the steam. One key issue for a given reservoir is the choice of the solvent used. The solvent must be chosen so that its phase behavior matches the reservoir system temperature and pressure and the operating temperature and pressure of the recovery process. In most cases, the preferred solvents are diluent (natural gas condensates) or propane or hexane [108,118]. Recently, other solvents, such as dimethyl ether, is being explored [119]. The phase behavior of the solvent must be tuned so that on injection it flows into the reservoir as a vapor, but readily dissolves into the bitumen or condenses at the edge of the depletion chamber. Furthermore, if in situ carbon rejection is desired (for depositing an asphaltene-rich phase within the reservoir containing the sulfur and metals), then a paraffinic (alkane) solvent is preferred. In these processes, the oil produced at the surface is an upgraded (solvent-de-asphalted) product that has greater value than the original reservoir oil given its lower sulfur, asphaltene, and metals content.

One key challenge for solvent-based recovery processes is the recovery and recycling of the solvent. The solvent often is more valuable than the heavy oil or bitumen being produced from the reservoir. This means that if losses of the solvent are significant, the economics of the recovery process suffers becoming at high losses uneconomic. For solvent-steam-based bitumen recovery processes, it is generally thought that more than 85%–90% of the solvent injected into the formation must be recovered for the process to be economic [29,107,108]. This means that the process itself must focus on the recovering both solvent and bitumen from the reservoir. Given their vastly different thermodynamic properties, this can be accomplished through pressure regulation, for example, pressure cycling where on the low pressure swing, the solvent is vaporized and produced from the reservoir. This could lead to other impacts, such as the cooling of the reservoir as the solvent vaporizes.

With respect to well design and arrangement, in most cases, for bitumen reservoirs, solvent-based recovery processes have adopted a well configuration similar to that of SAGD. This implies that the main drive force in the process is gravity drainage. In cyclic solvent-based recovery processes, such as the cyclic solvent process or cyclic solvent injection (being developed for postcold production heavy oil reservoirs), the solvent is injected in a manner similar to that of CSS.

Also, the completions for solvent injection if in vapor phase is considered to be the same as that of steam-based wells.

9.2.3 Conformance within the reservoir

For steam, air, or solvent-based recovery processes for heavy oil and bitumen reservoirs, the conformance of the agent (temperature or solvent) for reducing the viscosity of the oil within the reservoir is a critical consideration. If conformance is poor, that is, the agent is not spread uniformly and extensively within the reservoir, then the amount of heavy oil or bitumen contacted by the viscosity reduction agent is limited and consequently, the amount that is mobilized and then potentially moved to the production well is limited. Examples of the conformance of steam chambers in oil sands bitumen reservoirs are shown in Fig. 9.10. Thus in the design of the recovery process, it is essential that it attempts to establish and extend the conformance of the viscosity reduction agent as part of the process. In gravity drainage processes, such as SAGD, the steam releases its latent heat and the bitumen's temperature rises and its viscosity falls. At some point, under the action of gravity, the mobilized bitumen drains from its original pore space and then the steam flows in and replaces it in that pore space. This is referred to as a replacement process. In a solvent-based recovery process, a replacement process can also occur if the main drive force for movement of the mobilized bitumen is

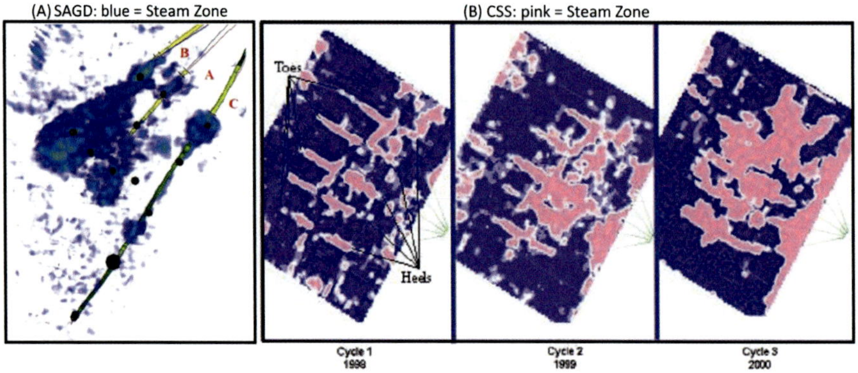

FIGURE 9.10

Examples of conformance of steam chambers (steam and gas push and cyclic steam stimulation) as determined from seismic interpretation: (A) SAGD operation and (B) CSS operation.

gravity. Otherwise, pressure-driven displacement processes can be a component of a solvent-based recovery process.

The conformance of the viscosity reduction agent is not only tied to the nature of the injected agent but also to the geological and oil property heterogeneity of the reservoir [57–61]. Although there do exist some reservoirs that are relatively uniform with respect to heterogeneity, most heavy oil and bitumen reservoirs are severely heterogeneous; many in the Athabasca oil sands deposit are in point bar depositional environments, as depicted in Fig. 9.11. These deposits have several rock types that have different storage (porosity) and permeability characteristics, which interfere with both steam injection and bitumen drainage. Examples are displayed in Fig. 9.12. This rock type heterogeneity results in less than ideal conformance with inefficiencies associated with the extent on how the injectant is mobilizing the oil and therefore the process performance and efficiencies (economic, energy, and emissions) are not ideal.

There have been technologies developed that attempt to deal with reservoir heterogeneity generally associated with the completions design [9–11]. For example, in traditional SAGD operations, the wells were completed with slotted liners, which were mainly designed to prevent sand ingress into the well. Nowadays, completions are designed using flow control devices (FCDs) with outflow control devices (OCDs) on the steam injection wells and inflow control devices (ICDs) on the fluid production wells. FCDs are generally designed to make the well flow resistances much larger than that between the reservoir and wellbore and consequently, the flow from or into the well tends to be more uniform no matter the heterogeneity of the reservoir [11]. Example simulations of the impact of FCDs on the conformance of the depletion chamber for SAGD are shown in Fig. 9.13. The results show that the FCDs do help to promote a more uniform depletion chamber along the horizontal well pair.

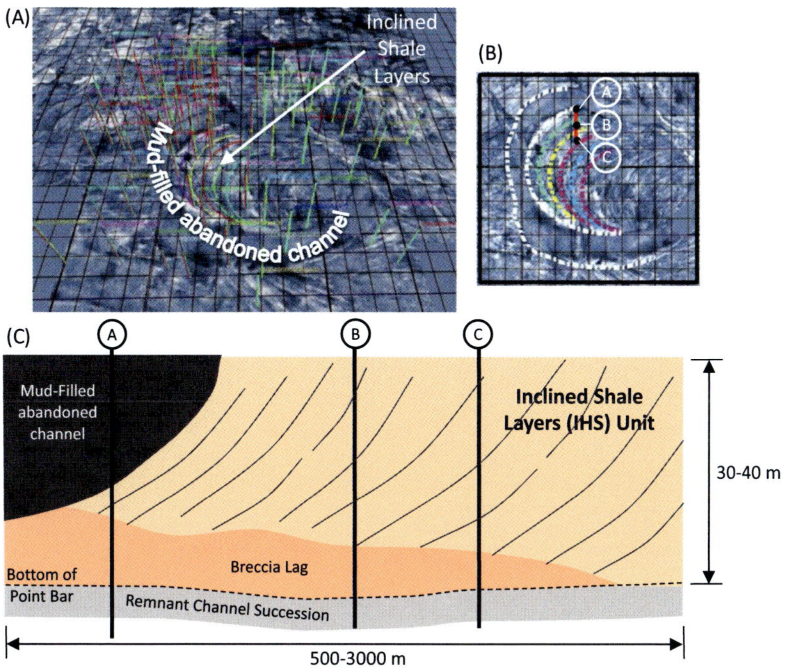

FIGURE 9.11

(A) Point bar model illustrating the heterogeneity encountered in oil sands reservoirs, (B) plan view with location of wells used to create the conceptual North–South stratigraphic cross section A–B–C displayed in (C). The cross section illustrates channel lag and inclined heterolithic strata geometries based on well logs.

Reproduced with modification from Y. Su, J.Y. Wang, I.D. Gates, SAGD well orientation in point bar oil sand deposit affects performance, Engineering Geology 157 (2013) 79–92.

One key technological advancement over the past 20 years has been in 4D seismic data acquisition, which has enabled direct visualization of the conformance zones within reservoirs [120]. In typical practice, the resolution of these methods is of order of tens of meters—this does not permit high resolution in the vertical direction in the reservoir given that the thickness of most heavy oil reservoirs is of order of a few tens of meters or lower (for cold produced heavy oil reservoirs in Western Canada, they tend to be between 3- and 10-m thick). However, in the horizontal directions, given that interwell spacing is of order of 80–120 m for many heavy oil and bitumen operations, a reasonable estimate of the conformance zone in the reservoir can be determined. Other technological innovations that have arisen over the past 20 years are improved pump designs, for example, pumps that can withstand sand production and multiphase fluid production [121,122], better instrumentation, for example, thermocouples to fiber

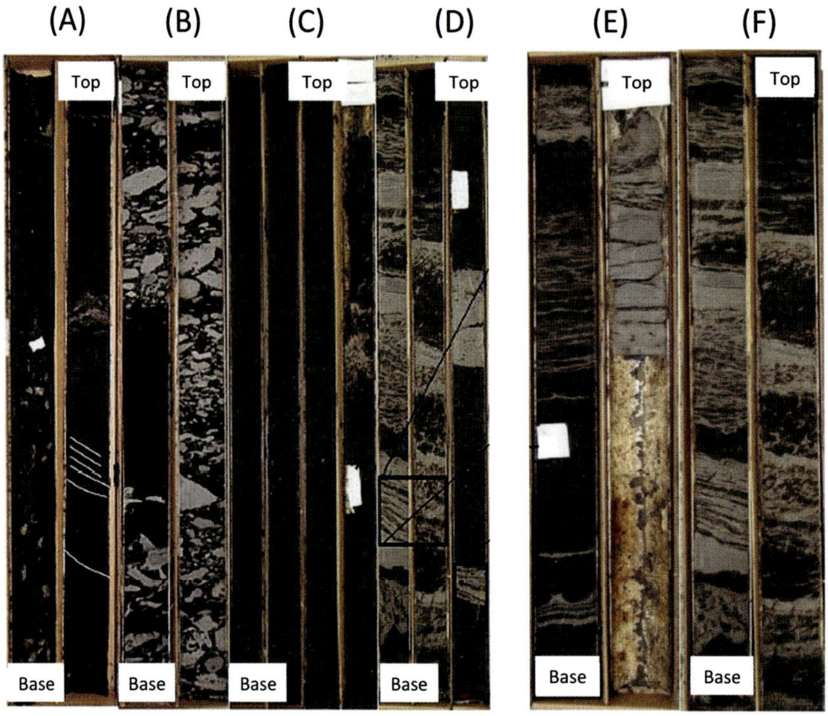

FIGURE 9.12

Examples of four facies found in oil sands reservoirs: (A) cross-stratified sands, (B) breccia lag—light material is siltstone in bitumen saturated sand, (C) massive to cross stratified very fine to fine grained sands, and (D) massive to bioturbated siltstone with sand interbeds used to model the point bar. Core boxes are 75-cm long. (E and F) Potential baffles to steam rise and oil drainage—these intervals can be up to several meters thick.

optic sensors [123–125], improved directional drilling [126,127], and new well configurations for heavy oil systems, such as fishbone wells [128].

9.3 Technology development for the future

Heavy oil systems, under cold production, are primary recovery processes with no injectants and are produced under solution-gas drive. This means that the amount of energy used in the recovery process is simply that of the pump used for producing the mobilized heavy oil to the surface. Thus its costs and emissions intensities are relatively low. However, given its low recovery factor (typically between 5% and 15%), follow-up recovery processes are required.

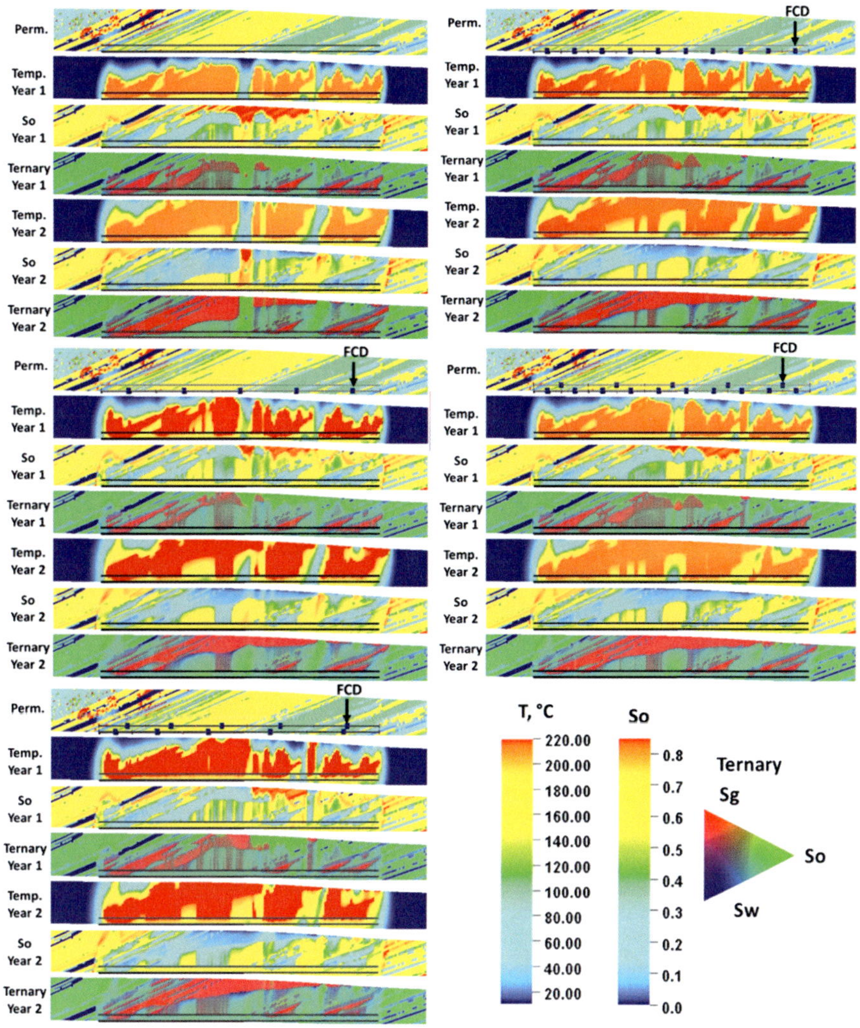

FIGURE 9.13

Effect of flow control device placement in the steam and gas push well pair on steam conformance in an Athabasca oil sands reservoir.

Reproduced with modification from Y. Su, Impact of point bar architecture on the performance of SAGD (PhD. dissertation), University of Calgary, Calgary, 2016.

Current options that are being explored all involve injection of agents into the reservoir to enable both mobilization of the heavy oil and the movement of the mobilized oil to the production well. With the imperative to lower emissions intensity, new technologies for producing heavy oil from postcold production will likely have to be solvent based or perhaps use carbon dioxide as part of the recovery process [14–17].

Bitumen systems require a lot of energy input and are thus costly to operate and generate relatively large emissions of greenhouse gases relative to conventional oil production [129]. Even with the technological advances of the past 30 years, given the requirements to lower the carbon intensity of hydrocarbon fuels, there remains a drastic need to lower the cost and emissions intensities of heavy oil and bitumen recovery processes [130–132]. Many government policies today require that emissions are brought to either net zero or near zero for energy sources by 2050 and many oil and gas companies have made commitments to doing this in their corporate policies [132]. This means that technological advancements in heavy oil and bitumen recovery processes must accelerate and be more rapidly deployed than what has happened over the past 30 years. This presents major challenges to heavy oil and bitumen production companies because technology development in this industry is typically capital intensive and it takes years to establish the benefits of the new technology [131]. However, the heavy oil and bitumen industry has been exploring new options to severely reduce their GHG emissions. For example, oil sands companies in Alberta, Canada, have been looking at options to use carbon capture and sequestration to reduce their overall GHG emissions to the atmosphere [133]. Other options include greater use of hydrogen as an energy vector within their operations, for example, the use of hydrogen for steam generation instead of natural gas [133]. Another option that is being explored is the use of small modular nuclear reactors for power and steam generation [133].

On the reservoir side, there has been a focus on additives coinjected with steam to improve the efficiency of heavy oil and bitumen recovery. These additives have focused on the use of surfactants, nanoparticles, and materials that induce surfactant generation within the reservoir itself [134–136]. However, there is a cost to the use of these additives and thus there is a desire that these additives are used at low concentrations to enhance recovery process performance. Another process that has been used is noncondensable gas addition to the reservoir. This is emerging as a cost-effective option for severely reducing greenhouse gas emissions and cost intensities of bitumen recovery processes [26–28].

On the drilling side, methods that could reduce the emissions intensity of drilling operations including natural gas, bi-fuel (natural gas and diesel), and electrical rigs [137,138]. Furthermore, more attention is being paid on how to optimize the drilling operation to reduce its time and expense. On the completion side, new and complex flow control devices are being designed and tested to enable more uniform and greater productivity and more uniform conformance of the depletion chamber within the reservoir [9–11]. These devices promote the production of oil over that of gas, thus enhancing the recovery of oil while preserving energy (in the form of gas pressure) within the reservoir. This leads to savings of steam injection, which consequently, lowers the GHG and cost intensities of the recovery process.

On the reservoir management side, given that many heavy oil and bitumen recovery operations have had over 10, and in some cases over 20 years of operation, much has been learned about how to operate multiple pads of wells to optimize the overall recovery of heavy oil or bitumen from the reservoir. This includes the use of interwell or interwell pair injection and production strategies. An example is the mega row strategy as used in cyclic steam stimulation operations in Cold Lake, Alberta [102,103]. Another related subject is how the early stages of the recovery process can affect the entire performance of the recovery process. As shown by Pinto et al. [139], the operations of SAGD well pairs in their first few months set the stage for their performance throughout their operating lives. This means that the start-up of the process could have significant impacts on the long-term economic and emissions performances of the well pairs.

Another option that is being developed by oil sands companies in Alberta, Canada, to achieve net zero by 2050 is through extensive use of carbon capture utilization and storage (CCUS) technology as well as electrification, fuel substitution (clean hydrogen), small modular nuclear reactors, and greater energy efficiency. These activities are directly tied to the nature, design, and operating strategies of the recovery processes themselves.

For ultra-low emission heavy oil and oil sands recovery processes, several options can be considered:

1. The recovery process does not use any hydrocarbon fuel for steam or hot water generation, thus obviating the generation of greenhouse gas. This provides options, such as hydrogen-based steam generation, nuclear-powered steam generation, renewably powered steam generation, or other nonhydrocarbon-based fuels (such as ammonia or other nitrogen fuels), providing the heat for steam or hot water generation.
2. The recovery process, if it uses steam or hot water, is coupled with CCUS to permanently remove the carbon dioxide. This option has significant cost and risk, and will require that the process of CCUS does not generate additional GHGs.
3. The recovery process does not use steam or hot water but rather uses solvent as the mobilization agent. If the solvent is to be heated, then it must be done by renewable (or nonhydrocarbon fuel) means or the generated GHGs should be sequestered. There is potential that the carbon dioxide generated could be sequestered in the reservoir as part of the recovery process if the amount of the GHG is sufficiently small.
4. The recovery process performs a chemical conversion of the heavy oil or bitumen within the reservoir to yield a low viscosity oil phase. This could be done through solvent deasphalting but energy will be required to move the low viscosity oil phase to the surface. If the energy required generates GHGs, then this could be used to displace the oil in the reservoir with consequent sequestration of the carbon dioxide within the reservoir (providing the amount of GHGs is sufficiently small).

5. Use air and steam injection and/or steam injection to produce hydrogen from in situ gasification of the heavy oil or bitumen within the reservoir. A fraction of the produced hydrogen is used to power the air compressor and steam generation units. The remainder of the hydrogen is the energy product from the reservoir.

9.4 Final remarks

Recovery technology for heavy oil and extra heavy oil (bitumen) systems has evolved for over 60 years with many attempts to find economically productive operations. Of the many options that have been discovered and reported in the academic, industry, and patent literature, only a few have survived and are now considered commercial technologies. The drive towards lower and near-zero emission intensity means that processes must produce oil but do so considering all sources of GHGs. For heavy oil systems that have undergone solution-gas drive, economic postcold production processes are required. For bitumen systems, currently, these include steam-assisted gravity drainage and cyclic steam stimulation. However, new technology or other approaches will be needed for producing heavy oil and bitumen reservoirs to move to lower and near-zero emission intensity. As of now, technology developments for recovering heavy oil and bitumen have focused on (1) hardware, for example, well placement and spacing, completions, injectants (e.g., steam-solvent or steam-additive systems), and surface facilities, and (2) operating strategy, for example, pressure, injection and production rates, and multiwell strategies. Moving to the future, heavy oil and bitumen recovery processes will need go beyond traditional reservoir and facilities technologies to include holistic accounting, handling, and treatment of GHGs as part of their operations.

Nomenclature

API	American Petroleum Institute
CCUS	carbon capture utilization and storage
cP	centipoises (viscosity unit)
cSOR	cumulative steam-to-oil ratio (steam expressed as cold water equivalent)
CSS	cyclic steam stimulation
DCSG	direct contact steam generators
DME	dimethyl ether
eMSAGP	enhanced modified steam and gas push
ES-SAGD	expanding solvent SAGD
FCD	flow control device
GHG	greenhouse gas
HRSG	heat recovery steam generator

ICD inflow control device
m meters
NCG noncondensable gas
NSolv warm solvent recovery process
OCD outflow control device
OTSG once-through steam generator
SAGD steam-assisted gravity drainage
SAGP steam and gas push
SA-SAGD solvent-assisted SAGD
SOR steam-to-oil ratio (steam expressed as cold water equivalent)
THAI toe-to-heel air injection
VAPEX VAPor EXtraction

References

[1] BP, Statistical Review of World Energy, BP, London, 2021.

[2] IEA, World Energy Outlook, International Energy Agency, Paris, 2020.

[3] Oil Sands Magazine, Western Canadian Select Explained, Alberta, 2021.

[4] AER, Alberta Energy Regulator Maps, Alberta <https://www.aer.ca/providing-information/data-and-reports/maps-mapviewers-and-shapefiles>, 2021.

[5] R.F. Meyer, E.D. Attanasi, Heavy oil and natural bitumen—strategic petroleum resources, Fact Sheet 70-03, U.S. Geological Survey, Reston, 2003.

[6] I.D. Gates, Basic Reservoir Engineering, Kendall-Hunt Inc, Dubuque, 2013.

[7] I.D. Gates, J. Wang, Evolution of in situ oil sands rccovcry tcchnology: what happened and what's new?, in: SPE-150686-MS, SPE Heavy Oil Conference and Exhibition, Kuwait City, Kuwait, December 12–14, 2011, Kuwait City, Kuwait.

[8] N. Ihejirika, Canada's Carbon Tax Hike and Strategic Implications for Oil & Gas Firms, Energy Insight 83, The Oxford Institute for Energy Studies, Oxford, 2021.

[9] Y. Su, I.D. Gates, Impact of flow control devices on SAGD performance from less heterogeneity to strongly heterogeneous reservoirs, in: SPE-174501-MS, SPE Canada Heavy Oil Technical Conference, June 9–11, 2015, Calgary, Alberta, Canada.

[10] S. Bannerjee, B. Hascakir, Design of flow control devices in steam-assisted gravity drainage (SAGD) completions, Journal of Petroleum Exploration and Production Technology 8 (2017) 785–797.

[11] G. Rosi, D. Zhu, I.D. Gates, et al., A review of passive flow control devices in petroleum extraction: implementation, well-design strategy, and physics of different flow regimes, Journal of Petroleum Science and Engineering (2022) (submitted).

[12] B. Tremblay, Cold production of heavy oil, in: J.J. Sheng (Ed.), Enhanced Oil Recovery Field Case Studies, Gulf Professional Publishing, 2013, pp. 615–666.

[13] C. Istchenko, I.D. Gates, Well-wormhole model of cold heavy oil production with sand, SPE Journal 19 (2014) 260–269.

[14] D. Zhao, I.D. Gates, On hot water flooding strategies for thin heavy oil reservoirs, Fuel 153 (2015) 559–568.

[15] D. Zhao, J. Wang, I.D. Gates, An evaluation of enhanced oil recovery strategies for a heavy oil reservoir after cold production with sand, International Journal of Energy Research 39 (2015) 1355–1365.

[16] A. Sharifi-Haddad, I.D. Gates, Modelling of cold heavy oil production with sand (CHOPS) using a fluidized sand algorithm, Fuel 158 (2015) 937–947.
[17] A. Sharifi-Haddad, I.D. Gates, CO_2-based heavy oil recovery processes for post-CHOPS reservoirs, Journal of CO_2 Utilization 19 (2017) 238–246.
[18] R.M. Butler, Thermal Recovery of Oil and Bitumen, GravDrain Inc., Alberta, 1997.
[19] I.D. Gates, S. Larter, Energy efficiency and emissions intensity of SAGD, Fuel 115 (2014) 706–713.
[20] A.D. Charpentier, J.A. Bergerson, H.L. MacLean, Understanding the Canadian oil sands industry's greenhouse gas emissions, Environmental Research Letters 4 (2009) 014005.
[21] J.A. Bergerson, O. Kofoworola, A.D. Charpentier, et al., Life cycle greenhouse gas emissions of current oil sands technologies: surface mining and in situ applications, Environmental Science & Technology 46 (2012) 7865–7874.
[22] B. Nimana, C. Canter, A. Kumar, Energy consumption and greenhouse gas emissions in the recovery and extraction of crude bitumen from Canada's oil sands, Applied Energy 143 (2015) 189–199.
[23] S.K. Das, R.M. Butler, Mechanism of the vapour extraction process for heavy oil and bitumen, Journal of Petroleum Science and Engineering 21 (1998) 43–59.
[24] S. Sleep, J.M. McKellar, J.A. Bergerson, et al., Expert assessments of emerging oil sands technologies, Journal of Cleaner Production 144 (2017) 90–99.
[25] V. Mohan, P. Neogi, B. Bai, Heavy oil and VAPEX process: a brief review, Improved Oil and Gas Recovery 5 (2021) 1–14.
[26] R.M. Butler, The behavior of non-condensable gas in SAGD—a rationalization, Journal of Canadian Petroleum Technology 43 (2004) 28–34.
[27] M. Austin-Adigio, I.D. Gates, Non-condensable gas co-injection with steam for oil sands recovery, Energy 179 (2019) 736–746.
[28] MEG Energy, Christina Lake regional project 2016/2017 performance—commercial scheme approval no. 10773, 2017, pp. 151–156.
[29] I.D. Gates, Oil phase viscosity behaviour in expanding-solvent steam-assisted gravity drainage, Journal of Petroleum Science and Engineering 59 (2007) 123–134.
[30] L.M. Dittaro, A.E. Jaafar, D.L. Perlau, et al., Findings from a solvent-assisted SAGD pilot at Cold Lake, in: SPE-165434-MS, SPE Heavy Oil Conference Canada, June 11–13, 2013, Calgary, Alberta, Canada.
[31] V. Mohan, Y. Su, J. Wang, I.D. Gates, Rich solvent—steam assisted gravity drainage (RS-SAGD): an option for clean oil sands recovery processes, Cleaner Engineering and Technology 8 (2022) 100463.
[32] D. Ji, T.G. Harding, Z. Chen, et al., Feasibility of electromagnetic heating for oil sand reservoirs, in: SPE-199910-MS, SPE Canada Heavy Oil Conference, September 28–October 2, 2020, Virtual.
[33] L. Hu, Combined electromagnetic heating and solvent injection for heavy oil recovery, PhD Dissertation, University of Alberta, Alberta, 2018.
[34] W. Wei, J. Wang, S. Afshordi, et al., Detailed analysis of toe-to-heel air injection for heavy oil production, Journal of Petroleum Science and Engineering 186 (2020) 106704.
[35] W. Wei, A. Rezazadeh, J. Wang, et al., An analysis of toe-to-heel air injection for heavy oil production using machine learning, Journal of Petroleum Science and Engineering 197 (2021) 108109.

[36] W. Wei, I.D. Gates, Lag times in toe-to-heel air injection (THAI) operations explains underlying heavy oil production mechanisms, Petroleum Science (2022) (in press).
[37] Alberta Energy Regulator, ST98-2020: Alberta Energy Outlook, Alberta, 2020.
[38] M.A. Carrigy, Geology of the McMurray Formation. Part III, General Geology of the McMurray Area, Research Council of Alberta, Memoir 1, 1959.
[39] I.D. Gates, N. Chakrabarty, Design of the steam and solvent injection strategy in expanding solvent steam-assisted gravity drainage, Journal of Canadian Petroleum Technology 47 (2008) 12–20.
[40] D.R. Gotawala, I.D. Gates, On the impact of permeability heterogeneity on SAGD steam chamber growth, Natural Resources Research 19 (2010) 151–164.
[41] F.J. Hein, Geology of bitumen and heavy oil: an overview, Journal of Petroleum Science and Engineering 154 (2017) 551–563.
[42] M.A.B. Khan, A.K. Mehrotra, W.Y. Svrcek, Viscosity models for gas-free athabasca bitumen, Journal of Canadian Petroleum Technology 23 (1984) 47–53.
[43] A.K. Mehrotra, W.Y. Svrcek, Viscosity of compressed Athabasca bitumen, Canadian Journal of Chemical Engineering 64 (1986) 844–847.
[44] M.J. Peacock, Athabasca oil sands: reservoir characterization and its impact on thermal and mining opportunities, in: Proceedings of the 7th Petroleum Geology Conference, 2009. March 30-April 2009, London, UK.
[45] T.W. Stone, D.H. Lau, W.J. Bailey, Control of reservoir heterogeneity in SAGD bitumen process, in: SPE-165388-MS, Heavy Oil Conference Canada, June 11–13, 2013, Calgary, Alberta, Canada.
[46] R. Strobl, Impact of reservoir quality on SAGD production-observation from the UTF phase B oil sand pilot, in: Proceedings of the Recovery-CSPG CSEG CWLS Convention, May 9–13, 2011, Calgary, Alberta, Canada.
[47] G. Yang, R.M. Butler, Effects of reservoir heterogeneities on heavy oil recovery by steam-assisted gravity drainage, Journal of Canadian Petroleum Technology 31 (1992) 37–43.
[48] Q. Chen, M. Gerritsen, A. Kovscek, Effects of reservoir heterogeneities on the steam-assisted gravity-drainage process, SPE Reservoir Evaluation & Engineering 11 (2008) 921–932.
[49] S. Chopra, L. Lines, D.R. Schmitt, et al., Heavy Oils: Reservoir Characterization and Production Monitoring, SEG Books, Tulsa, 2010.
[50] P.D. Flash, G.D. Mossop, Depositional environments of Lower Cretaceous McMurray Formation, Athabasca oil sands, Alberta, American Association of Petroleum Geologists Bulletin 69 (1985) 1195–1207.
[51] M. Fustic, L. Skulski, W. Hanson, et al., Geological mapping and reservoir characterization of oil sands reservoir by integrating 3D seismic, dipmeter, core descriptions, and analogs in the McMurray Formation, NE Alberta, Search and Discovery, article #40281, 2008.
[52] M. Hadavand, C.V. Deutsch, A practical methodology for integration of 4D seismic in steam-assisted-gravity-drainage reservoir characterization, SPE Reservoir Evaluation & Engineering 20 (2016) 353–362.
[53] S.M. Hubbard, S.G. Pemberton, E.A. Howard, Regional geology and sedimentology of the basal cretaceous Peace River oil sand deposit, north-central Alberta, Bulletin of Canadian Petroleum Geology 47 (1999) 270–297.
[54] Y. Ito, Oil Drainage characteristics during the SAGD process to explain observed field performance, in: SPE-170071-MS, SPE Heavy Oil Conference Canada, June 10–12, 2014, Calgary, Alberta, Canada.

[55] Y. Ito, S. Suzuki, Numerical simulation of the SAGD process in the Hangingstone oil sands reservoir, in: CIM 96-57, CIM Annual Technical Conference, June 10–12, 1996, Calgary, Alberta, Canada.

[56] P.A. Labrecque, J.L. Jensen, S.M. Hubbard, Cyclicity in Lower Cretaceous point bar deposits with implications for reservoir characterization, Athabasca Oil Sands, Alberta, Canada, Sedimentary Geology 242 (2011) 18–33.

[57] Y. Su, Impact of point bar architecture on the performance of SAGD, PhD Dissertation, University of Calgary, Calgary, 2016.

[58] Y. Su, J.Y. Wang, I.D. Gates, SAGD well orientation in point bar oil sand deposit affects performance, Engineering Geology 157 (2013) 79–92.

[59] Y. Su, J. Wang, I.D. Gates, Orientation of a pad of SAGD well pairs in an Athabasca point bar deposit affects performance, Marine and Petroleum Geology 54 (2014) 37–46.

[60] Y. Su, J. Wang, I.D. Gates, SAGD pad performance in a point bar deposit with a thick sandy base, Journal of Petroleum Science and Engineering 154 (2017) 442–456.

[61] R.S. Strobl, W.K. Muwais, D.M. Wightman, et al., Geological modeling of The McMurray Formation Reservoirs based on outcrop and subsurface analogues. Petroleum Geology of the Cretaceous Mannville Group, Western Canada, in: S.G. Pemberton, D.P. James (Eds.), Canadian Society of Petroleum Geologists, Memoir 18, 1997, pp. 292–312.

[62] R.S. Strobl, W.K. Muwais, D.M. Wightman, et al., Application of outcrop analogues and detailed reservoir characterization to the AOSTRA underground test facility, McMurray Formation, Northern Alberta. In: S.G. Pemberton, D.P. James (Eds.), Petroleum Geology of the Cretaceous Mannville Group, Western Canada, Canadian Society of Petroleum Geologists, Memoir 18, 1997, pp. 375–391.

[63] D.M. Wightman, S.G. Pemberton, The Lower Cretaceous (Aptian) McMurray Formation: an overview, in: S.G. Pemberton, D.P. James (Eds.), Petroleum Geology of the Cretaceous Mannville Group, Western Canada, Canadian Society of Petroleum Geologists, Memoir 18, 1997, pp. 312–344.

[64] R.M. Butler, A new approach to the modeling of steam-assisted gravity drainage, Journal of Canadian Petroleum Technology 24 (1985) 42–51.

[65] C.T.Q. Dang, N.T.B. Nguyen, W. Bae, et al., Investigation of SAGD recovery process in complex reservoir, in: SPE-133849-MS, SPE Asia Pacific Oil & Gas Conference and Exhibition, October 18–20, 1985, Brisbane, Queensland, Australia.

[66] Q. Lei, J.Y. Wang, I.D. Gates, Impact of oil-water relative permeability curves on SAGD behaviour, in: SPE-132185-MS, International Oil and Gas Conference and Exhibition, June 8–10, 2010, Beijing, China.

[67] J. Leung, Scaleup of effective mass transfer in vapour-extraction process accounting for field-scale reservoir heterogeneities, Journal of Canadian Petroleum Technology 53 (2014) 275–289.

[68] J. Wang, C.C. Ezeuko, I.D. Gates, Energy (heat) distribution and transformation in the SAGP process, in: SPE-157808-MS, SPE Heavy Oil Conference Canada, June 12–14, 2012, Calgary, Alberta, Canada.

[69] C. Wang, J.Y. Leung, Characterizing the effects of lean zones and shale distribution in steam-assisted-gravity-drainage recovery performance, SPE Reservoir Evaluation & Engineering 18 (2015) 329–345.

[70] J.B. Willis, H. Tang, Three-dimensional connectivity of point-bar deposits, Journal of Sedimentary Research 80 (2010) 440–454.
[71] M. Kim, In-situ visualization of steam assisted gravity drainage with chemical and solvent additives (Masters thesis), University of Toronto, Toronto, 2017.
[72] O. Mohammadzadeh, I. Chatzis, Pore-level investigation of heavy oil and bitumen recovery using steam assisted gravity drainage (SAGD), Oil & Gas Science and Technology – Rev. IFP Energies Nouvelles 65 (2010) 839–857.
[73] O. Mohammadzadeh, N. Rezaei, I. Chatzis, Pore-scale performance evaluation and mechanistic studies of the solvent-aided SAGD (SA-SAGD) process using visualization experiments, Transport in Porous Media 108 (2015) 437–480.
[74] A.H. Syed, N. Mosavat, J. Riordon, et al., A combined method for pore-scale optical and thermal characterization of SAGD, Journal of Petroleum Science and Engineering 146 (2016) 866–873.
[75] M. Ardali, M. Barrufet, D.D. Mamora, Experimental study of co-injection of potential solvents with steam to enhance SAGD process, in: SPE-144598-MS, SPE Western North American Region Meeting, May 7–11, 2011, Anchorage, Alaska, USA.
[76] M. Ardali, M. Barrufet, D.D. Mamora, Laboratory testing of addition of solvents to steam to improve SAGD process, in: SPE-146993-MS, SPE Heavy Oil Conference Canada, June 12–14, 2012, Calgary, Alberta, Canada.
[77] S. Esmaeili, H. Sarma, T. Harding, et al., Experimental study of the effect of solvent addition and temperature on two-phase bitumen/water relative permeability, SPE Reservoir Evaluation and Engineering 23 (2020) 1381–1402.
[78] L. Meng, A. Kordestany, B. Maini, et al., Experimental study of diffusion of vaporized solvent in bitumen at elevated temperatures, Fuel 280 (2020) 118595.
[79] H. Rahnema, M. Barrufet, D. Mamora, Experimental study of air injection in SAGD chamber, in: SPE-149195-MS, Canadian Unconventional Resources Conference, November 15–17, 2011, Calgary, Alberta, Canada.
[80] J.K. Donnelly, The best process for Cold Lake: CSS vs. SAGD, Journal of Canadian Petroleum Technology 39 (2000) 5–7.
[81] M. Saltuklaroglu, G.N. Wright, P.R. Conrad, et al., Mobil's SAGD experience at Celtic, Saskatchewan, Journal of Canadian Petroleum Technology 39 (2000) 45–51.
[82] G.E. Birrell, A.L. Aherne, D.J. Seleshanko, Cyclic SAGD—economic implications of manipulating steam injection rates in SAGD projects—re-examination of the Dover project, Journal of Canadian Petroleum Technology 44 (2005) 54–58.
[83] S.C. Gupta, S.D. Gittins, Christina Lake solvent aided process pilot, Journal of Canadian Petroleum Technology 45 (2006) 15–18.
[84] K.A. Miller, Y. Xiao, Lloydminster, Saskatchewan vertical well SAGD field test results, Journal of Canadian Petroleum Technology 49 (2010) 22–29.
[85] K.J. Jorshari, B. O'Hara, A new SAGD-well-pair placement: a field case review, Journal of Canadian Petroleum Technology 52 (2013) 12–19.
[86] Cenovus Energy Inc., 2015–2019 AER Annual Update: Christina Lake Project SAGD 8591, In-Situ Process Reports, Alberta Energy Regulator, Calgary.
[87] ConocoPhillips. 2006–2019 AER Annual Update: Surmont Project 9426, In-Situ Process Reports, Alberta Energy Regulator, Calgary.
[88] Husky Oil Operations Limited, 2015–2019 AER Annual Update: Sunrise SAGD 10419 and Tucker SAGD 9835, In-Situ Process Reports, Alberta Energy Regulator, Calgary.

[89] Imperial Oil Resources, 2013 AER Annual Update: EBRT and Cold Lake SA-SAGD Experimental Pilot, In-Situ Process Reports, Alberta Energy Regulator, Calgary.
[90] MEG Energy Corp, MEG Energy Regional Project Master, Calgary, 2018.
[91] Nexen CNOOC, 2015–2019 AER Annual Update: Long Lake SAGD 9485, In-Situ Process Reports, Alberta Energy Regulator, Calgary.
[92] Suncor Energy, 2015–2019 AER Annual Update: Firebag SAGD 8870, In-Situ Process Reports, Alberta Energy Regulator, Calgary.
[93] G. Giacchetta, M. Leporini, B. Marchetti, Economic and environmental analysis of a steam assisted gravity drainage (SAGD) facility for oil recovery from Canadian oil sands, Applied Energy 142 (2015) 1–9.
[94] M. Cokar, M.S. Kallos, I.D. Gates, Reservoir simulation of steam fracturing in early cycle cyclic steam stimulation, SPE Reservoir Evaluation & Engineering-Reservoir Engineering 15 (2012) 676–687.
[95] Y. Bao, J. Wang, I.D. Gates, On the physics of cyclic steam stimulation, Energy 115 (2016) 969–985.
[96] A.K. Tharanivasan, Asphaltene precipitation from crude oil blends, conventional oils, and oil with emulsified water, PhD dissertation, University of Calgary, Calgary, 2012.
[97] S. Ilyin, M. Arinina, M. Polyakova, et al., Asphaltenes in heavy crude oil: designation, precipitation, solutions, and effects on viscosity, Journal of Petroleum Science and Engineering 147 (2016) 211–217.
[98] J. Deldago-Linares, A. Cardenas, N. Morillo, et al., Asphaltene precipitation from Latin American heavy crude oils. Effects of solvent aromaticity and agitation on particle size reduction, Journal of Dispersion Science and Technology 40 (2019) 74–81.
[99] I.D. Gates, N. Chakrabarty, Optimization of steam-assisted gravity drainage (SAGD) in Ideal McMurray reservoir, Journal of Canadian Petroleum Technology 45 (2006) 54–62.
[100] I.D. Gates, C. Leskiw, Impact of steam trap control on performance of steam-assisted gravity drainage, Journal of Petroleum Science and Engineering 75 (2010) 215–222.
[101] J.P. Batycky, R.P. Leaute, B.A. Dawe, A mechanistic model of cyclic steam stimulation, in: SPE-37550-MS, SPE International Thermal Operations and Heavy Oil Symposium, February 10–12, 1997, Bakersfield, California, USA.
[102] R.J. Gallant, S.D. Stark, M.D. Taylor, Steaming and operating strategies at a midlife CSS operation, in: SPE-25794-MS, SPE International Thermal Operations and Heavy Oil Symposium, February 8–10, 1993, Bakersfield, California, USA.
[103] Z. Liu, S.D. Stark, Reservoir simulation modelling of the mature Cold Lake steam operations, in: SPE-160491-MS, SPE Heavy Oil Conference Canada, June 12–14, 2012, Calgary, Alberta, Canada.
[104] O. Akbilgic, D. Zhu, I.D. Gates, et al., Prediction of steam-assisted gravity drainage steam to oil ratio from reservoir characteristics, Energy 93 (2015) 1663–1670.
[105] A.A., Alturki, I.D. Gates, B.B. Maini, Co-injection of non-condensable gas improves ES-SAGD performance in shallow oil sands reservoirs with a small top water zone, in: SPE-137092-MS, Canadian Unconventional Resources and International Petroleum Conference, October 19–21, 2010, Calgary, Alberta, Canada.

[106] T. Guo, J. Wang, I.D. Gates, Pad-scale control improves SAGD performance, Petroleum (2018) 318–328.

[107] I.D. Gates, N. Chakrabarty, Design of the steam and solvent injection strategy in expanding-solvent steam-assisted gravity drainage, Journal of Canadian Petroleum Technology 47 (2008) 12–19.

[108] T.N. Nasr, B. Beaulieu, H. Golbeck, et al., Novel expanding solvent-SAGD process "ES-SAGD,", Journal of Canadian Petroleum Technology 42 (2003) 13–16.

[109] B. Orr, ES-SAGD: past, present and future, in: SPE-129518-MS, SPE Annual Technical Conference and Exhibition, October 4–7, 2009, New Orleans, Louisiana, USA.

[110] T. Xia, B. Greaves, M.S. Werfilli, et al., THAI process—effect of oil layer thickness on heavy oil recovery, in: PETSOC-2002-027, Canadian International Petroleum Conference, June 11–13, 2002, Calgary, Alberta, Canada.

[111] K.-W. Gwak, W.-S. Bae, A review of steam generation for in situ oil sands projects, Geosystem Engineering 13 (2010) 111–118.

[112] S.D. Herman, A.K. Mehrotra, A heat-transfer model for tube fouling in the radiant section of once-through steam generators, Canadian Journal of Chemical Engineering 99 (2021) 789–802.

[113] A. Ahmed, K.K. Esmaili, M.A. Irfan, et al., Design methodology of heat recovery steam generator in electric utility for waste heat recovery, International Journal of Low-Carbon Technologies 13 (2018) 369–379.

[114] B. Clements, High pressure direct contact oxy-fired steam generator, Patent US9512999, 2016.

[115] S. Ashoori, I.D. Gates, Carbon intensity of direct contact vs once-through steam generation for oil sands operations, Journal of Cleaner Production (2022) (submitted).

[116] R.G. Moore, C.J. Laureshen, J.D.M. Belgrave, et al., In situ combustion in Canadian heavy oil reservoirs, Fuel 74 (1995) 1169–1175.

[117] M.G. Ursenbach, R.G. Moore, S.A. Mehta, Air injection in heavy oil reservoirs – A process whose time has come (again), Journal of Canadian Petroleum Technology 49 (2010) 48–54.

[118] J. Sharma, I.D. Gates, Interfacial stability of in situ bitumen thermal-solvent recovery processes, SPE Journal 16 (2011) 55–64.

[119] A. Haddadnia, M. Zirrahi, H. Hassanzadeh, et al., Dimethylether—a promising solvent for ES-SAGD, in: SPE-189741-MS, SPE Canada Heavy Oil Technical Conference, March 13–14, 2018, Calgary, Alberta, Canada.

[120] W. Zhang, S. Youn, S., Q. Doan, Understanding reservoir architectures and steam-chamber growth at Christina Lake, Alberta, by using 4D seismic and cross well seismic imaging, SPE Reservoir Evaluation & Engineering 10 (2007) 446–452.

[121] M. Medina, C. Torres, J. Sanchez, et al., Real-time downhole monitoring of electrical submersible pumps rated to 250°C using fiber optic sensors: case study and data value in the Leismer SAGD project, in: SPE-153984-MS, SPE Western Regional Meeting, March 21–23, 2012, Bakersfield, California, USA.

[122] H. Mansir, M. Rimmer, L. Waldner, et al., Hybrid permanent magnet motor applications to electric submersible pumps in SAGD wells, in: SPE-204499-MS, SPE Gulf Coast Section Electric Submersible Pumps Symposium, October 4–8, 2021, Woodlands, Texas, USA.

[123] K. Bybee, Predicting flow distribution for SAGD producing wells by use of permanently installed fiber optics, Journal of Petroleum Technology 59 (2021) 57–59.

[124] M. Soroush, M. Mohammadtabar, M. Roostaei, et al., Fiber optics application for downhole monitoring and wellbore surveillance; SAGD monitoring, flow regime determination, and flow loop design, in: SPE-199938-MS, SPE Canada Heavy Oil Conference, September 28-October 2, 2020, virtual.

[125] W. MacPhail, J. Kirkpatrick, B. Banack, et al., SAGD production observations using fiber optic distributed acoustic and temperature sensing: SAGD-DAS – Listening to wells to improve understanding of inflow, in: SPE-180726-MS, SPE Canada Heavy Oil Conference, June 7–9, 2016, Calgary, Alberta, Canada.

[126] C. Zimmer, D. Richter, J. Person, et al., Drilling a better pair: new technologies in SAGD directional drilling, Journal of Petroleum Technology 51 (2012) 115–126.

[127] X. Chen, S. Sun, D. Tong, CT diagnostics of well trajectory and completion decision guidance for SAGD paired horizontal wells, in: SPE-193368-MS, SPE Thermal Well Integrity and Design Symposium, November 27–29, 2018, Banff, Alberta, Canada.

[128] M. Ozdemirtas, T. McGinn, J. Mejia, et al., Innovative fishbone SAGD well pair: an integrated approach to efficiently unlock the resource potential in Canadian oil sands play, in: SPE-173161-MS, SPE/IADC Drilling Conference and Exhibition, March 17–19, 2015, London, England, UK.

[129] A.R. Brandt, J. Englander, S. Bharadwaj, The energy efficiency of oil sands extraction: energy return ratios from 1970 to 2010, Energy 55 (2013) 693–702.

[130] Government of Canada, Energy and Greenhouse Gas Emissions (GHGs), 2019 https://www.nrcan.gc.ca/sites/www.nrcan.gc.ca/files/energy/Climate-change/pdf/CCCR_FULLREPORT-EN-FINAL.pdf. ISBN: 978-0-660-30222-5.

[131] M. Hannouf, G. Assefa, I.D. Gates, Policy insights to accelerate cleaner steam-assisted gravity drainage operations, Energies 15 (2022) 86.

[132] M. Hannouf, G. Assefa, I.D. Gates, Carbon intensity threshold for Canadian oil sands industry using planetary boundaries: is a sustainable carbon-negative industry possible? Renewable and Sustainable Energy Reviews 151 (2021) 111529.

[133] Oil Sands Pathways to Net Zero <https://www.oilsandspathways.ca>, 2022.

[134] R.G. Jonasson, M. Imran, K.D. Knorr, Screening surfactants for application in SAGD – Including a steam-phase coreflood test, in: SPE-189760-MS, SPE Canada Heavy Oil Technical Conference, March 13–14, 2018, Calgary, Alberta, Canada.

[135] L.A. Adetunji, A. Ben-Zvi, A. Filstein, Foam formulation for high temperature SAGD applications, in: SPE-198919-MS, SPE Thermal Well Integrity and Design Symposium, November 19–21, 2019, Banff, Alberta, Canada.

[136] R. Li, R. Etminan, Z. Chen, Chemical additives and foam to enhance SAGD performance, SPE-174489-MS, in: SPE Canada Heavy Oil Technical Conference, June 9–11, 2015, Calgary, Alberta, Canada.

[137] J. McEvers, B. Kent, Drilling with 100% natural gas engines, in: SPE-173072-MS, SPE/IADC Drilling Conference and Exhibition, March 17–19, 2015, London, England, UK.

[138] D.G. Hill, J. Johnson, J. Bell, et al., Natural gas as a fuel in drilling operations—analysis, testing, and implementation, in: SPE-140357-MS, SPE/IADC Drilling Conference and Exhibition, March 1–3, 2011, Amsterdam, The Netherlands.

[139] H. Pinto, X. Wang, I.D. Gates, On the ratio of energy produced to energy injected in SAGD: long-term consequences of early stage operational decisions, Journal of Petroleum Science and Engineering 199 (2021) 108271.

CHAPTER

Janus nanofluids for enhanced oil recovery

10

Wei Wang

Aramco Americas: Aramco Research Center-Boston, Cambridge, MA, United States

10.1 Introduction

Enhanced oil recovery (EOR) has been widely used to recover residual oil after the primary or secondary oil recovery processes. Conventional EOR techniques comprise chemical, thermal, microbial, miscible, and immiscible gas techniques. Among the conventional methods, chemical EOR has been widely used in oilfields with demonstrated high efficiency of oil displacement and low operational costs. Current chemical EOR techniques include the use of polymers, surfactants, or alkalis to improve sweep efficiency and oil displacement. The recovery percentage of conventional chemical EOR techniques, however, has reached a bottleneck, and technology innovation in EOR is needed for further improvement of oil production.

Nanofluids are aqueous fluids containing suspended nanoscale materials, usually colloidal nanoparticles (NPs), with or without the addition of other chemicals. The injection of nanofluids into the oil reservoir, that is, nanofluid flooding or nanoflooding has received extensive attention owing to its great potential as a novel chemical EOR technique. Compared with conventional chemicals used in water flooding, engineered nanomaterials could possess a variety of functions and properties by manipulating their morphologies and surface chemistry. The unique functions of nanomaterials allow nanofluids to recover those inaccessible oil reserves using conventional methods. Nanofluids have the advantages of low cost, high oil recovery efficiency, and wide applicability. Laboratory studies have demonstrated that nanofluids can alter the wettability of minerals, reduce interfacial tension (IFT) at oil/water interfaces, change the viscosity of fluids, and generate structural disjoining pressure at oil/rock interfaces. The applications of nanofluids in EOR have been summarized in literature reviews from the nanomaterials and petroleum engineering perspectives [1–12].

Janus particles, named after the Roman god "Janus" whose two faces look into the past and the future, have two different constituent materials and thus performances are distinctly compartmented on the same particle. Since Casagrande et al. [13] reported the first work related to the synthesis of Janus glass beads with one of the hemispheres hydrophilic and the other hydrophobic in 1988 and

Recovery Improvement. DOI: https://doi.org/10.1016/B978-0-12-823363-4.00010-8

Pierre-Gilles de Gennes [14] mentioned the concept of the "Janus Grain" in his Nobel lecture in 1991, over the past 30 years, a large number of nanoscaled Janus materials have been developed that have led to a wide range of new applications in macro/microstructures, electronic thin films, sensors, oil/water separation membranes, cell and drug encapsulations, and biomimetic interfaces [15–24], as well as in the field of EOR [25].

In this review, we highlighted the most recent advances with respect to the preparation of Janus nanomaterials by bulk phase processes, discussed the special features of Janus nanostructures, summarized different Janus nanomaterials that have been reported to be used for nanofluid flooding in the lab, and analyzed the functions of Janus nanoparticles (JNPs) affecting nanofluid flooding operations. We also discussed the mechanisms of JNPs in EOR, such as viscosity variation, IFT reduction, wettability alteration, and simulation studies of JNPs in EOR. Finally, the challenges and prospects and some suggestions for future studies to use JNPs in EOR processes are presented.

10.2 Distinct chemical and physical properties of Janus nanoparticles

The breaking symmetry of JNPs including 0D nanospheres, 1D nanorods, and 2D nanosheets brought distinct separated domains within one NP compared with uniform NPs, and typical Janus nanostructures are graphetically illustrated in Fig. 10.1. Each domain is capable of individual surface functionalization. Compared to homogenous NPs, JNPs exhibit fascinating amphiphilic, optical, magnetic, and catalytic properties, and the overall combinations resulted in plenty of JNPs for wide applications. For example, when JNPs are fabricated with asymmetric optical and magnetic properties, remote control of JNPs over an electrical or magnetic field is feasible. When separated compartments/surfaces are composed of polymers with

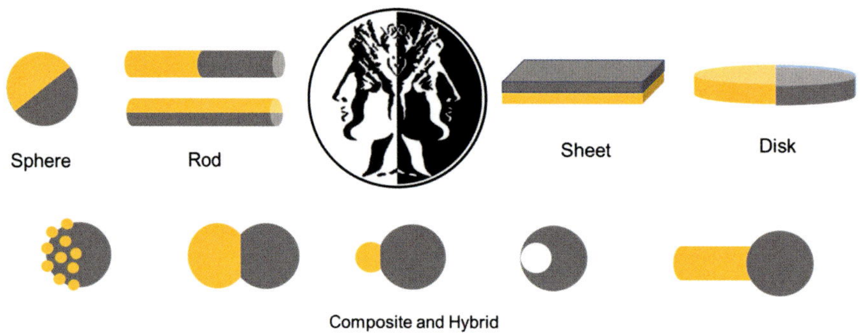

FIGURE 10.1

Typical types of Janus nanostructures.

different stimuli-responsive degradation properties, JNPs are capable of dual stimuli controlled degradation where the multistep controlled delivery and stimuli controlled release behavior become possible.

For application in EOR, amphiphilic JNPs with both hydrophobic and hydrophilic surfaces are important, since they could exhibit surfactant-like property and reflects properties from both NP and surfactant. The shape of the JNPs and the chemistry of functional groups on their surface have vital effects on the JNPs' interfacial behavior. JNPs usually exhibit higher interfacial activity than non-Janus NPs and thus are more effective in reducing the IFT in comparison to non-Janus NPs. Nonspherical JNPs in the shapes of rods, disk-like, sheet, wedge, etc. have been developed recently that showed a significant role at the oil–water interface to stabilize the emulsions. Nanofluids with JNPs have a higher viscosity than ones with NPs, which is a positive factor in the waterflooding process for oil movement. In addition, inorganic JNPs are stabler at high pressure and high-temperature reservoir condition than other chemicals, such as organic surfactants and polymers.

10.3 Synthesis and scale up production of Janus nanoparticles

Janus nanomaterials are of major importance for a wide variety of industrial applications ranging from particulate surfactants to medical treatments. Although various methods have been developed to produce JNPs, to date, major challenges facing the exploration and utilization of JNPs are facile synthetic routes, scale-up procedures, and ability to introduce a diverse set of functionalities. Especially, the scalability of synthesis must need to be addressed for JNPs to succeed in commercially driven applications. Generally, synthetic strategies for the fabrication of JNPs can be mainly grouped into three categories: masking, phase separation, and self-assembly [16], as presented in Fig. 10.2. Each of the strategies has some advantages and drawbacks.

JNPs can be made by either bottom-up, self-assembly-based techniques or top-down, micro/nanofabrication-based techniques. Both bottom-up and top-down fabrication techniques have their pros and cons, and they are complementary to each other. Together, these two distinctly different approaches will be able to generate a wide range of JNPs to fully achieve their potential in the application. For all these strategies, however, challenges remain in realizing controllable, scalable surface elective, and orthogonal functionalization.

10.3.1 Phase separation

The phase separation strategy is one of the most well-established bottom-up approach in the synthesis of JNPs. Two or more different chemical components in

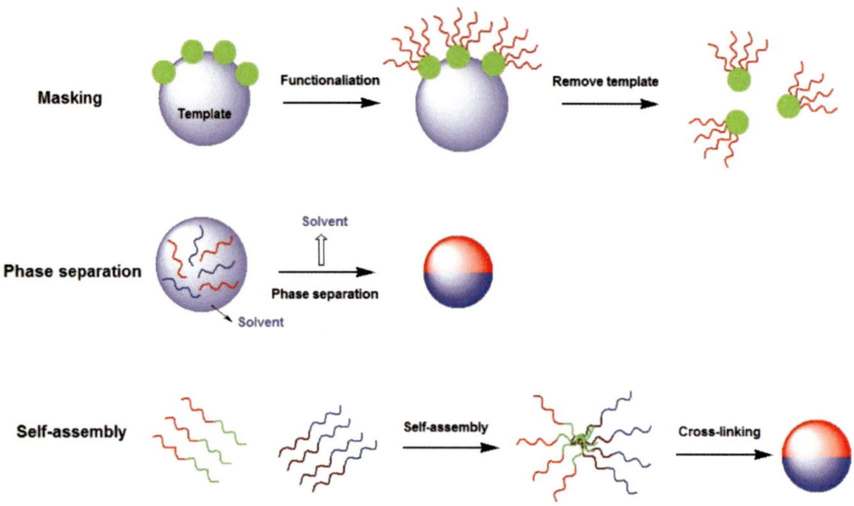

FIGURE 10.2

Schematic diagram illustrating the three main strategies for the preparation of Janus nanoparticles: masking, phase separation, and self-assembly.

Reproduced with permission from X. Fan, J. Yang, X.J. Loh, et al., Polymeric Janus nanoparticles: recent advances in synthetic strategies, materials properties, and applications. Macromolecular Rapid Communications 40 (2019) 1800203.

a mixed reaction can self-assemble in a controllable way to form Janus nanostructures. Although this technique has been widely exploited to fabricate JNPs comprising of inorganic materials and nanocrystals, phase separation has become the most practical method for synthesizing polymeric JNPs [23,26]. The mechanisms involved in the phase-separation reaction mainly the interaction between different polymers, and the interaction between polymers and other materials that are differed from polymers. Due to the unique behavior of fluids in the micrometer scale, the phase separation process can also be carefully controlled in microfluidic channels to form Janus droplets and solidify them into Janus particles by various types of solidification methods. The microfluidic method based on the phase separation strategy provides a unique platform to produce Janus particles on a larger scale in a continuous fashion [27,28], as demonstrated in Fig. 10.3. However, the resultant particles are typically rather large due to the relatively large size of fluidic channels. Thus the main problem is that the size of these formed particles ranges from one to hundreds of micrometers.

10.3.2 Self-assembly

The self-assembly strategy used to prepare JNPs relied on either the self-assembly of block copolymers or the competitive adsorption of incompatible ligands on the

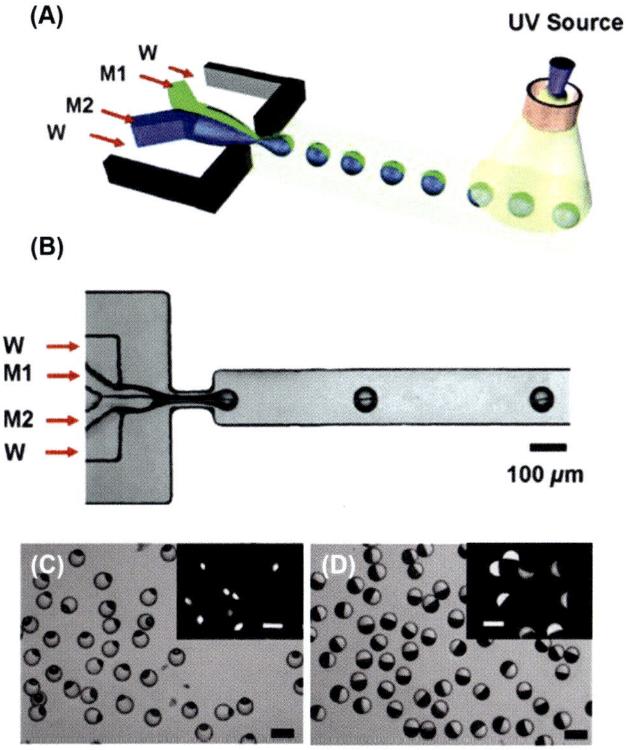

FIGURE 10.3

(A) Schematic representation of the generation of Janus droplets from immiscible monomers M1 and M2, emulsified in an aqueous solution of sodium dodecyl sulfate (W). The droplets are irradiated with ultraviolet light in the downstream channel. (B) Optical microscopy image of a formation of Janus droplets. (C and D) Optical microscopy images of Janus particles, bright and dark phases are polymers of M1 and M2, respectively.

Reproduced with permission from Z. Nie, W. Li, M. Seo, et al., Janus and ternary particles generated by microfluidic synthesis: design, synthesis, and self-assembly, Journal of the American Chemical Society 128 (2006) 9408–9412.

surfaces of preprepared homogeneous or core–shell particles. The self-assembly techniques of block copolymers in solution or bulk to prepare JNPs involve confining several individual polymers or blocks together simultaneously and is a flexible method to arrange different kinds of polymers orderly. Nanostructures and the geometric appearance of assembled JNPs are controlled through varied solubility of the polymers in solvent and their intrinsic physical properties. The synthesis of Janus particles by self-assembly via block copolymers was first described in 2001 by Erhardt et al. [29]. They produced a triblock polymer from polymethylacrylate, polystyrene, and low-molecular-weight polybutadiene.

The polystyrene and polymethylacrylate formed alternating layers in between which polybutadiene sat in nanosized spheres, and then, the blocks were cross-linked. After dissolving by THF and several washing steps, spherical Janus particles with polystyrene on one face and polymethylacrylate on the other, with a polybutadiene core are obtained. With this self-assembly strategy, the production of Janus spheres, cylinders, sheets, and ribbons is possible by adjusting the molecular weights of the blocks in the initial polymer and also the degree of cross-linking. However, the scalability of this strategy to prepare JNPs is somehow limited, since assembly of block copolymers at high concentrations is not always a well-controlled process and multistep treatments are needed, often bringing the purity as one of the major problems considered in resulted Jane structures.

10.3.3 Masking

The use of masking templates to prepare JNPs is probably the most flexible synthetic strategy, which is virtually applicable to any type of material and offers the possibility to modify the surface of NPs with a wide variety of functional groups. This strategy utilizes the asymmetric functionalization of NPs by exposing a portion of the surface to a particular environment, where the exposed surface of the NPs can be chemically modified toward a specific function, whereas the protected part of the surface is not modified. The masking process can be achieved through the top-down technique by depositing or immobilizing the NPs onto a solid substrate or through the bottom-up technique by assembling the particles at the interface of two fluid phases, then followed by selective modification on the unmasked surface. However, usually, only milligram level of particles can be produced in one batch although the functionalization of NPs deposited on solid substrates offers the broadest range of functionalization groups. This may limit its mass production and application at an industrial scale.

As another type of masking strategy, the Pickering emulsion method has been presented by Granick et al. [30] for the bottom-up synthesis of JNPs. In this method, Pickering emulsion templates are stabilized by solid NPs instead of molecular surfactants. During the synthesis, untreated solid NPs organized at liquid–liquid interface of emulsified molten wax and water to lower the overall surface energy of the Pickering emulsion system. The NPs are frozen in place when the wax solidifies. As a kind of masking technique, the exposed surfaces of the immobilized particles are modified chemically to achieve an asymmetric structure. Janus particles are obtained by dissolving the wax mask, and the inner surfaces can be further modified chemically. With this approach, Janus particles are able to be produced in gram-sized quantities because Pickering emulsions are easily prepared, and emulsion droplets normally have a large surface area [30].

Based on the Pickering emulsion approach, the syntheses of spherical JNPs, such as SiO_2, and 2D Janus nanosheets, such as graphene oxide (GO), have been demonstrated by several research groups. For example, by combining a Pickering-type emulsion and grafting-from polymerization via atom-transfer

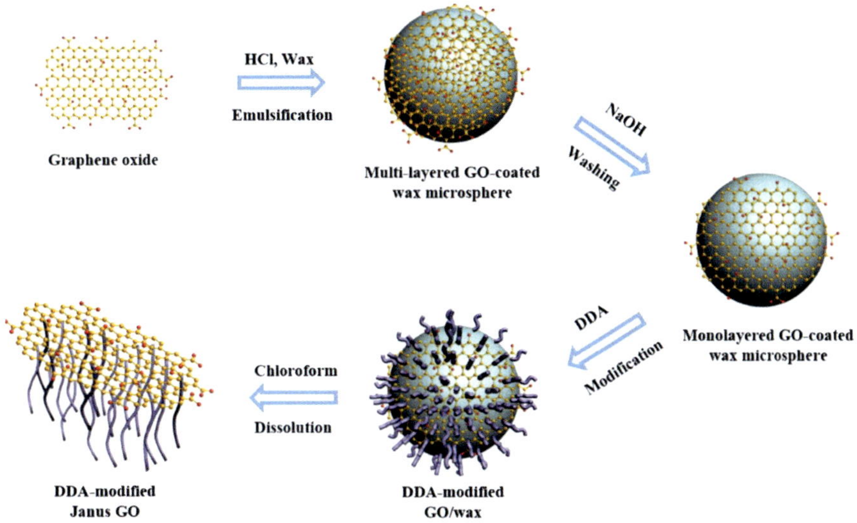

FIGURE 10.4

Schematic illustration of the preparation of asymmetric dodecylamine modified Janus graphene oxide nanosheets via Pickering emulsion template.

Reproduced with permission from H. Wu, W.Y. Yi, Z. Chen, et al., Janus graphene oxide nanosheets prepared via Pickering emulsion template, Carbon 93 (2015) 473–483.

radical polymerization, Janus graphene oxide (JGO) nanosheets bearing poly (methyl methacrylate) on one face as well as the symmetrically functionalized analog were prepared [31,32]. This work also demonstrated the direct visualization of polymer grown from the selective surface of GO. The asymmetric structure of JGO was also achieved by the graft of amino-containing chemicals on the selected surface in the wax-in-water Pickering emulsion template [33]. A typical procedure for the synthesis of dodecylamine modified JGO is schematically illustrated in Fig. 10.4.

However, this method is still not practical for scale-up in a reasonable manner to industrial application. Due to the limited interfacial area available in Pickering emulsions, it is technically difficult to produce JNPs at a larger than gram level in one batch. More efficient preparation of Pickering emulsions is required for the development of larger-scale production processes of JNPs.

10.3.4 Interfacial reaction

Probably the simplest and straightforward techniques used to prepare JNPs is through interfacial reaction in water–organic binary phase system. In literature reviews [16], this simple method was generally not considered one of the major strategies for the synthesis of Janus nanomaterials, due to its low efficiency with the little interfacial area available for reactions. By assembling the NPs onto the interface of water/organic phases, chemical reactions can occur in controllable and selective ways from the water phase side or the organic phase side and thus result in the formation of the asymmetric structure of NPs. For example, monodispersed $CsPbX_3/SiO_2$ and $CsPbBr_3/Ta_2O_5$ JNPs were successfully prepared by combining a water-triggered transformation process and a sol − gel process at water–hexane interface [34]. By treating Cs_4PbX_6 NCs dispersed in a nonpolar solvent with water, the formation of $CsPbX_3$ NCs was achieved at the water/hexane interface, which may provide an opportunity for selective NP surface modification via an effective sol–gel process to produce highly stable monodisperse $CsPbX_3/SiO_2$ and $CsPbBr_3/Ta_2O_5$ JNPs.

Langmuir-based methods have been adopted as an effective way for the production of JNPs that exhibit hydrophobic characters on one side and hydrophilic on the other, as depicted in Fig. 10.5. In the study by Pradhan et al. [35], *n*-hexanethiolatecapped hydrophobic gold NPs with average core diameter ~2 nm as the starting material were spread on water surface in Langmuir trough, then the laterally compressed dense gold nanoparticulate monolayer have a ligand-exchange reaction on the lower hemisphere from the water phase with hydrophilic 3-mercaptopropane-1,2-diol, which resulted amphiphilic JNPs.

Recently, Jeon et al. [36] have self-assembled functionalized GO monolayer at immiscible interfaces of water and ethyl acetate, enabling each surface to react with distinct chemical functions through Meisenheimer complex formation. JGO nanosheets with amphiphilic properties have been demonstrated, and typical synthesis procedure is graphically illustrated in Fig. 10.6. In principle, various functional pairs can be covalently bonded onto each surface of graphene to create a wide diversity of JGO. This approach provides general access to the creation of diverse surfactant materials and promising building blocks for 2D Janus nanomaterials.

Because the interfacial reactions only performed with a monolayer of materials at the interface in a biphasic system, resulting very limited Janus materials, and thus the major disadvantage for this method is limited potential for scaling up. To solve the problem, Wang et al. [37,38] recently developed an interfacial reaction system with stacked interfaces at nanometer scale, which hugely amplified interfacial area for chemical reactions in segregated domains. They proposed to use liquid crystal lamellar phases as templates to functionalize or grow Janus nanomaterials. The lyotropic liquid crystal phase medium consists of surfactant, water and oil to generate layered oil–water interfaces with adjustable intervals in ~5–100 nm [39,40], which can easily amplify the interfacial area by more than million times (at $>10^6$ orders), in comparison to the conventional biphasic solution method, as schematically illustrated in Fig. 10.7.

Using the lyotropic liquid crystal lamellar phase as "Interfacial Nano-Reactors" [37,38], the process for functionalizing 2D Janus graphene nanosheets is graphetically described in Fig. 10.8. In brief, hydrophilic GO nanosheets with –OH groups on their surfaces were first incorporated into the water layers of a

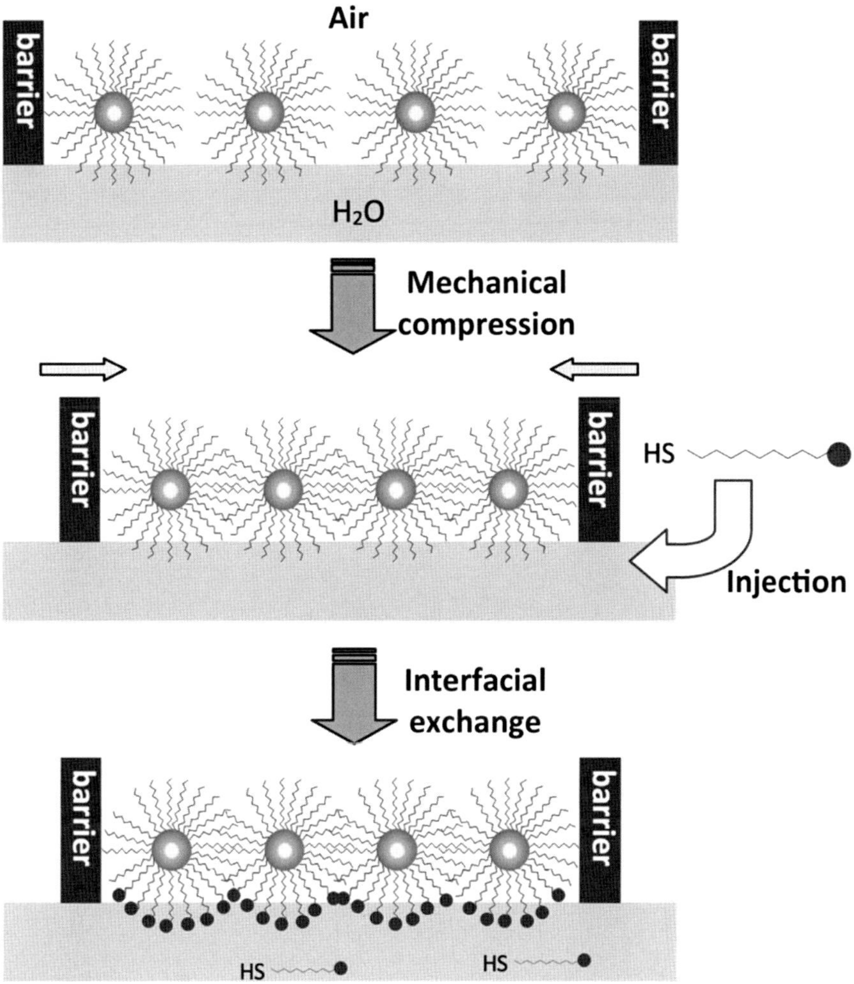

FIGURE 10.5

Schematic representation of the preparation of Janus nanoparticles based on the Langmuir technique [35].

Reproduced with permission from S. Pradhan, L.P. Xu, S. Chen, Janus nanoparticles by interfacial engineering, Advanced Functional Materials 17 (2007) 2385–2392.

lamellar phase, and then oil-soluble silane coupling agent with hydrophobic groups was introduced into the oil layers of the lamellar phase. By initiating the hydrolysis reaction with adding ammonia as catalyst to the water layer, the silane molecules were attracted onto the water/oil interface of water and oil to chemically binding with one side of the GO nanosheets, which resulted JGO nanosheets with hydrophilic–hydrophobic property. This method can be easily scaled up to

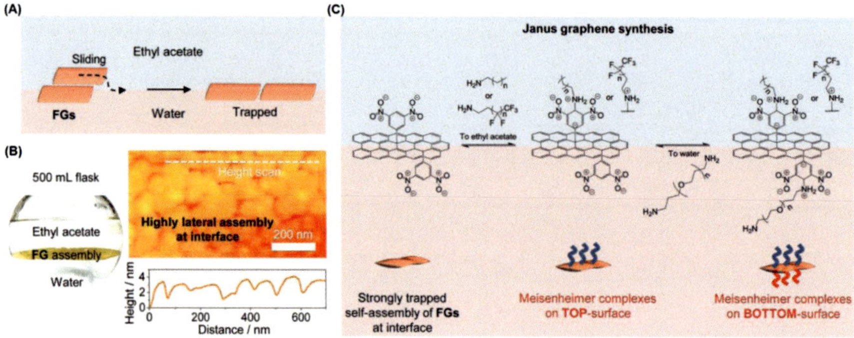

FIGURE 10.6

Self-assembly of functionalized graphene and Janus graphene synthesis at a liquid–liquid interface. (A) An assembly of individualized functionalized graphene sheets forms at the ethyl acetate–water interface, as a result of the high density of 3,5-dinitrophenyl groups on the π-conjugated surfaces that significantly reduce interlayer stacking. (B) Left: a photograph of the large area functionalized graphene assembly in a 500-mL flask. Right: atomic force microscopy topology of the functionalized graphene assembly that was lifted from the liquid interface onto a mica substrate. (C) Synthetic procedure for the preparation of 2D Janus graphene using orthogonal Meisenheimer complex formation on interfacially trapped functionalized graphene with 3,5-dinitrophenyl groups.

Reproduced with permission from I. Jeon, M.D. Peeks, S. Savagatrup, et al., Janus graphene: scalable self-assembly and solution-phase orthogonal functionalization. Advanced Materials 2019 1900438.

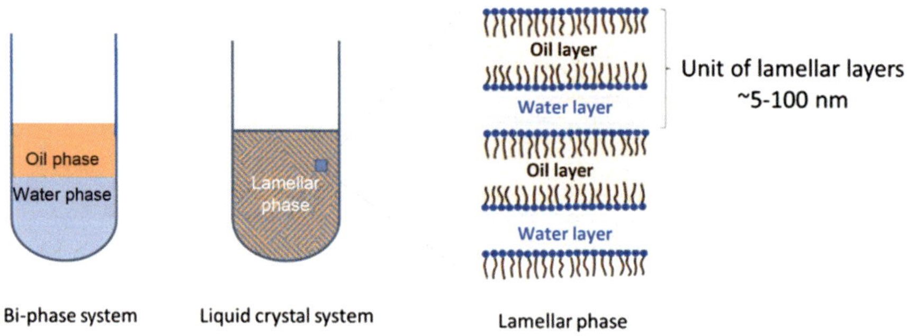

FIGURE 10.7

Graphic illustration of the biphase and the liquid crystal lamellar phase.

Reproduced from S. Chang, W. Wang, Bulk synthesis of Janus nanomaterials. US Patent application no. 16/818632, 2019.

produce the Janus nanomaterials at above gram level in a small reaction volume, which demonstrates high potential in larger quantity synthesis for industrial application. Interfacial activity of the synthesized JGO was characterized by surface pressure – area ($\pi - A$) isotherms with a Langmuir trough which evaluates the

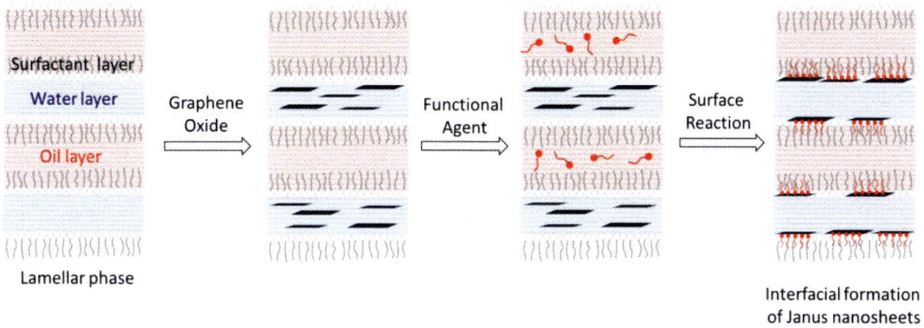

FIGURE 10.8

Schematic illustration of formation of Janus graphene nanosheets oil–water interfaces in lamellar phase.

Reproduced from S. Chang, W. Wang, Bulk synthesis of Janus nanomaterials. US Patent application no. 16/818632, 2019.

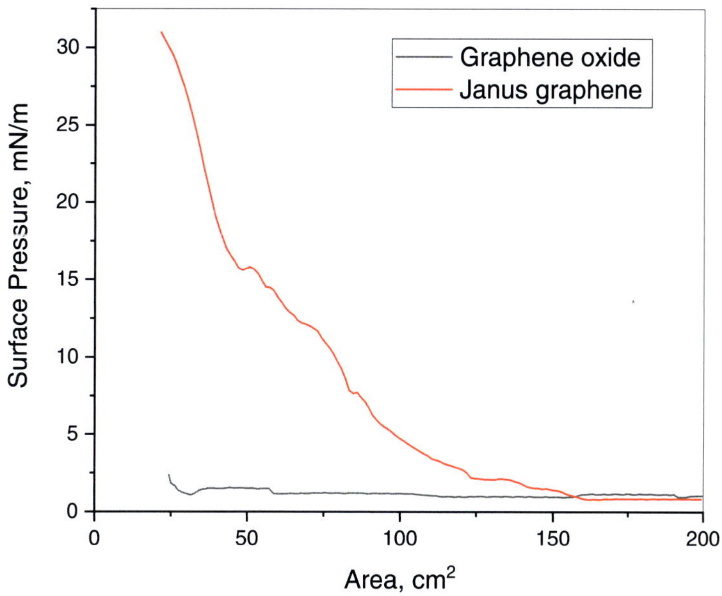

FIGURE 10.9

Surface-are isotherm of hydrophilic graphene oxide and amphiphilic Janus graphene oxide nanosheets. Graphene oxide and Janus graphene oxide suspensions were spread on water at the same concentration in Langmuir–Blodgett trough.

effects of compression on surface pressure. As shown in Fig. 10.9, the $\pi - A$ isotherm profile of JGO nanosheets exhibits a surfactant-like behavior clearly. This phenomenon is consistent with what have been observed for Janus nanosheets [32], as well Janus spherical particles [41] at the water – air interface.

10.4 Advantages of Janus nanofluids for enhanced oil recovery

Chemically EOR methods are utilized to increase the oil recovery by improving the mobility of fluids, altering the wettability of rocks, and/or lowering the IFT between brine and oil, and alkalis, surfactants and polymers have been used for this purpose. In the past decade, NPs have attracted much attention for application in oil recovery enhancement due to their unique colloidal and surface properties. Using NPs either alone or in combination with surfactants and/or polymers suspended in water or brine as nanofluids, a large number of NPs have been investigated. Studies in most of published papers indicate that NPs can be adsorbed on the oil/water interface, resulting in an alteration in the oil/water IFT, although some studies showed contradictory conclusions, as reviewed in literatures [1–12]. The shape, size, material, and concentration of NPs may govern the effect of nanofluid on EOR, however, the crucial roles are played by the surface functionalization of the NPs. With hydrophobic and hydrophilic groups or two different functional groups on the two sides, the JNPs hold advantages with both colloidal and surfactant-like properties. One of the most common applications of amphiphilic JNPs is stabilization of the oil − water interface by lowering of IFT. The non-homogeneous wetting distribution of JNPs can enhance their binding to liquid interfaces which improves the stability of the interfaces in foams and emulsions. Generally, the work done in this area of literature proved that functionalized JNPs have better surface activity than homogeneous normal NPs [42], exhibiting a great potential to achieve higher efficiency on EOR.

10.4.1 Oil recovery efficiency by Janus nanoparticles

Several types of JNPs have been explored for EOR by laboratory research. The experimental protocols in evaluation of EOR efficiency, such as core flooding and microfluidic channel flooding, have been well established. The mostly studied JNPs in EOR applications are silica-based NPs or graphene-based nanosheets, as summarized in Table 10.1. Rather than these two common types of JNPs, other types, such as amphiphilic molybdenum disulfide nanosheets and magnetic JNPs, have been also studied [50,51]. Laboratory studies have demonstrated that the JNPs can improve interfacial and rheological parameters in the oil–water–rock system, including fluids viscosity, IFT, and rock surface wettability.

10.4.1.1 SiO_2-based Janus nanofluids

Yin et al. [43] prepared composite SiO_2-based amphiphilic Janus nanosheets with one side containing hydrophilic carboxyl groups and the other side containing hydrophobic alkyl groups by a bottom-up synthesis strategy. The synthesized Janus nanosheets displayed a distinct ultrathin flake-like morphology with a lateral size of hundreds of nanometers. Core displacement experiments showed that

Table 10.1 Typical JNPs studied in oil recovery enhancement.

Type of JNPs	Method of synthesis	Morphology and functional group	Condition of core flooding	Increment of EOR%	Reference
SiO_2	$CaCO_3$ template/etching	Flakes Carboxyl/Alkyl (C_{18})	Nanofluid: 0.005 wt.% in DI water Core: berea sandstone Oil: crude	18.31%	Yin et al. [43]
SiO_2	Pickering emulsion template	Nanoparticles Si-OH/alkyl (C_{12})	Nanofluid: 0.005 wt.% in 5.0 wt.% NaCl Core: sandstone Oil: crude	13.46%	Wu et al. [44]
SiO_2	Chemical etching	Broken nanoparticles Si-OH/$-CH_3$ fragments	Nanofluid: 0.025–0.2 wt.% /surfactant 0.1–0.2 wt.% in DI water Core: packed quartz sand Oil: mixture of crude and diesel	7.4%–9.8%	Liu et al. [45,46]
NiO/SiO_2	Pickering emulsion template/ calcination	Nanoparticles NiO/SiO_2	100 mg/L in DI water Core: Ottawa sand micromodel Oil: light crude	25%–50%	Giraldo et al. [47]
Graphene	Pickering emulsion template	Nanosheets –OH/alkylamine	Nanofluid: 0.005 wt.%, 0.01 wt.% in 5% brine Core: man-made sandstone Oil: crude	15.2%	Luo et al. [31]
Graphene	Interfacial reaction	Amine/perfluoroalkyl	Nanofluid: 91 μg/mL in water Core: carbonate microfluidic model Oil: aramco crude	14%	Jeon et al. [36]
Graphene	Pickering emulsion template	Nanosheets –OH/amine (C_{12})	Nanofluid: 100 mg/L with surfactant 1000 mg/L in bine Core: artificial core Oil: heavy crude Temperature: 60°C	19.5%	Jia et al. [48]
Graphene	Interfacial nanoreactor template	Nanosheets –OH/alkyl (C_{18})	Nanofluid: 0.002 wt.% with surfactant in brine Core: artificial $CaCO_3$ micromodel Oil: aramco crude	10.5%	Wang et al. [49]

the Janus SiO_2 nanofluid could significantly increase the efficiency of oil recovery by ~18.31% even under an ultralow concentration of 0.005 wt.% and cause minimal impairment to the permeability.

Wu et al. [44] synthesized SiO_2-based amphiphilic JNPs by grafting organic hydrophobic chains onto one side of SiO_2 NPs using the Pickering emulsion template method. Experimental results showed that the oil–water IFT was reduced down to 2.28 mN/m when the concentration of the SiO_2 JNPs is 0.05 wt.% and the Janus nanofluids could effectively alter the wettability of the rock from oil wet to water wet. The displacement experiments with crude oil (viscosity of 3.43 mPa·s at 50°C) saturated low-permeability sandstone cores indicated that up to 15.74% increments of the oil recovery can be obtained by injecting 0.01 wt.% Janus nanofluids, while only 3.64% by unmodified SiO_2 nanofluid, as displayed in Fig. 10.10.

Liu et al. [45,46] have synthesized SiO_2 JNPs with size of 4 – 9 nm and specific surface area of up to 612.9 m^2/g by a surface-modification and a chemical etching combined route. The Janus nanofluid in deionized (DI) water could be tuned to achieve bidirectional regulation of surface wettability plus a reduction of the oil/water interface tension. Hence, a significant reduction (by 33~50%) of water injection pressure and an EOR (by 21.1% – 26.6%) were achieved.

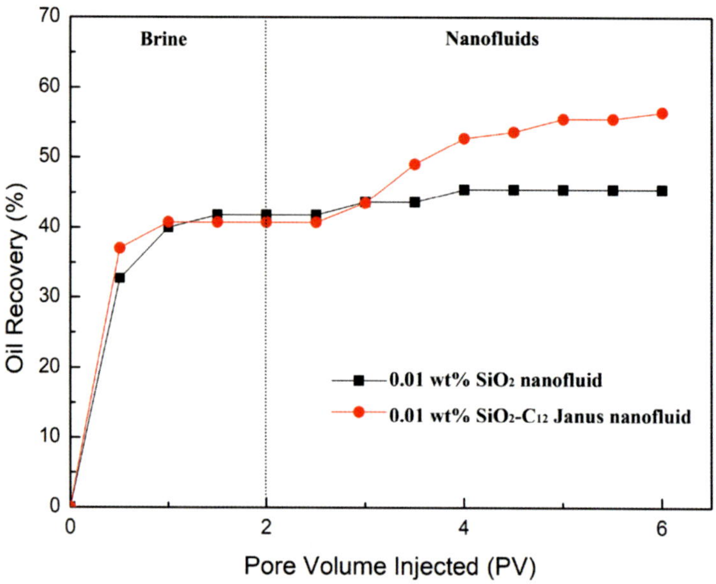

FIGURE 10.10

Oil recovery profiles using nanofluids flooding as tertiary recovery.

Source: Reproduced with permission from H. Wu, K. Gao, Y. Lu, et al., Silica-based amphiphilic Janus nanofluid with improved interfacial properties for enhanced oil recovery. Colloids and Surfaces A: Physicochemical and Engineering Aspects 586 (2020) 124162.

Apart from that, the SiO_2 JNPs are able to increase the viscosity of partially hydrolyzed polyacrylamide by 282.9% and significantly decrease its viscosity loss ratio in brine, causing an EOR of about 36.6%. They also have studied the utilization of SiO_2 JNPs/surfactant nanofluid without ultra-low IFT for improving oil recovery, and observed improved oil recovery by 7.4% − 9.8% with addition low percentage of SiO_2 JNPs in surfactants compared to pure surfactant in a sand-filled tube flooding. However, this evaluation of EOR was performed in DI water and did not consider the effects of ions in brine or connate water. In addition, according to the method for fabricating amphiphilic SiO_2 JNPs by a one-step NaOH etching in solution, the characterization did not provide solid evidence on the formation of Janus nanostructures, although the NPs exhibit amphiphilic properties.

10.4.1.2 Graphene-based Janus nanofluids

Luo et al. [31,52] produced the graphene nanosheets with an amphiphilic surface property by tuning the Janus balance of the GO with alkylamine, and performed core flooding experiment with the Janus nanofluids in crude oil saturated man-made sandstone rock. At very low concentration of 0.01 wt.%, the nanofluid of JGO nanosheets achieved about 15.2% enhancement in oil recovery, in comparison to those of conventional nanofluid flooding methods under similar conditions. Their results revealed that the nanosheets spontaneously approached the oil–water interface and reduced the IFT in a saline environment (4 wt.% NaCl and 1 wt.% $CaCl_2$), regardless of the solid surface wettability.

Jeon et al. [36] produced amphiphilic Janus graphene nanosheets through a dynamic covalent assembly process which directs the functionalization of graphene surfaces at liquid–liquid interface. To demonstrate usage of Janus nanofluids in EOR applications, a microfluidic-based reservoir micromodel has been used to understand oil–water–rock phase interactions and visualize the fluid transport processes in micropores. The micromodel developed by Wang et al. [49,53] for carbonate reservoir has excellent optical transparency, allowing to directly visualize the complex multiphase flows and geochemical fluid − calcite interactions by spectroscopic and microscopic imaging techniques. As showed by fluorescence imaging in Fig. 10.11, the oil distribution after flooding the micropores in the microfluidic chip using water and Janus nanofluid (at concentration of 91 μg/mL in DI water) reveals a 14% improvement in oil recovery by the Janus nanofluid.

Jia et al. [48] have modified GO nanosheets by dodecylamine at one side and evaluated interfacial activities of JGO in various oil/water systems. IFT measurement showed the JGO had greater interfacial activity than unmodified GO, and the IFT at heavy oil/water interface was significantly reduced to 0.12 mN/m at concentration of 100 mg/L of the JGO nanofluid. Furthermore, to achieve ultra-low level ($<10^{-2}$ mN/m) of IFT in heavy oil/water system, different types of surfactants, including dodecyltrimethylammonium bromide (DTAB), sodium dodecyl sulfate, and Triton X-100 (TX-100) were used to investigate their synergistic

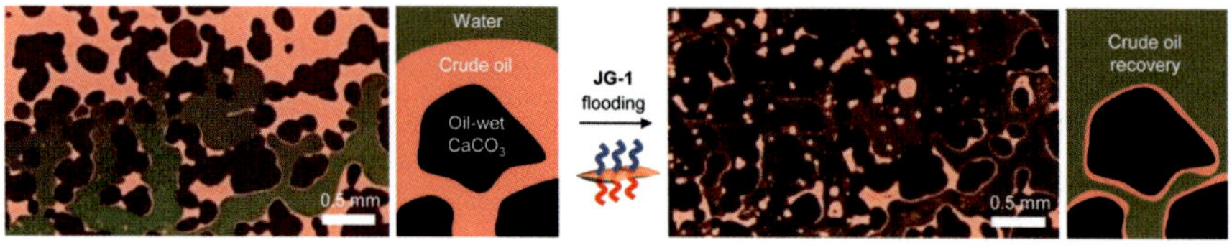

FIGURE 10.11

Janus graphene injection in a microfluidic system that emulates oil in a carbonate mineral formation. Left: after water (green-dyed) flooding, 39 ± 2% (determined by image processing) of crude oil (red) remained. Right: after a Janus graphene water solution flooding, 25 ± 1% of crude oil remained, indicating enhanced oil removal.

Reproduced with permission from I. Jeon, M.D. Peeks, S. Savagatrup, et al., Janus graphene: scalable self-assembly and solution-phase orthogonal functionalization. Advanced Materials 2019 1900438.

effects with JGO. The JGO/DTAB system exhibited excellent synergistic effects to dramatically decrease the heavy oil/water IFT to 8×10^{-3} mN/m, which was responsible for its remarkable ability to improve heavy oil recovery (19.5%) in core flooding tests.

Based on the "Interfacial Nano-Reactors" developed by Wang et al. [37,38] for Janus graphene synthesis, they further proposed to use petroleum surfactants, instead of expensive pure compounds as surfactant, and use crude oil as organic phase, instead of expensive pure organic solvents to build the "Interfacial Nano-Reactor." Upon the formation of Janus materials within the nanoreactors, no post-purification for the synthesized nanomaterial is necessary and all the components in the reactions can be used to play synergetic roles for enhance oil recovery process [49]. By evaluation with microfluidic micromodel of carbonate, additional enhancement of 10.5% in oil recovery was achieved, in comparison with using surfactant only.

10.4.1.3 Other types of Janus nanofluids

Giraldo et al. [47] prepared NiO/SiO_2 JNPs and the results showed a sharply increase of the capillary number (Nc) at a very low concentration of 100 mg/L of the JNPs, which mainly attributes to the decrease in the IFT and thus promotes the increase of the oil recovery. Oil displacement tests using a nanofluid based on the NiO/SiO_2 JNPs at the concentration of 100 mg/L showed an increase in oil recovery up to 50%. Raj et al. [50] have synthesized amphiphilic MoS_2 nanosheets and performed core flooding experiments in man-made core samples with various permeabilities in range of 2.5–2500 mD saturated with crude oil (100 cP) or model oil (25 cP). The brine with the salinity of 98,171 mg/L is used for secondary flooding and nanofluid containing 0.005 wt.% of MoS_2 nanosheets is used for tertiary flooding. Recovery factors of 21.18% and 18.25% of residual oil were observed in the tertiary flooding with 25 mD core samples saturated with model and crude oil respectively. However, the characterization did not provide solid evidence on the formation of Janus nanostructure. Glaser et al. [51] demonstrated that the self-assembly of synthesized Au/Fe_3O_4 Janus particles at the hexane–water interface results in a significant decrease in the IFT when compared to non-Janus particles. However, these JNPs systems are either not economic or complex in fabrication which limits their application at industrial scale.

10.4.2 Mechanisms of Janus nanofluids in enhanced oil recovery

The movement behaviors of NPs in porous media have been extensively investigated, focusing on the flow and rheological properties, and adsorption properties of the NPs on rocks. Mechanisms of oil displacement by nanofluids have been widely discussed in literatures. In general, the mechanisms mainly include disjoining pressure, pore channel plugging, increasing the viscosity of injection fluids, IFT reduction, wettability alteration, and preventing asphaltene precipitation, as schematically illustrated in Fig. 10.12. However, for the unique Janus nanofluids, some new phenomena have been observed during the EOR process, and none of these proposed mechanisms could be fully applicable to the experimental observation under different simulated hydrodynamic conditions.

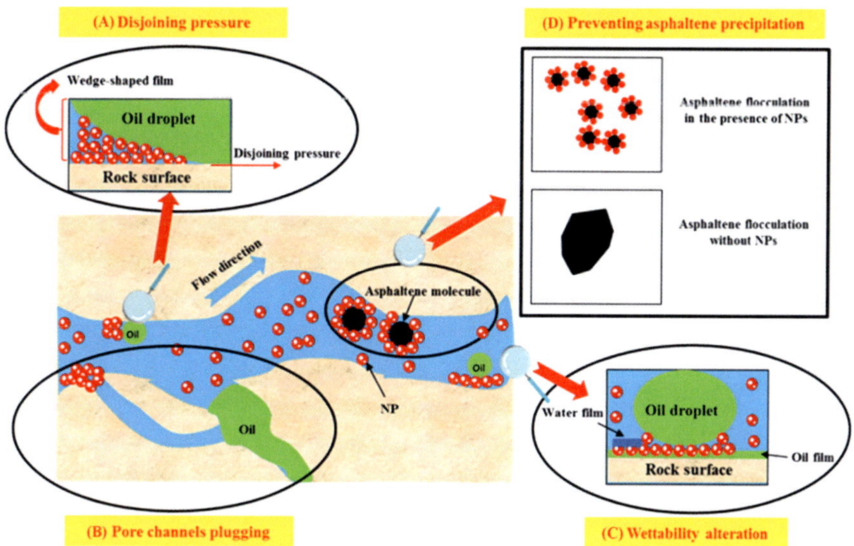

FIGURE 10.12

Schematic illustration of some proposed mechanisms for nanofluids enhanced oil recovery: (A) disjoining pressure, (B) pore channels plugging, (C) wettability alteration, and (D) preventing asphaltene precipitation.

Reproduced with permission from X. Sun, Y. Zhang, G. Chen, et al., Application of nanoparticles in enhanced oil recovery: a critical review of recent progress, Energies 10 (2017) 345.

An interesting phenomenon was the climbing film from JNPs. Lou et al. [31] described that the climbing film appeared and grew at moderate hydrodynamic condition to encapsulate the oil phase when applying the JGO nanosheets to an oil/brine system with increasing hydrodynamic power. With strong hydrodynamic power input, a solid-like interfacial film formed and was able to return to its original form even after being seriously disturbed. The film rapidly separated oil and water phases for slug-like oil displacement. The phenomena were confirmed by Wang et al. [49] at the water/crude oil interface with JGO. As displayed sample photos in Fig. 10.13, at ultralow concentration of 0.005 wt.%, the JGO nanosheets affect the water/crude oil interface significantly. When tilting the fluid sample (Fig. 10.13C), the water/crude oil interface becomes "rigid," implying the formation of a solid-like interfacial film of JGO which could drive oil movement in EOR process.

Jeon et al. [36] indicated that Janus graphene climbs the wall of a glass vial as a result of a high-energy interface created by a thin layer of water on the hydrophilic glass in contact with the hexane (oil phase) to minimizing the interfacial energy, as schematically illustrated in Fig. 10.14. They induced nonspherical emulsions with kinetically jammed Janus graphene by the restriction of equilibrium shapes. When the total area of Janus graphene surfactants oversaturates the area of the liquid interface ($\Sigma A_{JG} > A_{Interface}$), jammed assembly of Janus graphene arrests the nonequilibrium shapes of hexane in water. Climbing films as well as elastic oil–water interfaces made by JNPs can promote oil movement in EOR process.

To understand the high oil recovery enhanced by Janus nanofluid, two oil displacement mechanisms were proposed by Luo et al. [31], as illustrated in Fig. 10.15.

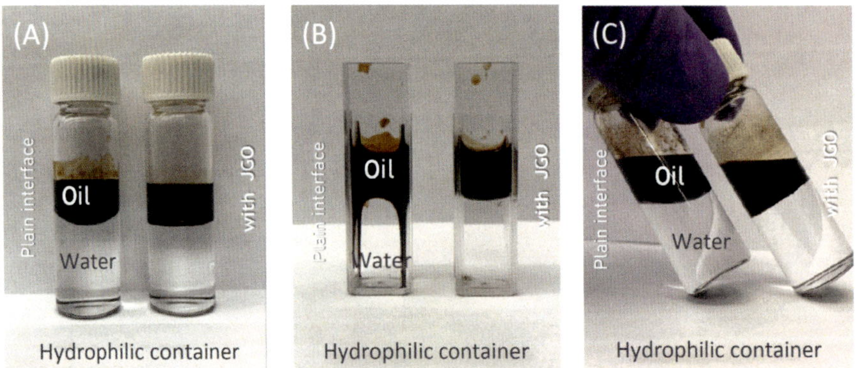

FIGURE 10.13

Photos of samples with and without Janus graphene oxide nanosheets in water-crude oil biphase: (A) in hydrophilic containers, (B) in hydrophobic containers, and (C) in hydrophilic containers with tilted angle of the containers.

Reproduced with permission from W. Wang, S. Chang, A.I. Abdel-Fattah, Enhanced oil recovery with Janus nanoparticles. US Patent application no. 17/381344, 2021.

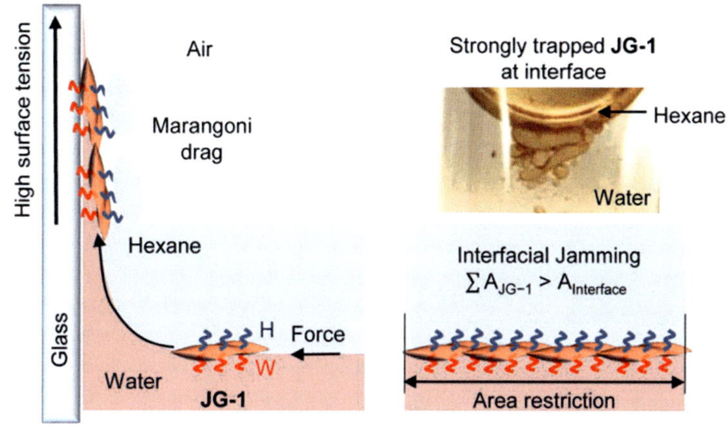

FIGURE 10.14

Left: Janus graphene climbs along the glass vial surface from the water–hexane interface to the air–glass interface with hexane through Marangoni flow. Right: the interfacial jamming of the Janus graphene surfactants is observed.

Reproduced with permission from I. Jeon, M.D. Peeks, S. Savagatrup, et al., Janus graphene: scalable self-assembly and solution-phase orthogonal functionalization. Advanced Materials 2019 1900438.

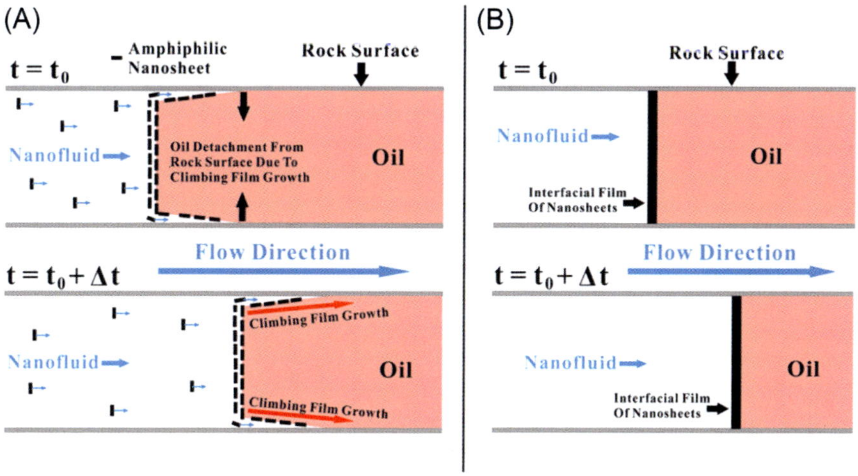

FIGURE 10.15

Schematic illustration of mechanisms of oil displacement by Janus graphene nanosheets. (A) Climbing film encapsulation mechanism for the water-wet surface. (B) Slug-like displacement mechanism.

Reproduced with permission from D. Luo, F. Wang, J.Y. Zhu, et al., Nanofluid of graphene-based amphiphilic Janus nanosheets for tertiary or enhanced oil recovery: high performance at low concentration, Proceedings of the National Academy of Sciences of the United States of America 113 (2016) 7711–7716

- *Mechanism I:*

The climbing film (a film of nanosheets along the solid surface) encapsulation for the water-wet surface. As shown in Fig. 10.15A, at $t = t_0$ the increased concentration of nanosheets due to the adsorption at the oil–water interface produces the concentration gradient leading to the transfer of nanosheets to the three-phase (nanofluid, oil, and rock solid) region, detaching and encapsulating oil from the rock surface. When flow continues under the gentle hydrodynamic condition from t_0 to $t_0 + \Delta t$, the film grows due to the ongoing supply of nanosheets from the nanofluid and carries the oil phase forward.

- *Mechanism II:*

Slug-like displacement by the interfacial film. As seen in Fig. 10.15B, at $t = t_0$ an elastic interfacial film forms at the oil–water interface at strong hydrodynamic power condition. The film can resist bending and also reform after being disrupted. As a result, at $t = t_0 + \Delta t$ oil is slug-like and displaced over a certain distance.

JNPs can be applied to heavy oil systems through adsorption mechanism. JNPs can adsorb on mineral surfaces altering the wettability of minerals from oil-wet to water-wet conditions, and adsorb on the oil/water surface altering the oil/water IFT and resulting in the formation of emulsions. Asphaltenes are also adsorbed on the surface of NPs, which reduces the asphaltene content in heavy oil, resulting in a decrease in the viscosity of oil, which helps in oil recovery. In the heavy oil/water system, it was noticed that both GO and JGO caused larger IFT reduction and the dramatic IFT reduction was closely related to the presence

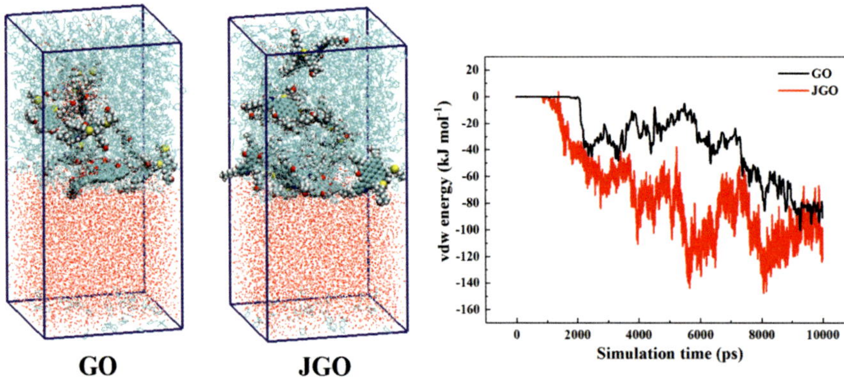

FIGURE 10.16

Simulation configuration of graphene oxide and Janus graphene oxide at the water (red)-simulated heavy oil (cyan) interface and the van der Waals energy of graphene oxide–asphaltene and Janus graphene oxide–asphaltene interaction.

Reproduced with permission from H. Jia, P. Huang, Y. Han, et al., Synergistic effects of Janus graphene oxide and surfactants on the heavy oil/water interfacial tension and their application to enhance heavy oil recovery. Journal of Molecular Liquids 314 (2020) 113791.

of asphaltene in the system [48]. The GO–asphaltene interaction could motivate the adsorption of GO at the heavy oil/water interface, and the additional hydrophobic interaction between the dodecylamine in JGO and asphaltene could cause further IFT reduction. Molecular dynamic simulations were employed to illustrate the interfacial interaction between GO/JGO and oil molecules and indicated that the more adsorbed asphaltene at the interface in the JGO system qualitatively indicated the more intensive JGO–asphaltene interaction. The calculation of van der Waals energy quantitatively reflected the stronger interaction between JGO and asphaltene, which might be attributed to the hydrophobic interaction between the hydrocarbon chain of dodecylamine in JGO and the branched alkyl in asphaltene (Fig. 10.16).

10.5 Conclusion and future perspective

The application of Janus nanomaterials in EOR has a high potential to improve oil productivity. Trying to increase the possibility of material combinations and the amounts of material production in synthesis, a wide range of processes has been developed to produce JNPs. Generally, Janus nanomaterials were prepared by bottom-up methods, including crosslinking of self-assembled block copolymers, selective chemical functionalization of freestanding films, and layer-by-layer self-assembly at air–water interfaces or on Pickering emulsion templates, or by top-down methods, such as masked etching on solid substrates. However, it is difficult to make mass Janus nanomaterials with precise control of size, morphology, and surface property by using such methods at an industrial scale, although some methods, such as the Pickering emulsion template and the "Interfacial Nano-Reactors," have some potential for scaling up. Therefore the bulk synthesis of such asymmetric Janus nanomaterials is of primary importance for their production at an industrial scale and helps to open up the field to commercial applications.

Successful Janus nanofluids floodings were only reported in laboratory research and there is not a report from field trials yet. We think the immaturity of Janus nanofluid flooding is due to a lack of understanding of the oil displacement mechanism and the behavior of NP under the downhole condition. Most of the laboratory investigations were performed at room temperature and ambient pressure, and also did not enough consider the effect of the extreme salinity of brine or connate water under harsh reservoir conditions. The chemical and colloidal stability of Janus nanofluids is crucial for their EOR applications since NPs in nanofluids can easily aggregate and sediment under the reservoir condition. The stability of nanofluids can be highly influenced by the morphology and properties of the NP and the fluid formula variables. More research efforts should be addressed toward the study of interactions between Janus nanofluids and rock surface and oil/water interface at a close real reservoir condition.

In addition, the type of reservoir rocks should be considered since the surface properties of different minerals may affect the interaction between rock and nanofluids, and the requirements for a successful nanofluids EOR can vary case by case. Almost all reported work in nanofluids EOR is based on sandstones, and very little experimental work is done on the carbonate rocks which is the major type of reservoir in the Middle East. The other factors limiting the wide application of nanomaterials in oilfield use include cost and environmental risk assessment.

The mechanisms of nanofluid flooding have generally been considered wettability alteration, reducing IFT, controllable viscosity, disjoint pressure for oil displacement, and stabilization for foam and emulsion, although some research results give controversial conclusions on which mechanism play a dominant role in different systems. Since these existing mechanisms could not explain enough some new phenomena related to Janus nanofluid flooding, two additional mechanisms have been proposed: the climbing film (a film of nanosheets along the solid surface) encapsulation for water-wet surface and the slug-like displacement by the interfacial film. To a fundamental understanding of JNPs behavior in porous media, proper selection of materials, and reaching optimum conditions for injection of JNPs, comprehensive simulation studies are necessary which remains a challenge. It is still a lack of suitable theoretical investigation and mathematical models to accurately describe these processes in EOR. Addressing these challenges can eventually lead to the development of high-efficient and cost-effective Janus nanofluids formulation for oilfield application and bring the nanofluids EOR technology to a promising future.

Nomenclature

AFM	Atomic force microscopy
ATRP	Atom-transfer radical polymerization
DI	Deionized
DTAB	Dodecyltrimethylammonium bromide
EOR	Enhanced oil recovery
FG	Functionalized graphene
GO	Graphene oxide
IFT	Interfacial tension
JG	Janus graphene
JGO	Janus graphene oxide
JNP	Janus nanoparticle
mD	Millidarcies
NP	Nanoparticle
SDS	Sodium dodecyl sulfate
THF	Tetrahydrofuran
vdW	van der Waals

References

[1] M.S. Kamal, A.A. Adewunmi, A.S. Sultan, et al., Recent advances in nanoparticles enhanced oil recovery: rheology, interfacial tension, oil recovery, and wettability alteration, Journal of Nanomaterials 2017 (2017) 2473175.

[2] K. Li, D. Wang, J. Jiang, Review on enhanced oil recovery by nanofluids, Oil & Gas Science and Technology 73 (2018) 37.

[3] B. Peng, L. Zhang, J. Luo, et al., A review of nanomaterials for nanofluid enhanced oil recovery, RSC Advance 7 (2017) 32246–32254.

[4] H. Panchal, H. Patel, J. Patel, et al., A systematic review on nanotechnology in enhanced oil recovery, Petroleum Research 6 (2021) 204–212.

[5] M.S. Alnarabiji, M.M. Husein, Application of bare nanoparticle-based nanofluids in enhanced oil recovery, Fuel 267 (2020) 117262.

[6] C. Negin, S. Ali, Q. Xie, Application of nanotechnology for enhancing oil recovery-a review, Petroleum 2 (2016) 324–333.

[7] M.Y. Rezk, N.K. Allam, Impact of nanotechnology on enhanced oil recovery: a minireview, Industrial & Engineering Chemistry Research 58 (2019) 16287–16295.

[8] P.J. Boul, P.M. Ajayan, Nanotechnology research and development in upstream oil and gas, Energy Technology 8 (2020) 1901216.

[9] Y. Sun, D. Yang, L. Shi, et al., Properties of nanofluids and their applications in enhanced oil recovery: a comprehensive review, Energy and Fuels 34 (2020) 1202–1218.

[10] J.A. Ali, K. Kolo, A.K. Manshad, et al., Recent advances in application of nanotechnology in chemical enhanced oil recovery: effects of nanoparticles on wettability alteration, interfacial tension reduction, and flooding, Egyptian Journal of Petroleum 27 (2018) 1371–1383.

[11] X. Sun, Y. Zhang, G. Chen, et al., Application of nanoparticles in enhanced oil recovery: a critical review of recent progress, Energies 10 (2017) 345.

[12] G. Cheraghian, L. Hendra, A review on applications of nanotechnology in the enhanced oil recovery part A: effects of nanoparticles on interfacial tension, International Nano Letters 6 (2016) 129–138.

[13] C. Casagrande, P. Fabre, E. Raphaël, et al., "Janus beads": realization and behaviour at water/oil interfaces, Europhysics Letters 9 (1989) 251–255.

[14] P.G. de Gennes, Soft matter, Science 256 (1992) 495–497.

[15] X. Zhang, Q. Fu, H. Duan, et al., Janus nanoparticles: from fabrication to (bio)applications, ACS Nano 15 (2021) 6147–6191.

[16] M. Lattuada, T.A. Hatton, Synthesis, properties and applications of Janus nanoparticles, Nano Today 6 (2011) 286–308.

[17] G. Agrawal, R. Agrawal, Janus nanoparticles: recent advances in their interfacial and biomedical applications, ACS Applied Nano Materials 2 (2019) 1738–1757.

[18] Y. Song, S. Chen, Janus nanoparticles: preparation, characterization, and applications, Chemistry: An Asian Journal 9 (2014) 418–430.

[19] Q. Yang, K. Loos, Janus nanoparticles inside polymeric materials: interfacial arrangement toward functional hybrid materials, Polymer Chemistry 8 (2017) 641–654.

[20] Y. Zhang, K. Huang, J. Lin, et al., Janus nanoparticles in cancer diagnosis, therapy and theranostics, Biomaterials Science 7 (2019) 1262–1275.

[21] J. Fu, D. An, Y. Song, et al., Janus nanoparticles for cellular delivery chemotherapy: recent advances and challenges, Coordination Chemistry Reviews 422 (2020) 213467.

[22] F. Liang, B. Liu, Z. Cao, et al., Janus colloids toward interfacial engineering, Langmuir 34 (2018) 4123–4131.

[23] X. Fan, J. Yang, X.J. Loh, et al., Polymeric Janus nanoparticles: recent advances in synthetic strategies, materials properties, and applications, Macromolecular Rapid Communications 40 (2019) 1800203.

[24] E. Poggi, J.F. Gohy, Janus particles: from synthesis to application, Colloid and Polymer Science 295 (2017) 2083–2108.

[25] Z. Tohidi, A. Teimouri, A. Jafari, et al., Application of Janus nanoparticles in enhanced oil recovery processes: current status and future opportunities, Journal of Petroleum Science and Engineering 208 (2022) 109602.

[26] W. Shi, D.A. Weitz, Polymer phase separation in a microcapsule shell, Macromolecules 50 (2017) 7681–7686.

[27] Z. Nie, W. Li, M. Seo, et al., Janus and ternary particles generated by microfluidic synthesis: design, synthesis, and self-assembly, Journal of the American Chemical Society 128 (2006) 9408–9412.

[28] S. Yang, F. Guo, B. Kiraly, et al., Microfluidic synthesis of multifunctional Janus particles for biomedical applications, Lab on a Chip 12 (2012) 2097–2102.

[29] R. Erhardt, A. Böker, H. Zettl, et al., Janus micelles, Macromolecules 34 (2001) 1069–1075.

[30] L. Hong, S. Jiang, S. Granick, Simple method to produce Janus colloidal particles in large quantity, Langmuir 22 (2006) 9495–9499.

[31] D. Luo, F. Wang, J.Y. Zhu, et al., Nanofluid of graphene-based amphiphilic Janus nanosheets for tertiary or enhanced oil recovery: high performance at low concentration, Proceedings of the National Academy of Sciences of the United States of America 113 (2016) 7711–7716.

[32] A. de Leon, B.J. Rodier, Q.M. Luo, et al., Distinct chemical and physical properties of Janus nanosheets, ACS Nano 11 (2017) 7485–7493.

[33] H. Wu, W.Y. Yi, Z. Chen, et al., Janus graphene oxide nanosheets prepared via pickering emulsion template, Carbon 93 (2015) 473–483.

[34] H.C. Hu, L.Z. Wu, Y.S. Tan, et al., Interfacial synthesis of highly stable CsPbX3/Oxide Janus nanoparticles, Journal of the American Chemical Society 140 (2018) 406–412.

[35] S. Pradhan, L.P. Xu, S. Chen, Janus nanoparticles by interfacial engineering, Advanced Functional Materials 17 (2007) 2385–2392.

[36] I. Jeon, M.D. Peeks, S. Savagatrup, et al., Janus graphene: scalable self-assembly and solution-phase orthogonal functionalization, Advanced Materials 31 (2019) 1900438.

[37] S. Chang, W. Wang, Bulk synthesis of Janus nanomaterials. US Patent application no. 16/818632, 2019.

[38] W. Wang, Synthesis of Janus nanomaterials. US Patent application no. 16/653367, 2019.

[39] W. Wang, S. Efrima, O. Regev, Directing silver nanoparticles into colloid-surfactant lyotropic lamellar systems, The Journal of Physical Chemistry B 103 (1999) 5613–5621.

[40] X. Chen, S. Efrima, O. Regev, et al., Doping silver nanoparticles in AOT lyotropic lamellar phases, Science in China Series B: Chemistry 44 (2001) 492–499.

[41] J. Lenis, S. Razavi, K.D. Cao, et al., Mechanical stability of polystyrene and Janus Particle monolayers at the air/water interface, Journal of the American Chemical Society 137 (2015) 15370–15373.

[42] B.P. Binks, P.D.I. Fletcher, Particles adsorbed at the oil – water interface: a theoretical comparison between spheres of uniform wettability and "Janus" particles, Langmuir 17 (2001) 4708–4710.

[43] T. Yin, Z. Yang, Z. Dong, et al., Physicochemical properties and potential applications of silica-based amphiphilic Janus nanosheets for enhanced oil recovery, Fuel 237 (2019) 344–351.

[44] H. Wu, K. Gao, Y. Lu, et al., Silica-based amphiphilic Janus nanofluid with improved interfacial properties for enhanced oil recovery, Colloids and Surfaces A: Physicochemical and Engineering Aspects 586 (2020) 124162.

[45] P. Liu, X. Li, H. Yu, et al., Functional Janus-SiO_2 nanoparticles prepared by a novel "cut the gordian knot" method and their potential application for enhanced oil recovery, ACS Appled Materials & Interfaces 12 (2020) 24201–24208.

[46] P. Liu, H. Yu, L. Niu, et al., Utilization of Janus-silica/surfactant nanofluid without ultra-low interfacial tension for improving oil recovery, Chemical Engineering Science 228 (2020) 115864.

[47] L.J. Giraldo, J. Gallego, J.P. Villegas, et al., Enhanced waterflooding with NiO/SiO_2 0-D Janus nanoparticles at low concentration, Journal of Petroleum Science and Engineering 174 (2019) 40–48.

[48] H. Jia, P. Huang, Y. Han, et al., Synergistic effects of Janus graphene oxide and surfactants on the heavy oil/water interfacial tension and their application to enhance heavy oil recovery, Journal of Molecular Liquids 314 (2020) 113791.

[49] W. Wang, S. Chang, A.I. Abdel-Fattah, Enhanced oil recovery with Janus nanoparticles. US Patent application no. 17/381344, 2021.

[50] I. Raj, M. Qu, L. Xiao, et al., Ultralow concentration of molybdenum disulfide nanosheets for enhanced oil recovery, Fuel 251 (2019) 514–522.

[51] N. Glaser, D.J. Adams, A. Böker, et al., Janus particles at liquid-liquid interfaces, Langmuir 22 (2006) 5227–5229.

[52] D. Luo, F. Wang, J.Y. Zhu, et al., Secondary oil recovery using graphene-based amphiphilic Janus nanosheet fluid at an ultralow concentration, Industrial & Engineering Chemistry Research 56 (2017) 11125–11132.

[53] W. Wang, S. Chang, A. Gizzatov, Toward reservoir-on-a-chip: fabricating reservoir micromodels by in situ growing calcium carbonate nanocrystals in microfluidic channels, ACS Applied Materials & Interfaces 9 (2017) 29380–29386.

Index

Note: Page numbers followed by "*f*" and "*t*" refer to figures and tables, respectively.

D

E

F

G

H

Printed in the United States
by Baker & Taylor Publisher Services